Food Science Text Series

The Food Science Text Series provides faculty with the leading teaching tools. The Editorial Board has outlined the most appropriate and complete content for each food science course in a typical food science program and has identified textbooks of the highest quality, written by the leading food science educators.

More information about this series at https://www.springer.com/series/5999

Richard Owusu-Apenten
Ernest R. Vieira

Elementary Food Science

Fifth Edition

Richard Owusu-Apenten
School of Biomedical Sciences
Faculty of Health and Life Sciences
University of Ulster
Coleraine, Londonderry, UK

Ernest R. Vieira
Georgetown, MA, USA

ISSN 1572-0330 ISSN 2214-7799 (electronic)
Food Science Text Series
ISBN 978-3-030-65431-3 ISBN 978-3-030-65433-7 (eBook)
https://doi.org/10.1007/978-3-030-65433-7

This Springer imprint is published by the registered company Springer Nature Switzerland AG
The registered company address is: Gewerbestrasse 11, 6330 Cham, Switzerland

Foreword

In the spring of 1970, I was a senior at the University of Rhode Island looking forward to graduation in a few months. My mentor at URI, Arthur Rand, called me in and told me of an opening at an agriculture school in Massachusetts for an instructor in food science. The school, The Essex Agricultural and Technical Institute, did not have a program at the time, so the position included setting up the new program which would offer an Associate of Applied Science degree in food science.

At the interview, the first question from the director of the school's college division was "what is food science?" I paused and answered, "how much time do you have?"

I would answer that question with the same response today, over 50 years later. Food science includes such a wide variety of areas, all of which require application of the basic sciences such as mathematics, biology, chemistry, and physics. How we as food scientists use these tools to create and maintain a food supply that is safe, nutritious, convenient, and affordable is surely a formidable task that is as important to the world health as any. The diverse positions my graduates have filled over the years clearly exemplifies this. Some of these professions include food processing, food product development, quality assurance, food safety, nutrition, public health, and education.

In the fifth edition of *Elementary Food Science*, Dr. Owusu-Apenten builds upon the prior work done by Nickerson, Ronsivalli, and me, and included are the many changes in food science and how they affect the world. As food scientists, we must share this information with young people so that our profession, so critical to the future health of our planet, will attract talented, enthusiastic, and creative students.

Since my retirement, I have volunteered at grammar and high schools and have had so much fun watching young people marvel at the wonders of food science. One of my grandsons put it best, "Grampi, food science is really magical!"

I agree.

Professor Emeritus
North Shore Community College
Danvers, MA, USA

Ernest R. Vieira

Preface to 5th Edition

The well-received book *Elementary Food Science* has been updated and revised, retaining many attractive features from the 4th edition, in particular the multipart structure.[1] In addition, the emphasis on practical aspects of food science and food handling remains evident in the 5th edition. However, the pace of change in food science has been rapid over the past 15 years. The "quantity" of food science has grown also as part of the wider information and digital revolution. Interestingly, the quality of the internet and online materials varies enormously, which can pose problems for the novice student or general reader. The twenty-first century emphasizes on student-centered instruction means that the student must demonstrate independent research skills, knowledge, understanding, critical thinking, and reflective writing. In view of such considerations, some notable changes in the book were:

1. Interrelated Food Science Topics (Part 1). This part has been shortened from nine to six chapters with enzymes and chemical reactions sections dispersed to other chapters. Two new expository chapters highlighting the food industry (Chap. 2) and the challenging problem of low consumer food literacy (Chap. 6) have been added to the 5th edition.
2. Food Safety and Sanitation (Part 2). Food safety is a new part, expanded from two to four chapters due to the trend for compulsory food-safety management.
3. Food Preservation and Processing (Part 3). This combines four chapters from the previous edition with two new chapters dealing with basic considerations in food processing (Chap. 11) and the quality of dried foods (Chap. 15).
4. Handling and Processing Food (Part 4). The topics fats and oils are dispersed in the text while updating and retaining eight out of nine chapters from the 4th edition.
5. Consumer expenditure for out of home eating now exceeds spending for home eating. A new Chap. 25 dealing with the foodservice and catering industries has been added, introducing this important topic for food science students.

[1]Vieira, E. R. (2013). Elementary Food Science (4th Edition), Springer Science & Business Media, 423 pp. Other editions appeared in 1976 and 1980, 1995, & 2013 (4th Edition).

6. The North American Industry Classification System (NAICS) is introduced throughout to emphasize strong links between science and the food industry.
7. Material for experimental food science and culinary arts is moved to the upcoming supplement.

Overall, employability and career paths in food science are more coherent than ever, from high schools all the way to section managers, product developers, or research scientist. Ironically, industry demand is growing for cross-expertise covering food science, human nutrition, culinary arts and foodservices. The 5th edition speaks also to some controversial public health issues, such as the suggested link between ultra-processed foods and chronic illness.

The core readership for *Elementary Food Science* remains unchanged, that is, those studying food science at secondary or post-secondary institutions (colleges and universities). Advanced candidates from other fields with low amounts of prior food science experience and those interested in the intersection of food science with the culinary arts, foodservices, nutrition, or public health areas will find sections of the book useful.

Richard Owusu-Apenten
Coleraine, Londonderry, UK

Contents

Part III Food Preservation and Processing

About the Authors

Richard Owusu-Apenten, (BSc. PhD. FIFST. FHEA, RNutr. RSci.) is an independent food scientist researcher. In the course of 30 years, he served as faculty in the Department of Food Science, Pennsylvania State University (USA), the Department of Food Science & Nutrition, University of Leeds (UK), and the School of Biomedical Sciences, Ulster University, Coleraine (UK). Dr. Apenten is a *Fellow* and Registered scientists (FIFST. RSci) of the Institute of Food Science & Technology (UK), a member of the Institute of Food Technology (IFT), a *Fellow* of the (UK) High Education Academy (FHEA) for teaching excellence and a Registered Nutritionist (RNutr) of the UK Association for Nutrition (AFN).

Ernest R. Vieira was chair of the Department of Food Science, Nutrition and Culinary Arts, Essex Agricultural and Technical Institute, Hathorne, Massachusetts, USA. He was also Professor of Chemistry and Food Science at the North Shore Community College, Danvers, Massachusetts, USA.

Part I

Interrelated Food Science Topics

1 Why Food Science?

Just as society has evolved over time, our food system has also evolved over centuries into a global system of immense size and complexity. The commitment of food science and technology professionals to advancing the science of food, ensuring a safe and abundant food supply, and contributing to healthier people everywhere is integral to that evolution – Philip E. Nelson, 2007 World Food Prize Laureate; Professor Emeritus, Food Science Dept., Purdue Univ.

1 Introduction

1.1 Food Technology and Food Science

The scientific study of food is one of human society's most important endeavors (Clausi et al. 2000; Skolnik 1968; Toops 2017). Food is necessary for growth, physical activity, and for maintaining health. Ensuring an uninterrupted supply of food, which is safe and affordable for a mostly urbanized global population, is one of the world's largest industries (Floros et al. 2010). According to data from the United States Department of Labor and Statistics there are about 15,010 professional food scientists and technologist in the United States, excluding those who identify as self-employed. The States of Minnesota, California, Texas, Wisconsin and Ohio have the highest concentrations of food Scientists and technologists (U.S. Bureau of Labor Statistics 2014). In addition, many millions of people (>10% of all US employees) work within the food industry in a variety of professional, technical, skilled, or semi-skilled occupations (cf. Chap. 2).

Ideally, everyone working in any area of food science and technology must possess some knowledge of basic and applied sciences. Before the 1920's, some of the topics covered by food science courses were taught as domestic science, home economics and nutrition (Dowd and Jameson 1918). Culinary and occupational home economics training for foodservice workers, initiated by the Smith-Hughes vocational education Act (1917), emphasized pre-college learning for work. Successive Carl D Perkins Acts modified vocational education gradually leading to the Career and Technical Education programs of today that prepare young people for work as well as college (VanLandingham 1995a, b; Bantang 2008; Zirkle 2012). Dieticians and nutrition students have always taken some college-courses in culinary arts and occupational home economics (Canter et al. 2007; Cooper et al. 2016). However, there is now growing interest in programs that combine culinary arts with food science or nutrition for the post-secondary sector (Tittl 2008). Culinary arts and food technology were joined recently to produce the Culinology™ degree endorsed by the Research Chefs Association (Bissett et al. 2010; Cheng et al. 2011; Cheng and Bosselman 2016). The interest in healthy eating is now leading to novel college course options, diplomas or BSc programs such as, "Culinary Nutrition" (Marcus 2013; Oregon State University 2019), "Culinary Food Science" (Ohio Department of Higher Education 2017; Iowa State University 2021) and also "Food, Nutrition and Culinary Science" (Temasek Polytechnic Singapore 2019).

Figure 1.1 shows how food science and food technology are defined by the Institute of Food

R. Owusu-Apenten, E. R. Vieira, *Elementary Food Science*, Food Science Text Series,
https://doi.org/10.1007/978-3-030-65433-7_1

Food Science is the discipline in which the engineering, biological, and physical sciences are used to study the nature of foods, the causes of deterioration, the principles underlying food processing, and the improvement of foods for the consuming public.
Food Technology is the application of food science to the selection, preservation, processing, packaging, distribution, and use of safe, nutritious, and wholesome food.
In practice, the terms food science and food technology are often used interchangeably

Fig. 1.1 Definition of foods science and food technology. (Institute of Food Technologists (IFT) 2014)

technologists (IFT). The training and education for food scientists and technologist will be discussed later in this chapter (Sect. 5).

1.2 Definitions for Food

Food is defined within international legislation (Table 1.1) as any substances intended for human or animal consumption, but excluding materials intended to be used *only* as drugs. Food includes drinks, chewing gum and water. From a legal point of view, all substances *intended* for human consumption are considered food, excluding tobacco and medicinal drugs.

Table 1.1 Some current definitions for food

Raw, cooked, or processed edible substance, ice, beverage, or ingredient used or intended for use or for sale in whole or in part for human consumption, or chewing gum – US food code 2013 (U.S. Food and Drug Administration 2013).
Articles used for food or drink for human or other animals, chewing gum, and articles used for components of any such article, – Federal Food, Drug, and Cosmetic Act, § 201(s) (U. S. Food and Drug Administration 2014)
Substance, whether processed, semi-processed or raw, which is intended for human consumption, and includes drink, chewing gum and any substance which has been used in the manufacture, preparation or treatment of "food" but does not include cosmetics or tobacco or substances used only as drugs. (Codex Alimentarius Commission 1997)

1.3 History of Food Technology and Processing

Food processing can be defined as any activity applied to foods prior to consumption (Horrocks 2003). Processed foods have been around in abundance since antiquity (Kim 2013). Moreover, the development of different of methods of food preparation and processing, occurred slowly by trial and error. Here and elsewhere, it is important to recall that much of human history occurred in a pre-scientific framework. A cultural or culinary focus of food processing has existed for much of recorded history.

Archeological evidence suggests that people of Ancient China, Egypt, Greece and Rome were all adept at preserving foods using vinegar, salt, spices, or honey (Selin 1997; Skolnik 1968). Sun drying probably predates agriculture. Some other traditional food processing methods that developed in pre-scientific times include, baking, cooking, curing, drying, fermentation, milling, pickling, salting and the preservation of meat using wood smoke (Horrocks 2003). Products like bread, wine, olive oil, noodles, cassava flour and maize flour were all developed without the aid of science.

Table 1.2 shows a tentative-timeline for processed foods for the past 2 million years. Other processed foods and recipes from 17,000 BCE to 2009 can be found elsewhere (Olver 2014).

A "scientific revolution" occurred in North West Europe, which led to unprecedented changes in human knowledge and understanding between the fifteenth and sixteenth century (Shapin 2008). Advances in biology, physics and chemistry unlocked some fundamental scientific laws of "nature". Vital scientific concepts (e.g. heat, energy, work, gas, liquids, molecules, microorganisms, mechanics, and electromagnetism) were discovered without which it was impossible "to do" science. The first steps towards the "scientific method" for enquiry emerged also requiring the testing of hypothesis by empirical experimentation.

Table 1.2 Time-line for some processed foods

1.8 million BCE, roasted meat
28,000 BCE, bread
7000 B.C., Persia, beer
5400 B.C., Persia, wine
5000 BCE, Poland, cheese
4500 BCE, Eastern Mediterranean, olive oil
3000 BCE, Ancient Egypt, palm oil
2400 BCE., Mesopotamia, pickles
2000 BCE., China, noodles
1900 BCE., Central America, chocolate
1500 BCE, China, bacon (salted pork)
1000 BCE., China, JIANG (fermented source)
500 BCE., India, sugar
AD. 700, Korea, kimchi
AD. 965, China, tofu
Seventh century, China/Persia, sugar refining
Tenth century, Basque Spain, salt cod
Fifteenth century Aztecs, peanut butter
Mid-fifteenth century, Yemen & Ethiopia, coffee
1767, England, carbonated water
1894, Michigan, USA, corn flakes
1908, Germany, glutamate and MSG
1926–1937, United States, SPAM
1950s, United States, chicken nuggets
1957, United States, high-fructose corn syrup

Source: Adapted from Kim (2013)

2 Origins of Food Science and Food Technology

2.1 Starting with Eighteenth and Nineteenth Centuries

The core natural sciences, which form the basis for food science and technology, did not emerge prior to the 1700s. We should recall that Benjamin Franklin's work on lightening as electricity dates from 1751, the discovery of oxygen occurred at about 1778, and Daltons atomic theory of matter did not emerge until 1805 (Wikipedia 2020b). Though, human societies demonstrated the capacity to organize and systemize knowledge, it is another thing entirely to talk of this as modern science. Some major scientific ideas such the "cell theory" of life, electromagnetism, electricity, germ theory of disease – were missing for much of human history. That the phrase "food science" was not widely used prior to 1938/World War II should not come as a great surprise (Vickery 1967). For reasons given above, it is safe to place the origins of food *science,* somewhere in the eighteenth century (Fig. 1.2).

2.2 Food Chemistry

The study of modern chemistry started with Antoine Lavoisier (1743–1793).[1] Historians consider Lavoisier as the father of the 1760 "chemical revolution" from France, which also laid the foundation for Nutrition. Lavoisier's approach to research is interesting and recognizable to modern readers; first, there was a thorough review of

Fig. 1.2 Portraits of Antoine Lavoisier (Left panel) and Benjamin Thompson (Count Rumford, b. 1753). (Photographs are in the public domain)

[1]Section 2 of this chapter contains biographic material from, Wikipedia (commons). Readers may refer to these sources for more detailed biographies. Additional references are cited as needed.

previous literature. Next Lavoisier developed hypothesis based on the past literature. Finally, there came empirical experiments designed to test his hypothesis (Partington 1989).

Lavoisier proved that flame combustion and cellular respiration were similar phenomena. The work on cell metabolism led ultimately to the measurements of metabolic heat using indirect calorimetry. According to the principles of indirect calorimetry, the total amount of energy produced when an animal metabolizes food (sugar) can be determined from the volume of carbon dioxide expired. Knowing how much energy is expended during physical activity is clearly important for working out energy balance. Indirect calorimetry is still one of the most convenient methods for determining human energy expenditure. Lavoisier's research was ultimately important also in the fields of agricultural chemistry, biological chemistry and nutrition (Partington 1989).[2]

Sir Benjamin Thompson (Count Rumford, b. 1753), is another important contributor to food chemistry. Thompson was active during the American war of independence but found time to invent the Rumford baking powder. Francois Megendie (1783–1855) is noted for his work on the chemical and nutritional value of gelatin.

Justus Von Liebig (1803–1873) developed the German food chemistry tradition that ultimately affected developments in the United States many decades later. For instance, Professor S.W. Johnson and his protégé, Professor Wilber Atwater all spent time in Germany during their doctoral and postdoctoral years. It was Atwater, who established the Agricultural Experimental Station framework in the United States. Professor Liebig is also credited with the invention of marmite and OXO cubes. Jean Baptist Dumas (1800–1884), apparently a great rival of Von Liebig, invented the Dumas protein assay (Partington 1989).

2.3 Food Microbiology

The invention of the microscope by Robert Hooke is considered the starting point for microbiology (Jay et al. 2005; Adams and Moss 2008; Ray and Bhunia 2013). However, Antoine Van Leeuwenhoek (b.1633–1723) was first to observe single cell organisms under a microscope – leading to the appellation father of microbiology.[3] Microbiology is defined as the study of microorganism and their ecology. Food microbiology is the application of microbiological principles to the study of food and the food system. In general, food microbiologists are concerned with three areas; (i) understanding the effect of microorganism in relation to food safety, (ii) the role of microbes in food spoilage and (iii) the use of microbes for food manufacturing (Jay et al. 2005; Adams and Moss 2008; Ray and Bhunia 2013).

It is believed that the first fermented products were produced round about 7000 BC or earlier (Oberg 2012). Examples of traditional fermented foods include, bread, beer, wine, pickles, kimchi, yogurt, and cheese.

Interestingly, the scientific basis for fermentation were unknown prior to the nineteenth century. Louis Pasteur (b.1822–1895) provided the experimental proof showing that microorganisms carried out fermentation. The germ theory (1857) developed by Pasteur helped to explain food spoilage and the causes of infectious human diseases. Pasteur demonstrated that heating liquid food (beer, wine and milk) protected it from spoilage if care was taken to avoid recontamination by microorganisms.

One half-century prior to Pasteur's discoveries, Nicolas Appert employed mild heat treatment 140–176 °F (60–80 °C) to extend the storage-life of foods. However, it is Pasteur's name (Pasteurization) which is now associated with thermal processing of beer and other liquid foods. By contrast, the term "Appertization" describes thermal processing using pressurized retorts to

[2]Lavoisier was executed during the French revolution and his death was considered premature by the scientific community.

[3]Terms such as "farther of" seem antiquated for modern readers, but were common currency in some of the literature sources cited.

Fig. 1.3 Louis Pasteur (1822–1895)

achieve temperatures above 100 °C so that the product is commercially sterile. Using metal containers for canning was developed by Girard, (cf. Sect. 2.4. below and Chap. 12) (Fig. 1.3).

The significance of pasteurization became evident for public health during the early 1900s. Mild heat treatment of milk turned out to be highly effective for the prevention of tuberculosis. Pasteurization is now widespread and is mandatory for many developed countries (Brink 1964). Pasteur's germ theory provided the scientific basis for improvements in food preservation and food safety (Gould 2000). Sterilization of medical equipment is another derivative technology that arose from the original developments in food technology.

2.4 Food Engineering and Processing

Canning is regarded one of the key milestones in the history of *modern* food processing (McLachlan 1975). Canning is generally credited to Mr. Appert (1749–1841) who worked as a confectioner in Paris. In 1810, Appert won a prize of 12,000 Francs for developing a preservation process, which enabled the French army to function without conscripting food from locals (Graham 1981). Appert's book "*L'art de conserver, pendant plsieurs années, toutes les substances animales et végétale*" ["The art of preserving animal and vegetable substances] shows that foods were preserved by heating inside of glass jars (Appert 1810). It seems the competition rules set out by the French Ministry of Interior required Appert to fund the publication of his book out of pocket as a condition for releasing his prize money. The first print of 200 copies of Appert's book were sold for 3 Francs each or 3.50 Francs with postage (Graham 1981; McLachlan 1975).

Following public disclosure, Appert's technology spread rapidly out of France. First, a French language second edition of the "L'art de conserver" was published in London in 1812. Then Longworth publishers of New York translated Appert's book into English; an electronic copy is available for preview from the University of Chicago Library (online). In hindsight, it seems Appert produced (inadvertently) the first text on the topic of food processing. Finally, he is credited with running the first "food preservation" business based on canning from about 1810 onwards.

Using a tin-plate for canning was probably developed entirely separately from the method invented by Appert (Graham 1981; Cowell 2007).[4] A compelling BBC article titled "*The story of how the tin can nearly wasn't*" describes an alternative history for metal canning (Geoghegan 2013). Shortly after Appert's prize became publicized "a foreign gentleman living abroad" submitted a patent application in the UK using tin cans for canning.[5] Mr. Philippe de Girard (1775–1845) was that French citizen who came to London (England) to patent tin/metal containers for canning food. Apparently, Girard was motivated to avoid bureaucracy or "red tape" common in France at the time (Geoghegan 2013).[6]

---

[4]Though glass and metal based canning technology probably arose in France at about the same time its not clear that Girard competed for the same prize as Nicolas Appert.

[5]The protracted nature of intellectual property (patenting) process suggests that the concept of metal canning developed, in parallel with Appert's glass based process.

[6]Appert's book describing canning technology was published in France a few months prior to the events described in the UK. The disclosure of glass based canning might have made it less likely that "similar" technology would be patentable in Paris. Similarly, the time required to develop an entirely new technology would be longer than 3 months. The French origin of commercial canning was possibly obscured by using Durand's name on the application for the patent in London.

Also according to the BBC documentary, Girard presented samples of tinned foods (milk, broth, roasted meat) for tasting at the Royal society of London. Photographs of letters from the Royal Society archives suggest that Mr. Peter Durand (patent agent for Girard) was recipient of a patent for canning using metal containers in London in 1811 (Geoghegan 2013). The patent for canning was sold for £1000 to Mr. Bryan Donkin an entrepreneur, inventor and mechanical engineer. Donkin worked on the canning technology for 2 years, and obtained a Royal charter from King George (III) of England allowing him to start commercial canning in 1813.

The first canning factory in Gt Britain was located in Southwark Park Road, inside the London borough of Bermondsey (UK). Apparently, Donkin stayed in the canning business until 1820 during which time, the British Admiralty were customers. However, by 1850 the quality of canned meat was considered inadequate. Microbiological problems affecting the quality of tinned food ("blow out" of tinned clams and lobster, souring of sweetcorn) were first tackled by William Lyman Underwood working in collaboration with Samuel C. Prescott at MIT (Prescott and Underwood 1897). The methods developed were later extended in the 1920–30's to research food safety, which led to increased confidence in the safety of commercially canned food (Fellers et al. 1928; Perkins 1964; Andress & Kuhn 1998).[7] The first canning factory in England was probably situated at or near the current location for Southwark Park Primary school.

There is some controversy about who invented canning partly because of the very short (3-month) time gap between Appert's prize award in Paris and the UK patent award (dated 28. Jan.1811). Moreover, Philippe de Girard invented a flax-spinning machine for another competition run by the French Interior Ministry (Wikipedia 2020a). It is partly because of this familiarity with Napoleonic prizes that lead some historians to regard Girard as the inventor of metal based canning (Robertson 2005; Cowell 2007).[8] There is at least one further twist in the origin story of the tinned food. By some accounts Mr. Appert, produced small quantiles of tinned sardines in France prior to publishing his bottle-jar method in 1810 (Graham 1981). It is not known whether Appert and Girard were directly acquainted with each other, in France or in England (Cowell 2007).

The introduction of canned food into North America is sometimes ascribed to Mr. William James Underwood but this too is contested for two reasons. First, home canning with glass jars was widely practiced owing to the publication of Appert procedures specifically for use by homemakers (Reber 2019). Secondly, Mr. Underwood, described as an English immigrant, opened a food preservation factory in Baltimore in 1822 (Graham, 1981). However, there seemed to be another food preservation factory in New York from about 1819 operated by Thomas Kensett & Ezra Daggett also originating from England. Moreover, Kensett and Co obtained a United States patent for tin metal cans from 1825. Both the Kensett and Underwood factory used glass jars to process their food but changed to more robust tinned cans from about 1840s. Tinned food, though regarded as expensive initially, was important for the California gold rush and the American civil war. Invention of the steam retort -autoclave by A.K. Shriver (1874) for high temperature processing is considered another highlight for the tinned food industry. A detailed historic account of the tin can industry in the US from 1810 to 1940 is available on the internet (Pearson 2016; Can Manufacturers Institute 2021).

2.5 Food Analysis

The chemical analysis of food substances and natural compounds contributed to the development of chemistry as a discipline (see 2.2). Early research involving food analysis include, those by Friedrich Christian Accum (1769–1838) who

[7]Some sources behind the BBC article – in particular material attributed to Dr. Norman Cowell's PhD thesis (1996) were not available for viewing by us.

[8]Under different historical circumstances, canning might rightly be called the process of "Girardization"

developed methods for the detection of adulterants thereby ensuring a safe and honest food supply. In 1820, Accum published a book titled, "A treatise on adulterations of food and culinary poisons exhibiting the fraudulent sophistication of bread, beer, wine, spirituous liquors, tea, coffee, cream, confectionery, vinegar, mustard, pepper, cheese, olive oils, pickles". The original text is worth viewing for a glimpse of the state of general food manufacturing 200 years ago (Accum 1820).[9]

3 Entering the Twentieth Century

Important advances in food technology were realized between1895 to 1938 (Institute of Food Technologists 2000; Skolnik 1968). As outlined previously, the technology of food canning came under scientific control; a further important development during this era was the use of artificial refrigeration equipment for storage and transportation of perishable foods, which was becoming common. Frozen foods reached retail outlets and home refrigerators began replacing iceboxes (Anonymous 1912; Bryce 1912; Pennington 1912). Some prominent people from the field of food science include, Samuel Prescott, William Lyman Underwood, Stephen Babcock, Edwin Hart, Clarence Birdseye, and Carl Fellers.

Samuel Cate Prescott, (1872–1962) was a microbiologist and food scientist who made significant advances in canning. Prescott based at MIT, collaborated with William Lyman Underwood (1864–1929) to solve the "blow out" issue that affected canned clams. The research showed that "blow-out" was due to microbiological activity, arising from germinating spores that survived the canning process for clams. Prescott identified that retorting at 121 °C for 10 min improves product quality. This finding helped ultimately to transform canning into a scientifically controlled, highly safe, technology of the present day. Research leading to heat inactivation conditions for *C. botulinum* spores were based on the experimental techniques developed by Prescott and Underwood twenty years earlier (Perkins 1964; Andress & Kuhn 1998). Prescott's interests in sanitation also led to his crusade in pasteurization and production of a sanitary milk supply. His interests widened to areas such as, food dehydration, refrigeration, food additives, and the chemical composition of foods. Prescott's work broke down many barriers between the basic sciences, enabling the development of interdisciplinary sciences such as food science.

Stephen Moulton Babcock (1843–1931) was a agricultural chemist, well known for developing the Babcock tests for dairy butter fat, in honor of whom Babcock Hall (the University of Wisconsin) is named. Like a number of his contemporaries, Professor Babcock trained in organic chemistry in Germany, returning thereafter to work in Cornel University and then as Chair of the University of Wisconsin food science department.

Edwin Bret Hart (1874–1953) was a pioneer nutritionist who conducted animal feeding studies alongside of Babcock and the nutritionist/biochemists Elmer Verner McCollum (1879–1967). The classic "single grain" experiment conducted with Babcock, McCullum and others, using different grain sources with equal calories, protein, carbohydrate, fat, and ash content showed that gross chemical constituents alone were not enough for a healthy diet. The investigations starting from 1906, led to improved understanding of importance of vitamins and minerals in food science.

Clarence Frank Birdseye II (1886–1956) is known for his contributions to the development of frozen foods. He was quite disappointed with the poor and unsanitary way in which fish were handled. These observations led to the development of several methods including quick-freezing to capture quality that would be lost if fish were handled improperly and frozen by slow methods. Birdseye was patient and worked many years in the 1920s and 1930s to get frozen foods accepted at the retail level. His discoveries, combined with advancements in freezing technologies, allowed substantial penetration into this market by 1939.

Carl Fellers showed interest in the nutritional quality of canned foods and other processed

[9]Available as a free e-book from Google Book Search.

foods. He served with the National Canners Association from 1921 to 1924 and his work led to the present high level of safety in canned foods. In 1926 he moved to the Massachusetts Agriculture College (now the University of Massachusetts), where he remained. Among his accomplishments are the invention of methods for pasteurizing dried fruits, canning Atlantic crab, the use of ascorbic acid as an antioxidant, the fortification of apple juice, and the use of cranberry culls to develop cranberry juice and other cranberry products that are so popular in today's market. Dr. Fellers was a founder member of the IFT and served as president in 1949–1950.[10]

3.1 The Interdisciplinary Years

Experts from many backgrounds influenced the development of food science in the twentieth century. Bernard Proctor (1901–1959) worked in food microbiology and studied the use of radiation for the preservation of foods. During World War II, he studied dehydrated foods and food packaging. Products were dehydrated and packaged in flexible and laminated containers in order to reduce weight, improve storage life and increase convenience. Many of the foods employed as soldiers rations, space foods, and as foods for hikers and campers are dehydrated and packaged using laminated foil (Farkas 1982; Kohn 1971)

Bernard L. Oser (1899–1995) was a biochemist, and a charter member of the IFT (Saxon 1995). Dr. Oser worked for 47 years with a commercial testing company (Food and Drugs Research Inc.) based in New York. The focus of Oser's research was food chemical safety and toxicology. His bioassays using test animals often gave different results from in-vitro "chemical" tests. The recognition of these differences led to the concepts of bioavailability, and the effect adsorption, distribution, metabolism and excretion (ADME) on the toxicology of chemical substances. Oser was associated with some controversial discussions on the health effects of cigarettes and the Delaney clause (Anonymous 2019; Saxon 1995). From 2000, the IFT supported the Bernard L. Oser Ingredient Safety Award open to nominees with 10 years' experience working in areas related to gererally recognized as safe (GRAS) status (Institute of Food Technologists (IFT) 2015).

Loren Sjostrom, from the consulting firm of Arthur D. Little (ADL), established a flavor laboratory and developed the flavor profile method for descriptive testing of food and beverages. The first paper describing the flavor profile method was presented at the IFT annual meeting of 1949. ADL trains people in the use of this method of sensory analysis, which is used widely.

Amihud Kramer (1913–1981) was a food scientist who qualified at the University Maryland in 1942. Professor Kramer is credited with promoting statistics in the field of food science. This application of statistical methods for quality control helped industry meet increasing quality and cost requirements. His 2-volume book, "Quality Control for the Food Industry" played a significant role in this influence (Kramer and Twigg 1970). He was active in devising procedures for production of desirable nutritional value and sensory quality at a reasonable cost.

3.2 Advances in Technology and Quality from 1960s to 1970s

Novel products such as freeze-dried coffee hit the market in the 1960s. Computer control in food processing plants was introduced, and product quality and efficiency of production increased. A variety of transparent, flexible and/or rigid plastic packaging and containers for food became common in the marketplace (Risch 2009); many new ingredients such as oil blends and flavorings were developed. Protein isolates from fish, whey, and soy were used to

[10] Many of the bibliographic materials for this section are available through Wikipedia, and the interested reader is referred to this source for more reading.

engineer and formulate new foods. Plant sanitation was improved by clean-in-place methods and foods with longer shelf lives resulted. Advances in dehydration technology improved dried milk flavor, and the use of enzymes produced unique products such as meat tenderizers. Single product plants for beer, milk, and baked products were automated resulting in increased production rates (US Food & Drugs Administration 2014a).

The 1970s brought a need for energy saving, and consumption of energy dropped by 25% in most food manufacturing plants. The era of health foods and organic foods was born during this decade. This gave the opportunity for many small-scale processors to become competitive with the bigger companies. By the end of the 1970s, smaller, quicker, user-friendly computers were introduced into food plants, contributing to more energy efficient plants producing higher quality products at a faster rate (US Food & Drugs Administration 2014a).

3.3 Progress from 1980s to 2000s

Along with the 1980s came great optimism on the part of food processors and increased consumer demand for food safety and quality. The "new generation" gourmet type product became more common. Advanced technologies in freezing systems, frozen distribution, retail and home freezers led to the development of new, high quality, frozen entrees. Aseptic packaging of products such as fruit juices using the "tetra Pak" box became common (Anonymous 1970). Microwave ovens invented in the 1950s began to appear in almost every home by the 1980s; industrial microwaves ovens were also applied for food manufacturing and foodservices applications (Chandrasekaran et al. 2013; Kalla and Devaraju 2017). Changing lifestyles in America led to more demand for minimally processed foods that retain fresh-like quality (Giovenzana et al. 2015).

Nutrition, health, convenience, and food safety continued to be priority concerns, for the twenty-first century. The Nutrition Labeling and Education Act (NLEA) of 1990 required the Department of Health and Human Services to propose new nutritional labeling regulations, and by 1994, these were in effect (U.S. Food and Drug Administration 2014). The US guidelines and food pyramid made it easier for the consumer to understand the importance of proper nutrition and a balanced diet (U.S. Department of Agriculture and U.S. Department of Health and Human Services 2010). The 1990s saw the introduction of genetically modified organisms (GMO) in the food system, which is growing in some sectors particularly related to insect pests or herbicide resistance crops in Agri-food production.

3.4 Computerization and the Internet

The period from 1990s and 2000 saw the ingress of mass computers into virtually all aspects of the food industry. Disciplines like food engineering and food analysis were probably first to deal with computerization, focusing on sensors, and the control of equipment. Currently, computer applications are found in all stages of the Agri-food system, including farm production; control of machinery, machine vision, logic controllers, equipment monitoring, process calculation (US Food & Drugs Administration 2014b). Another understated effect of computerization is the advent of the internet, which was for at time debated as a fundamental human right (Penney 2011). Artificial intelligence and robotics are predicted to affect large sections of economic activity, in particular foodservices and cooking (Frey and Osborne 2017).

3.5 Food Consumer Trends for the Twenty-First Century

Trends in food choice continue with increased demand for product variety. Currently an estimated 12,000 new grocery items are introduced into US market per year with a product attrition rate of 80–90%. The tendency to buy new food products (purchase diversity) increased with rising house hold expenditure on food, but was not strongly affected by factors such as, extent of college education, number of kids per household, age or race (Jekanowski and Binkley 2000).

Consumer demand remains for more fresh-like minimally processed foods. Other popular food groups included low additive – clean label foods, and food with improved food safety. Food and health continued to be important in the twenty-first century. Foods, with or without formal food labeling claims (Zink 1997) came into prominence including whole grains and other prebiotic or probiotic products intending to improve gut health (Rose 2014). Out of home eating and non-full service restaurants continued to increase in popularity. Food processors continue to respond to demand for niche products; demand for organic food continue to move towards mainstream. Ethnic foods, halal products, vegan and vegetarian offerings are increased also in popularity (USDA. Economic Research Services 2014).

4 Food Technology and Society

4.1 The Media and Food Literacy

How food is farmed, harvested, transported, processed, packaged, stored, prepared, and served is the subject of media scrutiny. Awareness of the underlying issues is important to promote consumer food literacy (Neff 2014).[11] However, the print media, internet and social media does not always present all the facts. Frequently, what is presented to the consumer are aspects that make sensational headlines and "sell newspapers" or else is liable to draw traffic to websites and blogs. The diverse nature of information sources available for food science and food technologists is also worth highlighting. Perhaps more than other disciples, it is difficult for the non-specialist public to judge the quality of food science information (Lupien and Lin 2004; Weaver et al. 2014).

Communication skills are necessary for food scientists to deal effectively with the different information sources (see Appendix 1). The technical reading materials for food science and food technology include, (i) journals, trade magazines, newspapers, (ii) Secondary publishing,(iii) patent information, (iv) data, and (v) grey literature from noncommercial organizations – meeting abstract, thesis, standards, government bulletins, market information, and newsletters etc., (vi) books, and (vii) trade publications (International Food Information Service 2005). In addition, food information sources can take the form of print, e-books or internet content.

Critical reading skills are needed, plus healthy skepticism towards expert opinion, to deal with the range of technical materials available in food science and technology. Consideration of possible "conflicts of interest" and sources of bias is essential in all discussions. The motivations of stakeholders or different pressure groups should be acknowledged if not celebrated. The reliability and trustworthiness of sources of information needs to be examined in order to make informed judgment.

Reliable information on food and nutrition issues can be accessed from the USDA National Agricultural library (https://www.nal.usda.gov/fnic). Other trustworthy sources include local boards of health and nonprofit agencies. Websites produced by Federal and state regulatory agencies are always a good bet when looking for information; In addition to the USDA, reliable information is provided by the Food and Drug Administration (FDA), State Departments of Public Health, and the Cooperative Extension Services. Professional associations such as the American Dietetic Association and IFT also provide additional sources of useful information. Many of these organizations are discussed further in Chap. 3.

Extension services run by State Universities generate reliable information and education opportunities for the public on a variety of topics including, nutrition, sanitation, and basic food processing; hygiene and food safety. The IFT and related professional organizations, provide accu-

[11] Food literacy is explored more comprehensively in Chap. 6.

rate information through the science communications programs regarding food related issues. The contact details for the IFT are:

Institute of Food Technologists
525 W. Van Buren, Ste 1000 Chicago, IL 60607
Phone: +1.312.782.8424
Fax: +1.312.782.8348
Email: info@ift.org
Web: https://www.ift.org/

Strategies for critical reading will help to identify reliable information sources on food legislation, labeling and food safety via the internet. Awareness of different levels of trustworthiness for different information sources is essential. Official government websites and noncommercial sources are also more likely to be trustworthy. Most internet browsers offer an "Advanced search" option, to allow filtering by date, type of document, or domain etc. Documents obtained from government (.gov) or college (.edu) websites are probably more reliable. Publications and technical articles that have undergone "peer" review should be sought out. Consumers should check the credentials of food experts and other writers about food and nutrition. The problem of poor food literacy for consumers is discussed further in Chap. 6.

4.2 Food Safety

Food can be vehicles for communicable and noninfective diseases. Foodborne disease agents are categorized as, biological hazards (pests, parasites, rodents), microbiological hazards (protozoa, bacteria, fungi, viruses), biochemical hazards (allergens, protein toxins), or chemical hazards (mycotoxins, pesticides, veterinary drug residues). In general, outbreaks of foodborne disease undergo constant surveillance in order to track new sources of bacteria at the State, Federal or international level (Centers for Disease Control and Prevention 2014).

Recent outbreaks of food-borne diseases were documented by OECD/WHO experts (OECD/WHO 2003; Rocourt et al. 2003). Animal-based foods were associated with serious epidemics including BSE in Europe, avian influenza viruses in Asia, and Ebola outbreaks of West Africa. These and similar zoonotic disease are transmitted from animal reservoirs to livestock or from wildlife hosts directly to the human consumer (Cutler et al. 2010). Ebola being one of the most alarming conditions is thought to be associated with the consumption of bush meat but the actual reservoir for the virus remains unknown (Pooley et al. 2015). It would be remise not to mention, the latest example of zoonosis which is the advent of severe acute respiratory syndromes (SARS) linked with coronavirus (SARS-Cov), including the Middle East SARS (MER-SARS) and Covid 19 pandemic that is thought be linked with wet markets of Wuhan, China (Contini et al. 2020).

Historically, the threat posed by milk-borne bovine tuberculosis was addressed via the adoption of mandatory pasteurization in Chicago in 1908; by 1924 the U.S. Public Health Service had developed voluntary sanitary standards for a so-called Grade a milk ordinance. The Canadian experience also confirms the health-benefits of pasteurization (Brink 1964).

Lapses of food safety occur inside of domestic settings and the foodservice sector (Sneed and Strohbehn 2008). It falls to State or Federal authorities to ensure "*minimum sanitary practices for food service providers and theatres that provide foods for the public*". The US Environmental Protection Agency monitors the importation of tainted foods through borders, airports and seaports. Federal and other regulatory agencies involved with food borne diseases are discussed in detail in Chap. 3.

4.3 Nutrition, Food and Well-being

The relationship between nutrition, food science and food technology is complex and intriguing (Mozaffarian 2016). Modern nutrition is about 100 years old (Mozaffarian et al. 2018). By contrast, food science developed after Second World War (Vickery 1967).[12] The discovery of vitamins led to new processing endpoints to minimize heating losses and food fortification. Other nutrition discoveries e.g. health effect of total/saturated fats, trans-fats, dietary fiber, and salt) led to the voluntary or compulsory reformulations of food products. Many serious chronic diet-related diseases, are leading to calls to examine, genetic differences and how these affect, diet and health relations, personalized nutrition and food technology (Lewis and Burton-Freeman 2010) (Mozaffarian et al. 2018).

4.4 Food and Public Health

Population studies show that what we eat is linked with wellbeing (Snowdon et al. 2013). For example, eating highly palatable, ultra-processed foods and alcohol consumption may contribute to non-communicable diseases such as, obesity, hypertension, stroke, diabetes, and cancer. A decrease in the rates of exercise, increased car-use and a more sedentary life-style have all been implicated in the development of metabolic conditions (Moodie et al. 2013). Supersizing, bigger food portion sizes, and product discounting are some of the other contributory factors for over-consumption. Some commentators accuse the food industry for "being in denial" about the negative impact of processed foods (Brownell and Warner 2009). Others think that the beneficial effects of processed foods are under-appreciated (Weaver et al. 2014).

Processed foods and foodservice purchases account for a significant proportion of foods eaten in developed countries (Dwyer et al. 2012). Therefore, promoting healthy eating should acknowledge the widespread use of processed foods. Other controversial considerations include the role of government-industry partnerships, industry self-regulation or independent legislation in promoting healthy eating (Moodie et al. 2013; Sharma et al. 2010). Punitive taxation (sugar tax) for high-calorie, energy-dense foods are being suggested to reduce the consumption of unhealthy food (Davey 2004). The public health dimensions of food science are now under increasing scrutiny (Ferruzzi et al. 2012; Nestle and Jacobson 2000; Rowe et al. 2011; Young and Nestle 2002).

Recent dietary guidelines is for lower intakes of, sugar, saturated fat, salt, and trans-fats. The food industry is developing new products and reformulating existing lines to produce more "healthier options" (Buttriss 2013). Recent reductions in hydrogenated vegetable oils and trans-fats in manufactured foods show what can be achieved. Food fortification programs may be initiated in response to perceived public health needs. Changing eating behavior is still seen as quite formidable undertaking (Krebs-Smith et al. 2010).

4.5 Food Agro-business and Development

The importance of food technology to industrialized countries is widely recognized (cf. Chap. 2). So enhancing the capacity for food technology in developing countries would be beneficial.[13] Traditional food-processing methods from developing countries (Brown and Pariser 1975) are not mechanized in part because of the lack of technical skills (Aworh 2008). Large food processing industries are concentrated in developed countries (Lahidji et al. 1998). Developing countries tend to export raw commodities, rather than added-value goods. However, the trade patterns may be chang-

[12] The chemical principles behind human nutrition is credited to Lavoisier in 1780s but modern nutrition began with the discovery of vitamins just over 100 years ago. Nutrition is associated with the dietetics. By contrast, food technology and foods science are part of the manufacturing sector.

[13] Currently, cconversations at the intersection of "food technology" and "international development" are not technical in nature and is categorized under "Agro-business" and "development". Exceptionally, Professor Aworh (Ibadan University) provides technical perspectives with his articles dealing with the need for mechanization/industrialization of post-harvest methods used for traditional food processing in order to promote development. Consideration of post-harvest losses led to the cachet material, filled under Agro-Industry and development, which emphasises economic, and policy issues.

ing, because developing countries are recognizing the need to increase their capacity for food processing in order to promote “export led growth”. Foreign direct investments and joint ventures are additional routes for enhancing food-processing capacity (Wilkinson 2004).

Technical requirements for industrializing traditional forms of processing in developed countries needs more attention (Aylward 1961). Past *technical* efforts to increase food security in developing countries focussed on agriculture production. Prevention of postharvest losses using new technology received limited attention (Affognon et al. 2015; Kitinoja et al. 2011). The possible impact of “introducing” food technology into the third world was discussed in relation to, food sovereignty, food resilience and food security (Edwards 2012).

The pivotal role of the Agro-Industries in developing countries was discussed in a high level conference “first global conference in agro-industries” (New Delhi (India; 8–11th April) organized by the Food And Agriculture Organization Of The United Nations (FAO), United Nations Industrial Development Organization (UNIDO), and the International Fund For Agricultural Development (IFAD) (FAO/UNIDO/IFAD 2008). The proceedings of the New Delhi meeting titled “Agro-Industry for development”, showed the discussions emphasised economic and policy issues (da Silva et al. 2009).

It helps to promote trans-disciplinary conversations, by recalling the definition of Agro-Industry (agribusiness, agro-processing) sector as “*businesses that process raw materials from agriculture, fisheries and forestry*”. The products of Agro-Industry include food, beverage, textiles, wood products such as, paper, and rubber products. Such manufactured products have added-value compared to farm-gate inputs, from which they derive. Interestingly, food and beverage processing is the most important (~50%) of the different agro-industries for high-income countries (Wilkinson 2004). Currently, the developing countries contribute to about 20% of the world trade in processed foods. There is general agreement that agro-industries can help with job creation in the developing world, particularly for the rural poor.

The Food and Agricultural Organization (FAO) of the United Nations produces free technical guides covering, food technology, processing, preservation, and small enterprises (FAO/WHO 2015). These and other initiatives are needed to promote a rapid diffusion of food processing technology to the developing countries in order to foster economic growth and improved food security (Haddad and Martorell 2002). So far, the refrigeration industry has highlighted its role in international development (International Institute of Refrigeration 2011). Perhaps one day, other food technology manufacturers will may also take on an international development remit.

4.6 Food Security

The 1996 FAO World summit declaration stated that food security exits. “when all people, at all times, have physical and economic access to sufficient safe and nutritious food that meets their dietary needs and food preferences for an active and healthy life”. According to common interpretations, food security has several main dimensions related to food availability, costs, preparation as well as temporal stability of these issues (Food and Agriculture Organization of the United Nations 2006) (Table 1.3).

Food security can be imperiled by a growing world population and climate change. The world’s population is projected to reach, 9 billion by 2050 resulting in an increased demand for food by 30–40%. It is further predicted that the world’s capacity for food production could decline owing to climate change (Cole et al. 2018). Under such circumstances, improvements of food security will require new ways to increase agricultural production, reduce food waste and enhance food

Table 1.3 The dimensions of food security (FAO)

1. AVAILABILITY – Supply issues, e.g. food production, stock levels and net trade
2. ACCESS – incomes, expenditure, markets and prices in achieving
3. UTILIZATION – Preparation method, biological uses
4. STABILITY – stability of dimension 1–3

preservation. A more industrialized food system may also help to increase the volume of food production. One lesson from developed countries is that, food technology can enhance availability, increase affordability, help to avoid seasonality, and promote greater consumer choice (Augustin et al. 2016; Cole et al. 2018; Weaver et al. 2014).

5 Food Science Education Outcomes

Upon completing of a program of study successfully, students will demonstrate skills, and qualities not possessed prior to their educational experience. Program outcomes for food science and food technology courses have been evolving for many decades (Hartel 2002; Iwaoka et al. 1966). Some of the characteristics of IFT certified food science and food technology programs at Universities and colleges are described below (Sects. 5.1, 5.2 and 5.3). Food and nutrition teaching at high schools up to K12 is covered alongside of consumer food literacy in Chap. 6.

5.1 IFT Certified Programs

Virtually all food science and food technology programs currently offered within the United States undergo review by the IFT (Hartel 2002).[14] Certification is organized through the higher education review board (HERB), which ensures that courses are adequately resourced. In addition, IFT certification ensures that essential learning outcomes, content, and learning activities, are consistent with pre-agreed standards (Table 1.4). A number of programs outside of the US have also achieved IFT certification. Aside from the formal certification route, food science and technology programs around the globe, including countries such as Nepal (Gartaula and Adhikari 2014, 2194), are adopting IFT educational standards. Guidelines for IFT certification are updated online periodically (Institute of Food technologists 2019).

Table 1.4 Areas of IFT certification (2018/2019)

Program requirements: Administrative and physical organization
Food science facilities
Undergraduate teaching faculty (teaching staff)
Program requirements: Curricular and educational
Foundational course descriptions
IFT program goals, standards, and essential learning outcomes (ELOS)

Source: Adapted from Institute of Food Technologists (2019)

Food science and food technology are normally based on a foundation of science, technology, engineering and mathematics (STEM) subject areas. Student intakes can be highly variable. However, by the end of their first year most students are expected to acquire a working familiarity with the natural sciences subjects such as, biology, chemistry, mathematics and physics. The preferred designation is "Food Science" (singular) as opposed to "food sciences.[15] The collection of STEM topics have each their own unique prestige and ethos (Schmidt et al. 2012).[16] However, the mixing the different disciplines together produces synergy so that food graduates exemplify excellent problem solving ability. There has always been opportunities also for students to take "minor" subjects alongside of food science and food technology; going on internships, study abroad, and doing self-directed research are also standard practices these days.

The training of food scientists involves multiple subject areas because real life problems seldom appear under neat little labels. Another advantage of multidisciplinary training is that thinking outside of the box comes naturally. Considering the structure of food science courses the whole is greater than the sum of its parts. Many areas previously grouped under the term "success skills" are

[14] There are approximately 220 food related Bachelor degree courses offered by colleges and universities in the United States. The vast majority are: full time BSc courses (142/221), of 4 years duration (166/221). Pre-Bachelor course were generally less than 2 years duration (21/221).

[15] However, one frequently encounters the term food "sciences", in connection with hybrid courses e.g. nutrition and food sciences.

[16] The acronym STEM is short for science, technology, engineering and mathematics. Nowadays, food science courses are also likely to offer opportunities for students to undertake modules in the social sciences, including topics from business, hospitality and tourism, palaeontology, and cultural studies.

now included in food science and food technology programs. There is widespread acknowledgement that written and oral communication skills are vital in the modern world. The professional skills includes attributes such was working effectively in groups (Bull and Clausen 2000). Developing leadership and professional skills also means improving ones awareness of diversity and equal rights including the need for reasonable adjustment for those that exhibit disability. A list of topics and studied as part of food science programs is shown in the Table below (Table 1.5).

Table 1.5 Food Science core disciplines

1. *Food chemistry (FC)*: The structure and properties of food components (water, carbohydrates, protein, lipids, other components and food additives); the chemistry of changes occurring during processing, storage, and utilization. [FC1–8]
2. *Food microbiology (FM)*: Microorganisms in food including beneficial, pathogenic, and spoilage; the influence of the food system on their growth, survival, and control. [FM1–6]
3. *Food safety (FS)*: Hazards (physical, chemical, biological) associated with foods and the food system; their transmission and control. [FS1–6][a]
4. *Food engineering and processing (FE)*: Food engineering principles; food preservation and processing; packaging materials and methods; cleaning and sanitation; water and waste management. [FE1–9]
5. *Sensory science (SS)*: Analytical and affective methods of assessing sensory properties of food. [SS1–3][a]
6. *Quality assurance (QA)*: Principles of food quality control and assurance. [QA1–4]
7. *Food laws and regulations (FL)*: Government regulations required for the manufacture and sale of food products. [FL1–4][a]
8. *Data and Statistical Analysis (DS)*: Collection, analysis, interpretation, and presentation of data. [DS1–3][a]
9. Critical thinking and problem solving (CT): Scientific reasoning through uncertainty in scientific and technical situations [CT1–5][a]
10. *Food Science Communication (CM)*: Oral and written communication. [CM1–3]
11. *Professionalism and leadership (PL)*: Organization and project management; skills necessary to work and interact with individuals from diverse backgrounds. [PL1–4][a]

Sources: Adapted from Institute of Food Technologists (2019)

[a]The number of essential learning outcome for each topic is shown in squared brackets [ ]. See Appendix 2

5.2 Characteristics of Food Science Programs

In addition to the natural science foundation topics, food science curricula also include more applied subjects such as, food biochemistry, food chemistry, food engineering, food microbiology, rheology, labeling, food packaging. A number of overarching themes are covered in food science programs including, Sensory characteristics, nutritional quality of foods, food safety management, and shelf life science or food preservation (Hartel 2002).

Modern food science curricular are evolving in the light of theories about how to promote adult learning. In ideal circumstances, instructional methods are diverse in order to cater for differences in learning styles (Table 1.6). Some students like more information whereas others learn best through a commonsense approach involving hands-on approach. All food science programs acknowledge the importance of experiential learning, and therefore include hands-on learning within "real-life" like situations. It is expected that courses will involve laboratory classes, some experiences using semi-pilot plant activity to demonstrate key principles related to topics such as, food dehydration, homogenization, packaging, and sealing foods. Alternatively, many courses will employ a teaching kitchen for small scale processing and a variety of cooking and foodservice scenarios.

Some food science programs incorporate business and marketing of food into the curriculum. This blending of business acumen and scientific training is indeed an interesting

Table 1.6 Learning and teaching activities in food science programs

1. Formal lectures
2. Group learning
3. Laboratory based classes
4. Pilot plant, experiential learning
5. Seminars and presentations
6. Problem based learning
7. Tutorials
8. Internship, work-based learning
9. Reflective portfolios
10. Journal club
11. Student lead research

Table 1.7 Employment data for agricultural and food scientists, 2018–28

Occupational title	SOC code	Employment, 2018	Projected employment, 2028	Change, 2018–28	
				Percent	Numeric
Agricultural and food scientists	19–1010	35,600	38,000	7	2300
Animal scientists	19–1011	2700	2800	7	200
Food scientists and technologists	19–1012	14,900	15,700	5	800
Soil and plant scientists	19–1013	18,000	19,400	8	1400

Adapted from Bureau of Labor Statistics U.S. Department of Labor (2020)

combination (see below). The food scientist must be a well-rounded applied scientist, able to work within many areas to solve the problems that develop in this ever-changing, exciting field.

5.3 Other Food Related Programs

The food science profession contains elements of "molecular gastronomy" though this term was not in use until recently. The field of culinology or culinary technology also combines elements of foods science with culinary arts with a focus on the art of cooking pertaining to specific cultures around the globe, regional cooking practices, ingredients choices, religious food laws and observance, ethnic foods, and food choice. However, graduate food scientists are no more likely to demonstrate culinary skills as compared to other graduates.[17]

Food science must be differentiated from food studies, which refers to the "*critical examination of food and its contexts within science, art, history, society, and other fields*". There is a definite humanities perspective in food studies including anthropological, cultural, and economic approaches to foods. Food social sciences are a thriving area for those wishing to focus on food policy. There are courses available leading to Undergraduate and postgraduate qualifications in food studies (Department of Nutrition and Food Studies (NYU) 2015).

[17]Joint food science programs are increasingly popular leading to degree certifications in, Food science & Business, Food and Nutrition, Food Science & Culinary art, Food Science & Hospitality

5.4 Careers in Food Science

Preparations for a career in food science require that a student complete relevant courses at high school and a Bachelor degree at a college or university. A few 2-year technical colleges offer Associate Degree programs, with graduates of these programs choosing to either obtain employment or to transfer their credits to universities or colleges to earn a Bachelor degree. Over 50 universities in the United States offer degrees in food science some through to the PhD level (USDA National Agricultural Library 2020).

Food scientists and food technologies are described as, persons who use "chemistry, microbiology, engineering, and other sciences to study the principles underlying the processing and deterioration of foods; analyze food content to determine levels of vitamins, fat, sugar, and protein; discover new food sources; research ways to make processed foods safe, palatable, and healthful; and apply food science knowledge to determine best ways to process, package, preserve, store, and distribute food" (U.S. Bureau of Labor Statistics 2014). Recently, the bureau of labor rebranded the food scientists more as "Agricultural and Food Scientists", which highlights the Agri-food system.

Annually there may be about 52,000 job openings for graduates in food and allied subjects (Hartel and Klawitter 2010). For 2028, it was estimated that there would be 38,000 job openings for agricultural and food scientists not including jobs in areas such as food retail (Bureau of Labor Statistics U.S. Department of Labor, 2020) (Table 1.7).

Appendices

A 1a: Sources of Food Science and Nutrition Information

Ask A Questior

Home **Topics** **Publications** **Collections** **Data** **Services** **About Us**

Food and Nutrition Information Center

https://www.nal.usda.gov/fnic

- DIETARY GUIDANCE
- LIFECYCLE NUTRITION
- DIET AND HEALTH
- SURVEYS, REPORTS AND RESEARCH
- FOOD COMPOSITION
- **PROFESSIONAL AND CAREER RESOURCES**
- FOOD SAFETY
- DIETARY SUPPLEMENTS
- FOOD LABELING
- NUTRITION ASSISTANCE PROGRAMS

▲Professional and Career Resources

- Resources
- **Academic Programs and Other Educational Opportunities**
- Art and Images
- Associations and Foundations
- Calendars and Conferences
- Dietetic Associations
- Ethnic and Cultural Resources
- Food Dictionaries and Encyclopedias
- Food Science and Technology
- Food Service
- Government Grant Information
- Journals and Newsletters
- Legislation and Policy
- Listservs, Blogs, and Mobile Apps
- Nutrition Education
- Online Learning
- Sources of Free or Low-Cost Materials
- Food Safety
- Dietary Supplements
- Food Labeling
- Nutrition Assistance Programs
- FNIC Frequently Asked Questions
- Quick Links for Educators, Health Professionals and Researchers

Source: Adapted from: USDA National Agricultural Library (2020). Food and Nutrition Information Center. Retrieved Jan, 2020, from https://www.nal.usda.gov/fnic

A 1b: Websites

Local extension websites
See: Land-Grant University website directory
https://nifa.usda.gov/land-grant-colleges-and-universities-partner-website-directory?state=All&type=Extension
USDA National Agricultural library;
https://www.nal.usda.gov/fnic
https://impact.extension.org/search/

A 2a: IFT Standards (Competencies) and Leaning Outcomes for Food Science and Food Technology Programs

Food chemistry (FC): *The structure and properties of food components (water, carbohydrates, protein, lipids, other components and food additives); the chemistry of changes occurring during processing, storage, and utilization*
Learning outcomes: Upon completion students will be able to:
FC.1. Discuss the major chemical reactions that limit shelf life of foods.
FC.2. Explain the chemistry underlying the properties and reactions of various food components.
FC.3. Apply food chemistry principles used to control reactions in foods.
FC.4. Demonstrate laboratory techniques common to basic and applied food chemistry.
FC.5. Demonstrate practical proficiency in a food analysis laboratory.
FC.6. Explain the principles behind analytical techniques associated with food.
FC.7. Evaluate the appropriate analytical technique when presented with a practical problem.
FC.8. Design an appropriate analytical approach to solve a practical problem
Food microbiology (FM): *Microorganisms in food including beneficial, pathogenic, and spoilage; the influence of the food system on their growth, survival, and control.*
Learning outcomes: Upon completion students will be able to:
FM.1. Identify relevant beneficial, pathogenic, and spoilage microorganisms in foods and the conditions under which they grow.
FM.2. Describe the conditions under which relevant pathogens are destroyed or controlled in foods.
FM.3. Apply laboratory techniques to identify microorganisms in foods.
FM.4. Explain the principles involved in food preservation via fermentation processes.
FM.5. Discuss the role and significance of adaptation and environmental factors (e.g., water activity, pH, temperature) on growth response and inactivation of microorganisms in various environments.
FM.6. Choose relevant laboratory techniques to identify microorganisms in foods.
Food safety (FS): *Hazards (physical, chemical, biological) associated with foods and the food system; their transmission and control.*
Learning outcomes: Upon completion students will be able to:
FS.1. Identify potential hazards and food safety issues in specific foods.
FS.2. Describe routes of physical, chemical, and biological contamination of foods.
FS.3. Discuss methods for controlling physical, chemical and biological hazards.
FS.4. Evaluate the conditions, including sanitation practices, under which relevant pathogenic microorganisms are commonly controlled in foods.
FS.5. Select appropriate environmental sampling techniques.
FS.6. Design a food safety plan for the manufacture of a specific food.
Food engineering and processing (FE): *Food engineering principles; food preservation and processing; packaging materials and methods; cleaning and sanitation; water and waste management.*
Learning outcomes: Upon completion students will be able to:
FE.1. Define principles of food engineering (mass and heat transfer, fluid flow, thermodynamics).
FE.2. Formulate mass and energy balances for a given food manufacturing process.
FE.3. Explain the source and variability of raw food materials and their impact on food processing operations.
FE.4. Design processing methods that make safe, high-quality foods.
FE.5. Use unit operations to produce a given food product in a laboratory or pilot plant.
FE.6. Explain the effects of preservation and processing methods on product quality.
FE.7. List properties and uses of various packaging materials and methods.
FE.8. Describe principles and practices of cleaning and sanitation in food processing facilities.
FE.9. Define principles and methods of water and waste management.

A 2b: IFT Standards (Competencies) and Leaning Outcomes

Sensory science (SS): *Analytical and affective methods of assessing sensory properties of food.*

Learning outcomes: Upon completion students will be able to:

SS.1. Discuss the physiological and psychological basis for sensory evaluation.

SS.2. Apply experimental designs and statistical methods to sensory studies.

SS.3. Select sensory methodologies to solve specific problems in food.

Quality assurance (QA): *Principles of food quality control and assurance.*

Learning outcomes: Upon completion students will be able to:

QA.1. Define food quality and food safety terms.

QA.2. Apply principles of quality assurance and control.

QA.3. Develop standards and specifications for a given food product.

QA.4. Evaluate food quality assessment systems (e.g. statistical process control).

Food laws and regulations (FL): *Government regulations required for the manufacture and sale of food products*

Learning outcomes: Upon completion students will be able to:

FL.1. Recall government regulatory frameworks required for the manufacture and sale of food products.

FL.2. Describe the processes involved in formulating food policy.

FL.3. Locate sources of food laws and regulations.

FL.4. Examine issues related to food laws and regulations.

Data and statistical analysis (DS): *Collection, analysis, interpretation, and presentation of data.*

Learning outcomes: Upon completion the students will be able to:

DS.1. Use statistical principles in food science applications.

DS.2. Employ appropriate data collection and analysis technologies.

DS.3. Construct visual representation of data.

Critical thinking and problem solving (CT): *Scientific reasoning through uncertainty in scientific and technical situations*

Learning outcomes: Upon completion and additional activities provided from the program, students will be able to:

CT.1. Locate evidence-based scientific information resources.

CT.2. Apply critical thinking skills to solve problems.

CT.3. Apply principles of food science in practical, real-world situations and problems.

CT.4. Select appropriate analytical techniques when presented with a practical problem.

CT.5. Evaluate scientific information.

Food science communication (CM): *Oral and written communication.*

Learning outcomes: Upon completion and additional activities provided from the program, students will be able to:

CM.1. Write relevant technical documents.

CM.2. Create oral presentations.

CM.3. Assemble food science information for a variety of audiences

Professionalism and leadership (PL): *Organization and project management; skills necessary to work and interact with individuals from diverse backgrounds.*

Learning outcomes: Upon completion and additional and leadership activities provided from the program, students will be able to:

PL.1. Demonstrate the ability to work independently and in teams.

PL.2. Discriminate tasks to achieve a given outcome.

PL.3. Describe social and cultural competence relative to diversity and inclusion.

PL.4. Discuss examples of ethical issues in food science.

Sources: Adapted from Institute of Food Technologists (2019). Essential learning outcomes are expressed as verbs based on Blooms taxonomy

A 3: Careers in Food Science

Advertising and communications	Plant supervisor
Analytical lab (manager)	Process researcher
Biochemist	Product and process developer
Cereal scientist	Product development
Consultant	Product development
Dairy products scientist	Project leader, technology
Environmental health inspector (food)	Project/product manager
FDA/USDA research scientist	Public health
Flavor chemist	Quality assurance
Flavor researcher	Quality assurance specialist
Food and drug inspector	Quality control supervisor
Food biochemist	Research and development
Food biotechnologist	Retail
Food chemist	Sales and marketing specialist
Food engineer	Sensory science
Food Industry R&D	Teacher (K12 schools)
Food ingredient sales	Teacher/academic (universities and colleges)
Food inspector	
Food marketing and sales	
Food microbiologist	
Food product consultant	
Food product development	
Food production manager	
Food quality and microbiology	
Food safety inspector	
Food safety/HACCP	
Food scientist	
Food technologist	
General manager, research	
Laboratory analyst	
Laboratory director	
Manager, meat applications	
Market researcher	
Meat scientist (research and development)	
Natural products researcher	
New technologies	
Packaging specialist (food)	
Plant manager	
Plant quality manager	

Sources: compiled from various US food science programme websites, with major contributions from, Angelo State University, Washing State University, Purdue University, Idaho State University, Penn State University and cf. Hartel and Klawitter, 2010

A 4: A Sample of Some Food Science Text Books

Borgstrom, G. (1968). Principles of food science.
1. Food technology and biochemistry.

Borgstrom, G. (1968). Principles of food science.
2. Food microbiology and biochemistry.

Pyke, M. (1968). Food science and technology.

Charley, H. (1970). Food science.

Pyke, M. (1970). Food science and technology.

Hall, C. W., A. W. Farrall, et al. (1971). Encyclopedia of food technology and food science series.

Potter, N. N. (1973). Food science.

Stewart, G. F. and A. A. Amerine (1973). Introduction to food science and technology.

Johnson, A. H. and M. S. Peterson (1974). Encyclopedia of food technology and food science series. Vol. 2. Encyclopedia of food technology.

Nickerson, J. T. R. and L. J. Ronsivalli (1976). Elementary food science.

Fox, B. A. and A. G. Cameron (1977). Food science – a chemical approach.

Birch, G. G., A. G. Cameron, et al. (1977). Food science.

Gaman, P. M. and K. B. Sherrington (1977). The science of food. An introduction to food science, nutrition and microbiology.

Potter, N. N. (1978). Food science.

Gaman, P. M. and K. B. Sherrington (1981). The science of food. An introduction to food science, nutrition and microbiology.

Charley, H. (1982). Food science.

Fox, B. A. and A. G. Cameron (1982). Food science – a chemical approach.

Potter, N. N. (1986). Food science.

Adrian, J., N. Adrian, et al. (1990). Dictionary of food science and industry.

Hui, Y. H. (1992). Encyclopedia of food science and technology.

Ronsivalli, L. J. and E. R. Vieira (1992). Elementary food science.

Charalambous, G. (1992). Food science and human nutrition.

Macrae, R., R. K. Robinson, et al. (1993). Encyclopaedia of food science, food technology and nutrition.

Potter, N. N. and J. H. Hotchkiss (1995). Food science.

Isaacs, N. S. (1998). High pressure food science, bioscience and chemistry.

Francis, F. J. (1999). Wiley encyclopedia of food science and technology.

Gutierrez-Lopez, G. F. and G. V. Barbosa-Canovas (2003). Food science and food biotechnology.

Chi-Fai, C. and K. Hoi-Shan (2005). A glossary of food science and technology.

Adrian, J. and N. Adrian (2006). Dictionary of food science and industry. (Dictionnaire agroalimentaire.).

Bateman, H., H. Sargeant, et al. (2006). Dictionary of food science and nutrition.

Hui, Y. H. (2006). Handbook of food science, technology, and engineering, 4 volume set.

Sunetra, R. (2007). Food science and nutrition.

Evans, J. A. (2008). Frozen food science and technology.

Hartel, R. W. and C. P. Klawitter (2008). Careers in food science. From undergraduate to professional.

Gallagher, E. (2009). Gluten-free food science and technology.

Campbell-Platt, G. (2009). Food science and technology.

References

Accum FC (1820) A treatise on adulterations of food and culinary poisons exhibiting the fraudulent sophistications of bread, beer, wine, spiritous liquors, tea, coffee, cream, confectionery, vinegar, mustard, pepper, cheese, olive oil, pickles. Retrieved from https://books.google.com.gh/books?id=YWAUAAAAQAAJ&dq=accum%20food%20adulteration&pg=PP1#v=onepage&q=accum%20food%20adulteration&f=false

Adams R, Moss O (2008) Food microbiology. RSC Publishing, 463 pp

Affognon H, Mutungi C, Sanginga P, Borgemeister C (2015) Unpacking postharvest losses in sub-Saharan Africa: a meta-analysis. World Dev 66:49–68

Andress EL, Kuhn GD (1998) Critical review of home preservation literature and current research. II. Early history of USDA home canning recommendations. Retrieved from https://nchfp.uga.edu/publications/usda/review/earlyhis.htm

Anonymous (1912) The romance of refrigeration. J Public Health 2(11):840–848. Retrieved from http://www.ncbi.nlm.nih.gov/pmc/articles/PMC1089468/

Anonymous (1970) Aseptic canning sells products. Low-volume items retain shelf life, flavour. Am Dairy Rev 32(8):42

Anonymous (2019) Bernard L Oser. Retrieved from https://www.sourcewatch.org/index.php/Bernard_L_Oser

Appert N (1810) L'art de conserver pendant plusieurs annÃ©es toutes les substances animales et vegetales. Retrieved from http://books.google.co.uk/books?id=3eF8N7gmcP4C

Augustin MA, Riley M, Stockmann R, Bennett L, Kahl A, Lockett T, Osmond M, Sanguansri P, Stonehouse W, Zajac I, Cobiac L (2016) Role of food processing in food and nutrition security. Trends Food Sci Technol 56:115–125

Aworh OC (2008) The role of traditional food processing technologies in national development: the West African experience. In: Robertson GL, Lupien JR (eds) Using food science and technology to improve nutrition and promote national development. International Union of Food Science & Technology, pp 31–49

Aylward F (1961) The place of food science and technology in the campaign against malnutrition. Some African problems. Proc Nutr Soc 20:112–117. https://doi.org/10.1079/pns19610031

Bantang SC (2008) A cook's tour: the journey to become a model culinary arts academy. Techniques: Connecting Education and Careers (J1) 83(3):14–19. https://eric.ed.gov/?id=EJ788422

Bissett RL, Cheng MSH, Brannan RG (2010) A quantitative assessment of the Research Chefs Association core competencies for the practicing culinologist. J Food Sci Educ 9(1):11–18. https://doi.org/10.1111/j.1541-4329.2009.00086.x

Brink GC (1964) How pasteurization of milk came to Ontario. Can Med Assoc J 91:972–972. Retrieved from http://www.ncbi.nlm.nih.gov/pmc/articles/PMC1927762/

Brown NL, Pariser ER (1975) Food science in developing countries. Science 188(4188):589–593. https://doi.org/10.1126/science.188.4188.589

Brownell, K. D., & Warner, K. E. (2009, March). The perils of ignoring history: big tobacco played dirty and millions died. How similar is big food? *Milbank Q, 87*(1): 259–294. Retrieved from http://www.ncbi.nlm.nih.gov/pmc/articles/PMC2879177/

Bryce PH (1912) The conservation of food products by refrigeration. Am J Public Health 2(5):325–328. Retrieved from http://www.ncbi.nlm.nih.gov/pubmed/18008662

Bull NH, Clausen JC (2000) Structured group learning in undergraduate and graduate courses. J Nat Resour Life Sci Educ 29:46–50

Bureau of Labor Statistics U.S. Department of Labor (2020) Occupational outlook handbook. Agricultural and Food Scientists: Job outlook. Retrieved from https://www.bls.gov/ooh/life-physical-and-social-science/agricultural-and-food-scientists.htm#tab-6

Buttriss JL (2013) Food reformulation: the challenges to the food industry. Proc Nutr Soc 72(1):61–69. https://doi.org/10.1017/S0029665112002868

Can Manufacturers Institute (2021) History of the can: an interactive timeline. Retrieved from https://www.can-central.com/can-stats/history-of-the-can

Canter DD, Etta MM, Boyce J (2007) The growing importance of food and culinary knowledge and skills in dietetics practice. Top Clin Nutr 22(4):313–322. https://krex.k-state.edu/dspace/bitstream/handle/2097/2604/CanterTCN2007.pdf?sequence=2091

Centers for Disease Control and Prevention (2014) Estimates of foodborne illness in the United States. Retrieved from http://www.cdc.gov/foodborneburden/index.html

Chandrasekaran S, Ramanathan S, Basak T (2013) Microwave food processing – a review. Food Res Int 52(1):243–261

Cheng M, Bosselman R (2016) An evaluation of the research chefs association's bachelor of science in culinology® core competencies. J Hosp Tour Educ 28(3):127–141. https://www.researchgate.net/profile/Michael_Cheng123/publication/305460731_An_Evaluation_of_the_Research_Chefs_Association

Cheng M, Ogbeide GCA, Hamouz FL (2011) The development of culinary arts and food science into a new academic Discipline—Culinology®. J Culin Sci Technol 9(1):17–26. https://www.smsu.edu/resources/webspaces/academics/programs/culinology/News/ChengOgbeideHamouz.pdf

Clausi A, Powers J, Francis K (2000) A century of food science. Retrieved from https://studylib.net/doc/8738024/a-century-of-food-science%2D%2D-institute-of-food-technologists

Codex Alimentarius Commission (1997) Definitions for the purposes of the codex alimentarius. Procedural manual (10th edn). Retrieved from http://www.fao.org/docrep/w5975e/w5975e00.htm#Contents

Cole MB, Augustin MA, Robertson MJ, Manners JM (2018) The science of food security. NPJ Sci Food 2:14. https://doi.org/10.1038/s41538-018-0021-9

Contini C et al (2020) The novel zoonotic COVID-19 pandemic: an expected global health concern. J Infect Dev Ctries 14(03):254–264

Cooper MJ, Mezzabotta L, Murphy J (2016) Food and culinary knowledge and skills: perceptions of undergraduate dietetic students. Can J Diet Pract Res 78(1):42–44. https://dcjournal.ca/doi/abs/10.3148/cjdpr-2016-3026

Cowell ND (2007) More light on the dawn of canning. Food Technol 61(5):5. https://www.ift.org/news-and-publications/food-technology-magazine/issues/2007/may/features/more-light-on-the-dawn-of-canning

Cutler SJ, Fooks AR, van der Poel WH (2010) Public health threat of new, reemerging, and neglected zoonoses in the industrialized world. Emerg Infect Dis 16(1):1–7. https://doi.org/10.3201/eid1601.081467

da Silva CA, Baker D, Shepherd AW, Jenane C, Miranda-da-Cruz S (2009) Agro-industries for development. FAO/IFAD, Cambridge, MA, 290pp. Retrieved from http://www.fao.org/family-farming/detail/en/c/292290/

Davey RC (2004) The obesity epidemic: too much food for thought? Br J Sports Med 38(3):360–363. discussion 363

Department of Nutrition and Food Studies (NYU) (2015) Food studies. Retrieved from http://steinhardt.nyu.edu/nutrition/food/ma/

Dowd MT, Jameson JD (1918) Food, its composition and preparation a textbook for classes in household science. Retrieved from https://archive.org/details/cu31924003551474

Dwyer JT, Fulgoni VL 3rd, Clemens RA, Schmidt DB, Freedman MR (2012) Is "processed" a four-letter word? The role of processed foods in achieving dietary guidelines and nutrient recommendations. Adv Nutr 3(4):536–548

Edwards L (2012) Food fight: the international assessment of agricultural knowledge, science, and technology for development. J Agrobiotechnol Manag Econ 15(1):70–76

FAO/UNIDO/IFAD (2008) Report of the global agro-industries forum improving competitiveness and development impact, New Delhi, India, 8–11 April 2008. Retrieved from http://www.fao.org/3/i0354e/i0354e00.htm

FAO/WHO (2015) Publications. Retrieved from http://www.fao.org/publications/en/

Farkas DF (1982) National research council advisory board on military personnel supplies national research council, planning committee on the workshop on dehydration compression of foods national research council, commission on engineering technical systems. Dehydration and compression of foods. Retrieved from https://books.google.co.uk/books?id=P0UrAAAAYAAJ

Fellers C, Prescott SC, Loomis HM, Geidel CD, Darling CA, Thom C (1928) Canned foods and the public health. Am J Public Health 18(7):893–896. https://ajph.aphapublications.org/doi/pdf/810.2105/AJPH.2118.2107.2893

Ferruzzi MG, Peterson DG, Singh RP, Schwartz SJ, Freedman MR (2012) Nutritional translation blended with food science: 21st century applications. Adv Nutr 3(6):813–819. https://doi.org/10.3945/an.112.003202

Floros JD, Newsome R, Fisher W, Barbosa-Canovas GV, Hongda C, Dunne CP, German JB, Hall RL, Heldman DR, Karwe MV, Knabel SJ, Labuza TP, Lund DB, Newell-Mcgloughlin M, Robinson JL, Sebranek JG, Shewfelt RL, Tracy WF, Weaver CM, Ziegler GR (2010) Feeding the world today and tomorrow: the importance of food science and technology. Compr Rev Food Sci Food Saf 9(5):572–599. Retrieved from http://www.ift.org/knowledge-center/read-ift-publications/science-reports/~/media/Knowledge%20Center/Science%20Reports/IFTScientificReview_feedingtheworld.pdf. https://onlinelibrary.wiley.com/doi/pdf/10.1111/j.1541-4337.2010.00127.x

Food and Agriculture Organization of the United Nations (2006) Policy brief: food security. Retrieved from http://www.fao.org/forestry/13128-0e6f36f27e0091055bec28ebe830f46b3.pdf

Frey CB, Osborne MA (2017) The future of employment: how susceptible are jobs to computerization? Technol Forecast Soc Chang 114:254–280

Gartaula G, Adhikari BM (2014) Challenges and prospects of food science and technology education: Nepal's perspective. Food Sci Nutr 2(6):623–627. https://doi.org/10.1002/fsn3.173

Geoghegan T (2013) The story of how the tin can nearly wasn't. Retrieved from http://www.bbc.co.uk/news/magazine-21689069

Giovenzana V, Beghi R, Civelli R, Guidetti R (2015) Optical techniques for rapid quality monitoring along minimally processed fruit and vegetable chain. Trends Food Sci Technol 46:2. https://doi.org/10.1016/j.tifs.2015.10.006

Gould GW (2000) Preservation: past, present and future. Br Med Bull 56(1):84–96

Graham JC (1981) The French connection in the early history of canning. J R Soc Med 74(5):374–381

Haddad L, Martorell R (2002) Feeding the world in the coming decades requires improvements in investment, technology and institutions. J Nutr 132(11):3435S–3436S. https://doi.org/10.1093/jn/132.11.3435S

Hartel RW (2002) Core competencies in food science: background information on the development of the IFT education standards. J Food Sci Educ 1(1):3–5

Hartel RW, Klawitter CP (2010) Careers in food science. From undergraduate to professional. Springer, 328 pp

Horrocks SM (2003) Food processing industry: technological change. In: Mokyr J (ed) The Oxford encyclopeodia of economic history, vol 3. Oxford University Press, pp 340–346

Institute of Food Technologists (2000) History of food science. Retrieved from http://www.ift.org/knowledge-center/learn-about-food-science/what-is-food-science/

Institute of Food Technologists (2019) IFT undergraduate program approval. Retrieved from http://www.ift.org/community/students/approved-undergrad-programs.aspx

Institute of Food Technologists (IFT) (2014) Approved undergraduate programs. Retrieved from http://www.ift.org/community/students/approved-undergrad-programs.aspx

Institute of Food Technologists (IFT) (2015) Bernard L. Oser food ingredient safety award. Retrieved from http://www.ift.org/Membership/Awards-and-Recognition/Achievement-Awards/Bernard-L-Oser-Food-Ingredient-Safety-Award.aspx

International Food Information Service (2005) Food science information discovery and dissemination. Retrieved from https://books.google.co.uk/books?id=7vifzZkFgoAC

International Institute of Refrigeration (2011) The cold chain, food security and economic development. Retrieved from http://www.iifiir.org/userfiles/file/publications/statements/Statement_2010_Cold_Chain.pdf

Iowa State University (2021) Culinary food science. Retrieved from https://www.cals.iastate.edu/majors/culinary-food-science

Iwaoka WT, Britten P, Dong FM (1966) The changing face of food science education. Trends Food Sci Technol 7(4):105–112

Jay JM, Loessner MJ, Golden DA (2005) Modern food microbiology. Springer, 790 pp

Jekanowski MD, Binkley JK (2000) Food purchase diversity across US markets. Agribusiness 16(4):417–433

Kalla AM, Devaraju R (2017) Microwave energy and its application in food industry: a reveiw. Asian J Dairy Food Res 36(1):34–47. Retrieved from http://www.arccjournals.com/uploads/articles/8DR1135.pdf

Kim E (2013) Processed food: a 2-million-year history. Sci Am 309(Issue 3) Retrieved from http://www.scientificamerican.com/article/processed-food-a-two-million-year-history/?page=1

Kitinoja L, Saran S, Roy SK, Kader AA (2011) Postharvest technology for developing countries: challenges and opportunities in research, outreach and advocacy. J Sci Food Agric 9(4):597–603

Kohn RM (1971) Space-food – a new branch of food science. Lebensm Wiss Technol 4(2):35–37

Kramer A, Twigg BA (1970) Quality control for the food industry. AVI, Westport

Krebs-Smith SM, Reedy J, Bosire C (2010) Healthfulness of the U.S. food supply: little improvement despite decades of dietary guidance. Am J Prev Med 38(5):472–477

Lahidji R, Michalski W, Stevens B (1998) The future of food: an overview of trends and key issues. In: Thefuture of food: long-term prospects for the agro-food sector. OECD Publications, Paris, pp 7–20. Retrieved from https://www.oecd-ilibrary.org/economics/the-future-of-food_9789264162006-en

Lewis KD, Burton-Freeman BM (2010) The role of innovation and technology in meeting individual nutritional needs. J Nutr 140(2):426S–436S. https://doi.org/10.3945/jn.109.114710

Lupien JR, Lin DX (2004) Contemporary food technology and its impact on cuisine. Asia Pac J Clin Nutr 13(2):156–161. Retrieved from http://www.ncbi.nlm.nih.gov/pubmed/15228982

Marcus JB (2013) Culinary nutrition: the science and practice of healthy cooking. Elsevier Science/Academic Press, 600pp. https://books.google.com.gh/books?id=p2j3v6ImcX0C

McLachlan T (1975) History of food processing. Prog Food Nutr Sci 1(7/8):461–491

Moodie R, Stuckler D, Monteiro C, Sheron N, Neal B, Thamarangsi T, Lincoln P, Casswell S (2013) Profits and pandemics: prevention of harmful effects of tobacco, alcohol, and ultra-processed food and drink industries. Lancet 381(9867):670–679

Mozaffarian D (2016) Dietary and policy priorities for cardiovascular disease, diabetes, and obesity: a comprehensive review. Circulation 133(2):187–225. https://doi.org/10.1161/circulationaha.115.018585

Mozaffarian D, Rosenberg I, Uauy R (2018) History of modern nutrition science – implications for current research, dietary guidelines, and food policy. Br Med J 361:k2392. Retrieved from https://www.bmj.com/content/bmj/361/bmj.k2392.full.pdf

Neff R (2014) Introduction to the U.S. food system. Public health, environment, and equity. John Wiley & Sons, San Francisco, 576pp. https://books.google.co.uk/books?id=At_hBQAAQBAJ&lpg=PR1&pg=PR1#v=onepage&q&f=false

Nestle M, Jacobson MF (2000) Halting the obesity epidemic: a public health policy approach. Public Health Rep 115(1):12–24

Oberg CJ (2012) History of food microbiology. Retrieved from http://faculty.weber.edu/coberg/class/3853/3853%20HistoryofFood.htm

OECD/WHO (2003) Foodborne disease in OECD countries: present state and economic costs. OECD Publishing. https://doi.org/10.1787/9789264105386-en. Retrieved from http://www.keepeek.com/Digital-Asset-Management/oecd/agriculture-and-food/foodborne-disease-in-oecd-countries_9789264105386-en#page1

Ohio Department of Higher Education (2017) Request for Proposal (RFP): bachelor of science in culinary and food science. From https://www.ohiohighered.org/content/request_proposal_rfp_bachelor_science_culinary_and_food_science

Olver L (2014) The food timeline. Retrieved from http://www.foodtimeline.org/

Oregon State University (2019) Nutrition and foodservice systems: undergraduate options. College of public health and human sciences. Retrieved from https://health.oregonstate.edu/nutrition/nutrition-and-foodservice-systems

Partington JR (1989) A short history of chemistry. Dover Publications, 415 pp

Pearson GS (2016) The democratization of food: tin cans and the growth of the American food processing industry, 1810–1940. Thesis and Dissertation 2756, Lehigh University, PhD, 448pp. https://asa.lib.lehigh.edu/Record/10673470

Penney JW (2011) Internet access rights: a brief history and intellectual origins. William Mitchell Law Rev 38:10. Retrieved from https://open.mitchellhamline.edu/cgi/viewcontent.cgi?article=1440&context=wmlr

Pennington ME (1912) The hygienic and economic results of refrigeration in the conservation of poultry and eggs. Am J Public Health 2(11):840–848. Retrieved from http://www.ncbi.nlm.nih.gov/pubmed/18008742

Perkins WE (1964) Prevention of botulism by thermal processing. In: Lewis KH, Cassel K et al (eds) Botulism: proceedings of a symposium. U. S. Department of Health, Education, and Welfare, Public Health Service, pp 187–201. https://books.google.com.gh/books?id=gHVrAAAAMAAJ

Pooley S, Fa JE, Nasi R (2015) No conservation silver lining to Ebola. Conserv Biol 29(3):965–967. https://doi.org/10.1111/cobi.12454

Prescott SC, Underwood WL (1897) Contributions to our knowledge of micro-organisms and sterilizing processes in the canning industries. Science 6(152):800–802. https://www.jstor.org/stable/1623441

Ray B, Bhunia A (2013) Fundamental food microbiology, 5th edn. Taylor & Francis, 624 pp

Reber PB (2019) Historic cookbooks 1800-civil war. Retrieved from https://www.angelfire.com/md3/openhearthcooking/aaCookbook1800.html

Risch SJ (2009) Food packaging history and innovations. J Agric Food Chem 57(18):8089–8092

Robertson GL (2005) Food packaging: principles and practice, 2nd edn. Taylor & Francis/CRC Press, 568 pages

Rocourt J, Moy G, Vierk K, Schlundt J (2003) The present state of foodborne disease in OECD countries. World Health Organization, Geneva, 43pp. Retrieved from https://www.who.int/foodsafety/publications/foodborne_disease/oecd_fbd.pdf

Rose DJ (2014) Impact of whole grains on the gut microbiota: the next frontier for oats? Br J Nutr 112(Suppl. S2):S44–S49

Rowe S, Alexander N, Almeida N, Black R, Burns R, Bush L et al (2011) Food science challenge: translating the dietary guidelines for Americans to bring about real behavior change. J Food Sci 76(1):R29–R37. https://doi.org/10.1111/j.1750-3841.2010.01973.x

Saxon W (1995) Dr. Bernard L. Oser, biochemist expert in food safety, dies at 95. New York Times. Retrieved from http://www.nytimes.com/1995/01/22/obituaries/dr-bernard-l-oser-biochemist-expert-in-food-safety-dies-at-95.html

Schmidt SJ, Bohn DM, Rasmussen AJ, Sutherland EA (2012) Using food science demonstrations to engage students of all ages in science, technology, engineering, and mathematics (STEM). J Food Sci Educ 11(2):16–22

Selin H (1997) Encyclopaedia of the history of science, technology, and medicine in Non-Westen cultures. Retrieved from https://books.google.co.uk/books?id=raKRY3KQspsC

Shapin S (2008) The scientific revolution. Retrieved from https://books.google.co.uk/books?id=6BIr19MTXAMC

Sharma LL, Teret SP, Brownell KD (2010) The food industry and self-regulation: standards to promote success and to avoid public health failures. Am J Public Health 100(2):240–246

Skolnik H (1968) History, evolution, and status of agriculture and food science and technology. J Chem Doc 8(2):95–98. https://doi.org/10.1021/c160029a011

Sneed J, Strohbehn CH (2008) Trends impacting food safety in retail foodservice: implications for dietetics practice. J Am Diet Assoc 108(7):1170–1177. https://doi.org/10.1016/j.jada.2008.04.009

Snowdon W, Raj A, Reeve E, Guerrero R, Fesaitu J, Cateine K, Guignet C (2013) Processed foods available in the Pacific Islands. Glob Health 9(53):7pp. Retrieved from https://www.ncbi.nlm.nih.gov/pmc/articles/PMC4016479/

Temasek Polytechnic Singapore (2019) Diploma in food, nutrition & culinary science. Retrieved Feb 16, 2019, from https://www.tp.edu.sg/schools/asc/food-nutrition-and-culinary-science

Tittl M (2008) Careers that combine culinary and food science. In: Hartel RW, Klawitter CP (eds) Careers in food science: from undergraduate to professional. Springer, pp 267–275. https://d261wqtxts261xzle267.cloudfront.net/32077814/Careers_in_Food_Science_From_Undergraduate_to_Professional.pdf

Toops D (2017) How did the food industry get (from there) to here? In: Beckley JH, Herzog LJ, Foley MM (eds) Accelerating new food product design and development, 2nd edn. Wiley/The Institute of Food Technologists, 387 pp

U. S. Food and Drug Administration (2014) Federal Food, Drug, and Cosmetic Act (FD&C Act). Sec. 201. [21 U.S.C. 321] Chapter II – Definitions 1. Retrieved from http://www.fda.gov/regulatoryinformation/legislation/federalfooddrugandcosmeticactfdcact/fdcactchaptersiandiishorttitleanddefinitions/ucm086297.htm

U.S. Bureau of Labor Statistics (2014) 19–1012 Food Scientists and Technologists. Retrieved from http://www.bls.gov/oes/current/oes191012.htm

U.S. Department of Agriculture and U.S. Department of Health and Human Services (2010) Dietary guidelines for Americans, 2010, 7th edn Retrieved from http://www.cnpp.usda.gov/DietaryGuidelines

U.S. Food and Drug Administration (2013) Food code (2013) Retrieved from http://www.fda.gov/Food/GuidanceRegulation/RetailFoodProtection/FoodCode/ucm374275.htm

U.S. Food and Drug Administration (2014) Guide to Nutrition Labeling And Education Act (NLEA)

requirements August 1994. Retrieved from http://www.fda.gov/iceci/inspections/inspectionguides/ucm074948.htm. #GUIDE FOR REVIEW OF NUTRITION

US Food & Drugs Administration (2014a) Computerized systems in food processing industry. Retrieved from https://www.fda.gov/inspections-compliance-enforcement-and-criminal-investigations/inspection-guides/computerized-systems-food-processing-industry

US Food & Drugs Administration (2014b) Computerized systems in food processing industry: guide to inspections of computerized systems in the food processing industry. Retrieved from https://www.fda.gov/inspections-compliance-enforcement-and-criminal-investigations/inspection-guides/computerized-systems-food-processing-industry

USDA National Agricultural Library (2020) College and university food science programs. Retrieved from https://www.nal.usda.gov/fnic/academic-programs-and-other-educational-opportunities

USDA. Economic Research Services (2014) Retailing & wholesaling: retail trends. Retrieved from http://www.ers.usda.gov/topics/food-markets-prices/retailing-wholesaling/retail-trends.aspx

VanLandingham PG (1995a) Postsecondary vocational programs vs. apprenticeships in American culinary arts. Educational Resources Information Centre (ERIC) ED387619. Retrieved from https://files.eric.ed.gov/fulltext/ED387619.pdf

VanLandingham PG (1995b) How has vocational culinary arts changed as a result of a redesign of the education system. Educational Resources Information Centre (ED387618). Retrieved from https://files.eric.ed.gov/fulltext/ED387618.pdf

Vickery JR (1967) The scope and status of food science. First international lecture of the Food Group of the Society of Chemical Industry. Chem Ind 3:109–114

Weaver CM, Dwyer J, Fulgoni VL 3rd., King JC, Leveille GA, MacDonald RS, Ordovas J, Schnakenberg D (2014) Processed foods: contributions to nutrition. Am J Clin Nutr 99(6):1525–1542

Wikipedia (2020a) Philippe de Girard. Retrieved from https://en.wikipedia.org/wiki/Philippe_de_Girard

Wikipedia (2020b) Timeline of scientific discoveries. Retrieved from https://en.wikipedia.org/wiki/Timeline_of_scientific_discoveries

Wilkinson J (2004) The food processing industry, globalization and developing countries. E-J Agric Dev Econ 1(2):184–201. Retrieved from http://ageconsearch.umn.edu/bitstream/11999/1/01020184.pdf

Young LR, Nestle M (2002) The contribution of expanding portion sizes to the US obesity epidemic. Am J Public Health 92(2):246–249

Zink DL (1997) The impact of consumer demands and trends on food processing. Emerg Infect Dis 3(4):467–469

Zirkle C (2012) The multitiered CTE/VET system in the United States—from high school to two-year colleges. In: Barabasch A, Rauner F (eds) Work and education in America. The art of integration. Springer, Dordrecht, pp 33–51. https://books.google.com.gh/books?id=kxlIO5578ecC

Food-Largest of All Industries

2

1 Introduction

1.1 Evidence That Food Is One Of the Largest of Industries

The food industry (Sect. 2.2) *contains within it* one of the largest manufacturing sectors within the United States (US) and the same is true for other developed countries. The formal and informal food sector are also of great importance for most developing countries, in particular when taken together with agriculture. The focus of this chapter is food manufacturing and other food related businesses. The top-10 food manufacturing regions of the world are listed in Table 2.1 citing total sales receipts unadjusted for expenses.[1] Several countries (Canada, New Zealand, Australia and South Korea) are noteworthy as global food manufacturing hubs out of proportion to their populations. The ASEAN countries are predicted to become food manufacturing giants in the near future. Two BRIC countries (Brazil and China) appear in the top-10 list for food manufacturing but few other developing countries occur. In most developed countries, earnings from food manufacturing is a significant part of the wealth created by the manufacturing sector.

Information from Table 2.1 is considered provisional for several reasons. First, the summary is from 2010 and is likely not an accurate reflection of current data. Secondly, revenue figures can vary because of differences about which businesses should be counted within the food industry. Thirdly, economic data is sometimes based on forecasting – but this fact is not always evident. To deal with sources of possible inaccuracy data were cross-referenced to two or more sources. Another safeguard to ensure data quality was to use information from government / official sources or the grey literature whenever possible. In summary, there are uncertainties about sources of data.[2] For instance, Table 2.1 results for US food manufacturing earnings are about 2-fold those provided by the USDA for roughly the same period (cf. Table 2.4). Likewise, the food manufacturing earnings for Brazil, China, Japan, Canada etc. are on the high side compared to data cited later in this chapter (Sect. 3). However, regardless of this caveat, it is feasible to present a snapshot of food manufacturing and allied businesses in the world at large.

[1]Revenue from industry is measured as Gross Value Added (GVA) or Gross Domestic Product (GDP). GVA is revenue adjusted for costs. GDP = GVA + Taxes – Subsidies received. (From https://www.investopedia.com/terms/g/gross-value-added.asp). This textbook cites both GDP and GVA values as given in original sources.

[2]Data from the grey literature show high degrees of “repetition”. Information appearing in one business report was often copied and propagated, without citations/accreditation.

R. Owusu-Apenten, E. R. Vieira, *Elementary Food Science*, Food Science Text Series,
https://doi.org/10.1007/978-3-030-65433-7_2

Table 2.1 Food and drinks manufacturing worldwide, 2010[a]

Region/ Country	Sales (billion US$)	% sales	Employees(x1000)
EU	1296	15	4248
US	650	13	1524
China	515	9	5827
Japan	350	10	1455
Brazil	141	18	1412
Canada	91	17	233
Mexico	80	24	390
Australia	67	20	215
South Korea	57	5	200
NZ	29	45	76

[a]Notes: EU = European Union, NZ = New Zealand; the value for sales not adjusted for expenses. % sales = percent manufacturing industry sales (Food and drink Europe 2012)

1.2 North American Industry Classification System (NAICS)

The US Federal government uses the North American Industry Classification system (NAICS) to track business activity. NAICS is not wholly infallible, but the system is useful for housekeeping, and for revealing which different businesses are (not) related. The NAICS codes apply throughout the North American trade zone covering Canada, Mexico and US (NAICS Association 2014).

An older system of standard industrial classification (SIC) codes are still used and sometimes both SIC and NAICS codes may occur side by side. In general, all businesses must self-identify using NAICS codes when filling tax returns (IRS 2014). The NAICS codes are used to help keep tabs on different types of food businesses throughout this book (Appendices 1 and 2).

1.3 Industry Classification Codes from Other Parts of the World

The United Nations (UN) uses the international standard industrial classification (ISIC) system for cataloging businesses. With ISIC each different business activity is designated a unique letter of the alphabet. Thereafter, each of 21 categories or letters (A, B, C…U) is followed by two digits, e.g. the ISIC code C10 is reserved for food manufacturing, C11 refers to beverage manufacturing and C12 is for tobacco (Appendix 2). The European Union replaced ISIC codes with NACE codes during the 1990s.[3] Starting from 2008 a new version NACE (NACE version 2) became enforceable for all European Union member States (Appendix 3).

2 The North American Food Industry

2.1 Classification of the Food Industry

The food industry is large and heterogeneous, which probably explains why there are multiple NAICS designations for different food-related businesses. cf. Appendices A1–A4.[4] For example, food manufacturers (NAICS 311), food packaging manufacturers (NAICS 322 & 326) and retailers (NAICS 445) have unique NAICS codes (Table 2.2). The designation NAICS 722 covers foodservices and out-of home eating,

There are limitations to conventional industry classifications. There are no NAICS codes for terms such as "food industry" and other phrases including "food processing industry", "farm to fork", "food chain", "food system", "food supply-chain", "food value chain" or "Agri-food system", which appear regularly in various technical documents.

[3]NACE is short for Nomenclature Statistique Des Activités Économiques dans la Communauté Européenne. English translation: Statistical Classification of Economic Activities in the European Community.

[4]An extended listing of NAICS codes for food related businesses appears in Appendices 1–6.

Table 2.2 Classification of food industry activities (North America)

NAICS	NAICS title
111, 112, 114	Agriculture (crop, animal farming) fish farming, fishing, hunting
311	Food manufacturing, feed for animals (excludes foods for cats & dogs)
322	Folding paperboard box, other paper board, converted paper board
325	Synthetic dye and pigment & fertilizer
326	Plastics bag, pouches, polystyrene, glass, p and pouch (food packaging)
33	Food machinery manufacturing, appliances, furniture, transportation equipment
423	Equipment merchant wholesalers
424	Wholesalers, paper, grocery, frozen food, dairy products, fresh fruit, farm supplies
445	Supermarkets and other grocery stores, convenience stores, specialty stores, food (health) supplements stores, gasoline stations stores, warehouse clubs, direct sales
541	Food testing laboratories, research and development
624210	Community foodservices
722	Full-Service restaurants, limited-service eating places, special food services, drinking places

[a]Adapted from (NAICS Association 2017)

2.2 The US Agri-Food Industry

The Agri-food industry spans several areas of business activity and has no single NAICS code. The related phrases "Agro-Industry" or "Agri-business" are used widely in discussions of the impact of food on international development (Krieger and Schiefer 2006; Hawkes 2008). Briefly, Agro-Industry refers to businesses that process raw materials deriving from farming, fishing and forestry. Agro-Industry products include food, beverage, textiles, wood products including paper, and rubber products. In all cases, Agri-food manufacturing products have greater added-value compared to the farm-gate / or port-side raw materials from which they derive. Some USDA economic data groups "agriculture and related industry" in a manner that correspond closely to the idea of Agri-food products (Fig. 2.1).

2.2.1 Definition of the Food Industry

Food was defined in Chap. 1 as items intended to be consumed. However, the "industry" part conjures up mechanization and the technologicalization of the processes associated with harvesting, preparing, storing, and selling food (Oddy and Drouard 2016). Large-scale food preparation and manufacturing uses machinery and appliances alongside of human intervention. The "touchless" kitchen or process plant predicted by some commentators is not yet a reality and the food industry will remain very people-centric for the near future.

The food industry is not well defined in most publications. For instance, google scholar lists 15,000 articles for the publication years 1989–2019, containing the title phrase "food industry". Using a sub-sample of 30 articles, we found two articles that provided a definition for the term "food industry" (Gurr 1989; Stern 2016). Another two papers described the food industry extensively, but without a portable definition (Sadiku et al. 2019; Parry 1999). Granted that a selection of 30 papers is not statistically "representative" of the 15,000 articles, but 28 "no definitions" out of 30 cases is perhaps noteworthy.

On occasion, the food industry has been defined, as "*all those commercial enterprises contributing to each stage of the food chain from primary agriculture, through food processors, wholesalers and retailers, either directly to the consumer or via the agency of various catering (i.e. foodservice) institutions*" (Gurr 1989). Another, more concise definition stated that the food industry is "…all *things going into the production, processing, distribution, and sale of foods*" (Stern 2016). Though technically correct, the second statement omits the foodservices.

In this text, the food industry will be defined as, "*a group of businesses involved with food production, converting food raw materials into ingredients or finished products or beverages,* foodservice (catering), retail and wholesale of food". The main activities of the food industry are, crop and livestock farming (agriculture),

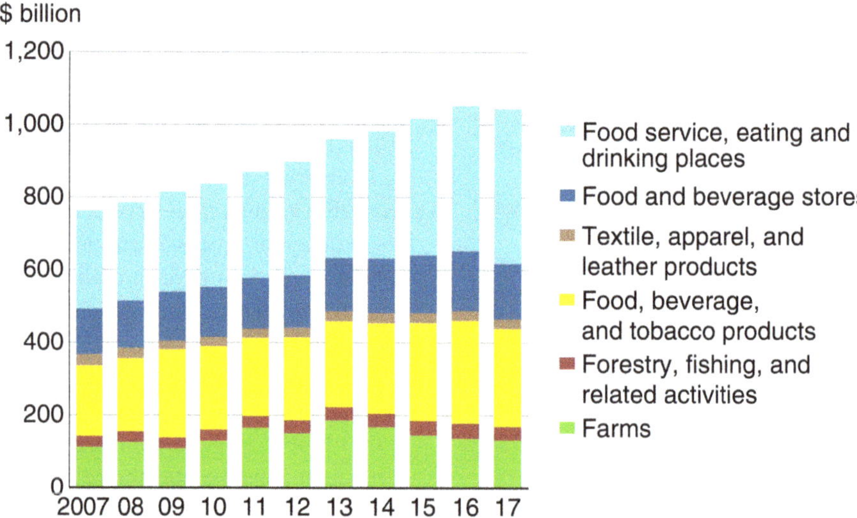

Fig. 2.1 Economic data aggregated for the Agri-food sector. From U.S. Department of Agriculture. (USDA Economic Research Service 2019)

Table 2.3 Number of businesses and employees in the US Agri-food industry

Sector[a]	NAICS	SIC	Businesses	Employees
Agriculture, fishing, hunting	111–112	0100–0200	67,934	749,400
Food manufacturing	311	2000	30,127	1,484,000
Beverage Manufacturing	312	2080	4841	131,438
Grocery & food wholesale	4244	5140–5180	33,051	756,070
Food & Beverage Stores	445[b]	5400	146,211	2,976,500
Foodservice	722	5810/12	587,457	10,488,800
Total	–	–	1,592,640	15,863,189

[a]Slightly different sector names occur for NAICS and SIC systems. F&B – food and beverage, NAICS – North American Industry Classification Scheme, SIC – Standard Industrial Classification. [b]Includes Supermarkets and grocery stores *(NAICS 445110), convenience stores (NAICS 44512), specialty food stores selling meat, fish, fruits and vegetables, baked goods etc. (NAICS 4452), and also beer, wine and liquor stores (NAICS 4453). Compiled by Authors. For a census date of ~2014, the total US workforce was approx. 148 million (U.S. Bureau of Labor Statistics 2014b; NAICS Association 2014)

seafood production (aquaculture and fishing), hunting, food manufacturing, commercial-scale food preparation (foodservice and catering), wholesale and retail of food (Table 2.3). Industrial fiber, textiles, wood products and tobacco will be excluded, though some of these items appear in official documents about food. Many experts use the terms Agri-food industry and food system, as alternative names for the food industry (Hawkes 2008; Krieger and Schiefer 2006; Wilkinson and Rocha 2009).

2.2.2 Characteristics of the US Agri-food industry

Recent official estimates from the US showed there were 1.5 million businesses and 15+ million

Table 2.4 Earnings of the US Agri-food sector (2012/13)

Sector	BillionUS$[a]	%GDP[b]
Agriculture, net farm income (111–112)	166.94	1.03
Food & Beverage manufacture (311 & 312)	233.34	1.44
Supermarket & Food & Beverage Store Sales	140.74	0.87
Food Service (722)	314.57	1.94
Non-traditional Food Store Sales	352.01	2.27
Convenience Stores Sales (not including gas)	166.94	1.03
Total	1374.54	8.46

[a]Official estimates of earnings of the US Agri-food sector (2012/13)
[b]Value excludes tobacco manufacturing and assumes GDP of US$16244.60 billion for 2014/2015 (U.S. Bureau of Economic Analysis (BEA) 2014). Currently (2020) the GDP of the US is estimated at US$20.0 trillion (Appendix 4)

Table 2.5 Key manufacturing sectors within the US

Sector	Manufacturing GDP %	Employees %
Chemical	14.6	6.8
Computer & Electronics	13.1	9.5
Food, Beverage & Tobacco[a]	12.4	13.7
Petroleum and Coal	9.8	1
Machinery	7.6	9.1
Fabricate metal	7.1	11.5
Other transportation equipment	5.4	5.7
Miscellaneous	4.6	4.8
Motor vehicles & trailers	4.4	6.3

[a]Corresponds to the food processing industry & tobacco (Manufacturing Institute 2014)

employees within the Agri-food industry (Table 2.3). For the 2014 fiscal year the Agri-food industry contributed 8.5 to 10% of the US total GDP (Table 2.4) and 1-in-10 of all jobs in the US (U.S. Bureau of Labor Statistics 2014b). Food industry enterprises can be small (one-employee) or large multinational businesses with many thousands of employees. On average there were 11 employees per US farming enterprise compared to 49 and 18 employees per food manufacturing or foodservices enterprise, respectively.

2.3 The Food Manufacturing Sector

Food manufacturing (NAICS 311) and beverage manufacturing (NAICS 312) are two of 21 industries forming the US manufacturing sector (Table 2.5; Appendix 2A).[5] All manufacturing industries "*transform raw materials, substances, or components into new products using mechanical, physical, or chemical methods*" (U.S. Bureau of Labor Statistics 2014c). For 2013, the greater manufacturing sector contributed an estimated US$$2.03 trillion (12.5% GDP) to the US economy alongside of 17.4 million jobs (11.2% total). The US$ earnings from manufacturing exceeded the GDP of any other national economy with the possible exception of China, Japan, Germany, India, United Kingdom, France, Italy, Brazil, Canada and Russia (Manufacturing Institute 2013).[6]

Food and beverage manufacturing businesses transform agricultural products (crops, livestock, fish and seafood) into food ingredients or finished products using mechanical, physical or chemical means (Ferman 2008). Interestingly, economic data for food, and beverage manufacturing (and tobacco manufacturing) are consistently reported together (U.S. Department of Commerce 2008). Earnings from transportation and distribution are not included in the metrics for the Agri-food sector possibly because haulage firms convey other goods besides food. The types of business and employees within the US food and beverage manufacturing industry are summarized in Tables 2.6, 2.7 and 2.8.

[5]A list of about 20 different manufacturing business groupings and their NAICS codes is available in Appendix A2A. The term "food processing" industry is largely absent from Official economic data, instead we find food and beverage manufacturing and the corresponding NAICS.

[6]The list corresponds to the top 10 largest economies in the world – not in order to size.

Table 2.6 The US food manufacturing sector (NAIC311) - employees and businesses

NAICS code	NAICS code description	Employees (March 2012)	Number of businesses
31	Manufacturing	10,984,361	295,643
311	Food manufacturing	1,427,283	25,543
3111	Animal food manufacturing	50,200	1738
31111	Animal food manufacturing	50,200	1738
311111	Dog and cat food manufacturing	22,422	282
311119	Other animal food manufacturing	27,778	1,456
3112	Grain and oilseed milling	53,140	788
31121	Flour milling and malt manufacturing	15,807	377
311211	Flour milling	11,081	288
311212	Rice milling	3778	67
311213	Malt manufacturing	948	22
31122	Starch and vegetable fats and oils	23,620	347
311221	Wet corn milling	6709	64
311222	Soybean processing	6174	114
311223	Other oilseed processing	g	49
311225	Fats and oils refining and blending	8963	120
31123	Breakfast cereal manufacturing	j	64
3113	Sugar and confectionery products	67,986	1666
31131	Sugar	13,250	78
311311	Sugarcane mills	3625	26
311312	Cane sugar refining	3273	23
311313	Beet sugar	6352	29
31132	Chocolate + confectionery from cacao	7112	179
31133	Confectionery from purchased chocolate	28,426	969
31134	Nonchocolate confectionery	19,198	440
3114	Fruit &,vegetable preserving, specialty food	155,163	1670
31141	Frozen food	85,424	659
311411	Frozen fruit, juice, and vegetable	29,754	225
311412	Frozen specialty food	55,670	434
31142	Canning, pickling, and drying	69,739	1011
311421	Fruit and vegetable canning	42,336	701
311422	Specialty canning	13,926	116
311423	Dried and dehydrated food	13,477	194

[a]Category names edited slightly from official titles (United States Census 2014)

The food, beverage (and tobacco) manufacturing industry ranks first in terms of numbers of manufacturing jobs and ranks third in terms of the contribution to US *manufacturing* GDP (Table 2.5). About 10% of all *manufacturing* jobs can be found in the Food and beverage sector (NAIC311 & 312) (Manufacturing Institute 2014). The US food and beverage manufacturing industry is summarized in Tables 2.6, 2.7 and 2.8. Likewise, food and beverage manufacturing provides ~10% of all manufacturing jobs in Canada and European Union.

2.4 The US Foodservice Sector

In 2008 the foodservice sector (out-of-home eating) accounted for 47 cents of each dollar spent on food.[7] Beginning from 2013, consumer spending for out-of-home eating in the US exceeded the spending for home-prepared food (Chap. 25);

[7]Two alternative spellings "foodservice" and "food service" (two words) are currently in use. The term "foodservice" is used in this book and chapter to reflect Industry sources. However, NAICS literature refers to the "food services" (two words). Readers are recommended to search both terms in further reading.

Table 2.7 The United States food-manufacturing sector (NAICS 3115 & 3117) - employees and businesses

NAICS code	NAICS code description	Employees (2012)	Number of businesses
3115	Dairy product	136,661	1587
31151	Dairy product (except frozen)	115,728	1186
311511	Fluid milk	54,833	420
311512	Creamery butter	1849	38
311513	Cheese	44,679	540
311514	Dry, condensed, evaporated dairy product	14,367	188
31152	Ice cream and frozen dessert	20,933	401
311520	Ice cream and frozen dessert	20,933	401
3116	Animal slaughtering and processing	488,848	3642
31161	Animal slaughtering and processing	488,848	3642
311611	Animal (except poultry) slaughtering	156,041	1494
311612	Meat processed from carcasses	101,442	1345
311613	Rendering and meat byproduct processing	8699	220
3117	Seafood product preparation and packaging	31,261	620
311711	Seafood canning	2103	92
311712	Fresh and frozen seafood processing	29,158	528
3118	Bakeries and tortilla	280,317	10,520
31181	Bread and bakery product	208,824	9336
311811	Retail bakeries	55,166	6400
311812	Commercial bakeries	134,389	2705
311813	Frozen cakes, pies, and other pastries	19,269	231
31182	Cookie, cracker, and pasta	55,043	822
311821	Cookie and cracker	31,519	380
311822	Flour mixes & dough from purchased flour	19,206	293
311823	Dry pasta	4318	149
31183	Tortilla	16,450	362

Data from March 2012 (United States Census 2014).

Table 2.8 The US food and beverage manufacturing sector (NAIC3119-3121) - employees and businesses

NAICS code	NAICS code description	Employees 2012	Number of Businesses
3119	Other food	163,707	3312
31191	Snack food	45,463	612
311911	Roasted nuts and peanut butter	15,554	246
311919	Other snack food	29,909	366
31192	Coffee and tea	14,224	433
31193	Flavoring syrup and concentrate	6966	168
31194	Seasoning and dressing	32,591	714
311941	Mayonnaise, dressing, and other prepared sauce	16,348	356
311942	Spice and extract	16,243	358
31199	All other food	64,463	1385
311991	Perishable prepared food	36,745	699
311999	All other miscellaneous food	27,718	686
312	Beverage and tobacco product	146,493	5059
3121	Beverage	131,438	4841
31211	Soft drink and ice	63,574	1207
312111	Soft drink	50,877	460
312112	Bottled water	8028	291
312113	Ice	4669	456
31212	Breweries	25,608	728
31213	Wineries	36,132	2694
Distilleries	Distilleries	6124	–

Data from March 2012 (United States Census 2014)

the foodservice and food retail businesses are the two main (competing) channels for selling food to the consumer (Tables 2.9 and 2.10). Interestingly, the consumer engages with the foodservice sector in a variety of forms: restaurants, fast-food enterprises, in-flight meals, hotels meals, hospital meals, schools, and/or drinking establishments (National Restaurant Association 2014). The scope of non-residential catering covers all out-of-homes eating aside from informal dinning. A further aspect of the catering industry deals with the provision of ready-to-eat meals purchased from retailers for home heating and consumption.

The foodservice sector (NAICS 722) is officially listed as a part of the Accommodation and foodservices sector (NAICS 72). Some 10 million+ employees work within the foodservice sector that includes businesses that prepare foods to-*order*, full-service restaurants, limited-service eating places, special foodservices, foodservice contractors, caterers, and mobile foodservices, and drinking places.

Table 2.9 The foodservices and drinking places (NAICS 7221-4)[a]

Prepare meals, snacks, beverages to customer orders
On premises, off-premises consumption
Offers food with/ without seating with/ without waiter/ waitress service
Full-Service Restaurants (7221)
Limited-Service Eating Places (7222)
Special Food Services (7223)
Drinking Places, Alcoholic Beverages (7224)

[a]Post 2017 literature refers to NAICS 722511 for full service restaurants and NAICS 722513 for Limited service eating places

Currently, there are well over 500,000 restaurant businesses and 980,000 restaurant locations in the US (National Restaurant Association 2014). For the 2013 fiscal year, the sector had sales of US$600 billion with 9% growth over the previous two years. For every dollar spent in the restaurant business another $2 may be generated in sales in other industry. Accordingly, the restaurant industry impact is estimated at US$1.8 trillion. Sales in the restaurant industry were estimated to account for 4% of US GDP (National Restaurant Association 2014). Another area outside of manufacturing which deserves mention is food wholesale and retail, including the supermarkets and other grocery stores with ~3 million employees (Fig. 2.2, Tables 2.9, and 2.10).

2.5 The Food Dollar Expenditure

Data from 2008 shows that 11–12 cents of every US consumer dollar spent for Agri-food businesses went to the farmers (Fig. 2.2). The share of the food-dollar going to food manufacturing,

Table 2.10 The United States foodservice sector - employees and businesses (NAICS 722)

NAICS code	NAICS code description	Employees/(2012)	Businesses[a]
72----	Accommodation and foodservices	11,556,285	649,011
721	Accommodation	1,864,708	63,398
7211	Traveler accommodation	1,815,117	54,126
7212	RV parks and recreational camps	38,273	7189
7213	Rooming and boarding houses	11,318	2083
722	Food services and drinking places	9,691,577	585,613
7221	Full-service restaurants	4,570,174	227,161
7222	Limited-service eating places	4,106,307	275,939
7223	Special foodservices	668,117	39,654
72231	Food service contractors	530,597	26,134
72232	Caterers	129,621	10,815
72233	Mobile foodservices	7899	2705
7224	Drinking places (alc. beverages)	346,979	42,859

[a]Number of businesses, not premises. (United States Census 2014)

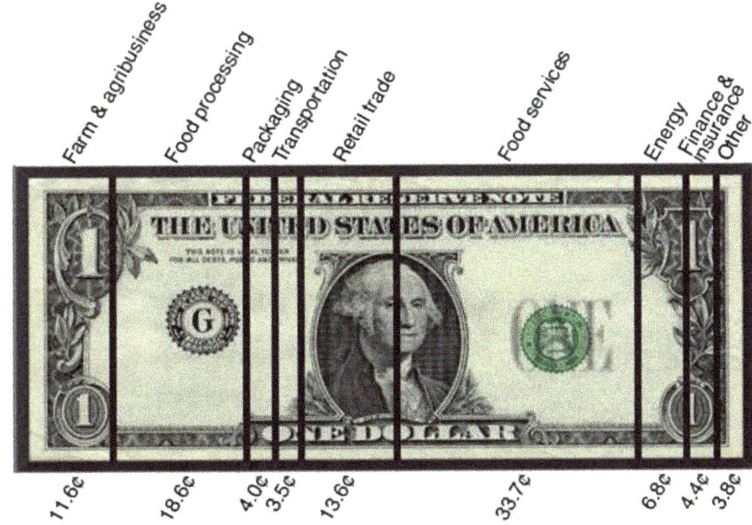

Fig. 2.2 The farm-share of the food-dollar bill (Canning 2011)

packaging and transportation was about 24% and for the foodservice sector 33.7% (Canning 2011). Historical data on food-dollar expenditure patterns is available from the USDA Economic Research Services (USDA-ERS) which tracks house-hold expenses.

The value of manufactured food products was about 8-times greater than the value of primary agricultural inputs (Canning 2011). Therefore, value is added to raw agricultural commodities as these are transformed into manufactured foods, and as foods transits to the foodservice sector. Processed foods have well-known enhancements compered to raw agricultural products, including a longer shelf life, improved food safety, increase in convenience, all-year round availability, and increased consumer choice. The foodservice value added to inputs ranges from culinary arts to intangibles such as the restaurant service (Chap. 25).

2.6 The US Food Retail Sector

Of the twelve sub-sectors within general retail (NAIC44) food and beverage retail (NAICS 445) is ranked first and second in terms of number of employees (2,872,426) or number of businesses (145624); these figures are estimates from Federal sources (U.S. Bureau of Labor Statistics 2014a). Total sales, industry trends and structure of the food retail were summarized by the USDA (USDA. Economic Research Services 2014). Total food retail sales reached US$ 571 billion in 2011 with 90% of receipts obtained from supermarkets (NAIC44511), 5.5% from convenience stores (NAIC44512) and 3.3% sales in other retail outlets. Tables 2.4 and 2.11 summarize key economic data for the food sales and retail sector.

In Table 2.4, the economic data for the retail sector (NAICS 44) do not include contributions from the foodservice and drinks sector (NAICS 722) though as noted previously, both are forms of retail. However, the 2009 economic summary from the National Retail Federation (NRF) combined food retail with foodservices and drinks – leading to an astonishing 28 million total jobs (1-in-6 jobs in the US), 3,617,486 Businesses (12% US total) and about US$1.6 trillion earnings or 8.5% national GDP. The NRF noted also that retail trade contributed indirectly to other industries, including manufacturing and transportation with an overall estimated direct and indirect GDP contribution of about US$ 2.6 trillion or 17.6% national GDP and 1-in-4 of all US jobs (National Retail Federation 2011).

Table 2.11 US Food & Beverage Retail Sector

NAICS code & descriptions	Employees	Number of businesses
445 Food and Beverage	2,872,426	145,624
4451 Grocery stores	2,582,042	91,530
44511, Supermarkets etc.	2,471,155	66,047
44512, Convenience Stores	110,887	25,483
4452 Specialty stores	139,449	21,767
44521, Meat Markets	33,782	5228
44522, Fish & Seafood	10,293	1957
44523, Fruit & Vegetable	20,332	2788
44529, Other Food Stores	75,042	11,794
445291, Baked Goods	19,560	3240
445292, Confectionery & Nut	19,520	3171
445299, All Other Specialty	35,962	5383
4453 Beer, Wine & Liquor	150,935	32,327

Table 2.12 The Canadian Agri-food sector

Industry or Sector (NAICS)	Businesses	$billiion CND
Agriculture, forestry & fishing (11)	179,833	?
Food Manufacturing (311)	7412	80.9
Beverages, manufacturing (3121)	1084	11.7
Food Wholesaler-Distributors (4131)	9669	98.9
Food and Beverage Stores (445)	35,013	107.3
Food Services and Drinking Places (722)	82,480	45.0

Table 2.13 Employment size category and region: December 2012 Food Manufacturing (NAIC311)

Classification	Micro	Small	Medium	Large
(Number of employees)	1–4	5–99	100–499	500+
Percent Distribution	25.40%	63.30%	10.40%	1.00%

Trends for the US food retail sector include supermarket consolidation. The 20-largest retailers currently account for US$449.3 billion worth (78.6% total) sales. At the same time, there was a reported increase in non-traditional food sales from a variety of outlets including, warehouse clubs, drug stores, and dollar stores. Another interesting trend is a growing interest in local foods (Zepeda and Li 2006; Martinez 2010). As noted above, food retail is traditionally seen as competing with the foodservice as a means of selling foods to the consumer.

2.7 Canadian Food Manufacturing

The Canadian Agri-food sector provides 12.5% of jobs nationally and 8.2% GDP (Tables 2.12–2.14). Canadian food manufacturing (NAICS 311) sector is enormous for a nation of 35 million people comprising 7412 business, and 236,237 employees (Canadian Industry Statistics (CIS) 2014). Approximately, 22% and 77% of food and beverage processing jobs were in the administrative and production grades, respectively. General economic indicators show food and beverage processing contributed ~10% of manufacturing jobs and approximately 2.4% of Canadian GDP. Manufacturing revenue was $67.8 billion in 2004 rising to $80.9 billion CAD ($72.9 billion US$) in 2011. By comparison, the drinks manufacturing (NAIC3121) sector comprised of about 1000 businesses with total revenues of about $11.7 billion CAD. The vast majority of food processing firms were classed as micro (25%) or small (65%) with 1–4 or 5–99 employees, respectively (Table 2.13). The major food manufacturing effort is concentrated in the provinces of Ontario > Quebec > British Columbia > Alberta (Table 2.14).

Per capita basis, Canada has a larger food industry compared to the US. Surprisingly also, the number of food and beverage stores are two-fold grater in Canada (per capita) as compared to the US. Similarly, the number of food manufacturing businesses also appear to be 50% greater in Canada on per capita basis (Table 2.14).

2.8 The North American Food Industry Summary

The US food industry is large, diverse and well documented. Overall, the US Agri-food industry contributed ~9% of the total GDP and 15.8 mil-

Table 2.14 Canadian food manufacturing by province: Number of Businesses (December 2012)

	Food manufacturing (NAICS 311)									
Province or territory	311	3111	3112	3113	3114	3115	3116	3117	3118	3119
Alberta	670	86	24	12	34	123	143	3	150	95
British Columbia	1069	76	20	61	71	70	110	137	303	221
Manitoba	295	41	21	5	19	44	54	9	68	34
New Brunswick	191	15	2	5	8	9	11	83	42	16
Newfoundland and Labrador	202	3	1	1	2	9	11	148	16	11
Northwest Territories	0	0	0	0	0	0	0	0	0	0
Nova Scotia	325	14	2	4	13	14	21	188	51	18
Nunavut	6	0	0	0	0	0	0	6	0	0
Ontario	2504	208	61	135	211	226	330	39	861	433
Prince Edward Island	79	5	3	0	6	10	4	35	10	6
Quebec	1765	171	48	119	118	150	247	73	583	256
Saskatchewan	302	41	22	3	9	82	71	1	56	17
Yukon Territory	4	0	0	0	0	0	0	0	3	1
Canada	7412	660	204	345	491	737	1002	722	2143	1108
Percent Distribution	100%	100%	100%	100%	100%	100%	100%	100%	100	100%

*3111 – Animal Food Manufacturing;	3116 – Meat Product Manufacturing
3112 – Grain and Oilseed Milling	3117 – Seafood Product Preparation and Packaging
3113 – Sugar & Confectionery Product	3118 – Bakeries and Tortilla Manufacturing
3114 – Fruit and Vegetable Preserving and Specialty Food	3119 – Other Food Manufacturing
3115 – Dairy Product	

Reference: Canadian Industry Statistics (Canadian Industry Statistics (CIS) 2014)

lion jobs (10.6% total) most of which were from the foodservice sector (~10.5 million; Table 2.3). Indeed, foodservice jobs exceeded the number of jobs from the US manufacturing sector (~9.6 million) minus food and beverage manufacturing (Table 2.6). The foodservices jobs also exceeded jobs from food and beverage manufacturing (1.6 million; Table 2.3) by 5.5-fold.

Most Agri-food industry earnings were from food sales through various channels (48% or US$658 billion, Table 2.4), followed by the foodservice (23%), food & beverage manufacturing (17%) and only then farming, fishing and hunting (12%).

Within Canada, some 8% of the national GDP and 10% of jobs were from the Agri-food sector. These are impressive stats by any standards. The food industry of the US is described as mature, with a strong international presence. Some 40% of the top 50 food-manufacturing firms have headquarters in the US. According to figures for 2000 the total exports from the US amounted to about US$30 billion as compared to $US37 billion imports. The main export foods include meat products, grain, fats and oils, fruit and vegetables. The major trading partners for the US include Japan, Canada, Mexico, Thailand, France and Italy. Interestingly, the value of US food industry involvement in foreign direct investments and partnerships (US$ 34 billion) exceeds the value of exports (Wilkinson and Rocha 2009). The value of internationally traded food (~US$60 billion) is about 10% of the total value of the food industry.

3 Global Food Industry

3.1 Global Food and Beverage Manufacturing

The global food manufacturing industry is outlined in this section for the period of 2008 to 2012. Economic statistics for the global food industry is available from market research firms at considerable expense.[8] The present discussion is based on

[8] Price considerations means that the current Market research data were unavailable for this project.

open access information from the USDA_ERS (USDA Economic Research Services 2012) and other sources.[9] The pattern of international trade by food manufacturing industries up until 2004 was summarized by Wilkinson from the Graduate Centre for Agricultural Development Federal Rural University, Rio de Janeiro (Wilkinson 2004). Globally about 90% of manufactured foods is for domestic consumption and 10% is traded internationally (Gehlhar and Regmi 2005).

The global food and beverage sector (farming, food production, distribution, retail and catering) was valued at between US$5.7 and US$7 trillion in 2014 with an annual growth rate of 3.5–5% (BusinessVibes 2014; Gehlhar and Regmi 2005; IMAP 2010; Research and Markets 2015). In 2008 the European Union had the largest food and beverage market. However, the US ranked first in terms of the food and beverage manufacturing size for a single sovereign state. Future growth of the food and beverage sector is anticipated to occur in the BRIC nations (Brazil, Russia, India and China) and also the Far East.

Worldwide, the beverage sector was valued at US$ 1.4 trillion, compared with US$ 3.3 trillion for food processing. Food retail was described as fragmentary, with the top 50 retail companies accounting for 20% market share (Gehlhar and Regmi 2005; Wilkinson 2004). Supermarket, hypermarkets, discounters, local stores, street vendors are all jostling for market share. Food manufacturing is likewise dispersed. The major manufactures are seeking to expand into developing markets via foreign direct investment and joint venture. In the remainder of this section, we will review the food industry in some other regions of the globe.

3.2 European Union Food Manufacturing

Food manufacturing (NACE 10) and the drinks manufacturing industry (NACE 11) form the largest manufacturing sub-sector within the European Union with revenue totaling US$1326.0 billion (€ 1017.0 billion) and 4.25 million employees. Table 2.15 shows the economic summary of European Union food and drinks manufacturing.

Table 2.15 The European Union food and drinks manufacturing industry (2011)

European Union Population	300 million
Turnover (US$ billion)	$1383.12
Employees (million)	4.25 (15%)
Exports (US$ billion)	$104
Imports (US$ billion)	$86
No of companies	287,000
World market share (Exports)	16.50%

Food and drink Europe (2012); US$ values converted from Euros @ exchange rate of 1 Euro = 1.36 US$

Table 2.16 shows a breakdown of food and drinks manufacturing across the European Union. Germany, France, Italy, United Kingdom and Spain are the top-5 food manufacturing countries within the European Union; the UK is in talks to exit from the EU. Table 2.17 lists manufactured food export / import trade for the EU (28) including the UK.

Information describing the food chain within the United kingdom (Table 2.18) is available, from the annual "Food statistics pocketbook" compiled by the Department of Environment and Rural Affairs or DEFRA (Department for Environment Food and Rural Affairs (UK) 2013). Sections of the DEFRA document have been required reading for many of my own courses in food science over the past 14 years.

The UK agri-food sector now comprises over 190,000 enterprises employing 3.7 million employees equivalent to 14% (1-in7) of all UK jobs and 7% of GDP. It is noteworthy that, food and drinks manufacturing adds 7400 enterprises, 400,000 employees and US$38.8 billion (UK£25 billion) to the UK national GDP. The effect seen with non-residential catering in the US is partially replicated in the United Kingdom in terms of jobs and revenue (Table 2.18).

3.3 Brazil

Brazil is currently the world's sixth largest economy (GDP of US$2.5 trillion) accounting for ~45% economic output from South America.

[9]Estimates from various sources differ partly because a proportion of the food and beverage manufacturing occur within an *informal* sector where concrete data is unavailable. In view of inconsistencies, readers are encouraged to compare data from multiple origins.

Table 2.16 European Union food and drinks manufacturing by country (2011)[a]

Country	Turnover US$ billion	Value added US$ billion	Employees (1000)	Businesses
Germany	222.088	15.64	500	5960
France	213.792	39.848	500	10,000
Italy	172.72	32.912	408	6300
United Kingdom	119.136	32.232	370	6500
Spain	113.968	27.2	446	30,000
Netherlands	80.512	19.448	131	4385
Poland	67.592	12.104	403	13,708
Belgium	60.52	9.112	89	4912
Denmark	34.544	4.352	55	1610
Ireland	29.92	8.16	43	689
Sweden	26.112	5.984	56	3400
Portugal	19.72	3.944	110	10,513
Austria	17.136	6.392	58	3921
Czech Republic	15.368	3.944	105	8360
Finland	15.368	3.4	33	1900
Greece	14.96	1.904	65	1180
Romania	14.28	2.992	186	8239
Hungary	11.288	2.72	97	6556
Bulgaria	6.392	1.088	99	5612
Slovakia	5.032	0.952	30	218
Lithuania	4.896	0.816	42	1205
Slovenia	2.992	0.68	16	1214
Latvia	2.176	0.408	25	788
Cyprus	2.04	0.544	13	863
Estonia	2.04	0.408	13	422
total	1274.592	237.184	3893	138,455

[a]US$ values were converted from Euros @ exchange rate of 1 Euro = 1.36 US$. Source from (Food and drink Europe 2012)

Table 2.17 Trade by European Union food and drinks manufacturing industry (2011)

2011 Exports	Million US$	%Inc2011/10	Imports	Million US$	Inc2011/10
USA	16681.39	10	Brazil	9517.28	17
Russia	10066.38	10	Argentina	7314.08	3
Switzerland	6503.81	8	USA	5558.32	11
Japan	5149.95	10	China	5527.04	13
China	4927.55	48	Switzerland	4837.52	12
Hong Kong	3873.93	45	Indonesia	4270.4	22
Norway	3768.29	19	Thailand	3848.8	18
Canada	3171.98	7	Turkey	2900.88	13
Australia	2394.97	15	Norway	2681.92	11
Saudi Arabia	2331.03	24	New Zealand	2394.96	15

Food and drink Europe (2012); US$ values converted from Euros @ exchange rate of 1 Euro = 1.36 US$

Table 2.18 Economic summary for the United Kingdom Agri-Sector (2013)

Sector	GVA (US$bn)	(%)	Employees/1000	%	No of Enterprises*
Ag. & Fishing	13.7	9	440	12	153,000
F&D Manufac.	38.8	27	400	11	7472
Wholesaling	14.5	10	210	6	15,115
Retailing	40.5	28	1192	32	53,641
NR catering	37.9	26	1499	41	115,177
Totals	145.4	100	3700	100	344,405

Notes: UK food industry sectors are, Agriculture and fishing (Ag & Fishing), Food and drink manufacturing (F & D Manufac), food wholesale (wholesale), food retail (Retail), and foodservice or Non-residential catering (NR catering). (Department for Environment Food and Rural Affairs (UK) (2013)

Table 2.19 Food and beverage manufacturing in Brazil (2011)

Population (million)	200.3
GDP (US$ billion)	2510
Manufacturing GDP (US$ billion)	297.6 (10% GDP)
Food & Beverage, GDP (US$ billion)	60.6 (18.2%mGDP)
Employees (million)	
No of Businesses	45,000
Food Exports(US$ billion)	10.0
Food imports(US$ billion)	1.2

Sources: Various (Food Export Association of the Midwest USA 2015; Jerger 2012; Macro Economic Meter 2014a)

Table 2.20 Food and beverage manufacturing in Russia (2011)

Population (million)	142
GDP (US$ billion)	2097
Manufacturing GDP (US$ billion)	195 (10% GDP)
Food & Beverage, GDP (US$ billion)	33.5 (17–20%mGDP)
Employees (million)	1.4
No of Businesses	53,000
Food Exports(US$ billion)	10.0
Food imports(US$ billion)	21.7

Sources: Various (Food and drink Europe 2012; Lubentsova 2011)

Food and beverage manufacturing is the largest manufacturing sector accounting for 18–20% of manufacturing GDP. The food and beverage sales from Brazil ranks in the top-10 countries (Table 2.1). The food industry of Brazil is described as advanced, with processing operations for the major food groups; bakery products and cakes (US$20 billion), chocolate (US$ 4 billion), and diary (US$ 7billion). In 2011, the beverage sector was valued at US$ 29 billion in Brazil. Currently 70% of processed food is for domestic consumption and the country has an export surplus of US$ 8 billion. The food retail sector in Brazil is large (current value ≈ US$ 126 billion) with an estimated one million employees (986,089). The foodservice sector (valued at US$122.6 billion) is also notable. More detailed information on the food industry in Brazil is available from trade associations and independent reports (Food Export Association of the Midwest USA 2015; Jerger 2012; Macro Economic Meter 2014a) (Table 2.19).

3.4 Russia

Russian food and beverage manufacturing sector is one of the most important of the BRIC countries and is currently valued at US$105 billion (2011) with a net contribution of US$33.4 billion towards manufacturing GDP (17–20% manufacturing GDP). The 2011 analysis from the US Department of Agriculture rated the Russian food industry as most able to withstand the 2008 financial crisis. The overall picture shows a vibrant modernizing sector, and discerning urbane customers with purchasing power estimated to be twice those of China. Features of the food and beverage sector in Russia are summarized in the Table 2.20 above.

To highlight the position as a leading BRIC country, Russia (2013) was a foremost market for EU food exports (US$ 10 billion) second only to the United States (US$ 16.6 billion) see Table 2.17. Russia accounted for 40% of European Union trade food exports to the BRIC countries (Food and drink Europe 2012; Lubentsova 2011). The value for Russian food production was estimated at US$105 billion, with profits of US$ 33.5 billion. Russia has several world-ranked food manufacturing sectors, e.g. diary sector (US$ 16–17 billion), juices (US$ 6 billion), confectionary sector probably the second largest worldwide (US$12 billion) and food ingredients (US$ 2billion). Within Russia both domestic and foreign companies operate in the food sector.

3.5 India

India is recognized as one of the world's leading producers of several raw food commodities including, cereals, grains, livestock, and fresh vegetables. However, the food and beverage manufacturing sector is thought to be small for a country of this size and the smallest of any of the BRIC countries with a global market share of ~1.4% (Table 2.21). Interestingly, India's total domestic spending on foods is sizeable and

Table 2.21 Food and beverage manufacturing in India (2014)

Population (million)	1241 (2012)
GDP (US$ billion)	1900
Manufacturing GDP (US$ billion)	194 (10%)
Food & Beverage, GVA (US$ billion)	17.5 (9% mGDP)
Employees, registered (million)	1.8
No of Businesses (registered)	36,871
No of Businesses (unregistered)	2,241,195
Food Exports (US$ billion)	31
Food imports (US$ billion)	19.8
World market share (Exports)	1.4%

mGDP – manufacturing GDP (Anon 2015; Macro Economic Meter 2014c; Invest India 2015; Mishra 2010)

estimated at US$181 billion (2009) or 21% of GDP (Mishra 2010). The *formal* food and beverage sector was valued at US$17.5 billion (1.4% GDP). A full discussion of the reasons accounting for the modest food and beverage sector in India is beyond the scope of this text, but commentators highlighted a distinct consumer preference for fresh foods in India as compared to other Asian countries, e.g. Malaysia. Another important consideration appears to be the large *unregistered* food and beverage sector in India estimated to be 60-fold larger than the registered sector.

The Indian authorities designated food and beverage manufacturing as a strategic sector for promoting growth in rural economies. It was recognized that growing the food and beverage manufacturing sector is important "to increase farm gate prices, reduce food waste, encourage value addition, promote crop diversification, promote employment and also increase export earnings". Food and beverage manufacturing was dubbed the "sunrise sector" by India's Ministry of Food Processing (Anon 2015) and the private-sector organization FICCI which is dedicated to developing the food processing sector (Federation of Indian Chambers of Commerce and Industry 2015). Several other sources provide insights and statistics about India's food industry (Macro Economic Meter 2014c; Invest India 2015; Mishra 2010).

Table 2.22 Food and beverage manufacturing in China (2011)

Population (million)	1363 (2012)
GDP (US$ billion)	7550
Manufacturing GDP (US$ billion)	2400 (32%)
Food Industry, GDP (US$ billion)	1100 (50% mGDP)
Employees (million)	6.74 (15% total)
No of Businesses	400,000
Food Exports(US$ billion)	50.5
Food imports(US$ billion)	28.8

Data adapted from Jun et al. (2013) and Bradbury (2012)

3.6 China

China's manufacturing sector served as the engine for the countries growth for 20–30 years. In 1992 manufacturing contributed 20% to China's GDP and is estimated at 30% (US$2400 trillion) in recent times. Food and beverage, and tobacco (ISIC31) contributed 50% mGDP (US$1100.0 billion; Table 2.22). Currently, China's food manufacturing sector is one of the largest worldwide though not well documented in western literature. More information is becoming available from Western Governments and private industry associations aiming to foster trade with China. Several reports published by the US Department of Agriculture (Jun et al. 2013), the Maine International Trade Centre (Bradbury 2012), New Zealand Trade and Enterprise (New Zealand Trade and Enterprise 2012), and also Agri-food Canada (Agri-food Canada 2010) provide useful coverage of the Chinese food industry. The Hong Kong Trade Development Council (HKTDC) website also hosts food industry articles focusing on China, Hong Kong, and other ASEAN countries (Hong Kong Trade Development Council 2013).

The food industry is divided into three sectors in China; (i) beverage manufacturing (~18%) covering alcoholic beverages- spirits, beer and wine, soft drinks and tea, (ii) food manufacturing (22%) -production of packaged food, pastries and confections, dairy, canned foods, fermented food, and condiments, and (iii) food processing (60%) referring to primary processing of agricultural commodities including, rice and, wheat milling, edible oil, sugar refining, slaughtering, production of salt, feed, and aquatic product processing. The food sectors in China correspond to North American NAIC311 (food manufacturing) and NAIC312 (beverage manufacturing).

Table 2.22 lists aggregate indicators for the food and beverage manufacturing and processing industries in China.

The processed food sector was valued at US$140 billion compared with US$16 billion for frozen foods, US$154 billion for fruit and vegetables, US$367 billion for foodservices and US$452–505 billion for food retailing. Therefore, the size of the Agri-food industry in China (excluding farming) is about US$1.16 trillion (Bradbury, 2012). The frozen food sector seemed to develop rapidly from a small number of "mum and pop" businesses to large national entities with the focus on Chinese traditional foods (vegetable and meat dumplings, sweet rice flour balls) with some diversification also into western style foods. Snack foods, prepared meat, dairy foods showed significant growth over the past 10 years or so.

The number of food manufacturing businesses in China is thought to have peaked at 41,000 in 2010 and declined to 31,000 firms in 2011 including 4874 beverage manufacturing firms, 6870 food manufacturing firms and 20,895 food processing firms (Bradbury 2012). The decline in numbers of businesses was thought to be the result of increasing industry consolidation and partly the result of the post 2008-economic downturn. The food industry as whole increased in value by 28% over the 2010–2011. A number of major driving forces for growth in the Chinese food industry were identified (Table 2.23). The major factors behind the growth of the food manufacturing in China is listed as, increasing urbanization and growing middle classes and incomes in first tier cities (Beijing, Shanghai, and Guangzhou), second tier cities (Chengdu, Shenyang, Wuhan, and Nanjing) as well as third tier cities (e.g. Changzhou). Some impediments to growth were noted in the Chinese food sector including, general rises in food costs and inputs, increasing labor costs, and rising food safety costs (Bradbury 2012).

Table 2.23 Factors influencing the food industry in China

Factors
Economic growth and increasing affluence (GDP increase 10%)
Urbanization, urban population to increase 607 to 822 million (2025 Est)
Food safety concerns
Health and Wellness
Demand for convenience foods
Improvements in infrastructure
Developments in retail

Source: Bradbury (2012)

Table 2.24 Food and beverage manufacturing in Hong Kong (2011)

Population (million)	6.6
GDP(US$ billion)	263
Manufacturing GVA (US$ billion)[a]	3.7
Food & Beverage, GVA (US$ million)	861 (23% mGDP)
No of Businesses	5400
Employees	49,000
Food Exports (US$ billion)	5.5
Food imports(US$ billion)	18.3

[a]Manufacturing, value added = Food, beverages and tobacco + Textiles and clothing + other manufacturing + Machinery and transport equipment + Chemicals, (Macro Economic Meter 2014b)

3.7 Hong Kong

In Hong Kong food and beverage manufacturing is about 1000-times smaller than in mainland China (Table 2.24). However, Hong Kong's food industry has attracted attention for a number of interesting reasons. First, 90% of food requirements is imported by Hong Kong. Second, the food, beverage and tobacco sector contributes 23.3% of manufacturing GDP about two-fold larger than other developed regions (Mecometer.com 2014). Thirdly, the Hong Kong food and beverage industry was differentiated from the Mainland China, by a greater dependence on re-exporting of goods (86% of businesses, 50% of total employees in food and beverage, and 90% of value addition etc.). The major manufacturing products for home consumption appear to be instant noodles and manufactured cereal products (macaroni, spaghetti), canned products, dairy products. The HK sector also benefits from the Chinese / Hong Kong Closer Economic Partnership Arrangement (CEPA) according to which goods manufactured in Hong Kong are eligible for duty free access to China's market (Hong Kong Trade Development Council 2013).

3.8 ASEAN Countries

The Association of Southeast Asian Nations (ASEAN) is an economic grouping of 10 nations from South Eastern Asia: Brunei, Myanmar, Cambodia, Indonesia, Laos, Malaysia, Singapore, Thailand, Philippines, and Vietnam. In 2015, ASEAN Economic Community (AEC) free-trade area is expected to begin operating. The ASEAN population comprises 600 million people with an apparent net GDP ranked eighth in the world. The food industry in a select few of the ASEAN countries is discussed in this section. Information concerning food manufacturing in ASEAN countries is available from the ASEAN Food and Beverage Alliance (AFBA) (Anonymous 2014). (Table 2.25)

Currently, the Agri-food sector contributes an average of 38% of jobs in ASEAN countries with net GDP contribution ranging from 50% for Myanmar to 33.4% for Cambodia. Food and beverage manufacturing contributed approximately 3.5% of manufacturing GDP for Singapore, and 11–20% GDP for Malaysia and Thailand (Anonymous 2012; Food Export Association of the Midwest USA 2011). As noted elsewhere estimates for economic data can vary considerable between different sources but the preceding figures agree broadly with those reported by others (Wilkinson 2004). A large number of consumers, and the rising middle classes are predicted to make the ASEAN countries especially attractive for future developments in the food industry. (Table 2.25)

The food manufacturing industry of Malaysia received attention recently from several international trade associations notably New Zealand, Australia, and the US (Ridley 2009; Food Export Association of the Midwest USA 2011; New Zealand Trade and Enterprise 2011; Anonymous 2012; Australian Trade Commission – Austrade 2014). Currently, food and beverage manufacturing produces 12% of manufacturing GDP from Malaysia (Table 2.26). Of the total of 37,861 registered manufacturing businesses some 3200 occur in the food and beverage sector, second only to the number of businesses for wearing apparels (24%); recent official estimates indicates that the number of food and beverage manufacturers in the formal sector may be of the

Table 2.25 Food industry data for ASEAN countries as per cent GDP

Country	m GDP (US$ billion)[a]	Food, beverage & tobacco GDP (US$ billion)	% mGDP[b]
Brunei	N/A	N/A	N/A
Burma (Myanmar)	N/A	N/A	N/A
Cambodia	0.585	0.038	6.6
Indonesia	142.21	32.40	22.8
Laos PDR	0.245	0.112	45.7
Malaysia	56.732	6.35	11.2
Singapore	38.168	1.43	3.8
Thailand	72.559	11.4	15.7
Philippines	28.867	6.45	22.3
Vietnam	5.786	1.747	11.8
Australia	74.252	13.98	11.8
Pakistan	22.93	5.153	22.5

[a]Data showing (ISIC 15-37) manufacturing GDP (mGDP) and GDP for food, beverage and tobacco (ISIC 31), [b]% mGDP = % manufacturing GDP derived from food, beverage and tobacco. Measures outputs minus costs of inputs data for 2012 (Mecometer.com 2014)

Table 2.26 Food and beverage manufacturing in Malaysia

Population (million)	30 (2014)
mGDP (US$ billion)[a]	60.7
Food, beverage & tobacco (US$ billion)	7.2–11
No of Businesses	3200 (5717)
%manufacturing GDP	12%
Employees	263,182
Food retail (US$ billion)	7.4
Food service (US$ billion)	6.0
Food Exports (US$ billion)	28.5
Food imports (US$ billion)	15–16

[a]Data showing manufacturing GDP (mGDP). A range of values shows estimates from different reports. Source of data: (Department of Statistics Malaysia Official Portal 2014; Food Export Association of the Midwest USA 2011; Macro Economic Meter 2014d; Anonymous 2012)

order 5717 (as opposed to 3200) with many more manufactures currently unregistered and in the informal sector. The foodservice sector comprises 142,918 registered Businesses (Department of Statistics Malaysia Official Portal 2014).

3.9 Australia & New Zealand

The Agri-food industry is divided into three components in Australia, (i) farms and fisheries, (ii) food and beverage processing, and (iii) retail food sales. As of 2010–2011, food and beverage manufacturing was the largest of the three sectors with 226,750 employers, and 21% of manufacturing sales and services revenues (Department of Agriculture Fisheries and Forestry 2011). These estimates do not include the contributions from the food retail sales which cover activities such as non-residential catering, restaurants and hotels. Food retailing in Australia is interesting according to the large contributions from supermarkets (64%) compared with restaurants and cafe sales (13%).

3.10 Global Food Industry Summary and Limitations

The preceding discussion covered the agri-food sector for a limited number of regions. South America was not covered in as much detail as warranted. Information on China was not as thorough as needed. The Middle East and the continent of Africa were not mentioned at all. These shortcomings represent limitations to our discussion.

Nevertheless, the evidence presented support the initial thesis of this chapter, which is that food is a major concern globally. Agriculture makes enormous contributions to the global economy. In developed countries, the value of raw agriculture products increase significantly due to processing. However, food processing and the food manufacturing industry, remains under-developed in many parts of the world. It is likely that the food industry will continue to grow in order to meet challenges of the growing world populations.

Appendices

Appendix 1

A1A: North American Industrial Codes for Agriculture, Fishing and Hunting Industries

NAICS	NAICS Title	Examples
1111	Oilseed and grain farming	Oil seed, grains, soybean, pea and bean, wheat corn, rice, other grain)
1112	Vegetable and melon farming	Melon, potato, other vegetables
1113	Fruit and tree nut farming	Oranges, apples, grape vineyards, strawberry, berry, tree nut
1114	Greenhouse, nursery, and floriculture production	Mushrooms, egg plants, other plants grown under cover
1121	Cattle ranching and farming	Beef cattle ranching and farming, cattle feedlots, dairy cattle, milk production
1122	Hog and pig farming	Hogs and pigs
1123	Poultry and egg production	Chicken, eggs, turkeys, hatcheries, other poultry
1124	Sheep and goat farming	Sheep farming, goat farming
1125	Aquaculture	Finfish farming, hatcheries, shellfish farming,
1129	Other animal production	Apiculture, horses and equine, furry animals, rabbit,
1141	Fishing	Fishing for finfish, shellfish, other seafood
1142	Hunting and trapping	Hunting and trapping
1151	Support activities for crop production	Soils preparation, crop harvesting, postharvest crop activities, farm labor, farm management, accountants

Excludes NAICS 113 for the timber, forestry and logging industries, adapted from https://www.naics.com/code-search/?naicstrms=111

A1B: North American Industrial Codes for food manufacturing related industries

NAICS	NAICS title	Examples
236210	Industrial building construction	Food processing plant construction
238290	Other building equipment contractors	Commercial kitchen food preparation equipment (e.g. mixers, ovens, stoves) installation
311111	Dog and cat food manufacturing	Dog food manufacturing
311119	Other animal food manufacturing	Pet food (except cat, dog) manufacturing
311211	Flour milling	Hominy grits (except breakfast food), manufacturing
311230	Breakfast cereal manufacturing	Corn breakfast foods manufacturing
311412	Frozen specialty food manufacturing	Nationality specialty foods, frozen, manufacturing
311422	Specialty canning	Italian foods canning
311423	Dried and dehydrated food manufacturing	Freeze-dried, or other dried fruits and vegetables
311514	Dry, condensed, and evaporated dairy product manufacturing	Dairy food canning
311612	Meat processed from carcasses	Meat canning (except baby food, pet food, poultry), from purchased carcasses
311615	Poultry processing	Canning poultry (except baby and pet food)
311919	Other snack food manufacturing	Potato sticks manufacturing
311942	Spice and extract manufacturing	Extracts, food (except coffee, meat), manufacturing
311991	Perishable prepared food manufacturing	Food, prepared, perishable, packaged for individual resale
311999	All other miscellaneous food manufacturing	Cocktail mixes, dry, manufacturing

A1C: North American Industrial Codes for Food Related Industries

NAICS	NAICS title	Examples
322212	Folding paperboard box manufacturing	Food containers, sanitary, folding, made from purchased paperboard
322219	Other paperboard container manufacturing	Sanitary food containers (except folding) made from purchased paper or paperboard
322299	All other converted paper product manufacturing	Molded pulp products (e.g., egg cartons, food containers, food trays) manufacturing
325130	Synthetic dye and pigment manufacturing	Food coloring, synthetic, manufacturing
325311	Nitrogenous fertilizer manufacturing	Plant foods, mixed, made in plants producing nitrogenous fertilizer materials
325312	Phosphatic fertilizer manufacturing	Plant foods, mixed, made in plants producing phosphatic fertilizer materials
326111	Plastics bag and pouch manufacturing	Food storage bags, plastics film, single wall or multiwall, manufacturing
326140	Polystyrene foam product manufacturing	Food containers, polystyrene foam, manufacturing
327213	Glass container manufacturing	Food packaging, glass, manufacturing
333241	Food product machinery manufacturing	Food choppers, grinders, mixers, and slicers (i.e., food manufacturing-type) manufacturing
333318	Other commercial and service industry machinery manufacturing	Food warming equipment, commercial-type, manufacturing
333993	Packaging machinery manufacturing	Food packaging machinery manufacturing
335210	Small electrical appliance manufacturing	Food mixers, household-type electric, manufacturing
336999	All other transportation equipment manufacturing	Food (vendor) carts on wheels manufacturing
337127	Institutional furniture manufacturing	Restaurant furniture (e.g., carts, chairs, foodwagons, tables) manufacturing

A1D. North American Industrial Codes for food merchant industries

NAICS	NAICS TITLE	Common keywords
423440	Other commercial equipment merchant wholesalers	Food service equipment (except refrigerated), commercial, merchant wholesalers
423830	Industrial machinery and equipment merchant wholesalers	Food processing machinery and equipment merchant wholesalers
424130	Industrial and personal service paper merchant wholesalers	Disposable plastics products (e.g., boxes, cups, cutlery, dishes, sanitary food containers) merchant wholesalers
424410	General line grocery merchant wholesalers	Food, general-line, merchant wholesalers
424420	Packaged frozen food merchant wholesalers	Frozen foods, packaged (except dairy products), merchant wholesalers
424430	Dairy product (except dried or canned) merchant wholesalers	Prepared foods, frozen dairy, merchant wholesalers
424480	Fresh fruit and vegetable merchant wholesalers	Health foods, fresh fruits and vegetables, merchant wholesalers
424490	Other grocery and related products merchant wholesalers	Baby foods, canned, merchant wholesalers
424690	Other chemical and allied products merchant wholesalers	Food additives, chemical, merchant wholesalers
424910	Farm supplies merchant wholesalers	Animal feeds (except pet food) merchant wholesalers

NAICS	NAICS TITLE	Common keywords
424990	Other miscellaneous nondurable goods merchant wholesalers	Pet supplies (except pet food) merchant wholesalers
445110	Supermarkets and other grocery (except convenience) stores	Food (i.e., groceries) stores
445120	Convenience stores	Convenience food stores
445299	All other specialty food stores	Specialty food stores
446191	Food (health) supplement stores	Nutrition (i.e., food supplement) stores
447110	Gasoline stations with convenience stores	Convenience food with gasoline stations
452311	Warehouse clubs and supercenters	Superstores (i.e., food and general merchandise)
454390	Other direct selling establishments	Street vendors (except food)

A1E. North American Industrial Codes for food related industries

NAICS	NAICS title	Example
541380	Testing laboratories	Food testing laboratories or services
541714	Research and development in biotechnology (except nanobiotechnology)	Biotechnology research and development laboratories or services in food science (except nanobiotechnology research and development)
541715	Research and development in the physical, engineering, and life sciences (except nanotechnology and biotechnology)	Food research and development laboratories or services (except biotechnology and nanotechnology research and development)
561439	Other business service centers (including copy shops)	Internet cafes (i.e., not serving food and beverages)
624210	Community foodservices	Food banks
713990	All other amusement and recreation industries	Hookah lounges (except primarily selling food and beverages)
722310	Food service contractors	Food service contractors, industrial
722330	Mobile foodservices	Mobile food stands
722513	Limited-service restaurants	Fast-food restaurants
811310	Commercial and industrial machinery and equipment (except automotive and electronic) repair and maintenance	Food machinery repair and maintenance services
923120	Administration of public health programs	Food service health inspections
923130	Administration of human resource programs (except education, public health, and veterans' affairs programs)	Food distribution program administration, government
926140	Regulation of agricultural marketing and commodities	Food inspection agencies

Appendix 2

A2A: North American Industrial Codes for Manufacturing Industries

NAICS code	Description
NAICS 311	Food manufacturing
NAICS 312	Beverage and tobacco product manufacturing
NAICS 313	Textile mills
NAICS 314	Textile product mills

NAICS code	Description
NAICS 315	Apparel manufacturing
NAICS 316	Leather and allied product manufacturing
NAICS 321	Wood product manufacturing
NAICS 322	Paper manufacturing
NAICS 323	Printing and related support activities
NAICS 324	Petroleum and coal products manufacturing
NAICS 325	Chemical manufacturing
NAICS 326	Plastics and rubber products manufacturing
NAICS 327	Nonmetallic mineral product manufacturing
NAICS 331	Primary metal manufacturing
NAICS 332	Fabricated metal product manufacturing
NAICS 333	Machinery manufacturing
NAICS 334	Computer and electronic product manufacturing
NAICS 335	Electrical equipment, appliance, and component manufacturing
NAICS 336	Transportation equipment manufacturing
NAICS 337	Furniture and related product manufacturing
NAICS 339	Miscellaneous manufacturing

A2B: Comparing NAICS and Other Codes for Food Manufacturing

NAICS	Title
311	Food manufacturing
3111	Animal Food Manufacturing
3112	Grain & Oilseed Milling
3113	Sugar & Confectionery Product Manufacturing
3114	Fruit & Vegetable Preserving & Specialty food manufacturing
3115	Dairy Product Manufacturing
3116	Animal Slaughtering & Processing
3117	Seafood Product Preparation & Packaging
3118	Bakeries & Tortilla Manufacturing
3119	Other Food Manufacturing
3121	Beverage Manufacturing
SIC	Title
SIC 2000	Food and Kindred Products
2010	Meat products
2020	Dairy Products
2030	Canned, Frozen, and Preserved Fruits, Vegetables, and Food Specialties
2040	Grain mill products
2050	Bakery Products
2060	Sugar and Confectionery Products
2070	Fats and Oils
2080	Beverages
2090	Miscellaneous Food Preparations and Kindred Products
ISIC	Title
C10	Manufacture of food products
C101	Processing and preserving of meat
C102	Processing and preserving of fish, crustaceans and mollusks
C103	Processing and preserving of fruit and vegetables
C104	Manufacture of vegetable and animal oils and fats
C105	Manufacture of dairy products

NAICS	Title
C106	Manufacture of grain mill products, starches and starch products
C107	Manufacture of other food products
C108	Manufacture of prepared animal feeds
C11	Manufacture of beverages
C1101	Distilling, rectifying and blending of spirits
C1102	Manufacture of wines
C1103	Manufacture of malt liquors and malt
C1104	Manufacture of soft drinks; production of mineral waters and other bottled waters

NAICS = North American industrial codes, SIC = standard industrial codes, International SIC (Advameg Inc 2014; NAICS Association 2014; U.S. Bureau of Labor Statistics 2014c; United Nations Statistics Division 2014)

Appendix 3

A3A: European Union NACES Codes for Food Manufacturing

Division	Group	Class	Description	ISIC rev.-4
10			Manufacture of food products	
	10.1		Processing and preserving of meat and meat products	
		10.11	Processing and preserving of meat	1010
		10.12	Processing and preserving of poultry meat	1010
		10.13	Production of meat and poultry meat products	1010
	10.2		Processing and preserving of fish, crustaceans and mollusks	
		10.20	Processing and preserving of fish, crustaceans and mollusks	1020
	10.3		Processing and preserving of fruit and vegetables	
		10.31	Processing and preserving of potatoes	1030
		10.32	Manufacture of fruit and vegetable juice	1030
		10.39	Other processing and preserving of fruit and vegetables	1030
	10.4		Manufacture of vegetable and animal oils and fats	
		10.41	Manufacture of oils and fats	1040
		10.42	Manufacture of margarine and similar edible fats	1040
	10.5		Manufacture of dairy products	
		10.51	Operation of dairies and cheese making	1050
		10.52	Manufacture of ice cream	1050
	10.6		Manufacture of grain mill products, starches and starch products	
		10.61	Manufacture of grain mill products	1061
		10.62	Manufacture of starches and starch products	1062
	10.7		Manufacture of bakery and farinaceous products	
		10.71	Manufacture of bread; manufacture of fresh pastry goods and cakes	1071
		10.72	Manufacture of rusks and biscuits; manufacture of preserved pastry goods and cakes	1071
		10.73	Manufacture of macaroni, noodles, couscous and similar farinaceous products	1074

Excerpt from Annex 1 of Regulation (EC) NO 1893/2006 of the European parliament Euro-Lex (2006)

A3B: European Union NACE Codes for Food Manufacturing

Division	Group	Class	Description	ISIC rev.-4
	10.8		Manufacture of other food products	
		10.81	Manufacture of sugar	1072
		10.82	Manufacture of cocoa, chocolate and sugar confectionery	1073
		10.83	Processing of tea and coffee	1079*
		10.84	Manufacture of condiments and seasonings	1079*
		10.85	Manufacture of prepared meals and dishes	1075
		10.86	Manufacture of homogenized food preparations and dietetic food	1079*
		10.89	Manufacture of other food products n.e.c.	1079*
	10.9		Manufacture of prepared animal feeds	
		10.91	Manufacture of prepared feeds for farm animals	1080*
		10.92	Manufacture of prepared pet foods	1080*
11			Manufacture of beverages	
	11.0		Manufacture of beverages	
		11.01	Distilling, rectifying and blending of spirits	1101
		11.02	Manufacture of wine from grape	1102*
		11.03	Manufacture of cider and other fruit wines	1102*
		11.04	Manufacture of other non-distilled fermented beverages	1102*
		11.05	Manufacture of beer	1103*
		11.06	Manufacture of malt	1103*
		11.07	Manufacture of soft drinks; production of mineral waters and other bottled waters	1104

Extracted from Annex 1 Regulation (EC) NO 1893/2006 of the European parliament (Euro-Lex 2006)

Appendix 4: United States Economic Indicators (2008–2018)

Year	GDP (US$)[a]	GDP per capita (US$)	GDP %-growth	Inflation (%)	Unemployed (%)	Government debt (in % of GDP)
2018	▲20,494.1	▲62,606	▲2.9%	▲2.4%	▼3.9%	▲77.8%
2017	▲19,485.4	▲59,895	▲2.4%	▲2.1%	▼4.4%	▼76.1%
2016	▲18,707.2	▲57,878	▲1.6%	▲1.3%	▼4.9%	▲76.4%
2015	▲18,219.3	▲56,770	▲2.9%	▲0.1%	▼5.3%	▼72.5%
2014	▲17,521.9	▲54,993	▲2.6%	▲1.6%	▼6.2%	▲73.7%
2013	▲16,691.5	▲52,737	▲1.8%	▲1.5%	▼7.4%	▲72.2%
2012	▲16,155.3	▲51,404	▲2.2%	▲2.1%	▼8.1%	▲70.3%
2011	▲15,517.9	▲49,736	▲1.6%	▲3.1%	▼8.9%	▲65.8%
2010	▲14,964.4	▲48,311	▲2.5%	▲1.6%	▲9.6%	▲60.8%
2009	▼14,418.7	▼46,909	▼−2.8%	▼−0.3%	▲9.3%	▲52.3%
2008	▲14,718.6	▲48,302	▼−0.3%	▲3.8%	▲5.8%	▲39.3%

Reference: adapted from Wikipedia (2020). [a]GDP in billion dollars

References

Advameg Inc (2014) Food & kindred products encyclopedia of business, 2nd ed. Retrieved from http://www.referenceforbusiness.com/industries/Food-Kindred-Products/index.html

Agri-Food Canada (2010) Agri-food past, present and future report China. Retrieved from http://www.ats-sea.agr.gc.ca/asi/3833-eng.htm

Anon (2015) Ministry of food processing – Government of India. Retrieved from http://www.mofpi.nic.in/

Anonymous (2012) "Market Watch 2012":The Malaysian food industry. Retrieved from http://www.malaysia.ahk.de/fileadmin/ahk_malaysia/Market_reports/The_Malaysian_Food_Industry.pdf

Anonymous (2014) ASEAN Food and Beverage Alliance (AFBA). Retrieved from http://foodindustry.asia/ASEAN-food-and-beverage-alliance

Australian Trade Commission – Austrade (2014) Food and beverage to Malaysia. Retrieved from http://www.austrade.gov.au/Export/Export-Markets/Countries/Malaysia/Industries/Food-and-beverage#.U3i8C00U9LN

Bradbury J (2012) China and Hong Kong: food opportunities for Maine. Retrieved from https://www.mitc.com/trade/FoodIndustryReportChina.pdf.pdf

BusinessVibes (2014) Facts and figures in global food and beverage industry. Retrieved from http://www.businessvibes.com/blog/facts-and-figures-Global-food-and-beverage-industry#sthash.vp9hYeB2.dpuf

Canadian Industry Statistics (CIS) (2014) Definition: food manufacturing – (NAICS 311) Retrieved from https://www.ic.gc.ca/app/scr/sbms/sbb/cis/definition.html?code=311

Canning P (2011) A revised and expanded food dollar series: a better understanding of our food costs, ERR-114, U.S. Department of Agriculture, Economic Research Service, February 2011. A revised and expanded food dollar series: a better understanding of our food costs. Retrieved from http://www.ers.usda.gov/media/131096/err114_reportsummary.pdf

Department for Environment Food and Rural Affairs (UK) (2013) Food statistics pocketbook 2013. Retrieved from https://www.gov.uk/government/uploads/system/uploads/attachment_data/file/243770/foodpocketbook-2013report-19sep13.pdf

Department of Agriculture Fisheries and Forestry (2011) Australian food statistics 2009–10. Retrieved from www.daff.gov.au/__data/assets/pdf_file/0011/.../food-stats2009-10.pdf

Department of Statistics Malaysia Official Portal (2014) Report on the survey of manufacturing industries. Retrieved from http://statistics.gov.my/portal/index.php?option=com_content&view=article&id=363&lang=en

Euro-Lex (2006) Regulation (EC) NO 1893/2006 of the European parliament and of the council of 20 December 2006 establishing the statistical classification of economic activities NACE Revision 2 and amending Council Regulation (EEC) No 3037/90 as well as certain EC Regulations on specific statistical domains. Retrieved from http://eur-lex.europa.eu/legal-content/EN/ALL/;ELX_SESSIONID=v4zSJGSQmHX11DpfShbyQvNhhpppmBXfhvXSqgLgxpkWkNX5dDWn!-415114057?uri=CELEX:32006R1893

Federation of Indian Chambers of Commerce and Industry (2015) Food processing. Retrieved from http://www.ficci.com/sector.asp?secid=15

Ferman J (2008) U.S. Department of Commerce Industry Report: Food Manufacturing NAICS 311. Retrieved from http://trade.gov/td/ocg/report08_processed-foods.pdf

Food & Drink Europe (2012) Data & trends of the European food and drink industry 2012. Retrieved from http://www.fooddrinkeurope.eu/uploads/publications_documents/Data__Trends_(interactive).pdf

Food Export Association of the Midwest USA (2011) Malaysia country profile. Retrieved from https://www.foodexport.org/Resources/CountryProfileDetail.cfm?ItemNumber=1029

Food Export Association of the Midwest USA (2015) Brazil country profile. Retrieved from https://www.foodexport.org/Resources/CountryProfileDetail.cfm?ItemNumber=1023

Gehlhar M, Regmi A (2005) Factors shaping global food markets new directions in global food markets, (United States Department of AgricultureAgriculture information bulletin ; no. 794). Retrieved from http://www.ers.usda.gov/media/872111/aib794_002.pdf

Hawkes C (2008) Agro-food industry growth and obesity in China: what role for regulating food advertising and promotion and nutrition labelling? Obes Rev 9(S1):151–161

Hong Kong Trade Development Council (2013) Processed food and beverages industry in Hong Kong Retrieved from http://product-industries-research.hktdc.com/business-news/article/Food-Beverages/Processed-Food-and-Beverages-Industry-in-Hong-Kong/hkip/en/1/1X3VQV4B/1X006S3L.htm

IMAP (2010) Food and beverage industry global report — 2010. Retrieved from http://www.proman.fi/sites/default/files/Food%20%26%20beverage%20Global%20report%202010_0.pdf

Invest India (2015) Food processing. Retrieved from http://www.makeinindia.com/sector/food-processing/

IRS (2014) Food industry overview – complete version. Retrieved from http://www.irs.gov/Businesses/Food-Industry-Overview%2D%2D-Complete-Version

Jerger C (2012) Brazil's food and beverage market report. Retrieved from http://www.s-ge.com/de/filefield-private/files/43077/field_blog_public_files/22183

Jun W, Tang C, Li J, Zhang S, Shen J, Zhang R (2013) China food manufacturing annual report. Retrieved from http://gain.fas.usda.gov/Recent%20GAIN%20Publications/China%20Food%20Manufacturing%20Annual%20Report_Beijing%20ATO_China%20-%20Peoples%20Republic%20of_1-31-2013.pdf

Krieger S, Schiefer G (2006) Quality systems in the Agri-food industry – implementation, cost, benefit and strategies. Retrieved from http://ageconsearch.umn.edu/bitstream/25795/1/pp060616.pdf

Lubentsova O (2011) Russian federation, food processing ingredients, the food processing sector in Russia. Retrieved from http://gain.fas.usda.gov/Recent%20GAIN%20Publications/Food%20Processing%20Ingredients_Moscow%20ATO_Russian%20Federation_12-8-2011.pdf

Macro Economic Meter (2014a) Brazil: manufacturing statistics. Retrieved from http://mecometer.com/infographic/brazil/manufacturing-statistics/

Macro Economic Meter (2014b) Hong Kong SAR, China : manufacturing statistics. Retrieved from http://mecometer.com/infographic/hong-kong-sar-china/manufacturing-statistics/

Macro Economic Meter (2014c) India : manufacturing statistics. Retrieved from http://mecometer.com/infographic/india/manufacturing-statistics/

Macro Economic Meter (2014d) Malaysia: manufacturing statistics. Retrieved from http://mecometer.com/infographic/malaysia/manufacturing-statistics/

Manufacturing Institute (2013) U.S. manufacturing is the world's tenth largest economy. Retrieved from http://www.themanufacturinginstitute.org/Research/Facts-About-Manufacturing/Economy-and-Jobs/10th-Largest-Economy/10th-Largest-Economy.aspx

Manufacturing Institute (2014) Manufacturing sectors by output and employment Retrieved from http://www.themanufacturinginstitute.org/Research/Facts-About-Manufacturing/Economy-and-Jobs/Sector-Output/Sector-Output.aspx

Martinez S (2010) Local food systems; concepts, impacts, and issues. DIANE Publishing Company, 87pp. https://books.google.com.gh/books?id=wVTjlY75WW8C

Mecometer.com (2014) China: manufacturing statistics. Retrieved from http://mecometer.com/infographic/china/manufacturing-statistics/

Mishra PK (2010) Bottlenecks in Indian food processing industry – survey 2010 Retrieved from http://www.ficci.com/sedocument/20073/Food-Processing-bottlenecks-study.pdf

NAICS Association (2014) NAICS identification tools. Retrieved from http://www.naics.com/search/

NAICS Association (2017) What is a NAICS code and why do I need one. Retrieved from https://www.naics.com/what-is-a-naics-code-why-do-i-need-one/

National Restaurant Association (2014) Industry impact. Retrieved from http://www.restaurant.org/Industry-Impact/Employing-America/Economic-Engine

National Retail Federation (2011) Retail powers the U.S. Economy. Retrieved from https://nrf.com/sites/default/files/Documents/Value%20of%20Retail.pdf

New Zealand Trade & Enterprise (2012) Exporter guide: food and beverage in China – market profile January 2012 Retrieved from http://www.med.govt.nz/sectors-industries/food-beverage/pdf-docs-library/information-project/market-profile-china.pdf

NewZealand Trade & Enterprise (2011) Food and beverage in Malaysia. Retrieved from https://www.nzte.govt.nz/en/export/market-research/food-and-beverage/food-and-beverage-market-in-malaysia/

Oddy DJ, Drouard A (2016) The food industries of Europe in the nineteenth and twentieth centuries. From http://www.ashgate.com/isbn/9781409454397

Parry DE (1999) Food industry costs, profits, and productivity. Food Cost Review, 1950–97 19. 2021. Retrieved from https://www.ers.usda.gov/webdocs/publications/41035/15330_aer780d_1_.pdf?v=0

Research and Markets (2015) Research and markets: global food and beverage retail industry 2012–2017: trends, profits and forecast analysis – market to reach US $5,776 billion in 2017. Retrieved from http://www.reuters.com/article/2012/07/13/idUS165413+13-Jul-2012+BW20120713 and http://www.researchandmarkets.com/research/kbvmpk/Global_food_and_be

Ridley L (2009) Food retailing in Asia. Retrieved from http://www.depi.vic.gov.au/agriculture-and-food/food-and-fibre-industries/market-access-and-competitiveness/target-markets/food retailing in asia

Sadiku MNO, Musa SM, Ashaolu TJ (2019) Food industry: an introduction. Inter J Trend Sci Res Develop 3(4):128–130. https://www.ijtsrd.com/papers/ijtsrd23638.pd

Stern A (2016) Food industry influence on dietary advice in the United States. In: Proceedings of the national conference on undergraduate research (NCUR) 2016, April 7–9, 2016. University of North Carolina, Asheville, NC. http://libjournals.unca.edu/ncur/wp-content/uploads/2021/2006/1706-Stern-Aaron-FINAL.pdf

U.S. Bureau of Economic Analysis (BEA) (2014) Gross-Domestic-Product-(GDP)-by-Industry Data. Retrieved from http://www.bea.gov/industry/gdpbyind_data.htm

U.S. Bureau of Labor Statistics (2014a) Food and beverage stores: NAICS 445. Retrieved from http://www.bls.gov/iag/tgs/iag445.htm

U.S. Bureau of Labor Statistics (2014b) Food services and drinking places: NAICS 722. Retrieved from http://www.bls.gov/iag/tgs/iag722.htm

U.S. Bureau of Labor Statistics (2014c) Manufacturing: NAICS 31-33. Retrieved from http://www.bls.gov/iag/tgs/iag31-33.htm#about

U.S. Department of Commerce (2008) U.S. Department of Commerce Industry Report: Food Manufacturing NAICS 311. Retrieved from http://trade.gov/td/ocg/report08_processedfoods.pdf

United Nations Statistics Division (2014) Detailed structure and explanatory notes ISIC Rev.4 (International Standard Industrial Classification of All Economic Activities, Rev.4). Retrieved from http://unstats.un.org/unsd/cr/registry/regcst.asp?Cl=27&Lg=1

United States Census (2014) 2011 County business patterns (NAICS). Retrieved from http://censtats.census.gov/cgi-bin/cbpnaic/cbpsect.pl or http://censtats.census.gov/cgi-bin/cbpnaic/cbpdetl.pl

USDA Economic Research Service (2019) Ag and food sectors and the economy. Retrieved from https://www.ers.usda.gov/data-products/ag-and-food-statistics-charting-the-essentials/ag-and-food-sectors-and-the-economy/

USDA Economic Research Services (2012) Global food markets. Retrieved from http://www.ers.usda.gov/topics/international-markets-trade/global-food-markets/global-food-industry.aspx

USDA. Economic Research Services (2014) Retailing & wholesaling: retail trends. Retrieved from http://www.ers.usda.gov/topics/food-markets-prices/retailing-wholesaling/retail-trends.aspx

Wikipedia (2020) Economy of the United States. Retrieved from https://en.wikipedia.org/wiki/Economy_of_the_United_States#21st_century

Wilkinson J (2004) The food processing industry, globalization and developing countries. E J Agri Devel Econ 1(2):184–201. Retrieved from http://ageconsearch.umn.edu/bitstream/11999/1/01020184.pdf

Wilkinson J, Flocha (2009) Agro-industry trends, patterns and development impacts. In: da Silva CA, Baker D, Shepherd AW, Jenane C, Miranda-da-Cruz S (eds) Agro-industries for development. The Food and Agriculture Organization of the United Nations and The United Nations Industrial Development Organization by arrangement with CAB International, Cambridge, MA

Wilkinson J, Rocha R (2009) Agro-industry trends, patterns and development impacts. In: da Silva CA, Baker D, Shepherd AW, Jenane C, Miranda-da-Cruz S (eds) Agro-industries for development. The Food and Agriculture Organization of the United Nations and The United Nations Industrial Development Organization by arrangement with CAB International, Cambridge, MA, pp 46–91. https://books.google.co.uk/books?id=GNXAggaNCd48C&lpg=PA46&ots=Sgiz-bXKY44&lr&pg=PA46#v=onepage&q&f=false

Zepeda L, Li J (2006) Who buys local food? J Food Dist Res 37(856-2016-56238):1–11. https://ageconsearch.umn.edu/record/7064/

Food Regulatory Agencies 3

1 Introduction

1.1 Overview of Agencies

The US has one of the safest Agri-food systems anywhere in the world (Buchanan 2000; Strauss 2011; Li et al. 2018). There are several federal agencies for regulating foods and food products, but only some have enforcement authority (Fig. 3.1); Food and Drug Administration (FDA), the Food Safety and Inspection Service (FSIS) section of the U.S. Department of Agriculture (USDA).[1] The main regulatory agencies associated with food safety are discussed in this chapter (Buchanan 2000; International Food Information Council 2014; Saxowsky 2010; USDA Food Safety and Inspection Service 2014b).

1.1.1 What Is Regulated by US Agencies

About 80% of the food supply is regulated by one US regulatory agency, the FDA (Table 3.1). The value of the US food trade is estimated at US$417 billion for domestic supplies and $49 billion in imported food excluding products, which are unregulated by the FDA (Food and Drug Administration 2007). Drawing up a plan which manages to sustain a culture of food safety is an enormous challenge.

1.1.2 Key Milestones in US Food Law

President Abraham Lincoln signed a bill during 1862, which led to the founding of the Department of Agriculture (now the USDA) headed by Isaac Newton. Charles M. Wetherill headed the Division of Chemistry, within the USDA. Later the division of Chemistry was renamed as Bureau of Chemistry. 1n 1883 Professor Harvey W. Wiley from Purdue University succeeded as Chief Chemist and head of the Bureau of Chemistry where he worked to raise awareness of food adulteration and other aspects of food safety. At about this time, many bills dealing with pure foods were debated in congress but were ultimately unsuccessful due to strong lobbying by various food interests. In 1927 the USDA Bureau of Chemistry was renamed the "Food, Drug, and Insecticides Administration" and in 1940 the name was changed to, Food and Drug Administration (FDA). The 1940 name change also coincided with the FDA moving from the USDA to the US Federal Security Agency, which was transformed, into the Department of Human Health Services (DHHS). It is noteworthy that the origins of the FDA is closely tied to the history of the USDA, which is in turn strongly tied to agriculture. Some of the major milestones in US food law are summarized in Table 3.2 and Table 3.3 (Merrill and Francer 2000; U. S. Food and Drug Administration, 2006a, 2009).

[1] In this text we reserve the abbreviation FDA to refer to the US Food & Drugs Administration (USFDA). Though other FDA's are operating around the world, most are patterned after the USFDA which is the first of its kind historically.

R. Owusu-Apenten, E. R. Vieira, *Elementary Food Science*, Food Science Text Series,
https://doi.org/10.1007/978-3-030-65433-7_3

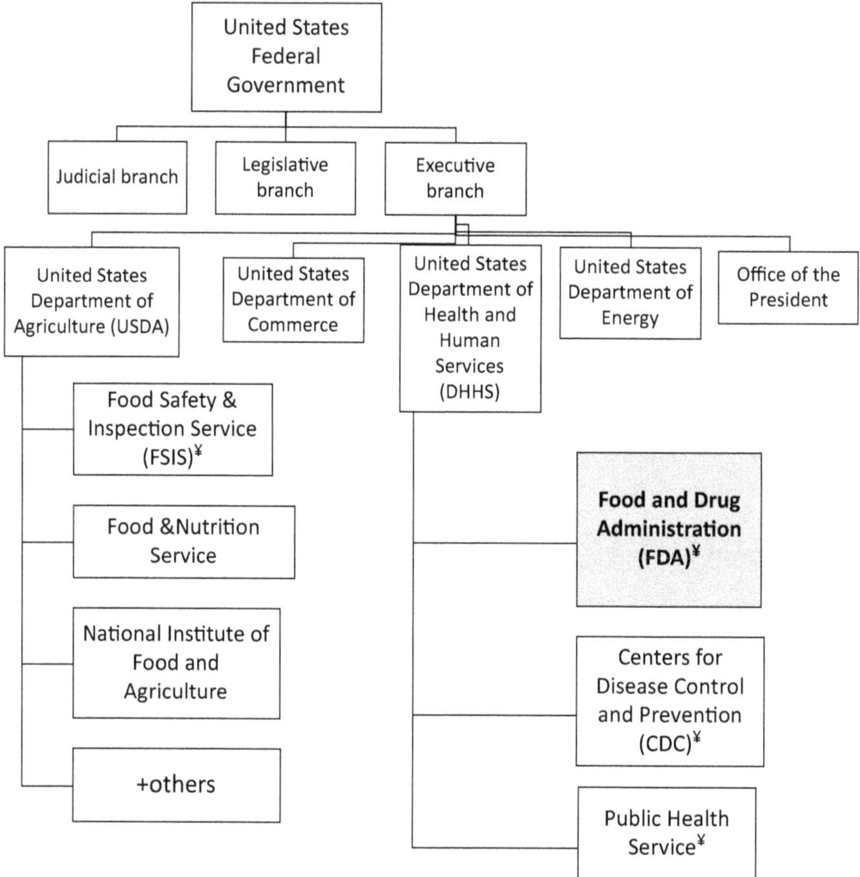

Fig. 3.1 US federal departments and agencies involved with food and public health regulatory affairs. Drawn using listings from (Wikipedia 2014c). ¥ Shows agencies with food regulatory function

Table 3.1 USDA regulation of registered domestic food facilities

Food manufacturers	(> 44,000
Warehouses, storage tanks, & grain elevators	113,000
Farms	200,0000
Restaurants and institutional food service establishments	935,000
Supermarkets, grocery stores, and other food outlets,	114,000
Registered foreign facilities producing for American markets	189,000

Regulatory agencies employ a range tools (*standards, regulations, ordinances,* and *codes*) to conduct enforcement. A *standard* can be described as "a measure" used to compare quality, whereas a *regulation* is a governmental control with the force of law (Table 3.4). An *ordinance, law,* or *statute* gives the legal right for governmental agencies to publicize and put regulations into effect. A *code* often equals a regulation or statue and is a collection of regulations, statutes, and procedures. One of the roles of regulation is to protect the public from foodborne illness. Food manufacturers, retailers, wholesalers, and foodservice establishments are required to meet standards for critical areas of their operations that require effective control measures.

1.2 Food and Drug Administration

The FDA is one of three agencies of the Department of Human Health Services, which is concerned with food safety. As relates to food

Table 3.2 Key milestones from US food law (1862–1962)

1862 – President Lincoln appoints Charles M. Wetherill, to the new Department of Agriculture
1880 – Peter Collier (Chief chemist USDA) petitions for a food and drug law
1883 – Dr. Harvey W. Wiley appointed Chief chemist, of the Bureau of Chemistry
1897 – Tea Importation Act for mandatory inspection of imported tea
1898 – Association of Official Agricultural Chemists (later AOAC International) forms a Committee on Food Standards chaired by Dr. Wiley
1902 – US Congress funds a study of chemical preservatives and colors by Bureau of chemistry
1906 – The first Food and Drugs Act prohibits sale of misbranded, adulterated foods and drinks
1907 – First Certified Color Regulations, approves seven food colors
1913 – Gould Amendment, early food labeling legislation
1933 – FDA proposes revision of 1906 Food and Drugs Act
1938 – The Federal Food, Drug, and Cosmetic (FDC) Act of 1938 comes into force
1939 – First Food Standards (canned tomatoes, tomato purée, and tomato paste).
1940 – FDA transferred from the Department of Agriculture to the Federal Security Agency
1949 – First FDA guidance "Procedures for the Appraisal of the Toxicity of Chemicals in Food"
1950 – Oleomargarine Act, clear labeling of margarine, to differentiate it from butter
1958 – Food Additives Amendment, introduces safety-testing, ban for carcinogenic additives.
1960 – Color Additive Amendment requires safety testing for new colors
1962 – Consumer Bill of Rights
1966 – Fair Packaging and Labeling Act
1968 – Animal Drug Amendments to the Food, Drug, and Cosmetic Act-Section 512
1969 – The White House Conference on Food, Nutrition, and Health; FDA of GRAS list

Source: FDA Consumer magazine (U.S. Food and Drug Administration 2006a) and (U. S. Food and Drug Administration 2009)

Table 3.3 Key milestones from US food law (1970–2010)

1970 – Environmental Protection Agency established; EPA assumes FDA oversight of pesticides
1973 – Low-acid food processing regulations; regulations regarding canning of low acid foods
1976 – Vitamins and minerals amendments stop vitamins and minerals regulations as drugs
1990 – Nutrition Labeling and Education Act (NLEA)
1992 – Nutrition facts, per-serving nutritional information, required under NLEA
1994 – Dietary Supplement Health and Education Act (DSHEA)
1997 – Food and Drug Administration Modernization Act
2000 – Dietary supplements structure function claims
2002 – Public Health Security and Bioterrorism Preparedness and Response Act
2003 – FDA mandatory labeling for trans-fat content
2004 – Food Allergy Labeling and Consumer Protection Act, mandatory labeling of food allergens 2009 – Country of Origin Labelling for meat, fish and shellfish; fruits and vegetables nuts
2010 – The Food Safety Modernization Act

Source: U. S. Food and Drug Administration (2009)

Table 3.4 What the FDA regulates

Animal and veterinary
Biologics
Cosmetics
Drugs
Food
Medical devices
Radiation-emitting products
Tobacco products
Vaccines, blood, and biologics

safety, the FDA works alongside of the Centre of Disease Control & Prevention (CDC) and the US Public health services (USPHS). The history of the FDA was summarized in the centenary article describing the life of Dr. Harvey (U.S. Food and Drug Administration 2006b).

1.2.1 What Products Does the FDA Regulate?

The FDA alone regulates about 80% of the domestic food supply. However, meat, poultry and, eggs are overseen by the USDA-FSIS (Sect. 1.5). In addition, the FDA is responsible also for regulating a host of non-food products including, drugs, cosmetics, tobacco products, dietary supplements, veterinary medicines and livestock feed (Table 3.4). Foodstuffs for companion animals (or pets) are also under the jurisdiction of the FDA (Appendix 1).

The FDA has two major offices concerned with foods (Table 3.5); (i) Centre for Food Safety and Applied Nutrition (CFSAN) and (ii) Office of Regulatory Affairs (ORA). The FDA's stated mission is, (a) ensuring a food supply which is safe, sanitary, wholesome, and honestly labeled and (b) protection of consumers and ensuring public health and compliance with FDA regulated products and minimizing risk associated with those products. FDA activities are outlined in Appendices 1–2 and discussed in the following sections (U.S. Food and Drug Administration 2014a). A useful overview of the FDA, in particular focusing on the organizational structure and recent changes in the physical facilities (buildings and offices), is available from Wikipedia (Wikipedia 2020).

1.2.2 Recalls, Outbreaks & Emergencies

The FDA hosts websites listing recall notices of foods, which are subject to misbranding, contamination or risks of food borne illness. Adverse events, hazardous products or outbreaks can be reported by members of the public, health professionals or by industry. Manufacturers as well as the FDA may post voluntary recall notices (U. S. Food and Drug Administration 2014b).

Table 3.5 The FDA food regulatory system

1) Department of Health and Human Services (DHHS)
2) Food and Drug Administration (FDA) See Retrieved from https://www.fda.gov/food
3) Center For Food Safety And Applied Nutrition (CFSAN)
4) Office of regulatory affairs
Food regulatory responsibilities:
– Enforce food safety laws for imported or domestically produce food, excluding meat, poultry, and eggs
– Regulations for 80% of U.S. food supply
– Nutrition labeling and standards
– Evaluate GRAS (generally recognized as safe) status for additives
– Inspect food establishments & warehouses
– HACCP systems
– Microbial standards
– Good Manufacturing Practices (GMPs) and Good Agricultural Practices (GAPs) standard
– Consumer education

Source: Adapted from Saxowsky (2010)

1.2.3 Guidance & Regulation

The Guidance and regulatory function of the FDA covers several important areas, for instance, (a) the food safety modernization act (FSMA; 2011) and (b) Food Facilities Registration (FFR), which is part of Bioterrorism act (2003). Briefly, FSMA is a new comprehensive food safety legislation signed into effect by President Obama in 2011 (Chap. 10). The FFR requires that food manufactures, packers and storage firms register their facilities in order that the authorities may determine their location rapidly in the event of a bioterrorism act. The responsibilities of the FDA in relations to guidance and regulation are summarized in Fig. 3.2.

1.2.4 Retail Food Protection: The Food Code

The FDA (CDC and FSIS) produces a food code setting out sanitation and food safety standards for State, City, Country, and tribal agencies, which have day-to-day responsibility for regulating food retail and foodservice operations (Chap. 25). Some 50 states, and 3 out of 6 territories have adopted the food code as a model for local food sanitation regulation. The first iteration of the "food code" appeared in 1993 but the standards and stipulations were based on food sanitation codes published beginning from 1934 (Tables 3.6 and 3.7). Other editions of the food code appeared from 1995 and thereafter at 1–2 year intervals. The food code provides a model for local agencies to develop their own codes. State and local food sanitation laws can be tighter, but not looser than the FDA code. Previous versions and the latest food code are available from the FDA website (U.S. Food and Drug Administration 2013).

1.2.5 Ingredients, Packaging and Labeling

The FDA oversees ingredients, packaging and labeling. The FDA works to ensure public understanding of important legal terms, as originally described by the FD&C Act (1938).

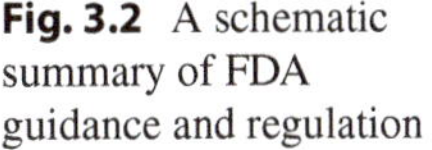

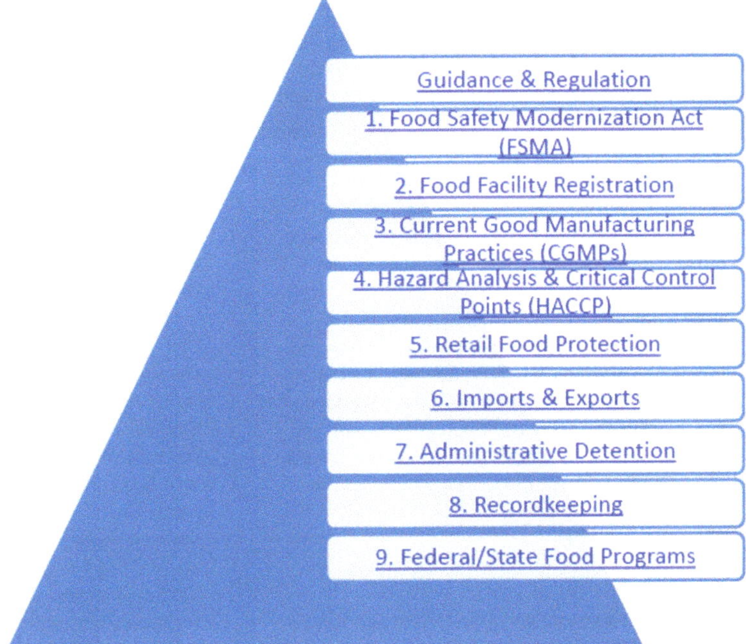

Fig. 3.2 A schematic summary of FDA guidance and regulation

Guidance and regulations are provided with respect to food allergens. For instance, the FDA established rulings in 2013 concerned with gluten-free labeling to meet the needs of an estimated three million celiac disease sufferers in the US. From August of 2014, the FDA set a new upper concentration limit for gluten-free labeling as 20 ppm (20 mg/kg food). Only foods containing less than 20-ppm gluten can be labelled legally gluten-free. Following the introduction of such new regulation, FDA sought to provide education and outreach to industry, in order to improve compliance (U.S. Food and Drug Administration 2014d) (Table 3.8).

1.2.6 Food Additives and Color Additives

The FDA has oversight regarding the safe use of indirect and direct additives. The indirect food additives refers to trace compounds or incidental additives such as substances that enter foods by leaching from packaging. The safety of additives is determined by considering a range of factors including, (a) composition and properties, (b) tolerable upper limits, (c) immediate or long-term effects and so-called (d) safety factors- numerical values imposed on intake limits to allow for possible differences in the physiological responses between humans as well as between humans and test animals. Additives perform a wide range of technological functions in foods including, shelf life extension (preservative), enhancement of nutritional value, improvements of sensory characteristics (texture, taste, appearance).

Some food substances require pre-approval to achieve GRAS (Generally Recognized as Safe) status. Acquiring GRAS status occur by petition (U. S. Food and Drug Administration 2018). Typically, a company wishing to introduce a new food substance must prepare a detailed dossier for the proposed chemical, containing a lot of information about its safety. It is after the GRAS petition has undergone scrutiny by FDA experts that the food substance is designated as GRAS. GRAS status is granted when the FDA declares that there are "no further questions" regarding an application. Information on GRAS status including an inventory of all compounds currently certified as GRAS is available from the FDA website (U.S. Food and Drug Administration 2014c). Food additives in common use before 1958 are presumed safe.[2]

[2]See 21 CFR Part 182 for details of the GRAS procedure.

Table 3.6 US food sanitations codes (1934–1982)

1934 – Restaurant sanitation regulations, proposed by the U.S. Public Health Service in cooperation with the conference of state and territorial health officers and the National Restaurant Code Authority
1935 – An Ordinance Regulating Food and Drink Establishments (Recommended by U.S. Public Health Service), December 1935, Mimeographed
1938 – Ordinance and Code Regulating Eating and Drinking Establishments, Recommended by the U.S. Public Health Service, March 1938, Mimeographed
1940 – Ordinance and Code Regulating Eating and Drinking Establishments, Recommended by the U.S. Public Health Service, June 1940, Mimeographed
1943 – Ordinance and Code Regulating Eating and Drinking Establishments, Recommended by the US Public Health Service, 1943, FSA, Public Health Bulletin No. 280 (Republished in 1955, DHEW, PHS Publication No. 37)
1957 – The Vending of Foods and Beverages -A Sanitation Ordinance and Code, 1957 Recommendations of the Public Health Service, DHEW, PHS Publication No. 546
1962 – Food Service Sanitation Manual Including A Model Food Service Sanitation Ordinance and Code, 1962 Recommendations of the Public Health Service, DHEW, PHS Publication No. 934
1965 – The Vending of Food and Beverages A Sanitation Ordinance and Code, 1965 Recommendations of the Public Health Service, DHEW, PHS Publication No. 546
1976 – Food Service Sanitation Manual Including A Model Food Service Sanitation Ordinance, 1976 Recommendations of the Food and Drug Administration, DHEW /PHS/FDA, DHEW Publication No. (FDA) 78-2091
1978 – The Vending of Food and Beverages Including A Model Sanitation Ordinance, 1978 Recommendations of the Food and Drug Administration, DHEW /PHS/FDA, DHEW Publication No. (FDA) 78-2091
1982 – Retail Food Store Sanitation Code, 1982 Recommendations of the Association of Food and Drug Officials and U.S. Department of Health and Human Services, Public Health Service, Food and Drug Administration, AFDO/HHS Publication

Table 3.7 US food code (1993–2013)

1993-food code, 1993 recommendations of the US Public Health Service, Food and Drug Administration, National Technical Information Service Publication PB94-113941
1995-food code, 1995 recommendations of the US Public Health Service, Food and Drug Administration, National Technical Information Service Publication PB95-265492
1997-food code, 1997 recommendations of the US Public Health Service, Food and Drug Administration, National Technical Information Service Publication PB97-133656
1999-food code, 1999 recommendations of the US Public Health Service, Food and Drug Administration, National Technical Information Service Publication PB99-115925
2001-food code, 2001Recommendations of the US Public Health Service, Food and Drug Administration, National Technical Information Service Publication PB2002-100819
2003-supplement to the 2001 food code, National Technical Information Service Publication PB2003–106843
2005-food code, 2005 recommendations of the US Public Health Service, Food and Drug Administration, National Technical Information Service Publication PB2005-102200
2007-supplement to the 2005 food code, National Technical Information Service Publication PB2007–112622
2009-food code, 2009 recommendations of the US Public Health Service, Food and Drug Administration, National Technical Information Service Publication PB2009-112613
2011-supplement to the 2009 food code, National Technical Information Service

Table 3.8 FDA oversight related to Ingredients & Packaging Definitions

Allergens
Food Additives & Ingredients
Generally Recognized as Safe (GRAS)
Packaging & Food Contact Substances (FCS)
Irradiated food & packaging
Labeling & nutrition
Environmental decisions

1.2.7 Packaging and Food Contact Substances

The FDA regulates food packaging materials and food contact substances (FCS), which are "*materials used for manufacturing, packing, packaging, transporting or holding foods*". Common FCS include plastic, paper, cardboard and associated compounds such as plasticizers, coatings, and adhesives. The FCS notification process established in 2000 enables any interested party to submit documentations for new FCS for evaluation by FDA experts. The overall safety determination is primarily based on a so-called threshold concept according to which the proposed new

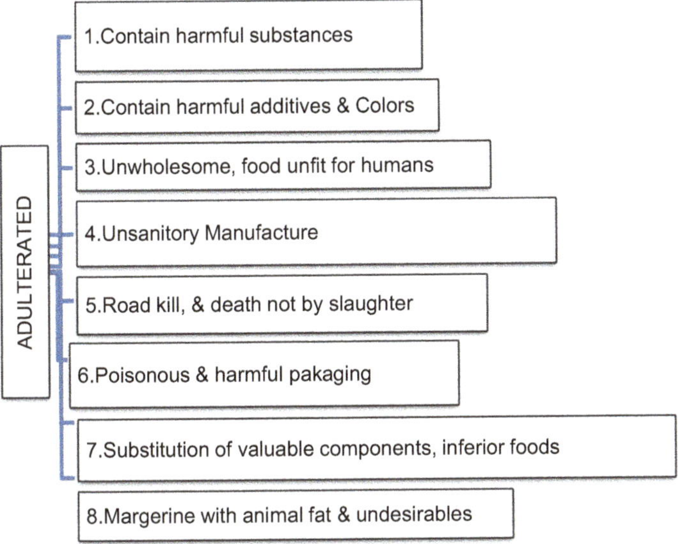

Fig. 3.3 Examples of food adulteration (21 CFR 601-Defintions)

usage should not push the cumulative exposure of a FCS above 3 mg per person per day (U.S. Food and Drug Administration 2014e).

1.2.8 Foodborne Illness

About 10 years ago there were an estimated 48 million cases of foodborne diseases in the US annually resulting in 128,000 hospitalizations and 3000 deaths (Chap. 8 Sect. 1.5) (Scallan et al. 2011). Foodborne diseases are caused by many hazards chief amongst which are microorganisms and viruses. The FDA oversight covers incidence and outbreaks and also assessment of exposure risk. There is an FDA website dedicated to consumer information and guidance on safe purchase, storage, handling and preparation of different food groups (U.S. Food and Drug Administration 2014b). Foodborne disease is discussed further in Chap. 8.

1.2.9 Food Adulteration

There are nine major forms of food adulteration defined in the *meat inspection* code 21 U.S. Code § 601 in rather precise terms (USDA Food Safety and Inspection Service 2014a).[3] Some food adulteration offences are not difficult to detect (Fig. 3.3 and Appendix 3). For example, there are tests to detect sources of contamination such as that from rodents (hairs, pellets, or urine), insects, dirt, and other detritus. Also, if a food is putrid, this can be detected by the ordinary human senses, a fact that is well known and accepted. However, the detection of decomposition is not easy as scientists may not agree on what constitutes decomposition of a particular food. Frequently, disputes based on the extent of decomposition of a food, therefore, have to be settled in court.

The presence of a variety of hazards in foods may constitute adulteration. However, it is not easy to enforce the section of the law prohibiting substances that may be deleterious to the health of the consumer. For example, pathogenic bacteria, such as the *Salmonella* organisms, can cause disease and even death. However, poultry and some other foods may contain these organisms, naturally. The testing for this group of organisms requires at least 2 days for results, another factor that serves to impede enforcement.

1.2.10 Misbranding

Misbranding is defined in the meat inspection code 21 U.S. Code § 601 (USDA Food Safety and Inspection Service 2014a). Misbranding refers to any one of 12 different offenses related to label-

[3]A similar description for adulteration appears for other foods (21 CFR 342). In all matters legal, the wording is usually exact. Readers affected by this issue should seek specialist help.

ing. Products may classed as misbranded for anyone of the reasons listed below (Table 3.9). If a standard of identity has been set up for particular food, the food can contain only the ingredients specified in the standard (21 CFR. 130). Where a food is found to contain ingredients or additives different from the food standard, the former cannot be traded using the same name as the food for which the standard exists. As an example, sulfur dioxide is allowed in some food and might be used to provide good color in ketchup. However, there exists a standard of identity for ketchup in which sulfur dioxide is not included. Therefore, where ketchup is found to contain sulfur dioxide, it might be seized and destroyed.[4]

Foods for which a standard of identity exists and which do not conform to the standard can be condemned on the basis that they have no identity. The FDA may opt to issue warning letters in cases of misbranding, or else instigate more drastic enforcement action leading to seizure and condemnation of the affected food. Either outright destruction, sale, reconditioning and or export may follow on from the condemnation of a food product (EveryCRSReport.com 2016).

Table 3.9 Some commons forms of food misbranding

(i) Labeling is false or misleading
(ii) Food is sold under the name of another food
(iii) There is an undeclared imitation of another food
(iv) There is improperly filled container or packaging, e.g. A container is oversized or under-filled
(v) Packaging is incorrectly labelled, has missing addresses and other details for manufacturer, packer, or distributor, inaccurate weight, or other measures of quantity
(vi) Is missing required labeling information
(vii) Fraudulently claims standardized composition
(viii) Claims standard fill but falls below the standard fill
(ix) Incorrectly labelled with the common or usual name of the food, – except for spices, flavorings, and colorings
(x) Claims special dietary uses, without information on nutrient and other dietary properties
(xi) Contains undeclared artificial flavoring, coloring, or chemical preservative
(xii) Is lacking information on proper handling or storage

Adapted from USDA Food Safety and Inspection Service (2014a)

Standards of identity have been set up for some bakery products, cacao products, cereal flours and related products, alimentary pastes, milk and cream, cheeses, processed cheese, some cheese foods and spreads, some canned fruits, fruit preserves and jellies, some canned shellfish, eggs and egg products, oleomargarine, some canned vegetables, canned tomatoes, tomato products, and other foods.[5] The FDA accepts all standards of identify of foods described by the Codex Alimentarious without modification (Codex Alimentarius Commission 2014).

Products wrongly labelled as to weight, portions, or ingredients are also classed as misbranded even if no standard of identity has been established. Confiscated foods provided they are deemed wholesome might not be destroyed due to misbranding. Instead, misbranded and confiscated products can be relabelled and re-sold. If the product fails to meet the "fill of container" requirement, it may be relabelled to rectify this fact and sold under the new label. Though the FDA can regulate food that are shipped interstate, a company may still come under the jurisdiction of the FDA if only a fraction of a food consignment crosses State lines.

1.2.11 Inspection of Food Manufacturing Facilities

The FDA has a responsibility to inspect any food processing plant, facility, or commercial food business that produces or prepares foods, ingredients or ready-to-eat meals for public consumption. The purpose of FDA inspections is to check for compliance with US food Safety laws. The FDA may close any premises it considers unsanitary or adulterating foods. For the period 2004–2009, the FDA inspected between 14,547 and 17,032 plants and premises annually. The FDA have not all the personnel necessary to inspect all

[4]Incidental addition of sulfite to foods with standards of identity is discussed in detail 21 CFR 130.8. Incidental addition of sulphites to products with standards of identity at >10 ppm addition must be declared on the product.

[5]Code of Federal regulations (21 CFR 130 to 21 CFR 169) lists standards of identity for some foods and food groups

plants directly. State Inspectors perform about 50% of facility inspections under contract with the FDA. The scope of inspections cover issues related to food safety – for example, chemical or microbiological hazards, and labelling (adulteration and misbranding) issues. Facility inspections also evaluate whether a food enterprise has an adequate food safety management plan (cf. Chaps. 9 & 10). General violations of food safety law result in one of four FDA responses; (a) official action indicated, (b) voluntary action indicated, (c) no action indicated or (d) refer to the State (Levinson 2011). Food plant inspections have become especially topical since the advent of FSMA. Food facility inspections are regarded as "a food safety audit" so food business owners requiring assistance would do well to seek professional advice (Stevens 2019; Food Industry Counsel LLC 2021).

1.3 Centre for Disease Control & Prevention (CDC)

The Centre for Disease Control and Prevention (CDC) located in Atlanta (GA) is an agency of the Department of Health and Human Services. Originally founded in 1946 the CDC is largely concerned with communicable diseases. The CDC has a wide remit that also covers foodborne illnesses, and control of such disease. Historically, experts from the CDC and the FDA (both agencies under the Department of Health and Human Services) and the Food Safety Inspection Services (FSIS) of the USDA worked together to produce the food code; The CDC is also well known for publishing statistical data and case studies in the *Morbidity and Mortality Weekly Report* (MMWR). CDC presents information on food safety via their highly well designed websites (Centers for Disease Control and Prevention 2012).

1.4 U.S. Public Health Service

The U.S. Public Health Service (USPHS) is an agency within the Department of Health and Human Services, which is also concerned with food regulation.[6] The USPHS ensures the sanitary quality of drinking water and foods served on interstate and international carriers (airlines, trains, etc.). This agency carries out research and surveillance on foodborne diseases (infections and intoxications) and sanitary processing and shipping of foods. USPHS works in conjunction with State agencies to set standards for coastal waters from which bivalve shellfish (oysters and clams) may be harvested for consumption. There are anti-pollution specifications also for waters from which bivalves can be harvested, subjected to purification procedures, and then relocated to approved sites.

Standards for the bacteriological quality of the bivalve shellfish and for waters in which shellfish are grown are set by the USPHS. Registered shellfish dealers are required by the USPHS to keep records indicating areas from which bivalves were harvested, from whom they were purchased, and to whom they were sold. The State and the Public Health authorities keep a list of approved shellfish dealers. It is up to State authorities to make sanitary surveys of bivalve growing areas and bacteriological tests of the waters of shellfish-growing areas and of the shellfish. If a particular dealer does not comply with regulations, he is taken off the approved list and any product that they ship interstate will be seized. Where the State program for shellfish sanitation does not meet the requirements of USPHS, all producers within the State are taken off the approved list and no bivalve product can be shipped interstate.

The Public Health Service works together with State authorities to set standards for milk and cream and to monitor diseases in dairy herds. The areas of responsibility extends to bacteriological quality of raw milk, raw cream, and certified milk.

1.5 Food Safety and Inspection Services (FSIS)

In 1886 the USDA acquired powers to regulate interstate food trade. In 1905 Upton Sinclair's famous book "The Jungle" detailing the poor sani-

[6]The USPHS is alternatively named, the Commissioned Corps of the U.S. Public Health Service

tary conditions in the meat facilities in Chicago was written, leading to public disquiet and support for the Pure Food Act of 1906. Another bill also passed in 1906 was the Federal Meat Inspection Act (FMIA) aiming to bring about a more stringent inspection of meat processing plants. The two bills were the responsibility of the Bureau of Chemistry and the Bureau of Animal Industry (BAI), respectively. Meanwhile the meat inspection function of the USDA was consolidated within the BAI which in 1912 gained additional powers for egg inspection. In 1906, the responsibility for meat, poultry and egg inspection passed from the BAI to the Agricultural Research Service (ARS) and from there to the Animal and Plant Health Service (APHIS). Finally, meat, poultry and egg inspection passed to Food Safety & Quality Services (FSQS) before that body was renamed as FSIS in 1981 (Table 3.10) (USDA Food Safety and Inspection Service, 2014b).

Table 3.10 Origins of the meat and poultry inspections within FSIS

1886	USDA to regulate interstate traded foods
1905	Upton Sinclair's book "The Jungle" describing Chicago meat facilities
1906	Pure Food Act – admin. Bureau of Chemistry
1906	Federal Meat Inspection Act (FMIA) – admin by BAI
1927	Bureau of Chemistry renamed Food, Drug & Insecticide Administration (FDIA)
1931	*FDIA renamed Food and Drug Administration (FDA)*
1938	Federal Food, Drug and Cosmetic Act – admin by FDA
1940	FDA moved from USDA to Department of Human Health Services
1953	BAI functions transferred to Agricultural Research Service (ARS)
1957	Poultry Products Inspection Act,
1958	Humane Methods of Slaughter Act
1965	ARS houses Meat Inspective Division and Poultry division
1971	Meat & Poultry Inspection goes to Animal and Plant Health Service (APHIS)
1977	Food Safety & Quality Services (FSQS) for meat and poultry grading
1981	*FSQS re-named Food Safety Inspection Service (FSIS)*
1996	FSIS launches Pathogen Reduction/HACCP Systems.

Source: Adapted from USDA Food Safety and Inspection Service (2014b). Key milestones from FDA history is also available from U.S. Food and Drug Administration (2006a)

1.5.1 The Meat Inspection Bureau

The Federal Meat Inspection Act of 1906, was administered by the US Department of Agriculture (USDA) through BAI, and then by the Meat Inspection Bureau (MIB) of the ARS and eventually by the FSIS. This law regulates the safety of meats (beef, pork, lamb) entering into interstate commerce. The Meat Inspection Bureau is also an enforcement agency. It differs from the FDA in that most of the regulation is carried out by inspectors stationed at the plant processing the food (Table 3.11).

Table 3.11 The FSIS food regulatory system

United States Department of Agriculture
Food safety and inspection service (FSIS)
Food regulatory responsibilities:
– Enforcement of food safety laws – Inspection of meat and poultry plants – Labeling requirements for meat – Food safety education

The MIB deals with any food that contains more than a small percentage of meat. When cattle, hogs, or sheep are slaughtered, one or more inspectors (who are veterinarians) must be stationed at the slaughtering plant if any parts of such products are to be shipped interstate. The animals may be inspected prior to slaughter, and if diseased, are destroyed. However, the main inspection comes after slaughter. As the animals are slaughtered, the carcass, entrails, and organs are tagged and identified with a particular carcass. The veterinarian inspectors test the viscera to determine whether or not the animals were diseased. Diseased animal carcasses including organs and entrails are covered with a dye and must be disposed of as tankage and not used for human consumption. (Tankage is slaughterhouse waste that is heat processed and dried and used as fertilizer.)

A group of "lay" inspectors are attached with the MIB. These inspectors have training but are not veterinarians. Lay inspectors deal primarily with those plants or areas of plants where meat is cut into portions for further processing or for shipment as fresh cuts; where sausages (fresh sausage, frankfurters, bologna and other cooked sausage, dried sausage, etc.) are produced; and where hams, shoulders, and bacon are cured, smoked, and so on. Lay inspectors determine that processing rooms and equipment are clean and sanitary before work is started, and if not satisfactory, the room is tagged and cannot be used for processing until cleaned and sanitized to the satisfaction of the inspector (Table 3.11).

The MIB maintains lists of approved ingredients for cured and processed products, and it is part of the inspectors' job to determine that nothing is added to processed products that is not on the approved list. Inspectors also check the amount of some approved materials added to processed products.

The MIB has laboratories that do some analyses. For instance, a certain amount of water can be added to some cooked sausages or to cured hams. The amount of water allowed is specified (e.g., 10% of the finished product in frankfurters). In frankfurters, the added water gets into the product as ice during the cutting of the meat, one of the reasons for adding water in this case being to keep the ingredients (the meat emulsion) cool during cutting, which improves the quality of the finished product. The inspector cannot tell exactly how much water is being added during the cutting, but if he or she becomes suspicious he or she will take samples of the finished product, ship them to a bureau laboratory, and have them analyzed for added water. In any case, samples for analysis may be taken periodically from processing plants. The same kind of inspection and analysis may be used for other processed meat products, including hams.

States also have inspectors who regulate the slaughter and processing of meats that are used intrastate but not shipped interstate. However, such inspection in the past has been far from adequate. In recent years federal authorities decided improvement in the inspection and regulation of local slaughtering and meat-processing plants was essential, and this improvement is presently being implemented.

The USDA periodically inspects (at least twice a year) "custom" slaughterhouses (those that slaughter but do not sell the meat). If the plants start selling meat wholesale or retail, the USDA will inspect more frequently. The inspections will be done for sanitation, adherence to USDA standards, and a USDA veterinarian will inspect the carcasses for disease.

The USDA also has a Total Quality Control (TQC) program in which a plant sets up its own regulations that are at least as stringent as those set by the USDA. The plant personnel conduct inspections and keep the records. USDA inspectors will inspect the plants regularly and will check all records closely.

1.5.2 The Poultry Inspection Service

The Poultry Inspection Service, similar in scope and operation to the MIB, is responsible for the inspection of poultry products. An agency within the FSIS, it ensures that all poultry sold in interstate commerce is processed in U.S. government-inspected plants and is wholesome. Although poultry inspection is mandatory, the grading of poultry products is voluntary on the part of the processor. Poultry is inspected prior to dressing, during evisceration, during packing, and after packing.

1.6 The Environmental Protection Agency (EPA)

The Environmental Protection Agency (EPA) authorizes and regulates the use of pesticides and other environmental contaminants, and it monitors compliance and provides technical assistance to the states. It sets standards for air and water quality and regulates the use of sanitizers and the handling of wastes. (Table 3.12)

Table 3.12 The EPA food safety responsibilities

Environmental Protection Agency (EPA)
– *Office of Prevention, Pesticides and Toxic Substances (OPPTS)*
Mission: Protect human health and the environment.
Food safety responsibilities:
– Ensure safe pesticide use in food production. – Determine tolerable upper limits for pesticide residue – Pest management techniques used on crops. – Regulations and standards for safe drinking water

2 Other US Regulatory and Inspection Agencies

There are inspection agencies that perform activities of a regulatory nature but not all are enforcement agencies. One such is the Voluntary Inspection Service of the U.S. Department of Commerce. If a fish processor so desires, he may have an inspector stationed in his plant on a permanent basis to assess the premises for sanitary conditions, determine that whole some ingredients are used, and grade the product. As previously stated, this is a voluntary service, and the processor must pay whatever costs are involved in such an inspection. With this type of inspection, the product is evaluated and graded as A, B, or C in quality, the last grade being the lowest acceptable quality. The packer can label the product accordingly.

Continuous inspection is available for certain canned and frozen fishery products. While the inspectors do not inspect every fish, they do inspect a sufficient number so that all lots of fish are sampled, and because each lot is relatively uniform in quality, unacceptable fish are bound to be identified. This is also true for canned fish. The inspection of frozen fish products does not permit the same degree of confidence, because frozen products are manufactured from blocks of frozen fish or shellfish, usually prepared and frozen outside the US. It is true that the inspection service can and does sample the finished product to determine quality, but this can only be a spot check that is not adequate to evaluate the quality of the whole pack. There is another reason why grading canned fish is more effective than grading frozen fish. Once the canned product has been processed, the quality does not change to any extent. This is not the case with frozen foods. Frozen fishery products are not indefinitely stable even at 0 °F (−17.8 °C). At higher temperatures they deteriorate at a much faster rate. It is unfortunate that while manufacturers tend to freeze and store fishery products to a temperature of 0 °F (−17.8 °C) or below, those who transport such products and especially retailers, during the handling and display of fishery product, do not. Thus, a product labeled grade A may actually be grade C or substandard by the time that it reaches the consumer.

The USDA has a grading service for fruits, fruit juices, and vegetables, both canned and frozen. Again, for canned fruits and vegetables, the grading is quite effective, but for frozen products the same problems exist as do for frozen fishery products. To make the grading of frozen foods effective, the temperatures at which these products are shipped, the temperatures at which they are handled and stored at the retail level, and especially the temperatures at which frozen foods are held during display at the retail level must be regulated.

2.1 The Internal Revenue Service (IRS)

The Internal Revenue Service (IRS) enforces, with FDA collaboration, the Federal Alcohol Administration Act and other pertinent regulations that control the commerce of alcoholic beverages (e.g., whiskey, wine, beer, brandy).

2.2 The National Bureau of Standards (NBS)

The National Bureau of Standards (NBS) is responsible for setting the official standards for units of weights and measures for all commercial products, including food. The Office of Technical Services (OTS) issues voluntary "Simplified Practices Recommendations" to limit types and sizes of packages, bags, and jars used as food containers.

2.3 The Federal Trade Commission (FTC)

The Federal Trade Commission (FTC) enforces the provisions of the FTC Act, which prevents unfair and deceptive trade practices. State and local authorities such as Boards of Health regulate inspections in establishments such as restaurants, cafeterias, and other foodservice establishments that con-duct local sales. Many state and local ordinances are patterned after the "Model food code" but some are different and many boards of health issue *guidelines* that explain the laws, making them easier to understand.

The regulation and standardization of food in international trade represent a prodigious and nearly impossible task. Yet the benefits to be derived are of such magnitude that they merit the necessary efforts to achieve them. The Codex Alimentarius Commission, an international organization, was formed by over 90 countries to establish food standards. Its importance is recognized when we are told that the U.S. FDA detains about 40% of the imported foods that it checks. What of the food shipments that escape inspection? What proportion of the imported foods is mislabeled? What proportion contains harmful substances or is contaminated or adulterated? It is because of these considerations that the FDA detains so much of the imported foods that it checks.

The member countries of the Codex Alimentarius send their experts to the international meetings to help formulate quality standards, which are stricter in some aspects than those of many individual countries. For example, the international standards require the listing of all ingredients in food formulations. The work done by this international body, slow though it is, should facilitate trade among the member countries, and the risks of foodborne illness and deceptive practices should be substantially reduced.

2.4 Trade and Professional Organizations

2.4.1 The Institute of Food Technologists (IFT)

The professional organization for food technologists and other food professionals, the IFT is a non-profit scientific society founded in 1939. Since 1947, the IFT has printed a monthly publication *(Food Technology)* that presents information regarding the development of new and improved food sources, products, and processes, their proper utilization by industry and the consumer, and their effective regulation by government agencies. The IFT also has a Communications Division that comments on and provides information on proposed regulations.

The IFT has a number of sections that have monthly meetings and the national IFT also has an annual meeting that is attended by about 20,000 people. The IFT also has a student organization and encourages the teaching of food science at various levels from elementary school through college.

2.4.2 Academy of Nutrition and Dietetics (American Dietetic Association)

The Academy of Nutrition and Dietetics (AND), formerly known as the American Dietetic Association (ADA), is the professional organization for dietitians and dietary technicians founded in 1917. The ADA has a registration program for qualified dietitians and dietary technicians who obtain education from an approved college or university program and successfully complete a certification examination. As of 2010, the AND had a membership of over 72,000 members. Local sections of the AND does much work with educational institutes to educate the public in proper nutrition and healthy life styles. Through regular position statements and publications, the AND provide authoritative commentary on current nutrition and public health discussions.

2.4.3 The National Restaurant Association (NRA)

The NRA is the national trade association for retail and institutional for food service establishments (Chap. 25). It promotes sanitation with governmental, scientific, commercial, and educational institutions.

2.4.4 The Educational Foundation of the National Restaurant Association (EF)

The EF provides educational services to the food industry which include certifications in areas such as foodservice sanitation and personnel administration. EF also offers a comprehensive risk management series called SERVSAFE that includes the sanitation certification.

2.4.5 The Center for Science in the Public Interest (CSPI)

The CSPI is a non-profit public interest membership organization that advocates improved health and nutritional policies. CSPI publishes a *Nutrition Action Newsletter* 10 times a year.

2.4.6 The American Council on Science and Health

The Council is a reputable consumer group that distributes reliable health and nutrition information. The ACSH is located in New York, and produces peer-reviewed reports on important health topics and publishes *Priorities,* a quarterly magazine.

2.4.7 The National Environmental Health Association (NEHA)

NEHA has a Registered Sanitarian program that has established a national standard based on education, experience, and testing for sanitarian qualifications. Some of its members are responsible for inspection services and environmental health programs.

2.4.8 The International Association of Milk, Food and Environmental Sanitarians (IAMFES)

IAMFES was a pioneer in the US milk sanitation program, and also publishes information on milk and food safety and generally accepted methods on how to investigate a foodborne illness.

2.4.9 The Frozen Food Industry Coordinating Committee

This organization has developed a code for *Recommended Practices for the Handling of Frozen Food* that describes handling of food from processing to retail food service.

2.4.10 The National Pest Control Association (NPCA)

The NPCA consists of licensed and certified pest control operators (PCOs) and provides guidelines and training materials for integrated pest management (IPM) and other topics related to safety and sanitation.

2.4.11 National Sanitation Foundation (NSF) International

The National Sanitation Foundation develops standards for equipment design, construction, and installation. Any manufacturer who wishes to have NSF International evaluate their equipment can do so and, if standards are met, the equipment will be included on a list and carry the NSF International seal.

2.4.12 Underwriters Laboratories, Inc. (UL)

UL lists all equipment that meets NSF International standards and also lists electrical equipment that passes U.L. safety requirements.

3 EU Food Regulatory Agencies

3.1 A Primer on the European Union

To understand food regulatory agencies within the European Union it will be useful to examine the Union itself. The European Union dates from the 1950's. Following the 2nd world war it became evident that closer economic ties between European States might allay future conflict. In 1951 the Treaty of Paris was signed to form the European Coal and Steel Community. On 25th March 1957 the Treaty of Rome established the European Economic Community (EEC) also known as the Common Market (Table 3.13). The original signatories (Belgium, Germany, France, Italy, Luxemburg & Netherlands) desired to promote better living and working standards by

Table 3.13 Core values of the European Economic Community (Article 3)

Elimination, of customs duties and all other measures with equivalent effect
Common customs tariff commercial policy towards third countries
Removal of the obstacles to free movement of persons, services and capital
A common agricultural policy
A common transport policy
Common market
Similar balances of payments

removing obstacles for balanced trade and fair competition (Centre for European Studies 2013). The 248 articles of the Rome Treaty set in motion an economic and political project that now embraces 28 nation states and 500 million consumers. Official data ranks the EU as the largest trading block in the world accounting for 16% of trade in manufactured goods. Controversially, the EU is also ranked as the number one economy in the world and at various times shares this ranking with the US. The EU is one of the largest food producing, food importing and exporting regions of the world; it is also the largest trading partner of the US (Europa 2014). (Table 3.13)

The EEC lasted from 1957 until February 1992 at which time the Maastricht Treaty (official name: Treaty for the formation of the European Community) was signed and the EEC was transformed into the European Community (EC). Whereas the EEC was about trade some feared or hoped that EC would ultimately transform into the "United States of Europe". The EC included economic as well as political and financial union. For instance, the EC brought in a common foreign policy, a Europe-wide passport and a single European driving license valid for all member States. From January 1, 1999 most EC states adopted a single currency called the Euro though some (UK and Denmark) opted out from that element of the Maastricht treaty. In 2007 The Lisbon treaty replaced the EC with the European Union (EU).

The EU is governed through three political institutions; (a) The EU commission –the law making body, (b) The EU parliament – representing European Citizens rights and the (c) The European Court. – Judiciary (Wikipedia 2014a). Policy is generally debated at the EU parliament and is ultimately translated into laws by the Commission. Now we will turn to food regulation within the EU.

3.2 The European Food Safety Authority (EFSA)

The European commission was initially guided in food policy by 9- Committees advising on food (Table 3.14). The situation prior to 2002 was that there was no single EU wide body responsible for food safety. Food safety legislation was enforced by national agencies with no uniformity of standards. Following several major food borne threats to public health in the 1990s a need was recognized for one regulatory agency to oversee food safety across the entire EU. A white paper describing the functions of the proposed European Food Authority (EFA) was published in January of 2000 and put out for consultation for 4 months. A final EU Commission proposal followed in September 2000 accompanied by enabling legislation in December 2001 but with the term "Safety" added to the title (Jukes 2002). The European Food Safety Authority (EFSA) started operations in 2002.

The function of EFSA is to advice on matters related to food safety usually following a request from the European Commission for the purpose of drafting legislation within the EU. EFSA can

Table 3.14 Pre 2002 food committees in the European Union

Scientific Steering Committee
Scientific Committee on Food
Scientific Committee on Animal Nutrition
Scientific Committee on Animal Health and Animal Welfare
Scientific Committee on Veterinary Measures relating to Public Health
Scientific Committee on Plants
Scientific Committee on Cosmetic Products and Non-Food Products intended for Consumers
Scientific Committee on Medicinal Products and Medical Devices
Scientific Committee on Toxicity, Eco-toxicity and the Environment

also respond to requests from industry (Table 3.15).

EFSA is concerned with (a) providing a comprehensive, integrated approach covering the food chain, from farm-to table, (b) food safety within and between EU states, (c) addressing the concerns of a wide range of stakeholders including, feed manufacturers, farmers, food manufacture/retail etc., (d) ensuring traceability of feeds and foods within the EU, (e) adopting risk assessment as a basic scientific principle for food safety assessments, (f) basing all decisions on the highest and best scientific advice, (g) adopting a precautionary principle; this means where there is insufficient information – then judgements will be based on risk avoidance until actual safety can be confirmed.

Comparing the FDA and EFSA shows up important differences in their functions. Recall that the FDA is charged with *enforcing* US federal law related to foods and drugs. In the EU, medicinal drugs are regulated through the European Medicines Agency (EMA); because EMA overseas veterinary drugs some cross talk with the EFSA panel on animal health can be anticipated. The FDA provides guidance to industry but also formulates binding regulations.

As part of its enforcement role, the FDA may carry out inspections to obtain evidence of violations. The FDA monitors also the food supply chain starting with farming, food manufacturing, to distribution, and eventually sales. By contrast, EFSA's role is advisory. EFSA committees and panels develop scientific opinions about a variety of food safety issues, from which food law may be made by the European Commission – a role not unlike the US congress (Table 3.15).

3.3 European General Food Law

The main EU legislation covering food safety is designated EC178/2002; which is the same legislation that introduced EFSA (Fig. 3.4). Article 14, 16, 18 and 19 in the general food law requires compliance by food businesses. The day to day enforcement of EC178/2002 is left

Table 3.15 EFSA panels areas of responsibility

EFSA Panel	Areas of concerns
Panels for risk assessment and scientific assistance	
Animal Health & Welfare (AHAW)	Animal diseases and animal welfare; food producing animals; fish and animal health; avian influenza; blue tongue; cloning
Biological Hazards(BIOHAZ)	Food safety and food-borne diseases; Campylobacter; Listeria; Salmonella; Bovine Spongiform Encephalopathy (BSE); Food-borne zoonotic diseases; meat inspection; parasites in food; qualified presumption of safety (QPS)
Contaminants (CONTAM Panel)	3-Monochloropropane-1,2 Diol Esters (3-MCPD); Acrylamide; Aflatoxins in food Contaminants in food and feed; Dioxins and PCBs; Meat inspection; Metals as contaminants in food; Mineral oil hydrocarbons; Mycotoxins
Plant Health (PLH)	Disease and threats to plant health Bee health; Invasive alien species; Plant pests affecting food crops
Panels for scientific evaluation of regulated products	
Additives, products or substances used in animal feed (FEEDAP)	Lenziaren; formaldehyde
Dietetic products, nutrition and allergies (NDA)	Dietary reference values; dietary guidelines; nutrition claims; health claims
Additives and nutrient sources added to Food (ANS)	Safety of additives; nutrient sources, substances deliberately added to foods; aspartame
Contact materials, enzymes, flavorings and processing aids (CEF)	Enzymes, flavorings, active and intelligent packaging substances; food contact materials;
Genetically modified organisms GMO (GMO panel)	Bee health; environmental risk assessment
Pesticides unit and the panel on plant protection products and their residues (PPR)	Consumer exposure to pesticides, pesticides risk assessment

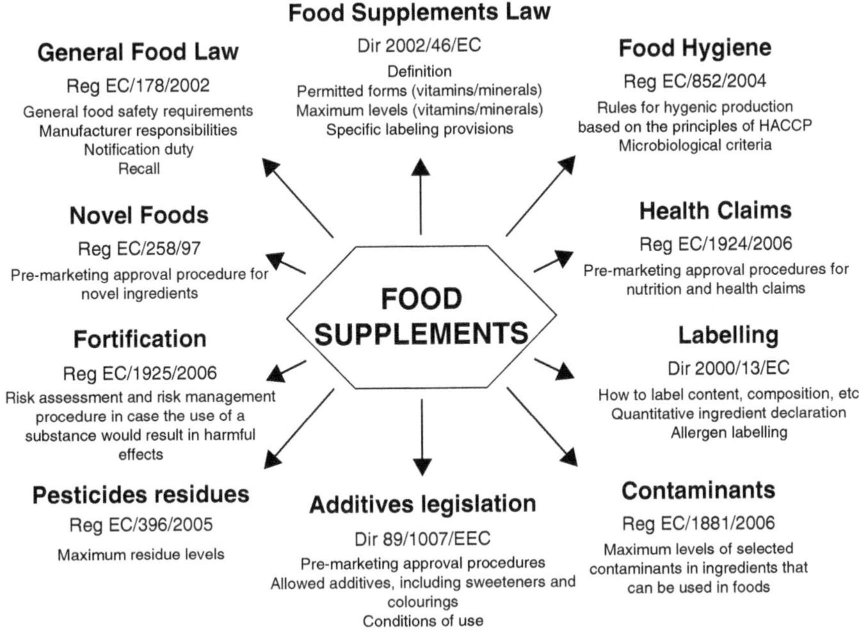

Fig. 3.4 Examples of major EU food legislation (with permission received from Author Dec 2014)

with Food Standards Agencies within each State. In general, EU legislation either reflects or supersedes existing State food safety legislation. For instance, the United Kingdom Food Safety Act of 1990 contains the main requirements for food safety for local businesses. The UK Food Standards Agency however, interprets and informs businesses of additional requirements encapsulated within the EC178/2002 and any later amendments. The general EU food law states that (a) food offered for sale must not be unsafe, (b) labeling on packaging should not be misleading, (c) food businesses must maintain traceability, which refers to accurate record keeping of B2B suppliers and finally (d) food businesses should have the capacity to recall foods that have been found to pose risk to the general population. (Fig. 3.4)

4 Food Regulation in China

4.1 Food Regulatory Agencies in China

Food legislation in the PR of China is a function of multiple agencies with oversight for different portions of the food supply chain from primary production, through to food processing and manufacturing, packaging and retail (Wikipedia 2014b). Food legislation is enacted as directives or regulations from either the State Council or from Departments of the State council (Table 3.16). The State Council is the equivalent of the US Congress, or the EU Commission. The majority of legislation however comes from Departments of the State Council –referring to specific ministries, commissions or agencies.

Table 3.16 Regulatory agencies with oversight for food safety in the PR of China

Agency	Area of responsibility
General Administration of Quality Supervision	Inspection, quarantine, imported and exported foods
Ministry of Agriculture	Food safety & agricultural production
Ministry of Commerce	Food trade, internal food markets
Ministry of Health*	Poisoning cases, food safety and & hygiene enforcement, & inspections; communicable disease prevention and treatment, food health; investigation of violations at local level; Food cosmetics and drug policy
Ministry of Science & Technology	Food manufacturing, processing and
National Institute of Nutrition & Food Safety	Food hygiene and nutrition research
State Administration for Industry & Commerce	Food authentication, advertising, labeling
State Drug Administration (SDA)	State drug policy
State Food and Drug Administration (SFDA)*	Food cosmetics and & drug policy

Source: Summarized from (Wikipedia 2014b), *food legislation transferred from Ministry of health to the SFDA in 2003

A listing of the agencies involved with food legislation is shown below. Some of the information in Table 3.16 is now dated. For instance, the State Drug Administration had a name change in April 16, 2003 when it became, the State Food and Drug Administration and also acquired responsibility for food legislation from the Ministry of health (Table 3.16).

It is generally supposed that both regulations and directives should be treated as mandatory (Table 3.17). China's food regulations and directives are available online but English versions of many documents are not yet uploaded.

Table 3.17 Food regulations and directives in the PR of China

Category	Regulations	Directives
All food	54	121
General	5	18
Manufacturing & Standards	5	9
Distribution & advertising	5	9
Import and export	6	21
Safety	1	13
Agricultural products	13	8
Aquatic Product & Seafood	0	1
Health Food	2	23
Packaging & labeling	3	1
Food additives	3	8
Others	11	10

Source: Adapted from http://www.chinafdc-law.com/laws/index_b_1.html

4.2 China's General Food Hygiene Laws

China's food hygiene law was enacted on 30th October 1995 at the 16th Session of the Standing Committee of the Eight National People's congress. The general food hygiene law contains 54 articles covering eight major areas (Table 3.18). The English version of the General food hygiene law is available online (China Laws and Regulations of Food Drug and Cosmetics 1995).

Article 54 from the Supplementary provisions of the general food hygiene law provides a glossary of terms according to which food is defined as: any finished product or raw food material provided for people to eat or drink, products traditionally served as *both* food and medicament, but excluding products used solely for medical purposes. A food additive is defined as natural or synthetic compounds added to food to improve quality. Utensils and equipment refers to machinery, piping, conveyor belts etc. which come into direct contact with foods. The definitions are on the whole similar to those used by the Codex Alimentarius Commission.

Of notable interest is Article 9 which covers prohibited foods: these being spoiled foods, toxic foods, and foods containing pathogens; uninspected meat products, poultry or other livestock

Table 3.18 Food Hygiene law of the Peoples Republic China

Section	Articles	Comments
I. General provisions	1–5	Preamble
II. Food hygiene	6–10	Environment, food facilities, utensils, personal hygiene
III. Hygiene of food additives	11	Hygiene requirements for food additives, materials
IV. Hygiene of containers, packaging, utensils and equipment used for food	12–13	Hygiene requirements for food packaging, containers, disinfection
V. Formulation of food hygiene standards and measures for food hygiene control	14–16	Food hygiene responsibility of Public health Departments, State Council; National and provincial hygiene standards
VI. Food hygiene control	17–31	Industry and hygiene, food production or marketing relations to safety, mandatory food safety personal, labeling and package claims, health claims, mandatory food hygiene, food hygiene licencing, imported/ export food inspection
VII. Food hygiene supervision	32–38	Food hygiene supervisors and inspectors at state, provincial levels.
VIII. Legal responsibility	39–53	Sanctions for violation of food hygiene laws
IX. Supplementary provisions	54	Key definitions

Summarized from: Wikipedia (2014b)

that died of disease, food containing unapproved food additives. Articles 39–53 describe a range of sanctions which may apply for violations of the food hygiene legislation. A food manufacturer responsible for a food poisoning outbreak is liable to a fine of 1000–50,000 Yuan and/ or fines of 1-5x the total illegal receipts. Food manufacturers may also have their food hygiene licences revoked without which they are barred from future food manufacturing. Foods processed without a hygiene licence is prohibited and manufacturers risk fines of 500–10,000 Yuan (82–1629 US$). The court fines for unsafe food additive use is about 5000 Yuan or US$815 (Wikipedia 2014b).

5 International Food Standards

5.1 Codex Alimentarius Commission

The Codex Alimentarius Commission (CAC) is an international food regulatory agency formed through the joint action of the Food and Agricultural Organization of the United Nations (FAO) and the World Health Organization (WHO). Apparently, the foundation of CAC in 1963 stems from the same post-WWII concerns that led to founding of the FAO in 1948. In part, the FAO was formed in a bid to facilitate agricultural and food development but this led also to concerns with ensuring international minimal standards for food safety, which is one of the main activities of CAC. The current membership of CAC consists of 160 nation states covering over 99% of the world populations. Codex holds regular meetings that are open to the general public. (Table 3.19)

CAC produces international food standards, regulations and codes based on advice received from three FAO / WHO experts committees – related to food additives, pesticide residues and microbial food safety. At any given moment several dozen other subject committees, commodity committees, task forces, and other WHO/FAO committees operate in support of CAC outputs. The CAC published standards are described as not mandatory for any nation or state. However, the standards for quality, microbial safety are accepted references points for negotiations and arbitrations by the World Trade Organization (WTO). Currently Codex standards for over 230 foodstuffs and processes are available online without charge (Codex Alimentarius Commission 2014).

Table 3.19 Milestones for the foundation of the Codex Alimentarius Commission (CAC)

1943 – Founding conference of the FAO is convened by President Franklin Roosevelt
1945 – Foundation of FAO of the United Nations
1948 – Foundation of the World Health Organization (WHO)
1955 – First Joint FAO/WHO Conference on Food Additives. Joint FAO/WHO Expert Committee on Food Additives (JECFA) sets out principles of food additive use.
1958 – The United Nations Economic Commission for Europe (UNECE) proposes Geneva Protocol for harmonized layout for food commodity standards
1958 – Council of the Codex Alimentarius Europaeus
1960 – First FAO Regional Conference for Europe, discussion of minimum food standards, labeling requirements, methods of analysis
1961 – FAO resolution passed forming Codex Alimentarius Commission (CAC).
1963 – Joint FAO/WHO Food Standards Programme initiated through the Codex Alimentarius Commission
1995 – CAC standards, and food safety guidelines, as reference for WTO Agreement
2011 – Session 34 of CAC attended by 625 delegates from 145 Member states

Summarized from a codex article. (Codex Alimentarius Commission (2014). "Codex timeline from 1945 to the present." Retrieved from http://www.codexalimentarius.org/about-codex/codex-timeline/en/)

5.2 International Organization of Standardization (ISO)

The international organization of standardization (ISO) is described on its website as a non-governmental membership organization, based in Geneva, Switzerland (International Organization for Standardization 2015). ISO is dedicated to formulating international voluntary standards for products, services and systems – in order to promote international trade. The ISO (abbreviated from *isos*, Gk for equality) has developed over 19,500 different standards including ~1000 standards covering elements of the food chain from farm to fork (Appendix 4).

6 Conclusion

Food regulations are not only the reserve of specialist but potentially affect everyone. The responsibility for ensuring that food is safe and wholesome for consumers' falls to governmental bodies, funded by public funds and taxation. The food and drugs administration (FDA) serves as the key regulatory body for foods in the US, supported in that function by the USDA and PHS. The sheer size, scope and complexity of the FDAs regulatory role cannot be underestimated.

Appendices

Appendix 1: Products Classes Regulated by the FDA

Products class	Examples
Biologics	Vaccines, blood and blood products, cellular and gene therapy products, tissue and tissue products, Allergenics
Cosmetics	Color additives for makeup and personal care products, skin moisturizers and cleansers, nail polish and perfume
Foods	All foods, dietary supplements, water, Food additives, infant formulas, other food products
Drugs	Prescription drugs (branded and generic),non-prescription (over-the-counter) drugs
Medical devices	Simple items (e.g. bedpans), complex technologies (e.g pacemakers),dental devices, surgical implants, prosthetics
Electronic products	Radiation emitting- microwave ovens, X-ray equipment, laser products, ultrasonic therapy equipment, mercury vapor lamps, sunlamps
Veterinary products	Livestock feeds, pet foods, veterinary drugs and devices
Tobacco products	Cigarettes, cigarette tobacco, roll-your-own tobacco, smokeless tobacco

Source: U.S. Food and Drug Administration (2014a)

Appendix 2

A2a: FDA Activities and Processes (2014)

1.	Recalls, Outbreaks & Emergencies	Deals with food recalls, safety alerts and advisories, outbreak investigations
2.	Guidance & Regulation	Guidance, FSMA, facility registration, CGMPs, HACCP, retail food, import/export
3	Ingredients, Packaging & Labeling	Deals with ingredients, additives, contact substances, GRAS, allergens, and nutrition labeling
4.	Foodborne Illness & Contaminants	Deals with preventing foodborne illness, information on pathogens, chemicals, pesticides, natural toxins, and metals
5	Dietary Supplements	Using dietary supplements and FDA's role in regulating supplement products and dietary ingredients
6	Compliance & Enforcement	Deals with Food Registry, warning/untitled letters, and inspection and compliance
7	International & Interagency Coordination	International outreach, trade and interagency agreements
8	Food Defense	Deals with malicious, criminal, or terrorist actions on the food supply

Source: U. S. Food and Drug Administration (2014a)

A2b: FDA Activities and Processes (2019)

1	Recalls, outbreaks & emergencies	Alerts, advisories & safety information, food safety during emergencies, outbreaks of foodborne illness, recalls of foods & dietary supplements
2	Industry guidance & regulation	Guidance, FSMA, facility registration, CGMPS, HACCP, retail food, import/export
3	Labeling & nutrition	Nutrition facts label, label claims, menu & vending machine labeling, gluten-free labeling
4	Ingredients & packaging	Food & color additives, GRAS, food allergens, food contact substances, new plant varieties
5	Dietary supplements	Products & ingredients, new dietary ingredients notification process, structure/function claim notifications
6	Compliance & enforcement	Reportable food registry, warning letters, inspections, compliance programs, adverse event reporting; recalls?
7	International & interagency coordination	International outreach, visitor's program, trade agreements, and interagency agreements
8	Food defense	Protecting the food supply, intentional adulteration, food defense plan builder
9	Science & research	Laboratory methods, whole genome sequencing, risk analysis, total diet study, consumer research
10	Chemicals, metals & pesticides in food	Acrylamide, arsenic, chemicals, metals, toxins & pesticides in food

Source: U. S. Food and Drug Administration (2020)

Appendix 3: Nine Forms of Food Adulteration

Adulterated foods
1) Contain added substances harmful to health 2) Contain substances harmful via feeding to livestock, e.g. feed or color additive 3) Consist of filthy, putrid, or decomposed substances or is unsound, unhealthful, unwholesome, or otherwise unfit for human food 4) Is manufactured or held under insanitary conditions 5) Is from an animal which has died otherwise than by slaughter 6) Is packaged in a container which is poisonous or may render the contents harmful 7) Has been intentionally subjected to radiation, other than as part of certified process 8) Has a valuable constituent abstracted, substituted, or damage or inferiority has been concealed; or substances have been added or packed to increase bulk or weight, or reduce its quality or strength, or make it appear better or of greater value than it is 9) Is declared as margarine but contains animal fat or any other filthy, putrid, or decomposed substance

Source: From 21 CFR 601

Appendix 4: Selected ISO Standards Related to Foods

ISO 22000:2005: Food safety management systems
ISO 22005:2007: Traceability in the feed and food chain – General principles and basic requirements for system design and implementation
ISO/TS 22002-1:2009: Prerequisite programmes on food safety – Part 1: Food manufacturing
ISO/TS 22002-2:2013: Prerequisite programmes on food safety – Part 2: Catering

References

Buchanan RL (2000) The development of science-based food safety regulations in the United States. Irish J Agric Food Res 39(2):331–342. https://www.jstor.org/stable/25562399

Centers for Disease Control and Prevention (2012) Foodborne illness and outbreak investigations: A farm to table overview. Retrieved from http://www.cdc.gov/foodsafety/investigations.html

Centre for European Studies (2013) Treaty establishing the European Economic Community (Rome, 25 March 1957). Retrieved from http://www.cvce.eu/obj/treaty_establishing_the_european_economic_community_rome_25_march_1957-en-cca6ba28-0bf3-4ce6-8a76-6b0b3252696e.html

China Laws and Regulations of Food Drug and Cosmetics (1995) Food hygiene law of the People's Republic of China. Retrieved from http://www.chinafdc-law.com/laws/detail_156.html

Codex Alimentarius Commission (2014, 01/2020) Codex standards. Retrieved from http://www.fao.org/fao-who-codexalimentarius/codex-texts/list-standards/en/

Europa (2014) EU position in world trade. Retrieved from http://ec.europa.eu/trade/policy/eu-position-in-world-trade/

EveryCRSReport.com (2016) Legal issues associated with FDA standards of identity: In brief. Retrieved from https://www.everycrsreport.com/reports/R44393.html

Food and Drug Administration (2007) Food Protection Plan of 2007. Retrieved from http://www.fda.gov/Food/GuidanceRegulation/FoodProtectionPlan2007/ucm132565.htm#

Food Industry Counsel LLC (2021) FDA inspection checklist. What to expect when they're inspecting. Retrieved from https://www.foodindustrycounsel.com/blog/fda-inspection-checklist-what-to-do-before-during-and-after-your-next-fda-inspection

International Food Information Council (2014) Ensuring a safe food supply: a concise guide to the U.S. food regulatory system. Retrieved from https://ucfoodsafety.ucdavis.edu/processing-distribution/regulations-processing-food

International Organization for Standardization (2015) About ISO. Retrieved from http://www.iso.org/iso/home/about.htm

Jukes D (2002) Food law: European Food Safety Authority (previously known as the 'European Food Authority'). From http://www.reading.ac.uk/foodlaw/eu/efa1.htm

Levinson DR (2011) Vulnerabilities in FDA'S oversight of state food facility inspections. Department of Health and Human Services, Office of Inspector General, 39pp. Retrieved from https://www.resolve.ngo/docs/hhs-ig-report-on-state-inspections.pdf

Li M, Baker CA, Danyluk MD, Belanger P, Boelaert F, Cressey P, Gheorghe M, Polkinghorne B, Toyofuku H, Havelaar AH (2018) Identification of biological hazards in produce consumed in industrialized countries: a review. J Food Prot 81(7):1171–1186. Retrieved from https://doi.org/1110.4315/0362-1028X.JFP-1117-1465

Merrill RA, Francer JK (2000) Organizing federal food safety regulation. Seton Hall L Rev 31:61. https://scholarship.shu.edu/cgi/viewcontent.cgi?article=1334&context=shlr

Saxowsky D (2010) US agencies involved with food safety. Retrieved from http://www.ndsu.edu/pubweb/~saxowsky/aglawtextbk/chapters/foodlaw/USagencies.html

Scallan E, Hoekstra RM, Angulo FJ, Tauxe RV, Widdowson MA, Roy SL, Jones JL, Griffin PM (2011) Foodborne illness acquired in the United States – major pathogens. Emerg Infect Dis 17(1):7–15. Retrieved from http://wwwnc.cdc.gov/eid/article/17/1/P1-1101_article

Stevens C (2019) Preparing for a food safety inspection: everything you need to know. Retrieved from https://modernrestaurantmanagement.com/preparing-for-a-food-safety-inspection-everything-you-need-to-know/

Strauss DM (2011) An analysis of the FDA food safety modernization act: protection for consumers and boon for business. Food Drug Law J 66(3):353–376. https://www.jstor.org/stable/26661214

U. S. Food and Drug Administration (2009, 01/31/2018) Milestones in U.S. Food and drug law history. Retrieved from https://www.fda.gov/about-fda/fdas-evolving-regulatory-powers/milestones-us-food-and-drug-law-history

U. S. Food and Drug Administration (2014a) Food. Retrieved from http://www.fda.gov/food/default.htm

U.S. Food and Drug Administration (2014, 2018) Recalls, outbreaks & emergencies. Retrieved from http://www.fda.gov/Food/RecallsOutbreaksEmergencies/default.htm

U. S. Food and Drug Administration (2014b, 2018) Recalls, outbreaks & emergencies. Retrieved from http://www.fda.gov/Food/RecallsOutbreaksEmergencies/default.htm

U. S. Food and Drug Administration (2018) Generally recognized as safe (GRAS). Retrieved from https://www.fda.gov/food/food-ingredients-packaging/generally-recognized-safe-gras

U.S. Food and Drugs Administration (2020) Inspection guides. Retrieved from https://www.fda.gov/

inspections-compliance-enforcement-and-criminal-investigations/inspection-references/inspection-guides
U.S. Food and Drug Administration (2006a) FDA milestones. *FDA Consumer magazine: 2006 Jan-Feb; 40(1):36–8.* Retrieved from http://permanent.access.gpo.gov/lps1609/www.fda.gov/fdac/features/2006/106_milestones.html
U.S. Food and Drug Administration (2006b) Harvey W. Wiley: Pioneer consumer activist. FDA Consumer Magazine 40(1):34–35. Retrieved from http://permanent.access.gpo.gov/lps1609/www.fda.gov/fdac/features/2006/106_milestones.html
U.S. Food and Drug Administration (2013) Food code (2013) Retrieved from http://www.fda.gov/Food/GuidanceRegulation/RetailFoodProtection/FoodCode/ucm374275.htm
U.S. Food and Drug Administration (2014a) About FDA: What does FDA regulate? Retrieved from http://www.fda.gov/AboutFDA/Transparency/Basics/ucm194879.htm
U.S. Food and Drug Administration (2014b) Foodborne illnesses: What you need to know. Retrieved from http://www.fda.gov/Food/FoodborneIllnessContaminants/FoodborneIllnessesNeedToKnow/default.htm
U.S. Food and Drug Administration (2014c) Generally recognized as safe (GRAS). Retrieved from http://www.fda.gov/Food/IngredientsPackagingLabeling/GRAS/default.htm
U.S. Food and Drug Administration (2014d) Gluten-free labeling of foods. Retrieved from http://www.fda.gov/Food/GuidanceRegulation/GuidanceDocumentsRegulatoryInformation/Allergens/ucm362510.htm
U.S. Food and Drug Administration (2014e) Regulatory report: FDA's food contact substance notification program. Retrieved from http://www.fda.gov/Food/IngredientsPackagingLabeling/PackagingFCS/ucm064161.htm
USDA Food Safety and Inspection Service (2014a) Federal Meat Inspection Act. Retrieved from http://www.fsis.usda.gov/wps/portal/fsis/topics/rulemaking/federal-meat-inspection-act
USDA Food Safety and Inspection Service (2014b) FSIS history. Retrieved from http://www.fsis.usda.gov/wps/portal/informational/aboutfsis/history
Wikipedia (2014a) European Union. Retrieved from http://en.wikipedia.org/wiki/European_Union#History
Wikipedia (2014b) Food safety China. Retrieved from http://en.wikipedia.org/wiki/Food_safety_in_China
Wikipedia (2014c) List of US federal agencies. Retrieved from http://en.wikipedia.org/wiki/List_of_United_States_federal_agencies#United_States_Department_of_Agriculture
Wikipedia (2020) Food and drug administration. Retrieved from https://en.wikipedia.org/wiki/Food_and_Drug_Administration#Food_and_dietary_supplements

Food Labels 4

1 Introduction

1.1 Background of Food Labeling Legislation

The historical records show that food laws existed in the American Colonies (Curtis et al. 2013). However, the first federal law to deal with food labeling specifically was the Federal Food & Drug Act (1906) also known as Pure Food & Drug Act (Meadows 2006). The 1906 Act made it a criminal offence to *manufacture* adulterated or misbranded food. Section 2 – prohibited the *introduction or shipment* of adulterated or misbranded food across State lines, or their import. Export of otherwise adulterated or misbranded food had to ensure the products conformed to the requirements of international customers (U. S. Food and Drug Administration).[1] The Gould Amendment (1913) required printed labels outside of packages indicating their contents (Meadows 2006). Tables 4.1 and 4.2 lists some of the major food laws from the US.

Sections of the Federal Food and Drug Act (1906) were superseded, repealed or transferred to section 341 of the Food, Drug & Cosmetic Act of 1938 (FD&C). Misleading labeling was one of many forms of misbranding addressed by the FD&C.[2] Some form of labeling had been in force since the Gould Amendment, but labeling legislation became stricter with the passing of the Fair Packaging & Labeling Act (1966) (FPLA) (Federal Trade Commission 2020).

The FPLA (1966) required the Federal Trade Commission and the FDA to produce legislation that ensured packaging for household consumer commodities is "labeled to disclose" three characteristics; (i) identity of the commodity, (ii) net content or weight, and (iii) the business name and address of the distributor, packer or manufacturer. Crucially, FPLA allowed for future additional legislation as needed *"to prevent consumer deception (or to facilitate value comparisons) with respect to descriptions of ingredients, slack fill of packages, use of"cents-off" or lower price labeling, or characterization of package sizes*" (Federal Trade Commission 2020). The two decades after the FPLA (1966) had few or no changes for labeling legislation. In 1990, the FDA introduced the Nutrition Labeling & Education Act, (NLEA) (Fig. 4.1).

US food labeling regulations can be retrieved from two repositories, Code Federal Regulations (CFR), Title 21, Food & Drugs, Subsections 100.1–100.108 (21CFR 101.1–101.108) contains labeling rules and regulations passed by the FDA, as mandated by the US congress. An electronic

[1]The 1906 Pure food and Drug act has been superseded (See Table 4.1).

[2]US Code, Chapter 9. Title 21, FOOD AND DRUGS, Chapter 9, Subchapter IV. SS 343 Misbranding

R. Owusu-Apenten, E. R. Vieira, *Elementary Food Science*, Food Science Text Series,
https://doi.org/10.1007/978-3-030-65433-7_4

Table 4.1 Examples of United States food legislations affecting food labeling

Federal Food and Drug Act, 1906 (21 USC. 1)
Federal Meat Inspection Act, 1906 (21 U.S.C. 12)
Food Inspection Decision (FID 76), 1907
The Gould Amendment, 1913
Food, Cosmetics & Drugs Act 1938 (21 USC. 9).
Food additive amendment, 1958
Color additive amendment, 1960
Fair Packaging & Labelling Act, 1966 (15 USC. 39).
Nutrition Labeling and Education Act 1990
Dietary Supplements Health and Education Act, 1990
Food Quality Protection Act, 1996
FDA Modernization Act, 1997
Food Allergy Labeling & Consumer Protection Act, 2004
Revised nutrition labeling section (Guide Chapter 7).

Source: Including Curtis et al. (2013), Meadows (2006) and U. S. Food and Drug Administration (2009d)

version of the CFR (e-CFR) is available online to anyone with an internet/DATA connection anywhere on the planet. The e-CFR is considered "unofficial" but is more current and updated every couple of days. The other repository for food labeling rules (and all other US laws) is the United States code (USC) – maintained by the Office of the Law Revision Counsel. The USC Title 21, chapter 9 contains FDA legislation and food regulations coming out of the USDA.[3] For historical interest, archival copies of the Federal Food & Drug Act (1906) is filed under USC title 21. Chapter 1 (Office of the Law Revision Counsel 2020).[4] USC Title 21, Chapter 9 – contains Food, Drugs & Cosmetics Act (1938) and amendments (Office of the Law Revision Counsel 2020). Finally, texts for CFR, USC and the US constitution are available from the "GovReg" website (GovRegs 2016).[5]

Table 4.2 Historic milestones in the US food law (1862–1962)

1862 – President Lincoln appoints Charles M. Wetherill, to the new Department of Agriculture
1880 – Peter Collier (Chief chemist USDA) petitions for a food and drug law
1883 – Dr. Harvey W. Wiley appointed Chief chemist, of the Bureau of Chemistry
1897 – Tea Importation Act for mandatory inspection of imported tea
1898 – Association of Official Agricultural Chemists (later AOAC International) forms a Committee on Food Standards chaired by Dr. Wiley.
1902 – US Congress funds a study of chemical preservatives and colors by Bureau of chemistry
1906 – The first Food and Drugs Act prohibits sale of misbranded, adulterated foods and drinks
1907 – First Certified Color Regulations, approves seven food colors
1913 – Gould Amendment, early food labeling legislation
1924 – Apple Cider Vinegar labeling case, the Supreme Court ruling. Food & Drugs Act (1906) applies to product's label that may mislead or deceive, even if technically true.
1933 – FDA proposes revision of 1906 Food and Drugs Act.
1938 – The Federal Food, Drug, and Cosmetic (FDC) Act of 1938 comes into force
1939 – First Food Standards (canned tomatoes, tomato purée, and tomato paste).
1950 – Oleomargarine Act, clear labeling of margarine, to differentiate it from butter.
1966 – Fair Packaging and Labeling Act
1968 – Animal Drug Amendments to the Food, Drug, and Cosmetic Act-Section 512
1969 – The White House Conference on Food, Nutrition, and Health; FDA of GRAS list
1990 – Nurition labeling and Education Act

Source: FDA Consumer magazine [7] and [6] (U. S. Food and Drug Administration 2009d)

1.2 Definition of Food Label and Labeling

There are four main considerations for food labeling: the food, container, labels and the act of labeling. Codex definitions of each these terms is reproduced below (Table 4.3). Briefly, a food label is usually a sectioned off part of food package that is intended to communicate key information for the consumer. Food labels can be stenciled directly on the product (cf. date mark-

[3]The FDA (as Bureau of Chemistry) started out as part of the USDA but was transferred to the Department of Public Health Services at round about 1930. About 80% of food regulations arises from the FDA whilst the USDA looks after meat products and related (cf. Chap. 4)

[4]The US Code (U.S.C) is a repository of all United States laws categorized under 54 Titles; each title is sub-divided into multiple "Chapters".

[5]The GovReg website is updated daily whilst formatting is also significantly improved.

ing eggs or meat). Other forms of labeling may entail tying a tag on a food product.

1.3 Why Food Labeling

The general aims of food labeling from the nineteenth century onwards were at least three fold; (i) to inform consumers and caterers what they purchased, (ii) to identify packaged foods with minimal confusion, and (iii) to combat food fraud. The need for labeling increased as the distances between, food producers, manufacturers and consumers grew. Longer food supply chains increase the need for food labels (Curtis et al. 2013). Currently, all packaged foods must be labeled (Table 4.4).

Food labels have an educational function as well as an advertising role (Frohlich 2010). The characteristics of food labels seem to influence nutrition outcomes (Roberto 2005). The extent to which consumers read food labels information

Electronic Code of Federal Regulations

e-CFR data is current as of January 23, 2020

Title	Volume	Chapter	Browse Parts	Regulatory Entity
Title 21 Food and Drugs	1	I	1-99	FOOD AND DRUG ADMINISTRATION, DEPARTMENT OF HEALTH AND HUMAN SERVICES
	2		100-169	
	3		170-199	
	4		200-299	
	5		300-499	
	6		500-599	
	7		600-799	
	8		800-1299	
	9	II	1300-1399	DRUG ENFORCEMENT ADMINISTRATION, DEPARTMENT OF JUSTICE
		III	1400-1499	OFFICE OF NATIONAL DRUG CONTROL POLICY

Need assistance?

Fig. 4.1 The electronic code of federal regulations is a repository of FDA rules

Table 4.3 Basic definitions of terms

Container – any package of food for delivery as a single item, whether by completely or partially enclosing the food and includes wrappers. A container may enclose several units or types of packages when such is offered to the consumer.
Food – any substance, processed, semi-processed or raw, which is intended for human consumption, and includes drinks, chewing gum and any substance which has been used in the manufacture, preparation or treatment of "food" but does not include cosmetics or tobacco or substances used only as drugs.
Label – any tag, brand, mark, pictorial or other descriptive matter, written, printed, stenciled, marked, embossed or impressed on, or attached to, a container of food.
Labelling includes any written, printed or graphic matter that is present on the label, accompanies the food, or is displayed near the food, including that for the purpose of promoting its sale or disposal,

Adapted from Codex Alimentarius Commission (1985a)

Table 4.4 Mandatory labeling items for pre-packaged foods

The name of the food
List of ingredients – except single ingredient foods, all ingredients are normally listed in descending order of weight (w/w).
List of additives
Net contents and drained weight
Name and address of manufacturer
Country of origin
Lot identification
Date marking and storage instructions
Best before date, – related to quality
Use by date – related to safety, food poisoning risk
Instructions for use

Adapted from Codex Alimentarius Commission (1985a)

affects the likelihood of healthy eating (Ollberding et al. 2011). Nutrition labeling, is considered one of several population-based approaches for improving public health (Mozaffarian et al. 2012).

1.4 Industry Guide for Food Labeling

The FDA produces a food labeling industry guide (September 1994), which undergoes periodic revisions (April 2008; October 2009; January 2013). The latest (2013) edition adopts a question and answer format. The guidebook references CFR in order that readers can view the details of each legislation (U. S. Food and Drug Administration 2010, 2013).

As a general principle, manufacturers' have a responsibility to keep up with the latest legislation for food labeling and future changes (U. S. Food and Drug Administration 2018c). The organization of this chapter was inspired by the FDA guide and CFR (Table 4.5).[6]

2 Labeling Requirements

2.1 General Principles

Food labels must bear the name of the product, the net contents (net weight or liquid measure in both common household and metric measures), and the name and address of the manufacturer, packer, or distributor. The net weight is different from the drained weight, for products packed in brine. The inclusion of net contents in a label is important for consumers, so that they can ensure that they make purchasing judgments related to unit costs (e.g., buying a large package may yield significant savings in unit cost). They can also use unit cost to select the most economical product from among different brands. Products must be labeled clearly, avoiding any confusion. Many foods are covered by a standard of identity, under authority of the FDA (Sec 2.2).

Table 4.5 The FDA industry guide to food labeling

Introduction
Background
General requirements for labeling
Name of food
Net quantity of food
Ingredients list
Nutrition labeling
Claims
Nutrient content claims
Health claims
Qualified health claims

Reference U.S. Food and Drug Administration (U. S. Food and Drug Administration 2010, 2013)

The FDA industry guidelines for labeling describe in very precise terms, the locations for labeling information on the package (U. S. Food and Drug Administration 2013). The principal display panel refers to the front (square) panel of the packaging, which is most easily visible to the consumer, under normal display conditions. The information panel is located *immediately* to right of the principal display panel. For a regular rectangular package, the size of the principal display panel can be calculated: height multiplied by width. For a cylindrical package, the principal display panel corresponds to 40% of the area calculated as height multiplied by circumference (Fig. 4.2).

The purpose of the information panel is to carry larger quantities of labeling information: ingredients lists, producer address, Country of origin, nutrition labeling etc. Some packages have two alternative information panels. For example, a rectangular package may have a "landscape" or "portrait" profile. There may be openings at the sides of the package immediately (right side) to the principal display panel. The information panel may need repositioning if the principal display panel occurs at the top of rectangular package. Complication may arise also if the boundary between the principal display panel and information panel becomes less clear e.g. irregular, round, or cylindrical packages and bottles (25 CFR Part 101. 2).

2.2 Standard of Identity for Food

Many foods have a "standard of identity" which refers to a detailed set of ingredients (21CFR

[6]Material in this chapter is introductory; seek legal advice before implementing food labelling.

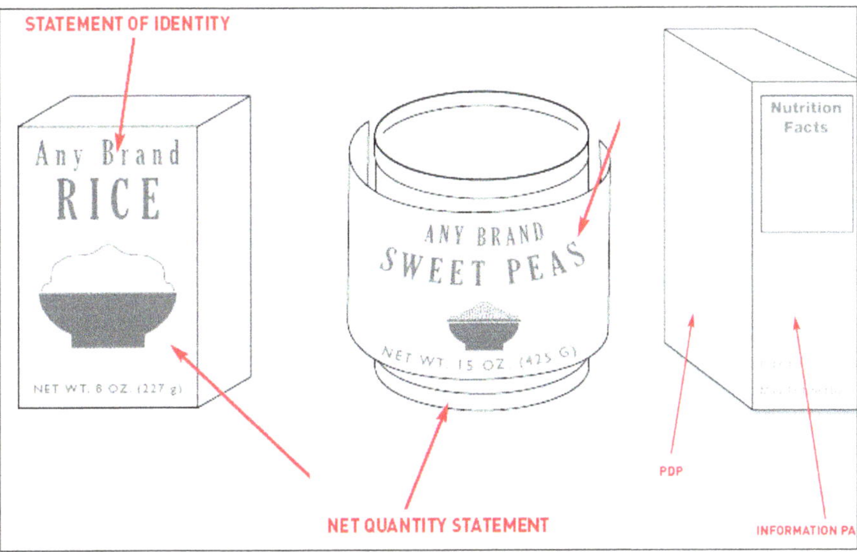

Fig. 4.2 Schematic diagram of some food packages showing the locations of the principal display panel and information panel. (Adapted from U. S. Food and Drug Administration 2013)

130.3). The FDA has been developing and enforcing food standards since 1938 (North Dakota State University 2020).[7] The idea behind food standards was to "*promote honesty and fair dealing of consumers*" (cf. 21 USC 343(g)). Standards of identity define compulsory ingredients, optional ingredients, minimal and maximum amounts allowed, and prohibited ingredients (EveryCRSReport.com 2016).

The standard of identity can be produced in response to a petition from one or more stakeholders or the FDA. Industry petitioners must undertake to appear for a public hearing to provide proof of evidence if required. The example for peanut butter illustrates that the process for establishing a standards of identity involves congressional hearings that may be attended by Consumer activists and others interested in safe foods. It was reported that the standard of identify for peanut butter took 12 years to be established (Boyce 2016).

[7]See Food, Drugs & Cosmetics Act (1938) subsection 401.

The FDA also reviews any food standards produced by Codex Alimentarious, and may adopt, review and amend any it sees fit (21CFR 130.5). Currently, there are at least 300 United States food standards published via the CFR (Table 4.6). The FSIS/USDA produces standards for identify for meat products (9 CFR 319) and poultry foods 9CFR 381 (155) and 9CFR 381 (177) (North Dakota State University 2020). A listing of 224 food standards are also available from Codex commission (Codex Alimentarious Commission 2020).

A product will be judged as not meeting a standard of identity where, (i) it contains any ingredient not listed in a standard, (ii) there are ingredients added but not listed in an the standard, and or (iii) the proportions of ingredients or manufacturing procedures do not match those depicted in an existing standard (21CFR130.8). However, NLEA (1990) allows selected food additives to be added standardized foods without changes to the standard (EveryCRSReport.com 2016).

2.2.1 The Statement of Identity

Declaring the name of a product is a basic requirement for labeling. The statement of iden-

Table 4.6 The FDA standard of identity for some foods[a]

21CFR130	130.3–130.20	Food standards- general
21CFR131	131.3–131.206	Milk and cream
21CFR133	133.3–133.196	Cheeses and related cheese products
21CFR135	135.3–135.160	Frozen desserts
21CFR136	136.3–136.180	Bakery products
21CFR137	137.105–137.350	Cereal flours and related products
21CFR139	139.110–139.180	Macaroni and noodle products
21CFR145	145.3–145.190	Canned fruits
21CFR146	146.3–146.187	Canned fruit juices
21CFR150	150.110–150.160	Fruit butters, jellies, preserves, and related products
21CFR152	152.126	Fruit pies
21CFR155	155.3–155.201	Canned vegetables
21CFR156	156.3–156.145	Vegetable juices
21CFR158	158.3–158.170	Frozen vegetables
21CFR160	160.100–160.190	Eggs and egg products
21CFR161	161.30–161.190	Fish and shellfish
21CFR163	163.5–163.155	Cacao products
21CFR164	164.110–164.150	Tree nut and peanut products
21CFR165	165.3–165.110	Beverages
21CFR166	166.40–166.110	Margarine
21CFR168	168.110–168.180	Sweeteners and table
21CFR169	169.3–169.182	Food dressings and flavorings

[a]Adapted from the Code of Federal Regulations (CFR). Currently 300 food standards are available

tity must be the *usual or common name* of the product, "as established by law or regulation" (21CFR 101.3 (2)). Adopting a traditional name signifies that a product does not differ substantially from the corresponding food standard. The statement of identity, must be printed on the principal display panel, using bold font 1/16th Inch minimum (21CFR 10.3). The physical state of the product should must also be reported (whether it is diced or sliced etc.) beside the statement of identify.

If there is no prior us*ual or common name*, the manufacturer may identify the product using a "*fanciful but descriptive*" name (21CFR 102. 5). The chosen name should describe a food product accurately and simply and avoid confusing it with other different food classes. The food name may include a characterizing ingredient – that is a component that affect product pricing and acceptability (North Dakota State University 2020).

2.2.2 Imitation Foods and Fake Foods

An imitation food is a product that resembles a traditional food, but is *nutritionally inferior* with above 2% difference compared to a relevant food standard (21CFR 101.3 (e)). The term "imitation" must appear alongside of the product name, "*if it is a substitute for and resembles another food but is nutritionally inferior to that food*". A nutritionally inferior food that carries the label "imitation" can be marketed under a traditional name. As an illustration, the name "imitation crabmeat" will avoid charges of "misbranding" because (i) crabmeat has a standard of identity, (ii) the imitation product is nutritionally inferior compared to the standard – contains no crabmeat, but (iii) the packaging is labelled "imitation". Currently, some imitation crabmeat sticks made from surimi is sold in the United States using a variety of statements of identity, e.g. "California roll" "Imitation crab" or "Simply surimi".[8] For reference, the codex standard for canned crabmeat applies to products that contain above 50% crabmeat.[9]

There appears to be no obligation for calling a product "imitation" if there is no prior usual or common name available. For a novel product, the manufacturer must adopt "*an appropriately descriptive term that is not false or*

[8]These names may be trademark terms.

[9]There does not appear to be a formal standard of identity for crabmeat in the United States, but manufacturers refer to their use of genuine fresh crab.

Table 4.7 Some imitation food products

Imitation food	Comment
Crabstick	Fish surimi starch mix
Meat analogue	Various e.g. extruded/precipitated plant protein isolate, lipid,
Artificial rice	Extruded rice, grain powder to resemble rice
Egg substitute	Various including tofu, chickpea flour
Plant milk	Water extracts from plant seeds e.g. almonds, coconut, etc.
Non-dairy creamers	Caseinate emulsions either liquid or powdered form for coffee whitening
Cheese analogues	Various gelled, savory products based on casein or starch
Custard (egg free)	Starchy egg-free replacement for traditional custards
Imitation shark fin	Gelling hydrocolloids to produce chewy texture

Source: various internet browsing and patent searches

misleading. The label may, in addition, bear a fanciful name which is not false or misleading" (21CFR 101.3 (3) (2)). In summary, manufactures can either designate a novel product as "imitation" or else give a new name. Each choice has disadvantages and advantages. The downside of choosing a completely new name is that the product may get less brand recognition initially. However, re-branding may avoid negative perceptions somtimes associated with the term "imitation".

Clearly, naming foods can be somewhat nuanced. The declared names for many plant based milk and meat substitutes seems to be proving controversial (Table 4.7). Interestingly, the term "imitation food" is sometimes confused with the more pejorative "fake food", even though the former is a legally recognized food category (Olmsted 2017).[10] The nutritional inferiority of imitation foods can be problematic also. For example, it is not clear whether imitation crabmeat is nutritionally inferior compared to conventional crabmeat (Fig. 4.3). The nutrition facts label for imitation crabmeat and conventional crabmeat does show only slight differences in the nutritional chacteristics despite the fact that one product is derived from fish surimi whilst the other is crustacean (Fig. 4.3). The imitation foods category seems to be a popular area for innovation and research.

[10] Artificial food, fake food.

2.3 Misbranding of Food

2.3.1 General Principles

Product misbranding may occur due to several circumstances. First, any feature that renders a food label misleading may count as misbranding. For example, the declared name on the package may be incorrect (see above). An imitation food, for which a normal or usual name exists, may be missing the term "imitation" – such a product will be classed as misbranded. Instances where a product name overemphasizes one ingredient out of proportion to multiple ingredients may count as misbranding. Errors related to geographic indication, where the manufacturers address is wrong also counts as misbranding (Table 4.8).

2.3.2 Penalties for Misbranding

Misbranding due to under filling and other causes, can lead to two broad responses from the FDA, (i) Warning letters. The FDA may send a warning letter to manufactures hoping they will take corrective action, (ii) Product seizure. The FDA is empowered to seize misbranded foods. The seizure is followed by a court hearing (condemnation hearing) following which the product may be destroyed, sold, or reconditioned for export (EveryCRSReport.com 2016).

2.4 Ingredients List

A listing of ingredients must be printed within the information panel, in order of descending amount (21CFR 101. 4). The specific names for ingredients should be listed and not generic names. Some generic designations are allowed for labeling (e.g. milk, skim milk, and egg yolk) where sub-components need not be specified.

Fig. 4.3 A nutrition facts label for imitation crabmeat (crabs sticks) compared with the nutrition facts label for crab (right panel). (Data compiled using data from Fitbit.com 2020)

Nutrition Facts

Serving Size 30 grams

Amount Per Serving	
Calories 35	Calories from Fat 0
	% Daily Value*
Total Fat 0.7g	1%
Saturated Fat 0.1g	1%
Trans Fat 0g	
Cholesterol 8mg	3%
Sodium 52mg	2%
Potassium 47mg	1%
Total Carbohydrate 2.1g	1%
Dietary Fiber 0.2g	1%
Sugars 0g	
Protein 4.8g	10%
Vitamin A 1% •	Vitamin C 0%
Calcium 2% •	Iron 2%

* Percent Daily Values are based on a 2,000 calorie diet. Your daily values may be higher or lower depending on your calorie needs.

Nutrition Facts

Serving Size 28 grams

Amount Per Serving	
Calories 29	Calories from Fat 0
	% Daily Value*
Total Fat 0.5g	1%
Saturated Fat 0.1g	0%
Trans Fat 0g	
Cholesterol 28.5mg	10%
Sodium 172.5mg	7%
Potassium 92.4mg	3%
Total Carbohydrate 0g	0%
Dietary Fiber 0g	0%
Sugars 0g	
Protein 5.8g	12%
Vitamin A 0% •	Vitamin C 1%
Calcium 3% •	Iron 1%

* Percent Daily Values are based on a 2,000 calorie diet. Your daily values may be higher or lower depending on your calorie needs.

Table 4.8 Forms of misbranding

1. Label is false or misleading
2. Uses another name
3. Non-declaration of imitation food
4. Misshapen or misfiled container
5. Missing labeling information
6. Information on label is difficult to read
7. Falsely claims meeting standard of identity
8. Does not meet standards of quality and fill of container

From USC 21. 9. 343

Added water must be listed per-weight basis similarly to other ingredients. Ingredients added at levels below 2% w/w must be listed last and indicated as "present at below 2%". Sources of food allergens must be listed. Significant amounts of sulfite (above 10 ppm), which can cause allergic responses, should be listed. Chemical preservatives (and additives) should be listed as per their common name, with a brief statement of their function. Special requirements for labeling natural or artificial flavors, and spices can be found in 21CFR 101.22.

2.5 Other Labeling Items

2.5.1 Name of Manufacture, Distributor or Packer

A food package must carry the name for the manufacturer, distributor or packer – together with their address: street, city, state and ZIP code. Streets, which are listed within a current city directory or telephone book, need no declaration. Products made by a third party can be labelled "Made for" or "Distributed by…". The manufacturers name and related information should appear on the information panel (21.CFR 101.5).

2.5.2 Net Quantity of Product

The weight of foodstuff within a package should appear on the principal display panel (Fig. 4.2). The net product weight (grams) must not include the weight of the packaging or filling brine for solid food; drained weight (i.e. weight of food minus any brine) must be reported. For liquid food or beverage, the product amount is expressed as a volume. To control for temperature effects, the product weight/ volume of frozen or refrigerated foods are measured at 4 °C or prior to melting, respectively. Other foods must be weighed at a temperature of 20 °C (21CFR 101.7).

The weight (or volume) of the product must appear within the lower 30% of the principal display panel. The font size used will vary depending on the display panel surface area; the minimum print size must be between 1/16th and one half of an inch depending on the panel size (21 CFT 101.7). The number expression for product quantity must be given to one decimal place (1 DP). Small deviations from the product weight may be tolerated owing to moisture gain or loss. For multiunit packaging, the weight for each unit and total package weight should be given, e.g. a six-pack might be described as "6–16 oz." package.

2.6 Vending Machine Food Labeling

A self-service device that dispenses food when money or currency is inserted without the need to refill between customers is called a vending machine. Food labeling requirements apply where a vending machine owner operates more than 20 units (21CFR 101.8). Vending machine operators (owner or a licensee) should provide nutrition information covering "Calorie" content for foods, serving size and servings per container. Food total energy must be expressed as 5-Calories increments (total food energy <50 Cal) or 10-Calories increments (for total food energy >50 Calories).

Exceptions to the vending machine labeling rules apply if an operator has below 20 machines. Additional labeling is not required also, if nutrition information is visible on any food packaging inside of the vending Machine "without obstruction". Labelling information can also be displayed in proximity to the vending machine, in prominent letters, and in a manner that allows viewing by customers prior to purchase (21CFR 101.8).

2.7 Below Standard Fill

Inaccurate weight or volume declarations may lead to charges of misbranding (cf. Sect. 2.3). In addition, a food (or consignment) will be judged "substandard" if there is an erroneous statement of quantity. Incorrect filling may arise due to accidental over-filling, under filling or incorrect labeling arising from equipment failure or human error.

A food product will be designated as "below standard fill" and unfit for general retail if the quantity of food is below the declared weight (21CFR 130.14(b)). Provided all other labeling requirements are met, misbranded foods can be sold to institutions run by federal, State or local government – including schools, prisons, and hospitals (21CFR 101.7). In addition, the receiver needs to acknowledge the voluntary purchase of a misbranded food consignment and indicate; "Product Mislabeled. Actual net weight (drained weight or volume where appropriate) may be as low as _% below labeled quantity".

3 Specific Requirements

3.1 Spices, Flavorings, Colorings and Preservatives

In broad terms, the labeling of "additives" is required without exception (21CFR 101.22).[11] However, the legislation is flexible, because items can be labelled using a specific name, or a generic name followed by the designation as "artificial" or "natural". Foods that contain preservatives should list the compound in a food label following by an explanatory phrase that explains the technological function; "preservative", or "helps to protect flavor" or "promotes color retention".

In the past, most colors did not have to be listed by name, but had to be listed within the

[11]Also see 21 CFR 201.20.

term "colorings." With the implementation of NLEA, this changed and colorings must be listed individually. Yellow Nos. 5 and 6, Red Nos. 40 and 3, Blue Nos. 1and 2, and Green No. 3 are all examples.

Protein hydrolysates are added to food for various reasons such as thickeners, flavorings, flavor enhancers, or nutrients. Because the law does not require all flavors to be identified by their common or usual names, some industries have included hydrolysates in the "flavorings" or "natural flavors" declaration. This has happened when the hydrolysates have been added as flavor enhancers, a use not exempt from common or usual name declaration. The FDA concluded that when hydrolysates are added as flavors, they also function as flavor enhancers and must be declared on the label by their common or usual name. In the past, terms such as "hydrolyzed vegetable protein," "hydrolyzed animal protein," or "hydrolyzed protein" were allowed. Now, the source of the protein, such as corn, soy, or casein, must be identified as in the declarations "hydrolyzed soy protein" or "hydrolyzed casein." In the case of casein (or caseinate), it must be identified as a milk derivative. This will help casein-sensitive people or people who, for religious reasons, avoid milk products, to eliminate products containing casein from their diet. Some "nondairy" creamers do contain caseinate, added to whiten effectively.

NLEA (1990) also has provisions concerning sweeteners, percentage ingredient information, sulfites, and requirements for juices. Voluntary inclusion of the food source in the name of sweeteners is now allowed. For example, "corn sugar monohydrate" may be used in addition to "dextrose" or "dextrose monohydrate," which is also permitted.

Manufacturers who decide to declare ingredients by percentage of formulation must do so by weight. They may use percentage declaration in as many or as few ingredients as they choose as long as the information is not misleading. They still must declare all ingredients in descending order by weight. Since 1986, sulfiting agents had to be listed in non-standardized foods. Now, standardized food labels must list them as well.

3.2 Juices and Concentrates

Specific labeling requirement apply to products that contain fruit or vegetable flavor or various proportion of juices (21CFR 101.22). Since 1990, "*a food that purports to be a beverage containing juice must declare the per cent of total juice on the information panel.*" This alone does not eliminate confusion, as some juice manufacturers use bland juices such as apple or white grape and prominently name the juice with some of the lesser used juice. As an example, the statement "strawberry or raspberry juice containing 100% juice" could lead a consumer to believe that the product contains only strawberry or raspberry juice when it actually may contain those juices in very small amounts. To correct this misconception, FDA has required manufacturers to either: state the beverage is flavored by the named juice such as "strawberry flavored juice drink" or Declare the amount of the named juice in a 5% range as "juice blend, 2–7% raspberry juice." Juice levels below 1% are not to be declared. Readers are recommended to refer to juice labeling regulations for more details (21CFR 101.22).

4 Food Allergens

4.1 Food Allergens Labeling and Consumer Protection Act

Food allergy affects 2% of children and 8% adults within the US annually (U. S. Food and Drug Administration 2009a). Consequences of allergy lead to 300,000 hospitalizations and 150–200 deaths per year. The main legislation relevant for food allergen labeling is the "Food allergen labeling & consumer protection Act, 2004 (FALCPA). Information and general guidelines to industry were published in 2006 (U. S. Food and Drug Administration 2006), 2009 (U. S. Food and Drug Administration 2009a) and most recently 2018 (U. S. Food and Drug Administration 2018a). The ultimate aim of the FALCPA is to reduce the risk of exposure to allergens (Thompson et al. 2006).

Table 4.9 A list of some major food allergen sources

Milk	Includes all dairy products
Egg	Liquid egg, dried egg, egg products
Fish	Flounder, cod or bass
Crustacean/ shell-fish	Shrimp, lobster, or crabs
Tree nuts	Almonds, pecans, or walnuts
Wheat	All wheat products, flour and ingredients
Peanut	Peanut butter, nut confectionery
Soybean	All

Source adapted from Thompson et al. (2006)

The legislation defines a "major food allergen" as any *ingredient* or food, which contains *protein* from one of eight food classes (Table 4.9). The 8-major allergen sources account for 90% of exposures; there are in fact nearly 160 different kinds of food allergen sources.

Effective from Jan 2006, it is mandatory to provide food allergen information on all food packaging regulated by the FDA. The types of allergens that need to be declared are interpreted broadly; for example, the designation "wheat" can be used to indicate that wheat derived ingredients or other cereals may be present. The current FDA listing of "tree nuts" contains over 15 different species.

Allergen labeling information may take two forms. First, labeling can be achieved if the name of one of the eight allergens appears anywhere in the information display panel. The name of the major allergen source may be given in parenthesis alongside an ingredient that is the source of the allergen, e.g. "casein (milk)" designates that the ingredient casein is from milk. Alternatively, the name(s) of allergen sources can be presented below the list of ingredients. In this case, a simple declaration can be using "Contains" statement. For example, a product suspected of containing egg or egg protein would be marked "Contains egg" (Thompson et al. 2006) (Table 4.10 and Fig. 4.4).

Unlike other labeling items, there is no lower threshold limit for declaring allergens. One may recall that a low level of sugar corresponding to 0.5 g per serving can be designated "zero sugar". By contrast, even trace quantities of protein antigens can elicit an allergic response.

Table 4.10 FDA listing of some tree nut allergens

Civilian name	Science name
Almond	*Prunus dulcis (Rosaceae)*
Beech nut	*Fagus spp. (Fagaceae)*
Brazil nut	*Bertholletia excelsa (Lecythidaceae)*
Butternut	*Juglans cinerea (Juglandaceae)*
Cashew	*Anacardium occidentale (Anacardiaceae)*
Chestnut	*Castanea spp. (Fagaceae)*
Chinquapin	*Castanea pumila (Fagaceae)*
Coconut	*Cocos nucifera L.*
Filbert/hazelnut	*Corylus spp. (Betulaceae)*
Ginko nut.	*Ginkgo biloba L. (Ginkgoaceae)*
Hickory nu	*Carya spp. (Juglandaceae)*
Lichee nut	*Litchi chinensis Sonn. (Sapindaceae)*
Macadamia nut/ bush nut	*Macadamia spp. (Proteaceae)*
Pecan	*Carya illinoensis (Juglandaceae)*

Source: adapted from U. S. Food and Drug Administration (2013)

4.2 Applicability of Allergen Labeling

The allergen labeling law extends to all food retailers and foodservice operations that package foods for human consumption. Starting from January 1, 2006 all packaged foods were required to be labelled, if the food belongs to a major food allergen group or where a packaged food "contains" ingredients, additives or colorings derived from one of the allergen groups (Table 4.9). The law covers conventional foods, dietary supplements, as well as foods for special dietary uses.

4.3 Exemptions from Allergen Labeling

The focus of food allergen legislation is "packaged" foods. Exposure risk arising from unpackaged foods are not covered. Exemptions from the food allergen labeling regulations apply for raw agricultural commodities (U. S. Food and Drug Administration 2013). Fresh foods, which means foods that have not been thermally processed or frozen, are not required to carry allergen label-

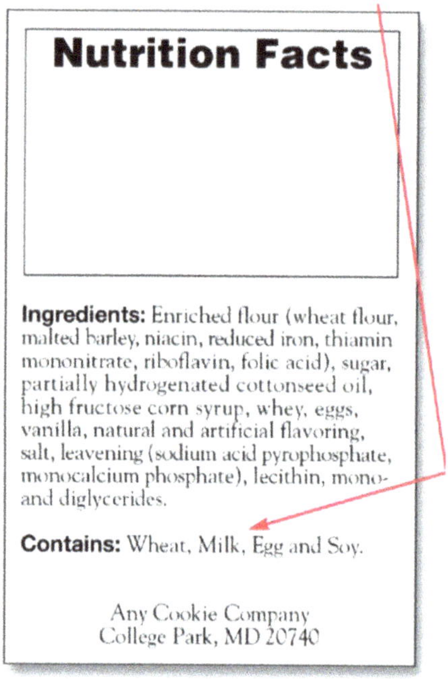

Nutrition Facts

Ingredients: Enriched flour (wheat flour, malted barley, niacin, reduced iron, thiamin mononitrate, riboflavin, folic acid), sugar, partially hydrogenated cottonseed oil, high fructose corn syrup, whey, eggs, vanilla, natural and artificial flavoring, salt, leavening (sodium acid pyrophosphate, monocalcium phosphate), lecithin, mono- and diglycerides.

Contains: Wheat, Milk, Egg and Soy.

Any Cookie Company
College Park, MD 20740

Fig. 4.4 The Information display panel, showing the nutrition facts panel and its relation to ingredients and allergens. (Adapted from U. S. Food and Drug Administration 2013)

ing. Refined oils of all kinds are another food group not required to carry food allergen labels. Currently, no "threshold" lower limit has been set for the amount of allergen necessary to trigger a requirement for labeling (U. S. Food and Drug Administration 2006, 2013). However, cross-contamination by food allergens does not require labeling (U. S. Food and Drug Administration 2006).

4.4 Precautionary Food Allergen Labeling

The food allergen labeling act FALCPA does not apply to products that "might" contain allergens due to unintentional addition. For example, an unintentional addition of allergen could arise due to cross-contamination of one food product by trace amounts of another. In principle, cross contact between foods could occur anywhere along the food supply chain: harvesting, storage, transportation, manufacturing or food preparation. However, manufacturers could consider using advisory or precautionary allergen labels informing the consumer where a product is "produced in facility that also uses allergens". Where the cross contact occurs at the food manufacturing stage, consumers might be informed that a food was "produced in a facility that also sues [allergen]" (U. S. Food and Drug Administration 2018b).

5 Nutrition Labeling

5.1 Characteristics of Nutrition Labeling

The Nutrition Labeling & Education Act 1990 (NLEA) made it compulsory for food packages to carry nutrition information. The aim of the nutrition label is to provide a description of nutrients in packaged foods so that the consumer can make informed choices about healthy eating (Codex Alimentarius Commission 1985b). As noted elsewhere, nutrition label information is considered one of several "population based" approaches for improving public health (Mozaffarian et al. 2012). Generally, nutrition labels have two components; a nutrient declaration consisting of a list nutrients and a separate section containing supplementary information (Codex Alimentarius Commission 1985b).

A nutrient is defined as any substance that is consumed as part of food and, which is used, by the human body for energy or for growth and maintenance. Some other food constituents where a deficit can lead to adverse physiological changes are also classed as nutrients (Codex Alimentarius Commission 1985b). The classical nutrients include protein, lipids, carbohydrates, vitamins and minerals, which are needed for wellbeing. Consideration is needed for components formerly classified as "non-nutrients" (e.g. dietary fiber) but which if missing from the diet can lead to chronic diseases. In the cases of fiber and certain other phytochemicals, the boundary between what is a nutrient or non-nutrients is beginning to erode (Codex Alimentarius Commission 1985b) (Table 4.11).

Nutrition labeling is similar to ingredient labeling in many respects (Sect. 2). For instance,

Table 4.11 Nutrition labeling sections from the Code of federal regulations

Subpart A	General Nutrition labeling
CFR 101.9	NL of food. → lists amounts of nutrients in a nutrition facts panel
CFR 101.10	Restaurant foods bearing nutrient content claims or health claims.
CFR 101.11	Standard menu items in covered establishments → list calorie values for menu items
CFR 101.12	Reference amounts customarily consumed per eating occasion → how to assess RACC
CFR 101.13	Nutrient content claims → claims food has high, low, or is free nutrients
Subpart C	Specific nutrition labeling requirements and guidelines
CFR 101.36	Dietary supplements.
CFR 101.42	Raw fruit, vegetables, and fish.
CFR 101.43	NL for the voluntary nutrition labeling of raw fruit, vegetables, and fish.
CFR 101.44	20 most frequently consumed raw fruits, vegetables, and fish in the US
CFR 101.45	Guidelines for raw fruits, vegetables, and fish (voluntary)

Source: Edited from 21CFR 101 Food labeling

Table 4.12 Revisions of the nutrition facts label

Item	Pre-2016	After 2016
1. Per serving	Busy statement	Simplified, total and number of servings per package
2. Food energy	Calories listed #3	Calories more prominent
3. Fat	Labels calories from fat	Omits Calories from total fat
4. Sugar	Labels as "sugar"	Labels as "total sugar"
		Separate declaration for "added sugar"
5. Micronutrients	Vit. D & Vit. A; iron and potassium	Vit. D and potassium, not vit C
6. DV footnote	More wordy	Simplified for 2000 Cal only

the nutrition label information should be "*not in any way false, misleading, deceptive or insignificant*" (Codex Alimentarius Commission 1985b). There is a need for nutrition labeling to be simple and easy to understand (Van den Wijngaart 2002). There is evidence showing that some key aspects of nutrition labeling (serving sizes) are not understood by the public who find it difficult to relate such facts to daily experiences (Cowburn and Stockley 2005).

5.2 Revised Nutrition Labeling Requirements

The requirements for NLEA legislation have undergone several amendments since 1993. For instance, the FDA introduced new recommended daily values (DV) in 1995. Nutrition labeling changes for lipid components (trans-fats, saturated fats and cholesterol) were initiated from 2003 and completed in 2006. Between 2003 and 2005, the FDA requested public comments on the planned changes to nutrition labeling, to reflect improved understanding of food and disease relations. Further revisions to the NLEA were completed in 2016 and communicated to the food manufacturing and foodservice industries in 2018 and 2019. Some of the latest alterations to the nutrition facts label are described below (Table 4.12 and Fig. 4.5)

5.3 Requirements for Nutrition Labeling

5.3.1 General Principles

Nutrition labeling rules (21CFR 101.9) apply to virtually all foods intended for human consumption.[12] Nutrition labeling must be printed on the food packaging or displayed separately at the point of sale. The quantities of various nutrients should be expressed per "serving", where one serving is defined as the amount of food customarily consumed by someone aged 4 years or older. Serving sizes can be determined using

[12]Some exemptions for nutrition labeling requirements (21CFR 101. 9 (paragraph J)) are discussed below (Sect 5.5.).

Fig. 4.5 Required items for nutrition labeling

carefully set out methodology (21CFR 101.12).[13] Discussions about labeling assume that a manufacturer has accurate food composition data and that their focus is how to navigate the labeling requirements.

The nutrition facts label must be a thickly outlined box positioned on the information panel. A mandatory list of nutrients must appear within a nutrition facts label (Table 4.13). For small packages (<40 sq. inches) the nutrition facts label can be moved to another panel. The quantity for each nutrient is expressed per serving, and per 100-g sample; equivalent values for percent-recommended daily value (RDV) should be also listed. An example nutrition facts labels can be seen in Fig. 4.3. Additional layouts, formatting and fonts may be be used (Figs. 4.5 and 4.6).

5.3.2 Nutrient Analysis

A nutrient declaration is an essential aspect of any product development, reformulation, or menu change. However, it is not always necessary to determine the types and amounts of nutrients present in a product by direct analysis. Nutrient profiles are available for a diverse range for specific products and branded products analyzed by the federal government's own scientists.

Table 4.13 A list of prescribed nutrients must be covered by nutrition labeling

Compulsory items	Voluntary items
Total calories	Calories from saturated fat
Total fat (g/serving)	Poly-unsaturated fats
Saturated fat	Mono-unsaturated fats
Trans fat or *Trans*	Fluoride
Cholesterol (mg/serving)	Soluble/insoluble fiber
Sodium (mg per serving)	Sugar alcohols
Carbohydrate, total (g/serving)	Vitamin C
Dietary fiber (g/serving)	
Total sugar (g/serving)	
Added sugar (g/serving)	
Protein (g/serving)	
Vitamins	
Vitamin A	
Potassium	

Summarized from 21CFR 101.9

[13] 21CFR 101.9 contains more than 100 occurrence of the terms or "serving" or "serving size in 29 pages.

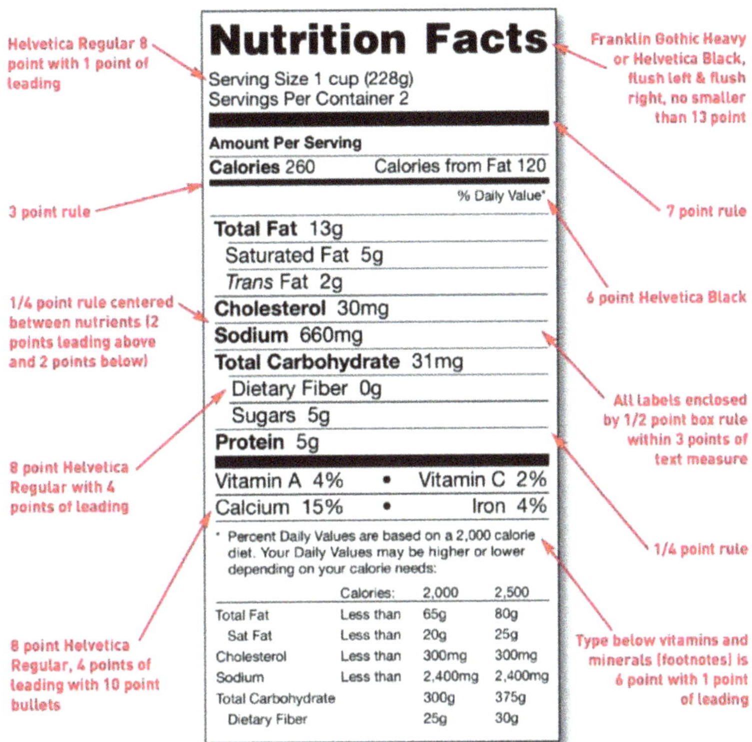

Fig. 4.6 Formatting and labeling of prepackaged foods. (Curtesy, US Food & Drugs Administration)

The nutrient composition of foods currently eaten in the United States is available as the USDA "nutrient database" compiled over several decades, and updated regularly. From October 2019, the USDA nutrient database was re-located to a "FoodData Central" access point. The new portal links general nutrient information with current research (USDA Agricultural Research Service 2019). The online USDA nutrient database supersedes data previously published by way of USDA Agricultural Handbooks.

Commercial nutritional analysis software are available to determine nutrient composition for menus and dishes. Such software normally incorporate USDA nutrient datasets, as the basis for menu or recipe analysis. Computerized nutritional analysis can be done in house by trained nutritionists. Alternatively, the nutritional analysis can be performed on a contract basis. Some of the more sophisticated nutrient analysis software will produce "ready to print" nutrient labels (cf. Fig. 4.6).

5.4 Nutrition Declaration

Current nutrition labeling regulations require a nutrient declaration, which is a list of key nutrients and the quantity of each expressed as weight per serving and as percent of daily value (Table 4.13 and Fig. 4.6).

5.4.1 Food Energy, Calories

The calorie value can be estimated from the proximate composition of the product (21CFR 101. 9 (c) (1)), protein, lipids, and total carbohydrate expressed as grams per 100 g: 4 Cal (x protein) + 4 Cal (x carbohydrate) + 9 Cal (lipids). Recall that 4.4.9 Cal values are standardized Atwater factors for converting gram-values for carbohydrate, protein and fat into Calories (kilocalories). The government's compilation of Atwater factors is taken from the classic USDA Agricultural handbook 74 published in 1950 and revised in 1973 (Merrill and Watt

Table 4.14 Atwater factors for some sugar alcohols (Cal per gram)

Isomalt	2.0
Lactitol	2.0
Xylitol	2.4
Maltitol	2.1
Sorbitol	2.6
Hydrogenated starch hydrolysates	3.0
Mannitol	1.6
Erythritol	0

1973).[14] Atwater factors do not themselves appear on food labels (Table 4.14). Atwater values for sugar alcohols (polyols) range from 1.5 to 4 Cal per gram (Livesey 1992, p77).

5.4.2 Total Fats, Saturated Fats

The total fats content (grams per serving) is measured as the sum of all fatty acids plus glycerol (21CFR 101. 9 (c) (2)). The values are cited as 0.5 g increments and a product containing 0.5 g or less of total fat per serving is generally assigned a value of "zero". The exact same rules apply for the declaration of saturated fats on food packaging. Currently, there is no compulsory requirement to report unsaturated fats content, presumably because this can be estimated as the difference between values for total fat and saturated fat.

5.4.3 Trans-Fats

As from 2006, food manufacturers are required to reduce the levels of trans-fatty acids in food (21CFR 101. 9 (c) (2) (II)). Labeling for trans-fats is also mandatory. The amount of trans-fats in food is determined from the weight of unsaturated fatty acids, that carry a one or more isolated double bond; the category excludes fatty acids with conjugated bonds. As above, levels of trans-fatty acids are expressed to the nearest 0.5 g. Values of trans-fats below 0.5 grams per serving are assigned a value of "zero".

[14]The book is an erudite account of food energy, which is highly relevant to contemporary discussions about obesity. This is recommended reading for anyone seriously interested in the relations between food intake and energy balance. The 109pp book is still available online from the USDA website.

The quantity of trans-fats present in any food or dietary supplement, can be quantified using one of two approved methods. The two official methods for measuring trans fatty acids for food labelling purposes are, gas liquid chromatography (GLC) or alternatively using an infra-red light absorption technique titled Attenuated Total Reflection– Fourier-Transform Infrared (ATR–FTIR) spectroscopy mostly abbreviated as FTIR (Mossoba et al 2009; Tyburczy et al. 2013)

5.4.4 Cholesterol

The levels of cholesterol in food must be declared as milligram per serving (mg/ serving) and recorded in intervals of 2 mg/serving. Values below 5 mg cholesterol per serving can be assigned a value of "zero" (21CFR 101. 9 (c) (3)).

5.4.5 Sodium

Sodium is a reflection of the amount of culinary salt present in foods. Sodium is declared as milligrams per serving (mg/serving). Values for sodium content are expressed as 5 mg intervals (1–140 mg/serving) or as 10 mg intervals (>140 mg/serving). Sodium values of 5 mg/serving is indicated as "zero" (21CFR 101. 9 (c) (4)).

5.4.6 Carbohydrates, Total

The total carbohydrate (carbohydrate, total) content must be expressed as grams per serving for all foods at 1 g-intervals. Products with less than 1-g per serving can be considered "zero" carbs and signified by the statement; "Contains less than 1 g carbohydrate". There is no formal recognition for the term "low carb" that is frequently applied to some forms of dieting. Total carbohydrate will include dietary fiber, total sugars and sugar alcohols (21CFR 101. 9 (c) (6)).

5.4.7 Dietary Fiber

Fiber is non-digestible carbohydrate from natural or synthetic sources, having three or greater sugar units. Dietary fiber should be declared as grams per servings basis, expressed as 1-g increments. The statement "Contains less than 1 g fiber" can be used where a product contains 1 g or less of fiber per serving (21CFR 101. 9 (c) (6) (i)).

5.4.8 Total Sugar and Added Sugar

For labeling purposes, natural sugar is defined the combination of free monosaccharides and disaccharides (glucose, fructose, sucrose) with values expressed as nearest 1 g. A product that contains less than 0.5 g is designated as "zero" value. From 0.5 g to 1 g sugar, the product can be designated "contains less than 1 g sugar" (21CFR 101. 9 (c) (6) (ii)),

Sugars added to food during processing must be declared separately. Added sugar covers free sugar as well syrup, honey and sugar from juice concentrates (21CFR 101. 9 (c) (6) (iii)). The expression for added sugar is "indented" under the value for total sugar with the phrase "includes X-g added sugar". As for total sugar, values for added sugars should be expressed to nearest 1 g; values of less than 0.5 added sugars can be designated as zero.

5.4.9 Protein

The quantity of protein (grams per serving) must be labeled to the nearest gram. Values of less than 0.5 g are cited as "zero". Crude protein can be determined from percentage N x 6.25 but other specific Kjeldahl/ Jones factors may apply for particular foods. Corrected protein concentrations must be reported, after multiplying the crude concentration by digestibility (scale 0–1.0) 'For adults, it is assumed that DV is 50 g/ day.

5.4.10 Vitamins and Minerals

Vitamin and mineral labeling must be age-specific and life cycle specific. Therefore, values for RDI must cited for infants (0–1 years), children (1–3 years), pregnant women and lactating women. Currently, only four vitamins and nutrients must be listed on packages; Vitamin D, Calcium, Iron and potassium with values expressed per serving along with the %RDI. Products with below 2% RDI for must be labeled as "not a significant source" of the appropriate vitamin/ mineral. Aside, from these categories no other vitamin or mineral needs to be listed unless it is a subject it is subject for nutrient content claim. Vitamins and minerals that are listed in standardized food declaration are not required to be listed. However, vitamins (e.g. folate) used for fortification must be listed in labels (21CFR 101.9 (c) (8)).[15]

5.4.11 Voluntary Declarations of Nutrients

Recent advances in nutrition knowledge are leading to changes in the nature of nutrient declarations. Some nutrient groups have been demoted as less important for labeling purposes. For example, the total fats content of foods is now considered more important compared with the proportion of unsaturated and saturated fats. The lists of food components for which labeling is now voluntary include; calories from saturated fat, polyunsaturated fat, monounsaturated fats, and fluoride. The voluntary items may need to be declared if manufacturers make specific nutrient content claims.

5.5 Exemptions and Exceptions from Nutrition Labeling

Food products that carry a nutrient content claim (cf. Sect. 6 of this chapter) must have nutrition labeling. Foods without a nutrient content claim may be exempt from the nutrition labeling depending on the, size of businesses, number of units of product sold annually, the nature of the food product and the extent of packaging (U. S. Food and Drug Administration 2007). Currently, there are at least 18 different exemptions from nutrient labeling (21CFR 9(j)). Restaurants that make a nutrient content claim (e.g. low fat) should when asked must provide written nutrition labeling information related to the nutrient of interest (21CFR 101.10).

5.5.1 Small Businesses Exemptions

Nutrition labeling exemptions apply for *retailers* with direct sales to consumers of sales less than US$500,000 or "gross annual food sales" below US$50,000.[16] Small businesses (manufacturers,

[15]Regulations for labelling vitamins and minerals have been revised recently (cf. 5.2)

[16]Note that the two FDA tests for what is a small retailer (annual sales of US$500,000 sales or US$50,000 "food"

Table 4.15 Possible reasons for nutrition labeling exemption

1	Retailers with direct sales less than US$500 k, Food sales < US$50 k 21 CFR 101.9(j)(1) & (18)
2	Restaurant foods – Except for menu calorie labeling (21CFR 101.11) (21 CFR 101.9(j)(2))
3	Ready for consumption foods, food services
4	Food with limited nutrients (e.g. coffee and spices) (21 CFR 101.9(j)(4))
6	Dietary supplements (21 CFR 101.9(j)(6))
8	Medical foods (21 CFR 101.9(j)(8)
9	Bulk food ingredients for food manufacturing 21 CFR 101.9(j)(9)
10	Raw fruit and vegetables
11	Packaged single ingredients, e.g. fish or game
12	Game meats (various) 21 CFR 101.9(a)(2), 21 CFR 101.9(j)(11)
13	Small packages with <12 sq. inch (77 sq.cm)
14	Shell eggs in cartons (21 CFR 101.9(j)(14))
15	Bakery products & delicatessen-type food (21 CFR 101.9(j)(3))
16	Infant formula, baby foods <4 years olds (21 CFR 101.9(j)(5) and 101.9(j)(7))
17	Donated food (21 CFR 101.9(a))
18	Unit packs from multipack foods, but multipack is labelled
19	Foods from bulk containers, if bulk container is labelled
20	Large package (40 sq. inch) with insufficient pdp or information panel space using other spaces
21	Low sales volume foods, <100 employers, selling <100, 000 units.

Adapted from (GovRegs 2016) and from 21CFR 101 (9) (j) (1) to 21CFR 9 (j) (18)

packers or distributors may exempt from nutrition labeling, if there are fewer than 100 full time workers (21CFR 9 (j) (1)(i)) and/or there are low volume sales (less than 100,000 units) of product annually in the United States (21CFR 101.9 (j) (18)). Small businesses meeting the above conditions must send a written notice to the FDA seeking exemptions every 12 months. Finally, there is an automatic exemption for those non-importing firms with less than 10 full time employees and sales of less than 10,000 total units annually (Table 4.15).

sales) are separated by the Boolean operator "OR" , which suggests that either condition may be applied independently in order to exclude/include a businesses from nutrition labelling requirements.

As noted elsewhere, there are no exemptions from nutrition labeling for any product with a nutrient content claim. Current instructions make it clear that eligible firms are responsible for filing their annual notice for exemption from nutrition labeling regulations, with no reminders sent to defaulters. Exemption from nutrition labeling regulation still requires that firms comply with general food labeling requirements (U. S. Food and Drug Administration 2007).

5.5.2 Ready-to-Eat Meals

Some exemptions from nutrition labeling apply for some ready to eat meals, e.g. foods delivered to care homes. Foods prepared by restaurants for immediate consumption do not require nutrient labeling. The exemption rules can be quite complicated and so manufacturers are advised to get professional legal advice; there may be a need for menu calorie labeling. Exceptions for nutrient labeling also apply to bakeries (NAICS 311811) and mobile vendors (NAICS 722330). The food service (NAICS 722310) sector is exempt from nutrition labeling including food from transporters (aircraft, trains), food service concessions at sports, entertainment & concert venues. The institutional food service providers (hospitals, schools, prisons) also appear to be exempt from nutrition labeling.

5.5.3 Raw Vegetables, Fruits and Fish

Raw produce are exempted from nutrition labeling, except if there is some sort of nutrient content claim. Interestingly there is also a provision for the voluntary labeling of raw vegetables, fruits and fish and the interested reader should refer to (21 CFR 101.45).

5.6 Menu Calorie Labeling

5.6.1 Basic Principles

About 2/3 of US adults are classed as overweight or obese. Excessive eating contributes to a positive energy balance, which leads to weight gain. Food expenditure outside of the home now exceeds that for home-prepared food. Consumers eating out need more information so that they can make more healthy food choices. It was predicted that consumers would reduce their calorie intake, once food energy values were made available at the point of sale (see below). Food manufacturers might reformulate foods once food energy values are publicized (Centre for Science in Public Interest 2010).

Statewide legislation for menu calorie labeling (21CFR 101.11) came into effect from May 7, 2018. The requirement to provide calorie values for menu items and restaurant foods came about 7 years after congress passed the relevant legislation as part of Obama's "Affordable healthcare Act". From 2010 to 2016 many States enacted menu labeling laws, including the States of New York, and California. Campaigners suggested widespread support for menu labeling but some from industry were not supportive.

Menu labeling is required for "*restaurants* or *retail food establishments*" that display three characteristics; (i) establishments do businesses under the "*same* name", (ii) the establishments have more than 20 "*location*", and (iii) establishments offer a "*substantially same menu*" – at the different outlets. A list of some businesses affected by menu labeling regulations include, grocery and convenience stores, pizza vendors, entertainment venues (cinema, sport arenas, amusement parks), cafeterias, coffee shops, salad bars, superstores, and some food service operators. The legislation covers partial-service restaurants, full-service chain restaurants and other chain providers serving prepared meals. Labelling must be provided, where food is consumed on the premises or else eaten shortly after purchase. Food sold by "self-service" restaurants "buffet line, cafeteria line, or similar self-service" is also covered by menu labeling legislation (Table 4.16).[17]

The typical menu label must declare food energy expressed as "Calories" or "Cal" per standard menu offered for sale. For instance, the total energy value for a whole "burger" roll plus filling can be reported. Additionally, information is required for different menu options, e.g., arising from different flavors, from the use of different toppings, or optional (regular versus large) sizes. A common design for menu labeling is that pricing and "Cal" information are placed side-by-side in adjacent columns. The "font size" used for menu labeling should be equal or greater than the font used for pricing, in order to ensure visibility. The calories information should be accompanied by a context statement indicating: "*2,000 calories a day is used for general nutrition advice, but calorie needs vary*" (21CFR.101.11).

[17]The italicized terms from this passage are each defined precisely in the menu labelling legislation (see 21CFR 101.11).

Table 4.16 Examples of standard menu items/ restaurant type foods

Combination meal; multicomponent meal listed on menu board as one item (e.g. burgers fries and drink
Variable menu item
Foods on display
Self-service foods
Beverages

5.6.2 Exemptions from Menu Calorie Labeling

Exemptions from the menu labeling regulation apply for restaurants chains with below 20 outlets. Food purchases classed as "*custom orders*", where food is prepared from non-standard menu for individual customers do not require menu labeling. Mobile vendors are also not required to provide menu labeling. Similarly, food from transport providers (e.g. trains, airplanes and ships) are exempt from menu labeling. Interestingly, schools are also exempt from menu labeling requirements, though the situation with other institutional food services is not certain at this time (21.CFR 101. 9(j)). Non-standard menu items, condiments, temporary menus or self-service foods sold for less than 60 days in the year also do not fall within the menu labeling regulation (U. S. Food and Drug Administration 2020a).

5.6.3 Discussions of Menu Labeling

The effectiveness of menu labeling policy is debated, even though the national policy is relatively new. Studies that compared the quantity of food calories ordered (calories per order) in restaurants before and after this policy showed mixed results (Long et al. 2015).[18] Another approach is to compare food ordered when test subjects are shown a menu with/ and without calorie labeling. Some evaluations combined the results from "prior" studies thereby generating a "systematic review"; roughly, a dozen systematic reviews on menu labeling are available with the most recent discussed here.

Some studies found that menu labeling could lead to reductions in the quantity of food energy purchased by consumers. For example, a study from Denmark found menu labeling produced ~100 Calories fall in food energy per order (Littlewood et al. 2016). The benefits of menu labeling were seemingly higher for neighborhoods within a higher socio-economic grouping (Sarink et al. 2016). Several investigations noted that menu labeling had mixed (positive or nil) effects depending on the context. The effect of menu calorie labeling was more positive for studies conducted in a laboratory setting, compared with real-life settings (Cantu-Jungles et al. 2017; VanEpps et al. 2016). A few studies showed that menu calorie labeling had no effect "whatsoever" regardless of the setting (Fernandes et al. 2016).

Pre-2011 evidence for the menu calorie policy is controversial (Allison 2011). As noted above, menu calorie labeling can lead to significant reductions in food energy purchased but there is no information about the amount of calories consumed. Few follow-up studies were done to see whether menu calorie labeling had long-term effects on study participants.

Past research suggests there is a negative association between the price of food and its energy content. High calorie (energy dense) foods are cheaper and more affordable compared with foods with a low energy density. People from low-income groups tend to choose cheaper higher calorie foods (Darmon and Drewnowski 2015; Drewnowski 2010). In real-life situations, pricing information is displayed next to menu calorie information on restaurant menu boards (Cantu-Jungles et al. 2017). There is little or no information available on the combined effect of menu calorie labeling and pricing on consumers. Studies are required under conditions where food pricing is not zero, or where pricing is artificially inflated in proportion with calorie content. Finally, not all experts agree that energy-dense foods are always cheaper (Carlson and Frazão 2012; Lipsky 2009).

Currently, many consumer groups support menu calorie labeling policy. The policy is also considered worthwhile because of the low cost to industry (Centre for Science in Public Interest 2015). Some public health experts support menu labeling, because other options may be limited (Allison 2011).

6 Nutrient Content Claims

6.1 The Gist of a Nutrient Content Claim

A nutrient content claim is a voluntary statement that highlights the level of specific nutrients in a food product. For instance, the manufacturer may wish to differentiate their product based on "a high" or "low" level claim for vitamin or fat, respectively.[19] The claim descriptors (e.g. high, low, free etc.) are each defined by legislation (cf. 21CFR 101.13, 21CFR105, and 21CFR107). With the exception of statements about minerals and vitamins, foods intended for young children (2–3 years) are not permitted to carry nutrient content claims (21CFR101.13b (3)).

The nutrient content claim refers directly to a specific nutrient. Alternatively, implied claims may refer obliquely to nutrients. Both expressed and implied claims are popular with manufactures. A statement such as "high in fiber" is a direct claim whereas "Contains oat bran" would be an example of an implied claim linked with fiber.

[18] A sustained drop of about 10% calorie intake is thought to be sufficient for weight loss.

[19] Nutrient Content Claims can be only for nutrients that are listed on the nutrition facts label.

6.2 Approved Nutrient Content Claims

Currently, there are approved nutrient content claims for food energy, sodium, lipids (total fat, cholesterol), and sodium (Table 4.17). Common descriptors for nutrient content claims are precisely defined in law; therefore identifying a food component as "low", "high", "more", "high potency" etc. is associated with known concentrations ranges of the food substance (Table 4.18).

6.3 Statements of Disclosure

Products carrying a nutrient content claim, and which have high levels of total fat, saturated fat, cholesterol or sodium must carry a "Statement of disclosure (21 CFR 101.13 (h))". The disclosure statement alerts consumers about the presence of health-risk nutrients, which are dietary agents linked with chronic disease. Currently, there is an association between several chronic diseases and dietary sodium, cholesterol, total fat and saturated fat. Health-risk nutrients occur mainly in "low nutrient density", high calorie, foods (Anderson and McMurray 1997).

Regular foods (carrying a nutrient content claim) with a threshold of 13 g total fat per serving require a statement of disclosure. By comparison, the threshold value is 19.5.g fat per serving of "main dish product" and 26 g per serving for a "meal product" (Table 4.19). A meal product refers to a breakfast, lunch or dinner offering that weighs at least 10 oz (>283 g). Furthermore, meal products contain 40 g-options from three of the following food groups; (i) breads, cereals and pasta etc., (ii) fruits and vegetables, (iii) milk, yogurt and dairy group, and (iv) high protein group – meat, fish and dry beans (21CFR 101.13 (l)). The "main meal dish" product (21CFR 101. 13 (m)) is similar to a meal product but weighs 6 oz (170 g).

Table 4.17 Describing the specific requirements for nutrient content claims

21CFR 101.54 – nutrient content (NC) claims for "good source," "high," "more," and "high potency."
21CFR 101.56 – NC claims for "light" or "lite."
21CFR 101.60 – NC claims for the calorie content of foods.
21CFR 101.61 – NC claims for the sodium content of foods.
21CFR 101.62 – NC claims for fat, fatty acid, and cholesterol content of foods.
21CFR 101.65 – implied NC claims and related label statements.
21CFR 101.67 – use of NC claims for butter.
21CFR 101.69 – petitions for NC claims.

Adapted from code of federal regulations (U. S. Food and Drug Administration 2010)

In order to require a statement of disclosure a product, (i) must exceed a threshold value for a health-risk nutrient, and (ii) must have a nutrient content claim. Currently, nutrient content claims are voluntary. Therefore, statements of disclosure are virtually voluntary (Palte 2020).

6.4 Front-of-Pack Labelling

As discussed previously, only food packages with nutrient content claim might also require a statement of disclosure for health-risk nutrients (Palte 2020). The US approach towards health-risk nutrients differs from other countries where statements of disclosure are mandatory by way of front-of-pack labelling. More than 20 front-of-pack labels were developed in the US on a voluntary basis (U.S. Food and Drug Administration 2009c, Wartella et al. 2010; Hawley et al. 2013; Dunford et al. 2017). By comparison, about 30 different countries have implemented compulsory front-of-pack labelling schemes (Jones et al. 2019).

The goal of front-of-pack labelling is to summarize nutrition information in a super-simplified form so that consumers can make healthy food choices at the point of purchase. A typical front-of-pack label (e.g. the multiple traffic light system in the UK) consists a series of logos (one symbol for each health-risk nutrient) plus numerical data. Each logo depicts whether a specific health-risk nutrient (e.g. salt) is low, medium or high. The front-of-pack logos can be multi-coloured or monochrome. In the former case, nutrient levels are usually color-coded as high (red), medium (amber) or low (green/ blue). The best front-of-pack labels provide an "at a glance"

Table 4.18 Descriptors for nutrition content claims

Component	Claim	Not more than	Per RACC[a]
Energy	Low	40 kcal per 100 g/20 kcal per 100 ml	40 kcal
	Free	4 kcal per 100 ml	5 kcal
	Reduced		<25% ref. food
Fat	Low	3 g per 100 g/1.5 g per 100 ml	
	Free	0.5 g per 100 g or 100 ml	0.5 g
	Reduced		<25% ref. food
Saturated fat	Low	1.5 g per 100 g / 0.75 g per 100 ml and 10% of energy	1.0 g
	Free	0.1 g per 100 g or 0.1 g per 100 ml	0.5 g
	Reduced		<25% ref. food
Cholesterol	Low	20 mg per 100 g/10 mg per 100 ml	20 mg
	Free	5 mg per 100 g/5 mg per 100 ml and, for both claims, less than: 1.5 g saturated fat per 100 g 0.75 g saturated fat per 100 ml and 10% of energy of saturated fat	2 mg
	Reduced		<25% ref. food
Sugars	Free	0.5 g per 100 g/0.5 g per 100 ml	0.5 g
	Low	NA	NA
Sodium	Low	120 mg per 100 g	140 mg
	Very low	40 mg per 100 g	
	Free	5 mg per 100 g	<5 mg
		NOT LESS THAN	
Protein	Source	10% of NRV per 100 g/5% of NRV per 100 ml or 5% of NRV per 100 kcal (12% of NRV per 1 MJ) or 10% of NRV per serving	
	High	2 times the values for "source"	
Vitamins and minerals	Source	15% of NRV per 100 g/7.5% of NRV per100 ml or 5% of NRV per 100 kcal (12% of NRV per 1 MJ) or 15% of NRV per serving	
	High	Times the value for "source"	

[a]*RACC* Reference Amounts Customarily Consumed from 21 CFR 101.54 (U.S. Food & Drug Administration 2021; Code of Federal Regulations 2021)

Table 4.19 Threshold values that trigger a statement of disclosure

Nutrient/food	All food	Main dish product	Meal product
Fat (g/serving)	13	19.5	26
Saturated fat (g/serving)	4	6.0	8
Cholesterol (mg/serving)	60	90	120
Sodium (mg/serving)	480	720	960
Code reference	21CFR 101.13	21CFR 101.13(m)	*21CFR 101.13(l)*

profile for the main health-risk nutrients for any population. In the US, the health-risk nutrients linked with chronic disease were cited as high intakes of, food calories, added sugar, salt, total fat and saturated fat (Institute of Medicine 2010, 2012; Dunford et al. 2017).

Many studies examined the effectiveness of different front-of-pack labelling styles, and how best to gather evidence for decision-making. Evidence shows that some types of front-of-pack labelling lead to increased purchasing of healthy food. Different countries/populations prefer different types of labels (Trieu et al. 2015; Croker et al. 2020). Mandatory front-of-pack labelling systems were introduced most recently in Chile and Mexico, which provide interesting test cases (Reyes et al. 2019; White and Barquera 2020).

As mentioned before, front-of-pack labelling is voluntary in the US though this might change in future. Many food manufacturers have developed "in-house" front-of-pack labels using varying styles, which consumers may find confusing. A committee of experts from the US reviewed many options for front-of-pack labelling to determine which ones most suited for applications nationwide (Institute of Medicine 2010). In 2012, it was recommended that USDA/FDA develop a front-of-pack labelling system consisting of logos for food energy (calories), salt, fat, saturated fat and added sugar (Institute of Medicine 2012; Dunford et al. 2017). Legal challenges were observed when of State of New York introduced front-of-pack label legislation based on a "first amendment" protection for commercial speech (Pomeranz 2015; Pomeranz et al. 2019). A detailed discussion of constitutional issues is best left to food lawyers. Nevertheless, it is worth noting that legal challenges affecting food-labelling legislation occur regularly in many countries (Pomeranz 2015; Pomeranz et al. 2019; Jones et al. 2019).

7 Health Claims

There is no doubt that food is required for maintaining health. The more serious point of contention is whether the nature of food consumed achieves additional benefits, such as better weight management, a lower fasting blood glucose levels, or high HDL cholesterol. Discovering whether a specific food substance is "anti-aging" or "anti-inflammatory" is a serious matter for all stakeholders. A successful health food is a win-win for all, from consumers to health insurance providers.

Nevertheless, the relations between diet and health is complicated and subject to misunderstanding. Indices such as high blood pressure, and high blood cholesterol can be confused with diseases (cardiovascular disease) though they are but *risk factors* for disease. Elevated fasting blood glucose indicates that a person is at high risk or is predisposed towards type-2 diabetes. It is widely agreed that life-style adjustments (including changes to the diet and increased physical activity) can alter the risk factors for disease and thereby promote long-term health.

7.1 General Health Claim Requirements

7.1.1 General Principles

A health claim is defined as "*a statement that characterizes the relations between a substance and the risk of disease or a health related conditions* (21 CFR 101.14 (a)). This sentence is misunderstood frequently, leading to unsuccessful health claims petitions by industry. To understand what a health claim is, it helps to break this concept down to consider the meaning of terms like "statement", "characterize", "a substance" etc. Notice also that health claims are concerned with "disease risk" and not about disease.

7.1.2 Forms of Statements

The health claim statement (21 CFR 101.14(a)(1)) can take many forms, including logos. FDA legislation describes other health claim statements as, referring to, "Third party" references, written statements including terms such "heart", symbols, logos or vignettes (21 CFR 101.14(a)(1)).

7.1.3 Substances

Substances are food components, which are the subject of a health claims; this includes conventional foods, single food components, or dietary supplements. The substance cited in a health claim is assumed to be safe or gererally recognized as safe (GRAS) at the concentrations intended to be consumed (21 CFR 101.14(a)(2)). In addition, a food substance for a claim must be relevant for the United States population or an identified subgroup. Finally, the health claim substance must contribute to the taste, aroma and/ or technical characteristic of a food. Claims that require a reduction of nutrient intake, must refer to specific nutrients (21CRF 101.14(b)) (Fig. 4.7).

7.1.4 Disease or Health Related Conditions

A health claim must relate to one specific ailment, "disease" or "health related condition". Moreover, diseases and health related conditions affecting the human body arise from malfunctioning organs, body parts, or system (e.g. circulatory,

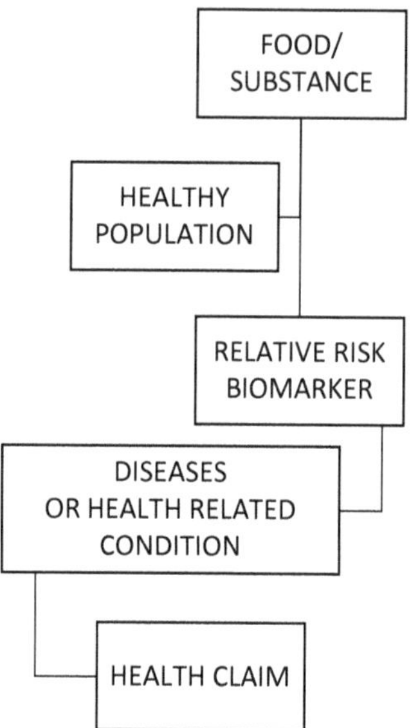

Fig. 4.7 Constituent elements of a health claim

respiratory, muscular system). Conventional wisdom suggests that dysfunction of the human body may arise due to damage or injury, or from wear and tear due to chronic factors. Nutrient deficiency diseases are excluded from health claims legislation (21 CFR 101.14(a) (5)).

7.1.5 Disease Risk and Biomarkers for Disease

Risk refers to the probability of harm.[20] For example, the risk of suffering a heart attack can be established by assessing the incidence of actual heart attacks per thousand individuals over a given timespan. The actual risk could be can ascertained using hospital records of numbers of persons treated. An alternative way to ascertain disease risk is by referring to certain diagnostic biological markers (biomarkers) of disease. Clinical biomarkers for disease risk are measured routinely by physicians' e.g. tests for glucose, blood pressure, serum cholesterol levels, etc. As an example, elevated fasting levels of glucose, is associated with increased risk of type 2 diabetes.

[20]The term "risk" appears briefly (twice) in 21CFR 101 (14). The reader is referred to Chapter 9 for further discussions on food safety risk.

Here T2D is the disease whilst blood glucose level is a risk factor for 2TD.

A health claim may refer to the fact that a food substance reduces blood glucose levels, but it may not state that a food reduces or treats or mitigates diabetes. In fact, *health claims must address healthy populations and avoid references to persons already suffering from a disease.* Claim statements that refer to diseases will be disqualified alongside of those dealing with disease mitigation, diagnosis or treatment. Any substance that is effective in treating a disease will be classed as a medicinal drug and legislated accordingly. Finally, the health claim must be based on information for which there is substantial scientific agreement (SSA) (U. S. Food and Drug Administration 2020b). This information must be about healthy people and refer to risk reduction or biomarkers for disease not to disease themselves.

7.1.6 Validity of Claims

The FDA determines the validity of claims based on the totality of evidence available, and examines whether there is significant scientific agreement (SSA) amongst trained experts. Newly proposed health claims are evaluated to establish that; (a) the claim communicates a clear disease-substance relationship, (ii) describes a specific disease and (iii) that the claim is truthful and not misleading and (iv) the claim has adequate information and details (Trumbo 2005; U. S. Food and Drug Administration 2020b; Yamini and Trumbo 2016).

7.1.7 Health Claim Application Process

The FDA is empowered by the NLEA Act (1990) to oversee pre-market approvals for health claims (U. S. Food and Drug Administration 2020b). From 2009, the FDA produced industry guidelines for constructing a petition for an authorized health claim for FDA evaluation (U. S. Food and Drug Administration 2009b). Companies wishing to deploy a health claim have two or three options; (i) repurpose an existing health claim for their needs. Intuitively, this seems the the most cost effective process, (ii) seek an entirely new health claim by petitioning the FDA, or (iii) seek a qualified health

Table 4.20 Criterion for validating health claim portfolios

Criterion 1	Explains food components relation to disease/health related condition
Criterion 2	Shows substance has either sensory, nutritive, or technical effect
Criterion 3	Component is safe and effective at proposed exposure concentration
Criterion 4	Adequate evidence, e.g. (i) Scientific supporting data, (ii) analytical data related to food, (iii) example health claim, (iv) computer literature searches, (v) copies of journal paper articles

Adapted from U. S. Food and Drug Administration (2013)

claim by petition. (21 CFR 101.70). A detailed description of portfolio structure is outside of the scope of this text. However, new petitions need to meet four general criteria (Table 4.20).

7.2 Examples of Current Health Claims

About one dozen FDA approved health claims are available covering several nutrients – disease risk relations. These are claims with substantive scientific agreement or highest grade of evidence. At the risk of repetition, it must be recalled that health claims cannot be applied to diseases. For example, evidence shows that eating fruit and vegetables may reduce the risk or prevent cancer, but there is "no" evidence showing that any food could in some way *treat* cancer.[21]

As noted above, health claim legislation does not cover disease "treatment, diagnosis or mitigation". In principle, a hypothetical food component which was found to be 100% effective for treating a specific diseases would not be allowed to carry a health claim, because that food would then be classed as a medicinal drug. Evidence of efficacy obtained using sick hospital patients (i.e. persons suffering from a disease) will be discounted because the results from sick people, *should not be extrapolated to the general healthy population*. This viewpoint led the industry to propose a new term "nutraceuticals" to refer to compounds that possess the characteristics of a nutrient as well as a pharmaceutical drug. Currently, the FDA does not recognize the term nutraceutical (Table 4.21).

Table 4.21 Example health claims and qualified health claims

21CFR101.72	Calcium, vitamin D, and osteoporosis.
21CFR101.73	Dietary lipids and cancer.
21CFR101.74	Sodium and hypertension.
21CFR101.75	Dietary saturated fat and cholesterol and risk of coronary heart disease.
21CFR101.76	Fiber-containing grain products, fruits, and vegetables and cancer.
21CFR101.77	Fruits, vegetables, grain products – fiber, soluble fiber, and risk of coronary heart disease.
21CFR101.78	Fruits, vegetables, and cancer.
21CFR101.79	Folate and neural tube defects.
21CFR101.80	Dietary noncariogenic carbohydrate sweeteners and dental caries.
21CFR101.81	Soluble fiber from certain foods and risk of coronary heart disease (CHD).
21CFR101.82	Soy protein and risk of coronary heart disease (CHD).
21CFR101.83	Plant sterol/stanol esters and risk of coronary heart disease (CHD).

Adapted from 21 CFR 101

Strict separation between foods and medicines is necessary, because otherwise manufacturers could be tempted to register new drug compounds as foods or nutrients, in order to avoid costly phase 1, phase 2 & phase 3 drug trials (U.S. Food & Drug Administration, 2021).

7.3 Qualified Health Claims

Qualified health claims are statements that the FDA determines have limited scientific evidence. For example, the qualified health claim for green tea and cancer reads: "*Green tea may reduce the risk of breast or prostate cancer although the FDA has concluded that there is very little scientific evidence for this claim*".

[21] The relations between health claims and diseases risk is worth reiterating. Many experts are liable to "let slip" by referring to diseases as opposed to disease risk. Actually, any substance which happened to treat cancer *symptoms* would be disqualified from the food-health claim process.

Table 4.22 Qualified health claim areas approved by the FDA

Health condition risk	Nutrients or food component
Cancer risk	Green tea, selenium, tomatoes, calcium, antioxidant vitamins
Cardiovascular disease risk	Oleic acid, B-vitamins, folic acid, nuts, omega-3FA, unsaturated-FA
Cognitive function decline risk	Phosphatidylserine
Diabetes (Type 2) risk	High amylose corn, Psyllium husk, whole grains, chromium picolate
Hypertension risk	EPA, DHA, calcium
Neural tube defect, risk	Folic acid
Peanut allergy, risk	Ground peanuts

Adapted from U. S. Food and Drug Administration (2019)

A qualified health claim usually means the totality of evidence available may be contradictory; for instance, some studies may show positive results whist others show no support. Alternatively, the type of studies presented by the manufacturer may be mostly laboratory based using isolated human cells or animal studies. A common form experiments called "observational studies" are considered to produce low quality evidence and are less reliable compared with evidence gathered by way of "gold standard" human randomized controlled trial (U. S. Food and Drug Administration, 2009b). The table above shows the main categories of qualified claims endorsed by the FDA (Table 4.22).

7.4 Organic Food Labeling

7.4.1 General Principles

The US organic food market was valued at US$28.4 billion in 2012, US$35 billion in 2014 and projected to reach US$70 billion in 2025. Currently organic foods sales represent about 4% of total food sales, involving nearly 75% of general grocery stores (USDA Economic Research Service 2019a). A national organic program and labeling process has been developed, as Organically Produced Products (OPP) transitioned from niche to mainstream over the past 30 years (USDA National Organic Program 2012). The organic certification is in part a "process certification". Therefore, achieving organic certification indicates farm production that addresses certain environmental and conservation issues.

General legislation covering both fresh and processed OPP is contained within "The Organic Food Act" (1990). The stated purpose of the organic food act was, (i) to establish clear standards applicable for fresh and processed OPP, (ii) to ensure consistent standards for OPP and, (iii) to promote interstate trade in OPP (7 USC 6501).[22] The USDA works in concert with state authorities to enforce rules for certifying organic producers (USDA Economic Research Service 2019a; USDA National Organic Program 2012).

Organic foods tend to be more expensive than conventional foods. The price premium for OPP compared to conventional food could provide economic incentive for misbranding. The United Nations cited several factors to explain the price premium for OPP. Firstly, the demand for OPP may be higher compared to supplies. Second, OPP have a higher production cost due to increased reliance on physical labor. Thirdly, postharvest handling cost for OPP may be increased by the legal requirement to handle these separately from conventional food, and finally, distribution channels for OPP tend to be smaller scale and less efficient (Food and Agriculture organization of the United Nations 1999, 2020). Experience from the US indicates that significant quantities of OPP are sold through local markets.

Comparative studies of the quality of OPP and conventional foods led to mixed results. Some studies indicated higher concentrations of nutrients for OPP (Brandt et al. 2011) whilst other researchers found no differences (Dangour et al. 2009). Organic dairy products were reported to have higher levels of omega-3 and omega-6 fatty acids (Palupi et al. 2012). Other studies found there was a lack of evidence for OPP being healthier than conventional foods, but the former

[22]The text for organic food legislation is available from US Code Title 7, Chapter 94 Organic Certification, subsections 6501 to 6524 (7 USC 6501 to 7 USC24).

were associated with lower dietary pesticides (Smith-Spangler et al. 2012). Organically produced meat had higher levels of essential lipids compared to conventionally produced meat, but such variations were difficult to separate from interspecies difference across of many studies (Srednicka-Tober et al. 2016). According to current research, the evidence for differentiating nutritional characteristics of OPP and conventional foods was weak (Vigar et al. 2019). In summary, qualitative differences between OPP and conventional foods remains controversial. Nevertheless, farming practices for OPP and conventional foods are clearly distinct (USDA Economic Research Service 2019b).

7.4.2 State Organic Production Certification of Nationwide Standards

Each state implements standards for OPP individually. Each state runs organic production inspection programme including annual certification and regular inspections by a state agent. In order to comply with OPP status, a farm must avoid certain prohibited practices related to soil and planting, fertilizers, and general management. For instance, GMO plants are to be avoided, alongside of synthetic fertilizers and natural poisons including arsenic and lead salts. The 2014 Farm Act offers financial support for farmers wishing to achieve organic certification (USDA Economic Research Service 2019b).

7.4.3 Grades of Organically Produced Products

The USDA certifies four grades of organic foods (Table 4.23). All levels of organic grading assume that the food products meet general pre-requisites; (i) foods are not produced using any "excluded methods". The use of bioengineered (genetically modified) organisms or synthetic fertilizers is forbidden for organic agriculture, as is night soil. Manufactured organic foods should not undergo irradiation, (ii) all food ingredients checks against a list of allowed and disallowed substances, and (iii) a USDA approved organic certification agent mediates on behalf for petitioners.

Compliance with the highest grade of organic certification (>95% of organic ingredients) will be indicated by placing a USDA approved organic seal on the principal display panel for a food package (Fig. 4.8). In addition, information related to specific organic ingredients will appear on the information panel alongside of the list of ingredients.

Table 4.23 USDA grades for organically produced products

Grade descriptor	Comments	USDA Seal/ Info Panel[a]
100% organic	Food from certified organic farm; manufactured with 100% certified organic ingredients; uses organic food additive or processing aides; names certifying agent	Yes/ yes
Organic	Product has 95% or greater organic ingredients excluding salt and water; uses 5% allowed ingredients; names certifying agent	Yes/ yes
Made with organic	Contains 70% or greater organic ingredients; uses 30% allowed ingredients; names certifying agent	NA/ yes
Organic ingredient	Identify specific organic ingredient	NA/ NA

Adapted from (USDA National Organic Program (2012)
[a]USDA Seal is displayed on Principal Display Panel and/ or "organic" ingredient is identified in Information Panel (cf. Fig. 1)

Fig. 4.8 The USDA seal carried by organically produced food products

7.5 Gluten Free Labeling

7.5.1 General Principles

Gluten is a protein associated with wheat and related cereals including rye and barley. Consuming foods that contain gluten leads to celiac disease in susceptible individuals. The symptoms of celiac disease arise from antibodies and inflammatory cells produced in response to gluten, which damage the human intestinal lining or microvilli. The damage to small intestines leads to malabsorption of nutrients, fatigue, skin complaints, and growth impairment. It is estimated that about three million people are affected by celiac disease in the United States ("Food labeling: gluten-free labeling of foods. Final rule," 2013).

In 2013, the FDA introduced legislation for "gluten free" labeling which came into effect on August 5th 2014 ("Food labeling: gluten-free labeling of foods. Final rule," 2013).[23] The legislation provides a clear definition of general terms like "gluten" and "gluten free". The new law also defines a threshold of 20 ppm gluten (i.e. 20 mg gluten per kilogram food) for gluten-free food. Food products must satisfy three requirements for gluten-free labeling (Table 4.24). However, labeling food products as "gluten free", "no gluten", "without gluten" is voluntary (21 CFR 101.9). Products that contains wheat as an ingredient cannot also claim to be gluten free – unless there is an additional explanation indicating how ingredients were refined in order to achieve gluten levels below 20 ppm. The FDA does not require gluten free foods to be processed on dedicated equipment in a bid to avoid incidental contamination, assuming the foods meet the gluten-free requirements.

Table 4.24 FDA pre-requirements for gluten free labeling

Avoid gluten-containing grains such as wheat, rye and barley.
Avoid food ingredients (e.g. flour) derived from a gluten containing grain
Excludes gluten depleted ingredients (e.g. starch) with above 20 ppm gluten
Avoids inherently gluten free food ingredients with a gluten level above 20 ppm

Adapted from (Food labeling 2013)

7.5.2 Hydrolyzed or Fermented Foods

The threshold value (<20 ppm) for deciding whether a food is gluten-free was derived using an "analytical based approach". Accordingly, the threshold value was set to reflect the levels of gluten, which is able to be determined by readily available ELISA methods. Though methods that are more sensitive exist for gluten detection, these were not yet affordable or practicable for everyday use. The alternative toxicological criterion for setting threshold values was not possible because there is insufficient human data for the No Observable Adverse Effect Level (NOAEL) for gluten.

The gluten-free legislation for beverages produced using gluten containing grains (e.g. barley) comes under the shared jurisdiction of the FDA and another organization, The Alcohol and Tobacco Tax and Trade Bureau (TTB). Beers brewed with and without hops are overseen by the TTB and FDA respectively. The TTB requires labeling to alert consumers about the potential gluten content in malted beers using stock statements: "Product fermented from grains containing gluten" or "This product was distilled from grains containing gluten". Malt-free beers are expected to follow broadly the same FDA requirements for other gluten-free foods (Table 4.24). The FDA adopted such "discretionary enforcement" for beers owing to the lack of reliable methods for gluten determination following fermentation and / or some forms of hydrolysis ("Food labeling: gluten-free labeling of foods. Final rule," 2013). The gluten-free labeling rule has been in force for less than a decade. However, current indications are that the gluten-free labeling is to be trusted (Thompson and Simpson 2015).

7.6 Conclusions

Food labeling legislation is a rapidly evolving subject. This and general space restrictions led to some otherwise very interesting topics being omitted from the current chapter. In particular labeling for

[23] Food Labeling; Gluten-Free Labeling of Foods 2013, 78 Fed. Reg. 47154.

genetic modified foods, seems likely to become mandatory. The country of manufacture has always been recorded for packaged foods. However, country of origin legislation (COOL) for agricultural produce is predicted to be very influential (Jones et al. 2009). Another area of continued interest is menu labelling and the possible addition of physical activity calorie equivalences (e.g. 100 Cal = approximately 1 mile walking) information at point of sale. The "food labeling guide" produced and updated at regular intervals by the FDA will continue to be an important reference point for industry (U. S. Food and Drug Administration 2013, 092018).

References

Allison DB (2011) Evidence, discourse and values in obesity-oriented policy: menu labeling as a conversation starter. Int J Obes (Lond) 35(4):464–471. https://doi.org/10.1038/ijo.2011.28

Anderson JJB, McMurray RG (1997) Overview: nutrition in sports and excercise. In: Wolinsky I (ed) Nutrition in exercise and sport, 3rd edn. Taylor & Francis, pp 49–63

ANNEX adopted in 2011. Revised in 2013, 2015, 2016 and 2017) Retrieved from http://www.codexalimentarius.net/input/download/standards/34/CXG_002e.pdf

Boyce AM (2016) When does it stop being peanut butter? FDA food standards of identity, Ruth Desmond, and the shifting politics of consumer activism, 1960s–1970s. Technol Cult 57(1):54–79. http://nrsharvardedu/urn-53:HULInstRepos:27022907

Brandt K, Leifert C, Sanderson R, Seal CJ (2011) Agroecosystem management and nutritional quality of plant foods: the case of organic fruits and vegetables. Crit Rev Plant Sci 30(1–2):177–197. https://doi.org/10.1080/07352689.2011.554417

Cantu-Jungles TM, McCormack LA, Slaven JE, Slebodnik M, Eicher-Miller HA (2017) A meta-analysis to determine the impact of restaurant menu labeling on calories and nutrients (ordered or consumed) in U.S. adults. Nutrients 9(10). https://doi.org/10.3390/nu9101088

Carlson A, Frazão E (2012) Are healthy foods really more expensive? It depends on how you measure the price. USDA-ERS Econ Inform Bull 96:50pp. https://doi.org/10.2139/ssrn.2199553

Centre for Science in Public Interest (2010) Menu labeling. Retrieved from https://cspinet.org/protecting-our-health/menu-labeling

Centre for Science in Public Interest (2015) Cost of menu labeling. Retrieved from https://cspinet.org/resource/cost-menu-labeling

Code of Federal Regulations (2021) 21 CFR 101. Subpart d – specific requirements for nutrient content claims. From https://www.ecfr.gov/current/title-21/chapter-I/subchapter-B/part-101/subpart-D

Codex Alimentarious Commission (2020) Standards. Retrieved from http://www.fao.org/fao-who-codexalimentarius/codex-texts/list-standards/en/

Codex Alimentarius Commission (1985a) General standard for the labeling of prepackaged foods. CXS 1–1985 (Adopted in 1985. Amended in 1991, 1999, 2001, 2003, 2005, 2008 and 2010. Revised in 2018). Retrieved from http://www.fao.org/fao-who-codexalimentarius/sh-proxy/en/?lnk=1&url=https%253A%252F%252Fworkspace.fao.org%252Fsites%252Fcodex%252FStandards%252FCXS%2B1-1985%252FCXS_001e.pdf

Codex Alimentarius Commission (1985b) Guidelines on nutrition labeling:CAC/GL 2–1985 (Adopted in 1985. Revised in 1993 and 2011. Amended in 2003, 2006, 2009, 2010, 2012, 2013, 2015, 2016 and 2017

Cowburn G, Stockley L (2005) Consumer understanding and use of nutrition labeling: a systematic review. Public Health Nutr 8(1):21–28

Croker H, Packer J, Russell SJ, Stansfield C, Viner RM (2020) Front of pack nutritional labelling schemes: a systematic review and meta-analysis of recent evidence relating to objectively measured consumption and purchasing. J Hum Nutr Diet 33(4):518–537. https://onlinelibrary.wiley.com/doi/510.1111/jhn.12758

Curtis PA, Steinberg EL, Parisi MA, Northcuttb JK (2013) How did we get where we are today? In: Curtis PA (ed) Guide to US food laws and regulations. Wiley, pp 23–53

Daley AJ, McGee E, Bayliss S, Coombe A, Parretti HM (2020) Effects of physical activity calorie equivalent food labelling to reduce food selection and consumption: systematic review and meta-analysis of randomized controlled studies. J Epidemiol Community Health 74(3):269–275. https://jech.bmj.com/content/274/263/269.abstract

Dangour AD, Dodhia SK, Hayter A, Allen E, Lock K, Uauy R (2009) Nutritional quality of organic foods: a systematic review. Am J Clin Nutr 90(3):680–685. https://doi.org/10.3945/ajcn.2009.28041

Darmon N, Drewnowski A (2015) Contribution of food prices and diet cost to socioeconomic disparities in diet quality and health: a systematic review and analysis. Nutr Rev 73(10):643–660. https://doi.org/10.1093/nutrit/nuv027

Drewnowski A (2010) The cost of US foods as related to their nutritive value. Am J Clin Nutr 92(5):1181–1188. https://doi.org/10.3945/ajcn.2010.29300

Dunford EK, Poti JM, Xavier D, Webster JL, Taillie LS (2017) Color-coded front-of-pack nutrition labels—an option for US packaged foods? Nutrients 9(5(480)):10pp. https://doi.org/10.3390/nu9050480

EveryCRSReport.com (2016) Legal issues associated with FDA standards of identity: in brief. Retrieved from https://www.everycrsreport.com/reports/R44393.html

Federal Trade Commission (2020) Fair Packaging and Labeling Act (1967). Retrieved from https://www.ftc.gov/enforcement/rules/rulemaking-regulatory-reform-proceedings/fair-packaging-labeling-act

Fernandes AC, Oliveira RC, Proenca RP, Curioni CC, Rodrigues VM, Fiates GM (2016) Influence of menu labeling on food choices in real-life settings: a systematic review. Nutr Rev 74(8):534–548. https://doi.org/10.1093/nutrit/nuw013

Fitbit.com (2020) NutritionFacts. Crabsticks. Retrieved from http://www.fitbit.com/foods/food?viewFood=on&foodId=537175351

Food and Agriculture organization of the United Nations (1999) Committee on agriculture. Fifteenth session rome, 25–29 January 1999, Red room. Organic Agriculture. Retrieved from http://www.fao.org/3/X0075e/X0075e.htm

Food and Agriculture Organization of the United Nations (2020) Organic agriculture: FAQ. Retrieved from http://www.fao.org/organicag/oa-faq/oa-faq5/en/

Food labeling (2013) Gluten-free labeling of foods. Final rule. Fed Regist 78(150):47154–47179

Frohlich X (2010) Buyer be-aware: the ethics of food labeling reform and 'mobilising the consumer'. In: Casabona CMR, Escajedo San Epifanio L, Ciri6n AE (eds) Global food security: ethical and legal challenges. EurSafe 2010 Bilbao, Spain 16–18 Septembe, vol 2020, pp 221–227

GovRegs (2016) U.S. Code of federal regulations. Retrieved from https://www.govregs.com/regulations

Hawley KL, Roberto CA, Bragg MA, Liu PJ, Schwartz MB, Brownell KD (2013) The science on front-of-package food labels. Public Health Nutr 16(3):430–439. https://doi.org/10.1017/S1368980012000754

Institute of Medicine (2010) Front-of-package nutrition rating systems and symbols: phase I report. National Academies Press, Washington, DC, 140pp. https://doi.org/10.17226/12957

Institute of Medicine (2012) Front-of-package nutrition rating systems and symbols: promoting healthier choices: phase II report. The National Academies Press, Washington, DC, 180pp. https://doi.org/10.17226/13221

Jones KG, Agapi S, Whitaker JB (2009) Country of origin labeling: evaluating the impacts on U.S. and world markets. Agric Econ Res Rev 38(3):397–405. https://citeseerx.ist.psu.edu/viewdoc/download?doi=310.391.391.975.8584&rep=rep8581&type=pdf

Jones A, Neal B, Reeve B, Ni Mhurchu C, Thow AM (2019) Front-of-pack nutrition labelling to promote healthier diets: current practice and opportunities to strengthen regulation worldwide. BMJ Glob Health 4(6):e001882. https://www.ncbi.nlm.nih.gov/pmc/articles/PMC6936575/pdf/bmjgh-6932019-6001882.pdf

Lipsky LM (2009) Are energy-dense foods really cheaper? Reexamining the relation between food price and energy density. Am J Clin Nutr 90(5):1397–1401. https://academic.oup.com/ajcn/article/1390/1395/1397/4598166

Littlewood JA, Lourenco S, Iversen CL, Hansen GL (2016) Menu labeling is effective in reducing energy ordered and consumed: a systematic review and meta-analysis of recent studies. Public Health Nutr 19(12):2106–2121. https://doi.org/10.1017/s1368980015003468

Livesey G (1992) The energy values of dietary fiber and sugar alcohols for man. Nutr Res Rev 5:61–84. From http://www.ncbi.nlm.nih.gov/pubmed/19094313

Long MW, Tobias DK, Cradock AL, Batchelder H, Gortmaker SL (2015) Systematic review and meta-analysis of the impact of restaurant menu calorie labeling. Am J Public Health 105(5):e11–e24. https://doi.org/10.2105/ajph.2015.302570

Meadows M (2006) A century of ensuring safe foods and cosmetics. FDA Consum 40(1):6–13. Retrieved from https://www.fda.gov/files/A-Century-of-Ensuring-Safe-Foods-and-Cosmetics.pdf

Merrill A, Watt BK (1973) Energy value of foods: basis and derivation. USDA Handbook No 74. Retrieved from https://www.ars.usda.gov/ARSUserFiles/80400525/Data/Classics/ah74.pdf

Mossoba MM, Moss J, Kramer JK (2009) Trans fat labeling and levels in U.S. foods: assessment of gas chromatographic and infrared spectroscopic techniques for regulatory compliance. J AOAC Int 92(5):1284–1300. https://doi.org/1210.1093/jaoac/1292.1285.1284

Mozaffarian D, Afshin A, Benowitz NL, Bittner V, Daniels SR, Franch HA, Jacobs DR Jr, Kraus WE, Kris-Etherton PM, Krummel DA, Popkin BM (2012) Population approaches to improve diet, physical activity, and smoking habits: a scientific statement from the American Heart Association. Circulation 126(1514–1563)

North Dakota State University (2020) Food law: standard of identity, petitions, food additives, food product claims, and organic food. Retrieved from https://www.ag.ndsu.edu/foodlaw/processingsector/standardofidentity

Office of the Law Revision Counsel (2020) US Code. Retrieved from https://uscode.house.gov/browse/prelim@title21&edition=prelim

Ollberding NJ, Wolf RL, Contento I (2011) Food label use and its relation to dietary intake among US adults. J Am Diet Assoc 111(5). https://doi.org/10.1016/j.jada.2011.03.009

Olmsted L (2017) Real food/fake food: why you don't know what you're eating and what you can do about it: Algonquin books, 336 pages

Palte GZ (2020) Be warned: a proposal to reform food product disclosure statements. Harv J Legis 56(2) Retrieved from https://harvardjol.com/2018/10/14/food-product-disclosure/

Palupi E, Jayanegara A, Ploeger A, Kahl J (2012) Comparison of nutritional quality between conventional and organic dairy products: a meta-analysis. J Sci Food Agric 92(14):2774–2781. https://doi.org/10.1002/jsfa.5639

Pomeranz JL, Wilde P, Mozaffarian D, Micha R (2019) Mandating front-of-package food labels in the U.S. – what are the first amendment obstacles? Food Policy 86:101722. https://www.ncbi.nlm.nih.gov/pmc/articles/PMC7441739/

Reyes M, Garmendia ML, Olivares S, Aqueveque C, Zacarías I, Corvalán C (2019) Development of the Chilean front-of-package food warning label. BMC Public Health 19(1):906. https://doi.org/910.1186/s12889-12019-17118-12881

Roberto CAKN (2005) Improving the design of nutrition labels to promote healthier food choices and reasonable portion sizes

Sarink D, Peeters A, Freak-Poli R, Beauchamp A, Woods J, Ball K, Backholer K (2016) The impact of menu energy labeling across socioeconomic groups: a systematic review. Appetite 99:59–75. https://doi.org/10.1016/j.appet.2015.12.022

Smith-Spangler C, Brandeau ML, Hunter GE, Bavinger JC, Pearson M, Eschbach PJ et al (2012) Are organic foods safer or healthier than conventional alternatives?: a systematic review. Ann Intern Med 157(5):348–366. https://doi.org/10.7326/0003-4819-157-5-201209040-00007

Srednicka-Tober D, Baranski M, Seal C, Sanderson R, Benbrook C, Steinshamn H et al (2016) Composition differences between organic and conventional meat: a systematic literature review and meta-analysis. Br J Nutr 115(6):994–1011. https://doi.org/10.1017/s0007114515005073

Thompson T, Simpson S (2015) A comparison of gluten levels in labeled gluten-free and certified gluten-free foods sold in the United States. Eur J Clin Nutr 69(2):143–146

Thompson T, Kane RR, Hager MH (2006) Food allergen labeling and consumer protection act of 2004 in effect. J Acad Nutr Diet 106(11):1742–1744

Trieu K, Neal B, Hawkes C, Dunford E, Campbell N, Rodriguez-Fernandez R, Legetic B, McLaren L, Barberio A, Webster J (2015) Salt reduction initiatives around the world – a systematic review of progress towards the global target. PLoS One 10(7):e0130247. https://www.ncbi.nlm.nih.gov/pmc/articles/PMC4511674/pdf/pone.0130247.pdf

Trumbo PR (2005) The level of evidence for permitting a qualified health claim: FDA's review of the evidence for selenium and cancer and Vitamin E and heart disease. J Nutr 135(2):354–356. https://doi.org/10.1093/jn/135.2.354

Tyburczy C, Mossoba MM, Rader JI (2013) Determination of trans fat in edible oils: current official methods and overview of recent developments. Anal Bioanal Chem 405(17):5759–5772. https://link.springer.com/article/5710.1007/s00216-00013-07005-z

U. S. Food and Drug Administration (2006) Guidance for industry: questions and answers regarding food allergens (Edition 4). Retrieved from https://www.fda.gov/media/117410/download

U. S. Food and Drug Administration (2007) Small business nutrition labeling exemption guide. Retrieved from https://www.fda.gov/food/labeling-nutrition-guidance-documents-regulatory-information/small-business-nutrition-labeling-exemption-guide

U. S. Food and Drug Administration (2009a) Food allergies: reducing the risks. Retrieved from https://www.fda.gov/consumers/consumer-updates/food-allergies-reducing-risks

U. S. Food and Drug Administration (2009b) Guidance for industry: evidence-based review system for the scientific evaluation of health claims. Retrieved from https://www.fda.gov/regulatory-information/search-fda-guidance-documents/guidance-industry-evidence-based-review-system-scientific-evaluation-health-claims

U. S. Food and Drug Administration (2009c, 09/20/2018) Guidance for industry: letter regarding point of purchase food labeling. Retrieved from https://www.fda.gov/regulatory-information/search-fda-guidance-documents/guidance-industry-letter-regarding-point-purchase-food-labeling

U. S. Food and Drug Administration (2009d, 01/31/2018). Milestones in U.S. Food and drug law history. Retrieved from https://www.fda.gov/about-fda/fdas-evolving-regulatory-powers/milestones-us-food-and-drug-law-history

U. S. Food and Drug Administration (2010) 21 CFR Part 101. Food labeling. Retrieved from https://www.accessdata.fda.gov/scripts/cdrh/cfdocs/cfcfr/CFRSearch.cfm?CFRPart=101

U. S. Food and Drug Administration (2013, 092018). Guidance for industry: Food labeling guide (Jan 2013). Retrieved from https://www.fda.gov/regulatory-information/search-fda-guidance-documents/guidance-industry-food-labeling-guide

U. S. Food and Drug Administration (2018a) Food allergen labeling and consumer protection act of 2004 (FALCPA). Retrieved from https://www.fda.gov/food/food-allergensgluten-free-guidance-documents-regulatory-information/food-allergen-labeling-and-consumer-protection-act-2004-falcpa

U. S. Food and Drug Administration (2018b) Food allergens. Retrieved from https://www.fda.gov/food/food-ingredients-packaging/food-allergens

U. S. Food and Drug Administration (2018c) Guidance for industry: food labeling guide. January 2013. Retrieved from https://www.fda.gov/regulatory-information/search-fda-guidance-documents/guidance-industry-food-labeling-guide

U. S. Food and Drug Administration (2018d) The history of FDA's fight for consumer protection and public health. Retrieved from https://www.fda.gov/about-fda/history-fdas-fight-consumer-protection-and-public-health

U. S. Food and Drug Administration (2019) Qualified health claims: letters of enforcement discretion. Retrieved from https://www.fda.gov/food/food-labeling-nutrition/qualified-health-claims

U. S. Food and Drug Administration (2020a) 21CFR101.11. Nutrition labeling of standard menu items in covered establishments. Retrieved from https://www.accessdata.fda.gov/scripts/cdrh/cfdocs/cfcfr/CFRSearch.cfm?fr=101.11

U. S. Food and Drug Administration (2020b) Authorized health claims that meet the significant scientific agreement (SSA) standard. Retrieved from https://www.fda.gov/food/food-labeling-nutrition/authorized-health-claims-meet-significant-scientific-agreement-ssa-standard

U.S. Food & Drug Administration (2021) The FDA's drug review process: ensuring drugs are safe and effective. Retrieved from https://www.fda.gov/drugs/information-consumers-and-patients-drugs/fdas-drug-review-process-ensuring-drugs-are-safe-and-effective

USDA Agricultural Research Service (2019) FoodData Central. Retrieved from https://fdc.nal.usda.gov/

USDA Economic Research Service (2019a) Oganic agriculture: market overview. Retrieved from https://www.ers.usda.gov/topics/natural-resources-environment/organic-agriculture/organic-market-overview/

USDA Economic Research Service (2019b) Oganic agriculture: the 2014 Farm Act. Retrieved from https://www.ers.usda.gov/agricultural-act-of-2014-highlights-and-implications/organic-agriculture/

USDA National Organic Program (2012) Organic labeling. Retrieved from https://www.ams.usda.gov/sites/default/files/media/Labeling%20Organic%20Products%20Fact%20Sheet.pdf

Van den Wijngaart AW (2002) Nutrition labeling: purpose, scientific issues and challenges. Asia Pac J Clin Nutr 11(2):S68–S71

VanEpps EM, Roberto CA, Park S, Economos CD, Bleich SN (2016) Restaurant menu labeling policy: review of evidence and controversies. Curr Obes Rep 5(1):72–80. https://doi.org/10.1007/s13679-016-0193-z

Vigar V, Myers S, Oliver C, Arellano J, Robinson S, Leifert C (2019) A systematic review of organic versus conventional food consumption: is there a measurable benefit on human health? Nutrients 12(1). https://doi.org/10.3390/nu12010007

Wartella EA, Lichtenstein AH, Boon CS (eds.) (2010) Front-of-package nutrition rating systems and symbols: phase I report. Institute of Medicine (US) Committee on Examination of Front-of-Package Nutrition Rating Systems and Symbols. Retrieved from https://www.ncbi.nlm.nih.gov/books/NBK209859/

White M, Barquera S (2020) Mexico adopts food warning labels, why now? Health Syst Reform 6(1):e1752063. https://doi.org/1752010.1751080/23288604.23282020.21752063

Yamini S, Trumbo PR (2016) Qualified health claim for whole-grain intake and risk of type 2 diabetes: an evidence-based review by the US Food and Drug Administration. Nutr Rev 74(10):601–611

Quality and Sensory Evaluation of Food

5

1 Introduction to Food Quality

1.1 Definitions of Quality

In marketing, sales and the advertising industries "quality" refers to desirable attributes associated with a product or service (Bremner 2000). The business and manufacturing industries, describe quality as "*a measure of excellence or a state of being free from defects, deficiencies and significant variations*" or ""*the totality of features and characteristics of a product or service that bears on its ability to satisfy stated or implied needs*" (BusinessDictionary 2015). Quality is discussed frequently in terms achieving consumer expectations, needs and their wants (Botta 1995). Several descriptions for food quality are listed in the Table below.

Establishing product specifications and their measurement is essential for monitoring quality (Bremner 2000). Such principles were adopted in food safety work in the 1960s that led to product specifications for food additives by JECFA (Food and Agriculture Organization of the United Nations 2013). As discussed in Chap. 3, internationally accepted standards of quality are available from the two major organizations, Codex Alimentarius Commission and the International Organizations of Standardizations (ISO).

1.2 Food Quality Attributes

The number quality dimensions currently being studied by food researchers is large (Bremner 2000). A recent survey showed over 7506 articles in the ISI web of science database featuring the phrase "food quality" in the title and nearly 300 reviews (March 2015). Allied terms included phrases such as "total quality management", quality control, and also quality assurance. Some current ideas surrounding the concept of food quality are considered below.

Quality foods will be affordable, wholesome, and safe in-line with the stated values and objectives of the Agri-food industry (Chap. 2). Food quality can be described also in relation to five dimensions related to sensory characteristics, functional properties, nutritional quality, storage-life and safety (SFNSS) – see Fig. 5.1 and Table 5.1.

Broadly speaking, sensory quality refers to appearance, texture, sound, taste and flavor of foods. Functional properties refer to the ability of some foods to support optimum health. Nutritional quality is dependent on the concentrations and bioavailability of vitamins, minerals, and certain beneficial non-nutrient components; the presence of toxic or anti-nutritional factors in foods also falls under the heading of nutritional quality. Storage life is determined by issues, which affect

R. Owusu-Apenten, E. R. Vieira, *Elementary Food Science*, Food Science Text Series,
https://doi.org/10.1007/978-3-030-65433-7_5

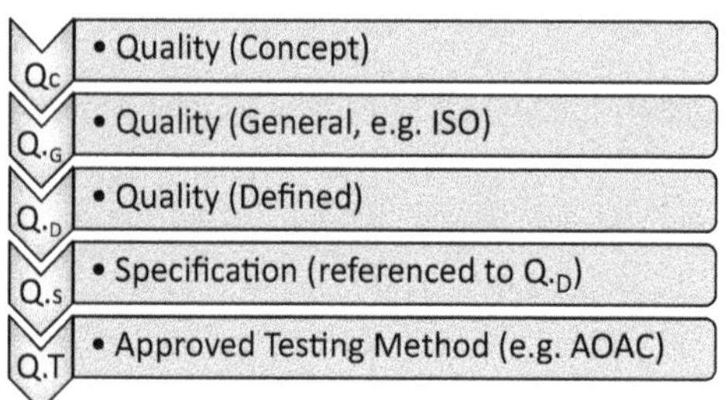

Fig. 5.1 The hierarchical approach to food quality. The diagram shows quality as a general concept and as a defined product standard / specification with an approved method for testing (Adapted from Bremner 2000)

Table 5.1 General definitions of quality

Consumer focussed
Extent to which a product fulfils consumer needs and wants, expectations
Combination of attributes that influence acceptability to a consumer
Product characteristics that satisfy consumer minimum demands / expectations
An aesthetic standard set by experienced users
Characteristics that determine degree of acceptability to consumers
A means for pleasing consumers not just protecting them from annoyances
Abstract quality
Factors (biological, physical) determining safety and nutritional value
Product characteristics used in a grading system
Product conformance to user requirements or pre-agreed standards

Adapted from (Botta 1995)

freshness, acceptability. Safety quality can be discussed in terms of an absence of microbiological, biological, and chemical, physical hazards for foods (Fig. 5.2).

Individual SFNSS quality characteristics (Fig. 5.2) are interrelated. As an example, storage-life covers a time beyond which there is an unacceptable loss of food safety or sensory quality. All quality dimensions are important but perhaps food safety is the predominant quality attribute (Botta 1995). Few would be happy with unwholesome food. Wider quality attributes. e.g. freshness quality; availability, convenience, and price quality seem to be derivatives of the SFNSS quality.

2 Food Sensory Quality

2.1 Sensory Characteristics

Sensory quality can be evaluated via the human senses. The appearance of food is probably the foremost attribute perceived by consumers that affects purchasing decisions (Moskowitz et al. 2012). It is also well known that the color, hue, and shape determines consumer acceptance of fresh fruits and vegetables (Barrett et al. 2010). Sensory properties such as taste, smell and texture contribute to food preferences, and hedonic ratings – but so far there is no evidence that it affects net food intake or general eating habits (Sorensen et al. 2003). However, some evidence suggests that high glycemic index foods and those high in fat content can produce behaviour characteristic of food addiction and cravings. (Schulte et al. 2015). Finally, an understanding of sensory properties is essential for new food product development (Lyon et al. 1992).

Food sensory evaluation is a multidisciplinary discipline, requiring a knowledge of food science, human physiology, and psychology (Lawless and Heymann 2010). Sensory evaluation can be performed using inanimate instruments or human test subjects. Considerable effort is often needed to establish correlations between human test results and those obtained using instrumental methods of sensory analysis.

Trained wine tasters can differentiate closely related products and may differentiate wines from different regions. Almost everyone can tell when a food contains too much or insufficient amounts of salt or when a tomato is ripened. Obviously, individuals will differ in their personal preferences and so some subjectivity is unavoidable. Nevertheless, factory personnel and sensory panel members can be trained to recognize specific elements of food quality. As an example, fresh produce can be hand-sorted by trained personnel as part of process-lines so that bad or damaged products are excluded. Actually, there are many similar examples of the use of human sensory evaluation as a basis for industrial processing.

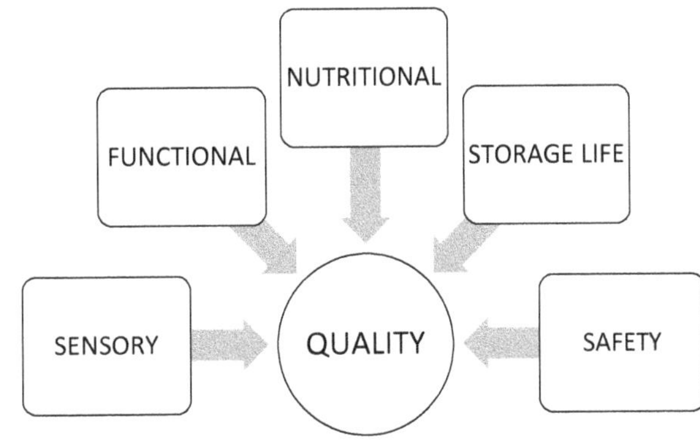

Fig. 5.2 Dimensions of food quality related to sensory, functional, nutritional, storage life and safety (SFNSS) characteristics

Attributes such as nutritional quality seem not to be overtly perceptible to the humans senses. Nevertheless, there are visual cues for nutritional quality such as ripening which is detectable.

2.2 Appearance

Many different characteristics add to or subtract from quality as perceived by the consumer. Examples include the brownness of a piecrust, the shape of a pickle, the wholeness of a product such as canned pineapple, for which the price decreases as the size changes from rings to chunks-same product, different shape. As noted previously, the list of factors which contribute to the appearance of food include, size, shape, surface texture, gloss, presence of dirt, blemishes, imperfections, indentations and the overall color. Appearance is determined by visual perception.

Color is perhaps the most recognizable of a food's visual attributes and one of the first levels of quality evaluations by the consumer. Color indicates freshness or spoilage of products such as fruits, and vegetables. It may also indicate the strength of a product as in the case of coffee and tea or completeness of cooking, as in the case of French fries or meats. For a food technologist, color can indicate many quality factors. The color of chocolate can give clues as to how it has been stored, and the color of a batter changes with density and can indicate an error in mixing. If too much oxygen is present in a package of dried foods such as tomatoes, a bleached color results.

2.3 Texture

The texture of food is felt by the tongue, palate, teeth, and even the fingers. Texture can be an indicator of freshness in many foods. With aging, many fruits and vegetables experience a breakdown of cell components resulting in a softer, soggy product that indicates low quality and spoilage. Other products, such as crackers and pretzels, must be protected from moisture gain by packaging in order not to soften.

Viscosity and consistency are characteristics that can have both visual and textural quality attributes. Syrups are visually more appealing if thicker, yet this increased viscosity also adds appeal to the feel of the syrup on the food. In some foods, such as ketchup, thickness is a major advertising and marketing feature.

2.4 Flavor

Flavor is a combination of taste and smell but is also influenced by a textural composite of sensations referred to as "mouth feel." Some foods, such as sauces or gravy that are thicker will often be judged to have the fuller or richer flavor. This perception may be either purely psychological or real. The change in texture may actually affect the volatility or solubility of a flavor compound, increasing the flavor intensity.

Color can also influence the perceived flavor in foods. Taste tests have been done in which the colors of fruit sherbets had been altered but the flavors were not changed. Only 43% of the flavors in the altered samples were identified correctly whereas just under 70% of the unaltered samples received correct responses. One tester even commented on how easy one sample was to identify because there were pieces of cherry in it. He actually tasted a sample of pineapple sherbet (with pieces of pineapple) that had been colored red!

Although color and texture have an effect, flavor is mainly controlled by the mixture of tastes (salt, sour, bitter, and sweet) and the numerous combinations of compounds that are responsible for aroma. The flavor of food is very complex and has not been completely described for many foods. Flavor measurement is largely subjective and thereby very difficult to measure. Differences of opinion on flavors and their contribution to quality result because of this. People have wide ranges of sensitivity to detect different tastes and odors and even when these are detected, individuals differ in their preference. The cultural and biological differences in people make flavor measurement quite difficult.

Some food companies utilize a method of flavor analysis developed in 1948 by Messrs. Sjostrom and Cairncross of Arthur D. Little, Inc. in Cambridge, Massachusetts. "The Flavor Profile" method is based on the concept that flavor consists of identifiable taste, odor, and chemical feeling factors (sensations in the mouth and throat, such as pepper bite) plus an underlying complex of sensory impressions not separately identifiable.

Trained panelists are needed for this method and they are selected based on their abilities to discriminate odor and flavor differences and communicate their perceptions. The panelists must take a series of tests to determine their abilities to notice flavor and odor differences and to identify basic tastes, rank, intensities, and identify common odorants.

After the panelists are selected, they must go through a training period of approximately 60h and 100h of practice sessions. The time of training varies depending on whether the panel will be testing any food or beverage or testing a specific food. Training for a single type of product requires less time. The training generally covers the nature of taste and smell, basic requirements for panel work, and techniques and procedures for reproducible odor and taste work. The development of sensory terminology through exercises and reference standards is also stressed. As the panelists advance, more difficult products are used and flavor situations of a more complex nature are covered. Interpretation and utilization of panel data are covered also.

When the panel is in operation, a quiet, well-lighted, odor-free environment removed from external distractions must be provided. All samples must be presented to the panel under identical conditions and should be analyzed and scored in an identical manner at the same time. A product's aroma is analyzed first, followed by the flavor, and finally the aftertaste.

Each panel has a leader and four or more members. The leader conducts a product orientation before the panel does the testing. In the orientation, the objectives of the product are outlined and samples to be tested are introduced with similar products to establish a framework for comparison. At this time, the panel will establish a list of character notes for the sample such as vanillin, peppermint, chalky, or buttery. Reference materials will also be chosen (pure compounds that display particular flavor and aroma notes). The panel then decides the best way to present and examine the samples. During the test, each panel member evaluates the sample independently using standardized Flavor Profile techniques and records the findings on a blank sheet. The panel leader, who is also a member of the panel, now has each panelist present his or her findings, records them, and leads a panel discussion to reach a consensus on each component of the description. The final profile may take three to five sessions and after this, the panel leader interprets and reports the results.

The components of the Flavor Profile method are:

1. Determine an overall impression of aroma and flavor, called amplitude. Amplitude measures balance (blend) and body (fullness) of flavor and aroma in the product. It is scored on a point system from 0 to 3 (in 1/2-point increments) with 0 representing no blend or fullness and 3 representing a high degree of blending and fullness.

2. Identification of perceptible aroma and flavor notes. These are called character notes and are objective rather than subjective (i.e., a note may be described as "vanilla" but not "good"). These include aromatics, basic tastes, and chemical sensations or feeling factors (cool, burn).
3. Intensity of each character note. The degree to which each character note is perceived is "intensity" and the scale is constant over all product categories. The point system such as the one used with amplitude is used, with 0 representing "not present" and 3 being "strong."
4. Order in which these character notes are perceived. The "order of appearance" is determined through standardized techniques for tasting and smelling because order of appearance will be influenced by the location of taste buds on the tongue, volatility of the components, and texture of the product.
5. Aftertaste. Aftertaste can include tastes, aromatics, and/or sensations. It is an important part of the flavor of a product. The sensations are noted at a specific time after tasting has been completed, usually 1 min. If aftertaste is a major factor, intensity ratings may be given. When color or texture is important to the product's description, it is also noted in the panel session.

When all data are collected and analyzed, a final test report, including complete identification of the sample(s) and the objectives and duration of the study is given. The methods used to prepare the product, the reference standards used, a summary of the amplitude ratings, and major character notes with their intensities are given. The presence of any off-notes should be mentioned and the order of appearance and aftertaste noted. Any observations of visual and textural qualities are important and should be described.

Another method is The Profile Attribute Analysis (PAA), which is an objective method of sensory analysis that uses an extensively trained panel to numerically describe the complete sensory experience through profile attributes. These are a limited set of characteristics that, when properly selected and defined, provide a complete description of the sensory characteristics of a sample with little descriptive information lost. PAA is based on the Flavor Profile Method in concept and implementation but by limiting the number of profile attributes the panelist measures, it is possible to evaluate four or more samples per session as compared with one per Flavor Profile session.

PAA relies on numerical measures to describe differences among samples, and "Flavor Maps" can be used to display summary statistics.

3 Sensory Evaluation

Because of differences in people's sensitivity, perceptions, and preferences, it is virtually impossible to create a food that everyone would choose as his or her favorite. The food scientist does have some methods of testing foods to get a consensus of what a majority of the prospective consumers will accept and therefore desire to purchase.

Testing product quality using human subjects is called sensory evaluation or sensory analysis. According to the IFT approved definition, sensory evaluation is "the *scientific method used to evoke, measure, analyze and interpret responses to products perceived through the senses of sight, smell touch, taste and heating*" (Lawless and Heymann 2010). This highly interesting area which blends psychology, biology, food science and statistics has grown enormously in the past 50 years to become an indispensable part of food product development and consumer science. Experts in sensory evaluation can be accessed on a commercial basis with laboratory facilities located in many of the State Universities, e.g. Washington State University and also Michigan State University. Most food science degree programs may offer training in this area.

There are three classes of sensory evaluation procedures; (i) discrimination tests or difference test are designed to establish when one product is different from a second product in some way, (ii) descriptive testing enables a determination of how much difference there is between products and, (iii) affective testing

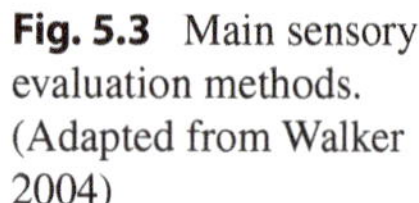

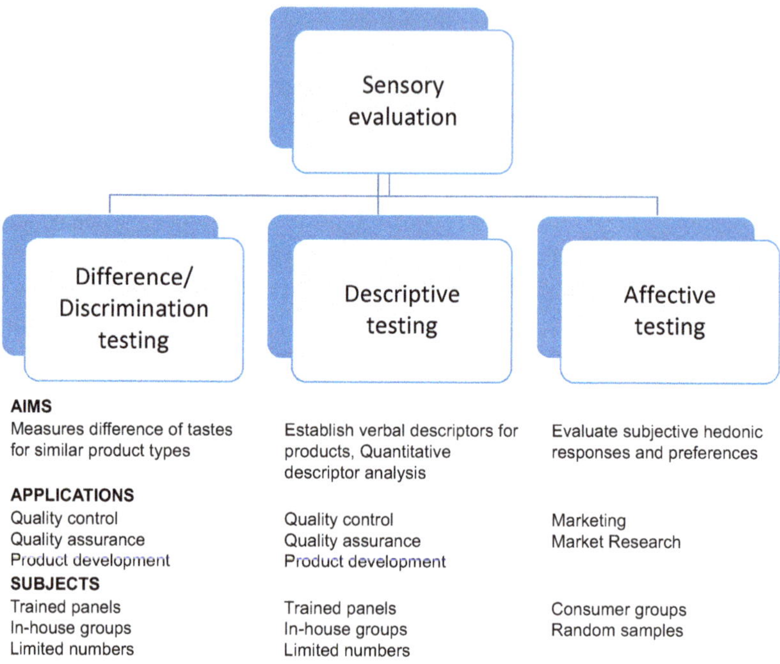

Fig. 5.3 Main sensory evaluation methods. (Adapted from Walker 2004)

determines how well a product is liked by the consumer (Fig. 5.3).

The idea behind sensory evaluation is that product characteristics can be evaluated objectively using the five human senses of sight, smell, touch, taste and hearing. Depending on the type of test, both untrained and trained persons may be used for sensory analysis with the former yielding data that is more representative of the general consumer. A brief outline of this highly specialized area of food science is provided below but the interested reader should refer to one of the many excellent textbooks devoted entirely to this area. (Kemp et al. 2009; Kilcast 2010; Lawless and Heymann 2010; Moskowitz et al. 2006; Stone et al. 2012).

3.1 Discrimination Tests

A discrimination text (difference text) is used to establish whether two products differ in some way. There are three sub-categories of difference tests depending on sample attributes and set of subsidiary information available (Table 5.2).

The duo test or paired-comparison test is applicable in situations where there is a simple well-defined attribute of interest (e.g. sweetness). A test subject is presented with two samples (AB, BA, AA, or BB) from which they must pick the sample with a stronger attribute (e.g. sweeter sample). Duo- trio testing is appropriate where there is no simple test attribute but *where* there is a reference standard available. For duo-trio testing, a subject is presented with the reference sample first, followed by a pair of food (AB) stimuli. The object of the duo-trio test is to pick the sample (A or B) most similar to the reference sample.

In the triangle-test (Fig. 5.5) a test subject is given three samples two of which are identical and asked to identify the odd sample. Assuming little or no difference between products, the "odd" sample would be identified correctly in 50% of instances or alternatively, only 50% of participants would identify the odd sample in any given trial. If the proportion of participants identifying the odd sample is different from 50% by a significant margin, it could be concluded that there was a detectable difference in the samples. The triangle procedure may be used to establish whether a particular product formulation is as acceptable as another is. Another use for difference tests is a situation in which a change in formulation is desired to reduce production costs.

3.2 Descriptive Testing

Descriptive testing refers to a series of related methods that evolved independently starting from the 1940s. The common feature of descriptive testing appears to be use of trained panelist. The aim of descriptive testing procedures is to evaluate products characteristics according to a pre-agreed intensity scale. A listing of some of the major descriptive tests (Table 5.3) and a detailed account of each is available (Lawless and Heymann 2010; Stone et al. 2012). Some descriptive testing, e.g. flavor and texture profiling are qualitative but QDA translates product attributes to numerical scales which can then be subjected to conventional analysis (Fig. 5.4).

Arguably, QDA is the most widely applied format for descriptive testing (Table 5.3). Skipping through the developmental aspects, the main elements of QDA are described as, (a) panel selection and training, (b) development and agreement of the sensory descriptors by panel members, (c) testing of products against the agreed descriptors and scales, and (d) robust statistical analysis typically involving ANOVA testing. The results of QDA is often presented in the familiar spider web plots, cobweb or radar plot (Fig. 5.4).

Table 5.2 Sensory evaluation facilities at Washington State University

Kitchen, food preparation, area, ovens, water baths, refrigerators, counter space
Sensory analysis booths
Data collection via lap top computer (Dell Dimension 2400) running Compusense® 5.0 software (Compusense, Inc.)
Sensory training room for descriptive analysis
Capability to train panelists
Conduct formal sensory evaluation on-site in individual booths
Food Processing Pilot Plant
Creamery Pilot Plant

Adapted from (School of Food Science, Washing State University, 2014)

Table 5.3 Descriptive analysis methods

Date	Method
1940	Flavor profiling
1963	Texture profiling
1974	Quantitative Descriptive Analysis (QDA)
1990	Spectrum descriptive analysis
1994	Free Choice Profiling

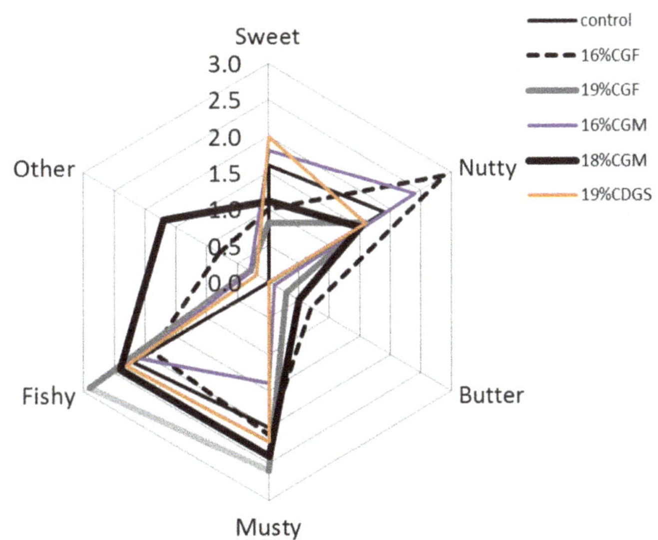

Fig. 5.4 A spider chart showing effect of diet on flavor descriptors for African tilapia fed varying amounts of corn gluten feed (CGF), corn gluten meal (CGM) or cord distillers grain soluble (CDGS) (Wu et al. 1996)

Two of the samples are identical and the other different. Please check the duplicate samples and score your preference for either the identical or odd samples.

- SCORING: Designate whether you prefer the pair or the odd sample by (P).
- Sample # Identical Samples (indicate by check mark) Score
- Plate #

Date: --------
Note: This test sheet was used in a triangle test done on two fish species prepared with identical recipes. To avoid problems with tendencies of testers to choose based on order of testing, samples were given in different sequences. This is shown below:

Plate #
1 AAB/ 2 ABB/ 3 BAB/ 4 BAA /5 BBA /6 BAA
Arrangement of Samples (left to right on plate)

Fig. 5.5 Triangle test score sheet

3.3 Affective Testing

Affective testing (hedonic testing) is intended to determine consumer acceptability and relative preferences for different products. The "acceptance" test usually requires a large number of individuals who represent the target group of potential consumers. For example, if you were testing a flavored cereal advertised by a cartoon character, you would use a panel of children. A gourmet-frozen entree would require a more mature panel. The panel here requires little training and is usually given a score sheet (Fig. 5.5) which rates different sensory attributes (appearance, odor, texture and flavor) from "excellent" to "inedible." Products may be scored using a "hedonic rating," which uses a numerical score of 9 for "excellent" and a score of 1 for "inedible." A mathematical average can now be determined for the food's acceptability. (Kemp et al. 2009; Kilcast 2010; Lawless and Heymann 2010; Moskowitz et al. 2006).

4 Instrumental Sensory Analysis

4.1 Objective Analysis of Food Quality

Instrumental methods of analysis can be deployed that provide surrogate measures of sensory quality. For instance, chemical odorants that determine the flavor and smell of foods can be measured using a gas chromatograph. Most dissolved chemicals can be measured instrumentally – to provide an index related to taste. Various physical attributes of foods, which affect their texture, can be measured using viscometers and rheological instruments.

Human testing can be costly for industry. Therefore, a lot of research is being conducted to develop instrumental methods for assessing food quality that could replace human sensory analysis. Methods that compare sensory and instrumental analysis of quality dealt with fatty foods (Waltking and Goetz 1983), apple cider and juice (Williams et al. 1983). Other reports deal with quality attributes for wine (Rapp 1998), cheese (Foegeding and Drake 2007), snack foods (Paula and Conti-Silva 2014), meat, lamb and poultry (American Meat Science Association 2015; Oltra et al. 2015; Luckett et al. 2014) and crispy foods (Roudaut et al. 2002). In the rest of this section, we describe some of the main instrumental method for assessing a range of food quality attributes (Otles 2008; Sliwinska et al. 2014).

4.2 Colorimetry of Foods

The color of liquid foodstuffs can be determined using colorimeters, which are instruments that measure light absorbance or transmittance when a sample is illuminated with visible radiation. When using transparent liquids, the degree of absorbance is related to color but such instruments are not routinely used for color measurement. For turbid samples, we can measure the transmittance or turbidity of a sample according to the degree of light transmitted. The major limitation associated with colorimetry is the need for transparent liquid foods devoid of particles of all types.

The color of solid foods can be readily measured using a reflector-meter. According to the tristimulus principle, the color of any object can be determined in terms of degree of by which the three primary colors (red, blue, green light) are reflected. The color of a solid object is measured according the XYZ scale, using values for X (red), Y (green) and Z (blue). The CIE color scheme was invented by the International Commission on Illumination (CIE; from the French, *Commission Internationale de l'Eclairage*) to try to mimic how humans perceive color via the rods and cones. The CIE color system is the basis for the reflector-meter, which is one of most popular instruments for measuring the color of solid foods, e.g. the Minolta Konica reflector-meter. The CIE L*a*b* and Hunter Lab. systems are the two alternative trismulus systems for measuring the color of solid objects (HunterLab 2008; 2012). The Hunter Lab reflectometer provides three numbers according to a degree of red-green (a) scale, (b) yellow-green scale and (L) lightness-darkness scale. A recent survey of the types of instrumentation employed for color measurement in the meat industry found that the majority of reports from the United States and the EU tended to employ a Minolta Konica reflector-meter, or Hunter Lab instruments (Tapp III et al. 2011). A recent development which seems likely grow in prominence is the use of Computer based vision and robotic systems for color measurement and for high-throughput sorting of food samples (Vyawahare et al. 2013).

4.3 Texture Monitoring Instruments

Food texture can be measured in a number of ways. Generally speaking, however, texture is the measure of resistance to force. Examples of the different forces are numerous. "Compression" is tested on foods that, when squeezed, remain in one piece, for example, bread. "Cutting'' is a force that goes through the food, thus dividing it, for example, an apple. A force applied away from a material resulting in tearing or pulling apart is termed "tensile strength," for example, donuts or breads. A force where one part of the food passes another is "shearing," as in the chewing of some candies and gum. When grading some foods, such as beefsteak, all of these factors can come into play. This branch of study is called food rheology.

Historically, the food technologist had many empirical tools to measure texture and control quality during processing. These include the "consistometer" for viscosity, the "squeeze-tester" for softness of bread, the "penetrometer" or "tenderometer" for tenderness in raw beef or baked goods or spreadability in margarine, and the "succulometer" which measures succulence by squeezing juice out of food. A range of many different types of equipment needed to measure various texture attributes can be found in the following reference (Kramer and Szczesniak 1973).

Modern commercial "all-in-one" texture analyzers can be fitted with a range of probes and accessories in order that one instrument may be used to measure multiple texture parameters. Examples of commercially available texture analyzers for use in the food industry settings include the TA.XTPlus texture analyzer from Stable Micro Systems Ltd. and distributed in North America by Texture Technologies (Texture technologies 2015). Two other instruments, the Texture Analyzer-TMS-PRO (Food Technology Corporation Ltd), and CT3 Texture analyzer (Brookfield Engineering Laboratories Inc 2015) are also widely used for a range of solid food product groups for a variety of technologically relevant measurements (Table 5.4).

As noted above, the texture of solid foods may be measured in a variety of ways but the technique called texture profile analysis (TPA) is deserving of special mention. Prior to the development of TPA in the 1960's several instruments had to be deployed when measuring different texture attributes; for instance, Kramer and Szczesniak 1973

Table 5.4 Applications of modern texture analyzers

Product types	Test methods	Attributes
Bakery products Confectionery Dairy Fruits and vegetables Gels and hydrocolloids Meat and fish Grains and snacks Pet food & animal feed	Break testing Bulk analysis Compression testing Penetration testing Shearing tests Tension	Adhesiveness Breaking point Chewiness Cohesiveness Elasticity Extensibility Fracture force Gel strength Hardness Resilience Springiness Tackiness

Partial listing compiled from (Brookfield Engineering Laboratories Inc. 2015; Food Technology Corporation 2015; Texture technologies 2015)

(p. 101) listed 40 or more in-house instruments for measuring food textures. The essence of TPA is the application of two successive compression forces to a solid food whilst measuring the structural resistance to deformation. Many textural attributes may be calculated based on the force-distance trace obtained from two-fold compression data. The TPA data provides information on a host of textural attributes e.g. hardness, cohesiveness, elasticity etc. (Fig. 5.6).

According to some accounts, TPA methods were first developed by Dr. Alina Surmacka Szczesniak, a sensory food scientist, then working at General Foods. The technique introduced five principal textural characteristics for solid foods (hardness, cohesiveness, adhesiveness, viscosity, and elasticity) as well as three other derived parameters (brittleness, chewiness, and gumminess) and how these could be measured using what was essentially one texture analysis machine (Friedman et al. 1963). Later TPA was adapted to the commercially available Instron Universal Testing machine by Professor Dr. Malcolm Bourne (Cornell University). TPA is currently one of the most common forms of texture analysis (Chen and Opara 2013; Rosenthal 2010).

4.4 Electronic Noses

In the mid-1980s, researchers at Britain's Warwick University used work done by the U.S. military in developing polymers that could conduct electricity to help planes evade enemy radar, and applied it to the development of the first electric nose (Mielle 1996). In their 1982 article which appeared in "Nature" magazine, Krishna Persaud and George Dodd patterned their new device on the mammalian olfaction system (Persaud and Dodd 1982). They realized that the human olfaction system contains many millions of sensory cells but that these could be separated into relatively few classes; according to current estimates there may be 1000 different genes coding for different odor receptors but 70% of these genes are non-functional meaning that there are approximately 200 different types of odor receptors.

On the other hand estimates for the number of different odors detected by humans range from 10,000 to one trillion. How then is the human nose able to distinguish between thousands of different smells? In olfaction, the 200 different receptors cells respond to vastly higher numbers of smell compounds perhaps depending on which sub-groups are stimulated, and the order in which these sensory cells become stimulated; one possible analogy is that many tunes may be composed using a fixed number of musical notes. The electric nose works on similar principles to detect very large numbers of odorants according to the overall patterns of responses generated via limited numbers of sensors. Advance statistical methods, are used to aid pattern recognition such as principal component analysis (PCA) and artificial neural networks (ANN).

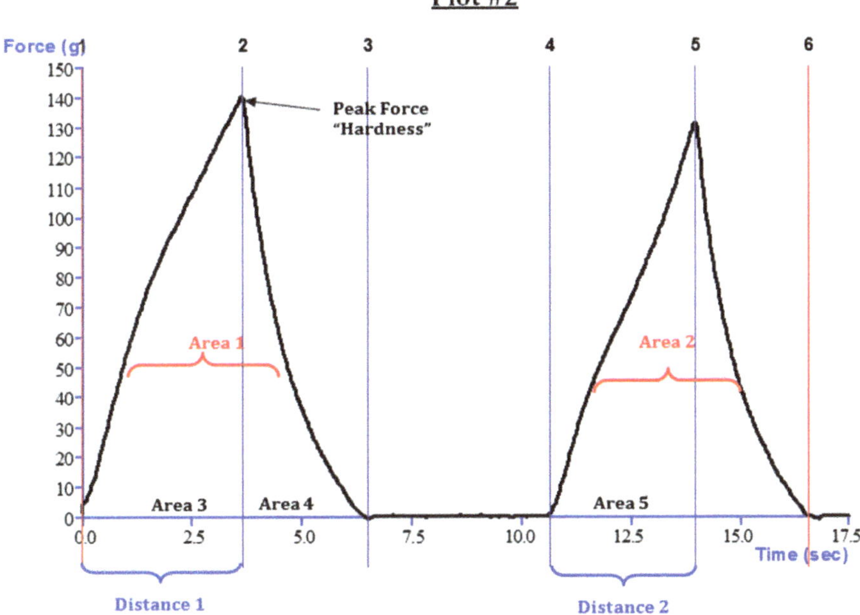

Fig. 5.6 A typical texture profile analysis derived from double sample compression (Courtesy, Stable Micro Systems, Ltd). Product characteristics are defined as Hardness = Maximum force at peak-1; Cohesiveness = Area peak-2/Area peak-1 or *Area 5/Area 4;* Springiness = Ratio of a product's original height. Distance 2/Distance 1 or Time 2/Time 1; Gumminess = Hardness x Cohesiveness = Hardness x (Area peak-2/Area peak-1); Chewiness (solids) = Hardness x Cohesiveness x Springiness; Resilience = Peak-1 (Area 4/Area 3)

According to current predictions electronic noses may find increasing applications in relation to food quality assessment, food spoilage detection, fermentation monitoring, biomedical applications and environmental monitoring (Gutierrez and Horrillo 2014; Loutfi et al. 2015). In medicine, for example, ailments such as diabetes and some liver and kidney diseases cause the production of certain odors that appear in the breath. Because the machines are more precise than the human nose in finding the specific odor components, scientists see its use as a diagnostic tool increasing in the future.

In the food industry, electronic noses could be used to detect specific aromas indicating spoilage, off quality, or use of a particular additive. This "technology of the future" is being used now in a number of applications and wherever the same component is being tested for in many samples, it is very useful. The use of this technology along with flavor panels gives the food scientist an additional tool to ensure a consistent, safe, high-quality food supply.

5 Processed foods quality and health

Economic experts propose that public health should be included as one measure for processed foods quality in addition to elements summarized in Fig 5.2 (food safety, sensory, functional, storage life). Moreover, the environment, climate change, and globalization are quality-related challenges facing the food processing industry. A central concern is the nutrition transition from traditional diets towards more ultra-processed foods that are known to be linked with some chronic diseases (Martinez-Perez et al. 2021). It is reasoned that, external oversight and regulations (e.g. labelling and taxation of added sugar) are needed to discourage unhealthy ultra-processed food products (Smith et al. 2011; Baum 2013; Qaim 2017; Hall 2018; Pedroza-Tobias et al. 2021). Some of these controversies are discussed in the sections on adult food literacy (Chap. 6) and ultra-processed foods (Chap. 11, Sect. 5) with the focus on food science and nutrition science.

References

American Meat Science Association (2015) Research guidelines for cookery, sensory evaluation, and instrumental tenderness measurements of meat. Retrieved from http://www.meatscience.org/docs/default-source/publications-resources/amsa-sensory-and-tenderness-evaluation-guidelines/research-guide/2015-amsa-sensory-guidelines-1-0.pdf?sfvrsn=6

Barrett DM, Beaulieu JC, Shewfelt R (2010) Color, flavor, texture, and nutritional quality of fresh-cut fruits and vegetables: desirable levels, instrumental and sensory measurement, and the effects of processing. Crit Rev Food Sci Nutr 50(5):369–389. https://doi.org/10.1080/10408391003626322

Baum CM (2013) Mass-produced food: the rise and fall of the promise of health and safety. Papers on Economics and Evolution, No. 1303, Jena, Max Planck Institute of Economics. 28pp. https://www.econstor.eu/bitstream/10419/88260/10411/742724263.pdf

Botta JR (1995) Evaluation of seafood freshness quality. Retrieved from https://books.google.co.uk/books?id=HmHgGL4e2w0C

Bremner HA (2000) Toward practical definitions of quality for food science. Crit Rev Food Sci Nutr 40(1):83–90. https://doi.org/10.1080/10408690091189284

Brookfield Engineering Laboratories Inc (2015) CT3 texture analyzer. Retrieved from http://www.brookfieldengineering.com/products/texture-analysis/ct3.asp

BusinessDictionary (2015) Quality. Retrieved from http://www.businessdictionary.com/definition/quality.html#ixzz3VfCcNa4I

Chen L, Opara UL (2013) Texture measurement approaches in fresh and processed foods – a review. Food Res Int 51(2):823–835. https://doi.org/10.1016/j.foodres.2013.01.046

Foegeding EA, Drake MA (2007) Invited review: sensory and mechanical properties of cheese texture. J Dairy Sci 90(4):1611–1624

Food and Agriculture Organization of the United Nations (2013) Online Edition: Combined compendium of food additive specifications. Retrieved from http://www.fao.org/food/food-safety-quality/scientific-advice/jecfa/jecfa-additives/en/

Food Technology Corporation (2015) Texture analyzers. Retrieved from http://www.foodtechcorp.com/texture-analyzers.html?gclid=CNqTsaX00MQCFQoXwwodCIIA7A

Friedman HH, Whitney JE, Szczesniak AS (1963) The Texturometer – a new instrument for objective texture measurement. J Food Sci 28:390–396

Gutierrez J, Horrillo MC (2014) Advances in artificial olfaction: sensors and applications. Talanta 124:95–105. https://doi.org/10.1016/j.talanta.2014.02.016

Hall KD (2018) Did the food environment cause the obesity epidemic? Obesity (Silver Spring) 26(1):11–13. https://www.ncbi.nlm.nih.gov/pmc/articles/PMC5769871/

Kemp S, Hollowood T, Hort J (2009) Sensory evaluation. A practical handbook. Wiley-Blackwell, Chichester/Ames

HunterLab (2008) Hunter L, a, b Versus CIE 1976 L*a*b*. Application Note 13(2):1–4. https://support.hunterlab.com/hc/en-us/articles/203993105-Hunter-L-a-b-Versus-CIE-203991976-L-a-b-an203993102-203993101

HunterLab (2012) Hunter L, a, b Versus CIE 1976 L*a*b*. Application Note. Retrieved from https://support.hunterlab.com/hc/en-us/articles/204137825-Measuring-Color-using-Hunter-L-a-b-versus-CIE-1976-L-a-b-AN-1005b

Kilcast D (2010) Sensory analysis for food and beverage quality control. A practical guide. Elsevier, Jordan Hill

Kramer A, Szczesniak AS (1973) Texture measurement of foods: psychophysical fundamentals; Sensory, mechanical, and chemical procedures, and their interrelationships. Retrieved from http://books.google.co.uk/books?id=IrNn75K_oe4C

Lawless HT, Heymann H (2010) Sensory evaluation of food: principles and practices, vol 2014. Springer, New York

Loutfi A, Coradeschi S, Mani GK, Shankar P, Rayappan JBB (2015) Electronic noses for food quality: a rcvicw. J Food Eng 144:103–111. https://doi.org/10.1016/j.jfoodeng.2014.07.019

Luckett CR, Kuttappan Vivek A, Johnson Lee G, Owens Casey M, Han-Seok S (2014) Comparison of three instrumental methods for predicting sensory texture attributes of poultry deli meat. J Sens Stud 29(3):171–181

Lyon DH, Francombe MA, Hasdell TA, Lawson K (1992) Guidelines for sensory analysis in food product development and quality control. Springer, Boston

Martinez-Perez C, San-Cristobal R, Guallar-Castillon P, Martínez-González MA, Salas-Salvadó J, Corella D, Castañer O, Martinez JA, Alonso-Gómez AM, Wärnberg J, Vioque J et al (2021) Use of different food classification systems to assess the association between ultra-processed food consumption and cardiometabolic health in an elderly population with metabolic syndrome (PREDIMED-plus cohort). Nutrients 13(7):2471. https://doi.org/2410.3390/nu13072471

Mielle P (1996) 'Electronic noses': towards the objective instrumental characterization of food aroma. Trends Food Sci Technol 7(12):432–438

Moskowitz HR, Beckley JH, Resurreccion AVA (2006) Sensory and consumer research in food product design and development. Blackwell Pub, Ames

Moskowitz HR, Beckley JH, Resurreccion AVA (2012) Sensory and consumer research in food product design and development, 2nd edn. Blackwell Pub, Ames

Oltra OR, Farmer LJ, Gordon AW, Moss BW, Birnie J, Devlin DJ, Tolland ELC, Tollerton IJ, Beattie AM, Kennedy JT, Farrell D (2015) Identification of sensory attributes, instrumental and chemical measurements important for consumer acceptability of grilled lamb Longissimus lumborum. Meat Sci 100:97–109

Otles S (2008) Handbook of food analysis instruments. CRC Press, Boca Raton

Paula AM, Conti-Silva AC (2014) Texture profile and correlation between sensory and instrumental analyses on extruded snacks. J Food Eng 121:9–14

Pedroza-Tobias A, Crosbie E, Mialon M, Carriedo A, Schmidt Laura A (2021) Food and beverage industry interference in science and policy: efforts to block soda tax implementation in Mexico and prevent international diffusion. BMJ Glob Health 6(8):e005662. https://gh.bmj.com/content/005666/005668/e005662.abstract

Persaud K, Dodd G (1982) Analysis of discrimination mechanisms in the mammalian olfactory system using a model nose. Nature 299(5881):352–355. https://doi.org/10.1038/299352a0

Qaim M (2017) Globalization of agrifood systems and sustainable nutrition. Proc Nutr Soc 76(1):12–21. https://doi.org/10.1017/S0029665116000598

Rapp A (1998) Volatile flavor of wine: correlation between instrumental analysis and sensory perception. Nahrung/Food 42(6):351–363

Rosenthal AJ (2010) Texture profile analysis – how important are the parameters? J Texture Stud 41(5):672–684. https://doi.org/10.1111/j.1745-4603.2010.00248.x

Roudaut G, Dacremont C, Pamies BV, Colas B, Le Meste M (2002) Crispness: a critical review on sensory and material science approaches. Trends Food Sci Technol 13(6–7):217–227

School of Food Science, Washing State University (2014) Sensory analysis: capabilities. Retrieved from http://sfs.wsu.edu/sensory/capabilities/

Schulte EM, Avena NM, Gearhardt AN (2015) Which foods may be addictive? The roles of processing, fat content, and glycemic load. PLoS One 10(2):e0117959. https://doi.org/10.1371/journal.pone.0117959

Sliwinska M, Wisniewska P, Dymerski T, Namiesnik J, Wardencki W (2014) Food analysis using artificial senses. J Agric Food Chem 62(7):1423–1448

Smith TG, Chouinard HH, Wandschneider PR (2011) Waiting for the invisible hand: novel products and the role of information in the modern market for food. Food Policy 36(2):239–249. https://www.otago.ac.nz/economics/otago087211.pdf

Sorensen LB, Moller P, Flint A, Martens M, Raben A (2003) Effect of sensory perception of foods on appetite and food intake: a review of studies on humans. Int J Obes 27(10):1152–1166. https://doi.org/10.1038/sj.ijo.0802391

Stone H, Bleibaum R, Thomas HA (2012) Sensory evaluation practices, 4th edn. Academic, San Diego

Tapp WN III, Yancey JWS, Apple JK (2011) How is the instrumental color of meat measured? Meat Sci 89(1):1–5. https://doi.org/10.1016/j.meatsci.2010.11.021

Texture technologies (2015) TA.XTPlus texture analyzer. Retrieved from http://www.texturetechnologies.com/texture-analyzers/TA-XTPlus-texture-analyzer.php

Vyawahare A, Jayaraj RK, Pagote C (2013) Computer vision system for colour measurement – fundamentals and applications in food industry: a review. Retrieved from http://www.rroij.com/jfpdt/index.php/jfdt/article/download/RRJFDT12/pdf

Walker G (2004) Introduction to sensory analysis. Retrieved from http://sst-web.tees.ac.uk/external/U0000504/Notes/Sensory/default.htm

Waltking AE, Goetz AG (1983) Instrumental determination of flavor stability of fatty foods and its correlation with sensory flavor responses. Crit Rev Food Sci Nutr 19(2):99–132

Williams AA, Langron SP, Arnold GM (1983) Objective and hedonic sensory assessment of ciders and apple juices. In: Williams AA, Atkin RK (eds) Sensory quality in foods and beverages: definition, measurement and control. E. Horwood, Chichester, pp 310–323

Wu YV, Warner K, Rosati R (1996) Sensory evaluation and composition of tilapia (Oreochromus niloticus) fed diets containing protein-rich ethanol by-products from corn. J Aquat Food Prod Technol 5(3):7–16. Retrieved from http://naldc.nal.usda.gov/download/25099/PDF

Consumer Food Literacy

6

1 Introduction

1.1 General Principles

Food literacy refers to the knowledge, skills and competencies, required to maintain healthy eating (Colatruglio and Slater 2016). As defined in Fig. 6.1, food literacy combines general literacy (reading, learning) with a knowledge of food. The food literate consumer will demonstrate basic knowledge of food and certain food-related competencies and behaviors (Fig. 6.1). Food literacy requires some understanding of the relations between eating and health, the influence of social and cultural factors on eating habits, health and nutrition; it is essential also to understand elements of legislation related to food labeling (Benn 2014). Alternative definitions for food literacy continue to be examined by experts (Amouzandeh et al. 2019; Cullen et al. 2015; Krause et al. 2018; Truman et al. 2017; Velardo 2015; Vidgen and Gallegos 2014; Yuen et al. 2018). Recent reviews found over 50-different detailed definitions for the concept of food literacy though there is a consensus emerging (Thompson et al. 2021). Currently a knowledge of food safety and hygiene is not included in discussions of food literacy.

Consumers with low levels of food literacy may be less able to assume full responsibility for healthy eating. Moreover, a big element of food literacy is the consumer's attitudes toward processed foods (Dwyer et al. 2012; Freedman 2012; Hsieh and Ofori 2007; Knorr and Watzke 2019; Weaver et al. 2014). Increasingly, the student of food science is required to be proactive in addressing consumer attitudes about their discipline. Therefore, the aim of this chapter is to introduce, a selection of topics at the intersection of food literacy and food science.

1.2 Critical Thinking, Food Literacy and the Press Media

The press media influences consumers, their knowledge of foods and their attitudes towards experts and food policy (Afshin et al. 2015; Ha 2018; Truswell 1998; Wellman et al. 1999; Yngve et al. 2006). Unfortunately, many newspapers and magazines tend to focus on so-called "good and bad" foods. By contrast, a position paper from the American Dietetic Association supported a "total diet" approach which emphasizes that "all food can fit into a healthy eating style" if consideration is given to the issue of portion size, and need for physical activity (Freeland-Graves and Nitzke 2002).[1] References to good and bad foods can be specious, particularly if the lists of bad foods

[1] A torrent of food stories include, the impact of food on chronic diseases and healthy ageing, food legislation, food miles, personalized nutrition, the emerging role of Artificial Intelligence (AI) on foodservice jobs, how information technology (food technology – *Sic*) may affect consumers interaction with food and food choices.

R. Owusu-Apenten, E. R. Vieira, *Elementary Food Science*, Food Science Text Series,
https://doi.org/10.1007/978-3-030-65433-7_6

Food literacy – definition
"a collection of inter-related knowledge, skills and behaviours required to plan, manage, select, prepare and eat foods to meet needs and determine food intake", as well as, "the scaffolding that empowers individuals, households, communities or nations to protect diet quality through change and support dietary resilience over time"

Fig. 6.1 Definition of food literacy. (Adapted from Colatruglio and Slater 2016)

keeps changing over time. Therefore, it seems to the consumer that "expert" recommendations about which food to eat and which to avoid, are conflicting and confusing (Dwyer et al. 2012; Freeland-Graves and Nitzke 2002).[2]

Recent discussions indicate a rising tendency to mistrust experts. In order to improve credibility, experts must declare relevant commercial interests when making recommendations, guidelines and undertaking discussions about food and nutrition. Experts should also authenticate their qualification and associated certification organization. Food and nutrition experts are being advised to develop consistent nutrition and food *messages, which are useable* by consumers (Anderson and Zlotkin 2000; Dwyer 2001; Truswell 1998; Wellman et al. 1999).

The consumers' responsibility for food literacy requires achieving basic critical thinking skills (Table 6.1). Readers should question routinely the technical proficiency of writers working for newspapers, magazines, television and bloggers from the internet. The designations "medical doctor" or "dietician" are protected professional tittles, reserved for trained and certified professionals. Unfortunately, anyone is free to operate as a "food expert" and to write, produce recipes, and *comment on food and nutrition issues*[3] (Table 6.1).

Currently, there appear to be few legal requirement for news and articles dealing with foods and eating to be "fact checked" or peer reviewed by certified food experts before publication. Sensationalized reporting (of food issues) tends to highlight novelty, controversy and preliminary scientific findings in order to gain audience share (Wellman et al. 1999).[4] Bias and a lack of transpar-

Table 6.1 Critical thinking aims and skills

Aims and characteristics
Aim1. Obtain knowledge of issues under consideration
Aim2. Assess amount and quality of evidence, identify unsubstantiated evidence
Aim3. Distinguish known factual information from and what is suspected
Aim4. Identify & acknowledge sources of bias and conflicts of interests
Aim5. Examine /re-examine assumptions
Aim6. Maintain objectivity,
Aim7. Reject conclusions prior to evaluation phase
Aim8. Avoid negativity or fault finding
Aim9. Identify personal views, interests, or wishes
Skills & processes
KU1. Identify underlying assumptions related to issues
KU2. Recognize the limitations of hypotheses
KU3. Ask questions that are clear and easily understood
KU4. Consider opposing views and opinions
KU5. Separate facts from opinions
KU6. Evaluate information sources for credibility
KU7. Organize and classify ideas and evidence
KU8. Evaluate and identify magnitude bias
KU9. Weigh all evidence
KU10. Separate relevant from irrelevant information
KU11. Use best available evidence
KU12. Draw logical conclusions based on evidence
KU13. Uses persuasive techniques
KU14. Acknowledge the variety of political, social, cultural, and personal perspectives

Adapted from Manitoba Government (2017)

[2]Paracelsus is credited with the statement that all substances are toxic – but that it is the dose consumed which is important.

[3]The Institute of Food Technology (IFT) is the main professional body for food scientists and technologists. Membership of the IFT should be the minimum requirement for food science and technology experts operating in the media domain. Likewise, a nutrition degree from a recognized tertiary education institution (College or University) might be minimum requirement for food experts.

[4]A sample of news headlines concerning processed foods from 2015–2018 include the following. Ultra-processed food linked to cancer, study says; 2018 / Ultra-processed food makes up majority of American's diets, despite calls for "clean" foods; 2016/ Processed meats rank alongside smoking as cancer causes – WHO; 2015 /Why processed food is like crack; 2015 (Search strategy-Google,

ency about "vested interest" are additional criticisms directed at the media (International Food Information Council Foundation 2010; Wellman et al. 1999). Lastly, training the new generation of food scientists must emphasize the need to engage with the public and media (Wellman et al. 1999).

In an article titled "is food a four letter word" Dwyer identified gaps in food literacy related to food processing and processed foods. Therefore, stereotypes and mischaracterization of processed foods could be undermining efforts to promote healthy eating (Dwyer et al. 2012). The term "processed" seemed to be being coopted as shorthand for high-sugar, high-salt and/ or high-fat food. There was a skewedness in attitudes expressed about processed foods in popular media, which needed to be addressed (Fig. 6.2).

The contribution of processed foods towards healthy eating also needed highlighting by disinterested experts and professionals (Dwyer et al. 2012; International Food Information Council Foundation 2010; Jones 2019; Knorr and Watzke 2019; Weaver et al. 2014).[5]

1.3 The Western Diet and Health

Investigations covering the "Western diet" and chronic diseases began in the 1950s and early 1960s.[6] At that time, medical doctors noted that several metabolic diseases were *correlated* positively with diets containing added sugar, fat, salt and low dietary fiber. In addition, the risk of non-communicable diseases declined for populations who ate traditional diets. New migrants that acculturated to the Western diet showed increased risk of chronic diseases (Dembert 1981; Trowell and Burkitt 1981).

Meanwhile, investigations from the United States showed that the risk of food-related chronic illnesses *increased* with income. Starting from a threshold of earnings, rises in household incomes were associated with an increasingly more westernized diet (Popkin 1994). Initially, households purchase greater volumes of food with rising income. However, further rises of income produced more luxury food purchases but no net increases in the total amount of food procured. There was increased consumption of fat, sugar and salt with rising income (Popkin 2001, 2015). Several factors unrelated to food and nutrition are also positively correlated with rising income, e.g. decreased personal energy expenditure, greater car-use, increasing TV access. Apparently, changes in disease risk were associated with a more affluent lifestyle (Yudkin 1964).

Influential commentary on the Western diet and associated ill effects reemphasizes the importance of food trends, which started quite a while back some 10,000 years ago, due to the transition from hunter gathering to an agricultural way of life. Foods that were unavailable to pre-agricultural societies were listed as, cereal grain, dairy food, refined sugar, alcohol, added salt, and refined vegetable oil. The reader should refer to the following literature for more discussions about "new foods" and current debates about the merits of a more Paleolithic diet (Cordain et al. 2005; Carrera-Bastos et al. 2011). To conclude this section, it is believed that adult

Advanced Search, Region = United States, "Processed food", News items).

[5]Positive guidelines on the use of processed foods in a healthy diet might be developed.

[6]The 1950s research compared locally produced traditional diets consumed in South East Africa with the "Westernized" foods produced commercially and available through organized food retail channels. The phrase "Westernized foods" has been retained in this discussion to reflect the original use from the 1950's medical literature.

> *"Many consumers do not realize that almost all foods currently consumed are processed.. that food processing has historically provided and will continue to provide a safe and abundant food supply that provides significant public health benefit. The mischaracterization of processed foods and food technology as unnatural, unsafe, and/or nutritionally inappropriate by some health professionals, advocacy organizations, and the media further makes the task of motivating consumers to eat more healthfully challenging".*

Fig. 6.2 Significance of processed foods (Dwyer et al. 2012)

education to improve food literacy will be helpful in addressing modern diet and health relations (Sumner 2013).

1.4 Nutrition Transitions and Income

Many developing countries are displaying nutrition transitions similar to those reported from East Africa and North America in the 1950's. As the average national income rises, there is increased consumption of vegetable oil (for frying) and added sugar. Eating beef, dairy products (dried milk, fluid milk, ice cream, and sweetened yogurt), eggs and pork all increase with rising income (Popkin 2001, 2015).

Both developing and developed countries showed similar trends where the diet was examined from 1960 projected to 2030 (Roser and Richie 2019). There was a rise in the net energy intake per person over the past half century for most countries (Roser and Richie 2019). In China, the ageing population, fewer dependent children and urbanization of migrant workers were major prompts for nutrition transition (Schmidhuber and Shetty 2005).

World trade liberalization and globalization of the food supply chain were also cited as facilitators for nutrition transitions (Schmidhuber and Shetty 2005). Food supermarkets emerged from about 1890 or about 150 years ago in Western countries (Toops 2017). Nowadays more international supermarket chains, food wholesalers and independent "cold stores" are spreading into smaller provincial towns within the 3rd World. The result is increasing consumer access to westernized foods. The marketing focus of the newly expanding supermarkets chains has also switched from middle-classes customers to the less affluent. Often cheaper more affordable products are developed for a less wealthy customer-base, including more snack foods, sugary drinks, and fast foods (Schmidhuber and Shetty 2005).

It is not certain that convergence towards the westernized diet is inevitable with rising income (Kearney 2010).[7] Recent investigations using Food Balance Sheets and Individual Dietary Surveys showed a 30% rise of total calorie intake from 1969 to 2050 (projected) with increasing calories from vegetable oils and animal products (Fig. 6.3). Calorie intake from eggs, milk, dairy products, meat products, white bread, and refined cereal products increased also. Interestingly, the intake of sugar/sweeteners were level globally over the same period (Kearney 2010).

1.5 Health and Processed Foods

Food is purchased from supermarkets, grocery stores or foodservice outlets usually in a processed state (International Food Information Council Foundation 2010).[8] Recent dietary guidelines for Americans (2015–2020) advised that consumers may cook using a combination of canned, frozen or fresh fruits and dark vegetables. Consumers were advised to review their overall food intake and not individual foods, which determine (health) outcomes.

The concept of "good and bad" is largely discredited since most foods can be eaten – in moderation (Freeland-Graves and Nitzke 2002). US dietary guidelines indicated that current intakes of processed meat and poultry, and salt intake should to be reduced (U.S. Department of Health and Human Services and U.S. Department of Agriculture 2015). Experts consider fresh food and processed foods as complimentary ingredients (Toops 2017).

The world supply of *fresh food* is expected to be insufficient to meet the needs of a growing population.[9] Also with increasing urbanization, those producing their own foods decreases. Consumers moving to urban centers will purchase rather than grow their produce (United States Department of Agriculture 2012). Unfortunately, food composition data relating to processed foods are only becoming available – so what we eat as processed foods is not known with great certainty (Eicher-Miller et al. 2012; Snowdon et al. 2014). The

[7]Nevertheless, there is an undercurrent also towards more localized food supply in some markets.

[8]The economic importance of the Agri-food industry is emphasized elsewhere (Chap. 2). Food is defined in Chap. 4. Some of the perceived advantages of processing are discussed in Chap. 11.

[9]Assuming all foods is "fresh" at the farm gate, the high rate of spoilage of unprocessed food limits availability.

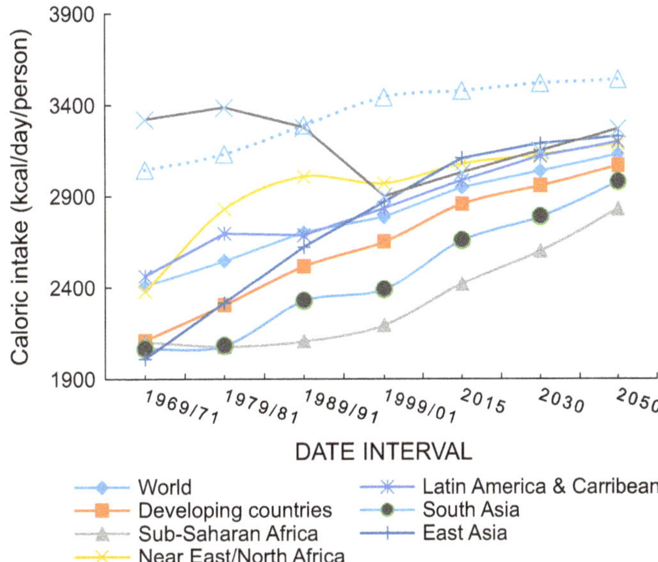

Fig. 6.3 Nutrition transitions occurring with time. Graph shows per capita food consumption (kcal per person per day). Drawn using data from (Kearney 2010)

recent designation of some foods as ultra-processed is further adding to the debate about processed foods (Baker and Friel 2016).

A textbook used in many nutrition & dietetics programs "*Understanding food: Principles and preparation*" describes the use of processed food within institutional foodservices (Brown 2007). For example, frozen fish can replace fresh fish; dried milk powder can replace fresh milk, frozen fruits and vegetables are useful replacements for fresh produce. Losses of nutrients from processed ingredient is rarely an issue because deficiency is rare from the standpoint of a *balanced diet*. Processed food ingredients can contribute to healthy eating but consumers should avoid added sugar and salt (Wolfram 2019).

Catering operations adopt processed food ingredients for cost control (Brown 2007). A preference for processed foods is more evident also during national holidays. Emergency food and food assistance programs, managing natural disasters (flooding, annual bush fires, the tornado season) and food banks for poverty alleviation tend to feature processed foods. Storing dehydrated or canned food is a common strategy for improving food security. Processed foods are better suited for international food assistance and relief compared to fresh foods (Marchione 2002; Poppendieck 1994).

Currently, all four major food and nutrition organizations within the United States acknowledge the significance of processed foods for healthy eating (Eicher-Miller et al. 2012).[10] However, the professions of food science and nutrition could cooperate more to promote public health. Exposing some common misunderstandings about processed foods may help consumers to make informed choices (Dwyer et al. 2012; Weaver et al. 2014).[11] Food and health relations are discussed further in Chap. 11 (Sect. 6).

2 Food Literacy and Cooking

2.1 Food skills

The skills for meal planning, ingredients selection, shopping, cooking and meal preparation seem to be decreasing over time (Weaver et al. 2014). The decline in food skills is important, because, research suggests that cooking at home is linked with healthier food choices and a higher overall healthy-eating index (Begley et al. 2019). A high frequency of home cooking

[10] See a joint statement from, the Academy of Nutrition & Dietetics (formerly the American Dietetic Association), American Association of Nutrition (ASN), the International Food Information Council (IFIC) and the Institute of Food Technology (IFT).

[11] Information packs containing useful facts information about processed foods for the general public were produced by International Food Information Council (IFIC) and the Institute of Food Technology (IFT)

was linked with a lower per capita food expenditure and fewer restaurant visits. Furthermore, those with higher cooking skills showed less reliance on pre-prepared foods and processed ingredients (Lam and Adams 2017; Stitt 1996). Cooking skills are essential for consumers to implement eating guidelines (Taillie 2018; Tiwari et al. 2017).

Not all the experts are convinced that cooking skill declined recently. Critics noted that cooking skills losses have not been measured properly (Lyon et al. 2003). Moreover, comparing cooking skills for modern consumers with those from the past could be problematic owing to recent improvements in cooking technology (e.g. gas ranges, electric cookers, microwave cookers, food mixers and blenders). Improved food technology could promote new cooking skills whilst older skills become redundant. The comparative poverty of the past (pre-world war II) may have led to higher cooking ingenuity. Additionally, the types of cooking skills demonstrated may change with the passage of time. For example, basic cooking skills could migrate from domestic kitchens to food factories. The modern consumer might possess more *creative* cooking- skills compared to a consumer from the past (Lyon 2003).

Teaching cooking skills was previously part of home economics classes at high school, but this declined from the 1950's (Lichtenstein and Ludwig 2010). A recent study suggested that cooking classes improve children's "food related" behaviors (Hersch et al. 2014). The meta-analysis of eight studies considered children (5–12 years), exposed up to 2 years regular teaching of cooking skills. The children had better cooking proficiency, increased cooking confidence, a higher consumption of fruits and vegetables, and increased dietary fiber intake. However, the authors noted only two high quality studies over the decade, and recommended further research to examine the effect of cooking classes on healthy eating (Hersch et al. 2014).

2.2 Frequency of Home Cooking

Recent trends suggests that home cooking decreased in the USA from 1960s to 2008 (Smith et al. 2013). There is a lack of consensus on this point; other research showed that home cooking increased in recent times (Taillie 2018) coinciding with the economic downtown of 2008. Prior to 2008, the tendency for home food preparation changes with gender, level of education, and race /ethnicity. For instance, the non-college educated male consumers were least likely to cook (33%) compared with college educated males (52%). College educated females (69%) cooked at home less frequently compared with non-college educated females (73%). Non-Hispanic black males were least likely to cook overall.

The rising numbers of women inside of paid employment and other societal changes may contribute to the decline of home cooking frequency. However, the effect of demographic factors were not consistent across all jurisdictions (Clifford Astbury et al. 2019). Females from a low socioeconomic grouping spent more time cooking compared to all other groups (Mejean et al. 2017).

2.3 Health Promotion and Food Literacy

Improvements of food literacy can improve eating habits (Anderson et al. 1998; Hersch et al. 2014; Wang et al. 2015). Interventions that increase food knowledge improve consumer attitudes towards healthy eating (Anderson et al. 1998; Hersch et al. 2014; Wang et al. 2015). Research showed that nutrition education increased knowledge and changed the behavior of consumers. Food and nutrition education also produced synergisms alongside of other interventions. For instance, programs to increase farm food production alone had low impact on health outcomes for very young children. However, adding nutrition education to the previous program, helped to promote breast-feeding, and the adoption of better hygiene (McNulty 2013). Therefore, health promotion techniques already used by nutrition experts successfully could be applied in order to improve awareness of food science and technology in the general population (Dwyer et al. 2012).

3 Food Literacy in Middle and High Schools

3.1 Nutrition Education in High Schools

Margaret Chaney, Chair of the home economics department at Connecticut College (Chaney 1945), discussed food and nutrition education in the K12 curriculum. Professor Chaney identified a need to provide support and opportunities for teachers to undertake continual professional development (CPD) in the area of food and nutrition education.[12] She recognized that food and nutrition teaching material needed to be age-specific, and tailored to different age groups. Cooperation between education supervisors, educators, high school nurses, foodservice operators and parents was also essential for program introduction (Chaney 1945). Such principles appear to be relevant today.

The American Dietetic Association support nutrition programmes for children and adolescents in schools (Bergman and Gordon 2010; Briggs et al. 2003; Pilant 2006; Stang and Bayerl 2003). Moreover, the ADA suggested that school food services departments should model good eating behavior for students inside as well as outside of schools. A range of initiatives were envisaged to promote wellness and optimal nutritional status including nutrition education programs, nutrition screening and assessment and strategies to correct undernutrition and over nutrition (Bergman and Gordon 2010; Briggs et al. 2003; Pilant 2006; Stang and Bayerl 2003). Current research suggests that nutrition education at school can have long lasting benefits on healthy eating habits. Food literacy interventions may be cost effective by reducing long-term health costs for treating obesity and other chronic conditions linked with the diet (Graziose et al. 2017). Promoting healthy eating habits at a young age is probably just as useful as correcting poor food habits and dietary recidivism in adults.

3.2 Standardized Food and Nutrition Education in Schools

Not so long ago, teaching in the areas of food and nutrition was described as tiny, scattered and inefficient for United States school's enrollment of 76 million children (Olson and Moats 2013; Smelkova 2015). A lack of timetabled space and low state priorities were the hindrances for school food and nutrition education programs (Hyunyi and Nadow 2004). Nowadays, all schools taking part in US Federal lunch programs must provide food and nutrition teaching.

Food and nutrition teaching occurs as part of the high school family and consumer Sciences (FCS) course or else within health education classes (Murimi et al. 2007).[13] However, there was a lack of standardized curriculum and food and nutrition topics were determined locally (Table 6.2). More recently, the FDA introduced a national standard curriculum and modules, dealing with foods and nutrition within "*Science and our food supply*" program (Table 6.3).

4 Adult Food Literacy

4.1 Influences and Correlations for Food Literacy

Food is a deceptively familiar topic, which can lead to a "false sense of security" for several reasons. First, consumers may find it just as difficult to

[12]The FDA has a website dedicated to professional development support for middle school and high school teachers covering food and nutrition topics in Health Sciences or Family, and Consumer Sciences (FSC) classes. For example see:www.fda.gov/teachsciencewithfood

[13]In the United States, Family & Consumer Sciences (FCS) covers most topics formerly taught under the title of Home economics.

Table 6.2 Some suggested topics for food and nutrition education in schools (K12)

Gardening (food growing),	Food choice systems (Food guide pyramids)
Cooking,	Nutrients in foods
Tasting sessions,	Snacks of fast foods
School lunchroom components	Health food choice
Trips to a farmers market	Consumer skills (Label reading)
Visits to foodservice or restaurant	Sanitation and food handling
Nutrition education lessons,	Preparation of healthy foods
Parent and community components,	Consumer skills, advertising
Role of packaging	Weight management
Food availability	Eating disorders
Taste –sensory education	Food choice, factors
Caloric requirements	Nutrition for sports
Sources of food, farm to table	Consumer skills
Budget principles	Ethnic foods

Source: Adapted from multiple sources (DeCosta et al. 2017; Murimi et al. 2007; Muzaffar et al. 2018)

Table 6.3 The FDA standards for schools: curriculum

Food safety training
Bacteria, including foodborne pathogens
Pasteurization technology
The science of cooking a hamburger
DNA fingerprinting
Outbreak analysis
1st Edition for high schools
Module 1; Serving Size and Calories
Module 2; Sugar in Beverages
Module 3; Sodium in Snack Foods
Module 4; Meal Planning
Module 5; Using the Nutrition Facts Label
Module 6; Healthy Eating Away from Home
2nd edition, for middle schools
Module 1: Introducing the nutrition facts label
Module 2: Nutrients to get less of
Module 3: Nutrients to get more of
Module 4: A closer look at fats
Module 5: healthy eating away from home

Notes: items for food safety and nutrition topics at high school, Adapted from Food and drug Administration (2016)

engage with food science topics compared to other topics (Tuorila and Monteleone 2009). Effort may be needed to ensure that the public is able to extract and use information such as those employed in food labeling (Cowburn and Stockley 2005), health claims, allergen advice (Gendel et al. 2013) or date labeling (Newsome et al. 2014). However, a recent meta-analysis found that population based approaches could be successfully applied to produce healthy eating (Mozaffarian et al. 2012).

Secondly, debates about food and nutrition occur "in private, unregulated space" and outside of a formal educational context. The first amendment allows free speech and prevents censorship of the media or any commercial speech (advertising) that is not *clearly* false (History.Com Editors 2017).[14] Thirdly, food is a multi-disciplinary profession. Therefore, maintaining relevance and keeping updated requires CPD through the workplace or leisure.

Fourthly, many food science *and* nutrition issues can be resolved effectively by a consensus involving opposing teams refereed using a set of pre-agreed rules and codes. Genuine professional conflicts may arise that require highly disciplined experts from agriculture, food science & technology, nutrition, marketing, food law etc. Any recommendations from expert panels needs to be communicated using simple language understandable by most consumers.

The fifth challenge is that competency as a frontline food professional requires concerted periods of training. Graduates can evaluate conflicting viewpoints and statistically *significant results.* Food and nutrition professionals are trained also to challenge simplistic narratives in the same way that medical doctors will confront fake medicine. There are areas of food and nutrition study where consensus expert advice is not yet available.

Finally, cultural issues appear to impact on food literacy. For instance, a study of young adults at university found that one major challenge for food literacy arose from the home environment. Firsthand involvement in food

[14] By comparison, the food manufacturing industry is regulated by the FDA and corresponding organizations worldwide (Chap. 3).

preparation was likely to promote food literacy (Colatruglio and Slater 2016).

4.2 Standards for Food Literacy

Educators have a significant role in promoting food literacy and informed choice. Currently, the public's understanding of processed foods is believed to be limited (Foegeding 2013; Hsieh and Ofori 2007; Kwon et al. 2004; Levine and Labuza 1990; Weaver et al. 2014). Discussions of processed food ingredients alongside of issues such as nutrient density, portion size control, will require a concentered public health effort of the type that deals with food safety issues. Unfortunately, there are no agreed standards for food literacy analogous to FDA nutrition education standards for schools (Sect. 3.2). Some of the main topics assumed to be important for food literacy are shown in (Fig. 6.4).

4.3 Food and Nutrition Controversies and Literacy

Controversies can detract from efforts to promote to healthy eating. Effective healthy eating messages should be perceived as consistent and non-conflicting over time (Freeland-Graves and Nitzke 2002). One persistent controversy is predicated on a bit of good news. Recently, the US Agri-food system produced 4000 Calories per head of population, which exceeds requirements (Nestle 2013). It was proposed that the bountiful farm-gate food supply incentivizes food corporations to promote overeating by increasing the (sweet, salt, fat) taste of food, whilst lowering cost, and increasing convenience.[15] It was claimed that food manufactures employ certain standard tactics or Corporate Political Activity Strategies (CAPS) (Table 6.4) to promote their commercial interests disproportionately compared to civic concerns (Jaichuen et al. 2018; Mialon and Mialon 2018). Economic experts discussing the recent rising obesity trend its possible link with processed foods, suggested a need more independent legislation, that is able to deal with "asymmetric information" (advertising), "regulatory capture" (tendency to influence regulations) and obfuscation strategies by big food corporations that can be traced back to nineteenth century (Smith et al. 2011; Smith and Tasnádi 2014). Similar viewpoints have been raised in connection with ultra-processed food (Chap. 11).

4.4 Adult Learning and Food Literacy

The principles of adult education or andragogy may be helpful for promoting food and nutrition education (InstructionalDesign.org 2019; Knowles 1980; Knowles et al. 2014). The majority of adults are independent learners, self-directed, and possessing prior experience of the subject matter and applications. According to Knowles, adult learning is likely to be successful also when instructors function as facilitators rather than top-down teachers (Knowles 1980). The range of opportunities and venues for promoting adult food literacy is wider than conversional school education (Sect. 5.1.). As noted above, the principles of nutrition health promotion should be adaptable for improving food literacy in the general population.

5 Food and Nutrition Education

Effort is needed to raise food and nutrition literacy, so that consumers can engage with healthy eating messages (see above). Some of the requirements for nutrition education should be aimed at improving consumer understanding of processed foods. In this section, we explore nutrition education and novel ways of thinking about food and nutrition.

[15] Enhancing food aaffordability, convenience and improved safety were goals for food science discipline, set by national governments in the post-world war 2 years.

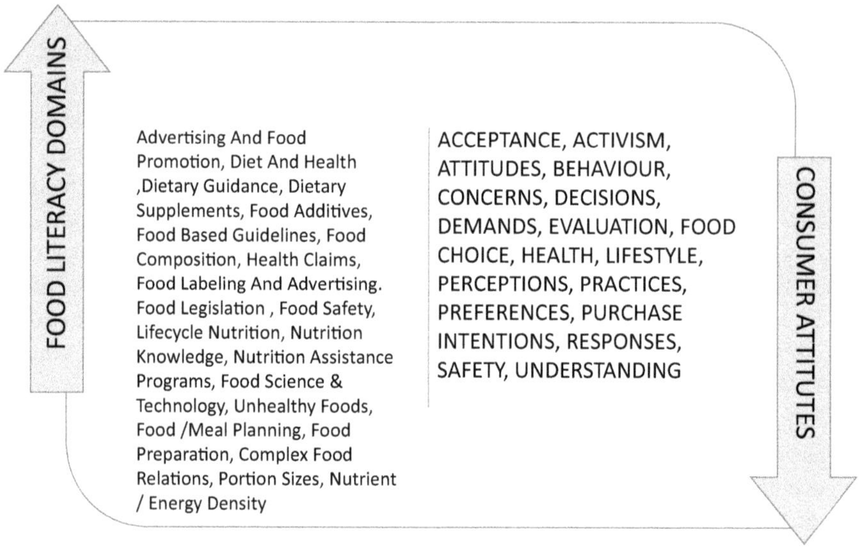

Fig. 6.4 Food education, knowledge, literacy and training needs. Adapted from various sources (Amouzandeh et al. 2019; Benn 2014; Cullen et al. 2015; Krause et al. 2018; Truman et al. 2017; Velardo 2015; Vidgen and Gallegos 2014; Yuen et al. 2018)

Table 6.4 Some criticisms directed at food companies

Encourage over eating, promote excess calorie intake by marketing, poor labeling,
Oppose dietary recommendation regulations
Lobbying activity to gain favor
Co-opt nutrition and food science professional through sponsorship
Act to win friends, disarm critics and play hard ball

Adapted from Nestle (2013).

Table 6.5 Current terms applied to nutrition education

Nutrition Education – US academic, official documents
Behavior Change Communication; UN, UK & developing countries
Health promotion – all nations, classical literature
Community nutrition – developing countries

Summarized from McNulty (2013)

5.1 The Nature of Food and Nutrition Education

Some alternative terms for nutrition education are listed in (Table 6.5) (McNulty 2013). The United Nations called nutrition education, – behavioral change communication (BCC). Another name for nutrition education is community nutrition. In North America, nutrition education is the preferred term (Table 6.5). In all cases, the basic concept for nutrition education is to motivate the audience and then change behavior related to healthy eating.

Nutrition education was defined by the ADA as "*... instruction or training intended to lead to acquired nutrition-related knowledge and/or nutrition-related skills...*" (McNulty 2013). Alternatively, nutrition education refers to "*any combination of educational strategies, accompanied by environmental supports, designed to facilitate voluntary adoption of food choices and other food and nutrition- related behaviors conducive to health and well-being; nutrition education is delivered through multiple venues and involves activities at the individual, community, and policy levels*" (Contento 2008). Less frequently, nutrition education refers to "nutrition guidelines and education" which are recommendations for a balanced diet (Anderson et al. 1998; Dwyer 2001) (Table 6.6).

Many international and national agencies are advocates for healthy eating including, UN agencies and some 50 different NGOs. However, there is a tendency for nutrition education programmes to take place independently within different sectors, e.g. health services, schools and education

Table 6.6 Two example definitions for nutrition education

.. Combination of educational strategies, accompanied by environmental supports, designed to facilitate voluntary adoption of food choices and other food- and nutrition-related behaviors conducive to health and well-being. Nutrition education is delivered through multiple venues and involves activities at the individual, community, and policy levels. Contento 2008
Community nutrition (Ndure et al. 1999) *A community nutrition program is a collection of activities aimed at solving the nutrition problems of a community with the full participation and cooperation of its members in a flexible, though systematic and well-researched manner. Community nutrition programs are implemented in both urban and rural settings.*

Adapted from Contento (2008) and Ndure et al. (1999)

systems. Health promotion/ nutrition education targets food choice, consumer believes and attitudes, motivations, and health-related behavior. Health promotion extends also to cover physical activity, smoking and alcohol use (McNulty 2013).

5.2 The Food and Nutrition System

Researchers from University of Queensland, Australia (Heywood and Lund-Adams 1991) provided an early account of the "food and nutrition system". The central idea is that all *activities* involved in food and nutrition can be joined to form a chain. A system is one way of expressing interrelatedness of different activities and structures. For example, there are eight major biological systems found in the body (circulatory system, digestive system etc.). The food and nutrition system is formed by interrelated subunits.

The food and nutrition system is an "open system" that exchanges energy and material inputs/ outputs with the environment. An open system is one where the boundaries are recognized as porous, allowing materials to enter and products to leave. The system model is important because it can be interrogated using different disciplines: agriculture, economics, food technology, management, medicine and sociology in addition to food and nutrition (Heywood and Lund-Adams 1991). Some potential constraints arising from political, social environment factors can be examined more clearly as they affect the food and nutrition system. Several different forms of the food system can be found in the following citation (Sobal et al. 1998).

The UN Universal declaration of human rights from 1948 recognized a person's" right to food.[16.] Growing, purchasing, preparing and consuming food is essential for a healthy life. Perhaps for this reason, every consumer should understand "where our food comes from". Awareness of the "food and nutrition systems" approach is vital in order to evaluate a wider scope of ideas and issues related to food and nutrition. For example, discussions about the diet and biodiversity become more manageable if we allow for the connectivity between nutrition, health, and policy (Fig. 6.5). The connections between farming practice, the natural / built environment, the food industry, nutrition and consumer issues can be visualized more clearly from a systems approach (Burlingame et al. 2009; Hall 2018; Malambo et al. 2016).

5.3 Kinds of Evidence for Food and Nutrition Policy

The intersection of food, and nutrition policy is a contested area (Sibbel 2012; Yach et al. 2010a, b). There exists a hierarchy of evidence – with differing degrees of reliability, for developing policy and recommendations (Anderson and Zlotkin 2000; Dwyer 2001; Truswell 1998; Wellman et al. 1999). The best quality evidence for food policy is derived from a type of study called, the randomized controlled trial (RCT), the systematic review or meta-analysis. Such "gold standard" evidence is used as the basis for food and nutrition policy. Unfortunately, the public is not sufficiently well informed about the significance of systematic reviews (Millen et al. 2016).

[16]The human right to foods.

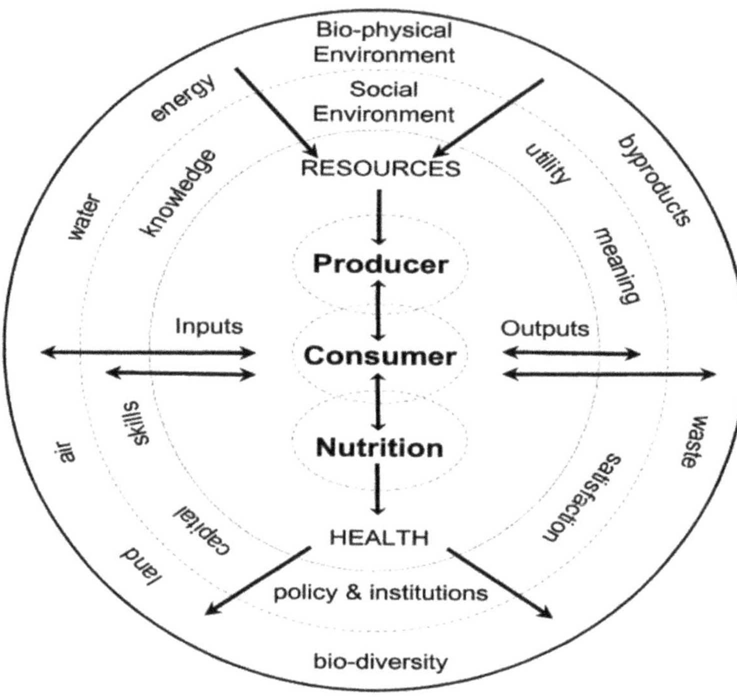

Fig. 6.5 An example of the food and nutrition system (from (Sobal et al. 1998) with permission from Elsevier

Many recent "findings" about the possible health effects of processed foods come from "observational studies" that reveal correlations between variables but not which is a "cause" and which is an effect. Observational studies are inexpensive to do but evidence from these so-called population studies are weak. Because of the phenomenon called "confounding", many outcomes from the observational studies are merely "suggestive". In addition, many risk factors for non-communicable diseases (e.g. obesity, cardiovascular disease) tend to be influenced by non-food variables including, household income, alcohol use, TV ownership and physical inactivity (Cavalli-Sforza et al. 1996).

Ideally, food policy should be using results from evidence-based practice, which follows a rigorously defined protocol based on the use of "PICO" items. PICO is an acronym for population (P), Intervention (I), Comparison (C) and Outcomes (O) (Millen et al. 2016).

5.4 Resolving Food and Nutrition Controversies

Genuine disagreements and competing viewpoints about food and nutrition issues should be resolved using evidence based practice. This requires an undertaking to review the totality of available evidence beginning with the highest-grade data.[17] For example, the health claim stating that consuming soybean protein lowers cholesterol is supported by evidence, from RCT performed with healthy humans, *who are seen as representative of the general population.* Typically, results from sick individuals are not permitted as evidence for a health claim.[18] Other areas for applying evidence based practice is in the development of dietary guidelines (Dwyer

[17]Definition: controversy (over/about/surrounding somebody/something) public discussion and argument about something that many people strongly disagree about, disapprove of, or are shocked by-The Oxford Learners Dictionary.

[18]Claims that refer to treating existing illnesses will be deemed as medicinal claims, which are related to drugs.

2001), derivation of dietary reference intake values and as stated above – assessing health claims.

Evidence based practice for nutrition issues require that all relevant RCT are located without bias, in accordance with pre-agreed criteria.[19] The results from separate RCT are then combined in order to produce a stronger overall measure. For instance, consider a policy request to establish the effect of salt intake on the average blood pressure for individuals described as Latina adults. In accordance with evidence-based practice, all relevant RCT evidence should be examined, in a way that enables valid recommendations about salt intake and health *in Latina adults*.

According the "PICO" approach (Table 6.7). We must define the population (P) of interest (Latina adults, age range); state the intervention (I) of interest (salt level), define clearly the types of outcomes (O) of interest (blood pressure). The selection of RCT using PICO approach must be "systematic" without bias, with clear "inclusion and exclusion" criterion. Publications in different languages should be evaluated if possible. The process for systematic reviews is increasingly accessible to nutrition and allied health graduates. Examples for systematic reviews, include those looking at the effect of different foods on diabetes risk (Murakami et al. 2005) (Sargeant et al. 2006), effect of pasteurization on milk vitamins and raw milk effects on health (Macdonald et al. 2011) and the effect of eating out on dietary quality (Lachat et al. 2012). Systematic reviews on diverse topics are usually freely available for downloading from the Cochrane library website.

Table 6.7 Framing food and health questions in line with evidence based practice

Populations¥	<u>Healthy subjects</u> Defined populations, Adults, Children, Ethnicities
"Intervention" (dietary variables)	Dietary and Commercially produced food <u>Carbohydrates</u> Caloric sweeteners, Fructose, Refined carbohydrates, added sugar,Sugar sweetened beverages. Dietary fiber <u>Fats and lipids</u> Total fat, Edible oils, Saturated fats, Trans-fatty acids <u>Animal products</u> Meat, Processed meat, Westernized diets <u>Food Additives</u> Salt, Emulsifier, Other
Comparisons / baseline¥	All healthy subjects High vegetables and fruit, high dietary fiber, low fat, low sugar, traditional diet, "hunter gatherers"
Outcomes¥	All healthy subjects Body weight / BMI, waist circumference, blood glucose, blood cholesterol, insulin resistance, blood pressure
Confounders	Ageing population, alcohol use, TV & media access, physical activity rate,Rising income, smoking, urbanization, motorized transport & Car ownership, existing illness

Notes: Adapted from (Popkin 2015). (¥) Only non-disease outcomes, biomarkers or risk factors for disease in otherwise healthy subjects. Exclude all those with overt diseases

6 Debating Fresh and Processed Food-Not

6.1 General Principles

Nowadays canned, dried or frozen fruits and vegetables provide about 50% of the daily intake of nutrients in the American diet (Dwyer et al. 2012). Arguably, most foods are processed to some degree prior to consumption.[20] Many professional chefs use processed food ingredients for convenience and to lower costs. However, many consumers lack the knowledge of how to use processed food ingredients for healthy meals (Hsieh and Ofori 2007; Knorr and Watzke 2019; Weaver et al. 2014). Food processing is a core subject within the food science discipline. There is a need for food literacy programs that improve consumer understanding of the merits and disadvantages of fresh versus processed foods. Moreover, the

[19] Often foreign language studies that should be included in a systematic review are not owing to difficulties with translation.

[20] Chapter 11 covers some general principles of processed foods.

responsibility for promoting this sort of food literacy falls with food science professionals.

It is well known that consumers appreciate the fresh-like quality achieved with minimally processed foods (MPF) (Rodgers 2016; Siddiqui and Rahman 2014). The manufacturing process for MPF such as freshly cut salad involves washing, cutting/ slicing, packaging and refrigerated storage.

Food additives and preservatives may be added to some processed foods in order to ensure food safety. For example, salt, nitrite or acidulants reduce the likelihood of *C. botulinum* sporulation, germination and toxin production (Chap. 12). Other additives may be added to deal with enzymatic browning, oxidation and general metabolism. The use of food grade sanitizers e.g. dilute solutions of hydrogen peroxide and chlorine, anti-browning agents, modified atmosphere packaging) are well documented. Refrigeration is another readily overlooked operational step that is legally required for MPF and certain packaged foods (Chap. 13, Sect. 2.3). Furthest removed from MPF and mildly processed foods we encounter a range of prepared foods produced using secondary food ingredients ranging from whole flour to isolated starch. Food additives are discussed in (Chap. 16).

Some plant foods subjected to "mild heating" also exhibit less heat damage to micronutrients (Shi and Le Maguer 2000). Different forms of processing affect (reduce / enhance) nutrient content according the temperature-time combinations used. For example, replacing boiling, baking or frying with steaming produce changes in product characteristics that are beneficial (Fan 2014; Miglio et al. 2008; Palermo et al. 2014). A continuum of processing intensity can be envisaged within conventional framework, for example by going from, "rare", "medium" to "well done" steaks.[21]

Discussing of the relative merits of processed *versus* fresh foods is contentious, but the stakes are high and many food scientists believe "the pendulum has swung too far". Many well-documented positive impacts of food processing on public health nutrition seem forgotten. Food science students should engage with the media more. (Baker and Friel 2014; Kelly and Jacoby 2018; Weaver et al. 2014). Current media focus on the effect of processing on single nutritional components (Mosha and Vicent 2005; Sanchez-Moreno et al. 2009). The rest of this section will consider why debating the merits of fresh and processed foods turn out less rewarding than expected.

6.2 Selection of baseline foods

A baseline food is a product that enables us to determine the effect of processing, vis a vis unprocessed foods. The baseline food is needed to evaluate the effect of processing on sensory, nutritional, storage, and safety characteristics. Unfortunately, there are usually no baseline foods to act as the basis for discussing processing. For example, baking bread requires secondary food ingredients be combined in the right proportions. The merits of baking *per se* is difficult to establish by comparing bread with its raw ingredients (flour, yeast, water). Studying the transformation of ingredients into bread will provide knowledge about baking technology and how this affects end-product quality. Discussing the merits of bread as a processed food is difficult because there is no such item as "unprocessed" bread for comparison. There are no baseline foods also for many so-called prepared foods, including cake, pasta and boiled rice. As mentioned with the Western diet (Sect 1.3) many processed foods being entirely the products of human invention do not appear in nature (Cordain et al. 2005; Carrera-Bastos et al. 2011).

Turning to dairy foods, the trick is to convert fluid milk into yogurt or a host of other products like cheese. Comparing milk with yogurt, cheese cream etc. provides knowledge about the ingredients, processing conditions and their impact on product quality. In dairy science, few will debate whether less processed milk is preferable to yogurt or cheese. There are debates about the

[21] The classification of different types of food processing according to their intensity is discussed in Chap. 11 (Sect. 1.1)

merits of raw milk versus pasteurized milk, but these are not part of the current discussions about "ultra-processing".

Another issue that prevents comparisons between baseline foods and the corresponding processed food is that the former are frequently unpalatable. Comparing a list of ingredients (e.g. cake ingredients) with a final processed food product provides few useful insights about merits of *unprocessed* foods even where the base-line food dilemma can be addressed. Such comments do not mean that metrics for food *quality* should not be expanded to include more health outcomes. The recent functional food revolution is evidence of consumer desire for health and wellness products (Rodgers 2016; Smith and Charter 2011).

6.3 Comparing Fresh with Minimal Processing Food

In principle, fruits and vegetables can be eaten with little or no processing. However, commercial supplies of fruits and vegetables require processing, even if it is washing and packaging. A MFP such as dried fruit is produced using a limited number of unit operations. It is conceivable to compare the merits of fresh peas with the MPF counterpart (frozen or dried peas) but the former is rarely available all year round. The MPF is also likely to be more convenient and less expensive.

6.4 Unpredictable Health and Safety Outcomes

The long-term health effects of processed foods are not known prior to product introduction. For example, the cardiovascular risks associated with trans-fats were discovered decades after the introduction of margarine. Saturated fatty acids remain controversial now as ever. Many jurisdictions are only now introducing taxes to control added sugar use (Afshin et al. 2017).

6.5 One Dimensional Focus on Food Merits

By one-dimensional debate, we mean discussions about the merits of fresh and unprocessed food that tend to focus on one issue, e.g. vitamin losses. Vitamin content is generally higher for unprocessed as compared to processed food. Nevertheless, vitamin deficiency is uncommon assuming that consumers achieve a balanced diet. Moreover, processing leads to trade-offs in terms of other sensory, nutritional, safety and storage attributes (See Chap. 11).

6.6 Confounding Effects of Different Food Groups

The different food group can confound the fresh versus processed food debate. For instance, consumption of *red* meat (food group) is association with some cancers, regardless of the extent of processing. However, increasing consumption of (all) meat and fish produced no changes in the risk of colorectal cancer and breast cancer. (Salehi et al. 2013; Zeraatkar et al. 2019a, b). Weather consuming processed meat affects the risks of all cancers and cancer mortality remains uncertain (Vernooij et al. 2019; Zeraatkar et al. 2019a, b).

6.7 Unprocessed Fresh Food Perishability

All foods will deteriorate with time, but the time-course for food spoilage will be shorter for unprocessed foods. The rate of food deterioration can be curtailed by processing. Increases of food stability affects availability, pricing, food security and environmental impact, which should be considered when discussing the relative merits of fresh and processed food products.

With such considerations in mind (Sect. 6.1 to 6.6) the loss of nutrients (vitamins C, B, E, carotenoids, phenols etc.) due to the storage of fresh fruits and vegetables were compared with losses due to freezing or canning. It was concluded nutrient losses during the fresh food storage were often underestimated, and that processed foods contribute significantly to dietary recommendations for Americans (Rickman et al. 2007a, 2007b).

6.8 The Total Diet Approach

The total-diets approach looks at the relations between human diet (dietary pattern) and health. Therefore, health is not related to single food items or nutrients but to the dietary pattern (Freeland-Graves, & Nitzke, 2002). The experts believe that processed foods and ingredients are compatible with a healthy diet. The possible link between diets containing too much ultra-processed foods and adverse health risk is discussed in Chapter 11.

7 Conclusions

Low food literacy is a relatively new issue for food science and technology. Different approaches will be required in order to educate consumers at different stages of the human lifecycle (infants, children, young adults, adults and elderly). The recent introduction of a US national standardized curriculum covering food and nutrition for school age children will help improve food literacy. The current shortfall in food literacy for the average adult consumer could be addressed using health promotion -type approaches outside of a formal educational setting, which is a strategy used for food safety education (Dwyer et al. 2012 2883; Hsieh and Ofori 2007; Knorr and Watzke 2019; Weaver et al. 2014).

References

Afshin A, Penalvo J, Del Gobbo L, Kashaf M, Micha R, Morrish K, Mozaffarian D (2015) CVD prevention through policy: a review of mass media, food/menu labeling, taxation/subsidies, built environment, school procurement, worksite wellness, and marketing standards to improve diet. Curr Cardiol Rep 17(11):98. https://doi.org/10.1007/s11886-015-0658-9

Afshin A, Penalvo JL, Del Gobbo L, Silva J, Michaelson M, O'Flaherty M, Mozaffarian D (2017) The prospective impact of food pricing on improving dietary consumption: a systematic review and meta-analysis. PLoS One 12(3):e0172277. https://doi.org/10.1371/journal.pone.0172277

Amouzandeh C, Fingland D, Vidgen HA (2019) A scoping review of the validity, reliability and conceptual alignment of food literacy measures for adults. Nutrients 11(4). https://doi.org/10.3390/nu11040801

Anderson GH, Zlotkin SH (2000) Developing and implementing food-based dietary guidance for fat in the diets of children. Am J Clin Nutr 72(5 Suppl):1404S–1409S

Anderson AS, Cox DN, McKellar S, Reynolds J, Lean ME, Mela DJ (1998) Take five, a nutrition education intervention to increase fruit and vegetable intakes: impact on attitudes towards dietary change. Br J Nutr 80(2):133–140

Baker P, Friel S (2014) Processed foods and the nutrition transition: evidence from Asia. Obes Rev 15(7):564–577

Baker P, Friel S (2016) Food systems transformations, ultra-processed food markets and the nutrition transition in Asia. Glob Health 12(1):80. https://doi.org/10.1186/s12992-016-0223-3

Begley A, Paynter E, Butcher LM, Dhaliwal SS (2019) Examining the association between food literacy and food insecurity. Nutrients 11(2). https://doi.org/10.3390/nu11020445

Benn J (2014) Food, nutrition or cooking literacy-a review of concepts and competencies regarding food education. Int J Home Econ 7(1):13

Bergman EA, Gordon RW (2010) Position of the American dietetic association: local support for nutrition integrity in schools. J Am Diet Assoc 110(8):1244–1254

Berry BW (1998) Color of cooked beef patties as influenced by formulation and final internal temperature. Food Res Int 30(7):473–478. https://doi.org/10.1016/S0963-9969(98)00023-4

Briggs M, Safaii S, Beall DL, American Dietetic, A., Society for Nutrition, E., & American School Food Service, A (2003) Position of the American dietetic association, society for nutrition education, and American school food service association--nutrition services: an essential component of comprehensive school health programs. J Am Diet Assoc 103(4):505–514

Brown A (2007) Understanding food: principles and preparation, 4th edn. Cengage Learning, Singapore, p 696

Burlingame B, Charrondiere R, Mouille B (2009) Food composition is fundamental to the cross-cutting initiative on biodiversity for food and nutrition. J Food Compos Anal 22(5). https://doi.org/10.1016/j.jfca.2009.05.003

Carrera-Bastos P, Fontes-Villalba M, O'Keefe JH, Lindeberg S, Cordain L (2011) The western diet, lifestyle, and diseases of civilization. Res Rep Clin Cardiol 2(1):15–35

Cavalli-Sforza LT, Rosman A, De Boer AS, Darnton-Hill I (1996) Nutritional aspects of changes in disease patterns in the Western Pacific region. Bull World Health Organ 74(3):307

Chaney MS (1945) Integrating nutrition education and activities in the school program. Am J Public Health Nations Health 35(7):728–731. https://doi.org/10.2105/ajph.35.7.728

Clifford Astbury C, Penney TL, Adams J (2019) Home-prepared food, dietary quality and socio-demographic factors: a cross-sectional analysis of the UK National Diet and nutrition survey 2008–16. Int J Behav Nutr Phys Act 16(1):82. https://doi.org/10.1186/s12966-019-0846-x

Colatruglio S, Slater J (2016) Challenges to acquiring and using food literacy: perspectives of young Canadian adults. Can Food Stud 3(1):96–118

Contento IR (2008) Nutrition education: linking research, theory, and practice. Asia Pac J Clin Nutr 17(S1):176–179

Cordain LES, Sebastian A, Mann N, Lindeberg S, Watkins BA, O'Keefe JH, Brand-Miller J (2005) Origins and evolution of the Western diet: health implications for the 21st century. Am J Clin Nutr 81(2):341–354. https://doi.org/310.1093/ajcn.1081.1092.1341

Cowburn G, Stockley L (2005) Consumer understanding and use of nutrition labeling: a systematic review. Public Health Nutr 8(1):21–28

Cullen T, Hatch J, Martin W, Higgins JW, Sheppard R (2015) Food literacy: definition and framework for action. Can J Diet Pract Res 76(3):140–145. https://doi.org/10.3148/cjdpr-2015-010

DeCosta P, Moller P, Frost MB, Olsen A (2017) Changing children's eating behaviour – a review of experimental research. Appetite 113:327–357. https://doi.org/10.1016/j.appet.2017.03.004

Dembert ML (1981) Western diseases: their emergence and prevention. Yale J Biol Med 54(6):504–505

Dwyer JT (2001) Nutrition guidelines and education of the public. J Nutr 131(11 Suppl):3074S–3077S

Dwyer JT, Fulgoni VL 3rd, Clemens RA, Schmidt DB, Freedman MR (2012) Is "processed" a four-letter word? The role of processed foods in achieving dietary guidelines and nutrient recommendations. Adv Nutr 3(4):536–548

Eicher-Miller HA, Fulgoni VL III, Keast DR (2012) Contributions of processed foods to dietary intake in the US from 2003–2008: a report of the Food and Nutrition Science Solutions Joint Task Force of the Academy of Nutrition and Dietetics, American Society for Nutrition, Institute of Food Technologists, and International Food Information Council. J Nutr 142(11):2065S–2072S

Fan Z (2014) Influence of ingredients and chemical components on the quality of Chinese steamed bread. Food Chem 163:154–162

Foegeding EA (2013) Obesity and food science being part of the solution. J Food Sci 78(9):IV. https://doi.org/10.1111/1750-3841.12261

Food and drug Administration (2016, 2019 April) Science and our food supply: using the nutrition facts label to make healthy food choices. Teachers guide for middle school classrooms. 2nd Edition Retrieved from https://wayback.archive-it.org/7993/20190423025420/https://www.fda.gov/Food/FoodScienceResearch/ToolsMaterials/ucm2006976.htm

Fox M (2012) Defining processed foods for the consumer. J Acad Nutr Diet 112(2):217

Freeland-Graves J, Nitzke S (2002) Position of the American dietetic association: total diet approach to communicating food and nutrition information. J Am Diet Assoc 102(1):100–108

Gendel SM, Khan N, Yajnik M (2013) A survey of food allergen control practices in the U.S. food industry. J Food Prot 76(2):302–306

Gibney M, Gibney MJ (2012) Something to chew on: Challenging controversies in food and health. University College Dublin Press, p 177

Goldstein MC, DMAGM (2009) Food and nutrition controversies today: a reference guide: ABC-CLIO, 242pp

Graziose MM, Koch PA, Wang YC, Gray HL, Contento IR (2017) Cost-effectiveness of a nutrition education curriculum intervention in elementary schools. J Nutr Educ Behav 49(8):684–691

Ha M (2018) Processed food linked to cancer? Uncritical media coverage ignores problems with study. Retrieved from https://theconversation.com/profiles/marie-ann-ha-328802

Hall KD (2018) Did the food environment cause the obesity epidemic? Obesity (Silver Spring) 26(1):11–13. https://doi.org/10.1002/oby.22073

Hersch D, Perdue L, Ambroz T, Boucher JL (2014) The impact of cooking classes on food-related preferences, attitudes, and behaviors of school-aged children: a systematic review of the evidence, 2003–2014. Prev Chronic Dis 11:E193. https://pubmed.ncbi.nlm.nih.gov/25376015/

Heywood P, Lund-Adams M (1991) The Australian food and nutrition system: a basis for policy formulation and analysis. Aust J Public Health 15(4):258–270. https://doi.org/10.1111/j.1753-6405.1991.tb00345.x

History.Com Editors (2017) First amendment. Retrieved from https://www.history.com/topics/united-states-constitution/first-amendment

Hsieh YH, Ofori JA (2007) Innovations in food technology for health. Asia Pac J Clin Nutr 16(Suppl 1):65–73

Hyunyi C, Nadow MZ (2004) Understanding barriers to implementing quality lunch and nutrition education. J Commun Health 29(5):421–435

InstructionalDesign.org (2019) Andragogy (Malcolm Knowles). Retrieved from https://www.instructionaldesign.org/theories/andragogy/

International Food Information Council Foundation (2010, 2014) What is a processed food? You might be surprised! Retrieved from https://foodinsight.org/wp-content/uploads/2014/07/IFIC_Handout1_high_res.pdf

Jaichuen N, Phulkerd S, Certthkrikul N, Sacks G, Tangcharoensathien V (2018) Corporate political activity of major food companies in Thailand: an assessment and policy recommendations. Glob Health 14(1):115. https://doi.org/10.1186/s12992-018-0432-z

Jones JM (2019) Food processing: criteria for dietary guidance and public health? Proc Nutr Soc 78(1):4–18. https://doi.org/10.1017/s0029665118002513

Kearney J (2010) Food consumption trends and drivers. Philos Trans R Soc B 365(1554):2793–2807

Kelly B, Jacoby E (2018) Public health nutrition special issue on ultra-processed foods. Public Health Nutr 21(1):1–4. https://doi.org/10.1017/s1368980017002853

Knorr D, Watzke H (2019) Food processing at a crossroad. Front Nutr 6:85. https://doi.org/10.3389/fnut.2019.00085

Knowles MS (1980) The modern practice of adult education. From pedagogy to andragogy. Cambridge Adult Education, Cambridge, p 400

Knowles MS, Holton EF, Swanson RA (2014) The adult learner: the definitive classic in adult education and human resource development. Taylor & Francis, p 402

Krause C, Sommerhalder K, Beer-Borst S, Abel T (2018) Just a subtle difference? Findings from a systematic review on definitions of nutrition literacy and food literacy. Health Promot Int 33(3):378–389. https://doi.org/10.1093/heapro/daw084

Kwon TW, Hong JH, Moon GS, Song YS, Kim JI, Kim JC, Kim MJ (2004) Food technology: challenge for health promotion. Biofactors 22(1–4):279–287

Lachat C, Nago E, Verstraeten R, Roberfroid D, Van Camp J, Kolsteren P (2012) Eating out of home and its association with dietary intake: a systematic review of the evidence. Obes Rev 13(4):329–346. https://doi.org/10.1111/j.1467-789X.2011.00953.x

Lam MCL, Adams J (2017) Association between home food preparation skills and behaviour, and consumption of ultra-processed foods: cross-sectional analysis of the UK National Diet and nutrition survey (2008–2009). Int J Behav Nutr Phys Act 14(1):68. https://doi.org/10.1186/s12966-017-0524-9

Levine AS, Labuza TP (1990) Food systems: the relationship between health and food science/technology. Environ Health Perspect 86:233–238. https://doi.org/10.1289/ehp.9086233

Lichtenstein AH, Ludwig DS (2010) Bring back home economics education. J Am Med Assoc 303(18) 1857-1858. https://doi.org/10.1001/jama.2010.592

Lyon P, Colquhoun A, Alexander E (2003) Deskilling the domestic kitchen: national tragedy or the making of a modern myth? Food Serv Technol 3(3–4):167–175

Macdonald LE, Brett J, Kelton D, Majowicz SE, Snedeker K, Sargeant JM (2011) A systematic review and meta-analysis of the effects of pasteurization on milk vitamins, and evidence for raw milk consumption and other health-related outcomes. J Food Prot 74(11):1814–1832

Malambo P, Kengne AP, De Villiers A, Lambert EV, Puoane T (2016) Built environment, selected risk factors and major cardiovascular disease outcomes: a systematic review. PLoS One 11(11):e0166846. https://doi.org/10.1371/journal.pone.0166846

Manitoba Government (2017) Social studies. Senior 2 social studies teacher notes. Critical thinking. Tn37. Retrieved from https://www.edu.gov.mb.ca/k12/cur/socstud/frame_found_sr2/tns/tn-34.pdf

Marchione TJ (2002) Foods provided through US government emergency food aid programs: policies and customs governing their formulation, selection and distribution. J Nutr 132(7):2104S–2111S

McNulty J (2013) Challenges and issues in nutrition education. Nutrition Education and Consumer Awareness Group, Food and Agriculture Organization of the United Nations, Rome. Retrieved from www.fao.org/ag/humannutrition/nutritioneducation/en/

Mejean C, Si Hassen W, Gojard S, Ducrot P, Lampure A, Brug H, Castetbon K (2017) Social disparities in food preparation behaviors: a DEDIPAC study. Nutr J 16(1):62. https://doi.org/10.1186/s12937-017-0281-2

Mialon M, Mialon J (2018) Analysis of corporate political activity strategies of the food industry: evidence from France. Public Health Nutr 21(18):3407–3421. https://doi.org/10.1017/s1368980018001763

Miglio C, Chiavaro E, Visconti A, Fogliano V, Pellegrini N (2008) Effects of different cooking methods on nutritional and physicochemical characteristics of selected vegetables. J Agric Food Chem 56(1):139–147

Millen BE, Abrams S, Adams-Campbell L, Anderson CA, Brenna JT, Campbell WW, Lichtenstein AH (2016) The 2015 dietary guidelines advisory committee scientific report: development and major conclusions. Adv Nutr 7(3):438–444. https://doi.org/10.3945/an.116.012120

Mosha TCE, Vicent MM (2005) Nutritional quality, storage stability and acceptability of home-processed ready-to-eat composite foods for rehabilitating undernourished preschool children in low-income countries. J Food Process Preserv 29(5–6):331–356. https://doi.org/10.1111/j.1745-4549.2005.00032.x

Mozaffarian D, Afshin A, Benowitz NL, Bittner V, Daniels SR, Franch HA, Jacobs DR Jr, Kraus WE, Kris-Etherton PM, Krummel DA, Popkin BM (2012) Population approaches to improve diet, physical activity, and smoking habits: a scientific statement from the American Heart Association. Circulation 126:1514–1563

Murakami K, Okubo H, Sasaki S (2005) Effect of dietary factors on incidence of type 2 diabetes: a systematic review of cohort studies. J Nutr Sci Vitaminol (Tokyo) 51(4):292–310

Murimi MW, Sample AD, Guthrie J, Landry D (2007) Nutrition education in team nutrition middle schools: teachers' perceptions of important topics to be taught and teaching curriculum used. J Child Nutr Manag 2(31):1–12

Muzaffar H, Metcalfe JJ, Fiese B (2018) Narrative review of culinary interventions with children in schools to promote healthy eating: directions for future research and practice. Curr Dev Nutr 2(6):nzy016. https://doi.org/10.1093/cdn/nzy016

Ndure KS, Sy MN, Ntiru M, Diène SM (1999) Best practices and lessons learned for sustainable community

nutrition programming. Academy. Retrieved from http://bvsper.paho.org/texcom/nutricion/Practices.pdf
Nestle M (2013) Food politics: how the food industry influences nutrition and health. University of California Press, Berkeley, p 534
Newsome R, Balestrini CG, Baum MD, Corby J, Fisher W, Goodburn K, Yiannas F (2014) Applications and perceptions of date labeling of food. Compr Rev Food Sci Food Saf 13(4):745–769. https://doi.org/10.1111/1541-4337.12086
Olson S, Moats M (2013) Nutrition education in the K-12 curriculum: the role of national standards: workshop summary. National Academies Press, Washington, DC, 3, the context for change. Retrieved from https://www.ncbi.nlm.nih.gov/books/NBK202126/
Palermo M, Pellegrini N, Fogliano V (2014) The effect of cooking on the phytochemical content of vegetables. J Sci Food Agric 94(6):1057–1070
Pilant VB (2006) Position of the American dietetic association: local support for nutrition integrity in schools. J Am Diet Assoc 106(1):122–133. https://doi.org/10.1016/j.jada.2005.11.006
Popkin BM (1994) The nutrition transition in low-income countries: an emerging crisis. Nutr Rev 52(9):285–298. https://doi.org/10.1111/j.1753-4887.1994.tb01460.x
Popkin BM (2001) The nutrition transition and obesity in the developing world. J Nutr 131(3):871s–873s. https://doi.org/10.1093/jn/131.3.871S
Popkin BM (2015) Nutrition transition and the global diabetes epidemic. Curr Diab Rep 15(9):64. https://doi.org/10.1007/s11892-015-0631-4
Poppendieck J (1994) Dilemmas of emergency food: a guide for the perplexed. Agric Hum Values 11(4):69–76
Rickman JC, Bruhn CM, Barrett DM (2007a) Nutritional comparison of fresh, frozen, and canned fruits and vegetables – II. Vitamin A and carotenoids, vitamin E, minerals and fiber. J Sci Food Agric 87(7):1185–1196. http://ucce.ucdavis.edu/files/datastore/1234-1778.pdf
Rickman JC, Barrett DM, Bruhn CM (2007b) Nutritional comparison of fresh, frozen and canned fruits and vegetables. Part 1. Vitamins C and B and phenolic compounds. J Sci Food Agric 87(6):930–944. https://www.cancentral.com/sites/cancentral.com/files/public-documents/UCDavis_Part931.2007.pdf
Rodgers S (2016) Minimally processed functional foods: technological and operational pathways. J Food Sci 81(10):R2309–R2319. https://doi.org/10.1111/1750-3841.13422
Roser M, Richie H (2019) Food per Person. Retrieved from https://ourworldindata.org/food-per-person
Salehi M, Moradi-Lakeh M, Salehi MH, Nojomi M, Kolahdooz F (2013) Meat, fish, and esophageal cancer risk: a systematic review and dose-response meta-analysis. Nutr Rev 71(5):257–267
Sanchez-Moreno C, de Ancos B, Plaza L, Elez-Martinez P, Cano MP (2009) Nutritional approaches and health-related properties of plant foods processed by high pressure and pulsed electric fields. Crit Rev Food Sci Nutr 49(6):552–576
Sargeant JM, Rajic A, Read S, Ohlsson A (2006) The process of systematic review and its application in agri-food public-health. Prev Vet Med 75(3–4):141–151. https://doi.org/10.1016/j.prevetmed.2006.03.002
Schmidhuber J, Shetty P (2005) The nutrition transition to 2030. Why developing countries are likely to bear the major burden. Acta Agriculturae Scandinavica C Food Econ 2(3–4):150–166
Shi J, Le Maguer M (2000) Lycopene in tomatoes: chemical and physical properties affected by food processing. Crit Rev Food Sci Nutr 40(1):1–42. https://doi.org/10.1080/10408690091189275
Sibbel A (2012) Public nutrition and the role of the food industry. Br Food J 114(6):784–797. https://doi.org/10.1108/00070701211234327
Siddiqui MW, Rahman MS (2014) Minimally processed foods: technologies for safety, quality, and convenience. Springer International Publishing 306 pages
Smelkova L (2015) Food education in America. Retrieved from https://d3n8a8pro7vhmx.cloudfront.net/foodday/pages/3852/attachments/original/1450802226/Food_education_report.pdf
Smith J, Charter E (2011) Functional food product development. Wiley, New York, p 528
Smith TG, Tasnádi A (2014) The economics of information, deep capture, and the obesity debate. Am J Agric Econ 96(2):533–541
Smith TG, Chouinard HH, Wandschneider PR (2011) Waiting for the invisible hand: novel products and the role of information in the modern market for food. Food Policy 36(2):239–249. https://www.otago.ac.nz/economics/otago087211.pdf
Smith LP, Ng SW, Popkin BM (2013) Trends in US home food preparation and consumption: analysis of national nutrition surveys and time use studies from 1965–1966 to 2007–2008. Nutr J 12:45. https://doi.org/10.1186/1475-2891-12-45
Snowdon W, Raj A, Reeve E, Guerrero R, Fesaitu J, Cateine K, Guignet C (2014) Processed foods available in the Pacific Islands. Global Health Globalization and Health 2013; 9:53; https://doi.org/10.1186/1744-8603-9-53. Retrieved from http://Globalizationandhealth.biomedcentral.com/articles/10.1186/1744-8603-9-53
Sobal J, Khan LK, Bisogni C (1998) A conceptual model of the food and nutrition system. Soc Sci Med 47(7):853–863
Stang J, Bayerl CT (2003) Position of the American dietetic association: child and adolescent food and nutrition programs. J Am Diet Assoc 103(7):887–893. https://doi.org/10.1053/jada.2003.50188
Stitt S (1996) An international perspective on food and cooking skills in education. Br Food J 98(10):27–34
Sumner J (2013) Food literacy and adult education: learning to read the world by eating. Can J Study Adult Educ 25(2):79–92. https://cjsae.library.dal.ca/index.php/cjsae/article/view/1410
Taillie LS (2018) Who's cooking? Trends in US home food preparation by gender, education, and race/ethnicity from 2003 to 2016. Nutr J 17(1):41. https://doi.org/10.1186/s12937-018-0347-9

Tiwari A, Aggarwal A, Tang W, Drewnowski A (2017) Cooking at home: a strategy to comply with U.S. dietary guidelines at no extra cost. Am J Prev Med 52(5):616–624. https://doi.org/10.1016/j.amepre.2017.01.017

Thompson C, Adams J, Vidgen HA (2021) Are we closer to international consensus on the term 'food literacy'? A systematic scoping review of its use in the academic literature (1998–2019). Nutrients 13(6):2006. https://doi.org/2010.3390/nu13062006

Toops D (2017) How did the food industry get (from there) to here? In: Beckley JH, Herzog LJ, Foley MM (eds) Accelerating new food product design and development, 2nd edn. Wiley/Institute of Food Technologists, Newark, p 387

Trowell HC, Burkitt DP (1981) Western diseases, their emergence and prevention. Harvard University Press, Cambridge, MA, p 456

Truman E, Lane D, Elliott C (2017) Defining food literacy: a scoping review. Appetite 116:365–371. https://doi.org/10.1016/j.appet.2017.05.007

Truswell AS (1998) Practical and realistic approaches to healthier diet modifications. Am J Clin Nutr 67(3 Suppl):583S–590S

Tuorila H, Monteleone E (2009) Sensory food science in the changing society: opportunities, needs, and challenges. Trends Food Sci Technol 20(2):54–62. https://doi.org/10.1016/j.tifs.2008.10.007

U.S. Department of Health and Human Services and U.S. Department of Agriculture (2015) 2015–2020 Dietary Guidelines for Americans. 8th Edition. December 2015. Retrieved from http://health.gov/dietaryguidelines/2015/guidelines/

United States Department of Agriculture (2012) Global food markets. Retrieved from http://www.ers.usda.gov/topics/international-markets-trade/Global-food-markets/Global-food-industry.aspx

Velardo S (2015) The nuances of health literacy, nutrition literacy, and food literacy. J Nutr Educ Behav 47(4):385–389.e381. https://doi.org/10.1016/j.jneb.2015.04.328

Vernooij RWM, Zeraatkar D, Han MA, El Dib R, Zworth M, Milio K, Johnston BC (2019) Patterns of red and processed meat consumption and risk for cardiometabolic and cancer outcomes: a systematic review and meta-analysis of cohort studies. Ann Intern Med. https://doi.org/10.7326/m19-1583

Vidgen HA, Gallegos D (2014) Defining food literacy and its components. Appetite 76:50–59. https://doi.org/10.1016/j.appet.2014.01.010

Wang D, Stewart D, Chang C, Shi Y (2015) Effect of a school-based nutrition education program on adolescents' nutrition-related knowledge, attitudes and behaviour in rural areas of China. Environ Health Prev Med 20(4):271–278. https://doi.org/10.1007/s12199-015-0456-4

Weaver CM, Dwyer J, Fulgoni VL 3rd, King JC, Leveille GA, MacDonald RS, Ordovas J, Schnakenberg D (2014) Processed foods: contributions to nutrition. Am J Clin Nutr 99(6):1525–1542

Wellman NS, Scarbrough FE, Ziegler RG, Lyle B (1999) Do we facilitate the scientific process and the development of dietary guidance when findings from single studies are publicized? An American Society for Nutritional Sciences controversy session report. Am J Clin Nutr 70(5):802–805. https://doi.org/10.1093/ajcn/70.5.802

Wolfram T (2019) Processed foods: what's ok and what to avoid. Retrieved from https://www.eatright.org/food/nutrition/nutrition-facts-and-food-labels/processed-foods-whats-ok-and-what-to-avoid

Yach D, Feldman ZA, Bradley DG, Khan M (2010a) Can the food industry help tackle the growing global burden of undernutrition? Am J Public Health. Retrieved from http://www.ncbi.nlm.nih.gov/pubmed/20395578

Yach D, Khan M, Bradley D, Hargrove R, Kehoe S, Mensah G (2010b) The role and challenges of the food industry in addressing chronic disease. Glob Health 6:10. https://doi.org/10.1186/1744-8603-6-10

Yngve A, Hambraeus L, Lissner L, Serra Majem L, Vaz de Almeida MD, Berg C, Kennedy N (2006) The women's health initiative. What is on trial: nutrition and chronic disease? Or misinterpreted science, media havoc and the sound of silence from peers? Public Health Nutr 9(2):269–272

Yudkin J (1964) Patterns and trends in carbohydrate consumption and their relation to disease. Proc Nutr Soc 23(2):149–162

Yuen EYN, Thomson M, Gardiner H (2018) Measuring nutrition and food literacy in adults: a systematic review and appraisal of existing measurement tools. Health Lit Res Pract 2(3):e134–e160. https://doi.org/10.3928/24748307-20180625-01

Zeraatkar D, Han MA, Guyatt GH, Vernooij RWM, El Dib R, Cheung K et al (2019a) Red and processed meat consumption and risk for all-cause mortality and cardiometabolic outcomes: a systematic review and meta-analysis of cohort studies. Ann Intern Med. https://doi.org/10.7326/m19-0655

Zeraatkar D, Johnston BC, Bartoszko J, Cheung K, Bala MM, Valli C, El Dib R (2019b) Effect of lower versus higher red meat intake on cardiometabolic and cancer outcomes: a systematic review of randomized trials. Ann Intern Med. https://doi.org/10.7326/m19-0622

Part II

Food Safety and Sanitation

Food Microbes, Quality and Fermentation 7

1 Introduction

1.1 Scope of Food Microbiology

Foods may be contaminated by microorganisms that cannot be seen with the naked eye and require a microscope to view. The study of micro-organism and their ecology (i.e. their interactions with the environment and each other) is called microbiology. Food microbiology is the application of microbiology principles to the entire food supply chain, starting, with agricultural production, manufacturing, foodservice, packaging, distribution, storage and retail (Adams and Moss 2008; Borgstrom 1968; Jay et al. 2005; Montville and Matthews 2008; Ray and Bhunia 2013).

Microorganisms that are associated with foods are no different from those found within the wider environment, e.g., soil, water, plants, animals and whatever else is employed as food and/or is used in the preparation, processing or packaging of food. Microorganisms include bacteria, yeasts, molds, algae, and protozoans. Some non-living microscopic biological entities such as viruses and prions, which cause disease, also come under the scope of food microbiology and veterinary science.

Food microbiology is concerned with three broad areas; (a) foodborne illness and food microbial safety, (b) microbe effects on spoilage and food quality, and (c) beneficial uses of micro-organisms in food manufacturing. Arguably, food microbiology has one of the best-integrated career paths in food science (McGrath 2008). The topic is covered in many classic texts which are recommended as further of further reading (Adams and Moss 2008; Jay et al. 2005; Montville and Matthews 2008; Ray and Bhunia 2013).

1.2 Sources of Microbes in Foods

The natural environment is a source of most of the microorganisms associated with foods. Microorganisms are found wherever there is liquid water. Indeed this is the rational for extraterrestrial searches for water, which is seen as pre-requisite for ET life. Besides, even where there is no liquid water, microbes may be present in a dormant state or as spores, which are reactivated by moisture. Transient extreme conditions employed in processing are no guarantee that food is completely safe from food-borne microbes. Many earthbound microorganisms live and survive under extremes of temperature, pressure, pH, and salt concentration (Beales 2004; Berry and Foegeding 1997). Environment sources of micro-organisms are outlined in the following section.

Air is a hostile environment for microbes because of the high concentration of damaging oxygen free radicals (oxy-radicals) and exposure to solar radiation within the visible, infrared and ultraviolet spectrum. The concentration

R. Owusu-Apenten, E. R. Vieira, *Elementary Food Science*, Food Science Text Series,
https://doi.org/10.1007/978-3-030-65433-7_7

of available nutrients in air is low be considered low. The atmosphere can have a low equilibrium relative humidity, and so airborne microorganisms must contend with varying degrees of desiccation. Dust, water droplets, aerosols, suspended oils and other detritus can be a source of microorganism. Gram-negative bacteria (e.g. E coli) are thought to perish in the atmosphere. Gram-positive bacteria survive. Some molds and fungi produce spores that survive air exposure.

Soil is probably the largest repository of bacteria. Some soil bacterial are industrially useful for producing pharmaceuticals. Food spoilage microorganisms from the soil include spore-forming clostridium and bacillus species.

Water is a potentially rich source of microorganisms within the food system. Water is used for irrigation of growing plants and for drinking by animals. Water is used in most forms of food processing, for washing fresh foods, and general cleaning and sanitation. Water is a popular medium for heat transfer during processes such as blanching and cooking. Water may be served with foods and or used for ice making. It is well appreciated that foods from aquatic (fresh water, marine) environments may be contaminated by psychrotrophic (cold-loving) bacterial. Shellfish – and other filter feeders may concentrate bacterial toxins from photosynthetic cyanobacteria (shell fish toxin). Coastal waters may be contaminated by (enteric) bacteria from farm and municipal sewage disposed of at sea. Lobster, shellfish/ muscle from coastal waters pose some risk of food safety. Detection of E-coli in sea/river water is an indicator of sewage contamination.

Living plants as well as animals are another source of microorganisms. Plant foods (fruit and vegetables) are major sources of bacterial contamination from e.g. fertilizer applications. Animal foods (e.g. meat) can be contaminated, e.g. poultry can be a source of salmonella. Abattoir contamination of meat is a concern.

As well as the main sources considered above, microorganism may also originate from food ingredients and from food contact surfaces, equipment and general packaging (Fig. 7.1).

1.3 Structure and Shape of Microbes

Bacteria, molds, and yeasts have rigid cell walls enclosing the cell materials and the cytoplasm, but they differ greatly in their other properties. The microbes most important in the spoilage and controlled changes in food have various shapes.

Many are rod-shaped and occur as either single cells, two adjoining cells, or short chains of cells. Other bacteria are spherical in shape (cocci). Some cocci exist mainly as a grapelike cluster of cells, the staphylococci, others as a cube like cluster of spherical cells, the sarcina. Some cocci occur in groups of two, the diplococci, or in chains, the streptococci. Some disease-causing bacteria are included in the cocci group.

Some bacteria are curved or comma-shaped rods, the vibrio, and others are long, slender, corkscrew-shaped cells, the spirochetes. Other bacterial groups either are not important to changes in food or are disease-causing types.

Molds are multicellular (bacteria and yeasts are single cells) and are made up of branched threads (hyphae), consisting of chains of cylindrical cells united end to end. Some of the hyphae serve to secure nutrients from the material in which they are growing while others produce the spores that provide for reproduction, or for new mold growth. Some molds produce a mycelium (mass of hyphae) that has cross walls (septa) while others do not have cross walls in the mycelium (Fig. 7.2).

Yeasts form single cells or chains of cells that may be spherical or of various shapes between the spherical and the cylindrical. One common characteristic of bacteria, yeasts, and molds is that they can exist either as active, vegetative cells or as spores. As vegetative cells, they metabolize, reproduce, cause food spoilage, and sometimes cause disease. They generally have a considerable effect on the environment. Their activity, however, is dependent on several environmental conditions, which are discussed later on. When conditions are unfavorable (such as high temperatures), the vegetative cells begin to die before they can reproduce, and soon all the

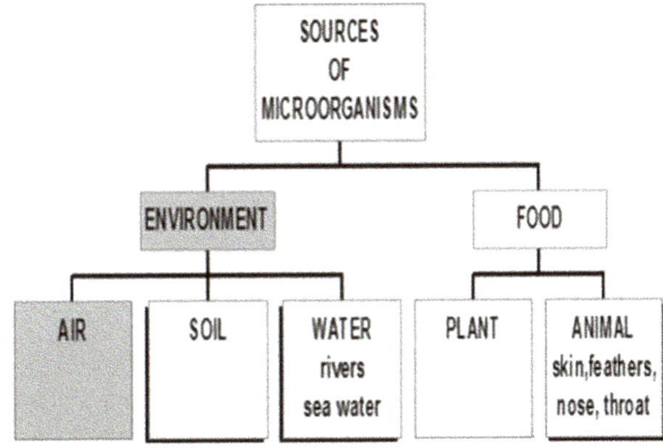

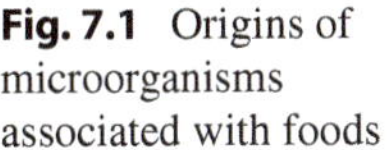
Fig. 7.1 Origins of microorganisms associated with foods

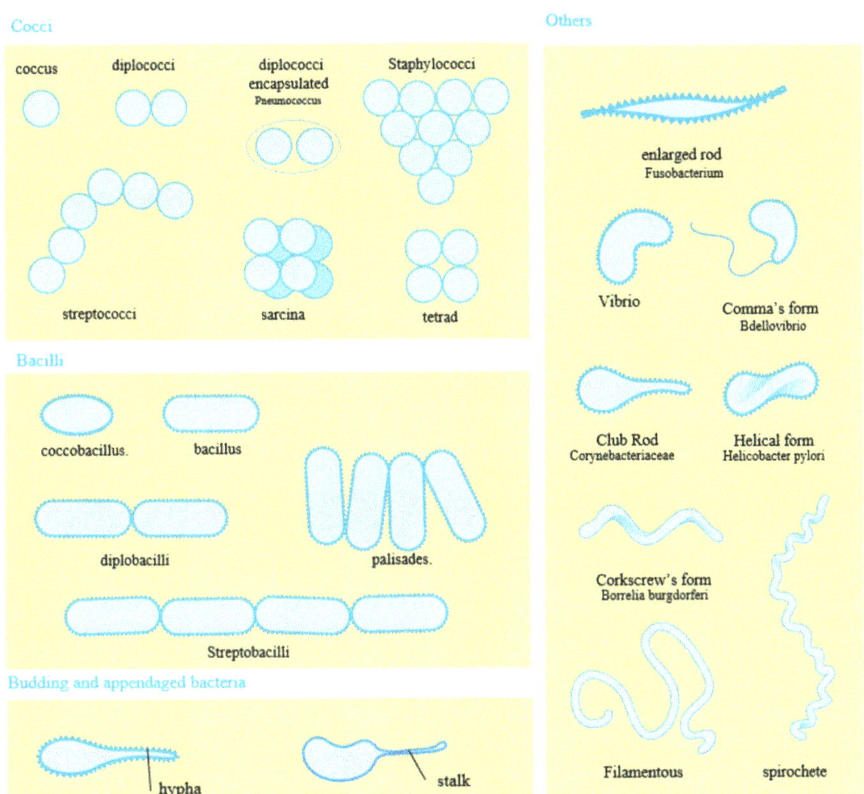

Fig. 7.2 Shapes of various microorganisms. (Source: Wikicommons)

vegetative cells die. Many microorganisms, however, also exist in the spore state, sometimes because it is a necessary step in reproduction, and sometimes because it is necessary for the microbe to survive unfavorable conditions. In other cases, certain microorganisms, such as some bacteria, exist in both the vegetative and the spore states and when conditions become unfavorable, the spores survive. In order for spores to survive the conditions that are destructive to their vegetative counterparts, they must have special properties. In general, spores are concentrated forms of their vegetative counterparts; thus, spores contain less water, and are more dense and smaller than the vegetative cells. The spore wall is thicker and harder than the cell wall.

1.4 Size of Microbes

Bacteria are comparatively small. The single cell of many bacteria is about 1140 millionth of an inch (1micron) in diameter but some bacteria may be 10 or 50 times larger than this. Because of their small size, most single bacteria are invisible to the naked eye, and when evidence of their growth can be seen, such as when slime forms on meat, the organisms will have multiplied to very large numbers, billions of cells to each square inch (6.5 cm^2) of the meat surface.[1,2]

The single cell of the mold, although not visible without magnification, is much larger than the single bacterial cell, and any significant growth of molds on foods can be seen. The visible mold comprises mycelia with or without spore heads. The spore head carries the spore that will give rise to more growth.

Yeasts vary in size from that of the more common spherical bacteria to a form several times larger. Like bacteria, when grown in solution, the individual cells cannot be seen, but eventually, cell numbers will accumulate to the point where the solution becomes turbid. There are instances in which groups of cells will form visible clumps (colonies) on foods or on the surfaces of solutions.

1.5 Motility in Microbes

Many types of bacteria are motile, that is, they are able to move about in the solutions in which they live by means of the movement of flagella, which are thin, protoplasmic, whip like projections of the cell. Some bacteria, such as the cocci, have no flagella and are not motile. Molds are not motile, but they spread in vine-like fashion by sending hyphae outward. Yeasts are not motile; consequently, they exist in relatively compact clusters, unless distributed by agitation or by some other external dispersing force.

[1] Bacteria form slimy extracellular polysaccharides, which are the major constituents of biofilms, along with some protein, lipids and DNA. Biofilm formation protects communities of bacterial embedded in it.

[2] Wikipedia lists, *Thiomargarita namibiensis (diameter 0.1–0.7 mm or 100–700 um) as the largest bacterial known and about visible to the naked eye. Cf.* https://en.wikipedia.org/wiki/Thiomargarita_namibiensis

2 Growth of Microbes

2.1 Bacteria Numbers and Its Significance

The number of disease causing bacteria present in any food determines whether the food is fit for consumption or not. The tendency for bacterial numbers to increase with time is termed "growth". Unlike higher organisms, bacterial growth is predominantly a matter of increasing numbers whilst individual cell size change by no more than two fold.

Bacteria reproduce by fission, a transverse division across the cell to form two new cells. There are some instances where sexual reproduction occurs in bacteria, one cell uniting with another before division occurs, but this is not the usual circumstance and need not be discussed further here.

Molds usually reproduce by means of spores, each organism producing spores in great numbers. Mold spores are of two types: sexual spores produced by the fusion of two sex cells, or asexual spores that arise from the fertile hyphae. Most molds produce asexual spores. Asexual spores are formed on the sides, end of hyphae threads (a conidium), or produced in a special spore case called a sporangium. Conidia contain many spores, and when a spore meets a suitable growth medium under favorable conditions, it grows to become the adult mold.

Yeasts reproduce by budding, by fission, or by spore formation. In budding, a projection is formed on the original cell that eventually breaks off to form a new cell. As is the case with bacteria, some yeast may undergo fission, the process in which one cell divides into two cells. Sometimes, yeasts reproduce by spore formations that may be sexual or asexual.

Bacteria grow by increasing their numbers. The growth pattern conforms to basic rules of exponential mathematics. The rate of increase in bacteria numbers or population is proportional to the *initial* bacterial load present in a food. The larger the initial load, population or number of bacteria (N_0), the faster the increase in observed bacterial numbers with time (N_t). Secondly, the *rate* of increase in bacteria numbers is dependent on a parameter, called "specific growth rate" (μ) which is the number of cell divisions per unit time. Consider the specific growth rate as being reflective of individual microbes and the quality of the environment, e.g. pH, temperature, nutrient availability etc.

Microorganism growth kinetics are briefly summarized by the familiar rate equation; Rate = μN_0. For the mathematically minded, the specific growth rate for a microbe is inversely related to the "doubling time" (t_D) which is the period required for cell fission; $t_D = 0.693/\ \mu$. Microorganisms with a high specific growth rate (smaller doubling time) grow more quickly, compared with microbes with a low specific growth rate (longer doubling time). According to the preceding principles, there are two fundamental ways to control microbe numbers in a food sample; first, we can reduce the initial numbers of bacteria in a food by good practice and by processing. Secondly, we can depress the doubling time for any remaining bacteria by secondary interventions, e.g. a low storage temperature.

Under favorable conditions, bacterial species multiply at an exceedingly rapid rate, with a doubling time requiring of about 20 min. From this standpoint, the *initial* number of cells with which a food is contaminated is very important, for if the number is high, only a few hours may be required for bacterial multiplication to reach a level that will cause food spoilage.

Under favorable conditions, reproduction will continue until there are billions of cells per ounce (29.6 ml) of solution or per square inch (6.5 cm^2) of surface.[3] In general, microorganisms reproduce exponentially. Therefore, one bacterial cell will become two, then four, eight, sixteen, thirty-two and so forth. Starting with 100 cells and allowing 5 generations of cell division will soon generate 100^5 – or 10,000,000,000 (10 trillion) cells.

[3]It should be noted that microorganisms must live in liquid material, and when they grow on food surfaces, they are actually living in the liquid available on or in the food.

2.2 Monitoring Growth of Bacteria

To recap, increases in the number of bacterial is referred to as growth. The degree of microbial population growth occurring within a fixed timed interval is called the *rate of microbial growth.* The rate of growth depends on the initial numbers of cells present. After adjusting for the initial number of cells, the rate of growth is now termed, the specific growth rate. Microbiologists are interested to measure how the specific growth is affected by the surrounding conditions such as temperature.

The presence of large number of microorganisms in a liquid food leads to turbidity and a decrease of light transmittance through a sample. Sample transmittance can be measured using a colorimeter set a light wavelength of 420–600 nm. Unfortunately, turbidity measurements are only possible for transparent foods, and is not very useful for milk, soups, food suspensions or solid foods. Also we cannot use transmittance measurements to determine bacteria numbers quantitatively, for surfaces and utensils.

2.2.1 Culture Methods

Small samples of solid foods and swabs from surfaces can be cultured in semi-transparent media and the turbidity measured. Thereafter, the rate of growth observed can be extrapolated backwards to estimate the initial numbers of bacteria sampled from the contaminated surface. Another well-established method for the detection of microorganism is by culturing with solid agar media and counting the colony spots formed by clumps of bacterial cells.

The culture plate method, used in many laboratories for monitoring bacterial numbers, is criticized as being slow and technically demanding. Alternative and quicker methods for monitoring microbial contamination measures the presence of adenosine triphosphate (ATP), DNA or antigens produced by the microorganism.

2.2.2 ATP Bioluminescence

The ATP bioluminescence method developed in the 1980s is now widely applied to monitor microbial contamination in the food industry. The main principle behind this method is that all living cells produce ATP. Therefore, the amount of ATP present in any environment (e.g. kitchen tabletops, utensils etc.) provides an indication of the number of living cell present and the degree of surface contamination. It is assumed that each bacteria cell contain a constant amount of ATP.

The ATP method is applied in the food and brewing industry (Hysert et al. 1976), for hygiene monitoring inside of milk and dairy plants (Kyriakides 1992) and for meat and minced meat analysis (Labots and Stekelenburg 1985). USDA researchers applied this technique to monitor meat carcass hygiene (Cutter et al. 1996; Siragusa et al. 1995). Investigators found that, ATP measurements were highly correlated with culture plate results, and could be used for monitoring potential fecal contamination on beef, chicken or pork carcasses.

The advantages of ATP assay for bacteria contamination include greater sensitivity and speed. For example, current ATP monitoring can be completed in a manner of minutes compared with a time of 36–48 h required for conventional culture methods. The advent of portable and lightweight ATP bioluminescence instruments has led to widespread uses of this method for hygiene and cleaning monitoring. One disadvantage worth noting is that ATP is found in all living tissues. Therefore, ATP luminescence monitoring is not specific for pathogenic microorganism.

2.2.3 DNA Polymerase Chain Reaction (PCR)

DNA polymerase chain reaction (PCR) is best suited for situations where there is a need to identify the DNA from a particular bacterial species. By analogy with forensic investigations, DNA is recovered from the food crime scene or suspect food. The DNA sample from an unknown pathogen is now is mixed with one only of 50 different synthetic DNA piece or primers each of which is "derived" from different pathogens. To this mixture is added the heat-stable enzyme (DNA polymerase), buffer, four DNA-bases, and the whole mixture (about 0.2 millimeters) is placed in a PCR thermal cycler. Here, the sample is subjected to a cyclical temperature regime which melts DNA leading to strand separation, followed by matching between the primer and crime-scene DNA, extension and amplification of the new DNA complex by about 1 billion-fold within about 3–4 h. The newly amplified DNA is now measured by agarose electrophoresis or using fluorescence dyes. The key advantages of PCR used in the field of investigative microbiology seems to be its speed, high sensitivity and precision (Ceuppens et al. 2014; Hanna et al. 2005; Ivey and Phister, 2011; Jany and Barbier 2008; Maurer 2011; Postollec et al. 2011).

The presence of pathogens can be confirmed within 4–5 h as the method is culture-independent, does not require culturing. PCR finds great use in tracking and identifying outbreaks. However, the PCR method has some limitations, including a possibility of false positive and negative results. The public health implications of PCR results need also to take account of the stable nature of DNA, which means that it may persist in the environment in the absence of live pathogen. Interestingly, PCR is particularly suited for hard-to-culture pathogens and identifying viruses. PCR services are commercially available in microbiology laboratories and the trajectory of this method is predicted to be upwards in coming years (Ceuppens et al. 2014; Hanna et al. 2005; Ivey and Phister 2011; Jany and Barbier 2008; Maurer 2011; Postollec et al. 2011).

3 Factors Influencing Microbial Growth

3.1 Extrinsic and Intrinsic Parameters

The growth of microorganisms is affected by a host of factors, which are classed as intrinsic to foods or else extrinsic and reliant on the environment surrounding the food (Table 7.1). Understanding growth factors is important in deciding whether a given food product as at risk

Table 7.1 Intrinsic and extrinsic factors for microbial growth

Intrinsic factors of food
Water availability (Aw)
pH and acidity
Redox potential (Eh)
Nutrient content
Antimicrobial substances
Physical properties/Biological structures
Extrinsic to foods, environmental factors
Temperature of storage
Relative humidity (RH) environment
Gases in environment
Other microbial activities
Microbe (implicit) factors

Adapted from U. S. Food and Drug Administration (2001) and IFT/FDA (2001)

Table 7.2 Water activity of selected American foods

Animal foods	A_w	Plant foods	A_w
Meat, Poultry, Fish	0.99–1.00	Fruits, Vegetables	0.97–1.00
Natural Cheeses	0.95–1.00	Bread	~0.96
Pudding	0.97–0.99	Bread, White	0.94–0.97
Eggs	0.97	Bread, Crust	0.30
Cured Meat	0.87–0.95	Baked Cake	0.90–0.94
Condensed Milk	0.83	Maple Syrup	0.85
Parmesan Cheese	0.68–0.76	Jam	0.75–0.80
Honey	0.75	Jellies	0.82–0.94
Dried Whole Egg	0.40	Uncooked Rice	0.80–0.87
Dried Whole Milk	0.20	Fruit Juice	0.79–0.84
		Fruit Cake	0.73–0.83
		Cake Icing	0.76–0.84
		Flour	0.67–0.87
		Dried Fruit	0.55–0.80
		Cereal	0.10–0.20
		Sugar	0.19
		Crackers	0.10

Adapted from U. S. Food and Drug Administration (2001) and IFT/FDA (2001)

of microbial contamination and which products should be subjected to so called temperature-time control for safety (TCS). We will see from these discussions that products such as flour may be stored at room temperature without TCS. By contrast perishable products such as, fresh meat and poultry (but not fresh fruits) should be subjected TCS to avoid microbial growth (cf. Sect. 3.9).

In the following sections, we discuss some of the variables known to affect bacterial growth in foods. This material is guided by the 2001 report on factors that affect bacterial growth provided to the FDA by the Institute of Food technologists (IFT) expert panel (IFT/FDA 2001).

3.2 Moisture

Microorganisms require the presence of moisture or available water in order to grow. The term water activity (A_w), expresses the availability of free water in foods. The A_w scale ranges from 0–1.0 and can be considered also to measure the energy of water in foods with 0 implying no free water and 1.0 referring to water which is fully able to perform the functions of this solvent (Table 7.1).

A_w can be measured from the moisture present within the vapor phase above a food, literally the vapor pressure (P) according to the familiar expression, $A_w = P_{(Food)}/P_{(solvent)}$, where $P_{(Food)}$ is the water vapor pressure in the atmosphere above a food and $P_{(Solvent)}$ is the water vapor pressure above a pure solvent, water. Some ordinary fresh-type foods have A_W of about 0.99 to 0.96 at ambient temperatures. Low water activity, which limits the growth of microorganisms in foods, may be brought about by dehydration achieved either with the addition of salt or sugar, or via the removal of water by drying. At low values of A_w, some water is tied up by binding to chemical compounds added to foods or to some food component, such as protein (Table 7.2).

Equilibrium relative humidity (ERH) describes the amount of moisture found in the atmosphere or air, which is in contact with food. The ERH is reached when the rate at which food loses water to

Table 7.3 Minimum water activity for microorganism growth

Organism	Minimum
Campylobacter spp.	0.98
Clostridium botulinum type E*	0.97
Shigella spp.	0.97
Yersinia enterocolitica	0.97
Vibrio vulnificus	0.96
Enterohemorrhagic Escherichia coli	0.95
Salmonella spp.	0.94
Vibrio parahaemolyticus	0.94
Bacillus cereus	0.93
Clostridium botulinum types A & B**	0.93
Clostridium perfringens	0.943
Listeria monocytogenes	0.92
Staphylococcus aureus growth	0.83
Staphylococcus aureus toxin	0.88

Adapted from U. S. Food and Drug Administration (2001) and IFT/FDA (2001)

Table 7.4 Minimum and maximum pH values for some foodborne microbes

Microorganism	pH (min)	pH (max)
Staphylococcus aureus growth	4.0	10.0
Yersinia enterocolitica	4.2	9.6
Salmonella spp.	4.2[1]	9.5
Listeria monocytogenes	4.39	9.4
Enterohemorrhagic *Escherichia coli*	4.4	9.0
Staphylococcus aureus toxin	4.5	9.6
Clostridium botulinum toxin	4.6	8.5
Clostridium botulinum growth	4.6	8.5
Vibrio parahaemolyticus	4.8	11.0
Bacillus cereus	4.9	8.8
Campylobacter spp.	4.9	9.0
Shigella spp.	4.9	9.3
Vibrio vulnificus	5.0	10.2
Clostridium perfringens	5.5	8.0

Adapted from U. S. Food and Drug Administration (2001) and IFT/FDA (2001)

its environment is equal to the rate at which it absorbs water from the environment. ERH is familiar as a meteorological concept and it is well to note also that ERH = A_w *100.

In general, microorganism cease to grow as A_w decreases. The molds and yeasts can grow in low moisture environments or until Aw falls below 0.65. Interestingly most pathogens are inhibited at a moderate Aw of about 0.93 with exception of S. aureus where growth ceases at A_w lower than 0.83 (Table 7.3).

3.3 The pH and Titrable Acidity of Food

The rate of growth of microbes is greatly affected by acidity.[4] Thus, microorganisms have an optimum pH at which they grow most rapidly and a pH range above or below which they will not grow at all (Table 7.4). Generally, molds and yeasts grow best at acidic (low) pH, as do some bacteria. At least part of the reason why fruits are usually spoiled by molds or yeasts and flesh-type foods (meats, fish, poultry, and eggs) are usually spoiled by bacterial growth is that fruits have a low (acidic) pH and flesh-type foods have a near-neutral pH. Many species of bacteria grow best at neutral or slightly alkaline pH. Some bacteria will grow at a pH as low as 4, whereas others can grow at a pH as high as 11. Interestingly, extremes of pH may be useful when used in combination with other interventions in a so-called hurdle technology.

In industry, the pH is not considered a wholly accurate measure of total acidity of a food because hydrogen ions in a non-dissociated state are not measured by a typical glass pH electrode. Typically, organic acids found in foods will exhibit weak acidity near pH 4 because ~50% of their hydrogen ions are in the non-dissociated state. Titrable acidity measured by titration with standard alkali determines the total available hydrogen ions in a food. Titrable acidity is considered a better measure of "true" acidity and microbial stability of foods compared with pH measurements.

3.4 Redox Potential (Eh)

Oxidation is a process where oxygen is added to other compounds, or hydrogen is removed.

[4]Acidity is routinely measured as pH (negative logarithm of hydrogen ion concentration. pH = −log [H+]. The pH scale ranges from 1 to 14. A pH value of 1–7 indicates acidic foods whereas pH 7–14 is alkaline. Neutrality is indicated by pH 7.0.

Oxidation is achievable also by the removal of electrons so that the "oxidation number" for an oxidized molecule increases correspondingly. The opposite of oxidation is "reduction" which is the addition of electrons or hydrogen to a compound. Normally, oxidation and reduction are two counter reactions, so as one molecule (Compound (I)) is oxidized, the electron released can be passed to a second compound (Compound II) which becomes reduced. Reduction and oxidation (Redox) reaction usually involve pairs of compounds or else one compound reacting with a metallic electrode.

Compound (I) → Compound $(I)_{Ox}$ + e

Compound (II) + e → Compound $(II)_{Red}$

The standard reduction-oxidation (redox) potential provides a measure of the tendency for electron to be transfer for any compound (E_h). By general agreement, if E_h is positive, then a compound is oxidizing. Reducing compounds have a negative E_h value. Hydrogen atom is assumed to hava a Eh value of zero. The environmental as whole possess a net reduction-oxidation potential (E_h) reflecting the balance of compounds present. Practical measurements for redox potential require a high impedance millivolt meter connected by wire to a bare platinum electrode and a standard silver/ silver chloride (Ag/ AgCl) electrode (Martin et al. 2013).[5]

A positive E_h shows that an environment is oxidizing (removes electrons) and has a tendency to accept electrons, or else oxygen rich and therefore aerobic. By contrast, a negative E_h value signifies a reducing environment with a tendency to release electrons, and/ hydrogen or which is anaerobic. Redox measurements are useful for monitoring water quality, soil sediments and general environmental monitoring. Values for E_h are affected by concentration of dissolved oxygen, pH and other components in the environment.

In general, aerobic microorganisms grow at E_h values between, +300 to +500 mV. The facultative anaerobes grow at E_h values between, −100 to +300 mV. Finally, the anaerobic microorganisms grow at Eh values of −250 to −100 mV. Table 7.5 lists redox potentials for some foods. The values for E_h should be taken as approximate, as they are affected by variety other intrinsic factors, e.g. pH of the food, temperature, exposure to air and of course the activity of microbes themselves. As noted above, bacterial are adapted to thrive at different ranges of E_h values. Moreover, the activity of certain microbes can lead to changes in the E_h value of foods (IFT/FDA 2001; Reichart et al. 2007).

Values for E_h can also affect the resistance of bacteria to heat inactivation (George et al. 1998; Riondet et al. 2000). The issues is important because, the redox potential of food changes with

Table 7.5 Approximate redox potentials for some foods (¥)

Food	+/− air	E_h (mV)	E_h(mV)
Milk	+	300	340
Cheese, Cheddar	+	−100	300
Cheese, Dutch	+	−310	−20
Butter serum	−	290	350
Egg (infertile after 14 d)	+	500	
Meats			
Liver, raw minced	−	−200	
Meat, Raw, post-rigor	−	−150	−60
Meat, Raw, minced	+	225	
Meat, Minced, cooked	+	300	
Cereals			
Wheat (whole grain)	−	−360	−320
Wheat (germ)	−	−470	
Barley (ground)	+	225	
Potato tuber	−	−150	
Plant juices			
Grape	−	409	
Lemon	−	383	
Pear	−	436	
Spinach	−	74	
Canned foods			
Canned foods, "Neutral"	−	−550	−130
Canned foods, "Acid"	−	−550	−440

Adapted from (U. S. Food and Drug Administration (2001) and IFT/FDA (2001)

¥Range of typical values at the natural pH of foods, in millivolts (mV) at platinum electrode vs. silver chloride reference

[5]This is the same devise used by electricians to check for live circuits, or by in laboratories to measure pH values. A new generation of digital millivolt meters are available that allow redox measurements for monitoring wastewater quality.

processing: souse vide products that are heated and vacuum packed, modified atmosphere packaging using hydrogen gas, addition of oxygen scrubbers, or addition of antioxidants increases the value for E_h (George et al. 1998; Riondet et al. 2000) (Fig. 7.3).

For example, E.coli, salmonella and listeria cultured at pH 7.0 showed increased heat resistance measured as the time for 6-decimal reductions in number, if redox potential was alterned from +380 mV to −348 mV by sparging with hydrogen (George et al. 1998). Interestingly, also when listeria were grown over wide range of starting E_h values, there was a decrease in redox potential towards more reducing conditions. Also the lag time for cell division increased with falling redox potential (Ignatova et al. 2010). The preceding results should be considered tentative and may be affected by the method of ajusting E_h values.

Investigations involving E_h values and food hygiene look likely to increase in the coming years, with the advent of more reproducible measurement systems. Some investigations suggest that hydrogen and other gases (Duried et al. 2003) or electrochemical reductions (Bolduc et al. 2006) might be used to adjust food redox potential to values where pathogens could be inhibited. Monitoring E_h was considered also as the basis for rapid electrochemical detection of surface contamination (Erdosi et al. 2011).

3.5 Nutrient Content

Microorganisms, especially bacteria, vary greatly in nutritional requirements from species to species. In the presence of particular inorganic salts, some bacteria can utilize the nitrogen in air to form proteins and utilize the carbon dioxide in air to obtain energy or to form compounds from which they can then obtain energy. Others can utilize simple inorganic salts, such as nitrates, as a source of nitrogen and relatively simple organic compounds, such as lactates, as a source of energy. Nearly all yeasts can derive all their nitrogen from lysine, an amino acid. Some bacteria require complex organic compounds for growth, including amino acids (primary units of proteins) and vitamins-especially those belonging to the B group-and traces of certain minerals.

In some cases, not only are trace minerals necessary, but their careful control is needed to sustain an optimum growth rate of microorganisms. There is some evidence that demonstrates the ability of at least some microbes to utilize substitute elements for required ones. Sometimes, one trace element may protect microbes from the toxic effects of the presence of other elements; for example, the presence of zinc has been reported to protect yeasts against cadmium poisoning.

Molds and yeasts, like bacteria, may require basic elements (carbon, hydrogen, nitrogen, phosphorus, potassium, sulfur, etc.) as well as vitamins and other organic compounds.

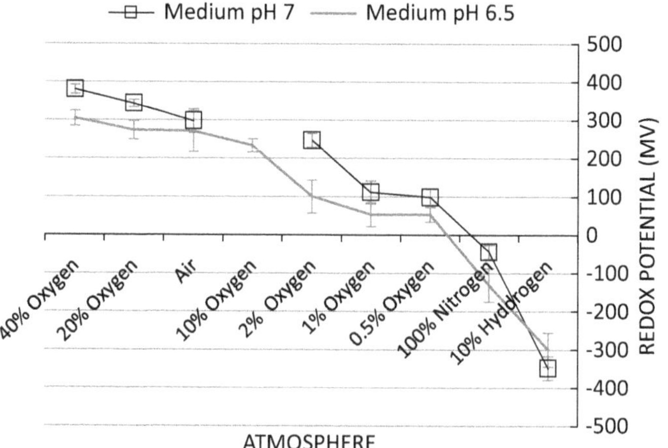

Fig. 7.3 Changes in redox potential for bacteria culture media under different atmospheres. (Drawn from George et al. 1998)

Although sugar is a nutrient important to microbes, some molds and yeasts can grow well in high concentrations that inhibit bacterial growth. In fact, yeasts grow extremely well in the presence of sugar.

3.6 Antimicrobial Agents and Preservatives

Foods may contain substances, which are not amicable for microbial growth. Examples include the enzyme lysozyme found in eggs, which can break down bacterial cell walls. Lacto-peroxidase, an enzyme found in milk will inactivate milk bacteria when supplemented with hydrogen peroxide and thiocyanate. Other bacteria produce natural antibiotics (e.g. the peptide nisin) which inhibit other microbes. High levels of citric and other organic acids present in the vacuoles of some plant foods will prevent microbial growth.

Food preservatives are a class of food additives used for extending shelf life by delaying food deterioration caused by microbes. The range of food additives used as preservatives includes substances with INS numbers 200–290. The classes of recognized preservatives include, organic acids (sorbic acid, benzoic acid, formic acid, lactic acid), and their salts, sulfur dioxide, sulfite, and nitrites (Codex Alimentarius Commission 1989).

3.7 Temperature

The growth rate of bacteria depends on temperature, as they do not grow above or below a specific range of temperatures (IFT/FDA 2001). Bacteria are classified according to the temperatures at which they grow. Psychrophilic bacteria grow fastest at about 68 to 77 °F (20 to 25 °C), but some can grow, although more slowly, at temperatures as low as 45 °F (7.2 °C), while others grow at temperatures as high as 86 °F (30 °C). Some will grow at temperatures as low as 19 °F (−7.2 °C), as long as the nutrient for growth is not frozen. Bacteria classified as mesophilic grow best at temperatures around 98 °F (36.7 °C), but some will grow at temperatures lower than 68 °F (20 °C) and others at temperatures as high as 110 °F (43.3 °C). Bacteria causing diseases of animals are mesophilic and mostly anaerobic. Bacteria classified as thermophiles grow best at temperatures as low as 113 °F (45 °C) and others at temperatures as high as 160 °F(71.1 °C) or slightly higher.

Although microorganisms grow within a given range of temperatures, their growth rate decreases greatly at the low or high temperatures within that range. When microbes are subjected to temperatures higher than those of their growth range, they are destroyed at rates that depend on the degree to which the temperature is raised above the growth range.

Some bacteria can produce spores whose degree of heat resistance varies with the species. Some types may be destroyed by raising the temperature (in the presence of moisture) to 212 °F (100 °C) over a period of several minutes while others can survive boiling many hours. Some may even survive temperatures as high as 250 °F (121.1 °C) for many minutes. Bacterial spores are some of the most heat-resistant living organisms. The significance of bacteria spores for the food canning industry is discussed in more detail later (cf. Chap 8, Sects. 3.2 & 3.4; Chap 12, Sects. 9.4 & 9.6).

In general, there are bacteria that can grow and survive under more extreme conditions compared to those tolerated by any mold or yeasts. Molds, as a class, can grow and survive under more extreme conditions than can yeasts. On the other hand, although the growth and survival of the many species of bacteria cover a broad range of conditions, each species is highly selective, and the range of conditions under which it can metabolize is generally narrower than that of molds and yeasts.

3.8 Oxygen Requirements

Some bacteria are aerobic, that is, they require oxygen for growth. Others grow when the concentration of oxygen is low (micro-aerophiles). Still others can grow in either the presence or absence of oxygen (facultative aerobes or facultative anaerobes). A number of bacterial species will not grow in the presence of oxygen because it is toxic to them, and these are called anaerobes. The anaerobes are interesting from a food safety point of view. *Most pathogenic, i.e. disease causing bacteria are anaerobes.*

It should be pointed out that although oxygen is toxic to anaerobic bacteria, they can grow sometimes under conditions that appear to include the presence of oxygen. The explanations for this is that organic materials (animal and vegetable matter) contain compounds that tie up the available oxygen, so-called oxygen scrubbers. In many instances, aerobic bacteria first grow and consume the oxygen and produce reducing compounds that also combine with available oxygen, leading to conditions suitable for anaerobic bacteria. Anaerobic conditions favor the survival of some pathogenic bacteria and renders them more resistant to heat. Therefore low oxygen gas mixtures used for packaging may affect food safety risk (George et al. 1998; Riondet et al. 2000).

Yeasts grow best in aerobic conditions, but some grow anaerobically (e.g., fermentative types) although more slowly. A drop in redox potential normally follows depletion of oxygen (cf. Sect. 3.4). In contrast to this, certain aerobic bacteria can utilize the oxygen in certain oxygen-bearing compounds, such as nitrates. In general, molds require oxygen for growth.

3.9 Potentially Hazardous Foods and Temperature Control for Safety

If environmental conditions within a food promotes the growth of microorganism, then such products will require refrigeration for safe storage. The FDA defines potentially hazardous foods (PHF) as those foods where water activity and pH are high ($A_W > 0.85$, pH >4.6) and which allow bacterial growth at room temperature.[6] In 2001, an IFT expert panel recommended replacing the tem "potentially hazardous" with the new phrase "time/temperature control for safety" (TCS), which may be defined as "*foods that require time/temperature control to limit pathogen growth or toxin formation that constitutes a threat to public health.*" Correct handling of TCS foods requires storing at temperatures outside of the so-called danger zone (IFT/FDA 2001).

4 Effects of Microbes on Foods

4.1 Undesirable Changes in Foods

The changes in food caused by microbial action can be classified as undesirable or desirable. Undesirable changes are further subdivided into (i) microbe induced undesirable changes that cause food spoilage, not usually associated with human disease, or (ii) undesirable changes that lead to food poisoning whether or not the food undergoes observable changes (Chap. 8). Bacteria, molds, and yeasts are the main causes of the spoilage of unpreserved foods, with bacteria playing the major role.

4.2 Food Spoilage

Several billion dollar (US$) worth of food is lost annually within the US due to microbiological spoilage (Buzby and Hyman 2012; Gunders 2012). Food spoilage refers to "*any changes in a food product that renders it unacceptable to the consumer from a sensory point of view*" (Gram et al. 2002). It is important to understand the basis of food spoilage in order that efficient forms of preservation can be devised (Gunders 2012)

Food spoilage can be detected organoleptically. That is, we can see, *smell, taste* or *feel* the spoilage and or experience combinations of the four sensations. Quite often, the evidence of microbial growth is easily visible as in the case of slime formation, a cotton-like network of mold growth, iridescence and greening in cold cuts and in cooked sausages, and even obvious large colonies of bacteria. In liquids, such as juices, microbial spoilage is often manifested by the development of a cloudy appearance or curd formation. The odors of spoiled protein foods are

[6]In 1962 the FDA introduced the term potentially hazardous food defined as: "*any perishable food which consists in whole or in part of milk or milk products, eggs, meat, poultry, fish, shellfish, or other ingredients capable of supporting rapid and progressive growth of infectious or toxigenic microorganisms.*"

very objectionable. Some of these obnoxious odors are common enough, and the compounds responsible for them are ammonia, various sulfides (as in the smell of clam flats), and hydrogen sulfide (the typical smell of rotten eggs).

The changes in the taste of spoiled foods arise arises partly from the loss of the characteristic good taste, and from the development of objectionable tastes. Thus, when a pear or orange spoils, the sweet characteristic taste is lost, and when milk spoils, it develops an acidic (sour), sometimes bitter, taste. The texture of spoiled foods sometimes feels slimy, while others may feel mushy.

Microbial food spoilage is due to the breakdown of food compounds by microorganisms. Protein is digested to form amino acids, and further broken down to odorous sulfur compounds, ammonia, and methylamines. Starch and sugars are converted to organic acids leading to sour taste (Gram et al. 2002). The breakdown of lipids releases short chain fatty acids, whilst fat oxidation leads to ketones and aldehydes; some bacteria also produce phenolic products. The exact type of spoilage depends on the nature of the food, type of microbial flora on it, as well as intrinsic factors (cf. Sect. 3.1). Details of food spoilage reactions were discussed recently with a focus on fish (Fraser and Sumar 1998; Ghaly et al. 2010), meat (Dave and Ghaly 2011) and wine (Bartowsky 2009)

5 Food Fermentation

5.1 Benefits and Advantages of Fermentation

Fermentation or the production of foods using microbial food culture dates from 10,000 BC (Wikipedia 2019). The scientific study of fermentation originated recently following the discoveries of Louis Pasteur in the nineteenth century. Fermentation is associated with many beneficial effects (Table 7.6). According to the FAO, fermentation is associated with improvements in food security, generation of income, improved nutritional characteristics, degradation of anti-nutritional factors and enhanced vitamin content (Battcock and Azam-Ali 1998b) (Fig. 7.4).

Fermented foods are disproportionately important for the 800 million (~11%) people currently suffering from world hunger. Globally, about 5–40% of caloric intake may be derived from fermented foods and beverages (Tamang and Kailasapathy 2010). Most of the world hungry live in Asia whereas, a higher percentage categorized as hungry may be in Africa. It is generally agreed, that fermented food improves food security through both affordability and access (World Food Programme 2019). In financially stressed communities the tendency for utilizing traditionally prepared fermented foods was higher and declined with increasing wealth. Fermented foods were noted as suitable for famine relief and for promoting survival. A lower cost, and improved affordability compared to canning or freezing accounts for the popularity of fermented foods in developing regions of the world including, Africa (Mokoena et al. 2016), South East Asia (Anal 2019; Azam et al. 2017), China (Liu et al. 2011) and Latin America (Eliza Romero-Luna et al. 2017).

Traditional fermented foods remain important also in parts of Mediterranean (Salameh et al. 2016), Eastern Europe (Soukand et al. 2015) Japan (Murooka and Yamshita 2008) and South Korea (Patra et al. 2016). In many developed countries the proportion of fermented foods in the diet may approach 30% (DFG Senate Commission on Food Safety (SKLM) 2010).

The FAO suggested that food fermentation is a useful strategy for preventing food waste, as various residues could be used for both submerged and solid state fermentation (Mokoena et al. 2016). Fermentable food wastes include some animal products (hides and bone) and process waste from plants (coconut press cake, banana leaves, pineapple peel).

Many fermented products have lower anti-nutritional factors, including lower levels of hydrogen cyanide in cassava and lower trimethylamine (TMA) in Icelandic fermented shark meat (Osimani et al. 2019). The mineral availability from many cereal foods can be significantly affected by a high phytic acid content (Gabaza et al. 2017). Investigations involving sorghum (5 varieties) and finger millet (4 varieties) grown in Kenya, revealed that 72 h traditional fermentation led to an average reduction in phytic acid

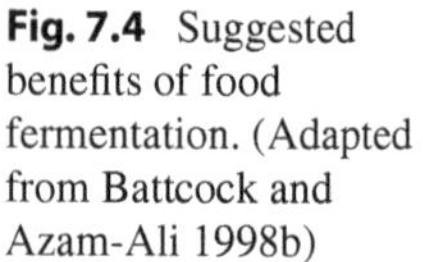

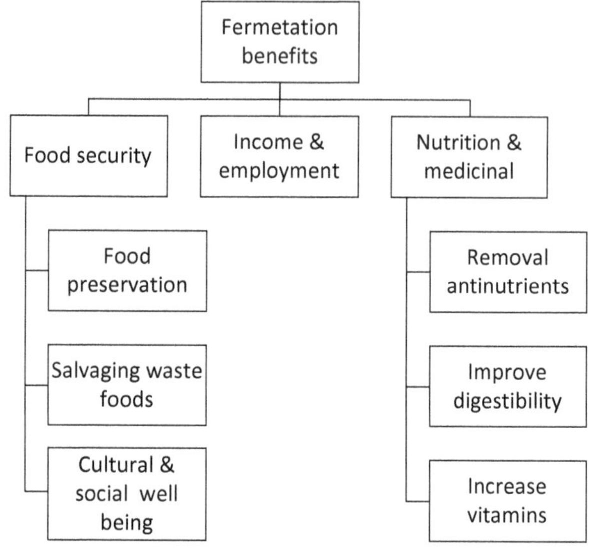

Fig. 7.4 Suggested benefits of food fermentation. (Adapted from Battcock and Azam-Ali 1998b)

Table 7.6 Benefits of food fermentation

Benefits (FAO)	Advantages (SKLM)
Improvements in food security,	Increased hygienic safety
Generation of income	Extended shelf life compared to the raw product.
Improvements in nutritional characteristics	
Improved digestibility,	Higher sensory quality
Removal of anti-nutritional factors	Degradation of toxic or harmful substances
Enhanced vitamin content	Minimal technology for production
	Low energy processing
	Classed as natural and organic food

Adapted from DFG Senate Commission on Food Safety (SKLM) (2010)

levels by 39% and 64%, respectively. Extension of the fermentation time to 96 h produced 54% and 79% reductions in phytic acid and increases in the bioavailability of calcium, iron and manganese (Makokha et al. 2002).

Fermentation of beans to create tempeh decreases flatulent causing carbohydrates, raffinose, stachyose, and verbascose by between 73–93% (Starzynska-Janiszewska et al. 2014). Further undesirable substances are degraded or partly degraded by fermentation such as, glucosinolates, goitrogens, lectins, phytic acid, proteinase inhibitors, oxalic acid, and indigestible carbohydrates associated with the development of flatulence.

The health benefits of fermented foods may derive partly from the action of microbial enzymes. Recently, attention was raised as to the potential probiotic benefits of lactic acid bacterial (Mokoena et al. 2016). Many authors reported potential health benefits of fermented foods as a source of probiotic lactic acid bacteria. (Baschali et al. 2017; Chaudhary et al. 2018; Eliza Romero-Luna et al., 2017; Todorov and Holzapfel 2015).

5.2 Food Fermentation for Preservation

Fermented foods are produced using, bacterial (lactic acid bacteria or acetic acid bacteria), molds or yeasts. Either mixed cultures are used, or else isolated and well-characterized "starter cultures" are used for fermentation. There are many traditional fermentation processes, where the isolation and characterization of microorganism have yet to be completed (Battcock and Azam-Ali 1998b; Tamang et al. 2016).

Virtually all food groups can be fermented for the purpose of preservation The sauerkraut process, which involves fermentation of salted cabbages, is taken as characteristic of other vegetable fermentation processes using lactic acid bacteria (Battcock and Azam-Ali 1998a). The flavor and texture of sauerkraut are due to the lactic acid and other metabolites produced by lactic acid bacteria in shredded cabbage to which salt (~2.5 g/100 g product) is added. The mixture is fermented by a three or more microbes; the starter culture contains a mixture of *Leuconostoc mesenteroides*, *Lactobacillus plantarum* and *Lactobacillus pentoaceticus* which work together at about 18–22 °C until a final acidity of 2.5%. During the fermentation, there occurs a succession *of* bacterial population as conditions in the food matrix changes (Battcock and Azam-Ali 1998a).

The flavor of dried-cured sausages (Italian salami, cervelat, etc.) is also produced by the growth of lactic acid bacteria in the ground meat substrate, within the casing during the period in which the sausages are stored on racks and allowed to dry. Many of the fermented foods acquire distinctive flavors but we must assume that the initial driving for food fermentation was preservation (Battcock and Azam-Ali 1998b).

According to one estimate, there are about 3500 fermented foods globally. Table 7.7 shows a selection of fermented foods from round the world (Campbell-Platt 1987; Tamang and Kailasapathy 2010). A few examples of fermented food groups are outlined in the rest of this section. Traditional fermentation products from round the world, maybe classified into 9 groups including some which are not considered mainstream (Table 7.8). Traditional fermented products are discussed further in Sect. 5.6).

Table 7.7 Some fermented fruit and vegetable products worldwide

Yeast fermentation	Lactic acid bacteria	Mixed fermentations
Fruit juices Banana beer Cashew wine Colonche Date wine Grape, fortified wines Grape, sparkling wine Grapes, red wine Grapes, white wine Jackfruit wine Tepache	Dry salted pickles Dry salted lime pickle Pickled cucumbers Pak-gard-dong (pickled leafy vegetable) Sauerkraut Tempoyak (pickled durian) Pickled beetroots Lamoun makbous (pickled lemons)	*Vinegars* Coconut water vinegar pineapple peel vinegar Palm wine vinegar Coconut toddy vinegar Nipa palm vinegar Quick process pickles
Plant saps Basi (Sugar cane wine) Muratina Palm wine Pulque Toddy Ulanzi (Bamboo Wine)	*Brined fruit or vegetable pickles* Green Mango Pickle Lime pickle (brined) Kimchi (pickled cabbage) Green olives Black olives Jackfruit pickle Pickled Radish Pickled cucumber Pickled leafy vegetables Other vegetables & fruits Sauerkraut	*Beverages* Cocoa products Cocoa powder Chocolate Coffee Other mixed fermentation products Vanilla Tabasco Tea
	Non-salted products Gundruk (pickled leafy vegetable) Kocho (pickled false banana) Sinki (pickled radish) Sunki Kanji Fermented tea leaves	

Adapted from Battcock and Azam-Ali (1998b) and Tamang et al. (2016)

5.3 Fermented Dairy Products

Milk and cream can be fermented to form a range of products including, soft and hard cheese, various types of yogurt, and sour cream (Law 1997; Robinson and Tamine 1993). The different products differ according to the type of microorganisms or so-called starter cultures used (Marshall 1987).

To obtain the curd in the manufacture of cheese, milk is cultured with bacteria that produce lactic in order to lower the pH. Thereafter calf rennet is added to precipitate the casein, although this can be accomplished in a different way. The particular flavors and textures of many cheeses are attributed to bacterial growth during or after curd formation. The distinctive flavor and texture of different cheese are attributed to a ripening phase that occurs during the aging where cheese is held in storage at a particular temperature for purposes of maturing.

Table 7.8 Classification of some fermented food groups and products

1. Alcoholic beverages
2. Cereals
3. Dried and smoked fish products
4. Legumes
5. Milk products
6. Miscellaneous products
7. Preserved meat products
8. Roots/tubers
9. Vegetables and bamboo shoots

Adapted from Battcock and Azam-Ali (1998b)

Cheeses, like Roquefort, Gorgonzola, blue, and Camembert, owe their particular flavor and texture to growth of molds. Under present-day manufacturing procedures, these cheeses are inoculated with molds of particular species, and incubated to allow the molds to grow and produce the desired textures and flavors.

The flavor of butter and buttermilk is the result of the formation of small amounts of a chemical compound from sugars or citrates by the growth of lactic acid bacteria in the cream prior to churning (Gettys and Davidson 1985). Pickles and olives are at least partially preserved by acid formed by bacteria when the raw materials are allowed to undergo a natural fermentation.

Yogurt is arguably the most popular fermented product, perhaps second only to Cheese. To produce yogurt, milk is heated to kill endogenous bacterial and then seeded a mixture microorganisms, *Lactobacillus acidophilus, Lactobacillus bulgaricus and Streptococcus thermophiles*. The lactic acid bacteria convert lactose to lactic acid thereby causing milk pH to drop sufficiently to coagulate some casein thereby producing a thickened mass. Basic yogurt may have fruits added (Marshall 1987; Robinson and Tamine 1993).

5.4 Alcoholic and Acetic Acid Fermentation

High alcohol beverages are produced by fermentations achieved using yeasts. Low alcohol products produced from fruit are called wines whilst those produced using cereals are beers or ales. Of course distilling off the alcohol from fermenting cereals, fruits, starch tubers and other starchy food leads to products such as gins, whiskeys and vodka identified with the US, Scotland and Russia. An extensive discussion of different types of alcoholic spirits is beyond the scope of this text. Liquid foods fermented with lactic acid bacteria produce no-alcohol or low alcohol beverages sometimes considered as functional foods (Baschali et al. 2017; Coskun 2017).

Acetic acid (vinegar) is produced from ethyl alcohol by the growth of Acetobacter (Fig. 7.5).

5.5 Fermented Baked Products

Various types of bread and other bakery products are leavened (raised) by yeast. The yeasts, in this case, not only produce carbon dioxide, a gas that causes the loaf to rise, but also produce materials that affect the gluten (protein) of flour, causing it to take on a form that is elastic. The elasticity of the gluten is necessary to retain the gas and to support the structure of the loaf. In summary, yeast fermentation has a large effect on the texture of bread dough (Courtin et al. 2014). Moreover, of the 150 or so flavor compounds associated with leavened bread, the majority is thought to be from yeast (Birch et al. 2014).

5.6 Traditional Fermented Products of the World

Traditional food technologies are usually small-scale, artisanal processes, with less reliance on equipment and investment capital (Chap. 11). Many small-scale processing methods also employ recipes less formerly codified, but rather passed down from one generation to the next. Accordingly, many fermentation processes from around the world are classed as traditional. Frequently, the microorganisms employed for traditional fermentation are not well characterized and the underlying processes are not optimized (Sanni 1993). Traditional fermentation uses mixed cultures of bacteria or else the so-called back-slop method, where a portion of a

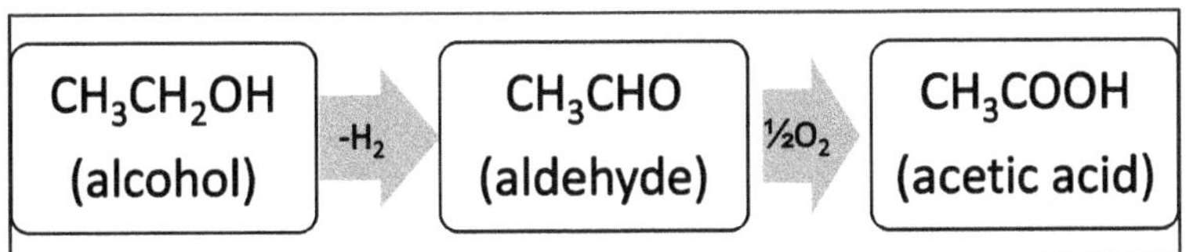

Fig. 7.5 Transformation of alcohol to acetic acid

previous fermentation product is used to initiate the next batch (Wikipedia 2019).

Present attitudes towards traditional fermentation processes are ambivalent. Some experts refer to traditional fermentation processes as cultural capital, representing a diversity of processes that remain acceptable to consumers (Oyewole 1997). However, critics of traditional fermentation processes noted that these involve primitive conditions, and craft based technology. Traditional fermentation were also considered labor intensive and likely to produce unstable products. Industrialization of fermentation processes were suggested, involving the adoption of Good Manufacturing Practices (GMP), the use of starter culture and treatment of products to achieve stabilization; e.g. pasteurization of traditional beers (Achi 2005a, b). Many experts are agreed on the need for more systematic study, and the assessment and development of microorganism, for commercialized production (Anal 2019; Luis Navarrete-Bolanos 2012; Ly et al. 2018; Tamang et al. 2016).

Traditional fermentation products from Chinese cuisine are now available in Western supermarkets, e.g. tempeh, miso, soysauce, natto, stinky tofu and fermented soymilk (Golbitz 1995; Li and Hsieh 2004). Chhang, a form of fermented rice wine popular in Nepal is produced, using cooked finger millet via a 12-day fermentation process. The traditional inoculum used for making chhang is similar those used for rice wine and believed to comprise of strains of lactobacillus and saccharomyces. Evaluation of the traditional chhang production method concluded there is good potential for modernization and larger-scale commercial production, perhaps by replacing the traditional inoculum with a well-defined mixture of microorganisms (Basappa 2002).

Table 7.9 Some traditional fermented products from Africa

Cereals	Burukutu, Pito, Kwunu, Ogi, Otika, Ndaleyi,
Fruits	Agadagidi, Cacao Wine, Eketeke, Ugba
Legumes	Daddawa, Iru, Ogiri, Ogboroti
Tubers	Abacha, Akara, Akpu, Elubo, Fufu, Gari, Loiloi, Kokobel
Tree Sap	Ogogoro, Palm Wine
Meat and Seafoods	Afonnama, Azu-Okpo, Nsiko, Oporo
Milk	Maishanu, Nono, Warankasi, Wara,

Adapted from Iwuoha and Eke (1996), Nkama et al. (1994), and Uzogara et al. (1990)

Traditional fermentation products from Africa were classified by the substrate food group fermented (Table 7.9) or classified by the different classes of product, e.g. non-alcoholic stiff porridges, non-alcoholic soft porridges and gruels or alcoholic fermentation products (Iwuoha and Eke 1996; Nkama et al. 1994). There is a drive towards industrialization, especially affecting the brewing sector and so commercial forms of pito and related beverages have been produced by brewers in East, Southern and West Africa (Amadou et al. 2011; Anonymous 1985).

Traditional fermented foods from diverse geographic regions are also receiving attention and resurgence. The rising interest in fermented food is attributed to a growing consumer desire for healthy and natural foods. Other traditional fermented foods include those from Turkey (Kabak and Dobson 2011), China (Liu et al. 2011), India (Kumar et al. 2013), Eastern Europe (Soukand et al. 2015) and the Mediterranean (Salameh et al. 2016). Industrializing traditional fermentations can be achieved via standard steps (Table 7.10).

Table 7.10 Proposed steps for upgrading traditional fermented foods

1. Isolate and identify microorganisms associated with the fermentations
2. Determine what microbes do in the product
3. Select, enrich and improve strains of bacterial, & cultivate starter culture.
4. Improve the process design, and manufacturing
5. Enhance the quality of starting raw materials
6. Do controlled laboratory studies of the fermentation process
7. Carry out pilot stage production.
8. Carry out industrial scale production trials and commercialize

Adapted from Haard et al. (1999)

5.7 Regulation of Microbial Food Cultures

Some microorganism are linked with foodborne infections, and so it is important to ensure that fermented foods, produced using bacterial or molds, are safe. It is also important to consider how fermented foods can be regulated effectively. Currently, there is a focus on identifying beneficial microorganism used for food fermentation. The expectation is that only non-pathogenic bacteria should be involved in food fermentations (Bourdichon et al. 2012; Laulund et al. 2017).

Legal regulation for the sector is difficult because many traditional processes involve so-called spontaneous fermentations with communities of microorganisms that are not fully characterized. The compositions of microbes can also change during the fermentation by a process of succession. Thirdly, the majority of fermented products claim a long history of safe use and could be (wrongly) assumed to be safe. Fourthly, there is no agreement whether microbes used for fermentation should be classed, as food additives, processing aides, or ingredients. Also microbes used for producing fermented products are sometimes regarded as different from those added only to achieve preservation (DFG Senate Commission on Food Safety (SKLM) 2010).

Fermented foods such as wine have a history dating back at least 8000 years or more, but the role of bacteria in fermentation was not known prior to the work of Louis Pasteur in 1859. The involvement of bacteria in milk fermentation was not established until 1877 (Bourdichon et al. 2012; DFG Senate Commission on Food Safety (SKLM) 2010). Significant progress towards the regulation of fermented foods, was made recently by the combined efforts of European Food and Feed Culture Association (EFFCA) and International Dairy Federation (IDF) which have compiled a list of all microorganisms documented in fermented foods by 2002 (Bourdichon et al. 2012; DFG Senate Commission on Food Safety (SKLM) 2010).

The groundwork for the effective regulation of fermented foods led to new basic definitions. For example, microbes used for food fermentation were designated as, microbial food cultures (MFC) defined as "*live bacteria, yeast or mold, used for food production*" but excluding microbes used entirely or solely for probiotic applications as food supplements. As of 2012, the list of documented food microorganism comprised 264 species, including 195 different bacteria and 69 different yeasts and molds; the most common bacteria for food fermentation appear to be 32 species of Actinobacteria and 94 species of Lactobacillus (Bourdichon et al. 2012).

5.8 United States Fermented Food Regulations

The US FDA regulates fermented foods in addition to diverse ingredients and microbes applied for farming use (Table 7.11). The regulated products include those for making food silage, for food manufacturing, or for animal feed. In general, the FDA regards (all) food ingredients as either classical food additives or as having GRAS (Gererally recognized as safe) status. Food additives can be used for food manufacture only after pre-market approval by the FDA (Chap. 16). On the other hand, GRAS status is awarded either, (i) by default to all food ingredients in historical use prior to 1958 or (ii) after a company notifies the FDA that an ingredient is GRAS as judged by an expert panel

Table 7.11 FDA regulated fermented foods, cultures and ingredients derived from microorganisms

Code federal regulations	Comments
21 CFR Part 172 and 173,	Food additives from microorganisms
21 CFR part 184	GRAS substances from microbes Organic acids Enzymes Agar Microbial enzymes Vitamins Butter starter
21 CFR Part 186	GRAS substances from microbes for indirect use
21 CFR Parts 131, 133, 136, and 137	Microbe derived foods for humans Acidified milk Yogurt Blue cheese Cheddar Cheese Enzymes
GRAS listed microorganisms	See GRAS list,

Adapted from Food and Drug Administration (2018)

References

Achi OK (2005a) The potential for upgrading traditional fermented foods through biotechnology. Afr J Biotechnol 4(5):375–380

Achi OK (2005b) Traditional fermented protein condiments in Nigeria. Afr J Biotechnol 4(13):1612–1621

Adams R, Moss O (2008) Food microbiology. RSC Publishing, 463 pp

Amadou I, Gbadamosi OS, Guo-Wei L (2011) Millet-based traditional processed foods and beverages – a review. Cereal Foods World 56(3):115–121

Anal AK (2019) Quality ingredients and safety concerns for traditional fermented foods and beverages from Asia: a review. Fermentation-Basel 5(1). https://doi.org/10.3390/fermentation5010008

Anonymous (1985) Major breakthrough in packaging sorghum beer. Food Rev 12(4)

Azam M, Mohsin M, Ijaz H, Tulain UR, Ashraf MA, Fayyaz A et al (2017) Lactic acid bacteria in traditional fermented Asian foods. Pak J Pharm Sci 30(5):1803–1814

Bartowsky EJ (2009) Bacterial spoilage of wine and approaches to minimize it. Lett Appl Microbiol 48(2):149–156

Basappa SC (2002) Investigations on chhang from finger millet (Eleucinecoracana Gaertn.) and its commercial prospects. Indian Food Industry 21(1):46–51

Baschali A, Tsakalidou E, Kyriacou A, Karavasiloglou N, Matalas A-L (2017) Traditional low-alcoholic and non-alcoholic fermented beverages consumed in European countries: a neglected food group. Nutr Res Rev 30(1):1–24. https://doi.org/10.1017/s0954422416000202

Battcock M, Azam-Ali S (1998a) Chapter 5. Bacterial fermentations. FAO Agricultural Services Bulletin No.134. Fermented fruits and vegetables: a global perspective. Retrieved from http://www.fao.org/3/x0560e/x0560e10.htm#5.6.2

Battcock M, Azam-Ali S (1998b) Fermented fruits and vegetables: a global perspective. FAO Agricultural Services Bulletin no.134. Retrieved from http://www.fao.org/docrep/x0560e/x0560e00.htm

Beales N (2004) Adaptation of microorganisms to cold temperatures, weak acid preservatives, low pH, and osmotic stress: a review. Compr Rev Food Sci Food Saf 3(1):1–20. https://doi.org/10.1111/j.1541-4337.2004.tb00057.x

Berry ED, Foegeding PM (1997) Cold temperature adaptation and growth of microorganisms. J Food Prot 60(12):1583–1594

Birch AN, Petersen MA, Hansen AS (2014) Aroma of wheat bread crumb. Cereal Chem 91(2):105–114

Bolduc MP, Bazinet L, Lessard J, Chapuzet JM, Vuillemard JC (2006) Electrochemical modification of the redox potential of pasteurized milk and its evolution during storage. J Agric Food Chem 54(13):4651–4657. https://doi.org/10.1021/jf052626k

Borgstrom G (1968) Principles of food science. 2. Food microbiology and biochemistry

Bourdichon, F., Casaregola, S., Farrokh, C., Frisvad, J. C., Gerds, M. L., Hammes, W. P., ... Hansen, E. B. (2012). Food fermentations: microorganisms with technological beneficial use. Int J Food Microbiol, 154(3), 87–97. doi:https://doi.org/10.1016/j.ijfoodmicro.2011.12.030

Buzby JC, Hyman J (2012) Total and per capita value of food loss in the United States. Food Policy 37(5):561–570

Campbell-Platt G (1987) Fermented foods of the world. A dictionary and guide. Butterworths, London

Ceuppens S, Li D, Uyttendaele M, Renault P, Ross P, Van Ranst M, Cocolin L, Donaghy J (2014) Molecular methods in food safety microbiology: interpretation and implications of nucleic acid detection. Compr Rev Food Sci Food Saf 13(4):551–577. https://doi.org/10.1111/1541-4337.12072

Chaudhary A, Sharma DK, Arora A (2018) Prospects of Indian traditional fermented food as functional foods. Indian J Agric Sci 88(10):1496–1501

Codex Alimentarius Commission (1989) Codex class names and the international numbering system for food additives CAC/GL 36-1989. Retrieved from http://www.codexalimentarius.org/input/download/standards/13341/CXG_036e.pdf

Coskun F (2017) A traditional turkish fermented non-alcoholic beverage, "Shalgam". Beverages 3(4). https://doi.org/10.3390/beverages3040049

Courtin CM, Rezaei MN, Jayaram VB (2014) Evaluating the impact of yeast fermentation on bread dough matrix rheology. Cereal Foods World, 59, (6S

Cutter CN, Dorsa WJ, Siragusa GR (1996) A rapid microbial ATP bioluminescence assay for meat carcasses. Dairy Food Environ Sanitation 16(11):726–736

Dave D, Ghaly AE (2011) Meat spoilage mechanisms and preservation techniques: a critical review. Am J Agric Biol Sci 6(4):486–510. Retrieved from https://thescipub.com/pdf/ajassp.2010.2859.2877.pdf

DFG Senate Commission on Food Safety (SKLM) (2010) Microbial food cultures. Retrieved from https://www.dfg.de/download/pdf/dfg_im_profil/reden_stellungnahmen/2010/sklm_mikrobielle_kulturen_101115_en.pdf

Duried A, Delbeau C, Divies C, Cachon R (2003) Use of redox potential modification by gas improves microbial quality, color retention, and ascorbic acid stability of pasteurized orange juice. Int J Food Microbiol 89(1):21–29. https://doi.org/10.1016/S0168-1605(03)00125-9

Eliza Romero-Luna H, Hernandez-Sanchez H, Davila-Ortiz G (2017) Traditional fermented beverages from Mexico as a potential probiotic source. Ann Microbiol 67(9):577–586. https://doi.org/10.1007/s13213-017-1290-2

Erdosi O, SzakmÃ¡r K, Reichart O, Szekely-Kormoczy P, Laczay P (2011) Application of the redox potential measurement-based rapid method in the microbial hygienic control. Acta Aliment 41(1):45–55. https://doi.org/10.1556/AAlim.2011.0005

Food and Drug Administration (2018) Microorganisms & microbial-derived ingredients used in food (partial list). Retrieved from https://www.fda.gov/food/generally-recognized-safe-gras/microorganisms-microbial-derived-ingredients-used-food-partial-list

Fraser OP, Sumar S (1998) Compositional changes and spoilage in fish (part II) -microbiological induced deterioration. Nut Food Sci 6:325–329

Gabaza M, Muchuweti M, Vandamme P, Raes K (2017) Can fermentation be used as a sustainable strategy to reduce iron and zinc binders in traditional African fermented cereal porridges or gruels? Food Rev Int 33(6):561–586. https://doi.org/10.1080/87559129.2016.1196491

George SM, Richardson LCC, Pol IE, Peck MW (1998) Effect of oxygen concentration and redox potential on recovery of sublethally heat-damaged cells of Escherichiacoli O157:H7, Salmonellaenteritidis and Listeriamonocytogenes. J Appl Microbiol 84(5):903–909. https://doi.org/10.1046/j.1365-2672.1998.00424.x

Gettys SC, Davidson PM (1985) A comparison of buttermilks made using fermentation, direct acidification, and a combination of both methods. J Dairy Sci 68(3):620–625

Ghaly AE, Dave D, Budge S, Brooks MS (2010) Fish spoilage mechanisms and preservation techniques. Am J Appl Sci 7(7):859

Golbitz P (1995) Traditional soyfoods: processing and products. J Nutr 125(3 Suppl):570S–572S

Gunders D (2012) Wasted: how America is losing up to 40 percent of its food from farm to fork to landfill. Nat Res Def Council 26:1–26

Gram L, Ravn L, Rasch M, Bruhn JB, Christensen AB, Givskov M (2002) Food spoilage—interactions between food spoilage bacteria. Int J Food Microbiol 78(1–2):79–97. http://www.innocua.net/web/download-3544/food-spoilage-interactions-between-food-spoilage-bacteria.pdf

Haard NF, Odunfa SA, Lee C-H, Quintero-Ramírez R, Lorence-Quiñones A, Wacher-Radarte C (1999) Chapter 2. Cereal fermentations in African countries. In fermented cereals: A global perspective. Retrieved from http://www.fao.org/3/x2184e/x2184e00.htm#con

Hanna SE, Connor CJ, Wang HH (2005) Real-time polymerase chain reaction for the food microbiologist: technologies, applications, and limitations. J Food Sci 70(3):R49–R53

Hysert DW, Kovecses F, Morrison NM (1976) A firefly bioluminescence ATP assay method for rapid detection and enumeration of brewery microorganisms. J Am Soc Brew Chem 34(4):145–150

IFT/FDA (2001) Evaluation & definition of potentially hazardous foods. A report of the institute of food technologists for the food and drug administration of the U.S. Department of Health and Human Services. Retrieved from https://www.fda.gov/downloads/Food/FoodborneIllnessContaminants/UCM545171.pdf

Ignatova M, Prevost H, Leguerinel I, Guillou S (2010) Growth and reducing capacity of listeria monocytogenes under different initial redox potential. J Appl Microbiol 108(1):256–265. https://doi.org/10.1111/j.1365-2672.2009.04426.x

Ivey ML, Phister TG (2011) Detection and identification of microorganisms in wine: a review of molecular techniques. J Ind Microbiol Biotechnol 38(10):1619–1634. https://doi.org/10.1007/s10295-011-1020-x

Iwuoha CI, Eke OS (1996) Nigerian indigenous fermented foods: their traditional process operation, inherent problems, improvements and current status. Food Res Int 29(5–6):527–540. https://doi.org/10.1016/0963-9969(95)00045-3

Jany J-L, Barbier G (2008) Culture-independent methods for identifying microbial communities in cheese. Food Microbiol 25(7):839–848. https://doi.org/10.1016/j.fm.2008.06.003

Jay JM, Loessner MJ, Golden DA (2005) Modern food microbiology. Springer, New York, p 790

Kabak B, Dobson ADW (2011) An introduction to the traditional fermented foods and beverages of Turkey. Crit Rev Food Sci Nutr 51(3):248–260. https://doi.org/10.1080/10408390903569640

Kumar RS, Kanmani P, Yuvaraj N, Paari KA, Pattukumar V, Arul V (2013) Traditional Indian fermented foods: a rich source of lactic acid bacteria. Int J Food Sci Nutr 64(4):415–428. https://doi.org/10.3109/09637486.2012.746288

Kyriakides A (1992) ATP bioluminescence applications for microbiological quality control in the dairy industry. J Soc Dairy Technol 45(4):91–93. https://doi.org/10.1111/j.1471-0307.1992.tb01790.x

Labots H, Stekelenburg FK (1985) ATP-bioluminescence: a rapid method for the estimation of microbial contamination in meat and meat products. Proc Eur Meet Meat Res Work 31(5):401–405

Laulund S, Wind A, Derkx PMF, Zuliani V (2017) Regulatory and safety requirements for food cultures. Microorganisms 5(2). https://doi.org/10.3390/microorganisms5020028

Law BA (1997) Microbiology and biochemistry of cheese and fermented milk

Li JR, Hsieh YH (2004) Traditional Chinese food technology and cuisine. Asia Pac J Clin Nutr 13(2):147–155

Liu S-N, Han Y, Zhou Z-J (2011) Lactic acid bacteria in traditional fermented Chinese foods. Food Res Int 44(3):643–651. https://doi.org/10.1016/j.foodres.2010.12.034

Luis Navarrete-Bolanos J (2012) Improving traditional fermented beverages: how to evolve from spontaneous to directed fermentation. Eng Life Sci 12(4):410–418. https://doi.org/10.1002/elsc.201100128

Ly D, Mayrhofer S, Domig KJ (2018) Significance of traditional fermented foods in the lower Mekong subregion: a focus on lactic acid bacteria. Food Biosci 26:113–125. https://doi.org/10.1016/j.fbio.2018.10.004

Makokha AO, Oniang'o RK, Njoroge SM, Kamar OK (2002) Effect of traditional fermentation and malting on phytic acid and mineral availability from sorghum (Sorghumbicolor) and finger millet (Eleusinecoracana) grain varieties grown in Kenya. Food Nutr Bull 23(3):241–245

Marshall VM (1987) Fermented milks and their future trends. I. Microbial aspects. J Dairy Res 54(4):559–574

Martin F, Ebel B, Rojas C, Gervais P, Cayot N, Cahon R (2013) Redox potential: monitoring and role in development of aroma compounds, rheological properties and survival of oxygen sensitive strains during the manufacture of fermented dairy products, lactic acid bacteria. R & D for Food, Health and Livestock Purposes. Retrieved from http://www.intechopen.com/books/lactic-acid-bacteria-r-d-for-food-health-and-livestock-purposes/redox-potential-monitoring-and-role-in-development-of-aroma-compounds-rheological-properties-and-sur

Maurer JJ (2011) Rapid detection and limitations of molecular techniques. In Annual review of food science and technology, vol 2 (vol 2, pp. 259–279)

McGrath M (2008) Overview of careers in the food science field. In: Hartel RW, Klawitter CP (eds) Careers in food science: from undergraduate to professional. Springer, New York, pp 23–42

Mokoena MP, Mutanda T, Olaniran AO (2016) Perspectives on the probiotic potential of lactic acid bacteria from African traditional fermented foods and beverages. Food Nutr Res 60. https://doi.org/10.3402/fnr.v60.29630

Montville TJ, Matthews KR (2008) Food microbiology: an introduction. ASM Press, Washington, DC

Murooka Y, Yamshita M (2008) Traditional healthful fermented products of Japan. J Ind Microbiol Biotechnol 35(8):791–798. https://doi.org/10.1007/s10295-008-0362-5

Nkama I, Abbo ES, Igene JO (1994) Traditional production and chemical composition of ndaleyi, a Nigerian fermented pearl millet food. Plant Foods Hum Nutr 46(2):109–116. https://doi.org/10.1007/BF01088762

Osimani, A., Ferrocino, I., Agnolucci, M., Cocolin, L., Giovannetti, M., Cristani, C., . . . Aquilanti, L. (2019). Unveiling hakarl: a study of the microbiota of the traditional Icelandic fermented fish. Food Microbiol, 82, 560–572. doi:https://doi.org/10.1016/j.fm.2019.03.027

Oyewole OB (1997) Lactic fermented foods in Africa and their benefits. Food Control 8(5–6):289–297

Patra JK, Das G, Paramithiotis S, Shin H-S (2016) Kimchi and other widely consumed traditional fermented foods of Korea: a review. Front Microbiol 7. https://doi.org/10.3389/fmicb.2016.01493

Postollec F, Falentin H, Pavan S, Combrisson J, Sohier D (2011) Recent advances in quantitative PCR (qPCR) applications in food microbiology. Food Microbiol 28(5):848–861. https://doi.org/10.1016/j.fm.2011.02.008

Ray B, Bhunia A (2013) Fundamental food microbiology, 5th edn. Taylor & Francis, 624 pp

Reichart O, Szakmar K, Jozwiak A, Felfoldi J, Baranyai L (2007) Redox potential measurement as a rapid method for microbiological testing and its validation for coliform determination. Int J Food Microbiol 114(2):143–148. https://doi.org/10.1016/j.ijfoodmicro.2006.08.016

Riondet C, Cachon R, Wache Y, Sunyol Bert E, Gbaguidi P, Alcaraz G, Divies C (2000) Combined action of redox potential and pH on heat resistance and growth recovery of sublethally heat-damaged Escherichiacoli. Appl Microbiol Biotechnol 53(4):476–479. https://doi.org/10.1007/s002530051644

Robinson RK, Tamine AY (1993) Manufacture of yoghurt and other fermented milk products. In: Robinson RK (ed) Modern dairy technology, volume 2: advances in milk products, vol 2. Springer, Cham, pp 1–48

Salameh C, Banon S, Hosri C, Scher J (2016) An overview of recent studies on the main traditional fermented milks and white cheeses in the Mediterranean region. Food Rev Int 32(3):256–279. https://doi.org/10.1080/87559129.2015.1075210

Sanni AI (1993) The need for process optimization of African fermented foods and beverages. Int J Food Microbiol 18(2):85–95

Siragusa GR, Cutter CN, Dorsa WJ, Warren J, Koohmaraie M (1995) Use of a rapid microbial ATP bioluminescence assay to detect contamination on beef and pork carcasses. J Food Prot 58(7):770–775

Soukand R, Pieroni A, Biro M, Denes A, Dogan Y, Hajdari A, Kalle R, Reade B, Mustafa B, Nedelcheva A, Quave CL, Luczaj L (2015) An ethnobotanical perspective on traditional fermented plant foods and beverages in Eastern Europe. J Ethnopharmacol 170:284–296. https://doi.org/10.1016/j.jep.2015.05.018

Starzynska-Janiszewska A, Stodolak B, Mickowska B (2014) Effect of controlled lactic acid fermentation on selected bioactive and nutritional parameters of tempeh obtained from unhulled common bean (Phaseolus vulgaris) seeds. J Sci Food Agric 94(2):359–366. https://doi.org/10.1002/jsfa.6385

Tamang JP, Kailasapathy K (2010) Fermented foods and beverages of the world. CRC Press, Boca Raton, p 460

Tamang JP, Watanabe K, Holzapfel WH (2016) Review: diversity of microorganisms in global fermented foods and beverages. Front Microbiol 7:377

Todorov SD, Holzapfel WH (2015) Traditional cereal fermented foods as sources of functional microorganisms. In: Holzapfel W (ed) Advances in fermented foods and beverages: improving quality, technologies and health benefits, vol 265, pp 123–153

U. S. Food and Drug Administration (2001 2015) Evaluation and definition of potentially hazardous foods – Chapter 3. Factors that influence microbial growth. A report of the Institute of Food Technologists for the Food and Drug Administration of the U.S. Department of Health and Human Services,December 31, 2001. Retrieved from http://www.fda.gov/Food/FoodScienceResearch/SafePracticesforFoodProcesses/ucm094141.htm

Uzogara SG, Agu LN, Uzogara EO (1990) A review of traditional fermented foods, condiments and beverages in Nigeria – their benefits and possible problems. Ecol Food Nutr 24(4):267–288. https://doi.org/10.1080/03670244.1990.9991145

Wikipedia (2019) Microbial food cultures. Retrieved from https://en.wikipedia.org/wiki/Microbial_food_cultures

World Food Programme (2019) Zero hunger. Retrieved from https://www1.wfp.org/zero-hunger

Microbial Foodborne Disease Outbreaks

8

1 Introduction

1.1 Food and Diseases

This is the first of three chapters (Chap 8-10) dealing with food safety. The aims from Chap 8 is to introduce the reader to a range of microbiological agents linked with foodborne disease and their significance for outbreaks. Knowing or *identifying* disease-causing agents (hazards) is usually the first step in the prevention of future outbreaks and epidemics. Another element required for food safety, is to control *exposure*. Various hygiene and sanitation methods for reducing exposure to foodborne microbes and other hazards are discussed in Chaps. 9 and 10, together with some approaches for food safety management.

There are several schemes for classifying human diseases, but the International Classification of Disease (ICD) is probably one of the most authoritative (World Health Organization 2019b). The WHO uses the ICD scheme for compiling mortality statistics for different causes of death. Other less formal classifications methods for human diseases are based on characteristics such as the duration of diseases (acute or chronic), mode of transmission (air, water, or foodborne etc.), type of causative agents (communicable versus non-communicable), the causes of diseases (injury, toxicity or deficiency), and the systems of the body affected (Fig. 8.1).

Foodborne diseases arise from the transmission of disease causing agents (hazards) to humans via the medium of food and/or the Agri-food system (Centers for Disease Control and Prevention 2014). The role of the Agri-food system in foodborne transmission is noteworthy in order not to restrict these discussions to finished food items. There are no restrictions for foodborne diseases, in terms duration but in general most tend to be acute, short terms conditions. The mode of transmission of food borne disease involve finished foods, but also living animals, food manufacturing premises, industrial equipment, utensils and foodservice workers. Foodborne diseases discussed in the rest of the chapter are those linked with microorganisms and "germs" and issues of hygiene and food safety. By contrast, food related public health concerns linked with chronic, non-communicable, metabolic diseases tend to be covered by nutrition professionals.

1.2 Foodborne Hazards

Any food-associated agent capable of causing harm is termed a food hazard. Foodborne hazards occur at all stages of the Agri-Food system from farming, to processing/manufacturing, retail and foodservices, and during home preparation.[1]

[1] Non-farm sources of food, e.g. wild meat trade has been

R. Owusu-Apenten, E. R. Vieira, *Elementary Food Science*, Food Science Text Series,
https://doi.org/10.1007/978-3-030-65433-7_8

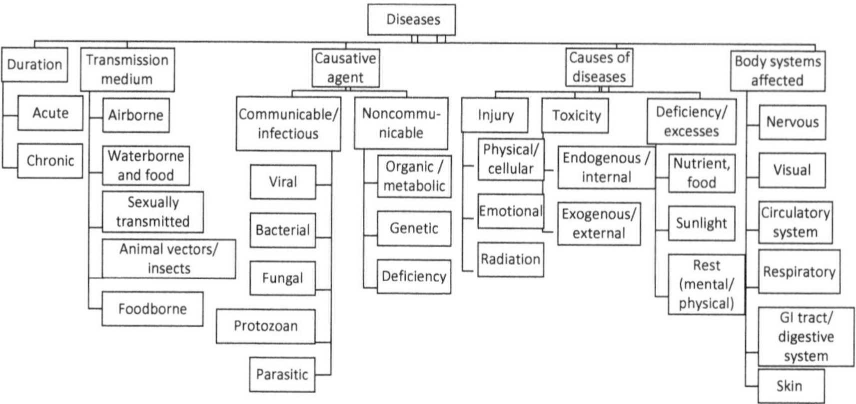

Fig. 8.1 Classification of some human disease. (Adapted from Wikipedia)

The main classes of foodborne hazards are chemical, physical, biological and microbiological agents. The chemical hazards include substances added to foods as processing aids, food additives or other substances, which function as unintentional contaminants, e.g. components may leach from food contact materials and packaging into foods. Other potential foodborne chemical substances include antibiotics and veterinary drugs used in the care of livestock.

Public health controversies concerning ground and surface water contamination caused by chemical spills are well publicized and have initiated regulatory action (Borchers et al. 2010; Jackson 2009). The effects of chemical hazards in foods are a topic of ongoing studies but there is keen interests these days concerning whether obesity can be partly blamed on the effect of hormone-like chemical agents unintentionally added to foods (Simmons et al. 2014).

Food grade industrial chemicals need handling carefully by all workers in the food system, and training in their proper use and storage must be stressed. There are concerns regarding chemicals to be controlled by foodservice managers and food processors covering the following: (i) Contamination with detergents and sanitizers; (ii) possible excessive use of preservatives, and spices; (iii) chemical action with containers such as acids with metal-lined containers; and (iv) toxic heavy metal contamination by lead within food production areas. It is necessary to avoid copper from water lines exposed to carbonated or other acidic beverages or contamination by cadmium from uncovered meats that come in extended contact with refrigerator shelves made from metal.

Physical hazards arise from particles or items that are not intended to be part of the food product. Chips of broken glass, metal fragments such as staples from packaging or metal curls from can openers, or even cherry pits in a "depitted" canned cherry are included in this group. Strict quality control and examination of raw and finished products are essential to control this hazard.

Microbiological hazards include the food poisoning resulting from the activity of microbes on the food. In many instances, grain used for animal feeds can become contaminated with mold growth during storage, the molds producing toxic materials that cause diseases in the animals to which the grain is fed.

Almost all the factors that contribute to foodborne diseases result from ignorance of proper food-handling procedures or the unwillingness of some individuals in the food industries (including food processors) to comply with the basic guidelines for proper food handling. Thus, foodborne diseases will continue to occur at unnecessarily high rates as long as (1) food handlers do not employ strict sanitation in both their personal hab-

a consistent source of microbial foodborne diseases ranging from tuberculosis, Ebola to Covid 19 viral epidemic of 2020.

its and in the maintenance of their work area and equipment, (2) foods are not properly refrigerated, (3) foods are not adequately processed, (4) cross-contamination is not avoided, and (5) management does not realize the importance of incorporating practices that prevent foodborne diseases.

Proper food handling procedures cover simple techniques, such as holding at specified temperatures, but they also include complex procedures, such as calculating processing times and predicting certain biochemical reactions that might result from processing modifications. Ordinarily, therefore at some level in the food-processing stage, the services of a professional food technologist are required. Once a food has been properly processed, the remaining handling steps require periodic quality control checks by personnel qualified in microbiology and sanitation.

Foodborne diseases are mostly caused by several species of bacteria, although viruses, parasites, amoebas, and other biological, as well as chemical, agents may be responsible. The disease-causing bacteria in foods, or their end products that cause disease, are transmitted by people consuming the food (Table 8.1). Foodborne diseases are classified as food infections and food intoxications, although the distinction between the two categories is not clear in some cases (e.g., *Bacillus cereus* and *Clostridium perfringens*).

Table 8.1 Organisms causing foodborne disease in the US (1990)

	Estimated	% Total
Bacterial		
Bacillus cereus	27,230	0.1
Botulism, foodborne	58	0.0
Brucella spp.	1554	0.0
Campylobacter spp.	2,453,926	6.4
Clostridium perfringens	248,520	0.6
E. coli O157:h7	73,480	0.2
E. coli O157:STEC	36,740	0.1
E. coli, enterotoxicgenic	79,420	0.2
E. coli (other)	79,420	0.2
Listeria monocytogenes	2518	0.0
Salmonella typhy	824	0.0
Salmonella, nontyphoidal	1,412,498	3.7
Shigella spp.	448,240	1.2
Streptococcus, food	50,920	0.1
Staphylococcus	185,060	0.5
Vibrio (other)	7880	0.0
Vibrio cholerae	54	0.0
Vibrio vulnificus	94	0.0
Yersinia enterocoltica	96,368	0.2
Bacteria subtotal	5,204,804	13.5
Parasitic		
Trichinella spiralis	52	0.0
Cyclospora cayetanensis	16,264	0.0
Toxoplasma gondii	225,000	0.6
Giardia lamblia	2,000,000	5.2
Cryptosporidium parvum	300,000	0.8
Parasite subtotal	*2,541,316*	*6.6*
Viral		
Hepatitis A	83,931	0.2
Rotavirus	3,900,000	10.1
Astrovirus	3,900,000	10.1
Norwalk-like	23,000,000	59.5
Viral subtotal	*30,883,931*	*79.9*
Grand total	38,630,051	100.0

Adapted from Mead et al. (1999)
Notes: Escherichia coli (ETEC) is Enterotoxigenic, E. Coli (STEC) O157 STEC is shiga toxin–producing, E. coli is other than STEC and ETEC, non-O157 diarrhea genic

1.3 Food Infections and Outbreaks

Foodborne disease refers to illnesses carried or transmitted to humans via the medium of food (Centers for Disease Control and Prevention 2014). Food infection refers to situations where a disease organism is carried through foods to the host (human or animal), where it actually invades the tissues and grows to numbers large enough to cause disease. A foodborne disease outbreak is defined by the International Association for Food Protection (IAFP)[2] as, "an incident where two or more people have the same disease, similar symptoms, or excrete the same poisonous microorganisms (pathogens)". A time, place, or person association between these people must be established. Elements of microbial foodborne diseases are discussed in the rest of this chapter.

[2]Formerly the International Association of Milk, Food and Environmental Sanitarians (IAMFES), the organization was renamed in 1999.

1.4 Epidemiology of Foodborne Disease

Readers will be broadly aware of classical epidemics, such as influenza, Ebola and other ominous diseases discussed in the media. Sadly, a smaller number of people will be familiar with foodborne disease epidemics that affect millions of people annually, and which are transmitted from person to person, via the medium of food (Altekruse et al. 1997; Tauxe 1997; Osterholm 2011; Scallan et al. 2011).

Tracking foodborne disease incidence is the job of epidemiologists. Within the United States, the Center for Disease Control and Prevention (CDC) has responsibility for monitoring foodborne disease. The CDC compiles information about foodborne disease outbreaks, mortality, types of microorganism implicated, and food practices associated with infections (Centers for Disease Control and Prevention 2001, 2003, 2004, 2006, 2009).

1.5 Economic Impact of Foodborne Disease

An influential report from 1999 indicated there were 76 million cases of routine foodborne illness annually resulting in 325,000 hospitalizations and 5000 extra deaths in the US (Mead et al. 1999). Foodborne illnesses cost US$26 billion annually due to numbers of workdays lost, and the cost of treatment.[3] A further reports from 2011 showed 47.8 million cases of foodborne illnesses, 127,000 hospitalizations and 3037 deaths per year (Scallan et al. 2011a, b).

The majority of microbes causing illness were unidentified. About one fifth (20%) of illnesses were ascribed to 31 named pathogens (Table 8.1). For outbreaks caused by known microorganisms, 50–80% were linked with viral agents (cf. Sect. 2).

Global impact estimates from the WHO showed 600 million cases of foodborne diseases annually (1 in 10 of the world population) with children under 5 years old disproportionately (40%) affected. A total of 420,000 adult deaths and 125,000 children's deaths arise from foodborne diseases per year. The most prolific causative agents for foodborne infections worldwide were noroviruses; bacterial causative agents included salmonella, campylobacter and E. coli. The WHO report noted listeria as a cause of premature births. The 550 million cases of foodborne diarrhea diseases produced 230,000 deaths (World Health Organization 2019a) (Table 8.1).

1.6 Under Reporting Foodborne Illnesses

Reported foodborne diseases represent a small percentage of actual cases. Under reporting is summarized using a food poisoning pyramid where the lower third represents cases of food poisoning that are unreported and for which no medical attention is sought. In most unreported cases of foodborne disease, the victims do not associate their symptoms with the consumption of a particular meal. The middle third of the pyramid represents cases where the sufferer visits their doctor but where stool specimens are not taken, no identification of microorganisms are made and no formal reporting is made to State or Federal authorities charged with monitoring the incidence of food poisoning. Finally, the final tip of pyramid represents the number of documented cases of food poisoning (Arendt et al. 2012).

Common reasons leading to underreporting of foodborne illness included the fact that subjects, did not know who to contact or were too ill to undertake reporting. Frequently also there was uncertainty regarding the causes of illness, and concerns whether reporting would be taken seriously by the authorities (Arendt et al. 2012).

1.7 Microbes Implicated in Foodborne Disease

Jones and Gerber from the Tennessee Department of Health showed considerable disparity between the identity of disease causing microbes as *per-*

[3] Adjusted for inflation US$26 billion i(1999) is equivalent to US$40 billion (2020)

ceived by health workers, and the identity of actual microorganisms implicated in foodborne illness (Jones and Gerber 2001). Survey results from public health workers implicated *salmonella, E. coli, staphylococcus,* and *shigella* as the four bacterial pathogens most involved in foodborne illness in the United States.

Actually the most frequent causes for foodborne disease in the United States were norovirus (~66%). For bacterial foodborne infections, the order of activity was campylobacter > salmonella > *Clostridium perfringens* > *E. coli,* >> *listeria* (Fig. 8.2).

The reasons for the disparity between perceived and actual causes of foodborne diseases remain puzzling. Perhaps those infections that produce severe symptoms or those that receive widespread news coverage make greater impact on public awareness (Jones and Gerber 2001). The pattern of infectivity in the US was replicated in other jurisdictions (Li et al. 2018). Viral foodborne diseases were a common cause of non-bacterial gastroenteritis worldwide (World Health Organization 2019a).

1.8 Attribution to Different Foods

Identifying foods associated with a high risk of foodborne disease is essential so that limited resources can be directed where they are most needed (Painter et al. 2013). A 10-year (1998–2008) study by CDC experts looked at foodborne diseases due to bacterial, viral, parasitic and chemical agents and identified specific food groups implicated in outbreaks. About 5000 separate outbreaks and 9.5 million illness were mapped with 17 food groups.

When all all-outbreaks are considered together 51% of cases were due to plant foods, 41% outbreaks were due to animal foods and 6% was due aquatic foods. The nature of hazards was distinct for different food groups. Aquatic foods tend to pose chemical and parasitic hazards, whist plant foods and animal foods were mainly associated with viral and bacterial agents (Fig. 8.3). Of 17 food commodity groups examined, the most risky in terms of percentage outbreaks were, 22% leafy vegetables, 17% dairy foods, 12% fruit and nuts and 10% poultry. (Painter et al. 2013;

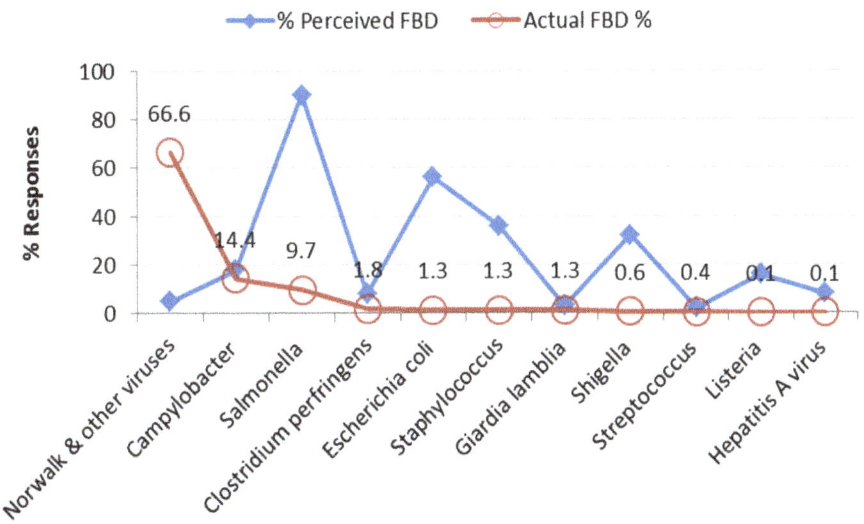

Fig. 8.2 Comparing actual and perceived causes of foodborne disease (FBD) in Tennessee (Jones and Gerber 2001)

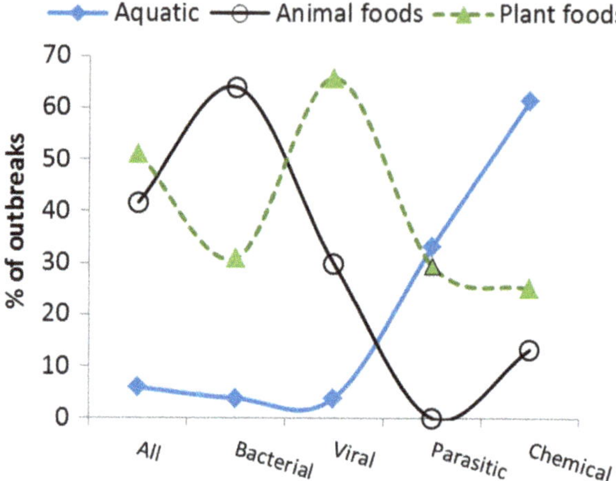

Fig. 8.3 Attribution of different types of foodborne illnesses to animal, plant or aquatic foods. (Drawn using data from Painter et al. 2013)

Centers for Disease Control and Prevention 2017). Based on earlier results (Bryan 1988) the main food vehicles for foodborne illness transmission might change depending on the range of risk factors prevailing at any time (cf. Chap. 9, Table 9.1 & Table 9.2).

Food linked frequently with the hospitalization of foodborne disease victims were (46%) animal foods, (41%) plant foods and (6%) aquatic food. Deaths from foodborne illness were mainly (43%) due to animal based foods, followed by (25%) plant foods and (6%) from aquatic. Poultry was identified as the single most hazardous commodity in terms of deaths from foodborne disease accounting for 19% of all deaths. Poultry tended to be infected by *Listeria monocytogenes* or *Salmonella spp*. at 63% or 26% of all occasions. (Painter 2013). The reservoirs, alternative environment sources, and transmission routes for these zoonotic pathogens continue to be investigated (Chlebicz and Slizewska 2018).

2 Foodborne Disease and Infections

2.1 FoodNet and Sources of Information

Information on foodborne disease has grown enormously the over past two-decades but CDC is the most reliable resource for anyone researching foodborne diseases (Sect. 1.2). Another comprehensive source of information on foodborne illness is FoodNet (Foodborne Diseases Active Surveillance Network).

FoodNet arose from a 1995 initiative involving the CDC and health departments from 10 states. The goal of FoodNet was to monitor and provide information on the foodborne disease burden for the whole of the United States. FoodNet now covers about 45 million people and provides the most current reporting on disease outbreaks FoodNet produces periodic reports and is also an authoritative source of information for some other influential publications on foodborne diseases (Chong et al. 2000; Lin et al. 2005; Mead et al. 1999; Scallan et al. 2011a, b).

A few of the notable foodborne disease agents are covered in the rest of this section, but its noteworthy that some of same agents can be contracted via water sources, and also as nosocomial or hospital acquired infections.

2.2 Norovirus (Norwalk Virus)

Noroviruses are some of the most prolific causes of foodborne disease and the least recognized. Noroviruses are associated with non-bacterial gastroenteritis, stomach flu or stomach bug with a 50% infection rate requiring about 18 virus particles (Matthews et al. 2012, Li et al. 2015). Norovirus is highly contagious, and can persist in the environment on surfaces, hands and clothing (Centers for Disease Control and Prevention 2015d). Sources of norovirus include food and water. An usually forthright description of norovirus infections, states "*you can get norovirus by accidentally getting the feces or vomit of an infected person in your mouth*". Norovirus infections occur from eating contaminated food or water. Norovirus infections were also associated with contaminated cutlery, and contact with a contaminated surface.

The incidence of norovirus disease was estimated at 19–21 million cases per year, between 56,000 and 71,000 hospitalizations and 570 to 800 deaths annually. According to conservative estimates, about 20% of all types of gastroenteritis and 50–60% of all foodborne illnesses were attributed to noroviruses (Belliot et al. 2014).

At present, norovirus infections occur most frequently within health care facilities (62.7%) compared with restaurants and banquets (22.5%), school and childcare facilities (6.1%) and other physical settings (7.2%). Cruise ships accounted for 1% of norovirus infections nationally; nevertheless about 90% of diarrhea infections on-board ship were attributed to norovirus from contaminated surfaces. Within foodservice establishments, the descending order of norovirus infections was sit-down restaurants (61%), catering and banquet (14%), institutional foodservice (3%), hospitals and nursing homes (2%) and places of worship (1%). Home dining accounted for 10% of infections (De et al. 2018; Centers for Disease Control & Prevention 2021). In nursing homes about 0.7% of outbreaks were foodborne compared with 28% from person-to-per contact (probably involving vomit) and 70% due to unidentified sources (Petrignani et al. 2015). Norovirus infections were a common reason for hospital closure. The incidence of norovirus infections in children (<5 years) was about fivefold higher than the incidence in adults (Baert et al. 2011; Belliot et al. 2014; Dicaprio et al. 2013).

Norovirus is reported to persist in the environment and on solid surfaces, which may explain the high infectivity of this agent. On the other hand, human noroviruses have proven very difficult if not impossible to culture outside of the human body (Barclay et al. 2014; Cromeans et al. 2014; Richards 2012). If non-human noroviruses strains from cats and mice are used as surrogates, they may not completely mimic their human counterparts. One consequence is that sanitation studies may not be transferable to actual behavior of norovirus in real life situations (Tung et al. 2013). Despite such reservations about surrogate viruses, the research is suggesting strategies for the control and prevention of norovirus transmission. Hand washing, using alcohol wipes, and disinfection of surfaces using quaternary ammonium sanitizers have all been proposed for controlling norovirus spread (Tung 2013, De et al. 2018).

2.3 Campylobacter, Campylobacteriosis

Campylobacter came to the attention of experts mostly from 1977 (Penner 1988). Initial research suggested that *Campylobacter jejuni* was implicated in 3–6% of diarrhea incidents in the United States (Peterson 1994). In the following two decades, it became evident that Campylobacter was responsible for about 14% of all incidences of foodborne illnesses (DuPont 2007). Campylobacter associated conditions have increased over the past 10 years in both developed and developing countries (Kaakoush et al. 2015). CDC data suggests that nearly 1-in-2 of all store-bought raw chickens is contaminated with campylobacter. Poultry is a major reservoir for campylobacter but cases are also associated with, water, shellfish and un-pasteurized milk (Allos 2001).

Campylobacter (14 + different types) grow optimally at the body temperature of poultry (41 °C) and most poultry appear able to carry this pathogen without ill effect. By contrast, about 500 CFU of bacteria are thought to be sufficient to cause illness in humans. Chopping boards, cross contamination from raw chicken, and undercooked barbecues are likely sources of campylobacter infections, which occur most frequently during the summer months; some evidence implicates the common housefly as an agent for the mechanical transmission of campylobacter (Ekdahl et al. 2005). Currently, campylobacter infections are the most frequent foodborne bacterial infection with an estimated 1.3 million cases annually. The symptoms of campylobacter infection, which appear within

2–5 days of contamination, include diarrhea, cramping and stomach pains and fever (Centers for Disease Control and Prevention 2015b).

2.4 Salmonellosis and Typhoid fever

Salmonellosis is caused by eating food contaminated with *Salmonella* bacteria. Every year there are an estimated 1.2 million cases, 19,000 hospitalizations and 380 deaths associated with of salmonellosis (Scallan et al. 2011a, b). Since only about 1% of acute digestive illnesses are reported in the United States, then the actual cases of salmonellosis may be far higher than official figures indicate (Centers for Disease Control and Prevention 2015h, i).

Typhoid fever, of which there are fewer cases than of salmonellosis, and hence fewer deaths, is caused by an organism belonging to the Salmonella species (S. Typhi), but this disease is usually not considered to be the ordinary salmonellosis for three reasons. Firstly, *Salmonella* can infect other animals as well as humans, whereas the typhoid germ is known to infect only humans. Secondly, typhoid fever is usually more severe than the ordinary salmonellosis; and thirdly, healthy adults must ingest an infective dose of about 100 non-typhoid *Salmonella* bacteria (cells) in order to cause salmonellosis, whilst about 500–1000 *Salmonella Typhi* cells are needed to cause typhoid fever (Hara-Kudo and Takatori 2011; U.S. Food and Drug Administration 2013; Kurtz et al. 2017).

New research is revealing how salmonellosis differs from typhoid fever. The bacteria agents responsible for salmollenosis and typhoid fever are closely related, but the former (Table 8.2) is efficiently arrested by intestinal cells leading to localized inflammation response (gastroenteritis) and diarrhea. By contrast, various slight genetic differences in the structure of *Salmonella Typhi* allows this bacterial strain to by-pass or evade the innate-immune system localized at intestinal cell lining. After gaining access to underlying tissue, *Salmonella Typhi* then passes into the circulation and multiplies rapidly. For people with HIV-AIDS and others with a weakened immune response, the non-typhoid and typhoid infections tend to be similarly dangerous (Raffatellu et al. 2008; Gal-Mor et al. 2014).

Table 8.2 Symptoms of salmonellosis

Intestinal symptoms
Abdominal cramps
Diarrhea
Fever
Headache
Nausea
Vomiting
Extra-intestinal, infections
Bacteremia (blood)
Meningitis (CNS membranes lining)
Osteomyelitis (bone)
Septic arthritis (joint)

Centers for Disease Control and Prevention (2015h, i)

The incidence of typhoid fever is low in the US. However, the chances of exposure may increase for long-haul travelers to areas with low food and water sanitation. Some international travelers are advised to consider typhoid fever vaccination. The occurrence of antibiotic resistance (multiple drug resistance) in *Salmonella Typhi* strains is another reason for the general concern with typhoid fever (Centers for Disease Control and Prevention 2011).

The ordinary symptoms of salmonellosis are abdominal pain, diarrhea, chills, and frequent vomiting. However, there are instances in which much more severe symptoms may be encountered. For example, the salmonella may pass to non-gastrointestinal sites leading to infections affecting the blood, muscles, brain outer membrane. The incubation period (time after ingesting the organisms until symptoms are evident) is 7–72 h. In typhoid fever, the incubation period is 7–14 days (Table 8.2).

Persons with salmonellosis can become carriers of the organism for a period after they have recovered from the disease. That is, they continue to discharge the organisms in their feces. Because of this, carriers often contaminate their hands with these organisms, which may not be removed completely after thorough washing. Hence, carriers may contaminate food with these bacteria, and in this manner, transmit the disease to others. In most salmonellosis cases, the carrier stage does not persist longer than 12 weeks after symptoms, and with typhoid fever, for shorter periods. However, there are isolated cases in which the carrier stage lasts much longer than 12 weeks, and 2% to 5% of those ill with typhoid fever may become permanent carriers.

The *Salmonella* bacteria are rod-shaped; they do not form spores, and thus they are not especially heat-resistant. They are motile (can move about in the water, in foods, or other materials in which they are found) and will grow either with or without air (oxygen). Now, more than 2000 types of *Salmonella* bacteria are known, all of which are considered infective to man. These *microorganisms are* widespread in nature, but there has been a great deal of attention spent on poultry and eggs salmonella (Gormley et al. 2010; Howard et al. 2012; Maciorowski et al. 2004; Samiullah 2013; Wight et al. 1996).

Whereas it is considered that many *Salmonella* bacteria must be consumed to cause the disease in a normal adult, it is known that the very old and especially the very young may contract the disease after ingesting only a few of these organisms. Therefore, any food, especially food eaten without cooking, should be kept essentially free of these bacteria.

Salmonella grow at temperatures near 95 °F (35.6 °C), but they will also grow more slowly at both higher and lower temperatures than this. It has been found that many foods are suitable for the growth of these organisms and that in some foods they will grow slowly at temperatures as low as 44 °F (6.7 °C) or as high as 114 °F (45.5 °C). Moreover, since the destruction of bacteria by heat is a matter that involves both time and temperature as well as the degree to which a food protects bacteria; temperatures as high as 140 °F (60 °C) may be required to bring about marked decreases in levels of these bacteria during the cooking of some foods.

Regarding the destruction of bacteria by heat, it should be explained that as they are heated and a temperature is reached at which they are destroyed, they are not all destroyed at once. For instance, if at about 140 °F (48.9 °C), 90% of the organisms would be destroyed in a period of 5 min, it would take 10 min to kill 99% of the organisms, 15 min to kill 99.9% of the organisms, and so on.

Table 8.3 USDA, FSIS regulation for the production of cooked beef, roasts beef, and cooked corned beef

Degrees Fahrenheit	Degrees Centigrade	Time (minutes)[a]
130	54.4	121
131	55.0	97
132	55.6	77
133	56.1	62
134	56.7	47
135	57.2	37
136	57.8	32
137	58.4	24
138	58.9	19
139	59.5	15
140	60.0	12
141	60.6	10
142	61.1	8
143	61.7	6
144	62.2	5
145	62.8	Instantly

[a]Processing time once temperature is reached

If some of these *Salmonella* organisms in foods survive whatever heating they receive during cooking, and the food is thereafter held at temperatures at which they will grow [44–110 °F (6.7–43.3 °C)], especially at room temperatures, the organisms may grow again to large numbers. Table 8.3 shows the USDA, FSIS regulation for the production of cooked beef, roast beef, and cooked corned beef. These times and temperatures will ensure destruction of *Salmonella* organisms:

Certain types of cooking are not sufficient to destroy all *Salmonella* bacteria that may be present in foods. Examples of cooked foods in which these organisms may survive are scrambled, boiled, or fried eggs and also meringue, turkey stuffing, oysters in oyster stew, steamed clams, and some meat dishes. Foods eaten raw or without further cooking, such as clams, oysters, milk powder, cooked crabmeat, and smoked fish, may be infective should they be contaminated with *Salmonella* bacteria, especially because they may be eaten by young people who are quite susceptible to infections of this kind.

Shellfish, egg products, prepared salads, and to some extent, raw and cooked meats have often been associated with the transmission of salmonellosis. Raw shellfish may be taken from waters contaminated with *Salmonella* bacteria and cooked shellfish meats may be contaminated by humans because they are usually removed from their shell by hand. Poultry of all types, beef cattle, and hogs may have salmonellosis or may be carriers of the organism causing the disease; hence under conditions of cooking in which these organisms are not destroyed, they may be transmitted to humans.

It has been reported that 2.6% to 7% of all fresh eggs are contaminated with *Salmonella.* The USDA recommends cooking all eggs until the yolks are solid to ensure destruction of all *Salmonellae.*

Pets, such as cats and dogs, can have salmonellosis and be carriers of the causative organism. As this is the case, children, especially the very young, who handle materials contaminated by pets and who are very susceptible to contracting the disease, may contract salmonellosis.

It may seem curious that animals can have salmonellosis. The reason appears to be that their feeds, especially fish meal, meat meal, and bone meal, fed to them as supplements, often contain *Salmonella* bacteria. This is also true of some of the dried types of food used for feeding pets. There are methods or procedures that would help to eliminate salmonellosis or greatly decrease the number of cases.

(a) Good sanitation methods and procedures in food a manufacturing plants, in restaurants and institutions serving food, and in the home are essential. This includes not only cleaning and sanitizing of equipment and utensils; elimination of insects and rodents; and cleaning and sanitizing of floors, walls, and so on, but also the personal cleanliness of workers who prepare or serve food. All personnel must wash and sanitize their hands prior to handling foods once having left their work stations for any reasons.
(b) To ensure safe food handling, all foods should be held at temperatures of 40 °F (4.4 °C) or below when not being cooked, prepared for cooking, or being served. This would not eliminate *Salmonella* organisms from foods, but it would prevent salmonellosis in healthy adults,
(c) Foods that can be eaten without further cooking should be produced under the best conditions of sanitation, and some foods of this type, such as milk powder, should be subjected to frequent bacteriological examination to determine that they are essentially free of *Salmonella* bacteria.
(d) Where possible, foods (poultry stuffing, etc.) should be cooked to temperatures of at least 150 °F (65.6 °C), at which it can be assumed that all *Salmonella* bacteria have been destroyed.
(e) Egg products (dried or frozen) should be pasteurized (heated to 140 °F (60 °C) for 3–4 min), then cooled prior to drying or freezing,
(f) Flocks of poultry known to have salmonellosis (this can be determined by testing) should be eliminated as egg producers. This can be done without economic loss because such poultry may be used as food,
(g) Any food, especially protein supplements, given to pets and other animals, should be treated to eliminate the *Salmonella* bacteria that infect these animals. This can be done, for instance, by pelletizing the food or supplements, a treatment that raises the temperatures sufficiently to destroy many of the bacteria that might be present,
(h) Animals that have died, from disease should not be eaten. (This rule is sometimes not observed on farms). Additional information on salmonella is available from, Centers for Disease Control and Prevention (2015h, 2015i) and website updates.

2.5 Shigellosis

Shigellosis also known as bacillary dysentery is a food and water-borne infection caused by bacteria of the genus (group) known as *Shigella.* Currently, the incidence of shigellosis is between 130,000 and 500,000 cases per year in the United States. International surveys suggest that cases of shigellosis are correlated positively with economic underdevelopment. Globally there was an estimated 80–165 million cases per year with 1.1 million deaths. About 99% of cases occur within developing countries (Bardhan et al. 2010; Ram et al. 2008). A study for 2010 estimated an annual mortality of 19,451 and 20,692 deaths due to Shigella in Africa and South East Asia regions (Lamberti et al. 2014).

The symptoms of shigellosis are diarrhea, abdominal cramps, and some fever. In severe cases, the symptoms are much more complex. There are thought to be five pathogenic species or types of *Shigella* bacteria as compared with more than 2000 for *Salmonella.* However, one of these organisms, *Shigella dysenteriae,* usually causes a much more severe disease than either *Salmonella* or other *Shigella* bacteria. The incubation period may be long as 7 days with an average of 4 days.

As in the case of salmonellosis, humans may become carriers of the *Shigella* organism. Person to person transmission of shigellosis has been noted and CDC data notes at risk groups to include, travelers, young children, non-heterosexual men, and HIV sufferers (Centers for Disease Control and Prevention 2015j).

It has been reported that the optimum growth temperature for *Shigella* bacteria is about 98.6 °F (37 °C) and that the range in which growth can occur is 50–104 °F (10–40 °C). They are relatively sensitive to heat and can tolerate up to 6% salt. Certain control methods are noted based on the knowledge of how food and drink should be handled;

(a) Since water is a known source of the organism causing shigellosis, all water used for drinking, for ice making, for adding to foods, or for the cleaning and sanitizing of equipment and eating utensils should be potable (drinkable). The portability of water is determined by sanitary surveys and by bacterial tests made on samples of the water. Any food-manufacturing or food-serving operation should use only water known to be potable. Municipal water supplies should be checked periodically to determine that they are not polluted. In some cases, water from deep wells from sources not connected with the municipal water supply is used for foods and cleaning. In addition, seawater is sometimes used for cleanup in food-serving or food-manufacturing operations. In such instances, management should arrange for bacterial tests to be made on such water supplies at frequent intervals to determine that they are not polluted,
(b) Because it is good practice in food handling to refrigerate foods not in use, another method of controlling shigellosis is to hold foods at 40 °F (4.4 °C) or below at all times when they are not being served or prepared for serving,
(c) Humans may become *Shigella* carriers; therefore, personnel known to have had an intestinal disease should be excluded from tasks that would bring them into direct contact with any food.

2.6 Vibriosis (Vibrio Illness)

Vibrio represents a large group of comma (curved rod) shaped bacteria found in salt-water and estuarine environments. CDC reports show that about six pathogenic forms including, *Vibrio vulnificus, Vibrio parahaemolyticus* and *vibrio cholera.* The first two strains produce vibrio illness, which features gastrointestinal symptoms occasioned by eating undercooked salt-water food, e.g. prawn, crab, oyster, squid etc. Current estimates show about 80,000 Vibriosis cases, 500 hospitalizations and 100 deaths in the US every year. There is some uncertainty in such data, due to under-reporting (Centers for Disease Control and Prevention 2015k).

Large outbreaks were reported 1997, 1998, and 2006 and since 2007, Vibriosis has been designated as a notifiable disease with data collated centrally by the CDC (Centers for Disease Control and Prevention 2015l). This disease is contracted almost solely from seafood, but direct infection through open wounds is also possible (Daniels 2011; Oliver 2005; Rose et al. 2001).

The symptoms of Vibriosis are abdominal pain, nausea, vomiting with diarrhea. A fever involving a 1–2 °F (0.55–1.1 °C) rise in body temperature is experienced in 60% to 70% of the cases. The period of incubation after ingestion of the organism is 24 h and the symptoms last for 1–2 days. It is questionable whether or not people who have had this disease become carriers.

The organism causing Vibriosis is a short, curved, rod-shaped, motile cell. It grows with or without oxygen and is believed to require 2% to 4% sodium chloride for growth. The organism is found naturally in the ocean, and, because it grows fastest at 86–104 °F (30–40 °C), is found in highest concentrations near the shoreline during the summer months. The likelihood of Vibriosis is higher in warmer months and 70% of cases occur from May to October.

In the United States, the food most often involved in the transmission of Vibriosis is oyster, but it is known to be present sometimes in cooked crab meat. Oyster beds may become affected in brackish and warm water. The control of Vibriosis involves some of the following precautions:

(i) Because *Vibrio* is found in high concentrations in seawater only during the period in which coastal waters are quite warm, it is preferable not to eat *raw mollusks* (squid, octopus, clams, and oysters) during the months of May to October when coastal waters are warmest. Some believe it is too risky to eat raw mollusks at any time!
(ii) Because *Vibrio parahaemolyticus* is quite heat-sensitive and would be destroyed by whatever cooking is necessary to remove crab meat from the shell, the presence of the organism in cooked crab meat must have been due to contamination after cooking. This indicates that good food plant sanitation is a method of controlling Vibriosis. The use of potable fresh water that has been adequately chlorinated for the cleaning and sanitizing of food plants and food-plant equipment is a method of controlling Vibriosis.

2.7 Foodborne Tuberculosis

An estimated 12.5 million cases of tuberculosis (TB) were reported of which 12% affected HIV positive individuals. Tuberculosis is caused by mycobacterium species in particular *Mycobacterium tuberculosis* which is a human pathogen. Person-to-person transmission of TB can occur via airborne aerosols in enclosed overcrowded places (de la Rua-Domenech 2006). The agent causing *bovine* TB is *Mycobacterium bovis*, which is associated with cows, small ruminants, domestic animals and wildlife. *Mycobacterium bovis* is transmissible to humans via raw milk obtained from infected cows, camels, goats, sheep and perhaps water buffalo (Boukary et al. 2012; Pesciaroli et al. 2014; Currier and Widness 2018). Bovine TB is a zoonotic disease (i.e. transmissible from animals to humans) and so raw or undercooked meat is another potential source of human infection (Etter et al. 2006).

Foodborne tuberculosis was common in the past but started to decline between 1920–1980 owing to the introduction of milk pasteurization (Fairchild and Oppenheimer 1998). A survey of foodborne illnesses associated with milk from 1980–1985 found many infections from Salmonella spp. and Campylobacter spp. with only a cursory mention of mycobacteria (Sharp 1987). However, there was concern that bovine TB might be resurgent in the US (Schneider and Castro 2003) and Europe (Torgerson and Torgerson 2010) since the 1980s.

Unpasteurized milk was a major transmission route for bovine foodborne TB in the late nineteenth and early twentieth century (Fitzgerald and Kaneene 2013). A noticeable resurgence of TB in cattle from the UK and elsewhere in Europe has been noted recently. However, there appears to be no evidence of TB infections from milk leading some experts to question efforts to eliminate TB in cattle by culling wildlife known to be carriers. Wildlife host for tuberculosis include white tailed deer and elk (north America), badgers and wild boar (Europe), brush-tailed possum (New Zealand), and the lechwe and wild buffalo from Southern Africa (Fitzgerald and Kaneene 2013).

2.8 Brucellosis

Brucellosis also known as undulant fever is another of the diseases that in the past was transmitted through raw milk. Brucellosis was controlled in recent years by milk pasteurization and the testing of dairy herds, which eliminated infected animals (Sharp 1987). *Brucella*

Table 8.4 Symptoms of human Brucellosis

Early symptoms	Later symptoms
Anorexia	Arthritis
Fatigue	Chronic fatigue
Fever	Depression
Headache	Meningitis
Malaise	Neurologic symptoms
Pain in muscles, joint, and/or back	Recurrent fevers
Sweats	Swelling of the heart (endocarditis)
	Swollen liver and/or spleen
	Swollen testicles and scrotum area

Adapted from Centers for Disease Control and Prevention (2015a)

melitensis can be transmitted through drinking or eating unpasteurized dairy products; brucellosis is also regarded as an occupational hazard for those working closely with animals, e.g. laboratory animal and zoo animal attendants. The symptoms of brucellosis are either acute fever or longer debilitating conditions (Table 8.4).

Internationally, brucellosis is described as a neglected and under recognized condition with high incidences (100 cases per 100,000 population) noted for Mediterranean, Middle East, and central Asia countries, e.g. Egypt, Iraq, Jordan, Saudi Arabia, Kyrgyzstan, and Azerbaijan. It is thought that brucellose may be endemic also in Mexico and parts of Latin America (13–26 cases per 100,000 of population). Brucellosis infections are associated with domesticated animals (including, camels, goats, sheep) - either living in close proximity with humans and or when products from such sources are undercooked (Dean et al. 2012; Mantur and Amarnath 2008; Rubach et al. 2013).

2.9 Other Food Infections

Only a limited number of the 31 pathogens identified as contributing frequently to cases of foodborne illness (Table 8.1) are described in Sects. 2.2, 2.3, 2.4, 2.5, 2.6, 2.7 and 2.8. Other pathogens that are known to cause severe symptoms include *Listeria monocytogenes and Escherichia coli* (serotype 0157:H7).

Listeria monocytogenes has attracted attention because of two cheese-related outbreaks that resulted in cases of meningitis, miscarriage, and perinatal septicemia. *L. monocytogenes* was implicated in well over 300 reported cases of listeriosis that resulted in approximately 100 deaths. These outbreaks also had a significant economic impact on the dairy industry, which suffered more than $66 million in losses due to product recalls (Ryser and Marth 1989). Listeria outbreaks for decade 1998–2008 between have been described (Cartwright et al. 2013). The ability of this organism to grow at refrigeration temperatures in combination with its appearance in raw and processed meats, poultry, vegetables, and seafood make it a serious threat to susceptible consumers and to the entire food industry. *L. Monocytogenes* has also survived the production conditions and storage of many types of cheeses (Ryser et al. 1985; Papageorgiou and Marth 1989; Piccinin and Shelef 1995).

*Yersinia enterocolitica (*Table 8.1*)* is a psychrotrophic organism that has been a causative foodborne pathogen in raw and pasteurized milk, and contaminated water. Unlike *L. monocytogenes,* it is destroyed by normal pasteurization. Yersiniosis usually results in enterocolitis, characterized by diarrhea, fever, and severe abdominal pain in the lower right quadrant. This illness sometimes has been incorrectly diagnosed as appendicitis and has resulted in unnecessary appendectomies. The organism is widely distributed and outbreaks have been traced to a wide variety of foods (Rahman et. al. 2011) such as milk that had contaminated chocolate syrup added to it after pasteurization (Black et al. 1978) and tofu that was packed in contaminated spring water (Tacket et al. 1985).

A strain of *Escherichia coli* (serotype 0157:H7) is considered the causative agent in hemorrhagic colitis, resulting in severe abdominal cramps and watery, grossly bloody diarrhea and hemolytic uremic syndrome, which is the leading cause of acute renal failure in children (Boyce et al. 1995). Vehicles of infection for this organism include raw or undercooked ground beef, undercooked ground beef sandwiches, raw milk, pork, lamb, and chicken. It is recommended to cook all ground beef to the well-done stage, which can be reached when the meat is held at 155 °F (68.3 °C) for at least 15 s. Prevention of this disease can be by proper cooking of ground beef and prevention of cross-contamination from raw foods and food handlers (Centers for Disease Control & Prevention 2021). Other food infections implicated *Bacillus cereus, Aeromonas hydrophilia, Arizona hinshawaii,* and *Plesimonas shigelloides* (Riemann and Cliver 2006).

3 Food Intoxications

Not all foodborne hazards are unnatural agents. Natural compounds formed by fungi, algae and some bacterial are highly potent toxicants. Food intoxications are conditions where a a living organism produces substances within the food that is toxic to humans and other animals. Typically, the toxicants remain even when the producing microorganism is destroyed.

3.1 "Red Tide" Shellfish Poisoning

Paralytic shellfish poisoning is caused by consuming shellfish, which has fed on toxic single cell algae called *Gonyaulax catenella* or *G. tamarensis.* These toxic dinoflagellates produce a toxin (saxitoxin) that can cause death in humans within 60–120 min after consuming contaminated shellfish. The amount of poison required to cause illness varies considerably. The toxicity is measured in mouse units (MU). One MU is the minimal amount of poison needed to kill a 20-g white mouse 15 min after intraperitoneal injection of an acidic extract of mussel (or other shellfish) tissue. Deaths in humans have resulted from a range of 3000 to >20,000 MU of the toxin.

When environmental conditions are right, concentrations greater than 20,000 organisms per milliliter can occur and the water becomes brownish red, hence the term *red tide.* Proper surveillance and control by governmental agencies is needed to ensure contaminated shellfish will not be harvested. One to three weeks after the "red tide" bloom has disappeared, mussels and clams gradually destroy or excrete the poison. For more insights about marine toxins and their role in foodborne diseases, the reader may refer to the following articles (Lipp and Rose 1997; Sobel and Painter 2005; James et al. 2010).

3.2 Staphylococcal Poisoning

The bacterium that causes staphylococcus poisoning is *Staphylococcus aureus,* the same bacterium that causes white-head pimples, infections, boils, and carbuncles (Mayo Clinic 2021). These cells are spherical or ovoid in shape, non-motile, and in liquid cultures arrange themselves in grapelike clusters, in small groups, in pairs, or in short chains. They grow best in the presence of air (oxygen), but they will also grow in its absence. They will grow in media or in foods that contain as much as 10% salt (sodium chloride). When these organisms grow in foods, they produce a toxin that can be filtered away from the food and from the bacterial cells. The toxin is not destroyed by normal cooking.

Foods that are most often at risk of staphylococcus infection may be listed in the order of declining frequency, ham> turkey meat > chicken/ chicken products > pork > roast beef > other meat. Non-meat products were of lower risk; potato salads >cream filled pastry > eggs/ egg salad > Mexican food (Bryan 1988). Almost any food (except acid products) seems to be suitable for the growth of *Staphylococcus aureus* leading to significant outbreaks in the past also involving cheddar cheese and other dairy products, bakery products and meat (Minor and Marth 1972a, b; Halpin-Dohnalek and Marth 1989). The overall rates of foodborne staphylococcus outbreaks has been declining over recent decades (Bergdoll and Lee Wong 2006).

The incubation time interval between eating and the first signs of illness ranges from 1 h to 7 h, peaking at about 3 h. The symptoms of staphylococcal oral infection are diarrhea (78%), vomiting (76%), weakness (70%), nausea (45%), headache (38%), and less frequently, stomach cramp (26%), fever (22%), bloody vomit (13%), and/ or bloody diarrhea (6%). The duration of staphylococcal infections is usually from 1 to 6 days (Denison 1936; Feig 1950).

The reason why ham products are frequently involved is that in their preparation, they may become contaminated with *Staphylococcus aureus,* and as this product contains 2–3% salt, other bacteria that might grow and inhibit the growth of staphylococci are themselves inhibited by the salt. Also, people who handle ham and its products are apt to believe that such foods are not perishable. However, ham and ham products are perishable and should always be held at 40 °F (4.4 °C) or below (Smith et al. 1983; Halpin-Dohnalek and Marth 1989).

The toxins produced by *Staphylococcus aureus* are not readily destroyed by heat though the organism itself is destroyed. Many foods and food extracts have been found to stabilize preformed S. aureus toxin during heating (Tatini 1976; Schwabe et al. 1990).

The toxins from *S. aureus* can damage host cells in a variety of ways. Firstly, some toxins (hemolysins) damage the membranes of red and white blood cells leading to lysis. Another group of toxin affect membrane function, altering transport of water and producing symptoms like diarrhea. Some of these toxins may also induce programmed cell death (apoptosis). A third wave of attach involves enzyme toxins, that digest vital host molecules like proteins needed for immune response, clotting, and healthy skin (Otto 2014).

Humans are the main reservoir of the organism causing staphylococcal poisoning (Smith et. al. 1983; Mayo Clinic 2021). Approximately 40% of normal human adults carry *Staphylococcus aureus* in their noses and throats and so the fingertips and hands often become contaminated with the organism. Infected cuts or abrasions (containing pus) is a definite source of the organism. Cows are a source of *Staphylococcus aureus,* particularly if the animals have mastitis (an infection of the udder).

In foods, *Staphylococcus aureus* will grow at temperatures as low as 44 °F (6.7 °C) and as high as 112 °F (44.4 °C). In some foods (turkey stuffing), temperatures as high as 130 °F (48.9 °C) have sometimes been necessary to destroy the organism. Staphylococcal poisoning may occur after a food has been held at temperatures that allow the organism to grow at relatively fast rates. Several methods were suggested for controlling staphylococcal food poisoning which are (Smith et al. 1983; Kadariya, et al. 2014):

(i) Hold all foods, when not being eaten or prepared for eating, at temperatures of 40 °F (4.4 °C) or below.
(ii) Prohibit all persons having boils, carbuncles, or pus abrasions or cuts on their hands from handling foods.
(iii) Require all personnel handling food in food manufacturing or food serving establishments to wash and sanitize their hands with solutions such as chlorine or one of the iodophors prior to performing their particular tasks.
(iv) Eliminate use of milk from cows with mastitis.

In Chap. 9, we present the requirements for food handling to avoid foodborne diseases in detail.

3.3 Botulism

Hundreds of species of clostridium occur in the environment (soil, foods, feces) with a few pathogenic ones causing serious diseases such as, (C. botulinum) botulism, (*C. difficle*) colitis, (*C. tetani*) tetanus, or (*C. difficile*) and antibiotic related diarrhea (Wells and Wilkins 1996; Rodriguez et al. 2016).

There are several types of botulisms, designated as food botulism, wound botulism, infant botulism, animal botulism or inhalation botulism (Caya et al. 2004; Centers for Disease Control and Prevention 2019; Dhaked et al. 2010; World Health Organization 2018). All forms of botulisms have the same basis. The botulin toxin functions as a specific nerve inhibitor by blocking the release of acetylcholine (Fig. 8.4). The symptoms of botulism arise primarily from muscle paralysis and therefore can be diverse: diffi-

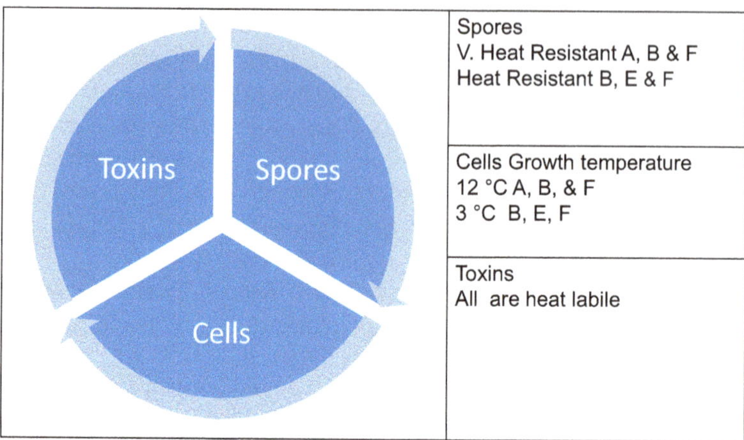

Fig. 8.4 Clostridium vegetative cells, toxins and spores – pose three forms of risk

culty of eye movement, double vision, slurred speech, abdominal distension, and a raw sore throat. In severe cases, and this disease is usually severe, the breathing mechanism and eventually heart action are affected, often resulting in death. The toxicant is the botulinum toxins, which functions as a neurotoxin or nerve blocking agent (Centers for Disease Control and Prevention 2019) (Fig. 8.4).

There are currently four main groups (Group I - IV) of *C. botulinum* bacteria distinguished by the different antibodies required to neutralize their toxins as types A, type B…Type G. (Peck 2006). Additional sub-types of toxins exist but these are less well studied (Poulain and Popoff 2019). The Group I (A, B toxin producers) and Group II (B, E, and F toxin producers) are human pathogens. Group III (C, D toxin producers) are causative agents for animal disease, and Group IV (G toxin producer) is nonpathogenic (Lindström et al. 2010). Type A toxin is most toxic with fatal doses cited as 0.1 μg (by injection), 1 μg (by inhalation) or 10 μg (via the oral route) but exact figures are somewhat difficult to come by (Gill 1982; Rossetto and Montecucco 2019). The purity of toxin preparations is about 100 "mouse units" per Nano gram with one mouse unit being sufficient to kill 50% of a group of mouse injected. Famously, *C. botulinum* toxic is the active component of "Botox" and related products used for cosmetic applications (reviewed by Koussoulakos 2009; Dhaked et al. 2010).

Group I (A, B toxin producers) consists *proteolytic C. botulinum*, which grow at temperatures above 10-12 °C (mesophilic). The spores from proteolytic *C. botulinum* are heat resistant and require temperatures of 120 °C to inactivate (see Chapter 12). The proteolytic *C. botulinum* are of concern for hermetically sealed or film-wrapped products that have low levels of oxygen (Lindström et al. 2010).

The Group II botulinum strains (B, E, F toxin producers) are *nonproteolytic* and psychrophilic. The cold-adapted bacteria is capable of growing at refrigeration temperatures as low as 38°F (3.3°C). The spores from nonproteolytic *C. botulinum* have low/ moderate heat resistance. Nonproteolytic *C. botulinum* are a concern for refrigerated foods packaged in a manner that reduces air exposure (Peck 2006; Lindström et al. 2010).

All botulinum organisms are anaerobic meaning they grow only in the absence of oxygen or under conditions in which oxygen is quickly taken up by chemical substances (reducing compounds) present in some foods. The *C. botulinum* toxins are produced in the food by the bacteria before the food is eaten.

Unlike the toxin produced by *Staphylococcus aureus*, toxins from Group II *Clostridium botulinum* are readily destroyed by heat, e.g., heating to the boiling point of water (212 °F 100 °C) or alternatively heating to 194 °F (90 °C) for 10 minutes (Peck 2006; Lindström et al. 2010). The US food Code provides practical advice on heating food adequately, the need for heating / cooling rapidly, and using proper holding temperatures, to avoid forming and germinating clostridium spores (U.S. Food and Drug Administration 2013). Moreover, the nonproteolytic *Clostridium botulinum* grow at temperatures as low as 38 °F (3.3 °C); current guidelines require that fleshy-foods (e.g. fish), and ready-to-eat foods with reduced oxygen packaging, be refrigerated at temperatures below 41 °F (5 °C) with strict shelf-life restrictions (U.S. Food and Drug Administration 2013; Peck 2006). Refrigeration requirements for ready-to-eat meals are discussed further in chapter 13.

Clostridium botulinum will not normally grow in canned acid foods (pH 4.5 or below), yet there have been some acid foods (pears, apricots, tomatoes) that have been involved in causing the disease. In such cases, some other organism (mold, yeast, or bacterium) has grown first and raised the pH of the food to the point where *C. botulinum* would grow (Huhtanen et al. 1976).

Most cases of botulism are caused by home-canning foods with low acidity. Therefore, one method of control would be to make sure that all vulnerable home-canned foods are pressure cooked at temperatures suitable to destroy all spores of *C. botulinum* that may be present. Pamphlets are available from the USDA that specify times and temperatures for home-canning different products using different size containers (National Center for Home Food Preservation 2015).

Heating home canned foods prior to consumption has no doubt saved many lives. A good control precaution is to boil all home-canned food using a saucepan for 10 minutes prior to serving (National Center for Home Food Preservation 2015).

Although rare, some commercially canned products have caused botulism. All manufacturers of such products should have technical advisors who can determine that the heat processes given their products are sufficient to destroy all spores of *Clostridium botulinum* that may be present.

To conclude, foodborne botulisms is a serious concern for low acid foods, with low air exposure due to sealing hermetically either by canning, in a jar, or by film wrapping and other packaging. The spores from proteolytic *C. botulinum* (which are far more heat resistant than the vegetative cells) are an issue for the canning industry (Chapter 12). The nonproteolytic *C. botulinum* strains are a problem for refrigerated foods with extended shelf life (Chapter 13). Overall, the requirements for controlling *C. botulinum* may vary according to whether the focus is on proteolytic or nonproteolytic strains, spores versus vegetative cells or if the concern is with the protein toxins (Fig 8.4). Strategies for controlling *C. botulinum* will be discussed in later sections of the book.

3.4 Clostridium Perfringens Poisoning

Perfringens poisoning is caused by *Clostridium perfringens*, which, like *Clostridium botulinum*, grows only in the absence of oxygen. There are five different *Clostridium perfringens* types (A-E) but the most linked with food poisoning is Type A (Wen and McClane 2004). In addition, pathogenic *C. perfringens* are spore-forming and toxin producers similarly to *Clostridium botulinum*. Even when the cells are killed by heating, some heat resistant spores survive and then germinate as happens with preheated foods (Sarker et al. 2000, Li and McClane 2006). The symptoms of perfringens poisoning are diarrhea and abdominal pain or colic. The illness occurs 8–22 h after the food has been eaten, and the symptoms are of short duration (1 day or less). The number of people involved in an outbreak is quite often comparatively large.

In *C. perfringens* poisoning, what ordinarily happens is that meat or poultry is cooked, then held at insufficiently high temperatures, then served. The spores survive the cooking process because they have some heat resistance; then when the meat or gravy is held at room temperature or on a steam table where the temperature is below 140 °F (60 °C), the spores grow and reproduce to large numbers, causing illness.

To control perfringens poisoning, meats and gravies should be refrigerated at temperatures of about 41 °F (5.0 °C) or lower below shortly after cooking if not immediately eaten, or should be held at temperatures not lower than 140 °F (60 °C) in preparation for serving (Schneider et al. 2020).

The food code contains further general advice regarding *C. perfringens* poisoning risk and food handling (U.S. Food and Drug Administration 2013). It is worth re-iterating that normal cooking does not inactivate C. perfringens vegetative cells or spores. During heating come-up time or during slow cooling there can be a rapid buildup of cells reaching a 1 million cells per gram of food needed to cause symptoms. Therefore, to reduce *C. perfringens risks* all foods subjected to normal cooking should not be stored at room temperature for longer than 4 hours. Hot foods should be cooled rapidly within 6 h from 135 °F (57.2 °C) to 41 °F (5.0 °C). In summary, Clostridium species are a key risk factor for cooked foods owing to their heat stability. Clostridia are also the main organisms for consideration during Appertization or canning heat process (cf. Chap. 17).

3.5 Other Food Intoxications

Other diseases may occur from the accidental accumulation of toxins in foods exposed to unusual environmental concentrations of chemical or biological toxins from polluted areas. In such cases, the toxins are not easily removed or destroyed; therefore, we must rely on our regulatory and public health agencies to make periodic inspections of foods that are suspect, as well as the area from which they are produced.

Bacillus cereus gastroenteritis is listed by some as food intoxication because it is believed that toxin is released in the food as a result of cell autolysis (enzymatic destruction of cells after death). Surveys of foods and ingredients have shown high percentages containing B. cereus. High numbers of the viable cells are required for symptoms to develop. These include abdominal cramps, watery diarrhea, and some vomiting. B. cereus is a spore former and can survive some cooking temperatures. Quick cooling of foods, proper holding temperatures for hot foods (140 °F [60 °C] or higher), and proper reheating to an internal temperature of 165 °F (74 °C) is necessary to prevent *B. cereus* foodborne infection. The temperatures for safe cooking, reheating, food-holding and storage are intended to manage anaerobic, spore-forming, bacillus species (*C. botulinum, C. perfringens, B. cereus*) in relation to protein rich foods, e.g. meat loaf, cooked poultry, and chicken casseroles (U.S. Food and Drug Administration 2013).

Mycotoxins are produced by molds. Some are highly toxic to many animals and potentially to humans. Aflatoxins are the most widely studied of the mycotoxins. As of 1980, 14 mycotoxins were recognized as carcinogens. The FDA has recognized the potential danger of mycotoxins and has established allowable action levels.

4 Foodborne Parasitic Diseases

4.1 Protozoa and Helminths

Wildlife, game and agricultural animals used as food sources carry parasitic organisms some of which are transmissible to humans (Centers for Disease Control and Prevention 2015c) (Table 8.5).

The control of foodborne parasitic agents is mostly by heating or freezing as per USDA advice. Other methods under investigation include parasite inactivation by chemical preservation, various food preparation measures (e.g. salting, fermentation, nitrite curing agents), high pressure processing and irradiation (Adams et al. 1997; Gamble 1997; Franssen et al. 2019).

The foodborne parasitic agents are protozoans (single cell animals) or helminths (liver fluke, tapeworms, roundworms etc.). Some of the helminths are not microorganisms by any stretch of this word but are worth mention nevertheless, by way of stressing the importance of good food hygiene. Foodborne parasitic agents are transmitted in undercooked seafood (meat, crabs, and fish), meat (red meat, chicken, venison, game etc.) or raw vegetables contaminated with faecal

Table 8.5 Parasites transmitted by food in the United States

Protozoa	Helminths
Cryptosporidium spp.	Trichinella spp. (mrw)
Giardia intestinalis	Anisakis spp. (mrw)
Cyclospora cayetanensis *Toxoplasma gondii*	Diphyllobothrium spp. (tw)
	Taenia spp. (tw)

Adapted from Centers for Disease Control and Prevention (2015g)
mrw microscopic round worms, *tw* tape worm

matter or manure. Parasitic infections are a subset of zoonotic conditions (disease transmissible from animals to man) which includes many of the bacterial and viral conditions discussed in earlier sections. A comprehensive glossary of parasitic diseases can be found at DPD-X the website of the Division of Parasitic Diseases and Malaria (DPDM) from the CDC (Centers for Disease Control and Prevention 2015c) (Table 8.5).

4.2 Trichinellosis or Trichinosis

Trichinellosis or Trichinosis is not a bacterial disease but caused by a microscopic roundworm, *Trichinella spiralis*. The incidence of trichinosis has declined steadily from about 400 cases per year in the 1940s to about 20 cases per year in 2007 (Centers for Disease Control and Prevention 2015f). There are also a few deaths attributed to this disease each year. Trichinosis is transmitted through the eating of pork, wild hog, bear and cougar meat. Trichinosis resulting from pork processed major packers is on the decline, whilst incidence from game meat is increasing (Hall et al. 2012; Kennedy et al. 2009; Wilson et al. 2015).

The symptoms of trichinosis vary with the number of organisms ingested. If large numbers of larvae (stage of the life cycle of the worm) are eaten, nausea, vomiting, and diarrhea develop 1–4 days. Aside from these gastrointestinal symptoms, further side effects arise within 1–2 weeks as roundworm larvae (produced by the adult worms) migrate from the intestines to the muscles. There may be intermittent fever as high as 104 °F (40 °C) that can last for a few weeks and a host of other outcomes. (Centers for Disease Control and Prevention 2015f) (Table 8.6).

Table 8.6 Later symptoms for roundworm

Chills
Constipation
Cough
Diarrhea
Fever
Headache
Itchy skin or rash
Muscle pain
Swelling of the face, particularly the eyes
Weakness or fatigue

Adapted from Centers for Disease Control and Prevention (2015f)

The upper eyelids may swell owing to the accumulation of fluids. Once the larvae have become embedded in areas between the muscle fibers, a cyst is formed that becomes calcified. In this state, the larvae remain dormant in the body over a period of years. Interestingly, the encysted stage is apparently able withstand freezing. The *Trichina* organisms in meat are present as the larval stage of a roundworm, and in the human intestine, develop into adult roundworms. The adult worms unite sexually and the females produce larvae that migrate from the intestine to the muscle tissues. The following is a description of a number of ways in which trichinosis can be controlled or prevented:

(a) *Trichina* larvae are destroyed by a temperature of 137 °F (58.3 °C) and so fresh pork should not be eaten unless all parts have been heated to this internal temperature or higher. As a general rule never to eat fresh pork that shows any pink or red color or that has not been heated to the point where no indication of uncooked meat fluids can be seen. The USDA has several rules governing the processing of pork sold as cured products,
   (i) The fresh pork shall have been frozen and held at 5 to −20 °F (−15 to −28.9 °C) for a period of 6–30 days, the time of holding depending on the temperature at which it is held and the size of the portion of pork;
   (ii) all parts of the cured product shall have been heated to at least 137 °F (58.3 °C) in ready-to-eat products;
   (iii) for dried, summer-type sausage (Italian salami, cervelat, etc.), the product shall have curing compounds added, after which it must be held for at least 40 days at temperatures not lower than 45 °F (7.2 °C),
   (iv) The cooking of garbage to be fed to hogs is considered a method useful for controlling trichinosis because

uncooked pork may be present therein. Such pork might contaminate the hog eating it, thus causing it to be a source of infection to humans.

In summary, roundworms are mostly associated with raw pork or wild hog but cases of human infection arising from cougar jerky, black bear, grizzly bear meat, and other wild game have also occurred in the United States (Wilson et al. 2015), Canada (Schellenberg et al. 2003), Japan, Korea (Kaewpitoon et al. 2008) and elsewhere. Trichinosis is perhaps a very good example of zoonotic disease, i.e. a disease that transmitted from livestock to human consumers.

Given the seriousness and long-term effects of trichinosis, "prevention is obviously the best policy". Following safe food handling strategies and avoiding under-cooked pork and game meat makes obvious sense but a 2008 outbreak at a California, social gathering resulted when attendees were served undercooked bear meat leading to 100% infection rate (Hall et al. 2012). In the serious event of suspected infection, medical expertise should be sought. Antithelmintic drugs have been used in the past to deal with parasitic worm infections.

4.3 Amebiasis

Amebiasis is caused by a single-celled animal, an amoeba, leading to dysentery in humans. The organism called *Entamoeba histolytica* has been known since about 1875 (Bruckner 1992). CDC reports that the Amebiasis is more common in the developing world and areas with a lower standard sanitation (Centers for Disease Control and Prevention 2015e). At any one time Amebiasis is thought to affect about 10–12% of the world's population with a mortality of 0.1% (Bruckner 1992; Verkerke et al. 2012).

About 10% of the infected individuals show symptoms these being, diarrhea, abdominal pain, fatigue, and fever are sometimes encountered. Incubation lasts from 2 days to several months, but is usually 3–4 weeks. Like other intestinal protozoans, there are several different stages in its life cycle, which explains how Amebiasis spreads. The infective stage of is a cyst form which survives in the general environment. When consumed, the cyst form transforms into an active motile organism called a trophozoites. These feed and reproduce asexually in the colon and then form further cysts before passed in the stool into the environment. The next generation of infection occurs when unsafe water, food or cutlery etc. contaminated with the *E. histolytica* cyst form is used. The control of Amebiasis is essentially a matter of good sanitary procedures focussion on the use of portable water (see Sect. 2.5);

(i) Water for drinking, ice making, adding to foods, for washing and sanitizing equipment/utensils and for irrigation should be potable,
(ii) Water taken from deep wells, lakes, or ponds, should be tested for bacteriological safety.
(iii) Persons known to have had amebic dysentery should be rigidly excluded from food handling of any kind,
(iv) During foreign travel avoid consuming non-bottled water, fountain drinks, drinks with ice-cubes (!) or fruits not peeled by yourself and dairy products of all kinds produced with non-pasteurized milk.

Amebiasis is not a bacterial disease, but it appears if bacteria indicative of pollution with human discharges are present. Carbonated water, carbonated soft drinks, and boiled tap water may be safe to drink (Centers for Disease Control and Prevention 2015e). Sharing cutlery, body towels, and eating salads have also been noted as risks factors for Amebiasis.

4.4 Giardia Lamblia

A waterborne disease, prevalent where surface waters serve as drinking water sources, has reached a high level of incidence in the United States. *Giardia lamblia,* a flagellate one-celled animal (protozoa), is the causative agent. Symptoms including abdominal cramps, diarrhea, fatigue, weight loss, and nausea may persist for 2–3 months if untreated. The organism may be spread through human feces, and animals such as beavers have been found to harbor the organism. Methods to prevent the transmission of the disease include appropriate disposal of fecal matter, wastewater treatment, and inclusion of a filtration step before chlorination in the preparation of drinking water from surface sources. Boiling the water will also kill the organism (Centers for Disease Control and Prevention 2015c).

4.5 Parasitic Tapeworms

Taenia solium (a tapeworm that may infest pork), *Taenia saginatta* (a tapeworm sometimes found in beef) and *Diphyllobothrium latum* (a tapeworm sometimes found in fish) can all cause illness in humans, but none of these pose any hazard when foods are thoroughly cooked. Small worms of the genus *Anisakis* may also infect fish and cause illness in humans when the infected fish is not properly cooked (Centers for Disease Control and Prevention 2015c).

References

Adams AM, Murrell KD, Cross JH (1997) Parasites of fish and risks to public health. Rev Sci Tech 16(2):652–660. https://europepmc.org/article/med/9501379

Allos BM (2001) Campylobacter jejuni infections: update on emerging issues and trends. Clin Infect Dis 32(8):1201–1206

Altekruse SF, Cohen ML, Swerdlow DL (1997) Emerging foodborne diseases. Emerg Infect Dis J 3(3):285–293. https://wwwnc.cdc.gov/eid/article/283/283/297-0304_article

Arendt S, Rajagopal L, Strohbehn C, Stokes N, Meyer J, Mandernach S (2012) Reporting of foodborne illness by U.S. consumers and healthcare professionals. Int J Environ Res Public Health 10(8):3684–3714

Baert L, Mattison K, Loisy-Hamon F, Harlow J, Martyres A, Lebeau B et al (2011) Review: norovirus prevalence in Belgian, Canadian and French fresh produce: a threat to human health? Int J Food Microbiol 151(3):261–269

Barclay L, Park GW, Vega E, Hall A, Parashar U, Vinje J, Lopman B (2014) Infection control for norovirus. Clin Microbiol Infect 20(8):731–740

Bardhan P, Faruque AS, Naheed A, Sack DA (2010) Decrease in shigellosis-related deaths without Shigella spp.-specific interventions, Asia. Emerg Infect Dis 16(11):1718–1723

Belliot G, Lopman BA, Ambert-Balay K, Pothier P (2014) The burden of norovirus gastroenteritis: an important foodborne and healthcare-related infection. Clin Microbiol Infect 20(8):724–730

Bergdoll MS, Lee Wong AC (2006) In: Riemann HP, Cliver DO (eds) Staphylococcal intoxications foodborne infections and intoxications, 3rd edn. Elsevier, Amsterdam, pp 523–562

Black RE, Jackson RJ, Tsai T, Medvesky M, Shayegani M, Feeley JC, MacLeod KI, Wakelee AM (1978) Epidemic Yersinia enterocolitica infection due to contaminated chocolate milk. N Engl J Med 298(2):76–79. https://www.nejm.org/doi/full/10.1056/NEJM197801122980204

Borchers A, Teuber SS, Keen CL, Gershwin ME (2010) Food safety. Clin Rev Allergy Immunol 39(2):95–141. https://doi.org/10.1007/s12016-009-8176-4

Boukary AR, Thys E, Rigouts L, Matthys F, Berkvens D, Mahamadou I, Yenikoye A, Saegerman C (2012) Risk factors associated with bovine tuberculosis and molecular characterization of Mycobacterium bovis strains in urban settings in Niger. Transbound Emerg Dis 59(6):490–502

Boyce TG, Swerdlow DL, Griffin PM (1995) Escherichia coli O157: H7 and the hemolytic–uremic syndrome. N Engl J Med 333(6):364–368. https://www.nejm.org/doi/full/310.1056/NEJM199508103330608

Bruckner DA (1992) Amebiasis. Clin Microbiol Rev 5(4):356–369

Bryan F (1988) Risks associated with vehicles of foodborne pathogens and toxins. J Food Prot 51(6):498–508. Retrieved from https://doi.org/410.4315/0362-4028X-4351.4316.4498

Cartwright EJ, Jackson KA, Johnson SD, Graves M, Silk BJ, Mahon BE (2013) Listeriosis outbreaks and associated food vehicles, United States, 1998-2008. Emerg Infect Dis 19(1):1–184. Retrieved from https://www.ncbi.nlm.nih.gov/pmc/articles/PMC3557980/

Caya JG, Agni R, Miller JE (2004) Clostridium botulinum and the clinical laboratorian: a detailed review of botulism, including biological warfare ramifications of botulinum toxin. Arch Pathol Lab Med 128(6):653–662

Centers for Disease Control and Prevention (2001) Preliminary FoodNet data on the incidence of foodborne illnesses – selected sites, United States, 2000. MMWR Morb Mortal Wkly Rep 50(13):241–246

Centers for Disease Control and Prevention (2003) Preliminary FoodNet data on the incidence of foodborne illnesses – selected sites, United States, 2002. MMWR Morb Mortal Wkly Rep 52(15):340–343

Centers for Disease Control and Prevention (2004) Preliminary FoodNet data on the incidence of infection with pathogens transmitted commonly through food – selected sites, United States, 2003. MMWR Morb Mortal Wkly Rep 53(16):338–343

Centers for Disease Control and Prevention (2006) Preliminary FoodNet data on the incidence of infection with pathogens transmitted commonly through food – 10 states, United States, 2005. MMWR Morb Mortal Wkly Rep 55(14):392–395

Centers for Disease Control and Prevention (2009) Surveillance for foodborne disease outbreaks – United States, 2006. MMWR Morb Mortal Wkly Rep 58(22):609–615

Centers for Disease Control and Prevention (2011) National typhoid and paratyphoid fever. Surveillance overview. Retrieved from https://www.cdc.gov/typhoid-fever/surveillance.html

Centers for Disease Control and Prevention (2014) Estimates of Foodborne Illness in the United States. Retrieved from http://www.cdc.gov/foodborneburden/index.html

Centers for Disease Control and Prevention (2015a) Brucellosis. Retrieved from http://www.cdc.gov/parasites/Amebiasis/general-info.html

Centers for Disease Control and Prevention (2015b) Campylobacter. Retrieved from http://www.cdc.gov/nczved/divisions/dfbmd/diseases/campylobacter/

Centers for Disease Control and Prevention (2015c) DPDx – Laboratory identification of parasitic diseases of public health concern. Retrieved from http://www.cdc.gov/dpdx/az.html

Centers for Disease Control and Prevention (2015d) Norovirus Retrieved from http://www.cdc.gov/norovirus/index.html

Centers for Disease Control and Prevention (2015e) Parasites – Amebiasis – Entamoeba histolytica Infection. Retrieved from http://www.cdc.gov/parasites/Amebiasis/general-info.html

Centers for Disease Control and Prevention (2015f) Parasites – Trichinellosis (also known as Trichinosis). Retrieved from http://www.cdc.gov/parasites/trichinellosis/

Centers for Disease Control and Prevention (2015g) Parasites: food. Retrieved from http://www.cdc.gov/parasites/food.html

Centers for Disease Control and Prevention (2015h) Salmonella. Retrieved from http://www.cdc.gov/salmonella/

Centers for Disease Control and Prevention (2015i, May 2019) Salmonella. Information for healthcare professionals and laboratories. Retrieved from https://www.cdc.gov/salmonella/general/technical.html

Centers for Disease Control and Prevention (2015j) Shigella – Shigellosis. Retrieved from http://www.cdc.gov/shigella/index.html

Centers for Disease Control and Prevention (2015k) Vibrio Illness (Vibriosis). Retrieved from http://www.cdc.gov/vibrio/index.html

Centers for Disease Control and Prevention (2015l) Vibrio parahaemolyticus. Retrieved from http://www.cdc.gov/vibrio/vibriop.html

Centers for Disease Control and Prevention (2017) Knowing which foods make us sick will help guide food safety regulations. Retrieved from https://tools.cdc.gov/medialibrary/index.aspx#/media/id/304694

Centers for Disease Control and Prevention (2019) Botulinism: resources & publications. Retrieved from https://www.cdc.gov/botulism/resources.html

Centers for Disease Control & Prevention (2021a) Common settings of norovirus outbreaks. Retrieved from https://www.cdc.gov/norovirus/trends-outbreaks/outbreaks.html

Centers for Disease Control & Prevention (2021b) E coli (Escherichia coli). Retrieved from https://www.cdc.gov/ecoli/index.html

Chlebicz A, Slizewska K (2018) Campylobacteriosis, salmonellosis, yersiniosis, and listeriosis as zoonotic foodborne diseases: a review. Int J Environ Res Public Health 15(5):29pp. Retrieved from https://www.mdpi.com/1660-4601/1615/1665/1863

Chong YY, Unklesbay N, Dowdy R (2000) Clinical nutrition and foodservice personnel in teaching hospitals have different perceptions of total quality management performance. J Am Diet Assoc 100(9):1044–1049. https://doi.org/10.1016/s0002-8223(00)00305-9

Cromeans T, Park GW, Costantini V, Lee D, Wang Q, Farkas T, Vinje J (2014) Comprehensive comparison of cultivable norovirus surrogates in response to different inactivation and disinfection treatments. Appl Environ Microbiol 80(18):5743–5751

Currier RW, Widness JA (2018) A brief history of milk hygiene and its impact on infant mortality from 1875 to 1925 and implications for today: a review. J Food Prot 81(10):1713–1722

Daniels NA (2011) Vibrio vulnificus oysters: pearls and perils. Clin Infect Dis 52(6):788–792

De J, Pabst CR, Montazeri N, Jones MK, Schneider RG, Schneider KR (2018) Preventing foodborne illness: norovirus. Retrieved from https://edis.ifas.ufl.edu/publication/FS129

de la Rua-Domenech R (2006) Human Mycobacterium bovis infection in the United Kingdom: incidence, risks, control measures and review of the zoonotic aspects of bovine tuberculosis. Tuberculosis (Edinb) 86(2):77–109

Dean AS, Crump L, Greter H, Hattendorf J, Schelling E, Zinsstag J (2012) Clinical manifestations of human brucellosis: a systematic review and meta-analysis. PLoS Negl Trop Dis 6(12):e1929

Denison GA (1936) Epidemiology and symptomatology of Staphylococcus food poisoning: a report of recent outbreaks. Am J Public Health Nations Health 26(12):1168–1175. Retrieved from https://www.ncbi.nlm.nih.gov/pmc/articles/PMC1562999/

Dhaked RK, Singh MK, Singh P, Gupta P (2010) Botulinum toxin: bioweapon & magic drug. Indian J Med Res 132:489–503

Dicaprio E, Ma Y, Hughes J, Li J (2013) Epidemiology, prevention, and control of the number one foodborne illness: human norovirus. Infect Dis Clin N Am 27(3):651–674

DuPont HL (2007) The growing threat of foodborne bacterial enteropathogens of animal origin. Clin Infect Dis 45(10):1353–1361

Ekdahl K, Normann B, Andersson Y (2005) Could flies explain the elusive epidemiology of campylobacteriosis? BMC Infect Dis 5:11

Etter E, Donado P, Jori F, Caron A, Goutard F, Roger F (2006) Risk analysis and bovine tuberculosis, a re-emerging zoonosis. Ann N Y Acad Sci 1081:61–73. https://doi.org/10.1196/annals.1373.006

Fairchild AL, Oppenheimer GM (1998) Public health nihilism vs pragmatism: history, politics, and the control of tuberculosis. Am J Public Health 88(7):1105–1117

Fitzgerald SD, Kaneene JB (2013) Wildlife reservoirs of bovine tuberculosis worldwide: hosts, pathology, surveillance, and control. Verterinary Pathol 50(3):488–499. Retrieved from http://www.ncbi.nlm.nih.gov/pubmed/23169912

Franssen FGC, Cozma-Petruţ A, Vieira-Pinto M, Jambrak AR, Rowan N, Paulsen P, Rozycki M, Tysnes K, Rodriguez-Lazaro D, Robertson L (2019) Inactivation of parasite transmission stages: efficacy of treatments on food of animal origin. Trends Food Sci Technol 83:114–128. https://doi.org/110.1016/j.tifs.2018.1011.1009

Gal-Mor O, Boyle EC, Grassl GA (2014) Same species, different diseases: how and why typhoidal and non-typhoidal Salmonella enterica serovars differ. Front Microbiol 5:10pp. Retrieved from https://pubmed.ncbi.nlm.nih.gov/25136336/

Gamble H (1997) Parasites associated with pork and pork products. Rev Sci Tech 16(2):496–506. https://www.greenme.com.br/wp-content/uploads/2015/2003/d9157.pdf

Gill DM (1982) Bacterial toxins: a table of lethal amounts. Microbiol Rev 46(1):86–94. https://journals.asm.org/doi/pdf/10.1128/mr.1146.1121.1186-1194.1982

Gormley FJ, Little CL, Murphy N, de Pinna E, McLauchlin J (2010) Pooling raw shell eggs: Salmonella contamination and high risk practices in the United Kingdom food service sector. J Food Prot 73(3):574–578

Hall RL, Lindsay A, Hammond C, Montgomery SP, Wilkins PP, da Silva AJ et al (2012) Outbreak of human trichinellosis in Northern California caused by Trichinella murrelli. Am J Trop Med Hyg 87(2):297–302

Halpin-Dohnalek MI, Marth EH (1989) Staphylococcus aureus: production of extracellular compounds and behavior in foods-A review. J Food Prot 52(4):267–282. Retrieved from https://doi.org/210.4315/0362-4028X-4352.4314.4267

Hara-Kudo Y, Takatori K (2011) Contamination level and ingestion dose of foodborne pathogens associated with infections. Epidemiol Infect 139(10):1505–1510

Howard ZR, O'Bryan CA, Crandall PG, Ricke SC (2012) Salmonella Enteritidis in shell eggs: current issues and prospects for control. Food Res Int 45(2):755–764. https://doi.org/10.1016/j.foodres.2011.04.030

Huhtanen CN, Naghski J, Custer CS, Russell RW (1976) Growth and toxin production by Clostridium botulinum in moldy tomato juice. Appl Environ Microbiol 32(5):711–715. https://www.ncbi.nlm.nih.gov/pmc/articles/PMC170388/

Jackson LS (2009) Chemical food safety issues in the United States: past, present, and future. J Agric Food Chem 57(18):8161–8170. https://doi.org/10.1021/jf900628u

James K, Carey B, O'halloran J, Škrabáková Z (2010) Shellfish toxicity: human health implications of marine algal toxins. Epidemiol Infect 138(7):927–940. https://doi.org/910.1017/S0950268810000853

Jones TF, Gerber DE (2001) Perceived etiology of food-borne illness among public health personnel. Emerg Infect Dis 7(5):904–905. Retrieved from http://www.ncbi.nlm.nih.gov/pmc/articles/PMC2631878/

Kaakoush NO, Castano-Rodriguez N, Mitchell HM, Man SM (2015) Global epidemiology of campylobacter infection. Clin Microbiol Rev 28(3):687–720. https://doi.org/10.1128/CMR.00006-15

Kadariya J, Smith TC, Thapaliya D (2014) *Staphylococcus aureus* and staphylococcal food-borne disease: an ongoing challenge in public health. Biomed Res Int(Article ID 827965): 9pp. Retrieved from. https://doi.org/10.1155/2014/827965

Kaewpitoon N, Kaewpitoon SJ, Pengsaa P (2008) Food-borne parasitic zoonosis: distribution of trichinosis in Thailand. World J Gastroenterol 14(22):3471–3475

Kennedy ED, Hall RL, Montgomery SP, Pyburn DG, Jones JL (2009) Trichinellosis surveillance – United States, 2002–2007. MMWR Surveill Summ 58(9):1–7

Koussoulakos S (2009) Botulinum neurotoxin: the ugly duckling. Eur Neurol 61(6):331–342. https://www.karger.com/Article/Pdf/210545

Kurtz JR, Goggins JA, McLachlan JB (2017) Salmonella infection: interplay between the bacteria and host immune system. Immunol Lett 190:42–50

Lamberti LM, Bourgeois AL, Fischer Walker CL, Black RE, Sack D (2014) Estimating diarrheal illness and deaths attributable to Shigellae and enterotoxigenic Escherichia coli among older children, adolescents, and adults in South Asia and Africa. PLoS Negl Trop Dis 8(2):e2705. https://doi.org/10.1371/journal.pntd.0002705

Li D, De Keuckelaere A, Uyttendaele M (2015) Fate of foodborne viruses in the "Farm to Fork" chain of fresh produce. Compr Rev Food Sci Food Saf 14(6):755–770. http://dx.doi.org/710.1111/1541-4337.12163

Li J, McClane BA (2006) Further comparison of temperature effects on growth and survival of Clostridium perfringens type A isolates carrying a chromosomal or plasmid-borne enterotoxin gene. Appl Environ Microbiol 72(7):4561–4568. https://www.ncbi.nlm.nih.gov/pmc/articles/PMC4920134/

Li M, Baker CA, Danyluk MD, Belanger P, Boelaert F, Cressey P, Gheorghe M, Polkinghorne B, Toyofuku H, Havelaar AH (2018) Identification of biological hazards in produce consumed in industrialized countries: a review. J Food Prot 81(7):1171–1186. Retrieved from https://doi.org/1110.4315/0362-1028X.JFP-1117-1465

Lin CTJ, Jensen KL, Yen ST (2005) Awareness of foodborne pathogens among US consumers. Food Qual Prefer 16(5):401–412. https://doi.org/10.1016/j.foodqual.2004.07.001

Lindström M, Myllykoski J, Sivelä S, Korkeala H (2010) Clostridium botulinum in cattle and dairy products. Crit Rev Food Sci Nutr 50(4):281–304. https://www.researchgate.net/publication/42346564_Clostridium_botulinum_in_Cattle_and_Dairy_Products

Lipp EK, Rose JB (1997) The role of seafood in foodborne diseases in the United States of America. Rev Sci Tech 16(2):620–640

Maciorowski KG, Jones FT, Pillai SD, Ricke SC (2004) Incidence, sources, and control of foodborne Salmonella spp. in poultry feeds. Worlds Poult Sci J 60(4):446–457. https://doi.org/10.1079/WPS200428

Mantur BG, Amarnath SK (2008) Brucellosis in India – a review. J Biosci 33(4):539–547

Matthews JE, Dickey BW, Miller RD, Felzer JR, Dawson BP, Lee AS, Rocks JJ, Kiel J, Montes JS, Moe CL, Eisenberg JN, Leon JS (2012) The epidemiology of published norovirus outbreaks: a review of risk factors associated with attack rate and genogroup. Epidemiol Infect 140(7):1161–1172

Mead PS, Slutsker L, Dietz V, McCaig LF, Bresee JS, Shapiro C et al (1999) Food-related illness and death in the United States. Emerg Infect Dis 5(5):607–625. Retrieved from http://stacks.cdc.gov/view/cdc/3262

Minor T, Marth E (1972a) Staphylococcus aureus and staphylococcal food intoxications. A review: III. Staphylococci in dairy foods. J Milk Food Technol 35(2):77–82. Retrieved from. https://doi.org/10.4315/0022-2747-4335.4312.4377

Minor T, Marth E (1972b) Staphylococcus aureus and staphylococcal food intoxications. A review: IV. Staphylococci in meat, bakery products, and other foods. J Milk Food Technol 35(4):228–241. Retrieved from https://doi.org/210.4315/0022-2747-4335.4314.4228

National Center for Home Food Preservation (2015) USDA complete guide to home canning, 2015 revision. Retrieved from https://nchfp.uga.edu/publications/publications_usda.html

Oliver JD (2005) Wound infections caused by Vibrio vulnificus and other marine bacteria. Epidemiol Infect 133(3):383–391

Osterholm MT (2011) Foodborne disease in 2011 – the rest of the story. N Engl J Med 364(10):889–891. https://doi.org/10.1056/NEJMp1010907

Otto M (2014) Staphylococcus aureus toxins. Curr Opin Microbiol 17:32–37. Retrieved from https://www.ncbi.nlm.nih.gov/pmc/articles/PMC3942668/

Painter JA, Hoekstra RM, Ayers T, Tauxe RV, Braden CR, Angulo FJ et al (2013) Attribution of foodborne illnesses, hospitalizations, and deaths to food commodities by using outbreak data, United States, 1998–2008. Retrieved from https://doi.org/10.3201/eid1903.111866

Papageorgiou DK, Marth EH (1989) Fate of Listeria monocytogenes during the manufacture, ripening and storage of feta cheese. J Food Prot 52(2):82–87. https://doi.org/10.4315/0362-4028X-4352.4312.4382

Peck M (2006) Clostridium botulinum and the safety of minimally heated, chilled foods: an emerging issue? J Appl Microbiol 101(3):556–570. https://sfamjournals.onlinelibrary.wiley.com/doi/full/510.1111/j.1365-2672.2006.02987.x

Penner JL (1988) The genus Campylobacter: a decade of progress. Clin Microbiol Rev 1(2):157–172

Pesciaroli M, Alvarez J, Boniotti MB, Cagiola M, Di Marco V, Marianelli C, Pacciarini M, Pasquali P (2014) Tuberculosis in domestic animal species. Res Vet Sci 97:S78–S85

Peterson MC (1994) Clinical aspects of Campylobacter jejuni infections in adults. West J Med 161(2):148–152

Petrignani M, van Beek J, Borsboom G, Richardus JH, Koopmans M (2015) Norovirus introduction routes into nursing homes and risk factors for spread: a systematic review and meta-analysis of observational studies. J Hosp Infect 89(3):163–178

Piccinin DM, Shelef LA (1995) Survival of Listeria monocytogenes in cottage cheese. J Food Prot 58(2):128–131. https://doi.org/110.4315/0362-4028X-4358.4312.4128

Poulain B, Popoff MR (2019) Why are botulinum neurotoxin-producing bacteria so diverse and botulinum neurotoxins so toxic? Toxins 11(34):18pp. https://www.mdpi.com/2072-6651/2011/2071/2034

Raffatellu M, Wilson RP, Winter SE, Bäumler AJ (2008) Clinical pathogenesis of typhoid fever. J Infect Dev Ctries 2(4):260–266. Retrieved from https://jidc.org/index.php/journal/article/view/19741286

Rahman A, Bonny TS, Stonsaovapak S, Ananchaipattana C (2011) *Yersinia enterocolitica*: epidemiological studies and outbreaks. J Pathog 2011:11pp. Retrieved from. https://doi.org/10.4061/2011/239391

Ram PK, Crump JA, Gupta SK, Miller MA, Mintz ED (2008) Part II. Analysis of data gaps pertaining to Shigella infections in low and medium human development index countries, 1984–2005. Epidemiol Infect 136(5):577–603. https://doi.org/10.1017/S0950268807009351

Richards GP (2012) Critical review of norovirus surrogates in food safety research: rationale for considering volunteer studies. Food Environ Virol 4(1):6–13

Riemann HP, Cliver DO (2006) Foodborne infections and intoxications, 3rd edn. Elsevier, Amsterdam, 922pp. Retrieved from https://books.google.com.gh/books?id=_OsZ9TcQTSsC

Rodriguez C, Taminiau B, Van Broeck J, Delmée M, Daube G (2016) Clostridium difficile in food and animals: a comprehensive review. Adv Exp Med Biol 932:65–92. https://link.springer.com/chapter/10.1007%1002F5584_2016_1027

Rossetto O, Montecucco C (2019) Tables of toxicity of botulinum and tetanus neurotoxins. Toxins 11(12):686. https://doi.org/610.3390/toxins11120686

Rose JB, Epstein PR, Lipp EK, Sherman BH, Bernard SM, Patz JA (2001) Climate variability and change in the United States: potential impacts on water- and foodborne diseases caused by microbiologic agents. Environ Health Perspect 109(Suppl 2):211–221. https://doi.org/10.1289/ehp.01109s2211

Rubach MP, Halliday JE, Cleaveland S, Crump JA (2013) Brucellosis in low-income and middle-income countries. Curr Opin Infect Dis 26(5):404–412. https://doi.org/10.1097/QCO.0b013e3283638104

Ryser ET, Marth EH, Doyle MP (1985) Survival of Listeria monocytogenes during manufacture and storage of cottage cheese. J Food Prot 48(9):746–750. https://doi.org/710.4315/0362-4028X-4348.4319.4746

Ryser ET, Marth EH (1989) New food-borne pathogens of public health significance. J Am Diet Assoc 89(7):948–954. https://doi.org/910.1016/S0002-8223(1021)02284-02287

Samiullah S (2013) Salmonella Infantis, a potential human pathogen has an association with table eggs. Int J Poult Sci 12(3):185–191

Sarker MR, Shivers RP, Sparks SG, Juneja VK, McClane BA (2000) Comparative experiments to examine the effects of heating on vegetative cells and spores of Clostridium perfringens isolates carrying plasmid genes versus chromosomal enterotoxin genes. Appl Environ Microbiol 66(8):3234–3240. https://www.ncbi.nlm.nih.gov/pmc/articles/PMC92139/

Scallan E, Hoekstra RM, Angulo FJ, Tauxe RV, Widdowson MA, Roy SL et al (2011a) Foodborne illness acquired in the United States – major pathogens. Emerg Infect Dis 17(1):7–15. https://doi.org/10.3201/eid1701.091101p1. Retrieved from http://wwwnc.cdc.gov/eid/article/17/1/P1-1101_article

Scallan E, Griffin PM, Angulo FJ, Tauxe RV, Hoekstra RM (2011b) Foodborne illness acquired in the United States – unspecified agents. Emerg Infect Dis 17(1):16. Retrieved from https://wwwnc.cdc.gov/eid/article/17/1/p2-1101_article

Schellenberg RS, Tan BJ, Irvine JD, Stockdale DR, Gajadhar AA, Serhir B et al (2003) An outbreak of trichinellosis due to consumption of bear meat infected with Trichinella nativa, in 2 northern Saskatchewan communities. J Infect Dis 188(6):835–843

Schneider E, Castro KG (2003) Tuberculosis trends in the United States, 1992–2001. Tuberculosis (Edinb) 83(1–3):21–29

Schneider K R, Goodrich-Schneider R, Kurdmongkoltham P, Bertoldi B (2020) Preventing foodborne illness associated with *Clostridium perfringens*. Retrieved June 5, 2021, from https://edis.ifas.ufl.edu/publication/FS101

Schwabe M, Notermans S, Boot R, Tatini SR, Krämer J (1990) Inactivation of staphylococcal enterotoxins by heat and reactivation by high pH treatment. Int J Food Microbiol 10(1):33–42

Sharp JC (1987) Infections associated with milk and dairy products in Europe and North America, 1980–85. Bull World Health Organ 65(3):397–406

Simmons AL, Schlezinger JJ, Corkey BE (2014) What are we putting in our food that is making us fat? Food additives, contaminants, and other putative contributors to obesity. Curr Obes Rep 3(2):273–285

Smith J, Buchanan R, Palumbo SA (1983) Effect of food environment on staphylococcal enterotoxin synthesis: a review. J Food Prot 46(6):545–555. Retrieved from https://doi.org/510.4315/0362-4028X-4346.4316.4545

Sobel J, Painter J (2005) Illnesses caused by marine toxins. Clin Infect Dis 41(9):1290–1296. https://academic.oup.com/cid/article/1241/1299/1290/278063?login=true

Tacket CO, Harris N, Allard J, Nolan C, Nissinen A, Quan T, Ml C (1985) An outbreak of Yersinia enterocolitica infections caused by contaminated tofu (soybean curd). Am J Epidemiol 121(5):705–711

Tatini S (1976) Thermal stability of enterotoxins in food. J Milk Food Technol 39(6):432–438. https://doi.org/410.4315/0022-2747-4339.4316.4432

Tauxe RV (1997) Emerging foodborne diseases: an evolving public health challenge. Emerg Infecti Dis J 3(4):425–434. https://wwwnc.cdc.gov/eid/article/423/424/497-0403_article

Torgerson PR, Torgerson DJ (2010) Public health and bovine tuberculosis: what's all the fuss about? Trends Microbiol 18(2):67–72

Tung G, Macinga D, Arbogast J, Jaykus LA (2013) Efficacy of commonly used disinfectants for inactivation of human noroviruses and their surrogates. J Food Prot 76(7):1210–1217

U.S. Food and Drug Administration (2013) food code (2013) Retrieved from http://www.fda.gov/Food/GuidanceRegulation/RetailFoodProtection/FoodCode/ucm374275.htm

U.S. Food and Drug Administration (2017) Food code (2017). Retrieved from https://www.fda.gov/media/110822/download

USDA Food Safety and Inspection Service (2013) Foodborne illness: what consumers need to know. Retrieved from http://www.fsis.usda.gov/wps/portal/fsis/topics/food-safety-education/get-answers/food-safety-fact-sheets/foodborne-illness-and-disease/foodborne-illness-what-consumers-need-to-know/CT_Index

Verkerke HP, Petri WA Jr, Marie CS (2012) The dynamic interdependence of Amebiasis, innate immunity, and undernutrition. Semin Immunopathol 34(6):771–785

Wells CL, Wilkins TD (1996) Chapter 18. Clostridia: sporeforming anaerobic bacilli. In: Baron S (ed) Medical microbiology, 4th edn. University of Texas Medical Branch at Galveston, Texas, 14pp. https://www.ncbi.nlm.nih.gov/books/NBK8219/

Wen Q, McClane BA (2004) Detection of enterotoxigenic *Clostridium perfringens* type A isolates in American retail foods. Appl Environ Microbiol 70(5):2685–2691. https://doi.org/10.1128/AEM.70.5.2685-2691.2004

Wight JP, Cornell J, Rhodes P, Colley S, Webster S, Ridley AM (1996) Four outbreaks of Salmonella enteritidis phage type 4 food poisoning linked to a single baker. CDR Wkly 6(8):R112–R115

Wilson NO, Hall RL, Montgomery SP, Jones JL (2015) Trichinellosis surveillance – United States, 2008–2012. MMWR Surveill Summ 64(Suppl 1):1–8

World Health Organization (2019) Botulism. Retrieved from https://www.who.int/news-room/fact-sheets/detail/botulism

World Health Organization (2019a) Food safety. Retrieved from https://www.who.int/news-room/fact-sheets/detail/food-safety

World Health Organization (2019b) International classification of diseases. Retrieved from https://www.who.int/health-topics/international-classification-of-diseases

9 Food Safety and Sanitation

1 Introduction

1.1 Safety and the Agri-Food Industry

Food safety is the responsibility for everyone involved with the Agri-food chain starting with farming, and moving to food processing and manufacturing, distribution, retail and foodservice (Fig. 9.1). The role of the consumer is also paramount; the "fight Bac!" initiative of the late 1990s aimed to improve food safety education for consumers (Partnership for Food Safety Education 2010).

Food produced and handled in a sanitary manner will prevent the growth of microorganisms. Safe food handling helps to avoid cross-contamination, which contributes to millions of cases of food poisoning each year in the United States. However, proper food handling requires personal hygiene and sanitary procedures that can be mastered with proper training (USDA Food Safety and Inspection Service 2015b).[1]

The goal of this chapter is to discuss some basic principles of food safety with a focus on prevention. We will also describe how safe food handling can help to reduce foodborne disease risk. The information will be relevant for safe food handling in the home and industry. At their work, all employees must be familiarized with the key facilities and equipment in the plant or commercial kitchen. It needs to be understood that effective cleaning and sanitization will reduce the risk of foodborne infections (U.S. Food and Drug Administration 2013) (Fig. 9.1).[2]

1.2 Food Safety Oversight by the FDA

The FDA oversees food safety on farms and all registered food premises, including processing plants, warehouses, storage tanks and grain elevators. In addition, the FDA oversees food safety, and the inspection of foodservice establishments (restaurants, institutional foodservices), supermarkets, grocery stores, and other food outlets. FDA guidelines for food safety are published as the "US Food Code" which is updated every two years for adoption by the State authorities (U.S. Food and Drug Administration 2013). Foreign

[1]Personal hygiene should be expanded to consider personnel (or employee) hygiene to more accurate capture the magnitude and generic nature of the underlying food safety issues. Cambridge Online dictionary defines personal as "relating or belonging to a single or particular person". https://dictionary.cambridge.org/dictionary/english/personal

[2]Food Safety is also discussed with a focus on foodservices in Chap. 30.

R. Owusu-Apenten, E. R. Vieira, *Elementary Food Science*, Food Science Text Series,
https://doi.org/10.1007/978-3-030-65433-7_9

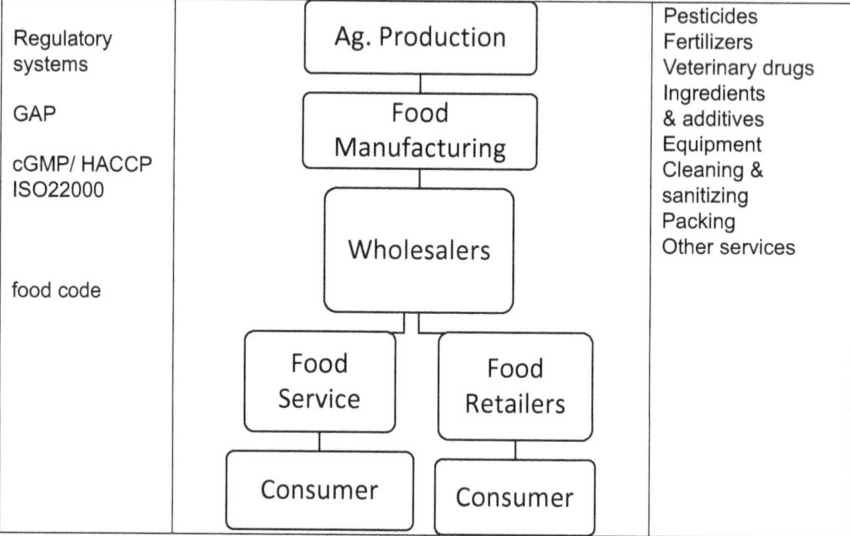

Fig. 9.1 Food safety management issues along the Agri-food system. Abbreviations: *GAP* Good Agricultural practice, *cGMP* current Good Management Practice, *HACCP* Hazard Analysis Critical Control Point

facilities registered within the US, which produce food for the American market, are also overseen by the FDA. As described in Chap 3, the FDA does not have jurisdiction over the meat, poultry and eggs industries (Food and Drug Administration, 2007).

Worldwide, the number of different foodborne diseases recognized exceeds 200. Overall, there were two million deaths from foodborne disease annually, mainly affecting the young, elderly, pregnant, and immune compromised persons. In 2015, the World health organization selected food safety as the main theme for World Health Day (World Health Organization 2015). Owing to the size of food safety commitments, regulatory agencies cannot achieve prevention of food borne disease without cooperation with industry. Foodborne illnesses are recognized as a major public health issue (Chap. 8).

1.3 Trending New Challenges for Food Safety

A number of recent changes in society may be contributing to increasing the risk of foodborne infections (Sneed and Strohbehn 2008). A more ageing population is leading to a demographic shift towards those with increasing susceptibility to foodborne disease. Availability of a greater range of prepared foods and ready-to-eat meals that are produced using centralized facilities means a greater risk of outbreaks spreading to a wider network of people (DuPont 2007; Lupien 2007; Zink 1997). Consumption of more fresh, minimally processed fruit and vegetables, prepared without a lethal cooking-step, adds to the risk of foodborne disease.

Another contributory factor for increasing foodborne disease risk is the increasing desire for foreign or imported exotic foods and the more globalized food supply. The FDA noted that foreign manufactured foods are not riskier as such. However, a greater diversity of food sources pose greater challenges for monitoring. Increasing urbanization, consumptions of foods on-the move, more away-from home eating and an increased frequency of international travel are all factors that contribute to the transmission of foodborne disease over longer distances than ever before.

Overall, the risk of foodborne disease seems to correlate with rising living standards; the possible relations between food hygiene, public health, extended supply-chains, international

Table 9.1 International and global factors that affect food safety

Changing dietary patterns
Changing food-handling patterns.
Expanding international/regional bodies and legal obligations
Greater range of food preparation preferences.
Human/animal interactions with potential for disease transmission
Increasing of food types and geographical sources
Increasing travel and tourism.
Increasing volume of international trade.
Intensification and industrialization of Agri-food System
New food and agricultural technologies.
New food processing methods.
Rising antibiotic resistance in microorganism.

Adapted from World Health Organization (2006)

trade and economics is a fertile area for discussions (Keener et al. 2014; Zylberman 2004). We should recall that correlation is not proof of causation. Owing to confounding factors, it would be hard to prove that a rise in living standards was a direct cause of foodborne disease.[3]

The FAO noted that many contributory factors for foodborne disease (Table 9.1) appear to be related to a rising volume and range of food types consumed (World Health Organization 2006). Some major food production hygiene issues are also notable at the international level, including resistance to antibiotics, and outbreaks of zoonotic conditions, e.g. BSE or mad cow disease. There are historical accounts of food infections and public health issues from the nineteenth and twentieth century France and EU, which show that International food safety concerns are not new (Zylberman 2004).

1.4 Sanitary and Phytosanitary Measures

The World Trade Organization (WTO) deals with differences of food safety standards between export and import countries that might lead to barriers to trade. Sanitary and phytosanitary measures (SPM) are a part of the General Agreement on Tariffs and Trade (GATT), which provide principles for addressing food safety issues related to world trade (Silverglade 2000; Zylberman 2004; WTO 2021) (Fig. 9.2).

The WTO recognizes that sovereign nations have the right to ensure that food trade items are safe for human, animal and plant health. However, there is a GATT requirement that food safety legislation should not be arbitrary, discriminatory, or protectionist for domestic producers. The SPM principles and food safety standards help to avoid disputes under WTO trade rules (World Trade Organization 2015b).

Therefore, SPM add to food safety discussions at the international level (World Trade Organization 2015a). Some recent global foodborne disease outbreaks that affected international trade include, BSE, avian influenza viruses, and also foot and mouth disease. There are also concerns that some private sector standards for quality and safety might adversely affect international trade. Discussions are underway to consider whether developing countries could apply less exacting (and costly) standards of safety. Whether an importing nation should be required to assist or compensate an exporting nation, where the former's genuinely higher food safety standard impedes international trade, is controversial (Silverglade 2000; World Trade Organization 2015a).

1.5 Risk Assessment for Foodborne Infections

The risk assessment process for foodborne illness is becoming influential particularly with international agencies (World Health Organization 2006). Key terms from risk assessment are passing to wider use and so familiarity with these is useful.

A foodborne hazard is defined as "….any *agent in, or condition of, food, with the potential to cause an adverse health effect*" (World Health Organization 2006). There are three classes of foodborne hazards - physical, chemical or bio-

[3]Increasingly wealthy consumers may eat out of home more often and be exposed to greater likelihood of foodborne illness.

Fig. 9.2 Sanitary and phytosanitary measures of the WTO aim to solve this riddle (WTO 2021)

How do you ensure that your country's consumers are being supplied with food that is safe to eat —"safe" by the standards you consider appropriate? In addition, at the same time, how can you ensure that strict health and safety regulations are not being used as an excuse for protecting domestic producers?

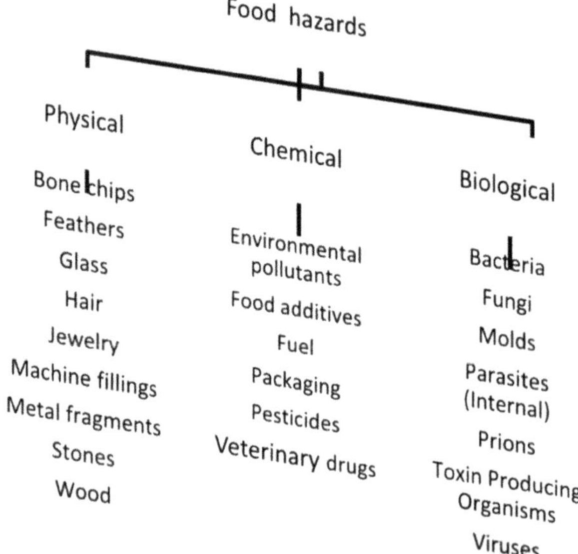

Fig. 9.3 Classification of foodborne hazards

logical (Fig. 9.3). Physical hazards are solid objects from the food that are difficult to masticate and that present a choking risk. Chemical hazards are either natural or artificial substances or contaminants that come into contact with a food, and which are harmful to the consumer at some dose. Biological hazards are harmful agents ultimately derived from living systems. There are potential borderline issues for the classification of foodborne hazards. For instance, aflatoxins produced by fungi, are still considered chemical hazards.

Risk analysis is applicable at any stage of the food system (Fig. 9.1). It is important to avoid compartmentalizing, because food hazards are transmitted along the food chain towards the consumer. The identities of foodborne hazards potentially located at all different stages of the food system should be considered. The requirement for traceability in food regulations is intended to deal with the transmission of foodborne hazards and to avoid the distribution of unsafe food (Aung and Chang 2015; Lupien 2005).

Another important food safety concept is that of *risk*, which is related to, (i) the probability of encountering harm and (ii) the severity or intensity of the harm. Food safety is frequently considered as an intricate set of activities designed to reduce risk. In theory, achieving a state of "zero risk" would ensure perfect food safety.

The technical approach for risk assessment involve up to four stages; (i) hazard identification, (ii) hazard characterization, (iii) exposure evaluation and (iv) risk characterization[4]. The aim of hazard *identification* is simply to list all possible classes and types of hazards associated with a food and/or food process. As noted above, the foodborne hazards can be classed as biological, chemical, or physical. To facilitate the hazards identification stage, the food safety management team must research all information

[4]Risk assessment is one of the three steps required for Risk Analysis; risk management (RM) and risk communication (RC) follow risk assessment. Some authorities see the three stages as interactive, rather than sequential.

(recipes, process literature, trade magazines) and produce a list of all *likely* hazards for the food (process) of interest including agricultural hazards if any (Fig. 9.1).

The aim of the hazard *characterization* is to establish a tentative dose-response characteristics for all significant hazards. Information is required about the symptoms of harm produced by different hazards and the concentrations, doses or numbers of bacteria required to produce different symptoms. The hazard characterization stage is usually completed by desktop research; alternatively, a small food company should seek external help from consultants. For some chemical hazards, some doses can be identified below which no toxicity occurs. For most physical hazards, the allowable dose will be zero.

For microbial pathogens, it is assumed there is no safe level of intake. Nevertheless, the *hazard* level associated with specific pathogens may be related to differences in the infective dose, the invading ability of each particle, the virulence or ability to cause disease, and the severity of the symptoms. For instance, it is thought that ingesting 18 particles of norovirus or 10–100 particles of Hepatitis A virus may be sufficient to cause disease. The infective dose for *E. coli* and other bacteria is estimated at 10 – 100 cells (U.S. Food and Drug Administration 2013).

The risk of foodborne infection is related to exposure frequency, or the likelihood of encountering hazards in any given time interval.

$$\text{Risk Magnitude} = \text{Hazard} \times \text{Exposure frequency} \quad (9.1)$$

The "risk equation" (Eq. 9.1) shows the type of "hazards" in any food environment and the "exposure frequency" are *independent* risk factors.[5] Therefore, lowering the exposure frequency to zero or near zero will promote food safety since the magnitude of risk will approach zero (Eq. 9.1). Many strategies for reducing the risk of foodborne illness reduce the frequency of exposure to disease causing food-borne microbes (Table 9.2).

[5]From basic algebra, 2 × 3 = 6, and so 2 and 3 are factors of 6. The multiplication rule from algebra means that any number multiplied by "zero" is zero.

Table 9.2 Risk factors for foodborne illnesses exposure

Contaminated equipment
Improper holding temperatures
Inadequate cooking (undercooking)
Poor personnel/personal hygiene
Unsafe food sources

Source: from US Food Code, p. 548 (U.S. Food and Drug Administration 2013)

2 Principles of Food Hygiene

2.1 Home Hygiene and Food Safety

Hands, fabrics, clothing, and contaminated surfaces are the main avenues for spreading infection in many settings, including, food production, food manufacturing, foodservices, hospital wards, care homes or everyday life situations (Bloomfield et al. 2017). Historical discussions of hygiene can seem fragmented by the differing focus on hand hygiene, laundering hygiene or surface hygiene. Nowadays, a more unified approach to hygiene is emerging that is applicable to homes and other situations (Bloomfield et al. 2017; Boyce 2016).

A survey from 1996 showed that infections in the modern home accounted for 20% of total foodborne diseases in the United States and 49–50% in Spain and New Zealand (Scott 1996). Food hygiene factors include, infrequent hand-washing following contact with raw chicken, and other cross contamination (Redmond and Griffith 2003). Home hygiene was influenced by tidiness, a sense of cleanliness, routine cleaning and an appreciation of good manners (Aunger et al. 2016; Godwin et al. 2006).[6] Studies focusing specifically on home hygiene, found infrequent hand washing after using the privy or, nappy changing significantly affected the spread of infection. Fecal contamination was detected at key sites and surfaces. A tendency for improved hygiene was also affected by a sense of aesthetics (Curtis et al.

[6]The topic of home hygiene is surprisingly under investigated in modern times and so evidence based recommendation are lacking. A PubMed search for the topic "home or domestic and hygiene yielded ~300 references, with most dealing with care homes.

2003). Other risky food handling behaviours contribute to foodborne disease in the home (Yang et al. 1998).

The home should be kept clean by periodic cleaning and by setting rules of conduct for members of the household, discouraging litter and smoking. Keep surfaces dust-free, by vacuum cleaning where possible. Enforce the habits that keep the house clean, prompting members of the household to help. The home should be kept free from rodents and other pests, such as roaches and flies, by a strict preventive program and by a continuing check for indications of the presence of these undesirable elements. When pests and insects are present, concerted efforts to eliminate them should not be spared. Using a private home or domestic kitchen for commercial food operations is prohibited. In addition, an area adjoining private sleeping quarters cannot be used for commercial food activity (U.S. Food and Drug Administration 2013)

2.2 Proper Food Handling

The principles of food hygiene that apply for domestic settings apply to foodservices, and food manufacturing (Al-Sakkaf 2013; Jacob and Powell 2009; Towns et al. 2006). This section introduces the "4C-concept" which is a cornerstone of proper food handling (Partnership for Food Safety Education 2010). Proper adherence to these principles is important to reduce the risk of foodborne disease. The four elements (4C's) refer to C1/cleaning, C2/cross-contamination, C3/cooking, and C4/cooling or chilling.

Cleaning: ensure proper washing of the hands, cutting boards, worktops and other contact surfaces for food.

Cross-contamination: implement steps to prevent the transfer of bacteria. Separate raw and cooked foods during preparations and storage; use different cutting boards for raw meat and fresh foods,

Cooking: food should be cooked and or heated to an appropriate internal temperature, to inactivate pathogenic microorganisms.

Cooling and chilling: foods should be chilled rapidly and then held and/or stored at sufficiently low refrigeration temperatures outside of the temperature danger zone that allows bacteria to multiply (Sect. 2.6).

2.3 Cleanliness and Cleaning

Tableware, knives, and pots and pans should have handles of metal or plastic rather than wood. Clean chopping boards, tableware and utensils soon after use rather than leaving for long periods in a sink or on the counter. Tableware and utensils should be washed in hot water and detergent and rinsed, then immersed in hot water (that has been heated to a minimum of 170 °F [76.7 °C]) for at least 30 s. Surfaces that contact foods should be easily cleanable material such as plastics and stainless steel (USDA Food Safety and Inspection Service 2015b).

Use automatic dishwashers when available because they clean effectively using temperatures higher than can be tolerated in hand washing. Keep the refrigerator and freezer clean and free from odors.

2.4 Cross Contamination Prevention

Transfer of bacterial from raw meats to cooked food is a risk factor for foodborne illness. To avoid cross contamination, separate raw meat from cooked product inside of a refrigerator. Use a separate chopping board and utensils for raw meats and another chopping board for and knife for vegetables and salads. By contrast, washing was not effective in combating cross contamination via chopping boards (Cogan et al. 1999; de Jong et al. 2008).

Cross-contaminations occurs via many routes. Bacterial, may spread from raw foods, staff, clothing, dirty hands, and poorly cleaned machinery. Infestation by pests should not be tolerated because living creatures are vectors for bacteria. Flies and rodents transfer bacteria from one food to another (Table 9.3). Deal with pest infestation by calling in the experts. For commercial prem-

Table 9.3 Pests that promote bacterial cross contamination

Ants
Beetles and weevils
Birds
Cockroaches
House flies and other flying insects
Insect larvae and maggots
Rats and mice

ises, frequent pest-checks, proper clearing, good initial quality controls for received materials are strategies worth adopting.

2.5 Cooking to High Internal Temperatures

To render food safe by conventional cooking, it is necessary to ensure that the center of the product reaches an internal temperature sufficient to inactivate any pathogenic bacteria lurking there (Sundberg and Carlin 1978). The main concerns with safe cooking deal with sources of animal proteins (meat, poultry, eggs and seafood) and potential contamination by four bacterial pathogens e.g. E. coli, salmonella, *staphylococcus aureus*, and *Listeria* (USDA Food Safety and Inspection Service 2014a).

It is possible to arrive at the temperature guidelines because of robust underlying scientific principles outlined below. The minimum temperature for inactivating bacteria can be estimated and adjusted for food matrix effects. Safe internal temperatures are available for different meat products, including whole joints, patties, meat loaf, chops and beefsteaks of known dimensions predetermined sizes (Table 9.4).

For most meat products, the internal temperature for safe cooking is 145 °F (62.8 °C). A higher safe internal temperature is adopted for cooking ground meat (160 °F/71.1 °C) to allow for the dispersal of bacterial within such comminuted products (USDA Food Safety and Inspection Service 2014a). For chicken and poultry, the safe internal temperature for cooking is 165 °F (73.9 °C).

Table 9.4 shows minimal internal temperatures for some animal protein food sources (USDA Food Safety and Inspection Service 2014b). The FSIS guidelines for safe internal cooking temperatures apply with few or no restrictions *irrespective* of different cooking apparatus or initial set temperatures.

Minimum internal temperatures for safe cooking practices were developed taking account of some basic food engineering principles for heating. As noted above the minimum internal temperatures required to render food safe, is independent of the type of cooking. According to the USDA/FIS advice, using a kitchen thermometer is the only sure way to ensure that the correct cooking temperature is reached. Different varieties of kitchen thermometers are available suited for different forms of cooking, e.g. oven cooking, deep oil frying, soups etc. (USDA Food Inspection Services 2011).

Direct measurement using a food thermometer removes a host of potential complications when ensuring safe food cooking. For instance, the type of appliances used influences the rate of heating for different foods. Gas, electric cooker or microwave cookers behave differently. The appliance temperature or power setting is also significantly different. Different types of cooking container, and the presence of packaging or foil wrapping can have an insulating effect preventing heating. The cooking medium, typically air, water or oil for baking, boiling or frying, respectively, affects the heating characteristics. Heating is dependent also on the food ingredients composition, and structural characteristics such as the proportion of air, water, lipids, carbohydrates or protein. Many dedicated food technology texts provide background information on heating (Karel and Lund 2003; Singh and Heldman 2008).

The internal temperature necessary to achieve safety must be considered paramount, before considering further endpoints for product quality, including palatability, cooking losses, color, juiciness, % moisture and tenderness (Cross et al. 1976; Lane et al. 1980; Parrish et al. 1973; Schmidt et al. 1970; Simmons et al. 1985). It may be desirable to minimize heating to control done-

Table 9.4 Recommended safe internal temperatures for cooking meat and poultry

Product	Minimum internal temperature [& rest time/min]
Beef, pork, veal & lamb, steaks, chops, roasts	145 °F /62.8 °C [3 min]
Ground meats	160 °F /71.1 °C
Ham, fresh or smoked (uncooked)	145 °F / 62.8 °C [3 min]
Fully cooked ham, from USDA-inspected plants	140 °F /60 °C
Fully cooked ham (to reheat)	165 °F /73.9 °C
All poultry[a]	165 °F /73.9 °C
Eggs	160 °F /71.1 °C
Fish & shellfish	145 °F /62.8 °C
Leftovers	165 °F /73.9 °C
Casseroles	165 °F /73.9 °C

[a]Including chicken breasts, whole bird, legs, thighs, wings, ground poultry, giblets, and stuffing. Adapted from USDA Internal temperature chart (USDA Food Safety and Inspection Service 2014a, 2016)

Table 9.5 Temperature zones for processing, and storing foods

Zone & temp (°F)	Comment (duration)	Temp (°C)
Zone A, 240–250	Commercial sterilization, low acid foods (pH >4.5)	115.6 to 121.1
Zone B, 212.0–240	Commercial sterilization, high acid foods (pH ≤4.5)	100 to 115.6
Zone C, 165 to 212.0	Most foods safe to eat	73.9 to 100
Zone D, 140 to 165	Storing hot foods hot	60 to 73.9
Zone Hz. 40 to 140	Danger zone for foods	4.4 to 60.0
Zone E, −32 to 40	Storing cold foods cold	0 to 4.4.
Zone F, 26 to −32	Storing super-chilled foods (2–3 Wks.)	−3 to −0
Zone one G, 0 to 26	Storing frozen foods (1–6 Mo.)	−17.8 to −3.3
Zone -I, −20 to 0	Storing frozen foods (6–24 Mo.)	−28.8 to −17.8
Zone -J, −40 to −20	Storing frozen foods (24 Mo.)	−40 to −28.8

ness but achieving safety requires using a food thermometer to avoid guesswork when aiming for the safe cooking temperature (USDA Food Inspection Services 2008).

The USDA/FIS define doneness as "as the condition of being cooked to the desired degree" (USDA Food Inspection Services 2008). Preparing a so-called rare, medium or well-done beefsteak involves different temperature-time treatments to achieve acceptable degrees of "doneness" according to subjective preferences of consumers. However, doneness is not always equivalent to "safe". If beef stake is undercooked then it may harbor pathogenic microbes. By contrast, over-done beefsteak could also pose a risk of chemical hazards such as the formation carcinogenic compounds.

Microwave cooking creates new challenges for safe cooking. Where food is heated from inside outwards, then a high-internal temperature will be no guarantee that the product surface is free from bacteria. For example, a study comparing chicken cooked using a conventional oven or microwave oven, found that the internal temperature was not a good predictor for microbiological safety in the former case (Lindsay et al. 1986).

In addition to heating to a safe internal temperature, it is essential to keep foods at a high temperature in order to prevent re-growing surviving bacteria. Hot holding is used in the food-services to keep food safe (Snyder and Matthews 1984) (Table 9.5).

Research spanning nearly two decades showed that many households may own a kitchen thermometer. However, routine measurements of minimal internal cooking temperatures were deficient. In addition, many influencers (celebrity chefs, internet bloggers) and also food recipes were not providing guidance or information related to minimum internal temperatures (Chambers et al. 2018; Medeiros et al. 2001; Morrison and Young 2019).

2.6 Cooling and Chilling

Keeping products out of the temperature "danger zone" within the entrie distribution chain is important for safe food handling. Products must be maintained at low temperatures during purchase, transport, preparation, and storage prior to consumption. The advice for chilling, cooling or refrigeration is applicable for individual consumers, foodservice operations or food manufacturers (Hui et al. 2004; Kennedy 2000; USDA Food Safety and Inspection Service 2015a, c).

When home shopping for foods, purchase perishable foods (e.g., milk, meat, and fish) last, and do not delay returning home once the shopping is completed. Bacteria, which spoil foods, can multiply to very large numbers in just hours while in a warm car. Upon arriving home, unpack perishable foods immediately and transfer quickly to a refrigerator or freezer. Maintain perishables at refrigerator temperatures of 32–38 °F (0–3.3 °C) without freezing. Frozen foods should be held at 0 °F (−17.8 °C) or below.

Refrigerators or freezers should be lightly loaded; otherwise, the temperature of bulk foods will not fall quickly enough. Slow cooling allows opportunity for the growth of bacterial pathogens. Reapportion bulk foods into smaller packs to encourage cooling. Large pots of stew, bulk hamburger, and large cuts of meat should put in small containers. Cooling food quickly helps to avoid spoilage (see Table 9.6).

Food should be packaged using impervious films in order to minimize freezer burn, oxidative deterioration, and dehydration. With the exception of ripe fruits and vegetables, cover all other foods before refrigeration. Cooked foods, meats, fish, cold cuts, bacon, and frankfurters should be packaged also before refrigeration. Nuts should be packaged and refrigerated to slow down the oxidation of their fats, which leads to rancidity. Greens (e.g., spinach) should be refrigerated unwashed because if water is retained they may spoil more quickly. They should, however, be washed before using (Hui et al. 2004; Kennedy 2000).

Table 9.6 Do's and don'ts for the safe cooling of foods

Do use shallow pans
Do cut it up to cool
Do use a thermometer
Do use ice to cool when possible
Don't use large containers
Don't use deep pans
Don't try to cool the whole roast
Don't guess at temperature
Don't always depend on the refrigerator to cool
Do divide the food mass into two or more smaller containers

Fresh foods should be used as soon as possible as most deteriorate rapidly. Baked goods, unripe fruit, and bananas can be held at room temperature for about 2 days, but longer storage requires freezing. Meats, eggs, poultry, fish, and other perishable foods should be held at safe holding temperatures below 40 °F (4.4 °C) or above 140 °F (60 °C) (Table 9.5).

2.7 Risky Food Handling

Meals consumed shortly following their preparation are least likely to cause food poisoning. Meals that are consumed hours after preparation at picnics and outings have increased risk of foodborne illness. Salads containing eggs, chicken, or turkey, or other products made from them pose specific hazards. If not eaten immediately, refrigerate foods and reheat just prior to use. Some products such as puddings, custards, and éclairs should be held under refrigeration at all times.

Home freezers are not of sufficient capacity to flash-freeze stuffed poultry, chicken, and especially turkey, and therefore poultry should not be stuffed prior to freezing; the FSIS advises against purchasing pre-stuffed poultry. Instead, poultry should be stuffed just prior to cooking.

Food leftovers should be used as soon as possible; this is especially true of salads, chicken, and other perishables. If, because of odor or other indicator, there is any doubt about the safety of a food, it is best to discard. As noted above, foods for picnic or outings are notably risky and requiring greater care during handling.

Precool all perishable foods, including ingredients for salads, preferably in a refrigerator. Heat all foods, to be served hot, to at least 165 °F (73.9 °C). Keep cold foods cold (40 °F [4.4 °C] or below) and hot foods hot (140 °F [60 °C] or above) until served. It is best to refrain from handling pets while preparing foods. Comprehensive advice on food handling for individuals as well those cooking for groups arc available from the USDA-FSIS (USDA Food Safety and Inspection Service 2015b)

Despite continued effort, a range of risky and unsafe food handling practices remain significant in the home. One USDA study involving 19,356 adults found (for the preceding 12 months) that sizeable numbers had, consumed undercooked eggs (50.2%), eaten home canned vegies (23.8%), eaten pink (undercooked) hamburgers (19.7%), eaten raw oysters (8%) or drank raw milk (1.4%). Many people frequently delayed washing cutting boards (19%) or washing their hands (18%) after handling raw meat (Yang et al. 1998). Apparently, many barriers remain to changing consumer behaviour, which goes beyond a lack of knowledge. Several experts noted that consumers are liable not to practice safe food handling even when in possession of appropriate information (Young and Waddell 2016).

Incidentally, the temperature boundary for safe food holding was revised to below or above 41-135 °F (5 – 57 °C) during the writing of this chapter (Food Code 2017.p574-581). The safe food handling temperature boundary (40°–140 °F (4.4–60°C) cited in the rest of this chapter appear in pre-2017 literature.

3 Retail Hygiene and Food Handling

3.1 Fresh Meat Handling

General food safety guidelines for retail outlets are based on the US food code, and are similar to those for food manufacturing plants (Buchanan 2000; Kambhampati et al. 2016; U.S. Food and Drug Administration 2013; U. S. Food and Drug Administration 2017). However, meat hygiene requirements are overseen by the FSIS. Many hygiene requirements are also applied globally (Food and Agriculture Organization of the United Nations 1991). In particular, fresh meat handling is of concern in a retail context.

Fresh uncut meats should be stored in a walk-in refrigerator with walls and ceilings of glazed tile to facilitate washing and cleaning. The floor should be constructed of unglazed tile and sloped to drains. The temperature of the meat storage room should be 32–37 °F (0–2.8 °C). Sawdust should not be used on the floor because this creates dust, which acts as a source of contamination.

The room where meat will be cut into retail portions should have the same wall, ceiling, and floor construction as the meat storage area. Benches used for cutting meat should have surfaces of stainless steel; cutting boards should be plastic or Teflon, which is impervious and easily cleaned and sanitized. The temperature of this room should be about 50 °F (10 °C), as personnel find it difficult to work at lower temperatures (Food and Agriculture Organization of the United Nations 1991).

A low relative humidity must be maintained in the cutting area otherwise, meat will condense moisture (sweat), which will facilitate the growth of spoilage bacteria. In addition, neither cut nor uncut meat should be allowed to accumulate in this area but should be moved back into the storage area or into display cases as soon as possible. The meat-cutting area should be cleaned and sanitized at least once per 8-h working period and should follow essentially the same methods as indicated earlier for food manufacturing plants.

When preparing ground meat or meat patty, use separate grinders (or grinder heads) for beef and pork. The reason for this is that pork may contain an infective roundworm (Trichinella species), which causes the disease trichinosis in humans. Therefore, pork should be cooked to high temperatures sufficient to destroy the worm cysts. However, beef burger (hamburger) is sometimes eaten undercooked or only lightly heated, and so it is appropriate to avoid contaminate the ground beef with pork.

Retail display cases for cut meats, chicken, cooked and fresh sausage products, bacon, and cold cuts should be maintained at a temperature of 32–38 °F (0–3.3 °C) and should be of the closable variety. Personnel in charge of display cases should make certain that products move in the order of "first in, first out" and that no item remains in the display case for long periods. Canned ham of the type requiring refrigeration should be maintained at temperatures of 38 °F (3.3 °C) or below and not displayed without refrigeration (Food and Agriculture Organization of the United Nations 1991). Extra information on meat handling practices can be found at the FSIS website and from the code of federal regulations (U.S. Government Publishing Office 2019).

3.2 Safe Handling Fresh Produce

Safe handling of fresh fruits and vegetables involve the whole food system (Fig 9.1). Contaminations from the farm environment (soil, water, livestock, and less commonly farm workers) and packing facilities pose significant levels of risk. Fresh fruit and vegetable are at risk also during minimal processing and packaging. Furthermore, many fruits and vegetables are eaten fresh without prior heat treatment, which is a lethal step for pathogens. (Beuchat and Ryu 1997; Murray et al. 2017; Li et al. 2015).

The main approaches for controlling food pathogens on fresh produce are, (i) by avoiding farm-based contamination, (ii) avoiding post-harvest cross contamination and (iii) washing with water and sanitizer (Beuchat and Ryu 1997; Li et al. 2015; Murray et al. 2017; Li et al. 2018). Worldwide, contamination of fresh produce by viruses and parasites were found alongside of bacteria. The overall prevalence of fresh food contamination was judged as low indicating that improvements in food safety are being made (Van Pelt et al. 2018).

Assuming food is safe when received at the retail outlet, fresh produce require proper handling to avoid safety and quality loss (Choi et al. 2016; Ramalho et al. 2015; Thaivalappil et al. 2018). The cells of plant foods continue to respire after harvest and the higher the temperature at which they are held the faster the rate of respiration. Respiration brings about chemical changes in fresh produce and a deterioration of quality. Loss of sweetness, loss of succulence, toughening, and the development of off-flavors are some of the changes that take place in fresh produce because of respiration.

Sweet corn on the cob, freshly picked and then stored at 35 °F (1.7 °C), is perfectly good in taste and texture after 15 days of storage. At high temperatures, the quality of sweet corn will be lost in a few hours. Storage areas for fresh produce should be clean and held at 32–37 °F (0–2.8 °C). Potatoes, turnips, and cabbages should be stored at about 50 °F (10 °C), as cabbage and turnips stand up well at this temperature, and potatoes convert starch to sugar at 40 °F (4.4 °C) or below-becoming sweet.

Most fresh produce should be displayed in an area or cabinet held at 32 to 37 °F (2.8 °C) or partially surrounded with ice, since such temperatures maintain quality. A possible exception is lettuce, which, if held in ice, may freeze, causing the leaves to wilt.

Fish and shellfish, such as shucked clams, oyster, scallops, and shrimp, should be presented in display case surrounded by ice. This provides a temperature of about 33 °F (0.6 °C) and is the best way to maintain this low temperature, without freezing, for products that are extremely perishable.

Milk, cream, sour cream, cheeses, and butter can be presented in an open-top display case, the temperature of which is held at 32–37 °F (0–2.8 °C).

3.3 Frozen Foods

Frozen foods, which are not indefinitely stable, are rarely handled under satisfactory conditions in retail stores. This topic has been extensively reviewed (Sun 2012; Erickson and Yen-Con 1997; Evans 2008; Kennedy 2000). The following is a brief summary of some key points.

Frozen foods deteriorate extremely slowly at −30 °F (−34.4 °C). Many foods have a storage life (no noticeable loss of quality) of at least 6 months at 0 °F (−17.8 °C), and some foods have a storage life of at least 1 year at this temperature. As the storage temperature is raised above 0 °F (−17.8 °C), for each 5 °F (2.8 °C), the rate of deterioration is approximately doubled.[7] Thus, a product that has a storage life of 6 months at 0 °F (−17.8 °C) will have a storage life of only 3 months at 5 °F (−15 °C). When frozen foods are delivered to the retail store, they should not stand on platforms at high temperature but should be immediately placed in the frozen storage room held at a temperature of 0 °F (−17.8 °C) or preferably at −20 °F (−28.9 °C).

Display cases used for holding frozen foods in the retail area should be of the open-top type, or of the enclosed shelf type. Shelf-type display cases for frozen foods are not suitable unless they are enclosed by doors. The reason for this is that cold air is heavier than warm air; hence, in the open-top case the cold air tends to remain in the area where the foods are held. Shelf-type frozen food cases must be refrigerated by blowing cold air out and over the product. Because this air is heavier than warm air, it tends to flow outward and downward into the room. In display cases of this kind it is therefore difficult to maintain temperatures of 0 °F (−17.8 °C) or below around the entire product. Those items in front where warm air has access are usually surrounded by air temperatures much higher than 0 °F (−17.8 °C), and hence are subjected to an accelerated rate of deterioration (Evans 2008; Kennedy 2000). Safe food handling guidelines for frozen foods require such products to be kept frozen hard at all times (US Food Code ss. 3.501.11 to 13). Inadvertent temperature rises, may lead to partially thawing the product and bringing it into the temperature danger-zone for growing food pathogens.

There are three FDA recommendations for thawing frozen foods safely; (i) thaw the frozen food under controlled low-temperature conditions using a refrigerator; (ii) thaw frozen food by submerging under water after sealing in a waterproof plastic bag to avoid contamination, (iii) cook the food product directly from the frozen state. In all cases, the objective is to ensure a proper "time/temperature control for food safety or TCS. (U.S. Department of Agriculture, 2013. p89). The TCS principle is also behind hot-holding or rapidly cooling foods to keep them out of the temperature danger zone (Table 9.5). Safe thawing is discussed further in Chap 13 (Sect 4.2)

3.4 Canned Foods

Most canned food are shelf-stable meaning that they can be stored without refrigeration. Shelf-stable food products do not require strict time/temperature control for food safety, which allows convenient storage (U.S. Food Code, p335). Other self-stable foods (e.g. dried food, food sterilized inside of flexible pouches, aseptically packed foods) can be stored at room temperature. Some canned foods are perishable (e.g. "keep refrigerated" canned ham) and must be refrigerated (USDA 2017).

In any event, canned foods for retail are heat processed for commercial sterility by FDA certified and properly trained personnel (Avena-Bustillos 2011). That means that the foods must

---

[7]The changes in rates of deterioration with temperature is often discussed in terms of the Q10, which the chages in rate observed with a 10 °C rise in temperature. For frozen foods the Q10 value ranges from 2–20 meaning a 10 °C rise in temperature may produce a 2–20 fold change in shelf-life depending on the product. See Erickson, M. C. and H. Yen-Con (1997). "Quality in frozen food." Pg. 385 for more discussions.

be heated to the point where all spores of *Clostridium botulinum* have been destroyed. Actually, to prevent spoilage (not disease) by other bacteria, canned foods should be heated beyond the point at which all disease-causing bacteria will have been destroyed (Derossi et al. 2011). The heating requirements for commercially canned foods are discussed further in Chap 12, Sect. 10.

A sizeable backlog of canned foods must be available to the retail store. This means that there must be a warehouse where canned foods can be stored. Such warehouses should be held at temperatures not above 75 °F (23.9 °C). For this reason, the temperature of the warehouse where canned foods are stored should be regulated as previously stated. Nor should the storage warehouse for canned foods be held at temperatures below 50 °F (10 °C), for if this is done, when higher temperatures are reached in this area or when the cans are placed in the retail outlet at higher temperatures, moisture may condense on the surface (the cans may sweat) and cause rusting of the outside surface, which discolors the label and is otherwise unsightly, and it may eventually weaken the can to permit microbial invasion of the contents.

Canned foods, both those stored in the warehouse and those held in retail stores, should be handled on a first-in, first-out basis. The reason for this is that although most canned foods are shelf-stable, they are not indefinitely stable. The usual type of deterioration after long periods of storage is due to internal corrosion of the container, which results in a swelled can or in leakage. Deterioration of this type is most often encountered in acid foods, such as tomato products, in which case internal corrosion produces hydrogen gas, causing the can to swell. The food in this case may be perfectly edible, but consumers would be running a risk to eat the food from a swelled can because they cannot be sure that the cause of the swelling was not due to gas produced by some disease-causing bacterium such as Clostridium botulinum. Swollen cans should always be returned to the supplier to be discarded properly.

Finally, noncommercial home canned products deserve a mention. Home canning involves the use of pressure-cooking or ambient pressure heating for high acid foods (Anon 1977). There is growing contingent of home canning.

3.5 Food Manufacturing Plant Sanitation

Food plant sanitation is necessary, first because it is a law (see the Food, Drug and Cosmetic Act), and second because it is ethical, economical, and expected. The requirements of food plant sanitation are covered by the principles of Good Manufacturing Practices (GMP) as outlined in Chap. 10.

4 Foodservice Hygiene

The significance of foodservice hygiene should not be underestimated. Recall that the foodservice and retail are the two main businesses that sell food directly to consumers. Virtually all risk factors for foodborne disease (Table 9.2) could be found within foodservice establishments. All over the United States, some 100 million meals are served each day, in over 950 thousand restaurants. Currently, over two-thirds of all reported food poisonings result from meals served inside of sit-down restaurants (Chap 25, Sect 5).

The actual numbers of outbreaks are uncertain because only a small percentage of all food poisonings may be reported. One thing is known, and that is the total number is very high, with some estimates indicating that there are several million foodborne disease incidences per year (U.S. Food and Drug Administration, 2013).[8] Unfortunately, the transmission of foodborne disease by occupational food handlers remains a problem (Insfran-Rivarola et al. 2020; Jacob and World Health Organization 1989; Ncube et al. 2019; Thaivalappil et al. 2018).

[8]The food safety guidelines for foodservices are based on the food code, and good manufacturing practice (GMP) guidelines detailed in Chap. 10.

4.1 Premises and Utensils

In general, the area where people are fed need not be special except that it can be cleaned easily and kept clean. The area should be pleasant; the floor must not be carpeted, but in any case, it must be easy to clean thoroughly.

It is required to have only potable (drinkable) water in foodservice establishments. If no potable water is used for cooling refrigeration units in walk-in refrigerators, for example, this water should not be connected with the potable water system, and the pipes should be painted with an identifying color.

In the food preparation and utensil-cleaning areas, the floors should be constructed of acid-resistant unglazed tile or of epoxy or polyester resin on a suitable base material.

Also, the floor should be sloped to drains to facilitate cleaning and prevent the accumulation of water. Drains should empty into lines separated from toilet sewer lines to a point outside of the building and should be so constructed as to prevent the possibility of backup into the building.

The walls of food preparation and utensil cleaning areas should be constructed of smooth glazed tile to distances equivalent to splash height, since this type of construction allows for easier and more effective cleaning. The junction of the walls and floor should be coved or curved, which also facilitates cleaning, there being no angled corners where food materials can lodge.

The surfaces of benches and tables in food preparation areas should be of stainless steel or plastic, because such materials are impervious, noncorrosive, and easy to clean. For the same reason, cooking utensils, including steam kettles, should be constructed of stainless steel.

Areas in which steam cooking, steam cleaning, or deep-fat frying is done should be provided with hoods and exhaust fans to the outside. Hoods should be constructed using stainless steel for ease in cleaning and should be equipped with traps to prevent condensed moisture from running back into foods being prepared.

Cutting boards should be made from plastic, preferably Teflon, as such material is easy to clean, does not absorb water, and does not foster bacterial growth.

4.2 Equipment and Facilities

Walk-in refrigerators should have floors constructed of unglazed tile that are sloped to drains to facilitate cleaning. The drains should not empty directly into the sewerage system. Instead, they should empty into a sink or other container that empties in turn into a drainage system, so there is no possibility of backup into the refrigerator. The walls and ceilings of walk-in refrigerators should be constructed of glazed tile to facilitate cleaning.

Wash basins, soap, hot and cold water, and a container of disinfectant (preferably a solution of one of the iodophors) together with paper towels should be present in the food preparation, utensil washing, and food exit areas to make certain that anyone having left his or her particular operation will wash, disinfect, and dry his or her hands prior to resuming work.

The managers of foodservice establishments, purchaser, or persons in charge, should obtain raw food materials only from reliable sources. It is essential that the water to be used for drinking, cooking, and cleaning is potable.

4.3 Food Preparation and Handling

The precautions for commercial premises are no different from those that have been given for household food safety (Sect. 2.5.) but if foods, including gravies, are to be held on steam tables, the temperature of all parts of the food should never fall below 140 °F (60 °C), and preferably not below 150 °F (65.6 °C). In addition, any food container held on steam tables should be emptied, removed, and replaced with a new container of the particular food instead of being partially emptied and refilled with more food.

If leftover cooked foods are to be refrigerated, they should be placed in covered impervious containers (plastic or metal) and labeled. A card catalog of leftovers should be maintained so that food may be thrown away if held for periods of more than 4 days at temperatures of 38 °F (3.3 °C) or above.

If cream-filled pastries, such as éclairs or pies, and salads, such as potato, tuna fish, crab meat, or chicken, are to be held in the refrigerator or in display cases, the temperature of such storage areas should be 38 °F (3.3 °C) or below.

4.4 Personnel Health and Personal Hygiene

Personnel with boils or pus-producing infections on their hands should not be allowed to handle foods or to clean utensils or equipment. Any personnel known to have had a recent intestinal ailment should be excused from work until such a period as it can be determined that he or she is not infective.

4.5 Waste Disposal

All garbage and waste materials should be held in leak-proof metal or plastic containers with tight-fitting covers when held on the premises of the foodservice establishment. Rubbish should also be held in this manner. After emptying, each container should be cleaned inside and outside, and the water used for such cleaning disposed of in the sewerage system. Such containers should be washed in an area well separated from that used for washing utensils; containers of garbage and trash should be stored in vermin proof rooms, well separated from the food preparation and serving areas.

The floors of such storage rooms, and the walls up to distances of splash heights, should have an impervious easy-to-clean surface. Storage areas of this type should be cleaned periodically. Care should be exercised to see that rodents and insects do not become established in waste storage areas.

4.6 Cleaning and Sanitization

At least two of five main risk factors for food-borne disease (Table 9.2) are directly impacted by cleaning (Etienne 2006. p1–3). Cleaning refers to the removal of dirt and/ or unwanted material by diverse approaches. Cleaning may involve chemical, physical cleaning, wet cleaning, dry-cleaning, or flame cleaning etc. The cleaned surface is then exposed to a sanitizer. It should be noted that cleaning and sanitization are two distinct processes.

Effective sanitization of wash-ware, utensils, food contact surfaces etc. is a legal requirement. In addition, it is known that sanitizer efficiency is affected by water temperature, and degree of water hardness (U.S. Food & Drug Administration 2017a).

In foodservice establishments, it is a violation of federal law to use a product as a chemical sanitizer unless it is Environmental Protection Agency (EPA) registered (Hwang et al. 2017; Prado-Silva et al. 2015; Wei et al. 2017). Some household bleach have EPA registration and the accuracy of all claims on the label has been proven. Many detergent soaps and cleaning agents have pesticidal (i.e. sanitizing) activity and these too require EPA registration (Environmental Protection Agency 2021). Follow the directions for sanitizing carefully. To sanitize dishes and other food utensils, the Food and Drug Administration (FDA) recommends a solution of 50 ppm available chlorine as a hypochlorite.

An effective method of sanitizing glasses, chinaware, and utensils is to wash in water at least at 120 °F (48.9 °C) with detergent followed by rinsing, then immersing in clean water at 170 °F (76.7 °C) for 30 s. Immersion for 10 s in a solution containing 50 ppm available chlorine or 30 s in a solution containing 12.5 ppm of an iodophor, in each case at a temperature not below 75 °F (23.9 °C), may be substituted for the hot-water dip for purposes of sanitizing utensils and tableware (Table 9.7).

After washing and rinsing, utensils must be submerged in the warm sanitizing solution (75 °F [24 °C] or higher) for 10 s. If other

Table 9.7 Commonly used FDA approved chemical sanitizers

Sanitizer (min. Conc)	Hypochlorite (Chlorine) (50 ppm)	Lodophor (Iodine) (12.5 ppm)	Quaternary Ammonium (200 ppm)
Temperature of solution	75 °F/23.9 °C+	75–120 °F/23.9–48.9 °C Note: Iodine leaves solution at 120 °F/48.9 °C and is no longer effective as a sanitizer.	75 °F/23.9 °C+
Time for sanitization	10 s, allow to air-dry after spraying on equipment	30 s for immersion, allow to air dry after spraying on equipment	≥30 s immersion, read instructions on labels
pH range, note detergent residues may raise pH	Below pH 10	Below pH 5.5	pH 7, varies with compounds, read label
Corrosiveness	Aluminum, silver, some plate utensils, stainless steel	Noncorrosive	Noncorrosive
Reaction with water biomass	Inactivated	Less effective	Unaffected
Reaction with hard water	Not affected	Not affected	Hardness >500 unsuitable for some quarts
Measurement of strength	Test kit or strip	Test kit or strip, detergent loses amber color as weakens	Test strip or kit

approved sanitizers such as iodophors or quaternary ammonium are used, 30 s of submersion is required. For counter tops, cutting boards, or equipment that cannot be submerged, spray or wipe with the proper strength solution and allow air-drying.

The sanitizers (Table 9.7) are chemical agents intended to disinfect inert objects and food contact surfaces (US food code. p393) excluding the human skin. From 2005, the phrase "hand sanitizer" was changed to "hand antiseptic" in the US food code. The hand antiseptics are regulated as, over either the counter drugs, food additives or GRAS substances. Since the Cov-19 outbreak of 2020, the classical or common meaning of the term hand sanitizers has become resurgent. Solutions of typically 70% alcohol is used as a topical disinfectant along with facemasks, which are mandatory. The medical literature and health scientists also retain the term "hand sanitizer" (Tamimi et al. 2014; Charbonneau et al. 2000).

4.7 Sources of Advice for Food Businesses

Food Safety requirements for food producers (farmers), manufacturers, and the foodservices are highly complex and constantly evolving. Therefore, food business proprietors are required to seek expert advice on legislation from local, or State extension experts covering basic hygiene requirements as outlined in Good Manufacturing Practices (GMP). Indeed, business will be required to hire, certified food safety champions on staff. For very small businesses, authoritative advice can be found from public health agents. Existing food business proprietors, aspirants, or others should seek advice on a possible requirement for "HACCP" for their businesses – this being one of the most exacting codes of practice. Material in this introductory chapter is not intended as legal advice on food safety. More details on GMP and HACCP will be found in Chap. 10.

References

Al-Sakkaf A (2013) Domestic food preparation practices: a reviewof the reasons for poor home hygiene practices. Health Promot Int 30(3). https://doi.org/10.1093/heapro/dat051. Retrieved from https://academic.oup.com/heapro/article/30/3/427/617876/Domestic-food-preparation-practices-a-review-of

Anon (1977) Home canning. A scientific status summary by the Institute of Food Technologist's Expert Panel on Food Safety & Nutrition and the Committee on Public Information. Food Technol 31(6):43–47

Aung MM, Chang YS (2015) Traceability in a food supply chain: Safety and quality perspectives. Food Control 39:172–184. https://doi.org/10.1016/j.foodcont.2013.11.007

Aunger R, Greenland K, Ploubidis G, Schmidt W, Oxford J, Curtis V (2016) The determinants of reported personal and household hygiene behaviour: a multi-country study. PLoS One 11(8):e0159551. https://doi.org/10.1371/journal.pone.0159551

Avena-Bustillos RJ (2011) Establishment of an FDA-approved international Better Process Control School for food canning supervisors. Prog Nutr 13(3):155–159

Beuchat LR, Ryu JH (1997) Produce handling and processing practices. Emerg Infect Dis 3(4):459–465

Bloomfield SF, Carling PC, Exner M (2017) A unified framework for developing effective hygiene procedures for hands, environmental surfaces and laundry in healthcare, domestic, food handling and other settings. GMS Hyg Infect Control 12 Doc08. https://doi.org/10.3205/dgkh000293

Boyce JM (2016) Modern technologies for improving cleaning and disinfection of environmental surfaces in hospitals. Antimicrob Resist Infect Control 5:10. https://doi.org/10.1186/s13756-016-0111-x

Buchanan RL (2000) The development of science-based food safety regulations in the United States. Irish J Agric Food Res 39(2):331–342. https://www.jstor.org/stable/25562399

Chambers E, Godwin S, Terry T (2018) Recipes for determining doneness in poultry do not provide appropriate information based on US government guidelines. Foods 7(8):126

Charbonneau DL, Ponte J M, Kochanowski BA (2000). A method of assessing the efficacy of hand sanitizers: use of real soil encountered in the food service industry. J Food Prot 63(4): 495-501. https://doi.org/10.4315/0362-4028X-4363.4314.4495

Choi J, Norwood H, Seo S, Sirsat SA, Neal J (2016) Evaluation of food safety related behaviors of retail and food service employees while handling fresh and fresh-cut leafy greens. Food Control 67:199–208. https://doi.org/10.1016/j.foodcont.2016.02.044

Cogan TA, Bloomfield SF, Humphrey TJ (1999) The effectiveness of hygiene procedures for prevention of cross-contamination from chicken carcases in the domestic kitchen. Lett Appl Microbiol 29(5):354–358

Cross HR, Stanfield MS, Koch EJ (1976) Beef palatability as affected by cooking rate and final internal temperature. J Anim Sci 43(1):114–121

Curtis V, Biran A, Deverell K, Hughes C, Bellamy K, Drasar B (2003) Hygiene in the home: relating bugs and behaviour. Soc Sci Med 57(4):657–672

de Jong AE, Verhoeff-Bakkenes L, Nauta MJ, de Jonge R (2008) Cross-contamination in the kitchen: effect of hygiene measures. J Appl Microbiol 105(2):615–624. https://doi.org/10.1111/j.1365-2672.2008.03778.x

Derossi A, Fiore AG, De Pilli T, Severini C (2011) A review on acidifying treatments for vegetable canned food. Crit Rev Food Sci Nutr 51(10):955–964. https://doi.org/10.1080/10408398.2010.491163

DuPont HL (2007) The growing threat of foodborne bacterial enteropathogens of animal origin. Clin Infect Dis 45(10):1353–1361

Environmental Protection Agency (2021) Determining if a cleaning product is a pesticide under FIFRA. Retrieved from https://www.epa.gov/pesticide-registration/determining-if-cleaning-product-pesticide-under-fifra

Erickson MC, Yen-Con H (1997) Quality in frozen food. xxi + 484pp

Evans JA (2008) Frozen food science and technology. x + 355pp

Food and Agriculture Organization of the United Nations (1991) Guidelines for slaughtering, meat cutting and further processing. Retrieved from http://www.fao.org/3/T0279E/T0279E00.htm#TOC

Food and Drug Administration (2007) Food Protection Plan of 2007. Retrieved from http://www.fda.gov/Food/GuidanceRegulation/FoodProtectionPlan2007/ucm132565.htm#

Godwin SL, Fur-Chi C, Coppings RJ (2006) Correlation of visual perceptions of cleanliness and reported cleaning practices with measures of microbial contamination in home refrigerators. Food Prot Trends 26(7):474–480

Hui YH, Cornillon P, Guerrero-Legaretta I, Lim MH, Murrell KD, Wai-Kit N (2004) Handbook of frozen foods. xi + 735pp

Hwang JH, Yoon JH, Bae YM, Choi MR, Lee SY, Park KH (2017) Effect of the precutting process on sanitizing treatments for reducing pathogens in vegetables. Food Sci Biotechnol 26(2):531–536. https://doi.org/10.1007/s10068-017-0073-7

Insfran-Rivarola A, Tlapa D, Limon-Romero J, Baez-Lopez Y, Miranda-Ackerman M, Ontiveros A-SK (2020) A systematic review and meta-analysis of the effects of food safety and hygiene training on food handlers. Foods 9(9):24pp. https://www.mdpi.com/2304-8158/2309/2309/1169

Jacob M, World Health Organization (1989) Safe food handling: a training guide for managers in food service establishments. World Health Organization, Geneva, 142pp. https://apps.who.int/iris/handle/10665/36870

Jacob CJ, Powell DA (2009) Where does foodborne illness happen-in the home, at foodservice, or elsewhere-and does it matter? Foodborne Pathog Dis 6(9):1121–1123. https://doi.org/10.1089/fpd.2008.0256

Kambhampati A, Shioda K, Gould LH, Sharp D, Brown LG, Parashar UD, Hall AJ (2016) A state-by-state assessment of food service regulations for prevention of norovirus outbreaks. J Food Prot 79(9):1527–1536. https://doi.org/10.4315/0362-028x.Jfp-16-088

Karel M, Lund DB (2003) Chatper 3: Heat transfer in foods. Physical principles of food preservation: revised and expanded. Retrieved from http://books.google.co.uk/books?id=FTJ4h-cMOYUC

Keener L, Nicholson-Keener SM, Koutchma T (2014) Harmonization of legislation and regulations to achieve food safety: US and Canada perspective. J Sci Food Agric 94(10):1947–1953. https://doi.org/10.1002/jsfa.6295

Kennedy CJ (2000) Managing frozen foods. Woodhead Publishing, Cambridge, 304pp

Lane RH, Muir WM, Mullins SG (1980) Correlation of minimum sensory doneness with internal temperature of deep fat fried chicken thighs. Poult Sci 59(4):719–723

Li D, De Keuckelaere A, Uyttendaele M (2015) Fate of foodborne viruses in the Farm to Fork chain of fresh produce. Compr Rev Food Sci Food Saf 14(6):755–770. http://dx.doi.org/710.1111/1541-4337.12163

Li M, Baker CA, Danyluk MD, Belanger P, Boelaert F, Cressey P, Gheorghe M, Polkinghorne B, Toyofuku H, Havelaar AH (2018) Identification of biological hazards in produce consumed in industrialized countries: a review. J Food Prot 81(7):1171–1186. https://doi.org/1110.4315/0362-1028X.JFP-1117-1465

Lindsay RE, Krissinger WA, Fields BF (1986) Microwave vs. conventional oven cooking of chicken: relationship of internal temperature to surface contamination by Salmonella typhimurium. J Am Diet Assoc 86(3):373–374

Lupien JR (2005) Food quality and safety: traceability and labeling. Crit Rev Food Sci Nutr 45(2):119–123. https://doi.org/10.1080/10408690490911774

Lupien JR (2007) Prevention and control of food safety risks: the role of governments, food producers, marketers, and academia. Asia Pac J Clin Nutr 16(Suppl 1):74–79

Medeiros LC, Hillers VN, Kendall PA, Mason A (2001) Food safety education: what should we be teaching to consumers? J Nutr Educ 33(2):108–113

Morrison E, Young I (2019) The missing ingredient: food safety messages on popular recipe blogs. Food Prot Trends 39(1):28–39

Murray K, Wu F, Shi J, Jun Xue S, Warriner K (2017) Challenges in the microbiological food safety of fresh produce: limitations of post-harvest washing and the need for alternative interventions. Food Qual Saf 1(4):289–301. https://doi.org/210.1093/fqsafe/fyx1027

Ncube F, Kanda A, Mpofu MW, Nyamugure T (2019) Factors associated with safe food handling practices in the food service sector. J Environ Health Sci Eng 17(2):1243–1255. https://www.ncbi.nlm.nih.gov/pmc/articles/PMC6985409/

Parrish FC Jr, Olson DG, Miner BE, Rust RE (1973) Effect of degree of marbling and internal temperature of doneness on beef rib steaks. J Anim Sci 37(2):430–434

Partnership for Food Safety Education (2010) Fight Bac! Retrieved from http://www.fightbac.org/

Prado-Silva L, Cadavez V, Gonzales-Barron U, Rezende AC, Sant'Ana AS (2015) Meta-analysis of the effects of sanitizing treatments on Salmonella, Escherichia coli O157:H7, and Listeria monocytogenes inactivation in fresh produce. Appl Environ Microbiol 81(23):8008–8021. https://doi.org/10.1128/aem.02216-15

Ramalho V, de Moura AP, Cunha LM (2015) Why do small business butcher shops fail to fully implement HACCP? Food Control 49:85–91

Redmond EC, Griffith CJ (2003) Consumer food handling in the home: a review of food safety studies. J Food Prot 66(1):130–161

Schmidt JG, Kline EA, Parrish FC Jr (1970) Effect of carcass maturity and internal temperature on bovine longissimus attributes. J Anim Sci 31(5):861–865

Scott E (1996) Foodborne disease and other hygiene issues in the home. J Appl Bacteriol 80(1):5–9

Silverglade BA (2000) The WTO agreement on sanitary and phytosanitary measures: weakening food safety regulations to facilitate trade? Food Drug Law J 55(4):517–524. https://www.jstor.org/stable/26660192

Simmons SL, Carr TR, McKeith FK (1985) Effects of internal temperature and thickness on palatability of pork loin chops. J Food Sci 50(2):313–315

Singh RP, Heldman DR (2008) Chapter 4 Heat transfer in food processing. In: Introduction to food engineering. Elsevier Science, Amsterdam, pp 247–401

Sneed J, Strohbehn CH (2008) Trends impacting food safety in retail foodservice: implications for dietetics practice. J Am Diet Assoc 108(7):1170–1177. https://doi.org/10.1016/j.jada.2008.04.009

Snyder PO, Matthews ME (1984) Microbiological quality of foodservice menu items produced and stored by cook/chill, cook/freeze, cook/hot-hold and heat/serve methods. J Food Prot 47(11):876–885

Sun W (ed) (2012) Handbook of frozen food processing and packaging, 2nd edn. CRC Press, New York, 936pp

Sundberg AD, Carlin AF (1978) Relation of final internal temperature to Clostridium perfringens destruction in beef loaves cooked in a crockery pot or conventional oven. J Food Sci 43(2):285–288. https://doi.org/10.1111/j.1365-2621.1978.tb02286.x

Tamimi AH, Carlino S, Edmonds S, Gerba CP (2014) Impact of an alcohol-based hand sanitizer intervention on the spread of viruses in homes. Food Environ Virol 6(2):140–144. https://www.ncbi.nlm.nih.gov/pmc/articles/PMC4032461/

Thaivalappil A, Waddell L, Greig J, Meldrum R, Young I (2018) A systematic review and thematic synthesis of qualitative research studies on factors affecting safe food handling at retail and food service. Food Control 89:97–107

Towns RE, Cullen RW, Memken JA, Nnakwe NE (2006) Food safety-related refrigeration and freezer practices and attitudes of consumers in Peoria and surrounding counties. J Food Prot 69(7):1640–1645

U.S. Food and Drug Administration (2013) Food code (2013) Retrieved from http://www.fda.gov/Food/GuidanceRegulation/RetailFoodProtection/FoodCode/ucm374275.htm

U.S. Food and Drug Administration (2017) Managing food safety: a manual for the voluntary use of HACCP principles for operators of food service and retail establishments. Retrieved from http://www.fda.gov/Food/GuidanceRegulation/HACCP/ucm2006810.htm

U.S. Food and Drug Administration (2017a) Annex 5. Conducting risk-based inspection. Food Code 2017:645–678. https://www.fda.gov/media/110822/download

US Food and Drug Administration (2017b) Food code (2017). Retrieved from https://www.fda.gov/media/110822/download.

U.S. Government Publishing Office (2019) Electronic Code of Federal Regulations. TITLE 9—Animals and animal products. CHAPTER III—Food Safety and Inspection Service, Department of Agriculture. SUBCHAPTER A—Agency organization and terminology; mandatory meat and poultry products inspection and voluntary inspection and certification. PART 416—SANITATION. Retrieved from https://www.ecfr.gov/cgi-bin/text-idx?SID=6a7ee18da35a7f2af8de2e936e0d904c&mc=true&node=pt9.2.416&rgn=div5

USDA (2017) Shelf-stable food safety. Retrieved from https://www.fsis.usda.gov/wps/wcm/connect/77ffde83-dc51-4fdf-93be-048110fe47d6/Shelf_Stable_Food_Safety.pdf?MOD=AJPERES

USDA Food Inspection Services (2008, 2013) Doneness versus safety. Retrieved from https://webcache.googleusercontent.com/search?q=cache:fBhdxIN-en4J:https://www.fsis.usda.gov/wps/wcm/connect/6cece12d-a1b9-43c3-9306-b163ea46e82b/Doneness_Versus_Safety.pdf%3FMOD%3DAJPERES+&cd=3&hl=en&ct=clnk&gl=us

USDA Food Inspection Services (2011) Kitchen thermometers. Retrieved from http://extension.umd.edu/sites/extension.umd.edu/files/_images/programs/FOODSAFETY/Kitchen_Thermometers.pdf

USDA Food Safety and Inspection Service (2014a) Beef from farm to table. Retrieved from http://www.fsis.usda.gov/wps/wcm/connect/c33b69fe-7041-4f50-9dd0-d098f11d1f13/Beef_from_Farm_to_Table.pdf?MOD=AJPERES

USDA Food Safety and Inspection Service (2014b) Chicken from farm to table. Retrieved from https://www.fsis.usda.gov/wps/wcm/connect/ad74bb8d-1dab-49c1-b05e-390a74ba7471/Chicken_from_Farm_to_Table.pdf?MOD=AJPERES

USDA Food Safety and Inspection Service (2015a) Basics for handling food safely. Retrieved from http://www.fsis.usda.gov/wps/portal/fsis/topics/food-safety-education/get-answers/food-safety-fact-sheets/safe-food-handling

USDA Food Safety and Inspection Service (2015b) Safe food handling fact sheets. Retrieved from http://www.fsis.usda.gov/wps/portal/fsis/topics/food-safety-education/get-answers/food-safety-fact-sheets/safe-food-handling

USDA Food Safety and Inspection Service (2015c) Safe food handling fact sheets/safe food handling/safe minimum internal temperature chart. Retrieved from https://www.fsis.usda.gov/wps/portal/fsis/topics/food-safety-education/get-answers/food-safety-fact-sheets/safe-food-handling/safe-minimum-internal-temperature-chart/ct_index

USDA Food Safety and Inspection Service (2016, Jan 15, 2015). Safe minimum internal temperature chart. Retrieved from http://www.fsis.usda.gov/wps/portal/fsis/topics/food-safety-education/get-answers/food-safety-fact-sheets/safe-food-handling/safe-minimum-internal-temperature-chart/ct_index

Van Pelt AE, Quiñones B, Lofgren HL, Bartz FE, Newman KL, Leon JS (2018) Low prevalence of human pathogens on fresh produce on farms and in packing facilities: a systematic review. Front Public Health 6(40):10pp. https://www.frontiersin.org/articles/10.3389/fpubh.2018.00040/full

Wei W, Wang X, Xie Z, Wang W, Xu J, Liu Y et al (2017) Evaluation of sanitizing methods for reducing microbial contamination on fresh strawberry, cherry tomato, and red bayberry. Front Microbiol 8:2397. https://doi.org/10.3389/fmicb.2017.02397

World Health Organization (2006) Food safety risk analysis: a guide for national food safety authorities. Retrieved from https://apps.who.int/iris/handle/10665/43718

World Health Organization (2015) Food safety; fact sheet. Retrieved from http://www.who.int/mediacentre/factsheets/fs399/en/

World Trade Organization (2015a) Current issues in SPS. Retrieved from https://www.wto.org/english/tratop_e/sps_e/sps_issues_e.htm

World Trade Organization (2015b) Sanitary and phytosanitary measures. Retrieved from https://www.wto.org/english/tratop_e/sps_e/sps_e.htm

WTO (2021) Sanitary and phytosanitary measures: introduction. Understanding the WTO agreement on sanitary and phytosanitary measures. Retrieved from https://www.wto.org/english/tratop_e/sps_e/spsund_e.htm

Yang S, Leff MG, McTague D, Horvath KA, Jackson-Thompson J, Murayi T, Boeselager GK, Melnik TA, Gildemaster MC, Ridings DL, Altekruse SF, Angulo FJ (1998) Multistate surveillance for food-handling, preparation, and consumption behaviors associated with foodborne diseases: 1995 and 1996 BRFSS food-safety questions. MMWR CDC Surveill Summ 47(4):33–57. https://www.cdc.gov/mmwr/preview/mmwrhtml/00054714.htm

Young I, Waddell L (2016) Barriers and facilitators to safe food handling among consumers: a systematic review and thematic synthesis of qualitative research studies. PLoS One 11(12):e0167695. https://www.ncbi.nlm.nih.gov/pmc/articles/PMC5132243/

Zink DL (1997) The impact of consumer demands and trends on food processing. Emerg Infect Dis 3(4):467–469

Zylberman P (2004) Making food safety an issue: internationalized food politics and French public health from the 1870s to the present. Med Hist 48(1):1–28

Food Safety Management, GMP & HACCP

10

1 Good Manufacturing Practices

1.1 Introduction

Regular inspection of food facilities enables, the detection of food safety violations, corrective action and enforcement (Phillips et al. 2006). Obviously, the threat of punitive action will help to enhance food safety. However, most experts believe that prevention (including inspection and monitoring) should be the first line of defense against foodborne diseases and deliberate forms of contamination (World Health Organization 2003). At present, the FDA supports active managerial control (AMC) for the foodservice and retail sectors. AMC is defined as, "*the development and implementation of food safety management practices to prevent, reduce and eliminate the risk factors for foodborne illness*". The hazard analysis critical control point or HACCP method (Sect. 5) is also built on the premise that prevention is a more effective form of control by contrast to the detection of foodborne disease.

The aims of this chapter is to introduce the main programs for managing food safety practices. The goal of Food safety management (FSM) is the prevention of foodborne disease. Good Agricultural Practices (GAP), Good Manufacturing Practices (GMP), and Good Hygienic Practices (GHP) deal with issues affecting the "environment" within which food is produced (farms), manufactured (food process facility) or prepared (commercial kitchen/ foodservice). The following information sources were particularly helpful for preparing Sections 1.1 to 1.5 (Codex Alimentarius Commission 2014; The CGMP Coalition 2011; U.S. Food and Drug Administration 2014; U.S. Government Publishing Office 2014; University of Nebraska Cooperative Extension 2015).

The principles of GMP were developed initially in the 1960's (Table 10.1). Major revisions of GPM occured in 1986 and then again 2002. Guidelines for GMP can be considered as a standalone food safety management strategy. In addition, the GMP scheme is also considered a compulsory pre-requisite program for HACCP. The US Food and Drugs Administration is responsible for GMP for all food groups, except poultry, eggs and meat that are overseen by USDA Food Safety Inspection Agency (U.S. Government Publishing Office 2014).

1.2 Basic Provisions for GMP

1.2.1 Personnel Disease Control and Hygiene

Infected food handlers and disease carriers are a significant route for the transmission of foodborne disease to consumers (Jacob and World Health Organization 1989; Hedberg et al. 2006; Thaivalappil et al. 2018; Ncube et al. 2019;

R. Owusu-Apenten, E. R. Vieira, *Elementary Food Science*, Food Science Text Series,
https://doi.org/10.1007/978-3-030-65433-7_10

Insfran-Rivarola et al 2020). The aim for GMP for employee disease control and hygiene is to reduce the risk of disease transmission from a food handler using strategies listed in Table 10.2.

Health. There is a requirement that food handlers should report health conditions likely to transmit to foods. Some of the diseases that are notifiable include, jaundice, diarrhea, vomiting, fever, sore throat with fever, infected skin wounds (boils, cuts), or discharges from the ear, eye or nose. Before returning to work, personnel reported previously as unwell must be certified by a physician as fit or recovered.

Cleanliness. Foodservice personnel are required to maintain physical cleanliness and good levels of hygiene, including daily changes of clothing prior to working with foods, food contact surfaces or packaging. Cleanliness also encompasses the wearing of suitable outer garment (typically white overall or frock), proper and regular hand washing, not wearing any potentially unsecured jewelry, proper care and maintenance of gloves, wearing appropriate hair restraints, and storing personal belongings and rations in a designated area away from the food processing area.

Fingernails must be trimmed short and clean with no nail polish; wearing of fingernail extensions is not allowed as this might serve as reservoirs for harboring bacterial. Without due care, fingernail extensions may be discarded accidentally into foods thereby posing a physical hazard. When possible, gloves should be worn when handling foods. With or without gloves, hands should be washed and dipped in a disinfectant prior to handling food. Similarly, hands should be washed after contact with any non-food contact surface including touching the mouth, nose, or other part of the body. The handwashing requirements for employees need to be stringent as the hands are thought to be involved in most instances of personal contamination of foods (USDA Food Safety and Inspection Service 2015a, b, c) (Table 10.2).

Hygiene practices. Food handlers should neither smoke, drink, nor eat in the work area. Pets and other animals do not belong in the food-processing area. Sneezing and coughing should be using the inner arm of the frock or inside collar. After using a handkerchief, the hands should be washed. Cloths should not be used for cleaning. Foods that appear to be unwholesome, or that may contain unacceptable contaminants, should not be handled (U.S. Government Publishing

Table 10.1 Good Manufacturing Practice in manufacturing, packing, or holding human Food (21 CFR 110)

Subpart A	GENERAL PROVISIONS
§110.3	Definitions
§110.5	Current good manufacturing practice.
§110.10	Personnel. Requirements for disease control, cleanliness, education and training, and supervision
§110.19	Exclusions.
Subpart B	BUILDINGS AND FACILITIES
§110.20	Plant and grounds. Construction and design to facilitate sanitary operations and maintenance
§110.35	Sanitary operations. Cleaning/sanitizing of physical facilities, utensils, and equipment; Storage of cleaning and sanitizing substances, pest control, sanitation of food contact surfaces etc.
§110.37	Sanitary facilities and controls. Water supply, plumbing, sewage disposal, toilet, hand-washing, rubbish and offal disposal
Subpart C	EQUIPMENT
§110.40	Equipment and utensils. Design, construction, and maintenance of equipment and utensils
Subpart D	[RESERVED]
Subpart E	PRODUCTION AND PROCESS CONTROLS
§110.80	Processes and controls. Covers raw materials and other ingredients; monitoring of manufacturing operations
§110.93	Warehousing and distribution. Transportation of food, protections from contamination and deterioration of the food and its container
Subpart F	[RESERVED]
Subpart G	DEFECT ACTION LEVELS
§110.110	Natural or unavoidable defects in food for human use that present no health hazard.

Source: Federal Register 51, 1986. Updated in Electronic federal Register updated 2019 (Office of the Federal Register (OFR) 2019)

Table 10.2 Some common personal hygiene rules for food handlers

Do not handle foods if suffering with a communicable conditions including skin infections
Ensure physical cleanliness. Wear clean (preferably white) uniforms. Avoid dangling jewellery.
Wear head covering, trim fingernails short and avoid nail polish
Wear gloves if possible.
Wash hands frequently with/without gloves and dipp in a disinfectant prior to handling food.
Rewash hands after touching a non-food contact surface, e.g. boxes, pallets, or parts of the body such as mouth, or nose.
Do not smoke, drink or eat in the work area.
Do not take pets and other animals to a food-processing area.
Confine sneezing and coughing to a handkerchief, tissue or inside the smock, or elbow joint.
Wash hands after using a handkerchief or tissue
Do not use cloths for cleaning.
Do not handle foods that seem unwholesome, or that may contain contaminants.

Office 2014) (Table 10.2). Finally, food handlers should keep their hair covered. Wearing hair restraint or head covering improves effectiveness, as well as serving to improve hygiene.

1.2.2 Education, Training and Supervision

Each food facility must have a manager dedicated to food safety compliance, called a "certified food protection manager". Revisions to the US food code (2017) noted that the Person In Charge "shall" be the certified food protection manager (U.S. Food and Drug Administration 2018). The certified food protection manager must be knowledgeable about food safety principles (including HACCP) as evidenced by showing "*proficiency of required information through passing a test that is part of an ACCREDITED PROGRAM*". Historically and still within some States, the certified food protection manager and the Person In Charge could be separate individuals.

The duties of the certified food protection manager is to, oversee food safety management overall, ensure adequate and ongoing employee training in food safety. There is a duty also for ongoing monitoring of temperatures for food hot holding or cold holding. Employee training should cover topics such as, the risks from bare hand contact with foods, handwashing requirements, fingernails and jewelry rules (US food code 2017, p. 70). All employee training in food safety should be documented and records of any certifications kept. Food facilities with a certified food protection manager (approximately 83% of the businesses sampled) were more likely to be compliant with the food code (Appling et al. 2018) and have a considerably lower chance of an outbreak (Hedberg et al. 2006).

1.3 Building and Facilities for GMP

1.3.1 Plant and Grounds Constructions

Construction of the food manufacturing plant should be from materials that do not hinder sanitation. A building constructed from steel, brick or concrete is preferable to wood, which is difficult to maintain in a clean condition and is more vulnerable to invasion by rodents, birds, and other pests. If a food process plant must be constructed of wood, the foundation should be "rat-stopped", which means constructed of cement to a distance of several feet below and above the ground (Codex Alimentarius Commission 2014).

The walls and roof-junction of the food plant should be weatherproof and impenetrable by insects. In food-processing or utensil-washing areas, the junction of the wall and floor should be curved and have no angled comers, to facilitate cleaning.

Other building details should help keep foods from contamination. Window ledges could be angled to prevent their use by personnel for storage. The floor should comprise acid-resistant unglazed tile or of epoxy resin material, an epoxy tile grout laid on cement, for instance.

Flooring should be sloped to drains so that water does not accumulate. A cement surface is undesirable, as it will become pitted, leaving areas where water and food scraps accumulate and where bacteria may grow to large numbers, thus becoming sources of contamination and putrid odors.

The internal walls should be covered with glazed tile at least up to distances of splash height, to facilitate cleaning.

Non-leaking walls (no openings, no doors) should separate areas. The boiler room must also be separate and closed off. Each entryway to the area producing the finished product should be equipped with a shallow pan containing a disinfectant, so that those who enter the area must step into the pan, thus disinfecting their shoes or boots.

1.3.2 Plant Exterior and Grounds

The surroundings for a food process plant should be neat, trim, and well landscaped (Codex Alimentarius Commission 2014). All parking spaces, roadways, and walks should be paved so that dust contamination of the air will be minimized, and contamination, such as animal droppings, will be washed away with each rain rather than be soaked into the ground to be airborne during dry spells.

The area surrounding a food plant, including platforms, should not be used for storing crates, boxes, or machinery, as these materials may harbor rodents that could eventually find their way into the plant. There should be no area around the plant where the landscaping allows potholes or depressions of any kind in which water may accumulate and become a breeding place for insects which then could become established within the plant.

Food materials, ensilage piles, or other organic wastes should not be present in any exposed area near the plant, as they attract and become breeding places for insects, especially flies, which are difficult to control in food plants, even under the best conditions. There should be no neighboring plants, such as chemical, sewage, poultry, or tanneries that may transfer bacteria or chemicals to the food plant.

1.3.3 Sanitary Operations

Food process facilities and commercial kitchens should be cleaned and sanitized thoroughly in accordance with a written cleaning scheduled for the floors, walls, benches and tables, conveyors, hoppers, fillers and kettles. Large food processing plants should have cleaning crews with a supervisor. Approved cleaning compounds should be used; as published by the U.S. Department of Agriculture in 1987, and it is available from that agency. The water used for cleaning should be potable and have a temperature of about 130–140 °F (54.4–60.0 °C).

The detergent selected should be suitable to remove the type of dirt encountered (Fig. 10.1).

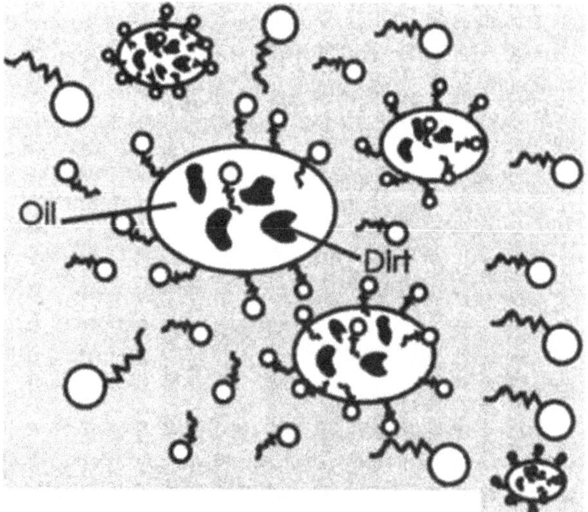

Fig. 10.1 Action of detergent soap molecules in removing soil and dirt

Depending on the application, detergents should have certain properties. In general, detergents should not be corrosive. When used in hard water, they should not form precipitates. They should have good wetting, and in many cases emulsifying, properties. All detergents contain surfactants (surface-active agents) that reduce tension where the detergent meets the soil surface, allowing the detergent to penetrate in order to loosen and disperse the soil. Most detergents used in the food industry are alkaline because most food soils are acidic in nature and will disperse more easily in the alkaline solutions. Detergents dislodge both organic and inorganic soils. They should saponify fats, have good dispersal and deflocculating properties, and not form residues or films on surfaces. High-pressure water and high-pressure steam can be used to flush hard to-reach places with detergent.

Detergents function as soap molecules, having an ionic and a nonionic end groups, which enables them to emulsify oil deposits effectively. The ionic or hydrophilic (water loving) "head" attracts water and the hydrophobic "tail" will "dissolve" in grease and oil so that the soil can be dispersed in the water medium (see Fig. 10.1).

Chemical agents used for controlling microbes include bacteriostats, which prevent the growth and spread of bacteria, and bactericides, which not only stop bacterial growth, but also destroy the bacteria. Some agents are bacteriostatic in small amounts and bactericidal in large amounts. Effective sanitation programs should be using mainly, if not completely, bactericides that include compounds in the following chemical classes: halogens, phenolics, quaternary ammonium compounds, alcohols, carbonyls, and miscellaneous others.

The oxidizing agents such as chlorine and iodine are considered to be the most important sanitizing agents known. Such common uses such as chlorination of drinking water and iodine treatment for cuts make these compounds familiar.

Phenol compounds, such as cresol, are considered to be also very good sanitizers; however, they have some disadvantages, among which are high irritation characteristics and relatively high cost.

Quaternary ammonium compounds, although effective in the control of algae and some bacteria, are relatively ineffective against a variety of microbes that are not resistant to the halogens and phenols.

Alcohols are not as effective as generally believed, and while some carbonyls, such as formaldehyde, are effective, they are also hazardous to use. There are other miscellaneous sanitizers, but they are generally low in effectiveness and high in cost. The three most commonly used sanitizers for food contact surfaces accepted by the FDA are chlorine, iodine, and quaternary ammonia. (cf. Chap. 9, Sect 4.5).

After cleaning and rinsing with hot water, equipment should be immersed for 30 s or longer and rinsed with a solution of iodophor containing 12.5 ppm of available iodine. Iodophors, or "tamed iodines," are combinations of iodine and surface-active agents that have the sanitizing advantages of iodine with minimized disadvantages.

Iodine by itself is not very soluble in water, has a high vapor pressure, is corrosive, and leaves a stain. In combination with surfactants, these undesirable properties are minimized. A solution containing 50 ppm of available chlorine at a temperature not below 75 °F (23.9 °C) may be substituted. All equipment so constructed as to hold liquid should be thoroughly drained after cleaning and sanitizing, and containers, such as pans, should not be nested after cleaning and sanitizing, as this prevents drainage and evaporation of moisture and thus provides moisture in which bacteria may grow. Detailed guidelines for cleaning and sanitizing practices appear in the US food code 2017 (sect. 4.6 and 4.7).

1.3.4 Sanitary Facilities and Controls

The range of sanitary facilities for GMP compliance include, water supply, adequate plumbing, sewage disposal, toilets facilities, hand washing and disposal of refuse (Table 10.3).

Table 10.3 GMP requirements for sanitary facilities and controls

Water supply	Adequate and safe quality
Plumbing	Adequate capacity, covey sewage, and floor drainage, avoid backflow or contamination
Sewage disposal	As above
Toilet facilities	Adequate, accessible, sanitary, in good repairs, self-closing doors
Handwashing stations	Adequate water at right temperature, provide towels, hand sanitizers, fixtures to avoid recontamination
Refuse and recyclable waste	Place receptacles at area generated

1.3.5 Water Supply and Plumbing

The water supply should be adequate for filling the plant's needs. All water that may contact either foods or surfaces that may be contacted by foods must be potable quality (safe for drinking). The water used for cleaning should be of adequate temperature and pressure and supplied via a plumbing system of adequate capacity and in conformance to building codes[1].The plumbing system for a food manufacturing plant should also have adequate capacity, convey all sewage to an appropriate disposal system without the possibility of backflow and contamination of food areas.

1.3.6 Sewage Disposal

Sewage disposal must be through a public sewerage system or through a system of equal effectiveness in carrying liquid disposable waste from the plant.[2] The sewerage system must conform to building codes and in no way be a source of contamination to the products, personnel, equipment, or plant. Drains must be sufficient to ensure the rapid and complete transfer of all wash water, spilled liquids, and so on, to the sewage system.

[1]Food Code 5-202.11 Approved System and Cleanable Fixtures.

[2]Food Code 5-402.10 Establishment Drainage System.

Table 10.4 Suggested lavatory facilities

Number of employees	No. of toilets	No. of wash basins
1–9	1	1
10–24	2	1
2FC 5–49	3	2
50–74	4	3
75–100	5	4

Notes: Add extra toilet for every 30 additional employees. Add 1 wash basic for every 50 employees above 100

1.3.7 Locker and Toilet Facilities

Locker rooms should be separate for male and female personnel, and a sufficient number of lockers should be available to provide one for each worker so that outside clothing may be stored. These locker rooms should be clean and tidy.

Separate toilet rooms with self-closing doors (that do not open directly into processing area) should be provided for men and women workers. Toilet rooms should have washbowls with hot and cold running water, soap dispensers, paper towels, and containers for refuse. The suggested minimum number of toilets and washbasins is given in the Table below. Urinals may be substituted for toilets in the men's rooms on a one-to-one basis, but the number of toilets must never be less than two-thirds of the recommended. The hand-washing units should be of foot- or electronically activated (Table 10.4).

Where personnel consume meals at work, a room separate from other rooms, including locker rooms, should be provided for this purpose. This room should be kept clean and sanitary. Drinking fountains should never be located in toilet rooms.

All personnel working in food-processing/preparation or utensil-cleaning areas should be provided with clean outer uniforms daily.

In processing or utensil-cleaning areas, there should be wash basins, a container of disinfectant (preferably a weak solution of an iodophor), and paper towels so that those in charge can make sure that workers wash and disinfect their hands before returning to work once having left their particular area of operation. This is of great importance to good food plant sanitation.

The food process plant or commercial kitchen must have adequate light to ensure that employ-

Table 10.5 Suggested minimum light requirements for food processing plants

Operation/rooms	Light Ft candles (Lux)[a]
Sorting, grading, inspection, Employee working with sharps, knives, grinders, cutting	50 (540 lux)
Processing, active storage Consumer self-service, buffets, under-counter refrigerators, toilets	20 (220 lux)
Instrument panels, switchboards	10 (100 lux)
Locker rooms etc., walk-in refrigeration, food storage & room;	

[a]Lighting of 100–150 ft. candles, is suited for a variety of food inspections (Food Code 6-303.11 Intensity)

ees can perform their duties effectively. Table 10.5 shows the minimum lighting requirements to carry out specific operations in the plant.

Smoking in food processing areas is prohibited and many plants now are "smoke free" environments. If a foodservice or food processing worker does smoke in designated areas, he or she must wash his or her hands thoroughly as one would after using the toilet. Signs should be conspicuously placed in the area to remind the workers to wash their hands.

Food processing plants should have good ventilation with filtered air, to reduce the atmospheric humidity, and hence the amount of condensation, which if allowed to occur, promotes growth of bacteria and molds on many surfaces including those of walls, ceilings, floors, equipment, utensils, and foods.[3] Minimal building requirements for food business premises can be seen for the State of Florida (Schmidt and Erickson 2008; Florida Department of Agriculture and Consumer Services (Division of Food Safety) (2021).

1.4 GMP and Equipment

1.4.1 Equipment and Utensils Design & Construction

GMP construction and design requirements for equipment and utensils apply regardless of size. Equipment and utensils, which come to contact with foods must be fabricated from non-toxic and durable material for ease of cleaning. Seems and joints should be smooth to avoid accumulation of residue. Freezers and cold storage rooms should be fitted with freezer/refrigeration thermometers. Contact surfaces and materials should not be sources of inadvertent contamination with a food hazard.

1.4.2 Maintenance of Equipment and Utensils

Equipment and utensils are to be constructed for ease of maintenance.

1.5 GMP and Production, Process Control

1.5.1 Processes and Controls

Current GMP requirements for processing control are wide-ranging, applicable to raw materials and all unit operations in the plant, with the aim of reducing food adulteration (Fig. 10.3)[4]. The food plant must have food safety protection manager, with overall charge for food safety. The food process plant is also assumed to have generalized quality assurance procedures and a process for product recall. Raw materials are required to be handled and inspected to assure low microbial numbers and decreasing risks of natural toxicants including aflatoxins and anti-nutritional factors. All major processes should be performed in accordance with standard operation procedures, that ensure protection from a variety of hazards.

1.5.2 Receiving and Storage

Raw materials and ingredients should be received, inspected, logged, and stored properly. Dry materials such as breading and flour, to be stored, should be held in a room constructed of materials, such as brick or cement, that do not allow the

[3]Food Code 4-301.14 Ventilation Hood Systems, Adequacy.

[4]In modern food regulations, adulteration refers to all forms of contaminations, whether intentional or unintended, including microbial contamination and classical economically motivated addition of inferior ingredients or withholding valuable components.

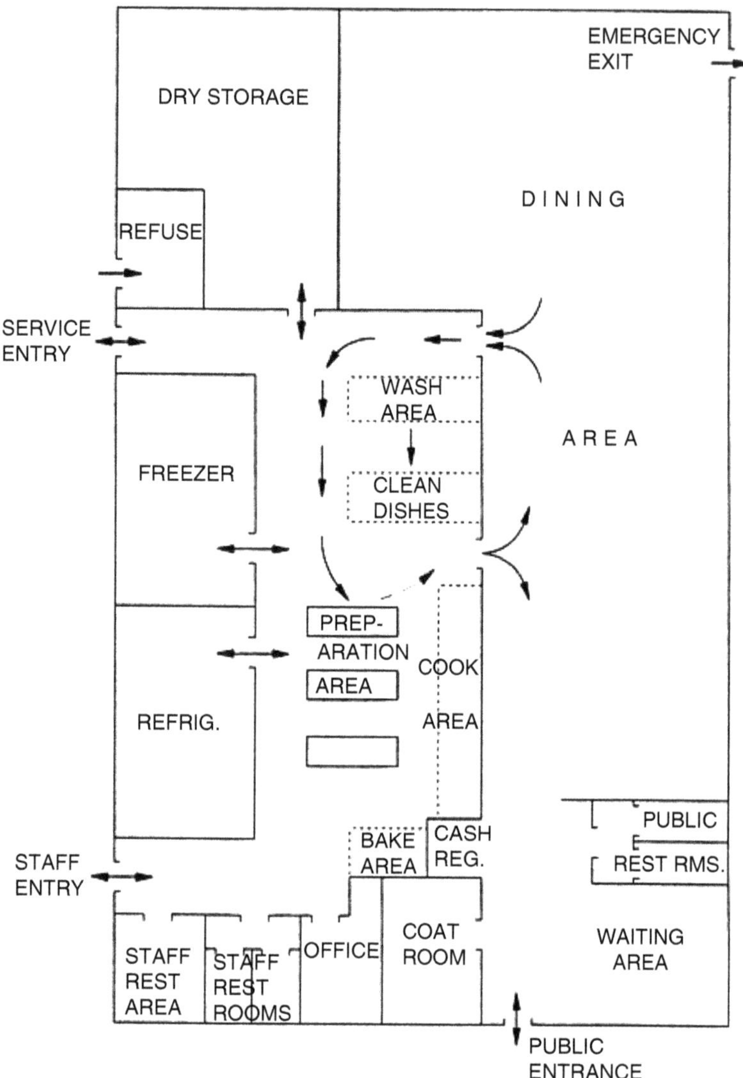

Fig. 10.2 Simplified floor plan for a foodservice operation

entrance of insects and rodents. Such rooms should be refrigerated to about 50 °F (10 °C) to prevent the hatching of eggs and the development into adult insects, should viable insect eggs be present.

An important part of plant sanitation is the establishment of a strict quality control over the incoming raw materials. No amount of plant sanitation can remain effective if incoming materials are allowed to bring in pests or contaminants.

In summary, the preceding sections deal with food safety management issues, affecting the environment for food manufacturing, commercial food preparation or food retail (buildings, facilities, personnel etc.). Food safety management practices for agriculture are considered next.

2 Good Agricultural Practices (GAP)

The good agricultural practices (GAP) guidelines came out of an address by President Bill Clinton[5] in 1997 over the safety of fresh fruit and vegetables (U.S. Food and Drug Administration 1998). The FDA developed GAP guidelines with the aim of

[5]President William Jefferson Clinton (42nd President of the United States)

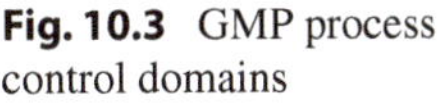

Fig. 10.3 GMP process control domains

reducing the risk of foodborne diseases from microbial contamination of fresh and minimally processed foods in line with several core principles;

- Foodborne disease prevention is the focus not mitigation
- The risk of foodborne disease exists for all parts of the Agri-food system
- GAP highlight supply chain steps important for fresh produce; agriculture cultivation and harvesting, storage, minimal processing, packaging, transportation.
- All water coming into contact with food presents a risk of contamination
- Animal manure is a recognized risk factor for microbial disease
- GAP should agree with existing food law and promote accountability

Some of the elements of GAP that deal with water, manure and compost risk, and personal hygiene issues are outlined below. It should be evident to readers that the requirements for GAP align closely with those of GMP (Sect. 1).

2.1 Water Use and GAP

According to GAP, water that comes into direct contact with fruits and vegetables, during growth or harvest, presents a risk of microbial contamination. Pathogenic microorganism as well other types of food hazards may occur within agricultural water, processing water, wash water and water used as cooling media. To reduce the risk of contaminations, all sources of water should be identified. Water wells need to be kept in good repair in order to reduce the risk of waterborne microbial contamination. Surface and ground water sources need to be checked for faecal contamination, which is possible in districts where manure is employed as fertilizer. Run-off water could bc casily contaminated with diesel and other soil residues. Steps should be taken also to protect the quality of water. Irrigation water should be considered as a particular source of foodborne diseases risk. Processing water that will come into close contact with foods, must be free of pathogens (U.S. Food and Drug Administration 1998).

2.2 Manure and Biosolids

Guidelines for GAP recognize that there is a legitimate use for animal manure as fertilizer especially if it is actively treated by composting to reduce microbial load. Composting at temperatures of 130–140 °F (54–60 °C) for 15-consecutive days reduces pathogens numbers in manure significantly. By comparison, passively treated (ageing) compost, and untreated manure poses risk of microbial contamination. Exposure to untreated manure or inadvertent contact with fresh food produce and growing plants should be avoided. Using untreated manure during a growing season is not recommended (U.S. Food and Drug Administration 1998).

2.3 GAP and Personal Hygiene

As with GMP (Sect. 3) employee personal hygiene, state of health and training is needed to avoid transmission of foodborne diseases to the consumer (U.S. Food and Drug Administration 1998).

3 Sanitation Performance Standards

3.1 Revised Sanitation Standard Operation Standards

The FDA is responsible for food safety regulation covering about 80% the food industry. However, poultry, egg and meat products are regulated by the USDA Food Safety Inspection Services (FSIS). Accordingly, the inspection of food manufacturing facilities is shared between the FDA and FSIS. The fact that food safety regulations are shared probably explains why the FDA developed GMP guidelines (21 CFR 110; Table 10.1) whilst the USDA-FSIS developed separate sanitation standards for eggs, poultry and meat process plants.

USDA-FSIS standards of sanitation (USDA Food Inspection Services 2016) are listed within the Code of Federal Regulations (9 CFR 416.1 to 9 CFR 416.17). The first half of USDA-FSIS s*anitation standard deals with* GMP. However, sanitation standard operating procedures (SSOPs) for animal food processing plants are not found within GMP. From 2016, FSIS introduced new Sanitation Performance Standards (SPS) alongside of SSOPs, which remain listed in Code of Federal Register. These programs probably serve as GMP-like prerequisites for HACCP for the meat, fish, poultry and eggs businesses (USDA Food Inspection Services 2016).

Sanitation requirements for meat products are complicated. The rules depend on the size of establishments with some types of businesses being exempt from inspection. Those businesses classed as small or very small seem to have different requirements and operators are advised always to seek professional advice. Restaurants, retail stores, and those processing meat on a small scale are exempt for inspection (USDA Food safety and Inspection Service 2019).

3.2 Construction Sanitary Performance Standards

The SPS requirement for a meat and poultry plants is to ensure effective cleaning similar to the requirement for GMP and the US food code;

- A food process plant should be constructed from durable materials
- Sanitation should not lead to contaminated food
- Constructions should enable the control of vermin
- Separate the spaces for food storage to avoid adulteration.

To ensure effective cleaning and sanitation, the walls and ceilings of the food process plant should be smooth, easily cleanable, lacking exposed piping; the floors should be durable, tiled, lacking or devoid of gaps wider than 1 mm, and fitted with graded drains. No carpets are allowed in food preparation areas and other floor coverings if present should be movable and washable[6]. A food premises with ill-fitting doors, gaps and holes in a window, flaking paintwork or evidence of mold growth are all deemed non-compliant with the sanitation standards (USDA Food Inspection Services 2016).

3.3 Lighting and Ventilation Sanitation Performance Standards

Good ventilation is a requirement for sanitation to avoid the buildup of odors, vapors and above all condensation. Transfer of odor or vapors count as adulteration as does dripping of condensate onto a product. Condensation is a risk whenever there are sudden changes in air temperature, leading to release of moisture. Steel production chutes, areas above open kettles or chill baths, refrigeration surfaces, loading docks are all areas with a high condensation risk due to temperature differentials.

[6]Food Code, 6-201.11 Floors, Walls, and Ceilings.

4 Other Hygiene Codes

4.1 The Food Code – United States

The United States federal food code dating from 1930s, and updated frequently since, provides a template for State and local legislators (U.S. Food and Drug Administration 2013). Many elements of the food code are incorporated into GMP. The food code also provides a basis for new SPS introduced in 2016 (Sect. 3.1).

4.2 Codex Alimentarius Food Hygiene Code

The international codes for food hygiene produced by the Codex Committee on Food Hygiene (CCFH) contain recommendations for maintaining food safety at all stages of in the food chain. The intentional food hygiene code was first published 1969 amended 1999, and revised 1997 and 2003. Currently, 21 further codes for specific food products have been developed based on the international food hygiene code (e.g., for dried fruit, tree nuts, low acid food, frog legs, milk products, egg products). The international code for food hygiene combines general principles of hygiene with GMP, thereby providing a foundation for the Hazard Analysis Critical Control Point (HACCP) systems for food safety (Table 10.6).

4.3 Food Hygiene in the European Union

Prior to 2004, the European Union had greater than 14 separate food hygiene directives for foods of animal origin (fish, meat, poultry, etc.).

Table 10.6 The international code of food hygiene, 1969 and, 2003

Primary production
Establishment: design and facilities
location, premises and rooms, equipment, facilities
Control of operation
control of food hazards, incoming material requirements, packaging, water, management and supervision, documentation and records, recall procedures
Establishment: maintenance and sanitation
maintenance and cleaning, cleaning programs, pest control systems, waste management, monitoring effectiveness
Establishment: personal hygiene,
health status, illness and injuries, personal cleanliness, personal behavior, visitors
Transportation;
General Requirements, Use and maintenance
Product information and consumer;
Training,
awareness and responsibilities, training programs, instruction and supervision, refresher training

Adapted from Codex Alimentarius Commission (1969)

From 2004, a single food hygiene code (EC directive 852/2004) for all foods was introduced, and came into force from 1 January 2006 (Eurolex 2015). In the new code, HACCP was made compulsory for all large food-manufacturing businesses. Also, imported foods were held to the same standards of food hygiene as food produced within the EU.[7]

The EU food hygiene code places the responsibility for compliance on the business operator. Furthermore, EC directive 852/2004 applies to the whole food chain including, where food is produced, processed, and marketed. Very small production for home consumption is excluded from the legislation, and so too are farm and direct supply retailers. Also excluded is any form of domestic food handling and preparation. Article 5 of the EC directive states that a 7-step HACCP process should be a permanent feature of all food businesses. Article 6 describes the need for registration and ongoing approval by competent national authorities (Table 10.7).

[7]Prior to EC directive 852/2004 the HACCP programme was recommended for businesses within the EU.

Table 10.7 Elements of the EU food hygiene directive (EC) 852/2004

Article 1: Scope; Applies to whole food chain. Highlights business operator has responsibility for compliance, excludes primary production for domestic use, domestic handling,,
Article 2: Definition of key terms
Article 3: Obligations of food business operator
Article 4: Specific hygiene requirements
Article 5: Requirement for HACCP
Article 6: Requirement for registration and approval of businesses
Article 7–9: State and National role in implementation
Article 10/11: Hygiene for imported and exported foods
Article 12–18: Plans for implementation, enactment from 2006 Jan

EC 852/2004 –Annex 1 incorporates GAP whereas Annex 2 covers GMP for food manufacturing. In many ways the EU food hygiene anticipates the principles illustrated by ISO22000 (see later). Finally, we note that national governments are responsible for local enforcement of the food code.

5 Hazard Analysis Critical Control Point

5.1 HACCP in the United States

Hazard Analysis Critical Control Point (HACCP) is a system for ensuring the safety of foods, which was developed in the late 1960s, and first presented to the public in 1971, due to the joint efforts by the U.S. Army Natick Laboratories, the National Aeronautics and Space Administration (NASA), and the Pillsbury Co (Penstate Food Science 2006; Surak 2009).

The status of HACCP implementation in the US food industry is summarized in Table 10.8. HACCP is currently mandatory for, seafood and Juice processing.

There are HACCP requirements also for several specialized food manufacturers, e.g. smoking food, curing meat, use of vinegar to acidy foods for packaging, live mollusk displays, and custom

Table 10.8 FDA guidance on HACCP for particular food product sectors

Date/ HACCP guidance	Title	References
1998; seafood	Procedures for the safe and sanitary processing and importing of fish and fishery products	Federal Register (1995), Food and drug Administration (2001b), and Oregon State University (2015)
2001 juice	Juice HACCP	Food and Drug Administration (2001a)
2005/foodservice and retail sectors	Managing food safety: A manual for the voluntary use of HACCP principles for operators of foodservice and retail establishments	Food and Drug Administration (2006b)
2006/dairy food processors	Hazards & controls guide for dairy foods HACCP	Food and Drug Administration (2006a)

preparation of animal foods (US food Code. ss. 3-502.11). The situation is complicated, as some States (e.g. California) have an extra level of scrutiny, requiring a HACCP plan with written approval – for TCS (time/temperature control for storage) foods packaged under reduced oxygen conditions (The Los Angeles County Department of Public Health 2020). Here and elsewhere, it is essential to refer to State versions of the US food code; always confer with local experts regarding food laws and its enforcement. The USDA-FIS also overseas HACCP for meat, fish, poultry and eggs (Kvenberg et al. 2000).

In all these cases, products not compliant with HACCP regulations are considered as adulterated in accordance with the 1938 Food, Drugs and Cosmetic Act (Food and drug Administration 2001a). By contrast, HACCP appears to be voluntary for foodservice and for food retail (Sect. 5.5).

5.2 Reasons for HACCP Non-compliance

Barriers for compliance with HACCP are widely discussed (Bas et al. 2007; Charalambous et al. 2015; Karaman et al. 2012; Macheka et al. 2013). Non-adoption of HACCP was explained by a series of factors. Several of the most important reasons for companies not to undertake HACCP were listed as (Bas et al. 2007; Milios et al. 2012).

- a lack of knowledge of the HACPP concept by many FBO,
- perception that the time required for HACCP implementation may be prohibitive,
- businesses with a high staff turnover were less likely to invest time and resources in HACCP,
- low management motivation,
- company size. Larger organizations were likely to devote resources to HACCP as compared to smaller food businesses (Bas et al. 2007; Milios et al. 2012).

Research from China showed that HACCP non-compliant businesses were more likely to be serving the domestic markets (Shaosheng et al. 2008). Obviously where the adoption of HACCP is mandatory then compliance tends to be wholesale regardless of the size of companies or any of the factors discussed above (Mensah and Julien 2011).

5.3 HACCP in Developing Countries

Food safety standards affect international trade (Chap. 9, Sect. 1.4). Adopting international standards of food safety in developing countries will open up markets, encourage trade and assists with much needed job creation (World Trade Organization 2015). However, it is believed that current standards of food safety in many developing countries are lacking. Achieving International standards seem to be hampered due to poor sanitary conditions, weak infrastructure, obsolescent or absent food safety legislation, dispersed food safety responsibility amongst different government ministries and organizations. A lack of laboratory equipment and expertise in low-income countries could also hinder the adoption of food safety standards. The FAO sought to raise awareness of policy makers about the importance of investing in food safety capacities as a means of promoting development goals (FAO Newroom 2005). Possible areas for improving food safety provision in developing countries include, food inspection enforcement, improved laboratory analysis and diagnoses of foodborne-disease, improved overall foodborne-disease surveillance, more certification of food safety management initiatives by producers and manufacturers (FAO Newroom 2005; Kirczicva et al. 2015; Kussaga et al. 2014; Wang et al. 2008).

5.4 Seven Stages of HACCP

There are seven basic stages for HACPP but there may be 10–14 different stages. Each step of a HACCP program is important and necessary for the assurance of a safe product (Food And Agriculture Organization of the United Nations 1998). HACCP is considered the best approach for assuring consumer safety because the method is preventative. The HACCP approach is a formal part of most certified FSMS. The seven stages of the HACCP are summarized in the Table below and outlined in the rest of this section (Fig. 10.4, Table 10.9).

5.4.1 Hazard Analysis

The hazards analysis will involve a review the recipe or ingredient list for potentially hazardous foods (PHF) which may be either separate items or ingredients (Table 10.10). Recall that, PHF are foods requiring time/temperature control for safety (Chap. 7, Sect. 3.9). A chart showing the path the food travels in your procedure will be needed. The extent of the risk associated with the product is determined partly by the number of PHFs you have in your product. The following factors should be considered in this risk assessment:

Customers with special needs-will they be old (nursing home), hospitalized and possibly immunodeficient, or have some other specific condition?

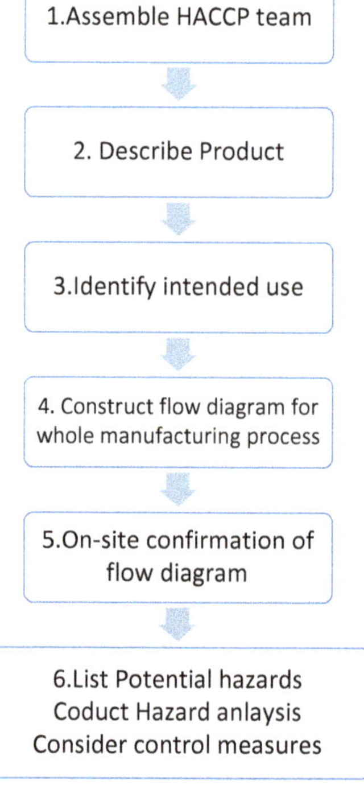

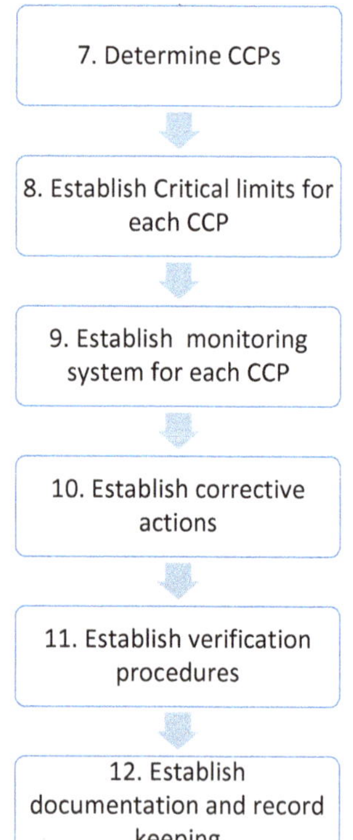

Fig. 10.4 Expanded steps for implementation of HACCP (Food and Agriculture Organization of the United Nations 2007)

Table 10.9 Seven principles of HACCP

Hazzard analysis to identify potential hazards in all stages of food process.
Determine critical control points (CCP) where controls are necessary to eliminate or reduce hazards.
Establish requirements to be met at each CCP.
Establish procedures to monitor each CCP.
Establish corrective actions when monitoring uncovers deviations from plan.
Establish record-keeping procedures.
Establish procedures to monitor effectiveness of HACCP.

Adapted from Food And Agriculture Organization of the United Nations (1998)

Table 10.10 Potentially Hazardous Foods (PHF)§

Any food that contains in whole or in part of:
Milk and milk products
Edible crustacean (shrimp lobster, crab)
Raw sprouts
Shell eggs
Baked or boiled potatoes
Sliced melons
Meats
Tofu or other soy products
Synthetic ingredients (e.g. textured soy proteins
Poultry
Fish
Plant foods that have been heat treated (e.g. beans)
Shellfish

Note: PHF does not include foods with a pH of 4.6 or below or a water activity (A_W). § The PHF are foods requiring temperature/time control for safety

Suppliers – assess if suppliers are reputable, listed with the FDA or an individual state as an approved shipper, as in the case of shellfish suppliers.

Size of operation – assess if your facility can handle the project.

Employees-assess whether employees need further training, equipment, or materials for the project.

5.4.2 Critical Control Points (CCPs)

Critical control points can now be added to the flow charts.

5.4.3 Establish Requirements to Be Met at Each CCP

Criteria or standards that must be met at each CCP need to be clearly established or agreed. These will be based on proven methods; research data or federal, state, or local food regulations. CCP requirements need to be specific e.g. related to temperatures, times, and recognized procedures whenever possible.

5.4.4 Monitoring Critical Control Points

Normally, CCP monitoring will involve employees responsible for the procedures. Having CCPs without monitoring is a waste of time (Table 10.10).

5.4.5 Taking Corrective Actions

Corrective action must be incorporated if the standards at any CCP are not being met. Corrective actions must be clear and based on the individual situation. For example, a corrective action for a food not being held at a proper hot holding temperature of 140 °F (60 °C) may be different depending on how long the food has been at the improper temperature. If it has been held improperly (less than 140 °F) for less than 2 h, reheat to 165 °F (73.9 °C) before serving. If it has been held at below 140 °F (60 °C) for more than 2 h, discard (US food Code 2017; p610).

5.4.6 Setting up a Record Keeping System

These should be simple, easy to do quickly, and should contain information needed to ensure safety. Checklists to examine all CCPs are useful here. Employees doing the job may be very helpful in setting up the form to be used.

5.4.7 Verify That the System Is Working

Verification that the HACCP plan is adequate may be done by laboratory analysis for microbiological or chemical hazards, depending upon the product. Internal quality control to check temperatures, texture, color, taste, absence of defects, or other quality and safety aspects are also useful.

5.5 HACCP for Foodservice and Food Retail

The FDA noted that the principles of HACCP as applied for food manufacturing cannot be applied to foodservice and food retail sector *without modification* (US food code, p549). The FDA backs AMC including a so-called "process-based HACCP" scheme. The purpose of AMC is "prevention of food borne disease, through a continuous system of monitoring and verification" (US food code, p. 549). Another intriguing feature of AMC is the focus on risk factors for foodborne illness, not the detection of diseases per se. Therefore, AMC requires the development and documentation of strategies for (i) effective equipment cleaning and decontamination, (ii) ensuring safe holding temperatures for food, (iii) avoiding undercooking food, (iv) maintaining high levels of personnel hygiene, and (v) elimination of unsafe food suppliers (cf. Table 9.2). The process-based HACCP for foodservice businesses involve foods being categorized into one of three TCS (time/ temperature control for safety) categories depending on process for preparation and holding. The HACCP plan is then developed for one of these three categories of product (US food code 2017, Annex 4).

6 Future of Food Safety Management

6.1 Integrated Food safety

All food business operators need to deploy a food safety system in order to ensure standards of hygiene and sanitation necessary to produce a food that is wholesome, free from adulteration and filth. The range of food safety management systems, include good agricultural practices (GAP) (Food and Agriculture Organization of the United Nations 2015), Good Management Practices (GMP) (The CGMP Coalition 2011; U.S. Government Publishing Office 2014) and/or a fully documented Hazard Analysis Critical Control Point (HACCP) system.

In particular, those food businesses with a worldwide reach are required to implement a HACCP plan with GMP as a prerequisite or baseline system (World Health Organization 2006). As noted above, it is a requirement for GMP that employees will be trained in basic food hygiene (Codex Alimentarius Commission 1969, 2014).

Table 10.11 Areas for Good Manufacturing Practices self-certification by food business in India (Food Safety and Standards Authority of India 2006)

1. Location and layout of food establishment
2. Equipment and fixtures
3. Storage systems
4. Personal hygiene
5. Water supply
6. Pest control system
7. Conveyance and transportation
8. Operational features
9. Documentation and records
11. Product information and consumer awareness
12. Training

Food safety management (FSM) is defined as "a set of interrelated or interacting elements to establish policy and objectives and to achieve those objectives, used to direct and control an organization with regard to food safety" (Crossley and Motarjemi 2011). Food Safety management can involve simple good hygiene programs, GMP alone or GMP in combination with HACCP (Crossley and Motarjemi 2011).

A practical example FSMS is provided by the *Food Safety and Standards Authority of India (FSSAI)* to food business operators (FBO) regardless of size (Food Safety and Standards Authority of India 2006). The basic process is meant to be achievable by all FBOs with support. The generic food safety plan starts with flow charting the food manufacturing process by the. The FSSAI process requires GMP self-certification against an agreed checklist provided by the FSSAI (Table 10.11).

6.2 FDA Food Safety Modernization Act (FSMA)

The FDA Food Safety Modernization Act (FSMA) signed into effect in 2011 by President Obama is described by as one of the most comprehensive changes to food safety management in the United States since the food codes in the 1930s (Koenig 2011). The history of the FSMA

Table 10.12 Seven new FDA Rules in the food safety modernization act

Accredited third-party certification
Current good manufacturing practice and Hazard analysis and risk-based preventive controls for human Food
Current good manufacturing practice and Hazard analysis and risk-based preventive controls for Food for animals
Foreign supplier verification programs (FSVP)
Mitigation strategies to protect Food against intentional adulteration
Sanitary transportation of human and animal Food
Standards for the growing, harvesting, packing, and holding of produce for human consumption
Voluntary qualified importer program (VQIP)

can be traced to the growing awareness of the food safety burden in the United States in the late 1990s, and the need to address potential bioterrorism threats to foods following the post the September 11 (9/11) atrocity of 2001 (Koenig 2011).

At the time of writing, over 60 new rulings and guidelines have originated from the FDA by virtue of the powers granted by the FSMA. Some major new powers (authorities) arising include, a requirement for mandatory GMP/HACCP for manufactures of human food and also those dealing with animal feed from 2015 (Table 10.12).

Interesting also, there is a requirement for importers to comply with United State food safety standards and certification by third parties (U.S. Food and Drug Administration 2019a). It seems likely that the FSMA will result in increasing applications of the HACCP principles in the food industry globally. At the time of writing, many of the directives of FSMA appear to be in the planning stage.

6.3 Threat Analysis Critical Control Points (TACCP) and Vulnerability Analysis Critical Control Points (VACCP)

The FSMA requires food businesses to develop food safety management strategies against intentional adulteration. The need for food defense arose from the realization that the Agri-food system could be vulnerable towards acts of terrorism including some by disgruntled employees. A threats and vulnerabilities analysis is analogous to HACCP, involving a systematic written plan for so-called Threat Analysis Critical Control Point (TACCP) or vulnerability analysis critical control point (VACCP) (Wareing and Hines 2016).

By analogy with HACCP, it is expected that a given TACCP or VACCP plan will be preventative. The written plan would comprise a flowchart of the process, consideration of critical control points for intentional adulteration, a procedure for food defense monitoring as opposed to direct detection of adulterants, which would be unknown in many cases. Steps for correction/mitigation action should be identified. At the time of writing, there are few examples of TACCP or VACCP plans in the public domain and one supposes these will be forthcoming in future (U.S. Food and Drug Administration 2019b).

6.4 Limitations

This chapter did not cover the inspection of food facilities, details of which are beyond the scope of this introductory text. Inspections are vital for detecting violations of food safety legislation and are (in any case) considered inevitable. The US food code 2017 (Annex 5) describes risk-based inspections. The certified food protection manager is mainly responsible for liaising with the public health inspector. Professional advice can be obtained also from State authorities who sometimes carry out inspections on behalf of the FDA.

References

Appling XS, Lee P, Hedberg CW (2018) Understanding the relation between establishment food safety management and risk factor violations cited during routine inspections. J Food Prot 81(12):1936–1940

Bas M, Yuksel M, Cavusoglu T (2007) Difficulties and barriers for the implementing of HACCP and food safety systems in food businesses in Turkey. Food Control 18(2):124–130. https://doi.org/10.1016/j.foodcont.2005.09.002

Charalambous M, Fryer PJ, Panayides S, Smith M (2015) Implementation of Food Safety Management

Systems in small food businesses in Cyprus. Food Control 57:70–75. https://doi.org/10.1016/j.foodcont.2015.04.004

Codex Alimentarius Commission (1969, 2003) General principles of food hygiene CAC/RCP 1–1969. Retrieved from http://www.codexalimentarius.org/input/download/standards/23/CXP_001e.pdf

Codex Alimentarius Commission (2014) Food hygiene, 5th edn. Retrieved from ftp://ftp.fao.org/codex/Publications/Booklets/Hygiene/FoodHygiene_2013e.pdf

Crossley S, Motarjemi Y (2011) Food safety management tools, 2nd edn. Report commissioned by the ILSI Europe Risk Analysis in Food Microbiology Task Force. ILSI Europe Report Series. Retrieved from http://ilsi.org/publication/food-safety-management-tools-2nd-edition/

Eurolex (2015) Food hygiene – basic legislation. Retrieved from http://ec.europa.eu/food/food/biosafety/hygienelegislation/comm_rules_en.htm

FAO Newroom (2005) Food safety and trade: the Codex reward. Meeting Codex standards can open doors to trade. Retrieved from http://www.fao.org/newsroom/en/focus/2005/104165/article_104287en.html

Federal Register (1995) 60 FR 65096 – procedures for the safe and sanitary processing and importing of fish and fishery products. Retrieved from http://www.gpo.gov/fdsys/granule/FR-1995-12-18/95-30332/content-detail.html

Florida Department of Agriculture and Consumer Services (Division of Food Safety) (2021) Food establishment minimum construction standards. From https://www.fdacs.gov/content/download/64038/file/food_establishment_minimum_construction_standards.PDF

Food And Agriculture Organization of the United Nations (1998) Food quality and safety systems – a training manual on food hygiene and the hazard analysis and critical control point (HACCP) system. Retrieved from http://www.fao.org/docrep/w8088e/w8088e00.htm#Contents

Food and Agriculture Organization of the United Nations (2007) FAO/WHO guidance to governments on the application of HACCP in small and/or less-developed food businesses. Food and Nutrition Paper 86. Retrieved from http://www.fao.org/3/a-a0799e.pdf

Food and Agriculture Organization of the United Nations (2015) Good agricultural practices Database Retrieved from http://www.fao.org/prods/gap/

Food and Drug Administration (2001a) Juice HACCP. Retrieved from http://www.fda.gov/Food/GuidanceRegulation/HACCP/ucm2006803.htm

Food and Drug Administration (2001b) Seafood HACCP. Retrieved from http://www.fda.gov/Food/GuidanceRegulation/HACCP/ucm2006764.htm

Food and drug Administration (2006) Managing food safety: a regulator's manual for applying HACCP principles to risk-based retail and food service inspections and evaluating voluntary food safety management systems. Retrieved from http://www.fda.gov/downloads/Food/GuidanceRegulation/UCM078159.pdf

Food and Drug Administration (2006a) Hazards & controls guide for dairy foods HACCP. Retrieved from http://www.fda.gov/Food/GuidanceRegulation/HACCP/ucm2007982.htm

Food and Drug Administration (2006b) Managing food safety: a manual for the voluntary use of HACCP principles for operators of food service and retail establishments

Food Safety and Standards Authority of India (2006) Manual of food safety management system, FSS Act, 2006 Retrieved from http://www.fssai.gov.in/Portals/0/Pdf/manual%20of%20food%20safety%20management%20system,%20fss%20act%202006.pdf

Hedberg CW, Smith SJ, Kirkland E, Radke V, Jones TF, Selman CA, EHS-Net Working Group (2006) Systematic environmental evaluations to identify food safety differences between outbreak and nonoutbreak restaurants. J Food Prot 69(11):2697–2702. https://doi.org/2610.4315/0362-2028X-2669.2611.2697

Insfran-Rivarola A, Tlapa D, Limon-Romero J, Baez-Lopez Y, Miranda-Ackerman M, Arredondo-Soto K, Ontiveros S (2020) A systematic review and meta-analysis of the effects of food safety and hygiene training on food handlers. Foods 9(9):24pp. https://www.mdpi.com/2304-8158/2309/2309/1169

Jacob M, World Health Organization (1989) Safe food handling: a training guide for managers in food service establishments. World Health Organization, Geneva, 142pp. https://apps.who.int/iris/handle/10665/36870

Karaman AD, Cobanoglu F, Tunalioglu R, Ova G (2012) Barriers and benefits of the implementation of food safety management systems among the Turkish dairy industry: a case study. Food Control 25(2):732–739. https://doi.org/10.1016/j.foodcont.2011.11.041

Kirezieva K, Luning PA, Jacxsens L, Allende A, Johannessen GS, Tondo EC et al (2015) Factors affecting the status of food safety management systems in the global fresh produce chain. Food Control 52:85–97

Koenig J (2011) The federal food safety modernization act: impacts in import and small-scale production sectors. Retrieved from http://nrs.harvard.edu/urn-3:HUL.InstRepos:8592051

Kussaga JB, Luning PA, Tiisekwa BPM, Jacxsens L (2014) Challenges in performance of food safety management systems: a case of fish processing companies in Tanzania. J Food Prot 77(4):621–630. https://doi.org/10.4315/0362-028X.JFP-13-254

Kvenberg J, Stolfa P, Stringfellow D, Garrett ES (2000) HACCP development and regulatory assessment in the United States of America. Food Control 11(5):387–401. https://www.semanticscholar.org/paper/HACCP-development-and-regulatory-assessment-in-the-Kvenberg-Stolfa/9304d9303ebca9308c9308a9316f9308fda9323b7495719f7495711c7495792ceca

Macheka L, Manditsera FA, Ngadze RT, Mubaiwa J, Nyanga LK (2013) Barriers, benefits and motivation factors for the implementation of food safety management system in the food sector in Harare Province, Zimbabwe. Food Control 34(1):126–131. https://doi.org/10.1016/j.foodcont.2013.04.019

Mensah LD, Julien D (2011) Implementation of food safety management systems in the UK. Food Control 22(8):1216–1225. https://doi.org/10.1016/j.foodcont.2011.01.021

Milios K, Drosinos EH, Zoiopoulos PE (2012) Factors influencing HACCP implementation in the food industry. J Hellenic Vet Med Soc 63(4):283–290

Ncube F, Kanda A, Mpofu MW, Nyamugure T (2019) Factors associated with safe food handling practices in the food service sector. J Environ Health Sci Eng 17(2):1243–1255. https://www.ncbi.nlm.nih.gov/pmc/articles/PMC6985409/

Office of the Federal Register (OFR) (2019, 4th April) Electronic Code of Federal Regulations. Title 21—Food and drugs. Chapter I—Food and Drug Administration, Department of Health and Human Services (continued). Subchapter B—Food for human consumption. PART 110—current good manufacturing practice in manufacturing, packing, or holding human food. Retrieved from https://www.ecfr.gov/cgi-bin/text-idx?SID=9fa6e3011dde222984586058c3a71c47&mc=true&tpl=/ecfrbrowse/Title21/21cfr110_main_02.tpl

Oregon State University (2015) The Seafood Network Information Center (SeafoodNIC). Retrieved from http://seafood.oregonstate.edu/index.html

Penstate Food Science (2006) HACCP historical timeline. Retrieved from http://sart.psu.edu/foodsafety/angel_stuff2/fssbook/docs/HACCP_history.pdf

Phillips ML, Elledge BL, Basara HG, Lynch RA, Boatright DT (2006) Recurrent critical violations of the food code in retail food service establishments. J Environ Health 68(10):24–30, 55. https://www.afdo.org/wp-content/uploads/2020/2009/Recurrent-critical-violations-of-the-food-code-in-retail-food-service-establishments-2021.pdf

Schmidt RH, Erickson DJ (2008) Sanitary design and construction of food processing and handling facilities. University of Florida IFAS Extension, 10pp. https://ucfoodsafety.ucdavis.edu/sites/g/files/dgvnsk7366/files/inline-files/26503.pdf

Shaosheng J, Jiehong Z, Juntao Y (2008) Adoption of HACCP system in the Chinese food industry: a comparative analysis. Food Control 19(8):823–828. https://doi.org/10.1016/j.foodcont.2008.01.008

State of Florida Department of Health (2009) Chapter 64e-11, Florida administrative code: food hygiene. From http://www.floridahealth.gov/%5C/environmental-health/food-safety-and-sanitation/_documents/64e-11.pdf

Surak JG (2009) The evolution of HACCP. Food Quality & Safety: Farm to fork safety. Retrieved from http://www.foodqualityandsafety.com/article/the-evolution-of-haccp/2/?singlepage=1

Thaivalappil A, Waddell L, Greig J, Meldrum R, Young I (2018) A systematic review and thematic synthesis of qualitative research studies on factors affecting safe food handling at retail and food service. Food Control 89:97–107

The CGMP Coalition (2011). Current food industry good manufacturing practices. Retrieved from http://www.affi.org/resources/current-food-industry-good-manufacturing-practices-aug-2011-prepared-the-cgmp-coalition

The Los Angeles County Department of Public Health (2020) California retail food code. From http://www.publichealth.lacounty.gov/eh/docs/specialized/cacode.pdf

USDA (2019) Allowed detergents and sanitizers for food contact surfaces and equipment in organic operations. Retrieved from https://www.ams.usda.gov/sites/default/files/media/8%20Cleaners%20and%20Sanitizers%20FINAL%20RGK%20V2.pdf

U.S. Food and Drug Administration (1998) Guidance for industry: guide to minimize microbial food safety hazards of fresh-cut fruits and vegetables. Retrieved from http://www.fda.gov/Food/GuidanceRegulation/GuidanceDocumentsRegulatoryInformation/ucm064574.htm

U.S. Food and Drug Administration (2013) Annex 5. Conducting risk based inspections. Food Code:587–620. http://www.fda.gov/Food/GuidanceRegulation/RetailFoodProtection/FoodCode/ucm374275.htm

U.S. Food and Drug Administration (2013) Food code (2013) Retrieved from http://www.fda.gov/Food/GuidanceRegulation/RetailFoodProtection/FoodCode/ucm374275.htm

U.S. Food and Drug Administration (2014) Retail & food service HACCP. Retrieved from http://www.fda.gov/Food/GuidanceRegulation/HACCP/ucm2006810.htm

U.S. Food and Drug Administration (2014) Good Manufacturing Practices (GMPs) for the 21st century – food processing. Retrieved from http://www.fda.gov/Food/GuidanceRegulation/CGMP/ucm110877.htm

U.S. Food and Drug Administration (2017) Managing food safety: a regulator's manual for applying HACCP principles to risk-based retail and food service inspections and evaluating voluntary food safety management systems. Retrieved from http://www.fda.gov/downloads/Food/GuidanceRegulation/UCM078159.pdf

U.S. Food and Drug Administration (2018) Summary of changes in the FDA food code 2017. From https://www.fda.gov/food/fda-food-code/summary-changes-fda-food-code-2017

U.S. Food and Drug Administration (2019a) Food Safety Modernization Act (FSMA). Retrieved from https://www.fda.gov/Food/GuidanceRegulation/FSMA/default.htm

U.S. Food and Drug Administration (2019b) FSMA final rule for mitigation strategies to protect food against intentional adulteration. Retrieved from https://www.fda.gov/Food/GuidanceRegulation/FSMA/ucm378628.htm#coverage

U.S. Government Publishing Office (2014) Electronic Code of Federal Regulations: Title 21: Food and Drugs, Part 110—Current good manufacturing practice in manufacturing, packing, or holding human food. Retrieved from http://www.ecfr.gov/cgi-bin/text-idx?SID=8d188e1ae469fb6b503a9849cd27bf3e&node=pt21.2.110&rgn=div5

University of Nebraska Cooperative Extension (2015) Good Manufacturing Practices (GMP's). Retrieved from http://www.foodsafety.unl.edu/haccp/prerequisites/gmp.html

USDA Food Inspection Services (2016) Sanitation performance standards compliance guide. Retrieved from https://www.fsis.usda.gov/wps/portal/fsis/topics/regulatory-compliance/compliance-guides-index/sanitation-performance-standards/sanitation-compliance-guide

USDA Food Safety and Inspection Service (2015a) Basics for handling food safely. Retrieved from http://www.fsis.usda.gov/wps/portal/fsis/topics/food-safety-education/get-answers/food-safety-fact-sheets/safe-food-handling

USDA Food Safety and Inspection Service (2015b) Safe food handling fact sheets. Retrieved from http://www.fsis.usda.gov/wps/portal/fsis/topics/food-safety-education/get-answers/food-safety-fact-sheets/safe-food-handling

USDA Food Safety and Inspection Service (2015c) Safe food handling fact sheets/safe food handling/safe minimum internal temperature chart. Retrieved from https://www.fsis.usda.gov/wps/portal/fsis/topics/food-safety-education/get-answers/food-safety-fact-sheets/safe-food-handling/safe-minimum-internal-temperature-chart/ct_index

USDA Food safety and Inspection Service (2019) HACCP and sanitation guidance. Retrieved from https://www.fsis.usda.gov/wps/portal/fsis/topics/regulatory-compliance/compliance-guides-index/haccp-guidance

Wang ZG, Mao Y, Gale F (2008) Chinese consumer demand for food safety attributes in milk products. Food Policy 33(1):27–36. https://doi.org/10.1016/j.foodpol.2007.05.006

Wareing P, Hines T (2016) Knowing your HACCP from your TACCP and VACCP. Retrieved from https://www.leatherheadfood.com/files/2016/08/White-Paper-Knowing-your-HACCP-from-your-TACCP-and-VACCP-FINAL1.0.pdf

World Health Organization (2003) Terrorist threats to food: guidance for establishing and strengthening prevention and response systems. Retrieved from http://apps.who.int/iris/handle/10665/42619

World Health Organization (2006) FAO/WHO guidance to governments on the application of HACCP in small and/or less-developed food businesses, FAO food and nutrition paper 86. Retrieved from http://www.fao.org/3/a0799e/a0799e00.htm

World Trade Organization (2015) Sanitary and phytosanitary measures. Retrieved from https://www.wto.org/english/tratop_e/sps_e/sps_e.htm

Part III

Food Preservation and Processing

Basic Considerations for Food Processing

11

1 Introduction

1.1 General Principles & Scope

Should all commercial food supplies be designated as processed? The late Dianne Toops, a prolific columnist for the magazine "Food Technology" implies as much (Toops 2017). Her article gives an historical overview of important developments in the American food industry, beginning from the time when the population numbered 60 million and 70% of people were farmers and self-sufficient in terms of their food needs. At the dawn of the twenty-first Century, the US population is over 360 million and primary food producers comprise about 1.6% of the total; even then, about 25% of the population live in rural areas (United States Department of Agriculture 2020).

Most US Americans purchase their foods from food retailers (supermarkets, grocery stores etc.) or from foodservice outlets. It seems reasonable to assume that all food purchased from stores are processed, including fresh produce. Food procured from foodservice outlets including fast-foods or sit down restaurants are described as "prepared" using industrial scale kitchens (Chap. 25),

Food processing technology is credited with transforming the commercial food supply over the past 150 years. Rapid advances in refrigeration and food freezing, canning, pasteurization, packaging, automated cooking, and food dehydration all helped to produce the food environment encountered by the modern consumer (Toops 2017). Ironically, the reputation of food technology and processed foods is becoming tarnished by association with certain public health concerns, in particular the rising obesity in industrialized countries (Cutler et al. 2003; Dwyer et al. 2012; Weaver et al. 2014).

This chapter provides an overview of basic issues including several alternative definitions for "food processing". Some common advantages and benefits of traditional and modern food of processing are described. Some perceived challenges arising from processing are described particularly, the notion of ultra-processed foods.

1.2 Definitions and Significance of Food Processing

Food processing is defined commonly as any activity applied to foodstuff or drink in order to prepare it for consumption (Singh and Heldman 2008; Smith and Hui 2008). The International Food Information Council (IFIC) described food processing as" *any deliberate change in a food that occurs before it is available for us to eat. It can be as simple as freezing or drying food to preserve nutrients and freshness, or as*

R. Owusu-Apenten, E. R. Vieira, *Elementary Food Science*, Food Science Text Series,
https://doi.org/10.1007/978-3-030-65433-7_11

complex as formulating a frozen meal with the right balance of nutrients and ingredients" (International Food Information Council Foundation 2010). With the exception of fruits and some vegetables (salad) very few foods are eaten raw, but even fresh produce usually undergo a form of processing and so the food groups are described as minimally processed. Commercial fresh produce undergo processing operations such as washing and cutting. As noted above, processed foods are prepared or manufactured by commercial (foodservice) kitchens or food process plants, respectively.

Excepting storage and transportation, most postharvest activities can be seen as forms of food processing (Fig. 11.1). However, the degree of processing varies for different foods, during or prior to these going from the farm-gate to the consumer. As discussed further below, most of the minimally processed food products are manufactured using a 5-stage process: Receiving → Sorting → Washing → Dewatering → Packaging (Shah and Nath 2006). Arguably, the lowest number of stages for a commercial food process is probably three: Receiving → Sorting → Packaging/Presentation. Manufacturing products like cheese and canned vegetables require a greater number of operational steps (Unit operations).

IFIC proposed the five categories of processed foods listed in Table 11.1 (International Food Information Council Foundation 2010).[1] We will consider ultra-processed foods later (Sect. 5) but this concept intersects with issues such as marketing. Processing is a core activity for food manufacturing (NAICS311& 312). The significance of processing can be seen from the central location of food manufacturing within commercial supply chains (Fig. 11.1).

Table 11.1 The IFIC classification of processed foods

Processed food class	Characteristics/examples
I. Minimally processed foods	Fresh like fruits and vegetables
II. Foods Processed to preserve	Canned, dehydrated, frozen, jarred
III. Food that Combine ingredients	Instant food, sources, mixes, gelatine
IV. Ready to eat meals	Require minimal or no home preparation
V. Foods packaged for freshness and convenience	Prepared deli foods, frozen meals

Adapted from International Food Information Council Foundation (2010, 2014)

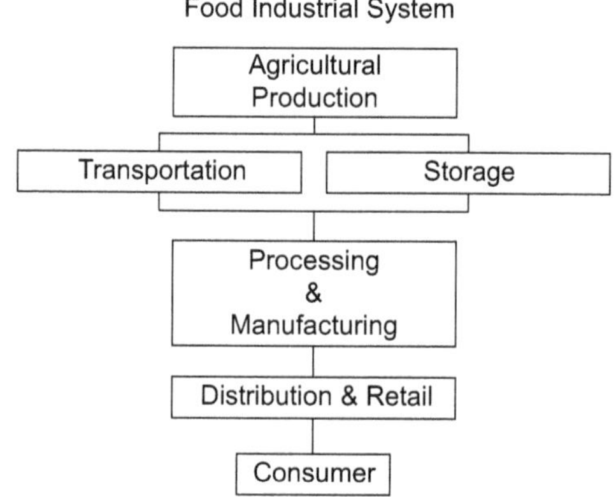

Fig. 11.1 The relation of processing and manufacturing to the rest of the food system

1.3 Aims and Benefits of Food Processing

1.3.1 Value Addition

Processed foods have higher perceived value compared to the raw commodities from which they derive. For instance, pasteurized milk costs more compared to milk at the farm gate. A large part of the "added value" for commercial foods arises from processing, packaging and or food service. Food processing and preparation within domestic settings is not monetized readily, but home "food labor" cost can affect how consumers behave (Davis and You 2010).

According to some estimates less than 50% of *the value* of food consumed by Americans is home prepared. The trajectory for home-prepared foods is likely to be downward over the coming years (Chap. 25). Due to the lower economies of scale, the dollar-value for home-food labor (cost per meal prepared) exceeds generally the labor cost for commercial food preparation. Economists believe the time-costs for home-food preparation affects the type of preprocessed food ingredient used (see below). In addition, a higher labor costs for home-food preparation also affects the overall

[1]Previously discussed in Chap. 1 (Sect. 4)

balance of time spent on housework, "market work" and childcare (Vernon 2005).

Food-labor cost per head decreases with increasing household size, i.e. the numbers of adults per household. Larger households can engage in bulk-food purchasing, preparing larger batches of food, and lower wastage (Deaton and Paxson 1998; Vernon 2005; Logan 2011). The inference is that smaller household sizes (e.g. single-adult households) benefit most from out-of home eating which matches with observations. Research shows that small households (single person households) have greater demand for out-of-home eating and manufactured foods (Kinsey 1990. p33; Umberger 2015). Moreover, out-of-home eating yields a time dividend or "convenience" for the consumer that is unable to spend "valuable time" for home-food preparation. To summarize, convenience or time saving is one of the values afforded by processed foods.

The price of minimally processed foods is generally higher compared to more processed (canned, dried or frozen) variety of fruits and vegetables. The higher price of fresh foods is due to their perishability, reduced availability and perceived quality. However, food processing leads to lower food prices overall. Research has shown that food pricing affects the amount and types of food consumed (Brinkman et al. 2010). The economic significance of foodservice (NAICS 722) and food manufacturing (NAICS 311) industries were discussed earlier (Chap. 2).

1.3.2 Preservation

Food processing can extend the storage life of products, which means food preservation (Rahman 2007). Indeed, the overlap between food processing and preservation was so great, that these terms were used interchangeably in the past. Clearly, not all processing operations extend shelf life. Peeling potatoes reduces the storage life unless the product is further processed in some way or frozen. Likewise, deshelling eggs will not extend shelf life, but liquid egg is useful as an ingredient for making angle cake, which is less subject to spoilage. Processing to inactivate enzymes found naturally in fresh foods extends shelf life. Some general benefits attributed to postharvest handling and processing and which contributes to the "added value" of processed foods are listed in Table 11.2 and discussed in Sect. 4.

Table 11.2 A list of attributes for processed foods that increases value

1. Increased food safety
2. Preservation and storage life extension
3. Conversion of raw food to edible products
4. Conversion of primary to secondary ingredients
5. Increase consumer benefits
6. Better packaging
7. Increased nutrition information

1.3.3 Safety

Incorporating a heating step during food processing enhances product safety. Heating represents a "lethality" step that reduces bacteria numbers. Some non-thermal methods of food processing, can also promote shelf life (Raso and Barbosa-Canovas 2003). To comply with HACCP, food process charts need to be analyzed on a case-by-case basis to incorporate strategies to ensure food safety.[2] Achieving preservation for minimally processed foods is particularly challenging (Tapia de Daza et al. 1996).

2 Types of Food Processing

2.1 Primary and Secondary Processing

IFIC suggested five types of processed foods, arising from a continuum of processing (Table 11.1). In addition to those broad categories, it useful to differentiate primary from secondary processing. Primary processing transforms raw foods into ingredients. Primary processing is applicable to all major food groups (meat & fish, milk, eggs, fruit and vegetables, cereals and legumes) thereby transforming these into ingredients. Usually, primary food ingredients are more stable and more convenient for preparing meals in the home or foodservice situation.

[2]HACCP is discussed more fully in Chap. 10.

Table 11.3 An example of some dairy ingredients from Canada[a]

Butter	Paneer
Buttermilk powder	Milk protein concentrate
Butter oil	Monterey Jack and Colby
Cheddar cheese	Mozzarella
Cottage cheese	Parmesan cheese
Cream	Quark cheese
Cream cheese	Ricotta cheese
Emmentaler and Swiss-style cheese	Romano cheese
Evaporated milk	Skim Milk powder
Fermented milk products	Sweet and condensed milk
Feta	Whey powder
Liquid whole milk	Whey protein concentrate
Mascarpone cheese	Whole milk powder
	Yogurt

[a]Adapted from Canadian Dairy Commission (2011)

Common examples of primary processing include the conversion of wheat to flour, transformation of shell eggs to liquid eggs, or the conversion of liquid milk to dried milk powder. More complex primary processing, include activities such as slaughtering and butchering to produce various cuts of meat. Most of the food commodities are also primary food ingredients.

Secondary processing refers to the transformation of primary ingredients into ready-to-eat products, including baked goods, ready meals, jam, pizza, cheese etc. Some foods ingredients can be consumed directly or used for more complicated dishes (Table 11.3). For instance, cheese can be eaten directly or used as toping for pizza. Sometimes, the distinction between primary and secondary processing is not clear-cut.

Interestingly, some Chefs and other foodservice operations may adopt secondary ingredients owing to their greater convenience. Using primary ingredients or "home recipes" for cooking leads to higher costs compared to "processed recipes" provided that "time-inclusive" prices are taken into account (Yang et al. 2015; Davis and You 2010).

2.2 Traditional Versus Modern Food Processing, and Overlaps

Traditional food processing refers to small scale, historical methods and practices; traditional food processing methods are less mechanized and also require limited technology and financial investment. Traditional processing methods dominated the food supply up until the eighteenth century when demand for preserved foods became acute because of a variety of issues, e.g. warfare,[3] population growth, urbanization and decreasing self-sufficiency in local food. The commercial food industry of sorts started about 1860s with the replacement of water-driven milling for wheat with mechanical power, invention of margarine as butter substitute, and some food preservation by canning and refrigerated storage (Oddy 2016).[4]

Modern food technologies started from Western Europe and spread to the new Colonies of Australia, North and South America (Farrer 2005). By contrast, techniques like food fermentation were performed by traditional cultures predating the industrial revolution (Brown and Pariser 1975; Farrer 2005; Oddy 2016). Some traditional food processing methods are listed below (Table 11.4).

The modernization of food processing required larger-scale production in order to benefit from economies of scale. Modern food processing makes increasing use of mechanization, automation, and greater levels of financial investment. As discussed elsewhere, the modern food process plant is usually purpose built, with specifically designed fixtures, utensils and machinery in line with Good Management Practice (GMP) requirements (U.S. Food and Drug Administration 2014). Modern food processing also requires a greater emphasize also on food safety and product quality. Transformations of

[3]The Napoleonic wars and two World Wars are all recognized as catalysts for food technologies such as canning, and food dehydration.

[4]The history of various food-processing technologies is discussed in chapters devoted to specific methods.

Table 11.4 Some traditional methods for food processing

Boiling
Brewing
Cheese-making
Cooling, or "refrigeration"
Drying
Fermentation
Germinating/malting
Milling
Honey as a preservative
Ice for preservation
Pickling
Pounding
Salting
Smoking
Soaking

Adapted from various sources including (Farrer 2005)

Table 11.5 Characteristics of traditional and modern forms of food processing

Traditional processing	Modern food processing
←Foodservice→[a]	
Small scale, cottage industry	Large scale
Home food preparation	Factory based
Low tech	Increased automation
Low capital investment	Greater capital req. (for machinery)
Primary focus on preservation	Industry consolidation
Historically important	
<u>*Emphasizes*</u>	<u>*Emphasizes*</u>
Artisanal & Cultural values	Efficient material handling
Customizable	Product design and development;
Non-formalized	Food safety (GMP, HACCP, traceability)
	Food packaging.

[a]Foodservice combines traditional and modern large-scale approaches

traditional to modern forms of food processing are currently underway within many developing countries, including China and India (Bawa 2007; Zhou et al. 2012).

Areas of potential overlap between traditional and modern forms of food processing are worth considering. Thus, foodservice operations employ commercial kitchens and equipment which have the characteristic "feel and appearance" of the home kitchen (Table 11.5). Nevertheless, food preparation in centralized locations require large-scale facilities and many modern food technologies. At the same time, foodservice operations also incorporate traditional or artisanal features such as the emphasize on, regional cuisines (Chap. 24).

There is overlap also between traditional and modern food processing, in areas of product development scale-up. For instance, many large-scale food production schemes will start with prototypical products developed on a small scale.

independent, and located within well defined spaces or locations within the manufacturing plant. Compliance with GMP requires that specific unit operations such as receiving (of raw materials) should take place within a designated area, e.g. provided with parking facilities for delivery vehicles and appropriate storage.

Different unit operations also require separate equipment or apparatus. The layout of unit operations will conform to the overall processing diagram. Therefore, materials flow from one machinery to a sequence that enables specific functions, e.g. sorting, washing, and packaging in the case of minimally processed foods. Food materials are conveyed efficiently from one unit operation to the next. The final unit operation for many cases is packaging and labeling (Table 11.6).

2.3 Unit Operations for Food Processing

Food manufacturing processes include many smaller, well-defined, steps called Unit operations or operational steps is self-contained,

2.4 Minimally and Moderately Processed Foods

The manufacturing operations for minimally processed foods include sorting, washing, and packaging for products such as fresh vegeta-

Table 11.6 Common unit operations for food processing

Material handing – vehicular transport, pumping, conveyor belting
Cleaning – washing, with water (+sanitizer), pealing to remove dirt
Separating – separation by size or density using sieving/centrifugation
Size reduction – grinding, meat slicer, homogenizing, blending
Pumping – Fluid or gas flow
Mixing – agitation to allow mass or heat transfer
Heat exchange – Heating, cooling, by freezers, ovens, pasteurizers
Concentrating – evaporation, reverse osmosis
Forming & shaping – By moulding
Controlling & monitoring – Sensors to monitor temperature, pressure, pH, time viscosity, weight, specific gravity, valves, humidity, time
Packaging – product placed inside glass, cardboard, plastic packing

bles, fruits and meat. Other minimally processed foods (e.g. salads, fruit pieces) incorporate a cutting operation before packaging (Shah and Nath 2006). As noted above Group (II) processed foods covers most food ingredients. Group (III) foods are normally processed ready-to-eat foods including bread, cookies, pizza and others (Moubarac et al. 2014). In contrast, the current designation "ultra-processed" food does not seem to imply a particular degree of processing (Sect. 5).

3 Benefits and Downsides of Food Processing

It is no accident that some processed foods cost more than "raw" ingredients. The extra costs arise from benefits (actual or perceived) for the consumer. Some of these issues such as increased convenience were described in relation to "value addition" (Sect. 1.3). Other benefits attached to processed foods are outlined here.

3.1 Improved Storage Life by Food Preservation

Food processing is associated with preservation and increased storage life. Indeed, processing was formerly synonymous with preservation but as discussed elsewhere this is not always true. The increase of storage-life achieved by processing is due to the lower number of spoilage microorganisms, and their decreased rate of growth, following processing. Processing also inactivates enzymes naturally present in raw foods, so these no longer promote undesirable changes that can affect colour, aroma, taste and the texture of food.

3.2 Improved Sensory Quality

Food processing can improve sensory quality. For instance, cooking improves the colour, taste, and the palatability of most foods. Improvements may arise due to starch granule gelatinization, and protein denaturation and gelation. Some improvements in nutritional characteristics arise also due to enhanced digestibility. Food borne allergens are notoriously heat stable but adequate heating can inactivate anti-nutritional proteins, e.g. trypsin inhibitors and lectins found in beans.

3.3 Nutrient Losses and Toxic Compounds

Undesirable changes can occur during food processing such as, the loss of nutrients. Hazardous compounds may arise with some high intensity long duration processes. Baking and frying may lead to cancer causing substances (carcinogens) called heterocyclic amines (HCAs) and polycyclic aromatic hydrocarbons (PAHs). Frying high-starch foods (French fries) promotes the formation of acrylamide a neurotoxin. Some other health-related changes affecting cooked food include the formation of Maillard reaction products and Advanced Lipid Oxidation Products formed by heat induced oxidation of unsaturated lipids. Leaching of toxic metals from cooking utensils e.g., aluminum, nickel and iron into processed foods is another area of concern (Kamerud et al. 2013).

4 Additional Consumer Benefits

4.1 Novelty and Innovation

Food companies introduce between 8000–12,000 new products in the United States every year, though some 90% of new products may fail (Jekanowski and Binkley 2000; Toops 2017). According to conventional wisdom, the high rate of food product innovations allows manufacturers to respond to, a constantly changing consumer needs, purchasing trends, a high pace of changes in lifestyle and also to remain competitive (Costa and Jongen 2006; Zink 1997).

4.2 Convenience

Some current trends affecting food-manufacturing industries include, a decline in the average time consumers spend on home food preparation and increased out-of-home eating – which are all driven by the need for convenience. There were reductions in the time for home preparation when using processed food ingredients as compared with home food recipes. The so-called "time-inclusive" price of food was lower when cooking with processed food ingredients (Yang et al. 2015; Davis and You 2010). Related factors that increase processed food consumption in low and middle-income countries include a rising urban population, increase in household incomes, and a growing middle class.

In summary, the benefits of processed foods are many and varied. Another less-recognized benefit is the greater nutrient stability of frozen foods as compared to fresh or unprocessed foods (Howard et al. 1996; abstract).

Table 11.7 Increased consumer benefits from processed foods[a]

All-year round variety
Decreased preparation time
Increased choice (less seasonality)
Greater convenience (prepared ready-to-eat)
Increase portability (weight decrease)
Increased food security
Increased food safety
Improved nutrient stability
Long storage allows bulk shopping
Lower costs

[a]Adapted from various sources

4.3 Processed Foods, the Environment and Development

Concerns about the environmental impact of the food system, led to increasing emphasis on locally produced food and organic food (Wuyang et al. 2012). Access to food processing technology was found to improve food security and well-being of rural poor in developing countries (Aworh 2015). The availability of small-scale technology enabled farmers to add-value to agricultural products, leading to wider benefits such as improvements in the diet and ability to meet school fees and other financial commitments (Silva-Barbeau et al. 2005).

5 Ultra-Processed Foods

5.1 Ultra Processed Food and Composition

A new classification for processed foods called the "NOVA" scheme is gaining traction with public health experts and nutrition professionals. For instance, the NOVA scheme was used by the World Health Organization as part of the decade of food and nutrition. However, the NOVA designation of some foods as ultra-processed has proven highly controversial, because the scheme did not reference "food processing" in the same

Table 11.8 Definition of ultra-processed foods

Ultra-processed foods are products "*...made from processed substances extracted or refined from whole foods – e.g. oils, hydrogenated oils and fats, flours and starches, variants of sugar, and cheap parts or remnants of animal foods – with little or no whole foods... include burgers, frozen pasta, pizza and pasta dishes, nuggets and sticks, crisps, biscuits, confectionery, cereal bars, carbonated and other sugared drinks, and various snack products... are made, advertised, and sold by large or transnational corporations ... very durable, palatable, and ready to consume, ...[Exhibit]... commercial advantage over fresh and perishable whole or minimally processed foods. ... are typically energy dense; have high glycemic load; ...low in dietary fiber, micronutrients, and phytochemicals; and are high in unhealthy ... dietary fat, free sugars, and sodium*."
Or
"*.. Industrial formulations manufactured from substances derived from foods or synthesized from other organic sources. They typically contain little or no whole foods, are ready-to-consume or heat up, and are fatty, salty or sugary,.. depleted in dietary fiber, protein, ... micronutrients and other bioactive compounds.... include: sweet, fatty or salty packaged snack products, ice cream, sugar-sweetened beverages, chocolates, confectionery, French fries, burgers and hot dogs, and poultry and fish nuggets*.'

Adapted from Monteiro et al. (2018)

manner defined by the IFIC (Table 11.1). At first, the designation ultra-processed referred to a miscellaneous group of products that contain high levels of salt, sugar, fat and or food additives (Monteiro et al. 2018). Later it was explained that food processing as defined by food experts was not the issue per se (Monteiro et al. 2018). Unfortunately, the subtleness in defining "ultra-processed" was not appreciated by social-influencers and even some senior food experts were confused (see below) (Table 11.8).

5.2 NOVA Classification of Processed Foods

The NOVA system underwent significant changes following a decade of discussions Carlos A. (Monteiro 2009). At its core, the NOVA classification proposes four groups or types of processed food (Group 1-IV) each with different

Table 11.9 The NOVA classification of foods by processing degree[a]

Group (I)	Unprocessed or Minimally processed foods. Fresh manufactured via a limited postharvest tasks, e.g. Cleaning, removing inedible fractions, portioning,
Group (II)	Processed culinary ingredients.
Group (III)	Processed foods (Canned foods. Made from Goup1 and 2 foods. Often edible on their own or eaten with other foods
Group (IV)	Ultra-processed foods (Group 4). Soft drinks, Savoury snacks, Sweet snacks and drinks; ready-to-consume formulations, prepared from ingredients

[a]Adapted from Monteiro et al. (2018)

kinds of ingredients used in their preparation (Table 11.9).

Group (I) minimally processed foods consist of products manufactured and/or containing unprocessed and minimally processed products or whole foods. Group (II) processed culinary ingredients refers to food products extracted from Group (I) foods by refining, milling or by hydrolysis. Many foods commodities could be regarded as culinary food ingredients such as, oils, fats, sucrose, flours, and starch. Items like fish, meat, liquid eggs, and meat products are also ingredients. Group (III) processed foods are prepared from Group (I) and (II) foods, and include products such as canned tuna, and canned vegetables. Finally, Group (IV) ultra-processed foods are distinguished by a lack of whole food ingredient.

According to the NOVA scheme ultra-processed foods are "reformulations of sources of energy, nutrients, fats, salts and which tend to be fatty and containing food additives". Group (IV) foods also have low amounts of dietary fiber, high glycemic index, and are also generally sugary and also quasi-addictive(Monteiro et al. 2018) (Table 11.9).

Ultra-processed foods (Table 11.8) are described as lacking in whole foods. High levels of salts, sugars, fats and additives are thought to be another distinguishing feature of ultra-processed food. Additionally, many ultra-processed foods have high energy density. Foods produced by multi-national corporations were over-represented in the NOVA group IV category. Ultra-processed foods are also considered hyper-palatable and habit forming which are properties that promote overeating. Interestingly, the ultra-processed foods concept extends to attributes such as the (aggressive) style of marketing and advertising (Mallarino et al. 2013).

The NOVA scheme excluded certain kinds of foods from consideration as ultra-processed foods. For instance, home-prepared foods and dishes *prepared* by full-service restaurants, and institutional foodservices outlets were excluded from consideration as ultra-processed foods. However, non-full service restaurant meals, take-outs, and other street foods, and fast foods were included as ultra-processed foods (Monteiro et al. 2018).

Currently, several governmental bodies and non-governmental bodies recognize the category of ultra-processed foods, including the Food and Agriculture Organization of the United Nations and also the Pan American Health Organization (Monteiro et al. 2018; Monteiro et al. 2013). Recent research showed that consumption of ultra-processed foods is increasing globally (Baker and Friel 2016; Juul and Hemmingsson 2015; Moubarac et al. 2014; Rauber et al. 2015). The potential link between ultra-processed foods and chronic disease (if true) could have a significant impact on how such products are used. Some evidence links ultra-processed foods and the risk of obesity and metabolic diseases in e.g. South America, North America, UK and Sweden (Juul and Hemmingsson 2015; Juul et al. 2018; Louzada et al. 2015; Moubarac et al. 2013; Rauber et al. 2018; Tavares et al. 2012).

Examples of ultra-processed foods were cited as, snacks, soft drinks, sweets / candy, chocolate, crisps, fries, ice-cream, and ready-to-eat meals (Costa et al. 2018). A study from Sweden listed examples from Group I-IV processed foods together with a listing of ultra-processed foods (Juul et al. 2015). The ultra-processed food group include, cookies (biscuits), cakes, pastries, ice cream, jam, confectionary (candy, chocolate), cereal bars; sugared drinks, flavored milk drinks, pizza and fish sticks. Another influential study organized ultra-processed foods into three main types; (A) frozen processed products including semi-ready oven potatoes, chicken nuggets, pizza etc., (B) Snacks like chips/ crisps, corn chips, pretzels, sweet snacks, salted nuts, confectionary, dairy products, and (C) Soft drinks –carbonated drinks, sweetened fruit and vegetable juices, ready to drink teas, and sports energy drinks (Monteiro et al. 2013). It may be noticed that the lists for ultra-processed foods include many "traditional" foods (snacks, soft drinks, sweets, chocolate, crisps, ice cream, pizza, pretzel, salted nuts etc.) alongside of new products like cereal bars and sports energy drinks.

Table 11.10 A critical appraisal of the NOVA categorizations of ultra-processed foods (UPF)

Criticism of NOVA	References
Ineffective at identifying high nutrient foods	Bleiweiss-Sande et al. (2019)
Lack of evidence the UPF are hyper-addictive	Gibney et al. (2017)
UPF classification is not consistent over time	Gibney (2019)
Uses non-traditional definition of "food processing" Low agreement with other classification methods, Lacks appeal – accessibility, practicality	Jones (2019)
UPF is linked with food composition not food processing	Ludwig et al. (2019)

5.3 Criticisms of Ultra-Processed Terminology

The "NOVA" scheme for ultra-processed foods was critiqued on a variety of grounds (Gibney 2019; Gibney et al. 2017). For instance, Professor Mike Gibney noted that the NOVA classifications is not superior to other foods classifications – that address issues of food composition. It was noted that the NOVA scheme lacked clearer cut-offs between foods, which are designated as "high" in sugar, or fat etc. The degree of *processing* implied in the terms "ultra" or "highly" processed were inconsistent.[5] The proposed link between globalization and low dietary quality suggested in the NOVA system seems tentative, because current world export/import trade accounts for about 10% of the food markets (Chap. 2).

Another criticism offered for the NOVA system is that, calling certain food groups hyperpalatable and/or habit-forming was questionable. Professor Mike Gibney doubted whether the NOVA classification alone could function as a way of documenting dietary patterns, and making guidelines for healthy eating (Gibney 2019; Gibney et al. 2017). Putting "aggressive" marketing and advertising as part of food/diet relations is also difficult when assessing ultra-processing (Moreira et al. 2015). Others have also criticized the NOVA scheme for ultra-processed foods, as the basis for analyzing diet and health relations (Table 11.10). It was noted that weight gain arises from eating too much or not exercising enough. Therefore linking obesity to specific nutrients or food is discredited. Finally, any association between ultra-processed food components and obesity is not proof of a cause and effect relation.[6]

5.4 How Ultra-Processed Food Affects the Diet

Monteiro and others discussed the relevance of the ultra-processed foods in relation to the WHO decade of nutrition (Monteiro et al. 2018).[7] The concept of ultra-processed foods, which originated from public health debates in 2009 about food and health, has evolved and passed into general use by the WHO and other major food policy groups. The main uses of the ultra-processed food paradigm is in the analysis of dietary patterns. According to current discussions ultra-processed food compromise about 58%, 48% and 29% of food consumed in the United States, Canada, and Brazil, respectively (Martins et al. 2013; Monteiro et al. 2013).

In summary, the concept of ultra-processed food though important is not entirely without criticism. Importantly, the designation "ultra-processed" depends on a products *composition* not the extent of culinary or manufacturing transformation. That is why sweetened tea qualifies as ultra-processed but sugar-free tea does not (footnote 5). It is not certain whether foods branded as "low sugar", "low-fat", "low-saturated fats", "low-salt" etc. should be excluded from consideration as ultra-processed foods (Scrinis and Monteiro 2018). Nutrient content claims are discussed further in Sect. 6 of Chap. 4.

6 Alternative Food and Health Relations

Many commentators from the social sciences presented alternative ideas linking food technology to public health. A few of the explanations showing how food technology might affect society at large are discussed briefly below. See Chap. 6 (Sect 1.3 – 1.5) for discussion about wellbeing and its relations to the "Western diet", income / nutrition transitions and processed foods.

6.1 Food Technology Access, Decreasing Labor and Rising Obesity

In the article titled *"Why have Americans become more obese?"* economics experts from Harvard University theorized that the rising rates of obe-

[5] The latest version of NOVA is not related to conventional processing concepts per se; for instance, sugary drinks can be produced by addition of sugars and not by any form of "over-processing".

[6] Compare with the five classes of processed foods currently recognized: (I) Minimally processed foods, (II) Foods processed for preservation, (III) Food as mixtures of combined ingredients, (IV) Ready-to-eat processed foods, and (V) Prepared foods/meals

[7] The ultra-processed food category originated round about 2009 and as from 2019, the PubMed database showed four narrative reviews, three randomized controlled studies dealing with ultra-processed foods in the title. Currently, more high-level evidence is needed for the concept of ultra-processed foods effects.

sity was correlated with food technology *access* and increased mass production of food over the past 20 years.[8] According to this thesis, greater access to food technology decreases the time-costs for food preparation. Moreover, easier faster cooking correlates with rising obesity even for individuals within the same nuclear family. Additional analysis suggested that rising levels of obesity were strongly associated with increasing caloric intake rather than decreasing, physical activity or energy utilization. *Snacking* was identified as the most likely mechanism for weight gain. By contrast, the total caloric intake at main meal times decreased over 20 years (Cutler et al. 2003).

International comparisons showed there were lower rates of obesity in low-income countries with low access to food technology. For example, the Mediterranean region showed lower access to food manufacturing technology. There were increased time-costs for food preparation in the Mediterranean countries and lower obesity rates compared with Northern European countries where access to food technology was greater (Cutler et al. 2003). The article did not mention traditional diets or traditional food processing per se, which are likely to require longer preparation times.

6.2 Work Patterns and Food Labor

Michelle Szabo (York University, Toronto) provided a feminist analysis of the "*socio-cultural and political-economic factors related to the contemporary demand for convenience in food provision*" (Szabo 2011). The article evaluated the significance of convenience foods for "*contemporary employer and gender relations*". One conclusion is that, adopting more fresh ingredients, homegrown produce, or sustainably grown foods is positive for health, but the costs for "social reproduction" were unaccounted for. The article covered other contemporary issues including the organization of food work, food work in the context of changing employment and household conditions, gender and food work.[9] The section titled "*Increasing the feasibility of reconnecting with foods*" contained general proposals for decreasing the reliance on convenience foods, by changing attitudes and conditions around paid and unpaid work. Apparently, the majority of males in North America (excluding Mexico) do not contribute to food work sufficiently to offset the extra labor-costs (unpaid) needed with traditional ingredients. Therefore, one possible gender-specific strategy for decreasing consumer reliance on convenience foods would be to promote cooking-culture amongst men and boys (Szabo 2011).

To summarize, using some pre-processed ingredients reduces the "time-costs" and general cost of *home-cooked food* (Yang et al. 2015; Davis and You 2010).

6.3 Policy and Healthy Food Environments

Food policy experts analyzed the failure to halt world obesity rises, in terms of a decreasingly healthy food environment, defined as "*the collective physical, economic, policy and sociocultural surroundings, opportunities and conditions that influence people's food and beverage choices and nutritional status*". Accordingly, weight gain and obesity should not be considered only as issues for individual responsibility. Rises in obesity will not be solved solely by consumer education. Government policies and regulations to improve the food environment should be considered. A combination of public – private (Government and food industry) partnerships might be useful for

[8]Correlation is no prove causation. Two variables may be correlated coincidentally, if both are linked with a third (unknown) variable. There is no simple explanation why greater access to food preparation technology should lead to weight gain for both genders given there is inequality of food labor. Somehow, easier food preparation using technology promote a positive energy balance (section 6.2)

[9]The term "food labor" is a more frequent description used by economic scholars. The terms in this text are are actual titles used by the author

achieving calorie reduction, but such approaches were thought to be insufficient and lacking in accountability. In addition, the recent globalization and trade liberalization trends may have reduced national governments ability to enact food legislations to reduce calorie intake. A need for strong legislation, monitoring and accountability were found to be essential to ensure healthy food environment (Swinburn et al. 2013, 2015; Hall 2018). These preceding discussions chime with the idea that mega-trending food issues, such as obesity and the impact of food production on global warming will require a systems approach for their resolution (Chap. 6; Sect. 5.2).

There is considerable interest in using fiscal policy (taxation and subsidies) to increase the prices of specific (processed) food ingredients e.g. sugar, or sugary drinks in order to promote healthy eating (Niebylski et al. 2015; abstract). Furthrmoore, subsidies could be used to lower the price of fruit and vegetables. The relations between food prices and diet quality is being actively studied (Andreyeva et al. 2010; Powell et al. 2013; Afshin et al. 2017).

7 Conclusions

Food processing is one of the oldest human activities. With few significant exceptions, most of the commercial food supply and food eaten in the home is processed to some extent. Interestingly, some consumers seem to be becoming alienated from processed foods. This chapter examined some basic considerations related to processed foods including, benefits and disadvantages of processed foods. The benefits of food processing is captured in terms of "added value", which explains why manufactured (processed) foods achieve higher prices compared to farm-gate prices. Some of the recent controversy surrounding ultra-processed foods are discussed. The role of economic policy (subsidies and taxation) for moderating food-eating patterns is mentioned. This chapter provides material on food processing that is particularly relevant for food literacy (Chap. 6).

References

Afshin A, Penalvo JL, Del Gobbo L, Silva J, Michaelson M, O'Flaherty M, Capewell S, Spiegelman D, Danaei G, Mozaffarian D (2017) The prospective impact of food pricing on improving dietary consumption: a systematic review and meta-analysis. PLoS One 12(3):e0172277

Andreyeva T, Long MW, Brownell KD (2010) The impact of food prices on consumption: a systematic review of research on the price elasticity of demand for food. Am J Public Health 100(2):216–222

Aworh OC (2015) Promoting food security and enhancing Nigeria's small farmers' income through value-added processing of lesser-known and under-utilized indigenous fruits and vegetables. Food Res Int 76(4):986–991. https://doi.org/10.1016/j.foodres.2015.06.003

Baker P, Friel S (2016) Food systems transformations, ultra-processed food markets and the nutrition transition in Asia. Glob Health 12(1):80. https://doi.org/10.1186/s12992-016-0223-3

Bawa AS (2007) Historical developments in food science and technology – Indian perspective. J Food Sci Technol 44(6):553–564

Bleiweiss-Sande R, Chui K, Evans EW, Goldberg J, Amin S, Sacheck J (2019) Robustness of food processing classification systems. Nutrients 11(6). https://doi.org/10.3390/nu11061344

Brinkman HJ, De Pee S, Sanogo I, Subran L, Bloem MW (2010) High food prices and the global financial crisis have reduced access to nutritious food and worsened nutritional status and health. J Nutr 140(1):153S–161S

Brown NL, Pariser ER (1975) Food science in developing countries. Science 188(4188):589–593. https://doi.org/10.1126/science.188.4188.589

Canadian Dairy Commission (2011) Dairy ingredient profiles. Retrieved from http://www.milkingredients.ca/index-eng.php?id=170

Costa AI, Jongen WMF (2006) New insights into consumer-led food product development. Trends Food Sci Technol 17(8):457–465

Costa CS, Del-Ponte B, Assuncao MCF, Santos IS (2018) Consumption of ultra-processed foods and body fat during childhood and adolescence: a systematic review. Public Health Nutr 21(1):148–159

Cutler DM, Glaeser EL, Shapiro JM (2003) Why have Americans become more obese? J Econ Perspect 17(3):93–118

Davis GC, You W (2010) The thrifty food plan is not thrifty when labor cost is considered. J Nutr 140(4):854–857. https://doi.org/810.3945/jn.3109.119594

Deaton A, Paxson C (1998) Economies of scale, household size, and the demand for food. J Polit Econ 106(5):897–930. https://www.jstor.org/stable/810.108 6/250035?origin=JSTOR-pdf

Dwyer JT, Fulgoni VL 3rd, Clemens RA, Schmidt DB, Freedman MR (2012) Is "processed" a four-letter word? The role of processed foods in achieving dietary guidelines and nutrient recommendations. Adv Nutr 3(4):536–548

Farrer K (2005) To feed a nation. A history of Australian food science and technology. Csiro Publishing, Collingwood, 230 pp

Gibney MJ (2019) Ultra-processed foods: definitions and policy issues. Curr Dev Nutr 3(2):nzy077. https://doi.org/10.1093/cdn/nzy077

Gibney MJ, Forde CG, Mullally D, Gibney ER (2017) Ultra-processed foods in human health: a critical appraisal. Am J Clin Nutr 106(3):717–724. https://doi.org/10.3945/ajcn.117.160440

Hall KD (2018) Did the food environment cause the obesity epidemic? Obesity (Silver Spring) 26(1):11–13. https://doi.org/10.1002/oby.22073

Howard L, Wong A, Perry A, Klein BP (1996) The stability of nutrients in fresh and processed vegetables. IFT Annual Meeting Abs. ISSN: 1082–1236

International Food Information Council Foundation (2010, 2014) What is a processed food? You might be surprised! Retrieved from https://foodinsight.org/wp-content/uploads/2014/07/IFIC_Handout1_high_res.pdf

Jekanowski MD, Binkley JK (2000) Food purchase diversity across US markets. Agribus Int J 16(4):417–433

Jones JM (2019) Food processing: criteria for dietary guidance and public health? Proc Nutr Soc 78(1):4–18. https://doi.org/10.1017/s0029665118002513

Juul F, Hemmingsson E (2015) Trends in consumption of ultra-processed foods and obesity in Sweden between 1960 and 2010. Public Health Nutr 18(17):3096–3107. https://doi.org/10.1017/s1368980015000506

Juul F, Martinez-Steele E, Parekh N, Monteiro CA, Chang VW (2018) Ultra-processed food consumption and excess weight among US adults. Br J Nutr 120(1):90–100. https://doi.org/10.1017/s0007114518001046

Kamerud KL, Hobbie KA, Anderson KA (2013) Stainless steel leaches nickel and chromium into foods during cooking. J Agric Food Chem 61(39):9495–9501

Kinsey JD (1990) US demographic trends and their relationship to food markets. Staff Paper P90-14. University of Minnesota, 105pp. https://ageconsearch.umn.edu/record/13264/

Logan TD (2011) Economies of scale in the household: puzzles and patterns from the American past. Econ Inq 49(4):1008–1028. https://www.nber.org/system/files/working_papers/w13869/w13869.pdf

Louzada ML, Baraldi LG, Steele EM, Martins AP, Canella DS, Moubarac JC et al (2015) Consumption of ultra-processed foods and obesity in Brazilian adolescents and adults. Prev Med 81:9–15. https://doi.org/10.1016/j.ypmed.2015.07.018

Ludwig DS, Astrup A, Bazzano LA, Ebbeling CB, Heymsfield SB, King JC, Willett WC (2019) Ultra-processed food and obesity: the pitfalls of extrapolation from short studies. Cell Metab 30(1):3–4. https://doi.org/10.1016/j.cmet.2019.06.004

Mallarino C, Gomez LF, Gonzalez-Zapata L, Cadena Y, Parra DC (2013) Advertising of ultra-processed foods and beverages: children as a vulnerable population. Revista de Saude Publica 47(5):1006–1010

Martins AP, Levy RB, Claro RM, Moubarac JC, Monteiro CA (2013) Increased contribution of ultra-processed food products in the Brazilian diet (1987-2009). Revista de Saude Publica 47(4):656–665. https://doi.org/10.1590/S0034-8910.2013047004968

Monteiro CA (2009) Nutrition and health. The issue is not food, nor nutrients, so much as processing. Public Health Nutr 12(5):729–731. 2009/05/01. Retrieved from https://doi.org/10.1017/S1368980009005291

Monteiro CA, Moubarac JC, Cannon G, Ng SW, Popkin B (2013) Ultra-processed products are becoming dominant in the global food system. Obes Rev 14(Suppl 2):21–28. https://doi.org/10.1111/obr.12107

Monteiro CA, Cannon G, Moubarac JC, Levy RB, Louzada MLC, Jaime PC (2018) The UN decade of nutrition, the NOVA food classification and the trouble with ultra-processing. Public Health Nutr 21(1):5–17. https://doi.org/10.1017/s1368980017000234

Moreira PV, Baraldi LG, Moubarac JC, Monteiro CA, Newton A, Capewell S, O'Flaherty M (2015) Comparing different policy scenarios to reduce the consumption of ultra-processed foods in UK: impact on cardiovascular disease mortality using a modelling approach. PLoS One 10(2):e0118353. https://doi.org/10.1371/journal.pone.0118353

Moubarac JC, Martins AP, Claro RM, Levy RB, Cannon G, Monteiro CA (2013) Consumption of ultra-processed foods and likely impact on human health. Evidence from Canada. Public Health Nutr 16(12):2240–2248. https://doi.org/10.1017/S1368980012005009

Moubarac JC, Batal M, Martins AP, Claro R, Levy RB, Cannon G, Monteiro C (2014) Processed and ultra-processed food products: consumption trends in Canada from 1938 to 2011. Can J Diet Pract Res 75(1):15–21. https://doi.org/10.3148/75.1.2014.15

Niebylski ML, Redburn KA, Duhaney T, Campbell NR (2015) Healthy food subsidies and unhealthy food taxation: a systematic review of the evidence. Nutrition 31(6):787–795

Oddy DJ (2016) Introduction. In: Drouard A, Oddy DJ (eds) The food industries of europe in the nineteenth and twentieth centuries. Routledge, London, pp 1–16

Powell LM, Chriqui JF, Khan T, Wada R, Chaloupka FJ (2013) Assessing the potential effectiveness of food and beverage taxes and subsidies for improving public health: a systematic review of prices, demand and body weight outcomes. Obes Rev 14(2):110–128

Raso J, Barbosa-Canovas GV (2003) Nonthermal preservation of foods using combined processing techniques. Critic Rev Food Sci Nutr 43(3):265–285

Rauber F, Campagnolo PD, Hoffman DJ, Vitolo MR (2015) Consumption of ultra-processed food products and its effects on children's lipid profiles: a longitudinal study. Nutr Metab Cardiovasc Dis 25(1):116–122. https://doi.org/10.1016/j.numecd.2014.08.001

Rauber F, da Costa Louzada ML, Steele EM, Millett C, Monteiro CA, Levy RB (2018) Ultra-processed food consumption and chronic non-communicable diseases-related dietary nutrient profile in the

UK (2008(−) 2014). Nutrients 10(5). https://doi.org/10.3390/nu10050587

Scrinis G, Monteiro CA (2018) Ultra-processed foods and the limits of product reformulation. Public Health Nutr 21(1):247–252

Shah NS, Nath N (2006) Minimally processed fruits and vegetables – freshness with convenience. J Food Sci Technol Mysore 43(6):561–570

Silva-Barbeau I, Hull SG, Prehm MS, Barbeau WE (2005) Women's access to food-processing technology at the household level is associated with improved diets at the pre-harvest lean season in The Gambia. Food Nutr Bull 26(3):297–308

Singh RP, Heldman DR (2008) Introduction to food engineering. Retrieved from https://books.google.co.uk/books?id=jebJgWHADi4C

Smith JS, Hui YH (2008) Food processing: principles and applications. Wiley, New York, 524 p

Swinburn B, Vandevijvere S, Kraak V, Sacks G, Snowdon W, Hawkes C, Barquera S, Friel S, Kelly B, Kumanyika S, L'Abbé M (2013) Monitoring and benchmarking government policies and actions to improve the healthiness of food environments: a proposed government healthy food environment policy index. Obes Rev 14:24–37

Swinburn B, Kraak V, Rutter H, Vandevijvere S, Lobstein T, Sacks G, Gomes F, Marsh T, Magnusson R (2015) Strengthening of accountability systems to create healthy food environments and reduce global obesity. Lancet 385(9986):2534–2545

Szabo M (2011) The challenges of "re-engaging with food" connecting employment, household patterns and gender relations to convenience food consumption in North America. Food Cult Soc 14(4):547–566

Tapia de Daza MS, Alzamora SM, Chanes JW (1996) Combination of preservation factors applied to minimal processing of foods. Crit Rev Food Sci Nutr 36(6):629–659

Tavares LF, Fonseca SC, Garcia Rosa ML, Yokoo EM (2012) Relationship between ultra-processed foods and metabolic syndrome in adolescents from a Brazilian Family Doctor Program. Public Health Nutr 15(1):82–87. https://doi.org/10.1017/s1368980011001571

Toops D (2017) How did the food industry get (from there) to here? In: Beckley JH, Herzog LJ, Foley MM (eds) Accelerating new food product design and development, 2nd edn. Wiley/The Institute of Food Technologists, Newark, 387 pp

U.S. Food and Drug Administration (2014) Good manufacturing practices (GMPs) for the 21st century – food processing. Retrieved from http://www.fda.gov/Food/GuidanceRegulation/CGMP/ucm110877.htm

Umberger WJ (2015) Demographic trends: implications for future food demand. Agricultural Symposium Federal Reserve Bank of Kansas City July 14–15, 2015, 16pp. https://www.kansascityfed.org/documents/7023/umberger-paper.pdf

United States Department of Agriculture (2020). Ag and food statistics: charting the essentials: farming and farm income. Retrieved from https://www.ers.usda.gov/data-products/ag-and-food-statistics-charting-the-essentials/farming-and-farm-income/

Vernon V (2005) Food expenditure, food preparation time and household economies of scale. SSRN. From https://ssrn.com/abstract=630862 or https://doi.org/10.2139/ssrn.630862

Weaver CM, Dwyer J, Fulgoni VL 3rd, King JC, Leveille GA, MacDonald RS, Ordovas J, Schnakenberg D (2014) Processed foods: contributions to nutrition. Am J Clin Nutr 99(6):1525–1542

Wuyang H, Batte MT, Woods T, Ernst S (2012) Consumer preferences for local production and other value-added label claims for a processed food product. Eur Rev Agric Econ 39(3):489–510. https://doi.org/10.1093/erae/jbr039

Yang Y, Davis GC, Muth MK (2015) Beyond the sticker price: including and excluding time in comparing food prices. Am J Clin Nutr 102(1):165–171

Zhou GH, Zhang WG, Xu XL (2012) China's meat industry revolution: challenges and opportunities for the future. Meat Sci 92(3):188–196. https://doi.org/10.1016/j.meatsci.2012.04.016

Zink DL (1997) The impact of consumer demands and trends on food processing. Emerg Infect Dis 3(4):467–469

12 Thermal Processing and Canning

1 Introduction

1.1 History of Thermal Processing

There are many forms of thermal processing. However, commercial sterilization applied to canning is one of the best known. The aim of this chapter is to introduce commercial canned food manufacturing using "post-fill thermal processing". Here, the food is sealed hermetically inside of cylindrical metal cans or glass jars before heating (Brody 2002, Pflug 2010). Other forms of thermal processing, e.g. heating followed by aseptic packaging, will be mentioned briefly. Home canning is not covered except that the principles described for acidic foods (Sect. 3.3) can be adapted for domestic canning.

In the book *"The Descent of Man, and Selection in Relation to Sex"*, Charles Darwin suggested that fire and language were probably the two most important discoveries in early human history. Darwin speculated that early knowledge of fire arose from natural wildfires and volcanic fires (Darwin, 1872). It is been supposed that human engagement with fire developed in three distinct stages. First, early hominids developed fire foraging (1.0–0.5 million years ago), by searching areas of landscape subjected to wildfires due to lightning and other natural causes. The taste and nutritional quality of heat damaged or cooked food would have been persuasive over time (Gowlett 2016).[1]

A second phase of human engagement with fires is said to have occurred within social and domestic circumstances at about 0.4 MYA. Humans learned to harvest, maintain and protect slowly burning wood, and animal dung to provide access to "non-seasonal" fires. Cooking with fire and corralling prey probably dates from this period. Fires also provided light, protection from nocturnal predators, warmth and the means of cooking.[2] The third-stage in the engagement with fire led to pyrotechnological processes such as, pottery kilns and metalworking; consideration should be given also to the recent development of the steam engine and internal combustion engine (Gowlett 2016). The commercial exploitation of heat for food preservation developed comparatively recently (Fig. 12.1).

[1] Cooking using hot springs or geothermal energy may have predated the use of fire for cooking. See: https://www.economist.com/science-and-technology/2019/05/25/did-cooking-in-hot-springs-make-humans-brainy. Currently, there are many other methods for heating food, without open flames – including, the use of steam, radiative/electromagnetic heating, and electrical methods.

[2] Currently many depictions of prehistoric pyro technology appear to be male dominated.

R. Owusu-Apenten, E. R. Vieira, *Elementary Food Science*, Food Science Text Series,
https://doi.org/10.1007/978-3-030-65433-7_12

Man in the rudest state in which he now exists is the most dominant animal that has ever appeared on the earth. … *He has discovered the art of making fire, by which hard and stringy roots can berendered digestible and poisonous roots or herbs innocuous. This last discovery, probably the greatest, excepting language, ever made by man, dates from before the dawn of history, pg. 132 Decent of Man*

Fig. 12.1 A quote from Darwin's "Descent of man, pg. 132" (Darwin 1872)

1.2 Canning Not as We Know It

The industrial process for destroying foodborne microorganisms and spores by heating is called commercial sterilization or canning. The development of canning is attributed to the French confectioner Nicholas Appert (Barbier 1989). In 1800 Appert developed the process of preserving foods by sealing within glass jars using cork and then heating with boiling water. Peter Durand of England is credited with developing the use of metal containers in 1810,which were fabricated and sealed by hand soldering, prior to heat treatment. In 1819, William James Underwood of the United States started the first civilian canning factory in Baltimore (Goldblith 1971; Graham 1981).[3] Another origin story for canned food is discussed in Chap. 1 (Sect. 2.4). Initially, preserving foods by heating using boiling water took too long, requiring about 6 h, so salt was added to the water bath, which increased the boiling temperature, thereby shortening the processing time. One problem with using salt to raise the process temperature was that saline corroded the cans, so the next innovation was to heat in steam under pressure. The higher the pressure, the higher the temperature at which water boils (see Table 12.1) and the shorter the processing time (Bitting 1912. p8). The invention of the autoclave-retort by A.K. Shriver (1874) is regarded as one of the highlights of canning technology (Pearson 2016).

Table 12.1 Boiling point of water at different atmospheric pressures[a]

Pressure (at sea level)		Boiling temperature/for water	
Psig	g/cm^2	°F	°C
0	0	212.0	100.0
1	70.4	215.4	101.9
2	140.8	218.5	103.6
3	211.2	221.5	105.3
4	281.6	224.4	106.9
5	352.0	227.1	108.4
6	422.4	229.6	109.8
7	492.8	232.3	111.2
8	563.2	234.7	112.6
9	633.6	237.0	113.8
10	704.0	239.4	115.2
11	774.4	241.5	116.4
12	844.8	243.7	117.6
13	915.2	245.8	118.8
14	985.6	247.8	119.9
15	1056.0	249.8	121.0
16	1126.4	251.6	122.0
17	1196.8	253.4	123.0
18	1267.2	255.4	124.1
19	1337.6	257.0	125.0
20	1408.0	258.8	126.0

[a]psig = pound per square inch, guage = 0 for atmospheric pressure or 14.7 psi

Older food technology literature appear to use the term "thermal processing" to mean solely canning (Stevenson 1992). Sealing a food product within a tin container and the subsequent thermal processing at higher than 248 °F (120 °C) were considered inseparable (Awuah et al. 2007, Holdsworth and Simpson 2008). More generally, thermal processing may involve any number of temperature-time treatments including those used for cooking at home (Holdsworth 1985). Nowadays, thermal processing refers to many temperature-time treatments; high-pressure and temperature conditions are used still for commercial canning of low acid foods with a pH above 4.5 (pH >4.5). However, high acid foods such as tomatoes can be treated to milder canning conditions using jars as containers as frequently applied for "home canning". Safety guidelines require that low acid foods should never be subjected to home canning without a pressurized cooker (Anon 1977; Zepp 2019; National Center for Home Food Preservation 2015; Schmutz et al. 2020).

[3]It is believed that the British Admiralty produced canned meat and other ships fare for the Navy, prior to the civilian use of this technology.

1.3 Economic significance of canned food

Canning involves glass jars or cylindrical metal cans mostly, though other shapes and materials can be used (Brody 2002). Also virtually, any food may be canned. The US metal canned food sector is spread across of several NAICS designations; (a) NAICS 311421 - Fruit and vegetable canning, (b) NAICS 311422 - Specialty canning, ready-to-eat foods, (c) NAICS 311711 - Seafood, seafood products, seafood soups, (d) NAICS 311514 - Dairy food canning, condensed milk, evaporated milk (e) NAICS 311612 - Meat and meat product canning and (f) NAICS 311615 - Canned poultry, poultry products.

There are five-large US canned food groups by economic value not all of which correspond with the NAICS designations. For instance, 2016/2017 data showed the following classes and earnings (US$billion): canned specialty foods including soups and ready meals (5.2 US$billion), canned vegetables (3.9 US$billion), canned meat and meat products (2.7 US$billion), canned fruit (2.0 US$billion), canned seafood (2.0 US$billion) and canned "other" food (0.8 US$billion) (Statistica 2019). The US canned food market for 2017 was valued at about US$16 billion (Grand View Research Inc. 2018) compared with a global canned food market worth US$92 billion that is projected to reach US$100.2 billion by 2020 (Fortune Business Insights 2020).

The Can Manufacturers Institute (US) reported a total shipment of cans per year (2019) for beverage (97.2 billion cans), food (25.3 billion cans), vegetables (8.59 billion cans), soups and miscellaneous foods (4.53 billion cans) and fruit (757 million cans). Other canned food (baby food, juices, seafood, meat and poultry) accounted for 3.7 billion cans (Can Manufacturers Institute 2021).

Some key benefits and advantages of canned foods were highlighted by the industry; accessibility, low cost and general affordability. Canned food may be used as part of the recommended dietary pattern for Americans, not to mention the general compatibility with large-scale catering (Can Manufacturers Institute 2021).

The environmental benefits of canned foods were rated highly. For instance, canned food being shelf-stable at room temperature requires no additional energy for preservation. Many canned foods also require less effort or little extra preparation thereby adding to energy savings. Also canning reduces food waste. Finally, aluminium packaging has high recycling ability compared to other forms of packaging (Can Manufacturers Institute 2021).

2 Commercial Sterilization or Canning

2.1 Definitions

The heat treatment associated with canning resembles that used to sterilize hospital equipment (Rutala et al. 2016). However, the thermal process accompanying canning produces a condition of commercial sterility rather than complete sterilization (Heldman and Hartel 1997a).

Commercial sterility is defined by FDA legislation (21 CFR 113.3) as, " *the condition achieved, (i) by the application of heat which renders the food free of, (a) microorganisms capable of reproducing in the food under normal nonrefrigerated conditions of storage and distribution; and (b) viable microorganisms (including spores) of public health significance; or (ii) by the control of water activity and the application of heat, which renders the food free of microorganisms capable of reproducing in the food under normal nonrefrigerated conditions of storage and distribution*" (Govinfo.Gov 2021). The origins of the term commercial sterility has been deliberated in the scientific literature (Pflug 1987, Anderson et al. 2011).

The goal of commercial sterilization is to destroy the spores of pathogenic bacteria. This fact is easy to overlook. Bacterial spores are more resistant than vegetative cells towards heat, chemical sanitizers, UV light and other inactivation agents. The reason bacteria undergo sporulation is in order to lie dormant during harsh conditions and to wait for favorable ones (Brown 2000; Setlow 2006, 2014). If not destroyed by the canning process, any remaining spores will germinate to form actively growing (vegetative) cells that will then form toxins.

Therefore, commercially sterile food has no vegetative (actively dividing) pathogenic bacterial cells, together with a very low *probability* (1-in-a-billion chance) of surviving (pathogen) spores. In addition, spores from highly thermostable organisms may exist within canned food but they will be unable to germinate/ grow under the storage temperature conditions (below 104 °F or 40 °C) recommended for canned food (Myrseth 1985). Spoilage of canned foods by thermophilic microbes is described in Sect. 3.

Food canning is a well-established and mature technology. The principles and equipment needs for canning are contained in many classical texts. (Binsted and Devey 1970; Gutterson 1972; Malik and Dhingra 1975; Roberts 1970). Many early discussions of canning principles appear to be out of print, and/or are difficult to access currently despite efforts to digitize such content recently (Bitting 1912, Lewis 1964). Canning operations for low acid foods (pH above 4.6) and associated GMP are regulated firmly (Govinfo. Gov 2021).

2.2 Structure of the Tin Can

By the early 1900s, machines did the manufacturing and sealing of tin cans (Oldring and Nehring 2007, Brambilla et al. 2016, Robertson 2016). The lid used for the cans contained a rim to which a plastic film gasket could be added, and the rim could be sealed tightly by machine by first crimping it over and under the flanged top of the can body and then pressing the two together by a second roller operation. The plastic gasket made the can hermetically sealed (airtight) by filling the tiny voids produced between the rim of the can cover and the flange of the can as they were brought together. This method is used currently for sealing tin-plate metal containers for canning.

Normally, cans are described in terms of the dimensions of the diameter (D) and the height (H) (Fig. 12.2) and each of the dimensions is given in numbers having three digits each. The first digit is in inches (1 in = 2.54 cm), and the next two digits are in 1/16 of an inch (0.16 cm). Thus, when a can is described as a 202 × 214 can, this means that its diameter is 2 (2/16th) in. (5.40 cm) and its height is 2 (14/16th) in. (7.30 cm). For those cans, which are not round, as in the case of sardine cans, three dimensions are given: the length, the width, and the height.

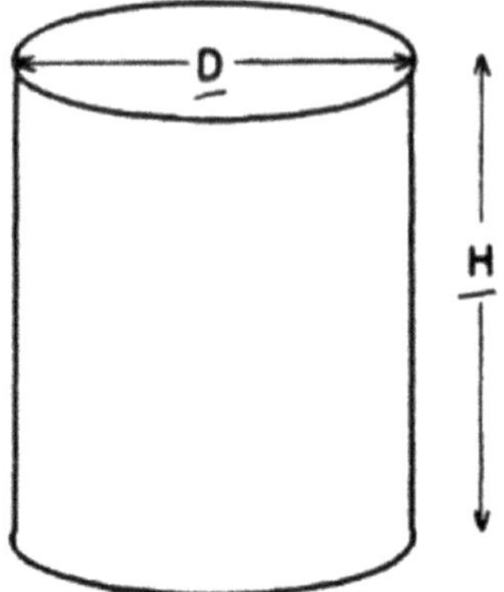

Fig. 12.2 Diameter (D) and height (H) of a metal container

Various lacquers were developed for lining cans, to prevent food discoloration that could occur because of interactions between the food and the can. Other innovations include, quick-opening cans with pull tabs, as are used for soft drinks, sardines, and nuts; cans that are opened with a slotted key that comes with the can, as are used for sardines and cured hams; pressurized cans that use a propellant to dispense whipped cream and cheese spreads; and extruded or drawn cans, as are used for beverages, tuna, and sardines. The extruded can has several advantages over the conventional can, a major one being the elimination of the bottom and side seams, which reduces the probability of seam failures, eliminates the use of lead (used in the side seam), and permits more stable stacking on shelves, requiring somewhat less vertical space.

2.3 Filling the Cans

Food is filled into cans before they are sealed by machine. For instance, with peas, a central hopper, which is kept filled with peas, is located above a rotating plate with openings through which the peas fall. When the plate rotates to the point at which the plate opening is just below the hopper, the opening is just above a container (below the plate), which when filled will hold a certain volume or weight of peas. The peas thus fall down and fill the bottom container, which then rotates to a point where the peas are released into a can. In the case of vegetables, after the food is added to the container, canner's brine is added automatically to fill the can, filling all voids and covering the vegetables. This liquid is usually added whilst hot, and because such foods are not packed under vacuum, the hot liquid provides whatever vacuum will be present in the container after processing and cooling. Canner's brine is held in a reservoir on the filling machine and is heated in this reservoir. As explained later, the temperature of the brine should be held at 170 °F (76.7 °C) or higher.

Semiliquid packs that heat by conduction, such as concentrated pea soup or mushroom soup, may be filled hot and can be filled volumetrically by automatic means. Solid packs such as tuna fish and corned beef may be filled by hand. A small amount of hot liquid (brine or oil) may be added to provide some vacuum, or the product is passed through a steam (exhaust) box to be heated and provided with some vacuum before the can is covered and sealed.

2.3.1 Liquid in Cans

Vegetables and fruits are usually packed in liquid. Canner's brine, a weak solution of sugar and salt, is ordinarily used for vegetables, and sugar solutions that may be as concentrated as 55% sugar or as dilute as 25% sugar are used for fruits. These liquids afford some protection against heat damage because they permit convection heating, which occurs at a faster rate than conduction heating.

2.3.2 Pretreatment of Foods

Foods undergo a pretreatment prior to the canning process, but these pretreatments differ depending on the foods, and no attempt is made to cover all of them here. Some pretreatments are applied to many different foods. One of these, usually applied to vegetables, is blanching (Fig. 12.3).

2.3.3 Vacuum in Cans

Canned foods are sealed under vacuum for several reasons. If canned foods were not under vacuum, the cans would swell should they be stored at higher temperatures or lower pressures than those at which they were packed. Thus, cans packed at sea level at ordinary temperatures would swell due to expansion of gas within the can if the cans were shipped to Denver, Colorado, which because of its high altitude (about 1 mi

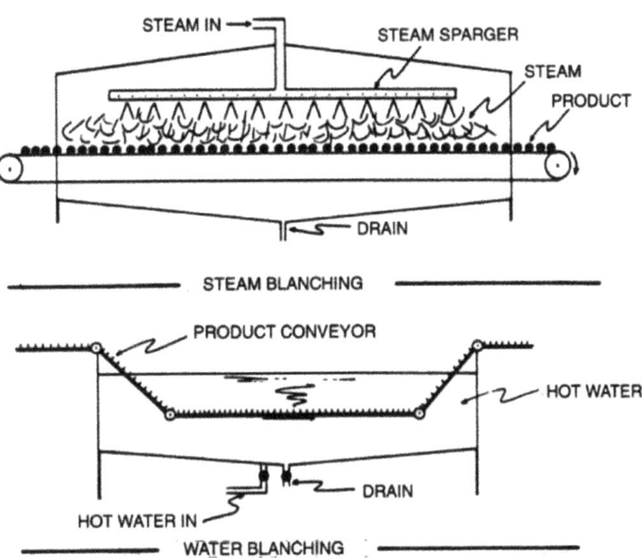

Fig. 12.3 Continuous blanching

[1.61 km] above sea level), is under reduced pressure, or if the cans were shipped to tropical areas or held in unusually hot places. When cans are in a swelled state, they are normally suspected of containing food that has spoiled, because when bacteria grow, many of them produce gases. Thus, a perfectly good canned food might be discarded because it could be suspected of being spoiled. Another reason for canning foods under vacuum is to remove oxygen. During heat processing, oxygen reacts with the food, causing undesirable changes in color and flavor of the product.

Finally, if at least some of the air is not removed from the container prior to sealing, then the product may have to be cooled under pressure (usually air pressure). Other wisc, thc can will buckle (a permanent distortion along the cover seam), and such cans will be discarded. The reason for the buckling can be explained. During the heating process, the retort is under pressure, for example, 20 psi (1.41 kg/cm^2). As the temperature inside the can reaches the high temperature in the retort, moisture in the food or brine vaporizes, exerting pressure against the inside of the can. In addition, the residual air in the can tends to expand as the internal temperature increases, and the expanding air exerts a pressure against the inside of the can. By the end of the process, the can is under equilibrated pressures from both the inside and the outside (Fig. 12.4). In this situation, the can is not stressed, because the internal pressure is counterbalanced by the outside pressure and vice versa.

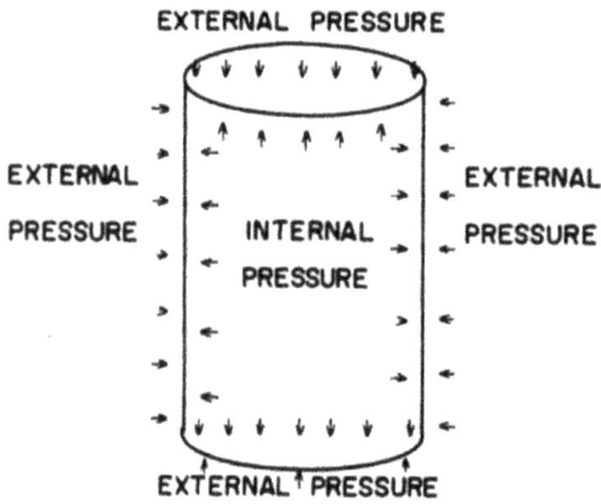

Fig. 12.4 Pressure equilibrium during thermal processing

At the end of the process, the external pressure, relative to the can, may be immediately released by shutting off the steam supply to the retort and opening the retort valve to the outside, allowing all its internal steam to escape. However, the internal pressure in the can cannot be released immediately because the can cannot be vented to the outside. Thus, the wall and lids of the can are under internal pressure, causing distortion, possibly even damage, to the seals (formed by the lids and the can body). This problem can be eliminated by replacing the hot steam in the retort with air under pressure. In this way, the heat source is removed without altering the pressure equilibrium. With the removal of the heat source, the contents of the can will cool down, condensing water vapor and cooling the residual air, returning it to its original volume, thus reducing the internal pressure until it no longer pushes against the inside of the can. Removing as much air as possible from the container prior to processing tends to minimize buckling. However, even with vacuum packing, cans of large diameter must be cooled under pressure to prevent distortion of the container.

Some foods, such as whole kernel corn, may be packed under a very high vacuum in order to provide for good heat penetration, when only a small amount of liquid is used with the foods. When this is done, a special beaded can with protruding ridges around the central part of the can body to strengthen the can may be needed. Otherwise, the container body may panel (become flattened) as the result of the difference in pressure between the inside of the can and atmospheric pressure on the outside.

2.3.4 Obtaining the Vacuum

The vacuum inside of canned foods may be obtained in several ways. One of the most common methods is to add hot food to the container. In this way, the residual air is removed, resulting in a partial vacuum. A second method is to add cold food to the container and to preheat it by passing it through a steam box uncovered, or only partially sealed, prior to sealing. In either of these techniques, the heat causes the product and the air in the headspace to expand, pushing air out of the container. In addition, the water vapor in the headspace displaces air, and trapped air in the food is driven out. In this condition, the can is sealed, and when the product is cooled, it will be under vacuum, as much of the air is removed from the container.

A vacuum in canned foods can also be obtained by subjecting the container to a mechanical vacuum just prior to sealing it in the chamber. A good vacuum can be obtained in this manner, but there are some limitations. For instance, if the product is packed in liquid, such as vegetables packed in canner's brine or fruit packed in syrup, when the vacuum is applied, much of the liquid may be flashed out. This is caused by the dissolved and occluded (trapped bubbles) air in the liquid, which comes out as a gas when a sudden vacuum of high intensity is applied. The sudden release of air causes some of the liquid to spatter out of the can. To avoid this, liquids must be subjected to vacuum treatment prior to filling into the can in order to remove dissolved and occluded air.

The third type of vacuum used in canning foods is called a steam jet vacuum. Just before the cover is placed on the can to be sealed, a jet of steam is forced over the contents of the can. This does not provide a high vacuum and removes only air from the headspace of the food in the can. It is used mainly for materials packed without liquid (Fig. 12.5).

2.4 Sealing the Cans

Cans are sealed automatically by machine. In sealing, the cover falls onto the top of the can automatically; the base plate of the sealer (upon which the can rests) raises the can with cover up tightly against the chuck. The edges of the can cover and the flanged body top are subjected to the action of two different rotating rollers. The first roller crimps the cover and body flange so that the edge of the cover is bent around and under the edge of the body flange; the second roller flattens and presses the top seam together so that it forms a double seam, which is a tight seal and a plastic gasket located in the outer rim of the cover (Fig. 12.6).

In the canning operation, it is imperative that the sealing machines provide a tight seal. This can be checked by removing the cover to observe the configurations of the cover hook and body hook (Fig. 12.7). The dimensions of the seal components may then be measured to ensure that the dimensions fall within certain limits. Tools for exposing the seam components and special micrometers for measuring their dimensions are available. A simple way to determine that the base plate, first operation (first roller), and second operation (second roller) of the sealing machine have been adjusted properly is to fill a can with boiling water, seal, cool, and then measure the vacuum in the can by means of a gauge that reads in inches of vacuum (1 in = 2.54 cm). The vacuum gauge has a sharp shaft that is

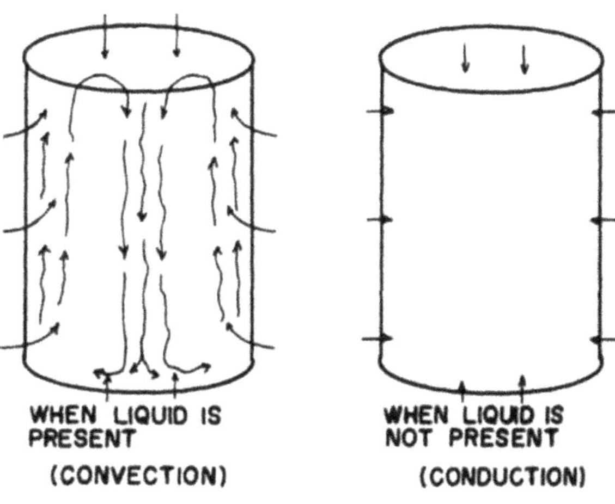

Fig. 12.5 Convection and conduction heating patterns in canned food

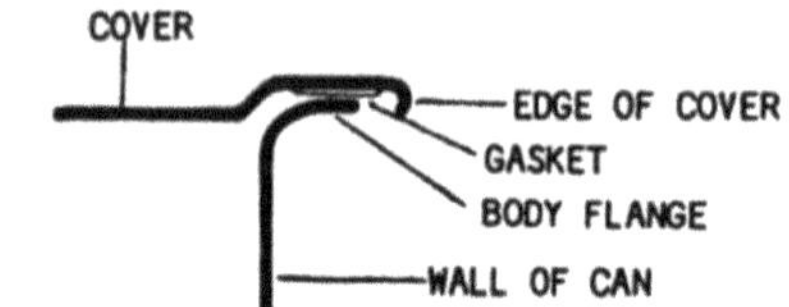

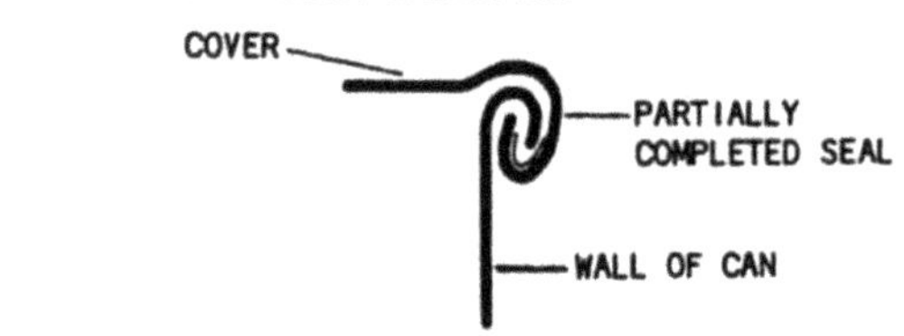

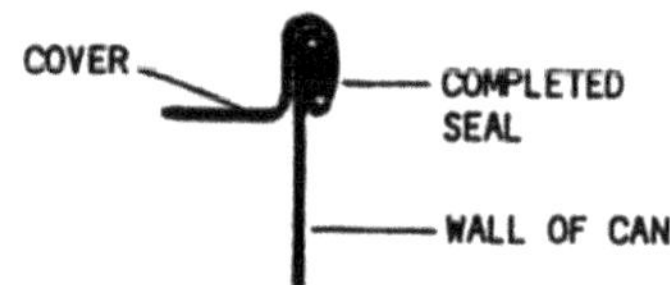

Fig. 12.6 Cross-sections of cover and body of can at three stages of sealing

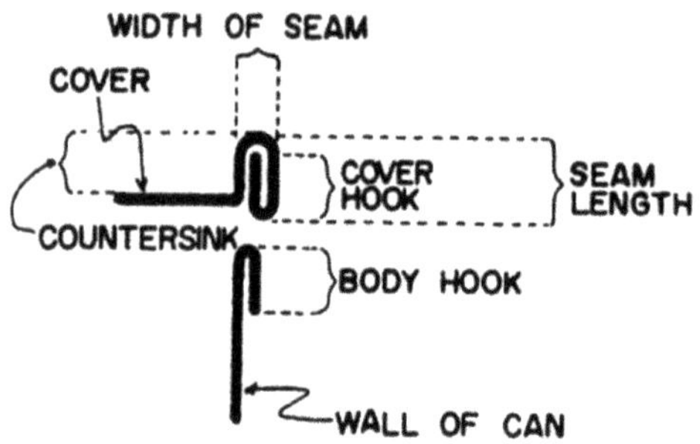

Fig. 12.7 Components of a tin can double seam in exploded view

pushed through the cover of the can, and a rubber gasket prevents air from entering the can. If a vacuum of 10 in (25.4 cm) or more is obtained, it can be assumed that the can is effectively sealed.

The sealing operation is checked at the start of the day's operation and periodically throughout the processing period. The importance of an effective seal cannot be overemphasized owing to the unfavorable economic impact that will certainly result from FDA seizure of product and product recalls from the market or from lawsuits if food poisonings should occur.

2.5 The Heat Process

Heat processing is a key operation in the food canning process (Food and Agriculture organization of the United Nations 1985). All canned foods must be exposed to heating for periods sufficient to ensure the destruction of microorganism that can cause sickness and, in some instances death. The sterilization step for canned foods is critical for the success of the overall process. A great deal of highly detailed preliminary work goes into establishing the precise temperature and time profiles for Appertization, which is the technical term for sterilizing canned foods.[4]

The principles of Appertization require that the death kinetics of a whole range of pathogenic bacteria are determined. In particular the decimal reduction time (D-value), which is the heating time required to reduce the initial population of bacteria by 10-fold, must be found (Adams and Moss 2008, p. 66–76). So the destruction at different times need to be recorded for the most heat

[4]Heat processing to stabilize foods against microbial spoilage was invented by Nicolas Appert, in honour of whom the term Appertization is named.

stable bacteria of concern. The other key parameter for thermal processing is the Z-value (°C), which is the temperature change required to alter the D-value by 10-fold.

Another preliminary consideration is to perform heat-penetration studies for specific configurations of cans; the object is to establish the actual heating profile at the geometric centre of the package in order to establish the time required to reach the set process temperature, allowing for the size, shape of the can, and the type of food filled. Each heat-penetration study will yield a unique temperature-time profile for one type of canned product. It is from the heat-penetration curve that the actual killing (lethality) effect must be established (Sect. 10). The heat penetration study is required by law. Further basic principles are summarized later in this chapter.

3 The Conventional Heat-Processing

3.1 Chamber Retort

The ordinary retort (Fig. 12.8) is the three-crate retort in which one crate of cans is placed on top of another in the retort (Teixeira 2013). When cans are placed in these crates, they may be allowed to fall into the crate haphazardly or they may be placed in the crate on end in an orderly manner. If the latter system is used, perforated metal separators must be placed between the layers of cans. This provides for an adequate circulation of steam around the cans necessary for optimum heat transfer to the product.

All retorts should be fitted with automatic steam valves that can be set to allow steam to flow into them to raise the processing temperatures only to the desired point. Retorts should also have a temperature sensor (inside of the retort) attached to a recorder on the outside and a chart showing temperature and time of processing for each batch of product. Processing charts should be identified with the code on the cans of each batch and should be kept on file. If the product is to be cooled in the retort after processing, an air pressure system with automatic inlet valve that can be set for a definite air pressure should be a part of the installation. In such installations, there should be an automatic pressure outlet valve attached. Otherwise, when cans with high vacuum are cooled, the outside pressure may become high enough to cause paneling of the cans (Teixeira 2013).

All retorts must be fitted with fast-opening valves to allow venting. When the steam is first turned on, the vent valve should be opened wide and left open until the temperature is raised to 220 °F (104 °C). This removes, from the retort, air that would otherwise cause cold spots around some of the cans and thus prevent adequate heating. The reason for this is that air is a poor heat conductor. In addition, during retorting, a small bleeder petcock valve at the top of the retort should be kept in the wide-open position. This removes air that may come in with the steam and provides for good circulation of steam. The requirements for canning low acid foods are legally binding. The legislation also requires a food safety trained and certified personnel in charge of the canning process (Govinfo.Gov 2021).

3.2 Cooling Heat-Processed Foods

Cans of food that have been heat processed may be cooled in the retort by allowing water to flow in after the steam has been turned off, in which case they may be cooled under a pressure of air or steam exerted over the water level in the retort. In other operations, the retort may be blown down and the crates of cans removed and moved slowly through a cooling canal. In either case, the cooling water should be potable (drinkable) and should not have a high bacterial count. The reason for this is that as the vacuum is forming in the can because of the cooling, the gasket in some cans may be soft enough to permit microscopic amounts of the cooling water to be sucked into the cans. If the cooling water is high in bacterial count, enough bacteria, which may cause spoilage or even disease, may thus enter the can and contaminate the product. Thus, it is desirable to chlorinate cooling water so that it contains a residual of 5 ppm available chlorine. This level of

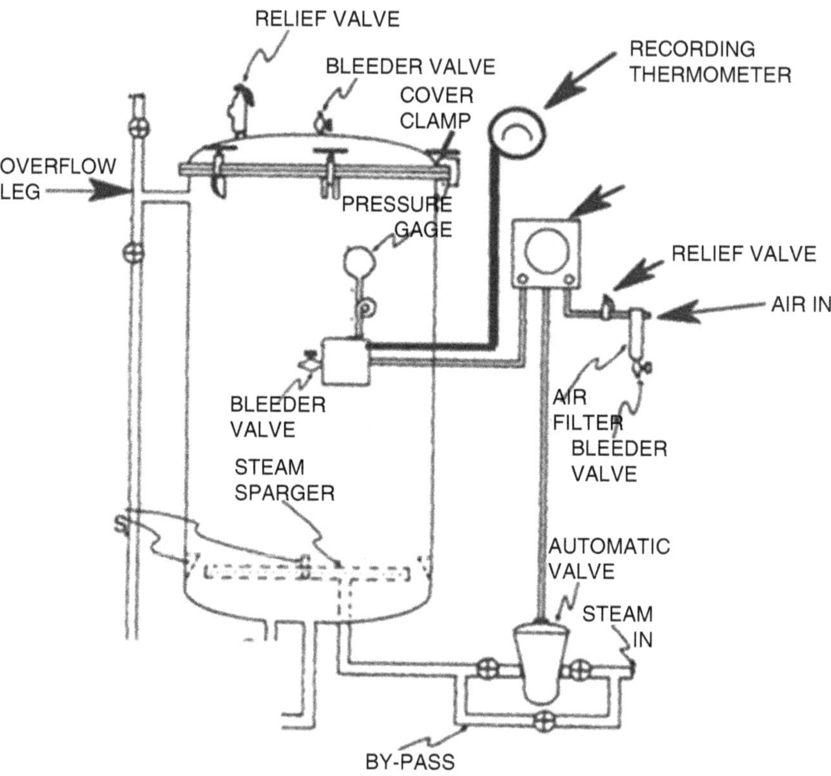

Fig. 12.8 Conventional retort with recording thermometer

chlorine is enough to keep the water relatively free from bacteria.

During cooling, the average temperature of the product should be brought to 95–100 °F (35–43.3 °C) as quickly as possible. The reason for this is twofold. First, if this temperature is maintained, the surface of the can will be warm enough to evaporate moisture remaining from the cold water bath. Should this water not be evaporated, it would cause rusting of the outside of the can, possibly spoiling the label and making the container unsightly, so that it would have to be rejected. Second, if the temperature is not lowered sufficiently, it may promote the growth of residual thermophilic bacteria, which would cause spoilage of the food within the can.

Thermophilic bacteria form spores and many types of thermophilic bacteria are unusually heat-resistant. Two spore forming thermophilic bacteria are mostly credited with spoiling low acid canned vegetables and ready meals stored at temperatures above 98.6 °F (37 °C); the spoilage thermophiles are *Geobacillus stearothermophilus* (formerly *Bacillus stearothermophilus*) and, *Moorella thermoacetica* (André et al. 2013). Therefore an attempt is made to keep as many thermophilic bacteria out of the product as possible. This can be accomplished by preventing the buildup of these bacteria at various points in the canning operation in the plant. The commercial process used today will reduce the number of thermophiles in foods to a minimum and only a few thermophilic spores will be present in the product; these will cause no problems, provided the product is held at temperatures below 110 °F (43.3 °C).

Cans must be washed to remove grease and food particles that get onto the outside of the can during the various procedures involved in the operation. This is done by passing the can through alkaline or detergent solutions, then through rinse water. If done after heat processing, the temperature of the rinsing solutions should be high enough to provide for evaporation of the water, thus preventing corrosion of the outside of the container.

After cooling, cans are usually packed into cases or stored in large masses in a warehouse under conditions in which they cool very slowly to room temperature. To prevent growth of thermophiles in the product prior to canning, canner's brine in the reservoir on the filling machine should be held at 170 °F (76.7 °C) or higher, a temperature at which no growth can take place.

3.3 Canning of Acid Foods

Foods that have a pH of 4.5 or lower are considered acidic foods. Canned acid foods need lower temperature processing to attain commercial sterilization compared with normal foods with a higher pH. The reason for this is that bacteria are more easily destroyed by heat when present in acid solutions. Moreover, spore-forming bacteria generally will not grow in foods having pH values of 4.5 or less. There are some exceptions to this; for instance *Bacillus coagulans* (formerly *Bacillus thermoacidurans*) can grow in tomato juice (maximum pH 4.5) and cause spoilage described as flat sour (Becker and Pederson 1950, York et al. 1975).

Acid foods are ordinarily processed by heating the cans in boiling water until all parts of the product have reached 180–210 °F (82.2–98.9 °C), and then cooling the cans. An exception is tomato juice, which is now often processed by flash heating to 250 °F (121.1 °C), holding at this temperature for 0.7 min, cooling to 200–210 °F (93.3–98.9 °C), filling into pre-sterilized can, sealing, and inverting the can so that the sterilizing effect of the heat (200–210 °F [93.3–98.9 °C]) at that pH will act on the can cover.

A partial listing of foods that have a pH of 4.5 or less, includes apples, apple juice, apricots, blackberries, blueberries, boysenberries, cherries, cherry juice, all citrus fruits and their juices, currants, gooseberries, loganberries, papaya juice, peaches, pears, pickles, pineapple in various forms and pineapple juice, plums, prune juice, raspberries, rhubarb, sauerkraut and sauerkraut juice, strawberries, tomatoes, tomato juice, and youngberries. Home canning acid foods has grown in popularity over recent years, based on the use of wide-mouthed glass Mason jars (National Center for Home Food Preservation 2015, Zepp 2019).

4 Other Methods of Heat Processing

4.1 Continuous Agitating Retort

Some processing retorts for canned foods are different from the conventional retort (Teixeira 2013). In the agitating retort, cans enter the retort continuously on a conveyor through a special inlet that prevents loss of steam. The cans are conveyed back and forth for the period of time necessary for sterilization, and also rotated around their long axes. Except for solid-packed foods, agitation of the product speeds up heat penetration. Cans exit from the continuous agitating retort through a special valve and enter the cooling system, which is set up much in the manner of the retort except that it is filled with cooling water.

In an "agitort" the cans or containers of cans are attached to a wheel that rotates during processing. Cans are rotated end over end so the air in the headspace travels along the sides of the can to mix the food, and back again (see Fig. 12.9).

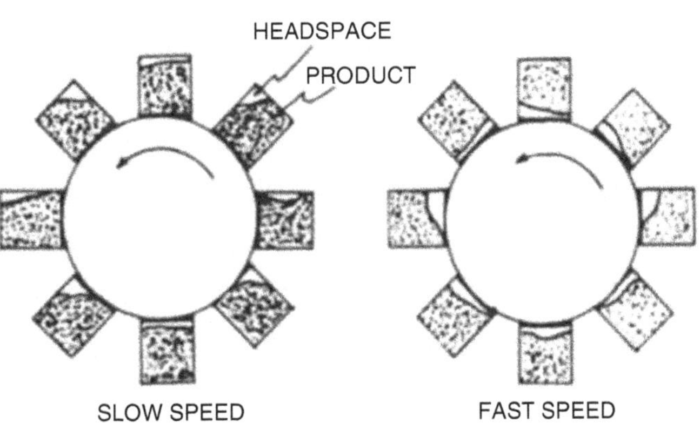

Fig. 12.9 Headspace patterns in fast-speed and slow-speed agitorts

The agitation of the contents hastens heat transfer even in semisolid packs. In this system, the cans are cooled within the agitort. Because the process depends on the rate at which the cans rotate, special equipment has been devised to count the rotations the cans undergo during the process.

4.2 The Hydrostatic Cooker

Hydrostatic cookers are used in Europe and to some extent in the United States (Fig. 12.10). In this system, there is a central chamber with a narrow entrance on one side and an exit chamber on the other side. The system is partially filled with water, and when steam is turned on in the central chamber, the water is forced up to higher levels in the entrance and exit legs (chambers).

Cans enter through the warm water of the entrance chamber on a conveyor, gradually entering warmer water as they approach the steam chamber. They are conveyed through the steam chamber at temperatures and for times that provide for commercial sterilization of the product. The cans are then conveyed out of the steam chamber, first through the warm water and eventually through the cold water of the exit chamber. The cans may be rotated around the long axis as they are carried along. Because the pressure in the hydrostatic cooker depends on the water head (height of water), this type of cooker is quite tall (about 40 ft [12.2 m]).

The agitort and the hydrostatic cooker are somewhat faster than other heat-processing methods, both because of the agitation that speeds up heat penetration and because in such conditions somewhat higher temperatures (up to 270 °F [132.2 °C]) may be used without causing excessive heat damage to the product. This allows for the canning of some semiliquid foods, such as mushroom soup or cream-style corn, in large cans. Because the labor involved and the cost of the container per unit weight of food are smaller as the can size is increased, it is economical to can foods in larger containers. This is desirable for institu-

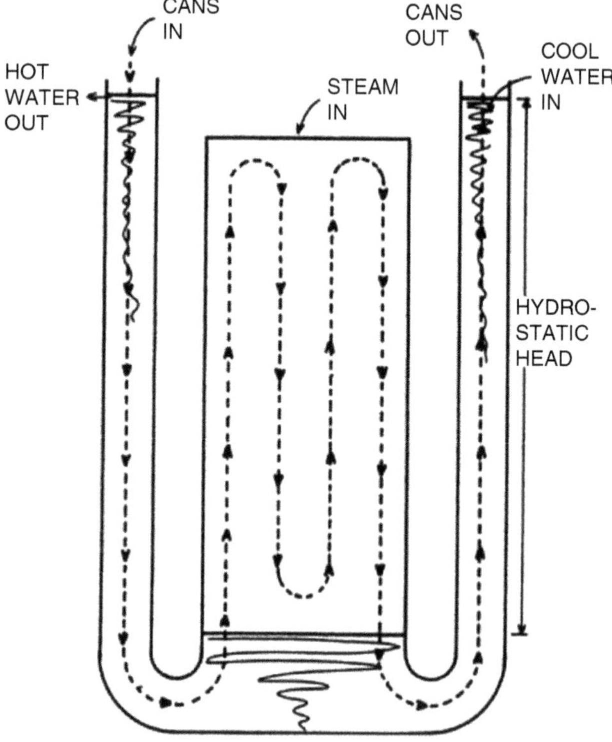

Fig. 12.10 Hydrostatic retort

tions feeding large numbers of people, because it costs them less to buy and handle large cans.

4.3 Aseptic Fill Method

The aseptic process was deemed by the Institute of Food Technologists to be the most important food science innovation during the 50-year period 1939 to 1989 (Brody 1971). In one system, the pureed material or liquid is passed in thin layers through a heating system wherein the temperature of the product is raised quickly. The product is then pumped to a holding chamber where it is held at high temperatures for 10–20 s, then pumped in thin layers through a cooling system wherein the temperature of the product is lowered quickly. It is next pumped to a filling system where it is forced into pre-sterilized containers, and the cans are sealed with pre-sterilized covers. Such systems can be pre-sterilized with very high temperature gas or steam before the operation is started. Flexible or plastic packages that will not stand the high temperatures of sterilization may be pre-sterilized by the use of lower temperatures in combination with chemicals or ultraviolet light. Hydrogen peroxide is often used in this process. The filling and sealing of the containers must be carried out under sterile conditions.

4.4 Cooking Under Pressure

Another method for sterilizing canned food is using pressure chambers. As the pressure of the air in a chamber is raised above atmospheric pressure, the temperature at which water boils is raised; and if suitable pressures are used, temperatures of 260 °F (127.0 °C) or higher may be attained. In these conditions, foods that contain discrete particles may be heated in open steam kettles in the pressure chamber for sufficient time to allow sterilization at the center of the food particles, then filled into cans and sealed. The cans are then inverted, allowed to stand for a few minutes, then cooled. It is not necessary to pre-sterilize the cans in this case because the high temperature of the product, when filled, will destroy whatever bacteria may be adhering to the inside. The containers are inverted to ensure contact between the bacteria adhering to the inside of the cover and the very hot product that would destroy them. Personnel working in chambers under pressure may need to be pressurized slowly in locks when entering and depressurized slowly in locks when leaving. Otherwise, they may be subject to the bends, a condition caused by nitrogen dissolving in the blood when humans are subjected to high pressures, and coming out as gas bubbles when the pressure is suddenly released. The "flash 18" canning process, which avoids overcooking and adverse flavour changes, is now being used for products such as chicken salad (Teixeira 2013; Potter 2013, p. 198).

4.5 The Sous Vide Process

This process was developed in France and used in various European countries. *Sous vide* is the French term for *under vacuum* and whenever food is packaged under vacuum, the potential for the growth of *Cl. botulinum* must be of primary concern. By this process, foods are packed in vacuum and heat-processed but not enough to provide for the destruction of botulinum bacteria. The products, reputed to be of superior quality, must be held under refrigeration until prepared for consumption; otherwise surviving bacterial spores can vegetate and produce toxin.

The U.S. Food and Drug Administration (FDA) describes sous vide packaging as meaning, "raw or partially cooked food is vacuum packaged in an impermeable bag, cooked in the bag, rapidly chilled, and refrigerated at temperatures that inhibit the growth of psychrotrophic pathogens" (US Food Code 2017, p. 18). Alternatively, sous vide process is where " raw or partially cooked food is placed in a hermetically sealed, impermeable bag, cooked in the bag, rapidly chilled, and refrigerated at temperatures that inhibit the growth of psychrotrophic pathogens" (US Food Code 2017, p. 625). The complete evacuation of the sous vide package seems not to be mandatory. Nevertheless sous vide products are grouped alongside of other "reduced oxygen products" (ROP) from a food safety viewpoint. Only licensed food, establishments are allowed to use the sous vide process that must be under a verified HACCP system. Like the cook/chill process and other ROP, sous vide food must be stored refrigerated at temperatures at or below 38 °F (3.3 °C) to avoid sporulation, growth and toxin production by *C. botulinum* (US food code 2017, p. 473). Sous vide products should be pasteurized adequately at 165 °F (74 °C) for 2 minutes before use (U.S. Food and Drug Administration 2017). In Europe, sous vide (under Vacume) refers to a method of cooking requiring an evacuated bag, and cooking under controlled low temperatures conditions using a water-bath at 131–194 °F (55–90 °C) for long periods (1–7hrs). The safe handling of sous vide and other ROP are discussed further in Chap. 13. In Europe, sous vide products must be consumed within 2 hours of cooking (Zavadlav, et al. 2020). Refrigerated storage requirements for sous vide and other food with reduced oxygen packaging are discussed further in Chap. 13.

4.6 Microwave Processing

Microwave energy, generated by a magnetron, is electromagnetic radiation with a frequency of 915 or 2450 megahertz. Both frequencies are authorized by the federal government for use with microwave ovens (Brody 2011; Tang 2015). Because water molecules are polar (have positive and negative ends), they oscillate as they align alternately between the positive and negative charges of the microwave energy. When frozen foods are exposed to microwave energy, the liquid water in them (all frozen foods contain some water in the liquid state) begins to heat due to the friction created by high-speed oscillations of the water molecules. Most of the water in frozen foods is in the ice state and is not affected by the microwave energy, until heat generated by the liquid water molecules is conducted to the ice, melting it, and thereby producing more liquid water to become involved in the heating process.

Applications of microwave devices for food manufacturing, foodservice and for domestic applications are well documented (Goldblith and Decareau 1973 (Abstract), Sale 1976, Chandrasekaran et al 2013, McHugh 2016, Hedeen et al. 2016, Kalla and Deveraju 2017, Guzik et al. 2021). The first uses of microwave energy in industry occurred during the 1970s when cooperative experiments between the Gloucester Laboratory, a USDC facility, and the Raytheon Co. resulted in the introduction of a continuous microwave tunnel in the seafood industry for thawing or tempering of shrimp and fish blocks (Bezanson et al. 1973). In tempering, the product is not thawed, yet lends itself to further processing such as cutting or coating with a breading.

Those early experiments also tested the feasibility of using microwave energy for shucking oysters. The innovation soon spread to the meat industry, and now microwave processing is extended to other applications including baking, pasteurization, and sterilization. Because the highest temperature attainable in conventional microwave units is 212 °F (100 °C), it is necessary to pressurize the processing chamber of the microwave unit in order to achieve the high temperature required to effect sterilization in a reasonable time period.

Microwaves have many uses in the food industry (Brody 2011, McHugh and Milczarek 2019) but special attention will be drawn towards canning applications. Two commercially viable microwave processing units, that satisfy (FDA) regulatory requirements for safe thermal processing, are known to be developed at the Washington State University; these are the microwave assisted thermal pasteurization system (MAPS) and microwave assisted thermal sterilization (MATS) (Tang 2015, Tang et al. 2018).

The operating conditions for MATS were proven adequate for inactivating *C. botulinum* spores. Since temperature-time profiles and lethality calculations are mandatory FDA requirements for safe thermal process, engineers had to develop new ways to measure the temperature at the slowest heating point for food packs during microwave treatment. There are expected to be many applications for the MATS and MAPS systems in the ready-to-eat food sector (Tang 2015, Tang et al. 2018). The scientific principles behind microwave sterilization and pasteurization have been described in detail (Ahmed and Ramaswamy 2007).

5 Containers

5.1 Cans

The speed with which cans are handled during processing and the structural protection they lend to the contents may make them still the most desirable overall container for heat-sterilized foods but a range of other packaging options have emerged – reviewed by Brody (Brody 2002).

5.2 Flexible Pouches

The materials used in pouches are composite laminates and quite strong. They include an outer layer of polyester, which has nearly the tensile strength of steel and resists wear and tear, an inner layer of nylon-11 or polypropylene that resists wear and tear and seals well, and aluminum foil as a middle layer for eliminating light from the contents as well as for making the pouch material impermeable to gases. The layers are bonded together with an adhesive. While flexible pouches can be processed in conventional retorts, special systems have been developed; in most cases, the systems are continuous (Table 12.2).

Table 12.2 Advantages and disadvantages of flexible pouches compared to metal or glass containers

1. Their shape adapts to available space.
2. They are lighter in weight.
3. When empty, they require less space.
4. When filled, their thickness dimension is small, permitting faster processing times and mitigating heat damage to foods.
5. Heat penetration is greater.
6. They are not subject to corrosion or breakage.
7. They are easier to open.
Flexible pouches have a few disadvantages:
1. Pouches cannot be filled as rapidly.
2. They are awkward to handle (e.g., stacking), especially the larger ones.
3. They give no structural support to fragile contents.
4. They are somewhat vulnerable to tearing and cutting.

5.3 Glass Containers

As a container material, glass has several desirable properties. It permits visibility of the product, imparts no taste, and is non-corrosive. The use of glass containers precedes by centuries the use of metal cans and, of course, flexible pouches. The advances in closures of glass containers are probably more dramatic than the advances in the design of the containers and, in fact, have forced the changes in container design. Cork closures have been perhaps the most important.

Cork is light, compressible, hermetically sealable, inexpensive, plentiful, and stable. Corks have long been in use for closing glass bottles, and still are. Ordinary cork closures (stoppers), however, are unable to withstand the buildup of internal pressure, except where the pressure is not too great and when they are used with bottles having small openings, such as in wine and champagne bottles. The diameters of the openings of glass jars used for preserving fruit and vegetables are so large that a fitted cork would be easily pushed out (for any given pressure, the total force increases as the area of the opening of the glass container increases). Even for small openings, it was necessary to develop hold-down devices, such as clamps and wire fasteners, when high pressures were expected, as in the bottling of beer. The crimped metal crown with cork liner was an effective closure for beer and soda bottles. However, aluminum screw-caps are now used in nearly all cases. Variations of screw-top, widemouth jars evolved to the modern Mason jar (named after its inventor). Tight-fitting, screw-type closures find a wide variety of uses in modern glass containers for commercially pasteurized and heat-processed foods as well As in home-canning operations.

However, it was not until more advanced jar closures were developed that canning in glass became significant. These closures include the

Phoenix cap, the Sure Seal cap, and the Vacuum Side Seal cap, types found in many tumbler-sized tapered jars such as those used for cheese spreads and jellies; and the Amerseal cap, a modified screw type cap such as found in capping jars for apple sauce, jellies, and wide-mouth fruit juice bottles.

Screw-type and pressed/pry-off-type lids and caps are currently in widespread use. Also, most containers are now tamperproof or to show when they have undergone tampering.

5.4 Microwavable Containers

The ease and speed with which microwavable entrees can be heated and served will accelerate an already growing demand for them. Not only are microwave ovens found in nearly every home, but they are also available in nearly every work place for use by employees. Packaging for microwavable meals includes trays made of high-gas barrier materials (such as polypropylene bonded to an ethylene vinyl alcohol polymer) that satisfy the criteria for such containers: (1) microwavable, (2) lightweight, (3) unbreakable, (4) easy to open, and (5) re-closable.

5.5 Plastic, Flexible Containers

The advances in aseptic packaging have allowed packing of heat-processed foods such as ketchup in "squeezable" bottles that are lighter, unbreakable, and very convenient in dispensing the product.

6 Warehouse Storage of Canned Foods

When canned foods have been heat-processed and cooled and the cans have been cleaned and dried, either they are stored in warehouses in bulk until labeled, cased, and shipped out, or they are labeled, cased, and stored in a warehouse until shipped out. Lithographed cans may be used, in which case labeling is not necessary.

Warehouses should be maintained so that their temperature does not rise much above 85 °F (29.4 °C) or fall below 50 °F (10 °C). Very high temperatures may promote the growth of thermophilic bacterial spores present in small numbers in the food. Very low temperatures may lower the temperature of the cans to the point that, should a sudden hot spell occur, cans will sweat (condense moisture), eventually causing external corrosion of the cans.

7 FDA Regulations and Safety

Good Manufacturing Practices (GMP) are approved by the FDA for both low-acid foods [those foods with a pH greater than 4.6 and a water activity (Aw) of greater than 0.85] and acidified foods. Use of the GMPs will help assure food safety and wholesomeness.

The Emergency Permit Control regulations help ensure safety by requiring all processors to register their production plants and methods with the FDA (Avena-Bustillos 2011). The manufacturers must also keep detailed records of all production and have them available for FDA inspection. These records are so very important if a problem is discovered and a recall of all potentially dangerous products must be made. The purpose of all these regulations is to ensure the destruction of all *Cl. botulinum* organisms and ensure that all canned food is safe for human consumption.

In foods for which a vacuum is not achieved and *Cl. botulinum* is not the target organism, pasteurization is used to eliminate pathogens or heat-resistant spoilage organisms. Some species of *Listeria* and *Salmonella* have been found to be more heat resistant than once thought. Use of the D-values for these organisms is crucial in developing new pasteurization times and temperatures that are effective in their destruction.

8 Basic Principles and Examples of Thermal Processing

8.1 Temperature and Heat Relations

Some basic principles behind heat processing (e.g. commercial sterilization/canning, pasteurization, and cooking) are presented in this section. The material is intended to be accessible for anyone with a working understanding of basic mathematics.

It is well to start with the nature of heat. In general, heat can be described as the total motions or kinetic energy of molecules making up an object. Heat is formally defined as a *form energy, which transfers between objects that have differences in temperature* (Henderson 2016). For instance, the molecules of water can be agitated *directly* using microwaves leading to a rising heat content. The molecules of a solid might vibrate about a fixed position, or undergo stretching and bending motions. The kinetic energy of gases and fluids derives from the velocity of constituent molecules, as they zip about within a container. A deficiency of heat is termed "cold".

The relations of heat and temperature can be described by analogy to some common electrical appliances such as a toasters or domestic kettles. A phenomenon called Joule heating occurs within a toaster when electric current (I) flows along a special piece of wiring "heating element" possessing a resistance (R). The magnitude of Joule heating is related to the size of the current and resistance, where, q* represents the heat energy, t = time, and $(V_A - V_B)$ is the electrical potential difference (Volts; V) across a resistive element. The electrical power rating, P (Watts) = q*/ t and consequently 12.1 applies.

$$P = I^2R \quad \text{or} \quad P = I \times (V_A - V_B) \tag{12.1}$$

Temperature and heat relate in the same way that voltage and current are also inter-related. It is found that current flows between any two points with unequal voltage.[5] By analogy, heat will transfer from a high temperature object to a low temperature object, if these are linked by a path with a low thermal resistance (or in other words) high *thermal conductivity* (Table 12.3).

Also, adding a fixed quantity of heat to an object – will produce a rise in its temperature. Contrary to one historic believe, the transfer of heat within a solid does not require a net movement of materials within the conducting material.[6] With the popular toy called Newton's cradle kinetic energy (from one suspended metal ball) is transferred without a net movement of adjacent spheres. Heat energy transfers by "*collisions*" between adjacent atoms but there is no requirement for the atoms making up the container, to enter the food.

Table 12.3 Thermal physical properties of some materials

Material	Thermal conductivity (W/m °K)	Specific heat (J/kg °K)
Aluminum (pure)	205	870
Water	0.59	4182
Ice	2.22	2093
Wood (oak)	0.17	2000
Copper	395	397
Steel	22–25	554
Silver	418	236
Tin	62.1	217
Iron	55	484
Plastic (PVC)	0.14–0.28	880–1170
Cardboard	0.048–0.064	1700

Some values adapted from National Physical Laboratory (2018). Values are at ~100–120 °C

[5] Electrical capacitors being an exception. Current flow generally occurs along metal wiring but lightening shows that electricity can flow through air.

[6] During cconvective heat transfer through fluid, the bulk material flows thereby assisting heat transfer. However, this is different from the classical caloric theory that considers heat as a "caloric fluid" that flows from one object to another.

Two otherwise identical materials, which differ in terms of temperature (ΔT) will also differ in many *physical* attributes including the net kinetic energy possessed by their molecules (heat content), volume, pressure, viscosity, texture, tensile strength, stretching, electrical resistance etc. Two bulk systems at unequal temperatures will differ also in terms of the rates of any chemical changes, taking place within them. A hotter object will deteriorate or degrade faster than a colder one. Comparing liquid water with ice illustrates how temperature affects the physical properties of materials.[7]

8.2 Heat Capacity

The relation between heat and temperature for otherwise identical objects depends on their material composition.[8] The ratio of heat quantity (q*) added compare to the temperature change (ΔT) is described as the "specific heat" capacity or specific heat (Cp).

By international agreement the quantity of heat needed to raise the temperature of 1-gram of water by 1 °C (the specifc heat of water) is called the *calorie*. The heat capacity for water is often reported also as 4.2 (Joule/°K/g). In other words a calorie = 4.2 (Joule).

$$\mathrm{Cp} = \mathrm{q}^* / \Delta T. \quad (12.2)$$

The *kilocalorie* (Calorie or 1000 calorie) is the quantity of heat required to raise the temperature of 1000 g (35.3 Ounces) of water by 1 °C.[9]

According to the preceding, a hot piece of metal has grater heat energy compared to a cold piece of metal. Cooling involves thermal energy loss. Moreover, hot tea possess greater heat energy compared to iced tea, as indicated by their respective temperatures. Fortunately, consuming hot beverage does not contribute to greater weight gain compared to consuming a cooled beverage. The human body and other living organisms utilize chemical energy rather than mechanical energy for repair and growth. Some typical values for the specific heat of different materials are given in Table 12.3.

8.3 Heat Transfer

Thermal processing involves the transfer of energy from a heat source (e.g., cooker, oven, fryer etc) to the food. Processes such as cooking, pasteurization, Appertization (commercial sterilization), food drying, refrigeration, freezing, microwave heating and the plate heat exchange are all dependent on heat transfer.

The preeminent goal for thermal processing is ensuring food safety and preservation (shelf-life extension). Thermal processing can also improve nutritional properties by inactivating enzyme inhibitors, plant toxins and other anti-nutritional factors in foods. Heating is found to improve the sensory characteristics of many foods. Heat energy removal is required also for manufacturing ice cream and other frozen desserts.

The heating medium (air, water, or cooking oil) during cooking affects the rate of temperature increase. The form of packaging has an obvious impact on the rate of heating. Conductive materials such as the tin-plate allows more effective heating as compared with cardboard and plastic, which are better insulators (Table 12.3).

Heating and cooling are based on the same scientific principles. However, heat transfer leading to a rise of temperature is treated as positive algebraically. During cooling (e.g. refrigeration) heat transfer is assigned a negative algebraic sign. As discussed above, heat transfer from a high to low temperature object occurs via one of three modes: conduction, convection and radiative thermal energy transfer (Fig. 12.5). Readers should consult one of the following food engineering texts for more details about heat transfer (Earle and Earle 1983; Heldman and Hartel 1997b; Singh and Heldman 2008). The following

[7]Heat transfer produces large changes in intermolecular forces and phenomena like vaporization/condensation and also melting/freezing. The KE possessed by particles is proportional to their temperature, Energy = kT (where k is called by chemists the Boltzmann constant).

[8]For simplicity, consider heat flow between two objects with identical composition.

[9]Notice this is the "Calorie" with a capital – C; therefore one kilocalorie = 1000 calorie = 1 Calorie (Cal)

section contains a short summary of the main forms of heat transfer.

8.3.1 Conduction

Conductive heat energy transfer occurs when molecules in a material vibrate against one another, and the vibrations are transferred through the material. The rate of heat conduction decreases in the order solid > liquids > gases due to the lower density of molecules in a gas (density is mass divided by volume). The rate of heat conduction through different materials (e.g. stainless steel, copper, or plastic packaging) is of practical importance if one wants to achieve efficient thermal processing on the one hand, or efficient cooling. The rate of conductive heat transfer is described by the Fourier relation (12.3);

$$Q = k\,A\,\Delta T / dX \qquad (12.3)$$

where Q is the quantity of heat transferred per unit time ($Q = q^*/t$) or Watt (W), k = Thermal conductivity of the material (W/ m °K), A = heat transfer area (m^2), ΔT = temperature change across the material and dx = material thickness (m).

From Table 12.3, the thermal conductivity *constant* (k) for copper and tin is about an order of magnitude greater than steel which allows for a more efficient heat transfer in the former (National Physical Laboratory 2018). Therefore, a copper-bottomed cooking pot will heat up more rapidly compared with a stainless steel cooking pot. A high thermal conductivity is also important for chilling food products.

8.3.2 Convection

Heat energy transfer by convection occurs due to bulk movement of a heated fluid, e.g. gas or liquid. During *natural convection,* a fluid closer to the heat source becomes less dense, rises and is replaced by cold fluid under of the influence of gravity (Fig. 12.5). In f*orced convection*, a mechanical device (pump or fan) is used facilitate fluid movement thereby increasing the efficiency of heat transfer. The rate of heat transfer by convection from a flat surface to a fluid is expressed by Newton's law of cooling.

$$Q = h\,A\;\Delta T \qquad (12.4)$$

where, h = the convection coefficient (W/ m^2 °K) and other symbols are as defined previously. As noted above, convective heat transfer occurs by bulk movement. From equation (12.4), the value of h changes with conditions such as, the degree of turbulence, geometry or shape of the container, type of fluid, and fluid velocity. Therefore, convective heat transfer can be improved by mixing. The temperature-time profile for convective heat transfer must be determined for each particular system.

8.3.3 Radiation Heat Transfer

During radiative heat transfer, thermal energy is carried by electromagnetic particle- waves through space. Radiative heat transfer is quite common, because this is how the suns heat energy reaches our planet. In addition to visible light, the sun radiates a great deal of other particles with a wavelength and energy outside of the those (wavelength 400–700 nm) seen by the human eye (violet, indigo, blue, green, yellow, orange, red). The visible light spectrum is part of a wider electromagnetic spectrum. The particle/waves with larger wavelengths (lower energies) compared to red light are infrared waves and then microwaves (Feng et al. 2012; Krishnamurthy et al. 2008) (Sect. 4.6).

8.3.4 Heat Transfer in Real Processing Situations

In many actual processing situations, heat transfer may involve an admixture of conduction, convection and radiation processes rather than a single mode. In practical situations, heat transfer can be summarized by the relation;

$$Q = \Phi\,A\;\Delta T \qquad (12.5)$$

where Φ = overall heat transfer coefficient (W/ m^2 °K) and other symbols are as defined before. This relation looks fairly similar to the Fourier and Newton equations. However, the heat transfer constant Φ should be considered a kind of hybrid constant possessing some of the characteristics of k or h. Engineers consider that various elements in the heat transfer system present their own resistance to heat transfer that are combined. To allow for the complicated nature of heat transfer it is necessary to determine heat transfer characteristics directly using highly robust thermometers, temperature probes or thermistors fixed to tin-plate cans or other food containers (Sect. 10.4 and Fig. 12.14).

8.3.5 Forms of Heating

Traditional forms of heat generation involve the combustion of fuels, such as wood, charcoal and fossil fuels (gasoline, petroleum oil, coal and natural gas). Electromagnetic radiation (microwave and infrared heating) is also used for cooking. Microwave and infra-red heating involves the excitation of water molecules, thereby increasing their vibrations within the food and resulting in a temperature rise.

8.4 Cooking and Cooking Methods

8.4.1 Cooking methods

Cooking refers to a collection of thermal processes involved in the preparation of foods for consumption (Brown 2007). The majority of cooking methods involve either dry heating or moist heating (Table 12.4). Different cooking methods form the basis of different cuisines. Aesthetic appeal provides a rational for choosing one method for cooking over another.

Irrespective of the cooking method, it is essential to ensure that the food interior reaches a *minimum internal temperature* required to inactivate pathogenic bacteria of public health concern. This requirement for minimum internal temperatures is enforceable legally. Minimal internal temperatures for safe cooking have been tabulated for foods that require time-temperature controls for food safety (cf. Chap. 9, Sect. 2.5; Chap. 17, Sect. 2.5).

Food prepared by different cooking methods (Table 8.4) experience varying time-temperature histories prior to serving. The rate of temperature rise, the maximum temperature and overall cooking time may be different. The heat transfer media (oil, air, water) will be different also where cooking is by boiling, baking, deep fat frying, shallow frying, stir-frying etc. The severity of heat treatment for any cooking method can be evaluated based on the time-temperature profile record. The "Cook Value" is a parameter for estimating the severity of thermal processing during cooking (Sect. 10.6).

Table 12.4 Common cooking methods involving moist or dry heat

Moist heating	Dry heating
Blanching	Baking
Boiling	Roasting
Braising	Grilling
Poaching	Deep frying
Scalding	Pan frying
Simmering	Sweating
Steaming	Basting
Parboiling	Dry roasting
	Barbecue
	Stir frying
	Torching, flambe

Adapted from Brown (2007)

8.4.2 Cooking and food quality

Some early discussions of the science of cooking from a culinary perspective appeared nearly 40 years ago (Robins 1980). Interesting studies covering cooking methods for dried beans and vegetables were completed in the 1920s (Masters 1918; Masters and Garbutt 1920). Earlier investigations by the FDA were mentioned in secondary sources later (Masters 1918; Masters and Garbutt 1920). The effect of cooking on digestibility and other bulk food properties were reported in the 1930s (Clifford 1930; Sinoda and Kodera 1932; Sinoda et al. 1931; Tinkler 1931; Tinkler and Soar 1920). The effect of cooking on cooking losses, palatability and nutritive changes appeared between 1940 and 1970s (Batchelder 1943; Fenton et al. 1943; Cross et al. 1976).

8.4.3 Nutritional effects of cooking

Classic investigations into the effect of different cooking methods on nutrients losses occurred in the 1940s (Stewart 1946), with attention focused on water soluble vitamins particularly vitamin C (Allen and Burgess 1950; Holmes 1958; Kylen et al. 1961; Pasricha 1967), Vitamin B6 (Lushbough et al. 1959) and iodine (Goindi et al. 1995); generalized cooking and storage effects were also studied (Rumm-Kreuter and Demmel

1990; Severi et al. 1997). Cooking foods from frozen was recommended for preventing vitamin loss (Nursal and Yucecan 2000). The scientific basis of cooking was covered (McGee 2007), along with molecular gastronomy and cuisine (Cassi 2011). The degree of nutrient retention following various cooking methods are available on-line. cf. *USDA Table of Nutrient Retention Factors Release 6*. https://www.ars.usda.gov/ARSUserFiles/80400525/Data/retn/retn06.pdf.

The focus on cooking studies turned to health during the 1980s. There was concern that highly digestible starches could contribute to diabetes risk and studies evaluated glycemic index of different foods (Bornet et al. 1990; Collings et al. 1981). The potential effect of high temperature cooked (grilled, fried, roated, barbequed) meat on the risk of pancreatic, breast, and colorectal cancer was noted and linked with the formation of substances called heterocyclic amines and polycyclic aromatic hydrocarbons (Anderson et al. 2002; Santarelli et al. 2008; Sinha et al. 2005).

Fumes from stir frying meat were flagged up as a cancer risk also apparently linked with the inhalation of heterocyclic amines in cooking fumes. Evidence from one case control study suggested that daily exposure to stir-frying fumes raised the risk of adenocarcinoma by 4–5-fold compared with controls. There was significantly greater risks for cooking fume related cancer for subjects already noted as smokers compared with non-smokers (Seow et al. 2000).

Table 12.5 Decimal reduction time for some vegetative microorganism found in foods

Vegetative organisms	D_{65} (min)
Mycobacterium tuberculosis	0.8
Salmonella sp.	0.02–0.25
Salmonella senftenberg	0.8–1.0
Staphylococcus aureus	0.2–2.0
Escherichia coli	0.1
Yeasts and molds	0.5–3.0
Listeria monocytogenes (D_{60})	5.0–8.3
Campylobacter jejuni (D_{55})	1.1

D-values are for 65 °C (149 °F) unless otherwise indicated, Z-value = 5.1 °C

Adapted from Adams and Moss (2008) and Sung and Collins (1998)

Table 12.6 Examples of pasteurized liquid foods

Beer and other beverages
Cold blended chilled dressings and chutneys
Fruit juice
Glazes and drizzles
Human milk
Liquid egg
Marinades
Milk
Mixes and salsas
Pestos and dips
Savory sauces (base sauces, meat sauces, pasta sauces, dessert sauces)
Soups
Vinegar
Water

8.5 Pasteurization of Liquid Foods

Pasteurization was defined narrowly by the World Health Organization as "*a microbiocidal heat treatment aimed at reducing the number of any pathogenic microorganisms in milk and liquid milk products, if present, to a level at which they do not constitute a significant health hazard*" (Table 12.5). In addition, pasteurization conditions were selected "*to effectively destroy the organisms Mycobacterium tuberculosis and Coxiella burnetii* "(Codex Alimentarius Commission 2004, 2009). More generally, traditional pasteurization methods involve "*the application of heat to liquid foods to inactivate vegetative cells of disease producing microorganisms*". Table 12.6 shows a list of some pasteurized liquid foods and semi-liquid food products.

Three equivalent pasteurization conditions are currently in use for milk, according to the different temperature –time combinations used; vat pasteurization also known as holder pasteurization (145.4 °F/63 °C, 30 min), flash pasteurization or high temperature short time (HTST) pasteurization (163 °F/73 °C, 15 s) and Ultra High Temperature (UHT) pasteurization (302 °F/150 °C, 2–3 s). All three forms of pasteurization have similar effects on microorganism, but HTST is more effective for vitamin retention (Table 12.7).

Table 12.7 Temperature and time combination for three major pasteurization processes

Name	Temperature °F (°C)	Time
Holder pasteurization	145.4 (63)	30 min
High temperature short time (HTST)	163.4 (73)	15 s
Flash pasteurization/flash heating	163.4 (73)	15 s
Ultra high temperature (UHT)	302.0 (150)	2–3 s

Cf. Sect. 9.5 for additional pasteurization times

The temperature-time treatments for milk pasteurization (Table 12.7) are intended to reduce initial pathogen levels by up to one million fold (log 6 reduction). *Coxiella burnetii* is described as the most heat stable pathogen likely to occur in milk (Cerf and Condron 2006, National Advisory Committee on Microbiological Criteria For Foods (NACMCF) 2006). The Z- and D_{63}-value for Coxiella burnetii is 4.3 °C and 3.7 min (respectively) showing the heat resistance is similar to *Mycobacterium tuberculosis* (D_{65} = 0.78 min) but this bacteria is a human pathogen (Sung & Collins, 1998).

However, the D_{65} was 0.1 min for *Mycobacterium bovis* and *Mycobacterium caprae* (Hammer et al. 2015). These microbial pathogens transmit bovine and goat tuberculosis to consumers of raw cow milk and goats milk, respectively (de la Rua-Domenech, 2006, Pesciaroli et al. 2014). The heat resistance of *M. bovis* and *M. caprae* is below that of Coxiella ensuring that vat pasteurization following current guidelines produces a safe product. In fact, the pasteurization temperature for milk was revised from 135 °F (61 °C) in 1924 to 145.4 °F (63 °C) in the 1950's based on the believe that *C. burnetii* is the most heat resistant pathogen associated with milk (Cerf and Condron 2006, National Advisory Committee on Microbiological Criteria For Foods (NACMCF) 2006). Further details about milk pasteurization can be found in Sect. 9 and in Chap. 18.

Aside from milk, other pasteurization conditions are product and pathogen specific. For example, crabmeat was pasteurized by heating to 185 °F (85 °C) for between 4min and 31 min to kill non-proteolytic *C. botulinum spores*, depending on the type of container used as packaging. Liquid egg white was pasteurized by heating at 132 or 134 °F (55.6 or 56.7 °C) for 6.2 or 3.5 min to kill salmonella (Bean et al. 2012, National Advisory Committee on Microbiological Criteria For Foods (NACMCF) 2006, p.1214 & p.1215). Egg white pasteurization is discussed further in Chap 19.

8.6 Blanching of Fresh Fruits and Vegetables

Blanching can be achieved by immersing fruits or vegetables in boiling or nearly boiling water for 5 min, which is not sufficient time to bring about cooking. Heating by steam at one atmosphere pressure is also possible until the temperature of the food is brought up to 180–190 °F (82.2–87.8 °C). The immediate effect of blanching is to inactivate enzymes present naturally in fresh foods and which if left unchecked will catalyze deterioration. Blanching or parboiling is highly recommended before home freezing. In the food industry, blanching is applied prior to canning or freezing.

9 Bacterial Death Kinetics

9.1 Rational and Kinetic Considerations

For canning to be successful, we must know both the temperature and heating time necessary to render food safe (Perkins 1964). The temperature-time combination used for processing must also avoid over-processing which leads to a loss of quality.[10] Detailed descriptions of bacterial

[10]As discussed elsewhere food quality attributes include

death kinetics can be found from food engineering texts used for IFT approved courses (Fellows, 2016; Heldman & Hartel, 1997b; Karel, et al. 1975). Shorter useful descriptions can be found in the following (Perkins 1964, Holdsworth 1985, Myrseth 1985, Al-Baali, 2006; Singh & Heldman, 2008).

As discussed in Chap 1, the study by Prescott & Underwood (circa 1896) related to "blow out" of tinned clams provided the scientific basis for canning. The studies that led to the prevention of *food spoilage* for canned food involved bacteriological examination of samples and establishing the minimum temperature-time treatment at the coldest-point of a can. It was some twenty years later in the 1920s that similar canning studies were conducted, but with the focus on food safety and the inactivation of *C. botulinum* spores (Perkins 1964; Andress & Kuhn 1998). What follows is a short summary of some basic concepts from food engineering and kinetics.

The time-course or kinetics for bacterial inactivation is described using standard notations: D-value, Z-value and F-value (Food and Drug Administration 2000; U. S. Food and Drug Administration 2014). The principles of chemical reaction kinetics apply to bacterial death kinetics with minor changes (Table 12.8).

The focus of chemical kinetics is on the initial concentration of substances (C_0) measured as mol per liter (mol/l). For microbial death kinetics, the "concentration" refers to the number of living bacterial in the food (N_0) measured as coliform units per ml (CFU/ml). For both chemical and microbial kinetics, the aim is to establish how the initial quantity (C_0 or N_0) changes with time – and how this *change* is then affected by temperature, pH etc. It is always important to summarize microbial cell death kinetics using simple equations, which can be used later for predictions and process design. Predicting bacterial death kinetics is important inorder that engineers

Table 12.8 Comparing rate parameters for chemical kinetics and microbial inactivation kinetics

Chemical kinetics		Inactivation kinetics	
Initial concentration	C_0	Live bacterial initially	N_0
Concentration later	C_t	Live bacterial later	N_t
Rate =	$k\,[C_0]^P$	$k\,[N_0]^P$	
Rate constant (min^{-1})	k	k	
Time constant (min)	1/k	D-value	
Activation energy	E (Joule/mol)	Z-value (°C)	

Notes: the units of Activation and Z-value do not map directly to each other

or the plant operator does not rely on a "trial and error" approach. However, the prediction element can only occur once basic kinetic parameters (D-value, Z-value and F-value) are determined painstakingly.

9.2 First Order Bacterial Inactivation Kinetics

The death of bacteria can be described by an exponential relationship with time. So the rate of destruction is proportional to the initial numbers of bacteria (N_0). The higher the starting population of bacteria the more rapid is their rate of destruction. The first order model implies also that injury to "one critical" cell component is sufficient to cause bacterial death. In reality, cell injury probably affects many components and so deviation from the exponential killing curve occurs, particularly at the beginning and the end of the thermal destruction process. However, the simplicity of the first order model is so great that slight losses of accuracy are considered a worthwhile tradeoff.

The exponential destruction curve for bacteria can be transformed into a simple straight-line graph, e.g. Y = mx + c, where Y is the *logarithm* of the number of live bacterial cells after a given heating time (log. N_t) and "x" is the heating time (t-minutes); the "y-intercept" constant-c represents the initial number of contaminating bacteria

those related to, sensory (S), functional (F)nutritional (N) and storage (S) characteristics. Considering the effect of thermal processing on quality of various food commodities is a key concern in food science.

(log. N_0). The graph gradient "m" is called the rate constant (k). The rate constant has units or dimensions of inverse time (min^{-1}). The time for deactivating 90% of bacterial (D-value) relates to the rate constant (D-value = 2.3/k). To apply the first order model, it is necessary to record the number of living bacterial for several time points whilst a food is being heated at a single fixed temperature.

9.3 Decimal Reduction Time (D-Value)

It is important to estimate the holding time required to reduce the initial population of bacteria by 10 - fold or 90%. The D-value is determined from a semi log graph of live bacteria cell numbers (Log N_t) versus holding time (t). The resulting straight-line graph has a negative slope equal to, k (Fig. 12.11). Where Log N_t- Log N_0 = 1, then t = 2.3/k = D-value. Therefore, the D-value is inversely related to the "rate constant" from chemical kinetics. Heating for a timed duration (t) equal to 1D, 2D or 3D (min) will produce, 10-fold, 100-fold, or 1000-fold decline in the initial number of live bacterial cells. Most foods are heated for a time equal to 12D minutes in order to provide a wide margin of safety (U. S. Food and Drug Administration 2014).

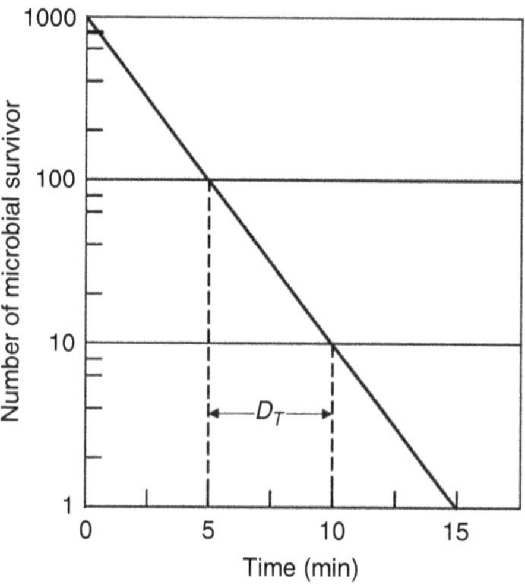

Fig. 12.11 A graph showing the thermal inactivation of microorganisms as a function of heating time at a fixed temperature (T). The decimal reduction time (D_T) is shown as 5 min. (Adapted from Al-Baali and Farid 2006)

$$D = \frac{t_1 - t_0}{\log N_0 - \log N_1} = \frac{2.3}{k} \quad (12.6)$$

There are many different bacteria in foods, each possessing their own D-value. Moreover, the D-value may change with the properties of the food, e.g. levels of salt and acidity (pH). To design a thermal process effectively, it is essential to establish the D-value for the most stable "target" microorganism. A list of some D-values recorded at 63 °C for significant foodborne pathogens is shown in Table 12.5. In practice, average D-values collected from diverse literature sources may be examined in order to derive average values for process design.

9.4 The Z-Value for Processing

The D-value decreases exponentially with rising temperature (Fig. 12.12). Higher temperatures lead to more rapid killing of bacteria and high values for k (≅1/D-value). So plotting a semi-log graph (log-D vs. Tempertaure) enables us to determine the Z-value, which is the *temperature change* required to produce a 10-fold change in D-value (log D_1-log D_2 = 1). The equation bellow shows how to calculate the Z-value from D-values recorded at two temepratures (T_1 and T_2). For increased accuracy, the Z-value can be determined from a graph of log D-value plotted against temperature.

$$Z = \frac{T_2 - T_1}{\log D_1 - \log D_2} \quad (12.7)$$

The D-value and Z-value are unique for each type of microorganisms. However, differences that are more notable occur in the average Z-value and D-value for spores and vegetative bacterial cells. As mentioned before, dormant spores may be 100s to thousands of times more heat resistant compared with the actively growing vegetative cells (Brown 2000; Setlow 2006, 2014). The

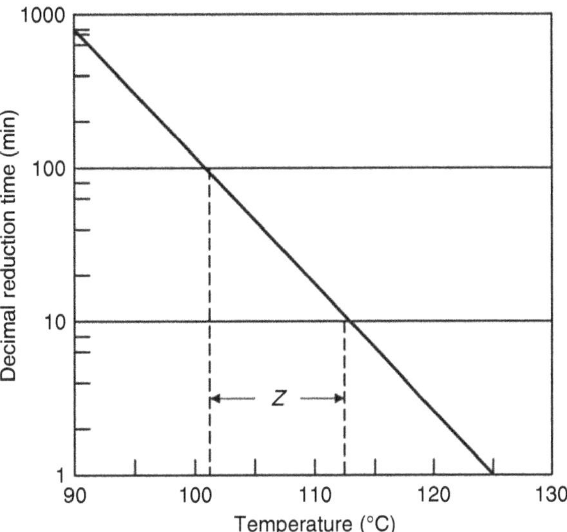

Fig. 12.12 A graph D-values plotted versus temperature gives a straight like. Y-axis shows D-values, X-axis is temperature. (Adapted from Al-Baali and Farid 2006)

Z-value (5 °C) for cells shows there is greater sensitivity to temperature change as compared with spores (Z= 10 °C) (see Sect 9.6).[11]

Obtaining the Z-value from two D-values recorded at two temperatures, enables prediction of D-values for a wide range of temperatures. Recall also that the dimensions of the D-value is time (minutes) so we can determine the process time corresponding to 12D (12 times D-value) (Fig. 12.12).

9.5 Prediction of D-Values and Z-Values for Pasteurization

Obtaining the D-value and Z-value for a target pathogenic microorganism enables the design of different pasteurization processes. For example, food engineers have established that a temperature of 63 °C is adequate for vat pasteurization leading to a decimal reduction time at 63 °C (D_{63} value) not exceeding 2.5 min for most pathogens and a Z-value ≅5.1 °C. In non-technical language, heating food to a temperature of 145.5 °F (63 °C) will kill 90% of the starting bacteria after 2.5 min. Additional implications of such data require unpacking, but the results are surprisingly rewarding (Fig. 12.13)

Recall that the D-value is the time to reduce bacterial numbers by 10-fold (one log cycle, because log (10) = 1). So heating a food sample at 63 °C for 12D values (12 x 2.5 min = 30 min) will reduce pathogen numbers by log-12 cycles (10 × 10 × 10 etc. or 10 raised to the power 12, 10^{12}). Therefore, the starting number of microbes will be reduced by 1 million, million fold.

Also, the Z-value shows the temperature-dependence of log. D_T and is analogous to activation energy (E_{act}) from chemical kinetics. The Z-value, allows the prediction of D_T over a wide range of temperatures. For example, consider the effect of raising the temperature of pasteurization from 63 °C by about 10 °C on the process parameters (Figure 12.13). Since the Z-value is 5.1 °C then D_{73} is predicted to be (2.5 min/ 91.34 =) 0.027 min and 12 × D_{73} = 0.324 min or about 19 s (see Table 12.7 and 12.9). Flash pasteurization, which is a form of HTST pasteurization, involves

[11]The analogy between Z-value and Activation energy from chemical kinetics is frequently highlighted.

$$z = \frac{T_2 - T_1}{Log\ D_1 - Log\ D_2}$$

$$\log D_1 - \log D_2 = \frac{(T_2 - T_1)}{z}$$

$$\frac{D_1}{D_2} = 10^{(T_2 - T_1)/z}$$

$$D_2 = D_1 . 10^{(T_1 - T_2)/z}$$

$$D_2 = 2.5\ min. 10^{(63-73)/5.1}$$

$$D_2 = 2.5\ min. 10^{(-10)/5.1}$$

$$D_2 = 2.5\ min. 10^{(-1.96)}$$

$$D_2 = 0.027\ min$$

$$D_2 = 1.6\ seconds$$

$$12x\ D_2 = 19.7\ second$$

Fig. 12.13 Estimation of process time following a 10 °C rise in temperature (see the text for details)

heating liquids (e.g. milk, beverages, juices etc) to ~73 °C for 20 s, which has the same effect as heating at 63 °C/30 min.[12] Additional equivalent pasteurization temperatures prescribed by the international Dairy Food Association can be seen in Table 12.9.

The current US pasteurization guidelines is for HTST milk processing at 72 °C for 15 s, in accordance with the grade A Pasteurized Milk Ordinance (Ranieri et al. 2009). Depending on the fat content of milk, HTST process can be performed at 72–75 °C for 15 s. Recent investigations showed pasteurization conditions established in the 1960s were exceeded using modern equipment to inactivate *Mycobacterium bovis* (Hammer et al. 2005). Interestingly, raising the pasteurization temperature above 72 °C to about 76 °C produced a counter-intuitive increase in the survivability of milk spoilage organisms following sample refrigeration. A plausible explanation is that spores found in raw milk are activated by heating (72–85 °C) thereby enhanc-

Table 12.9 A list of pasteurization temperatures and times

Temperature	Time	Pasteurization type
145 °F/63 °C	30 min	Vat or "holder" pasteurization
161 °F/72 °C[a]	15 s	High temperature short time (HTST)
191 °F/89 °C	1.0 s	Higher-heat shorter time (HHST)
194 °F/90 °C	0.5 s	Higher-heat shorter time (HHST)
201 °F/94 °C	0.1 s	Higher-heat shorter time (HHST)
204 °F/96 °C	0.05 s	Higher-heat shorter time (HHST)
212 °F/100 °C	0.01 s	Higher-heat shorter time (HHST)
280 °F/138 °C	2.0 s	Ultra pasteurization (UP)

Adapted from International Dairy Food Association (2019)

[a]Operators are advised to increase temperatures by 3 °C for high fat milk

[12]The equivalence of holder pasteurization and flash pasteurization does not hold for other quality outcomes. For example, vitamin retention is lower following holder pasteurization compared to flash pasteurization.

ing the growth of bacteria that spoil refrigerated HTST pasteurized milk (Ranieri et al. 2009, Martin et al. 2012).

Raising the process temperature from 63 °C to about 90 °C (Z-value = 5.1 °C) decreases the D-value by a factor of ten raised to the power ((90-63)/5.1) which is $10^{(-5.29)}$ or 100,000-fold. Therefore, the pasteurization time is 1800 s at 63 °C and 0.02 s at 90 °C, which is virtually instantaneous. The boiling point of milk is round 100.2 °C slightly above water (Luisa 1995. p.83).

One interesting application is the so-called flash-heating method for pasteurizing breast milk, water or weaning foods using rudimentary equipment (Israel-Ballard et al. 2007). Development workers in the third world noted that boiling *briefly* and then cooling, provides a simple but robust approach for rendering baby feeds and drinking water safe from foodborne illness.

Flash-heat pasteurization of human milk was proposed as an interim measure to reduce maternal transmission of HIV to infants (Israel-Ballard et al, 2007, World Health Organization 2010). Therefore, mothers should express own milk, heat to 100 °C or until the milk boils ever so briefly, cool without re-contamination, and use this for feeding. To a rough approximation, the temperature-time profile required for flash-heat pasteurization is similar to the ones described as Higher-Heat Shorter Time (HHST) pasteurization (Table 12.7).

9.6 The Reference F-Value and Lethality

The F-value or "process lethality" is defined as a process time (minutes) that is necessary to destroy an agreed number of viable cells. Commercial sterilization is designed to ensure the destruction of *Clostridium botulinum* spores that would germinate to form vegetative cells, which can produce toxins.[13] The industry standard for ensuring the destruction of *C. botulinum spores* involves heating to 249.8 °F ((121 °C) for 12D minutes. The "F_{121} value" is the nominal process time necessary to achieve microbial destruction where a sample is heated to 121.1 °C (249.8 °F) *instantaneously*. Using previous estimations the "complete" heat destruction of proteolytic *C. botulinum* spores suspended in phosphate buffer at 248 °F (120 °C) required 4 min (Perkins 1964). Using real foods spiked with *C. botulinum* spores, the average D_{121} value was 0.2 min with a Z value of 10 °C (Brown 2000; Bull et al. 2009).

Comparing D_{121} values shows there are entities in foods that are more heat resistant compared to *C. botulinum* spores (Table 12.10 & Table 12.11). Nevertheless, the *C. botulinum* spores pose a great public health threat. The 12D process for destroying *C. botulinum* spores is (12 × 0.2 min) about 2.4 min. Therefore, the lethality-value for *C. botulinum* spores is frequently cited as (F_{121}-value =) 2.4 min. However, European Union experts adopted the average D_{121} value of 0.25 min for *C. botulinum* spores and consequently the sterilization value (F =12D) is equal to 3 min.

Table 12.10 Heat stability of some bacterial spores cited as D-value at 121.1 °C

Bacterial endospores	$D_{121.1}$ (minutes)
B. stearothermophilus	4–5
C. thermosaccharolyticum	3–4
Desulfotomaculum nigrificans	2–3
B. coagulans	0.01–0.07
C. botulinum types (A and B)	0.1–0.2
C. sporogenes (P.A. 3679)	0.1–1.5
C. botulinum type E D_{80}	0.1–3.0
C. botulinum type E D_{110}	0.1 second
Z-value = 10 °C	

Adapted from Adams and Moss (2008) and Myrseth (1985)

[13] The toxin from CB is reportedly one of the most poisonous substances for humans though it can be used for cosmetic procedures with appropriate handling.

Table 12.11 Examples of D and Z values of spore forming organism and vegetative cells*

Microorganisms	T_{Ref} (°C)[a]	Z-value (°C) Mean (range)	Log D-value (range)	D $_{Ref.}$ (min)
Spore formers	121.1	10 (7–12)	0.2 to 0.69	0.01–5
Vegetative cells	70	5 (4–7)	−1.62 to 1.04	0.03–11
C. botulinum spores (A, B F)	121.1	10.2	−0.78 (−1.24 to 0.32)	0.17 (0.058–0.48)
L. monocytogenes	70	7.0	−1.06 (−1.84 to 0.28)	0.87 (0.014 to 0.52)

Adapted from Bean (2012)

[a]$T_{Ref.}$ is the reference temperatures for pasteurization (~70–75 °C) or Appertization (120–121.1 °C)

$$F = D_{121}\left(\log N_0 - \mathrm{Log}N_t\right)$$

And (12.8)

$$F_{121} = 12 \times D_{121}$$

Heating a food product to 121 °C (249.8 °F) instantaneously, then holding for 2.4 min. before cooling to room temperature immediately, will produce a 12-log cycle reduction in the number of *C. botulinum* spores. An initial contamination of some 10^{12} spores "per can" will be reduced to 1 spore per can by a 12D process. However, an initial contamination of 10 spores per can will be reduced to a final contamination risk of, 1 spore per 100 million cans.

The F_{121} value is unattainable directly because heating real food samples does not occur instantaneously. Cooling hot food also takes time. Normally, heat transfer depends on the physical attributes of the food and also the size and shape of the container (Myrseth 1985). To estimate the actual processing time, requires that the lag-time encountered during heating and cooling is taken into account.

10 Estimating Processing Times

To determine the overall F-value (designated F_0), it is necessary to find the temperature-profile at the "slowest heating point" for the can (see Fig. 12.14). Recall that heating occurs through convection or conduction leading to different "cold points". Figure 12.14 shows different locations of the cold point that can occur in foods of different consistencies and moisture levels. Several strategies used by food engineers to determine process times for canning are considered below. First, we present an intuitive method for estimating process times. Other process time calculations require advanced concepts from mathematics.

10.1 Intuitive Estimation of Process Times

To ensure that a tinned food is processed for a sufficient time and temperature to achieve a 12 log cycle (10^{12}-fold) reduction of heat-resistant *C. botulinum* spores, we must identify the slowest heating point in the product (Fig. 12.14). Then the "come up" time it takes to heat this "cold point" to the set temperature is determined. The come-up time is added to "the theoretical" time for a 12D process.

For example, Bigelow and co-workers reported, for No. 2 cans filled with 20 ounces (20 oz) of corn, that it takes 25 min for the slowest heating point to reach a maximum temperature of 214 °F (101 °C) using a 250 °F (121.1 °C) retort (cited by Perkins 1964). Assuming the average $D_{121.1}$ for *C. botulinum* spores is 0.2 min the food must be processed for a total time of: 25 + (0.2 × 12) or 27.4 min. Heating the sample for 27.4 min will ensure the 10^{12} fold reduction in *C. botulinum* spores. Notice from this example that the *theoretical* $F_{121.1}$ value

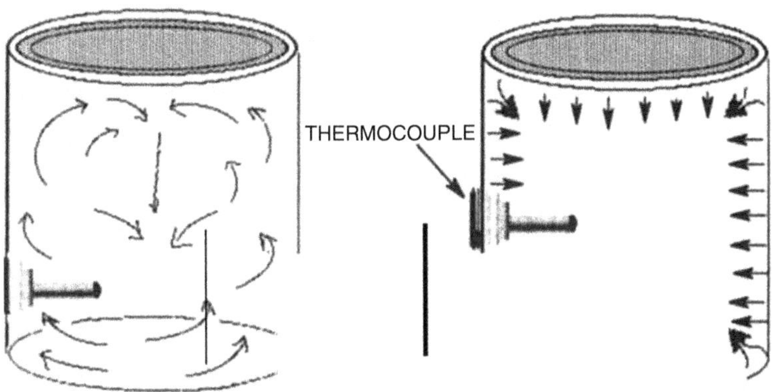

Fig. 12.14 Cold points in cans with convection and conduction heating

remains (12 × 0.2) equal to 2.4 min. The extra 25 min actual process time is necessary to allow for the slow heat transfer in real canned products. The cooling time was not considered in this estimation.

The intuitive approach for estimating F_0 can lead to several problems. First, it is implied that the canned product is heated to the set temperature of121.1 °C. We show below that this is not a real requirement; indeed heating to 121.1 °C is likely to lead to over-processing. Second, as noted above the preceding analysis did not take account of the time required to cool the food sample. Finally, the come-up times can be highly variable depending on product characteristics, e.g. the shape of can, type of food contained and whether the canning equipment (retort) allows sample mixing.

10.2 Instantaneous F-Value and Lethality Calculations

A more refined process calculation is available that acknowledges the target temperature of 250 °F (121.1 °C) might not be attained at all. Additional refinements for estimating the F_0-value take account of non-instantaneous temperature changes in the sample during heating. The "unit of lethality" (L_R) is defined, as the heat kill equivalent to 1 min process at 121 °C against an organism of a given Z-value; L_R is defined by Eq. 12.9 based on first principles (Myrseth 1985).

$$L_R = D_{121.1} / D_T \tag{12.9}$$

$$L_R = 10^{(T-121.1)/Z} \tag{12.10}$$

Assuming the food is processed at 121.1 °C, we obtain the idealized D_{121} value. By contrast, if a food is processed at another temperature (T) the decimal reduction times takes on a value D_T. So using a process temperature (T) together with a process time equal to D_T(min) will produce the same killing effect as processing at 121 °C for time equal to D_{121} (Myrseth 1985).

10.3 Total Lethality Calculations

Recall that the Z-value can be used to predict D_T over a wide range of temperatures (T). At temperatures below 121.1 °C, the magnitude of D_T increases relative to D_{121}. The *duration* of any 12D_T process increases as the process temperature decreases. The link between process lethality and temperature is as shown from Eq. 12.9 and 12.10. Accordingly, it is possible to figure-out the overall F-value (F_0) by taking note of "minute to minute" changes in the sample temperature throughout any given canning thermal process. The F_0 calculation involves three steps (Singh 2016, Simpson et al. 2020).

1. Conduct a heat-penetration study to find the temperature-time profile

Table 12.12 Example total lethality calculation

Time (min)	Temp (°C)	L_R	Sum-L	Fo (min)
0	24	0		
5	24.5	0	0.000	
10	34	0	0.000	
15	54	0	0.000	
20	72.5	0	0.000	
25	87	0	0.000	
30	98	0.005	0.005	
35	105	0.025	0.030	
40	110.5	0.087	0.117	
45	114.5	0.219	0.336	
50	117	0.389	0.725	
55	119	0.617	1.342	
60	120	0.776	**2.118**	8.6
65	120	0.776	2.894	
70	106	0.031	2.925	
75	88	0	2.925	14.6

2. Calculate the instantaneous lethality values for each temperature point
3. Add all fractional lethality values to find the total lethality

The stages 1–3 of the lethality calculation are outlined in the following section.

10.4 Heat Penetration Studies

The purpose of heat penetration studies is to determine the temperature-time profile for the slowest heating point for a canned product (Table 12.12). Several identical tin-plate cans with the same dimensions as those used for the food are fitted with a thermocouple located in the center. The cans are filled with a food product sealed and then subjected to several thermal process with well-defined set-temperatures and times. For each proposed thermal process, temperatures are recorded continuously or at fixed time intervals. Typically, twelve cans must be processed and the average data recorded.

10.5 Total Lethality Estimation

During thermal processing the sample temperature increases to a maximum value, then decreases in line with set points for process controls (Fig. 12.15). Tracing the temperature change at the slowest heating point provides the temperature-time profile. It is necessary to assign a lethality value for each temperature "point" recorded.

The "general method" for lethality calculation is illustrated in Table 12.12 and Figure 12.15 for canned fish ((Myrseth 1985). First, the product time-temperature profile was recorded at 5 min intervals (Fig. 12.14) and plotted as a graph (Myrseth 1985). Second, the fractional lethality estimate for each temperature-point (Fig. 12.15 right-hand side y-axis) was calculated using Eq. 12.10 with 121.1 °C as the reference temperature and using the published Z-value of 10 °C for proteolytic *C. botulinum* spores. Third, the area under the lethality-time curve (Fig 12.15) is taken as a measure of overall severity of heat processing or the killing effect referenced to 250 °F (121.1 °C).

The area under the curve was determined as 14.6 min using the trapezoid rule. All fractional lethality values were summed and multiplied by 5 (2.925 × 5 = 14.6 min). Evidently, the 75 min-retort process (Fig. 12.15) had a lethality effect equivalent to an idealized thermal process assuming the slowest heating point within the canned fish product had been heated to 250 °F (121.1 °C) for 14.6 min (Myrseth 1985).

In theory, the product was over-processed compared with a 12 × D_{121} (F_{121}= 2.4min) standard for commercial sterilization. In practice, processing beyond the 12 × D_{121} standard occurs frequently in order to inactivate spores from thermophilic spoilage microbes (Table 12.10) which may germinate if cans are not stored at low ambient temperature, e.g. under tropical conditions (Ashton 1981, André et al. 2013, Provost et al. 2019). The spores from *Geobacillus stearothermophilus* and the other thermophilic food spoilage bacteria are many-times more stable than spores from *C. botulinum* (Brown 2000; Setlow 2014). The canning industry also employs 2–3 fold margin of safety for thermal processing to allow for unforeseen circumstances, e.g. an unexpectedly high initial level of microbial contami-

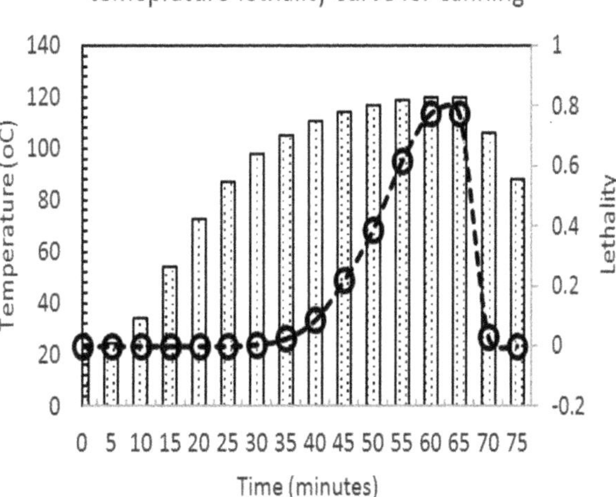

Fig. 12.15 An example total lethality calculation – from (Myrseth 1985). Left-hand axis (bars) shows the temperature-time profile. The lethality value for each temperature (circles) is plotted on the Right-hand y-axis. Sample data is shown in Table 12.12

nation (Pflug et al. 1981). As noted previously, FDA oversight over commercial canning is strict including, the requirement for expert operators (21CFR113.83) and regular inspections (Govinfo. Gov (2021). Nowadays, total lethality calculations are done using computers.

10.6 Cook value

Different cooking methods (Table 12.4) involve distinctive time-temperature treatments. The cook value may be defined as "*the heating time (t-minutes) after which the sample receives a heat treatment equivalent to one minute at* 100 °C (Gossett et al. 1984; Holdsworth 1985). Therefore, the cook value is analogous to the (F_0) value (previous section) except the reference temperature is 100 °C. Here the goal is to establish the severity of cooking – involving different temperature-time treatments and cooking methods. The cook value can be determined from, (i) the temperature-time profile recorded from the coolest point within a food sample during cooking, and (ii) a knowledge of the Z-value for quality loss (Z_Q). The mathematical formula for the cook value is identical to that used for lethality (F_0) calculations with minor changes (Luechapattanaporn et al. 2005; Kassama and Ngadi 2016).

Past researchers showed that US army rations consisting of pre-cooked scrambled egg (packaged using polymer-trays and covered with foil) could be sterilized by Radio Frequency heating or a conventional retort. The first requirement of both thermal processes was to achieve a >4 log (or 10,000-fold) reduction in *Clostridium sporogenes* spores. The Radio Frequency heating time of 20 min was needed as compared with 60 min for the conventional retort. Based on the individual temperature-time data and assuming Z_Q = 33 °C for quality loss, the cook value was estimated as 69 min for Radio Frequency heating and 153 min for the conventional retort. Product quality (color and texture) attributes were better using Radio Frequency heating, presumably because retorting led to overcooking (Luechapattanaporn et al. 2005).

Another study showed that fresh potato was "fully" cooked by boiling when the texture decreased to 6/100th of the initial value. Based on the recorded temperature-time profile during the boiling process for potatoes, the cook value was established as 7 min. Furthermore, the actual cooking time for boiling potatoes was 30–40 min due to the sluggish rate of heating (Lee et al. 1982, fig 3). Other reports showed the cook value was 3.7 min for baked potatoes compared with a cook value of 3.0 min for steamed potatoes. However, the actual cooking time (to reach a core temperature of 95 °C) was 45 min for oven cooking compared with 18 min during steaming, (Chiavaro et al. 2006).

Frying deboned chicken breast for optimum texture required a cook value of 450 sec, achieved by heating at 170, 180 or 190 °C for 300 seconds. In all cases, the internal cooking temperature did not rise above 108 °C. Cooking beyond 300sec increased moisture loss and oil uptake (Kassama and Ngadi 2016). Therefore, frying at higher than 170 °C was not essential if texture was the foremost concern. More generally, using high-temperature and shorter time cooking methods can reduce the cook value leading to improved quality (Holdsworth 1985; Ling et al. 2015).

References

Adams R, Moss O (2008) Food Microbiology. RSC Publishing, 463 pp

Ahmed J, Ramaswamy HS (2007) Microwave pasteurization and sterilization of foods. In: Rahman MS (ed) Handbook of food preservation, 2nd edn. CRC Press, pp 691–712. https://www.semanticscholar.org/paper/Microwave-Pasteurization-and-Sterilization-of-Foods-Ahmed-Ramaswamy/699b666f695859ee695344e695874a626554a695491c695804c698519e695858

Al-Baali AAG, Farid MM (2006) Principles of thermal sterilization. In: Sterilization of food in retort pouches, vol 2019. Springer, Boston, pp 25–32. https://doi.org/10.1007/1000-1387-31129-31127_31123

Allen MA, Burgess SG (1950) The losses of ascorbic acid during the large-scale cooking of green vegetables by different methods. Br J Nutr 4(2–3):95–100

Anderson KE, Sinha R, Kulldorff M, Gross M, Lang NP, Barber C, Kadlubar FF (2002) Meat intake and cooking techniques: associations with pancreatic cancer. Mutat Res 506-507:225–231

Anderson N, Larkin JW, Cole MB, Skinner GE, Whiting RC, Gorris LGM, Rodriguez A, Buchanan R, Stewart CM, Hanlin JH (2011) Food safety objective approach for controlling Clostridium botulinum growth and toxin production in commercially sterile foods. J Food Prot 74(11):1956–1989

Andress EL, Kuhn GD (1998) Critical review of home preservation literature and current research. II. Early history of USDA home canning recommendations. From https://nchfp.uga.edu/publications/usda/review/earlyhis.htm

André S, Zuber F, Remize F (2013) Thermophilic spore-forming bacteria isolated from spoiled canned food and their heat resistance Results of a French ten-year survey. Int J Food Microbiol 165(2):134–143

Anon. (1977) Home canning. A scientific status summary by the Institute of Food Technologist's Expert Panel on Food Safety & Nutrition and the Committee on Public Information. Food Technol 31(6):43–47

Ashton DH (1981) Thermophilic organisms involved in food spoilage: thermophilic anaerobes not producing hydrogen sulfide. J Food Prot 44(2):146–148. https://doi.org/110.4315/0362-4028X-4344.4312.4146

Avena-Bustillos RJ (2011) Establishment of an FDA-approved international Better Process Control School for food canning supervisors. Prog Nutr 13(3):155–159

Awuah G, Ramaswamy HS, Economides A (2007) Thermal processing and quality: principles and overview. Chem Eng Process 46(6):584–602

Barbier JP (1989) Appert, Nicolas, father of canning + food preservation. Historia 514:90–92

Batchelder EL (1943) Home drying methods and their effect on the palatability, cooking quality, and nutritive value of foods. Am J Public Health Nations Health 33(8):941–947

Bean D, Bourdichon F, Bresnahan D, Davies A, Geeraerd A, Jackson T, Membré J-M, Pourkomailian B, Richardson P, Stringer M, Uyttendaele M, Zwietering M (2012) Risk assessment approaches to setting thermal processes in food manufacture, International Life Sciences Institute (ILSI) Europe, 44pp. https://ilsi.eu/publication/risk-assessment-approaches-to-setting-thermal-processes-in-food-manufacture/

Becker ME, Pederson CS (1950) The physiological characters of *Bacillus coagulans (Bacillus thermoacidurans)*. J Bacteriol 59(6):717–725

Bezanson A, Learson R, Teich W (1973) Defrosting shrimp with microwaves. Proceedings of the Gulf and Caribbean Fisheries Institute 25:44–55. http://aquaticcommons.org/id/eprint/11932

Binsted R, Devey JD (1970) Soup manufacture. Canning, dehydration and quick-freezing. 3rd edn, xi+260pp

Bitting AW (1912) The canning of foods: a description of the methods followed in commercial canning, US Government Printing Office, 77pp. https://booksgooglecomgh/books?id=lT5DAAAAIAAJ

Bornet FR, Cloarec D, Barry JL, Colonna P, Gouilloud S, Laval JD, Galmiche JP (1990) Pasta cooking time: influence on starch digestion and plasma glucose and insulin responses in healthy subjects. Am J Clin Nutr 51(3):421–427

Brambilla L, Michel A, Bertholon R (2016) Condition of cans in collections: a challenge in conservation. In: Proceedings of the interim meeting of the ICOM-CC metals working group. 26N30 September 2016, New Delhi, India. 9pp. https://www.researchgate.net/publication/308983983_Condition_of_Cans_in_Collections_A_Challenge_in_Conservation

Brody AL (1971) Food canning in rigid and flexible packages. CRC Crit Rev Food Technol 2(2):187–243

Brody AL (2002) Food canning in the 21st century. Food Technol 56(3):12pp. https://www.ift.org/news-and-publications/food-technology-magazine/issues/2002/march/columns/packaging

Brody AL (2011) Advances in microwave pasteurization and sterilization. Food Technol 65(2):6pp. https://www.ift.org/news-and-publications/food-

technology-magazine/issues/2011/february/columns/packaging
Brown A (2007) Understanding food: principles and preparation, 4th edn. Cengage Learning, 696 pp
Brown K (2000) Control of bacterial spores. Br Med Bull 56(1):158–171. https://doi.org/110.1258/0007142001902860
Bull M, Olivier SA, Van Diepenbeek RJ, Kormelink F, Chapman B (2009) Synergistic inactivation of spores of proteolytic *Clostridium botulinum* strains by high pressure and heat is strain and product dependent. Appl Environ Microbiol 75(2):434–445. https://journals.asm.org/doi/full/410.1128/AEM.01426-01408
Canning food at home can be dangerous (1975) JAMA 231(9):904–904
Cassi D (2011) Science and cooking: the era of molecular cuisine. EMBO Rep 12(3):191–196
Cerf O, Condron R (2006) Coxiella burnetii and milk pasteurization: an early application of the precautionary principle? Epidemiol Infect 134:946–951. http://www.ncbi.nlm.nih.gov/pmc/articles/PMC2870484/
Chandrasekaran S, Ramanathan S, Basak T (2013) Microwave food processing—a review. Food Res Int 52(1):243–261. https://asset-pdfscinapseio/prod/2063436182/2063436182pdf
Chiavaro E, Davide B, Elena V, Roberto M (2006) The effect of different cooking methods on the instrumental quality of potatoes (cv. Agata). J Food Eng 77(1):169–178
Clifford WM (1930) The effect of cooking on the digestibility of meat. Biochem J 24(6):1728–1733
Codex Alimentarius Commission (2004, 2009) Code of hygienic practice for milk and milk products: CAC/RCP 57-2004. From http://www.fao.org/fao-who-codexalimentarius/standards/list-of-standards/en/
Collings P, Williams C, MacDonald I (1981) Effects of cooking on serum glucose and insulin responses to starch. Br Med J (Clin Res Ed) 282(6269):1032
Cross HR, Stanfield MS, Koch EJ (1976) Beef palatability as affected by cooking rate and final internal temperature. J Anim Sci 43(1):114–121
Darwin C (1872) The descent of man, and selection in relation to sex, vol 1
De la Rua-Domenech R (2006) Human Mycobacterium bovis infection in the United Kingdom: incidence, risks, control measures and review of the zoonotic aspects of bovine tuberculosis. Tuberculosis (Edinb) 86(2):77–109
Earle RL, Earle M (1983, May 2019) Unit operations in food processing: Web edition. Retrieved from https://nzifst.org.nz/resources/unitoperations/index.htm
Fellows PJ (2016) Food processing technology: principles and practice, 4th edn. Woodhead Publishing, 1152 pages
Feng H, Yin Y, Tang J (2012) Microwave drying of food and agricultural materials: basics and heat and mass transfer modeling. Food Eng Rev 4(2):89–106. https://doi.org/10.1007/s12393-012-9048-x
Fenton F, Barnes B, Moyer JC, Wheeler KA, Tressler DK (1943) Losses of vitamins which may occur during the cooking of dehydrated vegetables. Am J Public Health Nations Health 33(7):799–806
Food and Agriculture Organization of the United Nations (1985) FAO Fisheries Circular No. 784.Planning and engineering data 2. Fish Canning. (1). Canning principles. Retrieved from http://www.fao.org/docrep/003/r6918e/R6918E00.htm#Contents
Food and Drug Administration (2000) Kinetics of microbial inactivation for alternative food processing technologies. FDA Center for Food Safety and Applied Nutrition report–A report of the IFT for the FDA of the US Department of Health and Human Services. June, 2. Retrieved from https://www.fda.gov/files/food/published/Kinetics-of-Microbial-Inactivation-for-Alternative-Food-Processing-Technologies.pdf
Fortune Business Insights (2020) Canned food market size, share & covid-19 impact analysis, by type (canned seafood, canned fruits and vegetables, canned meat, and others), distribution channel (supermarkets/hypermarkets, specialty stores, and online retail), and regional forecast, 2020–2027. From https://www.fortunebusinessinsights.com/infographics/canned-food-market-103258
Goindi G, Karmarkar MG, Kapil U, Jagannathan J (1995) Estimation of losses of iodine during different cooking procedures. Asia Pac J Clin Nutr 4(2):225–227
Goldblith SA (1971) A condensed history of the science and technology of thermal processing [canning]. I. Food Technol 25(12):1256–1258
Goldblith SA, Decareau RV (1973) An annotated bibliography on microwaves their properties production and applications to food processing. MIT Press, Cambridge, MA, xi+356pp
Gossett PW, Rizvi SSH, Baker RC (1984) Quantitative analysis of gelation in egg protein systems. Food Technol 38(5):67–96
Govinfo.Gov (2021) 21 CFR PART 113 – thermally processed low acid foods packaged in hermetically sealed containers Retrieved from https://wwwecfrgov/cgi-bin/text-idx?SID=fcbf83657l9b6cf05ebd64075a1c8a6e&mc=true&node=pt212113&rgn=div5
Gowlett JAJ (2016) The discovery of fire by humans: a long and convoluted process. Retrieved from https://doi.org/10.1098/rstb.2015.0164
Graham JC (1981) The French connection in the early history of canning. J R Soc Med 74(5):374–381
Grand View Research Inc (2018) U.S. canned foods market size, share & trends analysis report, by type (seafood, vegetables, meat products, fruits, ready meals), competitive landscape, and segment forecasts, 2018–2025. From https://www.grandviewresearch.com/industry-analysis/us-canned-foods-market
Gutterson M (1972) Food canning techniques. Food Process Rev. Noyes Data Corporation:26
Guzik P, Kulawik P, Zając M, Migdał W (2021) Microwave applications in the food industry: an overview of recent developments. Crit Rev Food Sci Nutr:1–20. https://www.researchgate.net/profile/Paulina-Guzik/publication/351477101

Hammer P, Richter E, Rusch-Gerdes S, Walte HG, Matzen S, Kiesner C (2015) Inactivation of Mycobacterium bovis ssp. caprae in high-temperature, short-term pasteurized pilot-plant milk. J Dairy Sci 98(3):1634–1639. https://doi.org/10.3168/jds.2014-8939

Hedeen N, Reimann D, Everstine K (2016) Microwave cooking practices in Minnesota food service establishments. J Food Prot 79(3):507–511. https://doi.org/510.4315/0362-4028X.JFP-4315-4333

Heldman DR, Hartel RW (1997a) Commercial sterilization. In: Principles of food processing. Springer, New York, pp 55–82

Heldman DR, Hartel RW (1997b) Principles of food processing. Springer, New York

Henderson T (2016) Thermal physics – lesson 1 – heat and temperature: what is heat? Retrieved from http://www.physicsclassroom.com/class/thermalP/Lesson-1/What-is-Heat

Holdsworth S (1985) Optimization of thermal processing—a review. J Food Eng 4(2):89–116. https://d111wqtxts111xzle117.cloudfront.net/35128732/Optimisation_of_thermal_processing-a_review

Holdsworth SD, Simpson R (2008) Thermal processing of packaged foods, 3rd edn. Springer, New York, 525pp

Holmes EG (1958) A note on the water content of certain cereals as prepared for cooking in African households. East Afr Med J 35(9):549–550

Institute for Thermal Processing Specialists (2014) Guidelines for conducting thermal processing studies. From http://iftps.org/wp-content/uploads/2017/12/Retort-Processing-Guidelines-02-13-14.pdf

International Dairy Food Association (2019) Pasteurization. Retrieved from https://www.idfa.org/news-views/media-kits/milk/pasteurization

Israel-Ballard K, Donovan R, Chantry C, Coutsoudis A, Sheppard H, Sibeko L, Abrams B (2007) Flash-heat inactivation of HIV-1 in human milk: a potential method to reduce postnatal transmission in developing countries. JAIDS 45(3):318–323

Kalla AM, Devaraju R (2017) Microwave energy and its application in food industry: a review. Asian J Dairy Food Res 36(1):34–47. Retrieved from http://wwwarccjournalscom/uploads/articles/8DR1135pdf

Karel M, Fennema OR, Lund DB (1975) Principles of food science. Part II. Physical principles of food preservation

Kassama LS, Ngadi M (2016a) Relationship between oil uptake and moisture loss during deep fat frying of deboned chicken breast meat. Adv Chem Eng Sci 6(4):324–334. http://dx.doi.org/310.4236/aces.2016.64033

Kassama LS, Ngadi M (2016b) Relationship between oil uptake and moisture loss during deep fat frying of deboned chicken breast meat. Adv Chem Eng Sci 6(4):324–334. http://dx.doi.org/310.4236/aces.2016.64033

Krishnamurthy K, Khurana HK, Soojin J, Irudayaraj J, Demirci A (2008) Infrared heating in food processing: an overview. Compr Rev Food Sci Food Saf 7(1):2–13

Kylen AM, Charles VR, McGrath BH, Schleter JM, West LC, Van Duyne FO (1961) Microwave cooking of vegetables. Ascorbic acid retention and palatability. J Am Diet Assoc 39:321–326

Lee D, Pyun YR, Kwon YJ, Shin DH (1982) Changes in texture during the boiling process of potatoes. Korean J Food Sci Technol 14(1):16–20

Lewis KH, Cassel K (1964) botulism: proceedings of a symposium, U. S. Department of Health, Education, and Welfare, Public Health Service, 327pp. https://books.google.com.gh/books?id=gHVrAAAAMAAJ

Ling B, Tang J, Kong F, Mitcham EJ, Wang S (2015) Kinetics of food quality changes during thermal processing: a review. Food Bioprocess Technol 8(2):343–358

Luechapattanaporn KWY, Wang J, Tang J, Hallberg LM, Dunne CP (2005) Sterilization of scrambled eggs in military polymeric trays by radio frequency energy. J Food Sci 70(4):E288–E294

Luisa BG (1995) Handbook of milk composition. Academic Press, San Diego, 919pp. https://books.google.com.gh/books?id=MAbQjY0ppgQC

Lushbough CH, Weichman JM, Schweigert BS (1959) The retention of vitamin B6 meat during cooking. J Nutr 67(3):451–459

Malik RK, Dhingra KC (1975) Canning and preservation of fruits and vegetables, 140pp

Martin NH, Ranieri ML, Wiedmann M, Boor KJ (2012) Reduction of pasteurization temperature leads to lower bacterial outgrowth in pasteurized fluid milk during refrigerated storage: a case study. J Dairy Sci 95(1):471–475. https://doi.org/10.3168/jds.2011-4820

Masters H (1918) An investigation of the methods employed for cooking vegetables, with special reference to the losses incurred. Part I. Dried legumes. Biochem J 12(3):231–247

Masters H, Garbutt P (1920) An investigation of the methods employed for cooking vegetables, with special reference to the losses incurred: Part II. Green vegetables. Biochem J 14(2):75–90

McGee H (2007) On food and cooking: the science and lore of the kitchen. Scribner

McHugh T (2016) Microwave processing heats up. Food Technol 70(10):6pp. https://wwwiftorg/news-and-publications/food-technology-magazine/issues/2016/october/columns/processing-microwave-heating-applications-in-the-food-industry

McHugh T, Milczarek R (2019) MATS and MAPS advance food preservation. Food Technol 73(6):3pp. https://www.ift.org/news-and-publications/food-technology-magazine/issues/2019/june/columns/processing-mats-and-maps-advance-food-preservation

Myrseth A (1985) 1. Canning principles. Planning and engineering data. 2. Fish canning. FAO Fish Circ., (784): 1985 77 p. Retrieved from http://www.fao.org/docrep/003/R6918E/R6918E02.HTM

NAICS Association (2018) Canning. From https://www.naics.com/code-search/?naicstrms=canned,fish

National Advisory Committee on Microbiological Criteria for Foods (NACMCF) (2006) Requisite

scientific parameters for establishing the equivalence of alternative methods of pasteurization. J Food Prot 69(5):1190–1216. https://doi.org/1110.4315/0362-1028X-1169.1195.1190
National Center for Home Food Preservation (2015) Guide 01: principles of home canning. USDA complete guide to home canning, 2015 revision, 40pp. https://nchfp.uga.edu/publications/publications_usda.html
National Physical Laboratory (2018). 2.3.7 Thermal conductivities. Retrieved from http://www.kayelaby.npl.co.uk/general_physics/2_3/2_3_7.html
Nursal B, Yucecan S (2000) Vitamin C losses in some frozen vegetables due to various cooking methods. Nahrung 44(6):451–453
Oldring PKT, Nehring, U (2007) Packaging materials 7. Metal packaging for foodstuffs. International Lifesciences Institute Europe, 44pp. https://www.pac.gr/bcm/uploads/47-metal-packaging-for-foodstuffs.pdf
Pasricha S (1967) Effect of different methods of cooking and storage on the ascorbic acid content of vegetables. Indian J Med Res 55(7):779–784
Perkins WE (1964) Prevention of botulism by thermal processing. In: Lewis KH, Cassel K et al (eds) Botulism: proceedings of a symposium. U. S. Department of Health, Education, and Welfare, Public Health Service, pp 187–201. https://books.google.com.gh/books?id=gHVrAAAAMAAJ
Pesciaroli M, Alvarez J, Boniotti MB, Cagiola M, Di Marco V, Marianelli C, Pacciarini M, Pasquali P (2014) Tuberculosis in domestic animal species. Res Vet Sci 97:S78–S85
Pflug I (1987) Factors important in determining the heat process value, FT, for low-acid canned foods. J Food Prot 50(6):528–533. https://doi.org/510.4315/0362-4028X-4350.4316.4528
Pflug IJ (2010) Science, practice, and human errors in controlling Clostridium botulinum in heat-preserved food in hermetic containers. J Food Prot 73(5):993–1002. https://doi.org/1010.4315/0362-1028X-1073.1005.1993
Pflug I, Davidson PM, Holcomb RG (1981) Incidence of canned food spoilage at the retail level. J Food Prot 44(9):682–685. https://doi.org/610.4315/0362-4028X-4344.4319.4682
Potter NN (2013) Food science. Springer, 765 pp
Prevost S, André S, Remize F (2010) PCR detection of thermophilic spore-forming bacteria involved in canned food spoilage. Curr Microbiol 61(6):525–533
Rahman MS (2007) Handbook of food preservation. CRC Press, Boca Raton, 1008pp. https://booksgooglecomgh/books?id=sKgtq62GB_gC
Ranieri ML, Huck JR, Sonnen M, Barbano DM, Boor KJ (2009) High temperature, short time pasteurization temperatures inversely affect bacterial numbers during refrigerated storage of pasteurized fluid milk. J Dairy Sci 92(10):4823–4832. https://doi.org/10.3168/jds.2009-2144
Roberts AO (1970) Design and equipment of canning plant for beans in tomato sauce. Food Trade Rev 40(6):43–48
Robertson GL (2016) Chapter 7. Metal packaging materials. In: Food packaging: principles and practice, 3rd edn. CRC Press, pp 189–228. https://books.google.com.gh/books?id=BizOBQAAQBAJ
Robins GV (1980) Food science in catering
Rumm-Kreuter D, Demmel I (1990) Comparison of vitamin losses in vegetables due to various cooking methods. J Nutr Sci Vitaminol (Tokyo) 36(Suppl 1):S7–S14. discussion S14–15
Sale AJH (1976) A review of microwaves for food processing. J Food Technol 11(4):319–329
Santarelli RL, Pierre F, Corpet DE (2008) Processed meat and colorectal cancer: a review of epidemiologic and experimental evidence. Nutr Cancer Int J 60(2):131–144. https://doi.org/10.1080/01635580701684872
Schmutz P, Hoyle EH, McKay S (2020) Canning at home. Factsheet HGIC 3040. Home and Garden Information Centre, Clemson University, 6pp. https://hgic.clemson.edu/factsheet/canning-foods-at-home/
Seow A, Poh WT, Teh M, Eng P, Wang YT, Tan WC, Lee HP (2000) Fumes from meat cooking and lung cancer risk in Chinese women. Cancer Epidemiol Biomark Prev 9(11):1215–1221
Setlow P (2006) Spores of Bacillus subtilis: their resistance to and killing by radiation, heat and chemicals. J Appl Microbiol 101(3):514–525. https://doi.org/510.1111/j.1365-2672.2005.02736.x
Setlow P (2014) Spore resistance properties. Microbiol Spectr 2(5):TBS-0003-2012. https://doi.org/0010.1128/microbiolspec.TBS-0003-2012
Severi S, Bedogni G, Manzieri AM, Poli M, Battistini N (1997) Effects of cooking and storage methods on the micronutrient content of foods. Eur J Cancer Prev 6:S21–S24. https://doi.org/10.1097/00008469-199703001-00005
Simpson R, Nuñez H, Ramírez C (2020) Principles of thermal processing of packaged foods. In: Holden NM, Wolfe ML, Ogejo JA, Cummins EJ (eds) Introduction to biosystems engineering. Virginia Tech, 23pp. https://vtechworks.lib.vt.edu/bitstream/handle/10919/93254/Food_Thermal_Processing.pdf?sequence=10932&isAllowed=y
Singh P (2016) Thermal process calculations – lethality, thermal death time. Retrieved from https://www.youtube.com/watch?v=WaeeNrjTtIg
Singh RP, Heldman DR (2008) Chapter 4: Heat transfer in food processing. In: Introduction to food engineering. Elsevier Science, pp 247–401
Sinha R, Peters U, Cross AJ, Kulldorff M, Weissfeld JL, Pinsky PF, Hayes RB (2005) Meat, meat cooking methods and preservation, and risk for colorectal adenoma. Cancer Res 65(17):8034–8041
Sinoda O, Kodera S (1932) The chemistry of cooking: on the critical temperature in cooking the sweet potato. Biochem J 26(3):650–657
Sinoda O, Kodera S, Oya C (1931) The chemistry of cooking: the chemical changes of carbohydrates in the sweet potato according to various methods of cooking. Biochem J 25(6):1973–1976
Stevenson K (1992) Thermal processing food canning. In Hui YH, pp 2561–2566

Stewart CP (1946) Loss of nutrients in cooking. Proc Nutr Soc 4(2):164–171

Sung N, Collins MT (1998) Thermal tolerance of Mycobacterium paratuberculosis. Appl Environ Microbiol 64(3):999–1005

Tang J, Hong Y, Inanoglu S, Liu F (2018) Microwave pasteurization for ready-to-eat meals. Curr Opin Food Sci 23:133–141. https://s133.wp.wsu.edu/uploads/sites/1254/2018/1211/1322-Tang-2018-COIFS.pdf

Tang J (2015) Unlocking potentials of microwaves for food safety and quality. J Food Sci 80(8):E1776–E1793. https://onlinelibrary.wiley.com/doi/full/1710.1111/1750-3841.12959

Teixeira AA (2013) Thermal processing for food sterilization and preservation. Handbook of farm, dairy and food machinery engineering. M. Kutz, Elsevier, pp 441–467

Teixeira AA (2014) Thermal processing for food sterilization and preservation. In: Kutz M (ed) Handbook of farm, dairy, and food machinery engineering. Credo Reference, New York, pp 441–466. https://booksgooglecomgh/books?id=HUFmswEACAAJ

Tinkler CK (1931) The blackening of potatoes after cooking. Biochem J 25(3):773–776

Tinkler CK, Soar MC (1920) The formation of ferrous sulphide in eggs during cooking. Biochem J 14(2):114–119

U. S. Food and Drug Administration (2014) Sterilizing Symbols (D, Z, F). Retrieved from https://www.fda.gov/inspections-compliance-enforcement-and-criminal-investigations/inspection-guides/sterilizing-symbols-d-z-f

U.S. Food and Drug Administration (2017) Food code (2017). Retrieved from https://www.fda.gov/media/110822/download

World Health Organization (2010) Guidelines on HIV and infant feeding 2010: principles and recommendations for infant feeding in the context of HIV and a summary of evidence. Retrieved Aug, 2020, from https://apps.who.int/iris/bitstream/handle/10665/44345/9789241599535_eng.pdf

York GK, Heil JR, Marsh GL, Ansar A, Merson RL, Wolcott T, Leonard S (1975) Thermobacteriology of canned whole peeled tomatoes. J Food Sci 40(4):764–769

Zavadlav S, Blažić M, Van de Velde F, Vignatti C, Fenoglio C, Piagentini AM, Pirovani ME, Perotti CM, Bursać Kovačević D, Putnik P (2020) Sous-vide as a technique for preparing healthy and high-quality vegetable and seafood products. Foods 9(11):1537. https://wwwncbinlmnihgov/pmc/articles/PMC7693970/

Zepp M (2019) Approved canning methods: types of canners. Retrieved from https://extension.psu.edu/approved-canning-methods-types-of-canners

13 Low-Temperature Preservation

1 Introduction

1.1 Significance and Benefits of Refrigeration

Chilling and then storing foods at low temperatures without forming ice is called cold storage. However, chilling and holding foods at temperatures sufficient for ice formation is called freezing. Both chilling and freezing are forms of refrigeration, which is the deliberate reduction of temperature for the purposes of preservation. In the past, chilling and freezing were performed with little or no equipment. Mechanized refrigeration processes started from the nineteenth century (Rees 2013; Weightman 2012). The aims of this chapter is to present some basic features of food chilling and freezing. Consumer interest in food chilling and freezing is obvious since most homes own or aspire to a refrigerator.

Historically, refrigeration had a big impact on the American way of life. For instance, the coming of mechanical refrigeration allowed new patterns of production and settlement in the United States (US). Access to refrigerated transport (rail cars, ships, vans etc.) meant cold-food chains could connect otherwise distant parts of continental US and beyond. In the old-world, shipment of chilled foods and frozen foods from the antipodes became feasible as illustrated by the importation of meat from New Zealand and Australia to Europe.

Refrigeration is no less important in the current period for globalization. Efficient food supply cold-chains enable agricultural and fishery products (mainly seafood and chicken) to be processed in low-income jurisdictions to reduce costs (Hueston 2012; Yeong 2005). The potential transfer of food processing jobs due to globalization is raising concern in some quarters – by analogy with the globalization effects on clothing, electronics and car industries in recent times. Another interesting fact is the continuing spread of cold stores, extending secure-cold chains to many parts of the third word. The impact of refrigeration on the global food system and international trade is likely to increase.

Some common benefits of food refrigeration are listed in Table 13.1. Most obviously, cold storage extends shelf life and convenience. Longer storage life allows for bulk shopping by consumers, preparation and storage of whole meals at home and for cook-chill, cook freeze catering operations. The impact of refrigeration on the prevention of food borne disease has been extensively discussed (Marth 1998).

1.2 Refrigeration and Public Health

The introduction of commercial food cold stores and frozen foods in the nineteenth century was important for the transportation of fruits, milk,

R. Owusu-Apenten, E. R. Vieira, *Elementary Food Science*, Food Science Text Series,
https://doi.org/10.1007/978-3-030-65433-7_13

Table 13.1 General benefits of food refrigeration

Improves food safety
Maintains food quality
Extends food supply chains
Allows bulk shopping
Convenience
Avoids food waste
Novel product applications, e.g. cook-chill food services
Novel product lines (chilled tea)
Enhances food security

and meat to urban populations. The public health benefits of refrigeration were first discussed in a series of articles from the American Journal of Public Health (1912–1946) which are available online (Adams 1946; Anonymous 1912; Bryce 1912a, b; Pease 1912; Pennington 1912). From 1912, the extended cold-supply chains established by "the great packing houses" were newsworthy scientific items. Meat held at 32 °F (0 °C) could retain a wholesome state for 2–4 months (Bryce 1912a).

Food refrigeration improves the nutrition of urban populations also because chilled or frozen foods (by contrast with canning) retain much of the nutrients found in fresh foods. It was claimed that that chilling and freezing were instrumental in averting food shortages in 1912 even though the US population was quite modest by today's standards[1] (Bryce 1912a). Efficient cold-chains allowed rail transport of meat from the mid-West to the American East coast and then further still onwards to Victorian England. The successful shipments of frozen sheep meat from Sydney (Australia) to London were also noteworthy (Bryce 1912a).

By 1912 the benefits of large-scale food chilling and freezing had been realized for almost 30 years with many varieties of fresh foods (fruits from Californian, vegetables from Florida, sea food from the North West and meat from Mid-west) becoming shipped across the US and internationally (Enochian and Woolrich 2012). The low degree of microbial hazards associated with properly chilled and frozen food is another factor that affects health (James and James 2014).

1.3 History of the Pre-mechanized Chilling Industry

Domestic cold storage of food probably developed close to the Arctic Circle. Indigenous populations living in Norway, Sweden, Greenland, Alaska, Northern Canada, and parts of Siberia store excess meat in the natural cold. Iceland, just south of thc Arctic Circle is a historically significant source of stockfish, which is cod fish dried in the open cold air. Icelandic exports of stockfish to the Mediterranean were known in the middle Ages (Knútsson and Gestsson 2001). More recently, Nigeria is a major export destination for Icelandic stockfish (Icelandic Fisheries 2017). Coincidentally, ambient Arctic Circle temperature of 50 °F (10 °C) to −22 °F (−30 °C) are similar to conditions frequently used for food refrigeration and freezing (Fig. 13.1).

From the mid-eighteenth century, wealthier households kept winter ice in cellars from December through to June and sometimes August. Commercial food chilling developed in the nineteenth century, using ice recovered from lakes, large ponds and rivers frozen over during the winter. Large scale harvesting, storage, distribution, wholesale and retail of ice started round about 1807 and lasted right up to 1900 (Weightman 2012).

Naturally harvested ice from New England was exported to other parts of continental US and to the Caribbean, Calcutta (India), China and Australia (Weightman 2012). North American ice was exported to Europe also, and was later replaced by cheaper ice from Norway (Rees 2013). At first, lake ice was used for the preservation of fish and meat and then conveyed via railway to centres for consumption. The use of natural ice for food processing was at its peak right up to the early 1930s (Weightman 2012).

[1]Retrospective data shows the US population was 76.2 million (39.6% urbanized) and 92.2 million (45% urbanized) in 1900 and,1910, respectively (from https://en.wikipedia.org/wiki/Urbanization_in_the_United_States#Historical_statistics, Accessed May 2017).

Fig. 13.1 Air-dried stockfish produce from Iceland and Norway

1.4 Modern Refrigeration Capacity

Domestic fridges and freezers are the "must-have" appliances of the twenty-first century alongside of TV's and mobile phones (James et al. 2017; The Economist 2014). Domestic refrigerator ownership started from the 1950s in most high-income countries. Currently, about one quarter of American homes now contain two refrigerators. Globally, the number of home fridges and freezers is estimated at 1.4 billion units and predicted to rise by about 27% by 2020 and 60% by 2030 (Barthel and Götz 2012; The Economist 2014). The figures are corroborated by the International Institute for Refrigeration (Table 13.2).

From 1994 to 2014, refrigerator ownership increased from 24% to 80% in China (James et al. 2017; The Economist 2014). Turning to commercial refrigeration units, it is believed that only 15% of fresh produce that *could* benefit from low temperature storage is refrigerated and world-wide postharvest losses remain unacceptably high (James et al. 2017).

Food refrigeration accounts for 50% of the refrigeration industry, in terms of equipment use. The global sales in refrigerated produce was estimated at US$300 billion, and apparently greater than the total revenue for automobiles sales. Globally the refrigeration industry accounts for four in 1000 employees (12 million). However, there is a global lack of capacity in refrigeration capacity. Overall its thought that only about 6% of the global food production is subjected to refrigerated storage, compared to 4% food refrigeration in India and 90% fresh food

Table 13.2 Estimated number of refrigeration systems worldwide

The applications sectors	Equipment	Number of units
		Thousands
Refrigeration and food/domestic	Refrigerators and Freezers	1500,000
Refrigeration and food/commercial	Stand-alone, and centralized systems, condensers	95,000
Refrigeration and food/refrigerated transport	Vans, trucks, semi-trailers, trailers	40,000
Refrigeration and food/refrigerated transport	Refrigerated containers («reefers»)	
Air conditioning	Air-cooled systems	600,000
Air conditioning	Water cooled	2800
Mobile air-con systems	Air-conditioned vehicles (cars, vehicles and buses)	700,000
Refrigeration and health/medicine	Magnetic resonance imaging (MRI) machines	250
Refrigeration in industry/liquefied natural gas (LNG)	LNG receiving terminals, Liquefaction trains and tankers	0.625
Heat pump equipment	Residential, commercial or Industrial and reversible air-to-air conditioners	160,000
Leisure and sports	Ice rinks	13.5

Adapted from International Institute of Refrigeration (2015)

refrigeration in the UK. Interestingly, about 40% of the electricity supplied to supermarkets was utilized for refrigeration (International Institute of Refrigeration 2015).

1.5 Refrigeration and Food Security

The FAO states "Food security exists when all people, at all times, have physical, social and economic access to sufficient, safe and nutritious food which meets their dietary needs and food preferences for an active and healthy life" (Food and Agriculture Organization of the United Nations 2008; Grace 2016). The dimensions of food security include food availability, access, utilization, and stability of supply. It is becoming more evident that food-processing technology can improve food security in low in-come developing countries (Aworh 2015; United Nations Economic and Social Council 2017).

The role of refrigeration in supporting food security is considered important, against a backdrop of an ever-increasing world population that that is predicted to reach 9 billion by 2050. Most developing countries, where 89% of the world population reside, have serious deficits in refrigeration capacity leading to high levels of postharvest losses. The wider availability of refrigeration and more robust cold-supply chains is expected to reduce postharvest losses, improve food safety, and lead to improved food supply to cities – which is important given the growing trend towards urbanization. According to the International Institute of Refrigeration (Paris), increasing the refrigeration capacity in developing countries will help to support international trade. (International Institute of Refrigeration 1996, 2009, 2011, 2015).

1.6 Elementary Aspects of Low Tempreratue Storage

A brief discussion of the nature of heat, cold and temperature that will hopefully promote a deeper understand of cold storage can be found in Chap. 12.[2] Many fresh foods contain 60–80% of their weight as water. Therefore, it is important to consider how water behaves during chilling and freezing at low temperatures. Over a wide temperature range, water transforms from vapor to liquid and then to ice as it is cooled. Lipids undergo freezing/melting type phase-transitions leading to changes in their physical properties. Oils, which are liquids, turn into solids when the ambient temperature is decreased. The lipid membranes surrounding cells solidifies at below about 41 °F (5 °C) so that bacteria divide less efficiently. Psychrophilic microbes adapted to cold-temperatures will be less affected by refrigeration.

At low temperature the rate of deterioration from chemical, physical, enzymatic, or microbial changes all decrease. In general, lowering a food products temperature by 50 °F (10 °C) produces a twofold reduction in the rate of deterioration. Transferring a product from 77 °F (25 °C) to 59 °F (15 °C) and then to 41 °F (5 °C) produces a net fourfold extension of storage life. Further drops in temperature, to -5 °C and then −15 °C could produce an overall (2^4)16-fold extension in shelf life; as indicated above many food spoilage microorganisms may cease growing at below 5 °C leading to further shelf life gains from chilling and freezing (James and James 2014).

2 Cold Storage

2.1 Historic Perspectives

Cold storage involves chilling food to low temperatures between 59 °F (15 °C) and 28 °F (−2 °C) and holding or storing foods at these temperatures. Refrigeration is currently performed using mechanical units e.g., refrigerated warehouses, rail cars, trucks, refrigerated trailers, and refrigerated shop displays (Rees 2013). However, cold storage predates the industrial revolution. Historically, cold supply chains supported by mechanical refrigeration equipment allowed farm

[2] The principles of heat transfer were covered in Chap. 12 and readers are recommended to review that material when considering the present Chapter.

collection of fluid milk and opened up new market opportunities for the US meat packing industry and fruit and vegetable growers in California and Florida, and shipment of poultry and eggs by rail (Rees 2013). The great cities like New York could source food from wider afield. Freezing provided even greater impact compared with cold storage, which enabled international shipments of meat for the first time. A time-line for the development of mechanical refrigeration is summarized in Table 13.3.

The impact of mechanical refrigeration on international trade is well illustrated by the New Zealand meat and wool industry in the 1880s (Ministry of Culture & Heritage 2015; McLintock 2015). Prior to the advent of mechanical refrigeration, the New Zealand sheep industry was focussed on wool and tallow obtained by rendering or boiling down meat carcases. Wool production generated abundant amounts of live sheep that exceeded domestic demand for meat. This situation would change after the first trial shipments of frozen meat from Argentina and Australia round about 1878/9.

Table 13.3 Time-line for the development of mechanical refrigeration

1748 – William Cullen, invented cooling by vacuum evaporation
1800 – Michael Faraday, ammonia as refrigerant
1834 – Jacob Perkins, first refrigeration machine
1844 – John Carrie, first ice making machine
1851 – T. C. Easton & Co, New York chilled beef shipment to London
1861 – Thomas Sutcliffe Mort, first freezing works in Australia
1864 – Ferdinand Carrie patent for ammonia compression machine
1865 – First Cold storage house, New York
1868 – New York to Glasgow shipment of frozen meat
1878 – Argentina to England shipment of frozen meat
1879 – Australia to England shipment of frozen meat
1882 – Australia, New Zealand to London shipment of frozen meat
1882 – First plate freezer
1928 – Thomas Migley, invention of chlorofluorocarbons (CFCs)
1929 – Clarence Birdseye, fast freezing

Source: Compiled from several sources (Anonymous 1912; Bryce 1912a; Evans 2008b; Heldman and Nesvadba 2003; Pennington 1912)

The vessel SS Dunedin sailed from New Zealand on 15th February 1882 carrying 4000–5000 frozen carcases (130 tons of meat), which arrived in excellent state after a journey of 13,750 miles (Fig. 13.2). The shipment of meat arrived after roughly 3 months in-transit in London (England) on 24th May 1882.[3] This successful transport of frozen meat to UK allowed a rapid expansion of New Zealand meat exports with many more ships being fitted with freezers. Later, shipments of frozen milk, cheese, and butter also followed from New Zealand (Ministry of Culture & Heritage 2015; McLintock 2015).

The first household refrigerators were ice boxes in which ice was placed in a chamber at the top in order to cool a lower chamber where food was kept. Water from melting ice was allowed to run down to a container beneath the icebox and this had to be emptied periodically.

Little was known about the theory and principles of refrigeration in the early days. Much research was required to develop effective systems for circulating air in cold storage rooms. Actually, in fishing boats, a pile of ice in one corner of the pen was used to refrigerate fish instead of being applied directly to the fish as it is today.

Various mechanical refrigeration systems were developed in the past, such as vapor compression refrigerators, vapor absorption refrigerators and thermoelectric (Peltier cooling) refrigerators (Wikipedia 2017). A basic compression refrigerator consists of an insulated compartment or room (the refrigerator) and a continuous, closed system consisting of a refrigerant, expansion pipes or radiator-type evaporator located in the refrigerator, pump or compressor, and a condenser (Fig. 13.3). Refrigerants used with mechanical fridges included, ammonia and the chlorofluorocarbons (CFCs) or Freon. From 1987, CFCs were banned by international agreement owing concerns that these help to deplete the ozone layer; CFCs were replaced by hydro-fluorocarbons (HFCs), which are less damaging for atmospheric ozone. However, HFCs appear to posse other environ-

[3] The years for SS Dunedin meat shipments have been given by a variety of sources as 1882–1887.

Fig. 13.2 The SS Dunedin shipped the first cargo of frozen meat from New Zealand to England foreshadowing an export industry that is still vibrant to this day (Meat and Wool New Zealand 2015). Public domain picture published before 1923

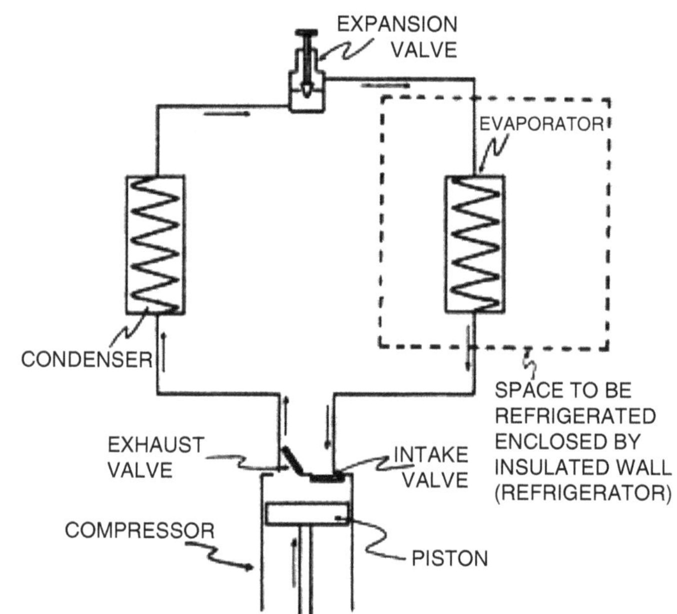

Fig. 13.3 Principle of mechanical refrigeration

mental hazards due their role as greenhouse effect promoting agents.

In most cases, the compressor and condenser are located outside the refrigerator compartment. The refrigerant flows into expansion pipes as a liquid. Here, it evaporates to a gas, and in changing from the liquid to the gaseous state, it absorbs heat. The gas is pulled into the compressor by the suction action of the pump and is then compressed into a smaller volume. The latter action causes the gas to heat up. This heat must be removed, which is accomplished by passing the condensed gas through a system of pipes or radiators usually cooled by water, sometimes by forced air. Cooling the condensed gas liquefies it, whereupon it is returned to the evaporator in the refrigerator. The conversion of the gas to a liquid also produces heat that is transferred to the water or air of the condenser. Special valves at both ends of the evaporator allow the required flow of liquid refrigerant in, and of gas out, of the expansion system in the refrigerator.

There are a number of ways in which refrigeration may be applied to the insulated area to be cooled. Expansion pipes where the refrigerant is evaporated may be located along the walls of the refrigerator. In this case, natural circulation of air (the cold air being heavier) may be depended on to refrigerate areas within the room away from the expansion pipes, or some type of forced-air circulation may be used. In some instances, radiation-type evaporation units are used. A fan that blows air through the radiator fins provides circulation of cold air throughout the refrigerator. An indirect method of cooling refrigerators depends on the use of refrigerated brine in an outside container, cooled (usually below the freezing point of water), then sprayed into a chamber within the refrigerator and returned to the outside cooler. The cold brine spray refrigerates and humidifies the air within the chamber and causes it to circulate outside of the chamber throughout the refrigerator. Refrigerated brine circulating through pipes may also be used.

2.2 Superchilling and Supercooling

Two separate technologies are described as superchilling. Classical superchilling developed in the 1920's involved chilling seafood onboard vessels using slurries of ice. By contrast, modern forms of superchilling require holding samples at very precise temperatures at the border between chilling and freezing (Kaale et al. 2011; Stonehouse and Evans 2015).

Food materials may be supercooled rapidly to temperatures below 32 °F (0 °C) without freezing (Fig. 13.6B). Superchilling allows partial ice crystallization whereas supercooling avoid ice formation altogether. These temperatures for supercooling are obviously at the limits of cold storage just prior to food freezing. With supercooling ice-formation is avoided thereby reducing process times. Supercooling is also believed to retain the fresh attributes of foods to a greater degree compared with food freezing. Some partial ice formation that occurs with superchilling and may function as a coldness reservoir during food distribution. Superchilled chicken and fish had 50% longer shelf life compared with the chilled products (Claussen 2011). Further research is ongoing in order to avoid excess ice formation during superchilling (Kaale and Eikevik 2014; Kaale et al. 2011; Stonehouse and Evans 2015).

Formerly, chilling foods like seafood involved using brine (salt solutions of varying concentrations) as a cold dip or spray. Chilling with brine is faster and more uniform than chilling in ice and the practice is still used today. Because the temperature of the brine can be reduced to well below freezing (to −6 °F [−21.1 °C] when sodium chloride is used and down to −67 °F [−55 °C] when calcium chloride is used), brines have also been used for freezing and, to a degree, are still used today.

Chilled seawater (CSW) is used to chill fish at sea. CSW consists of seawater and a predetermined amount of ice (usually in crushed form) sufficient to cool the catch and to hold it at the temperature of melting ice, which in seawater is somewhat below 32 °F (0 °C). *As* with brine, CSW cools more rapidly and more uniformly than ice alone. Refrigerated seawater (RSW) also used to chill fish at sea, is similar to CSW except that cooling is accomplished with mechanical refrigeration, which permits control of the RSW temperature. Actually, the salt concentration of RSW can be increased if desired, to operate at temperatures below the freezing point of seawater (Medina et al. 2009).

2.3 Refrigeration of Processed Foods

Many processed foods require refrigeration and these must be handled with care. The Reduced Oxygen Packaged (ROP) products are required to be refrigerated due to the risk of *C. botulinum or Listeria monocytogenes* growing under low oxygen (anaerobic) conditions. The ROP products include foods that are subjected to, modified atmosphere packaging, controlled atmosphere packaging, vacuumed packaging, sous vide and also, cook-chill processing (US food code, 2017. p.18). Refrigeration is mandatory also for ready-to-eat foods that include fresh cooked pasta, meat, and vegetable salads (Peck et al. 2018).

Processed foods requiring refrigeration have generally been cooked, pasteurized or else heat processed in some manner that is *insufficient* to produce commercial sterility (Chap. 12, Sect. 2.1). Proteolytic *C. botulinum* (Type A & B) spores can survive normal cooking, and may germinate once food is held within the temperature danger zone. Refrigeration represents a hurdle for proteolytic *C. botulinum*, which is mesophilic with a minimum growth temperature of about (10–12 °C). The spores of *C. botulinum* apparently do not germinate below 10 °C but guidelines are to keep foods at 34 °F (3 °C) for reasons discussed below.

The nonproteolytic strain *C. botulinum* (*Type E*) is a psychrophilic microbe, with a minimum growth temperature cited as 36.5–37.4 °F (2.5–3 °C). Nonproteolytic *C. botulinum* can grow and produce toxin under refrigeration conditions unless this is prevented by using time/ shelf-life restrictions (Peck 2006; Peck et al. 2018). The US food code recommends storage temperature/ shelf-life restrictions as follows; storage at 34 °F (1 °C) is for 31 days or storage at 41 °F (5 °C) is for 7 days. Alternatively, refrigeration can be used in conjunction with one other hurdle, to inhibit *C. botulinum* spores from germinating efficiently; low pH (below pH 4.6), low water activity (Aw < 0.91), nitrite (160–300ppm) or salt concentration of 4–5% (U.S. Food and Drug Administration 2017. p472). The effect of psychrotrophic microbes on food safety and the spoilage of refrigerated foods has been reviewed (IFT/FDA 2001; Kraft 1992).

2.4 Refrigerated Storage of Fruit and Vegetables

Chilled storage of fruit- and vegetables is one of the most important uses of refrigeration. Storing fruits and vegetables at low temperature may also involve modified atmosphere storage that means replacing oxygen with different percentages of more inert gasses such as nitrogen or carbon dioxide. This storage is used to slow down the ripening process in products such as apples, which are harvested in a seasonal, short period but are distributed as fresh product throughout the year. Modified atmosphere storage can prevent financial losses if severe winter storms stall deliveries, which will substantially affect shelf life.

Table 13.4 Considerations for refrigerated storage of fresh produce

Considerations
Postharvest biology and technology
Precooling and storage facilities
Heat load calculation
Controlled atmosphere storage
Temperature preconditioning
Modified atmosphere packaging[a]
Wholesale distribution center storage
Grocery store display storage
Chilling and freezing injury
Respiratory metabolism
Ethylene effects
1-methylcyclopropane (mcp)
Texture
Postharvest pathology
Flavor
Food safety
Nutritional quality

Source: Adapted from Gross et al. (2014)
[a]Includes other reduced oxygen packaging (ROP)

Conditions for refrigerated storage for fruits and vegetables can be exacting, being dependent on a number of physiological characteristics of different products. Fortunately, advice on the best practices and conditions for refrigerating 90 or so commercially important vegetables and fruit is available from UDA-ARS. The USDA handbook on fruit and vegetables has been published at least since the 1990s (Gast 1991) and been updated periodically (Gross et al. 2014) (Table 13.4).

2.5 Domestic Refrigeration Practices and Food Safety

The "4C-principle" for food hygiene describes how food safety can be promoted by proper food handling, involving adequate, Cleaning, Cooling & Chilling, and Cross Contamination avoidance, (cf. Chap. 17 and Sect. 17.2). We return to the notion of chilling and cooling and the importance of refrigeration practices for food safety.

Studies of consumer behavior and general knowledge surrounding refrigerator storages practices have been reported for specific localities, e.g., United Kingdom (Evans and Redmond 2015), Sweden (Marklinder et al. 2004), Portugal, Peoria – Illinois USA (Towns et al. 2006), all-US (Kosa et al. 2007), France (Lagendijk et al. 2008). However, until recently, consumer refrigerator practices had not been documented extensively within food journals. Investigations of refrigeration practices, (Carpentier et al. 2012; Evans and Redmond 2016; Godwin et al. 2006; Kosa et al. 2007; Lagendijk et al. 2008; Rohit et al. 2010) suggest that consumers needed support with good refrigerator practices (Table 13.5).

Research showed thermometer usage (~10%) in home refrigerators was infrequent. The vast majority of domestic refrigerators (~90% of appliances) were also operated at above 41 °F (5 °C) recommended by the USDA (Evans and Redmond 2016). However, results from France suggested that 37% of refrigerator users employ operating temperatures below 41 °F (Lagendijk et al. 2008). Some of the common deficiencies for domestic refrigerator practices were highlighted (cf. Table 13.5). A particularly significant issue seems to be the infrequent use of thermostats with domestic refrigerators (Evans and Redmond 2016; Kosa et al. 2007; Marklinder et al. 2004; Towns et al. 2006). Certain aspects of domestic refrigerator practices had not improved or changed over the past 40 years (James et al. 2017).

Table 13.5 Good refrigerator practice and hygiene

Low operating temperatures
Cleaning and disinfection frequency
Avoidance of condensation
Refrigeration thermometer use
Degree of fullness of fridges
Organization of fridge contents
Avoidance of food spills
Infrequent refrigerator door opening

From several sources. (Carpentier et al. 2012; Evans and Redmond 2016; Godwin et al. 2006; Kosa et al. 2007; Lagendijk et al. 2008; Rohit et al. 2010)

2.6 Refrigeration Practices in Industry

Guidelines for industry refrigeration practices appear in the US food code, which also covers aspects of GMP, e.g. refrigerator equipment design, maintenance, sanitizing, cleaning and staff training (U.S. Food and Drug Administration 2013). One of the key aims of refrigerated storage is to avoid growing *Listeria monocytogenes* and *C. botulinum*.

It is recommended that raw and cooked food products should be separated in commercial refrigeration and chilling plants. Raw foods (fish, meats, poultry, vegetables, and fruits) should be held in a different refrigerator from cooked foods. If cooked and raw foods must be stored together in the same refrigerator, the cooked foods are placed in covered containers, stored above, and separated from the raw foods. This precaution is necessary to prevent cross contamination where bacteria on raw food is transferred to the cooked foods.

The refrigeration capacity when holding foods at temperatures above freezing should be sufficient to take care of peak loads. That is, when comparatively large amounts of material are to be placed in the refrigerator within a short period, the refrigerator capacity must be adequate to cool all parts of all products down to the desired holding temperature within 1to 3 h. Often this is not possible and some alternative must be used. The usual alternative is to precool the product to or near to the desirable holding temperature before it is placed in the refrigerator. The product can be precooled in an insulated area where refrigerated air can be blown through the product, or in a chamber where the product can be placed under vacuum. (The basis of vacuum cooling is the evaporation of water from the product under low pressure.) Heat exchangers and thin film evaporators may be used for liquid or semiliquid materials.

Products should not be placed in the refrigerator in large bulk form unless they have been precooled. Individual units may be placed on trays and the trays placed on racks to facilitate cooling

Table 13.6 Selected guidelines for food Service and industry refrigeration practices

Food hygiene guidelines[a]	Comments
Time/temperature control ([b]3–501.12)	Refrigerate at <41 °F (5 °C) Or at any temperature if food remains frozen
Thawing ([b]3–501.13)	Thaw under refrigeration or running water (<70 °F/21 °C) with all food parts below 5 °C at all times
Cooling foods ([b]3–501.14	Temperature from 57 °C to 35 °C in 2 h or cooled from 57 °C to 5 °C in 6 h
Holding times	Time at 41 °F (5 °C) is 7 days, no restrictions if frozen
Refrigerated temperature logger	Fridge thermometers mandatory, inspection schedule twice/day
Cold transfer	As above
Cold holding records	Record as for HACCP, retain 6 months records
Labelling	Provide information on refrigerated conditions and use by date

[a]From the US food hygiene code (U.S. Food and Drug Administration 2013)
[b]Shows specified sections

of the product within the refrigerator. Once cooled, the individual units may be stacked closer together to conserve space. Liquids to be held in large quantities must be precooled before placing in the refrigerator, or they can be poured into small containers to facilitate cooling (Table 13.6).

Large commercial refrigerators should not open directly to the outside but rather to an anteroom that in turn opens to the outside, to reduce the amount of heat entering the refrigerator from the outside. The doors should be of the swinging, self-closing type. Fresh fish, meats, and poultry should be held at temperatures above, but as near to 32 °F (0 °C) as possible, to maintain good edible quality and prevent spoilage because of excessive enzyme action or excessive bacterial growth. Because these products generally freeze below 31 °F (−0.56 °C), a good temperature range for them is 31–35 °F (−0.56 to 1.67 °C). Such a temperature range is attainable in the well-regulated refrigerator. Safe refrigeration practices for food retail and the food services are outlined by the FDA (U.S. Food and Drug Administration 2013).

2.7 Refrigeration Plant Protection and Quarantine

In an earlier section, we discussed refrigerated cargo ship transport for produce in the late nineteenth century. Current practice requires that conditions for refrigerated cargo transport be clearly defined, to ensure that food quality is maintained. Food safety requires adequate refrigeration plant protection and quarantine (PPQ) procedures (Carefully to Carry Committee 2006). USDA guidelines for *cold treatment* of imported fresh fruit and vegetables are available that help to avoid the transmittance of foreign insect pests (adults, larvae, eggs) in cold shipments. The use of appropriate refrigeration conditions for shipment is seen as part of several phytosanitary measures (USDA 2013).

3 Food Freezing and Frozen Foods

3.1 Frozen Food Industry and Markets

3.1.1 The US Frozen Food Industry

Freezing is one of three key methods for preserving fruits and vegetable (NAIC3114) commercially, alongside canning (NAIC311421) and food drying (NAICS 311423). The frozen food industry also extends to specialty foods (NAICS 311412). A listing of the main frozen foods groups are shown the Tables 13.7 and 13.8.

The scope of the freezing industry is wide because virtually all food groups can be frozen. Economic information related to some of these industries can be found in the chapters covering the specific food groups.

In addition to preservation, freezing technology is used in other areas, e.g. manufacturing ice cream, frozen desserts, and for freeze concentration of juices. In the cases of ice cream and desserts an element of freezing preservation is needed also because such products have a limited storage life at room temperature.

The narrower frozen foods sector (NAICS 31141; see Table 13.7) is estimated at some 659 enterprises with a combined workforce of about 85,000 people and annual receipts amounting to US$50 billion or about 20% of the world total (Statistica 2015). The main product groupings for US frozen foods included in this appraisal were, specialty foods, fruits and vegetables, dinners and entrees, ice cream, pizza, breakfast foods, snacks and appetizers. Figures 13.4 and 13.5 compares frozen food sales by food categories for the US and the European Union. The sector is recognized as being one of the most innovative in food manufacturing (Tables 13.7 and 13.8).

Table 13.7 Classification of major frozen foods sectors (US)

Frozen fruit, juice, and vegetables (NAICS 311411)	Frozen specialty food[a] (NAICS 311412)
Ades drinks and cocktail mixes, blast freezing on contract basis, citrus pulp, concentrates, fruit juice, concentrates, vegetable juice, French fries pre-cooked, fruit, fruit juice concentrates, juices, fruit or vegetable, vegetables	Chop suey, chow mein, frozen dinners, food entrees, French toast, Macaroni, Mexican foods, nationality specialty foods, pancakes, pizza, pot pies, rice dishes, side dishes, soups (not seafood-based), waffles, whipped topping

[a]Excludes sea food

3.1.2 Global Frozen Food Markets

The global frozen food sector is worth 260 US$ billion per year (MarketsandMarkets 2014). The sector is currently valued in excess of US$26 billion for the European Union with the major segments being frozen ready meals, frozen fish and ice cream (Fig. 13.5). The benefits of freezing as compared to canning and drying are apparently, that the former retains more of the original quality of fresh foodstuffs. The processing time for food freezing is short compared other preservation methods (Barbosa-Cánovas et al. 2005).

3.2 Basic Principles of Food Freezing

3.2.1 The Preservation Effect of Freezing

Food is frozen and then maintained at lower temperatures compared with chilled foods. Differences in storage temperature aside, the longer shelf life of frozen products, (about 24 months) compared with chilled foods (max 1 month), is also due a water activity (A_W) difference. For frozen food at −20 °C, the value A_W was approximately 0.82 (Miyawaki 2018). Freezing will remove water from the bulk food in the form of ice, which can be physically separated from the liquid phase leaving behind a concentrate.

Table 13.8 The wider frozen food industry across different sectors[a]

3111	Animal Food (dog and cat food, other animal food manufacturing)
3112	Grain and Oilseed Milling
3113	Sugar and Confectionery Product
3114[a]	Fruit and vegetable preserving
31141[a]	Frozen food
311411[a]	Frozen fruit, juice, and vegetable
311412[a]	Frozen specialty food
3115	Dairy Product
31151	Dairy Product (not frozen) butter, cream, cheese
31152[a]	Ice Cream and Frozen Dessert
3116[a]	Animal Slaughtering and Processing
31161[a]	Frozen meat products–cf. animal slaughtering and processing
311612[a]	*Meat* Processed from Carcasses, includes fresh and frozen
311613	Rendering and Meat Byproduct Processing
311615[a]	Poultry Processing
3117[a]	Seafood Product Preparation and Packaging
3118[a]	Bakeries and Tortilla
311812[a]	Commercial Bakeries, fresh and frozen bread and bread-type rolls + other
311813	Frozen Cakes, Pies, and Other Pastries-frozen bakery products cakes, pies, and doughnuts. (excludes frozen bread)

[a]Some sectors with a refrigeration or freezing emphasis

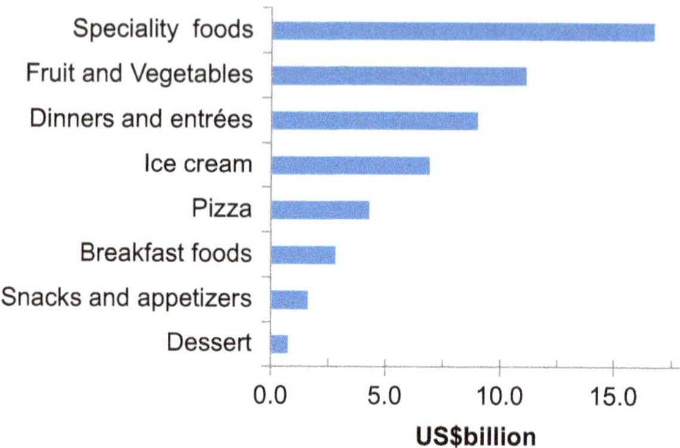

Fig. 13.4 A snapshot of frozen food sales in the US (2012) by category

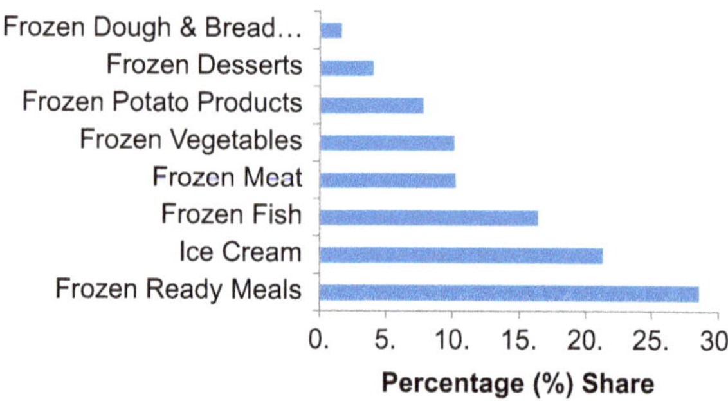

Fig. 13.5 Frozen food sales in the European Union (2013/14)

Cells within a freeze concentrated system will also experience dehydration stress due to the rising osmotic pressure. In summary the preservation action of freezing may be explained by multiple factors including, a decrease of temperature decreased A_W, formation of a glassy state (see below), declining rate of diffusion, and changes to cell structure. Figure 13.6A below shows the freezing of water at 32 °F (0 °C) to about 0 °F (−18 °C). The water in foods does not freeze completely at 32 °F (0 °C) because of dissolved substances in it.

3.2.2 Food Freezing Curves

The physical changes displayed by water during the freezing process can be summarized graphically (Fig. 13. 6B). For convenience the cooling curve is divided into four distinct stages; (1) removal of sensible heat and the cooling of liquid water, (2) freezing of the liquid water and removal of insensible heat also known as the latent heat for crystallization. During ice formation the temperature remains constant, (3) cooling of the solid ice mass, and (4) formation of a freeze concentrated phase (Fig. 13.6B).

The second stage of the freezing process is crystallization, which means ice is formed due to changes in the physical state of water from liquid to solid. Ice crystallization requires nucleation in a process similar to the formation of organic peals around particles of grit inside an oyster. Ice is believed to form when slowly moving water molecules form accretions around virtual nuclei present in solution. The presence of multiple ice nucleation sites leads to many small ice crystals and foods with a finer texture. For reasons that are not entirely clear, one of the benefits of rapid freezing is to ensure the formation of small multiple ice crystals. By contrast, slowed freezing leads to the formation of lower numbers of larger ice crystals associated with a coarser food texture.

The formation of ice will leave behind a freeze-concentrated solution of sugars, salts and other dissolved substances, which eventually forms a eutectic mixture. Continued reduction in temperature eventually freezes the eutectic mixture causing it to vitrify to produce a glassy solid at a so-called glass transition temperature (Tg). Glassy solids possess extremely high viscosity that

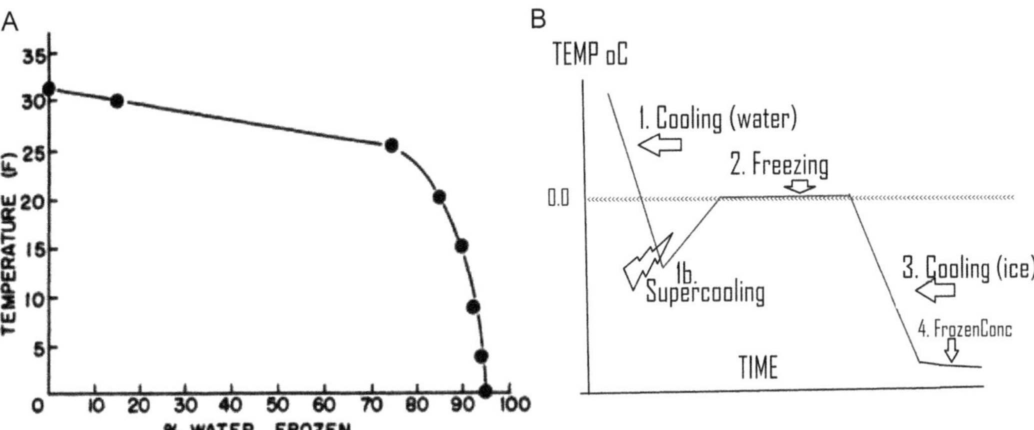

Fig. 13.6 (**A**) A freezing curve for water in foods at 32 °F to 0.5 °F (0–17.5 °C). (**B**) Freezing curve for pure water. Diagram shows (1) cooling of liquid water (1b) supercooling, (2) freezing & phase transition to form ice, (3) cooling of solid ice and (4) formation of freeze concentrate

extend the shelf life of foods more than expected from low temperature exposure alone (Le Meste et al. 2002). As illustrated by frequent occurrences of fossilized insects in amber, entrapment within a high viscosity glassy matrix can achieve unparalleled levels of preservation. Scientists are casting doubt on the survival of DNA for many million years in resin amber (Stankiewicz et al. 1998). Interestingly for food science, the benefits of forming a frozen glassy state appear to be similar to those associated with the glassy state during food dehydration. Examples of freezing curves are shown in Fig. 13.6A, B.

A glassy state formed within frozen and dehydrated foods has very high viscosity and is able to hinder the diffusion of reactants and nutrients needed for microbial growth and chemical changes (Le Meste et al. 2002). In reality, some unfrozen-liquid may remain even at temperatures corresponding to the Tg and perfect preservation is not fully realized. Nevertheless, there is considerable interest in the relations between the Tg of food materials and their quality.

The food polymer model explaining how water affects food stability suggests that the formation of glasses is really important for preservation and other physical chemical changes such as the collapse of food structures, phase changes, caking of powders etc. (Roos 2010)

3.2.3 Rapid Versus Slow Freezing

When foods are allowed to freeze slowly, water molecules migrate from smaller ice crystals to large ice crystals. This process called Ostwald ripening produces a more heterogeneous mass of ice, which impairs product quality. By comparison, when foods are frozen rapidly, sluggish water molecules do not have enough time to migrate to ice crystals at far distances and instead are "frozen in their tracks," so to speak, forming relatively small ice crystals. To maintain top quality, small ice crystals are more desirable than large ice crystals, and rapid freezing methods are employed. Small ice crystals contribute to the formation of a finer texture food and creates less damage to cellular foods. Accommodating small crystals within biological structure avoids membrane damage with less changes of color and texture when foods are thawed; allied issues apply when considering dehydrofreezing (see below).

A high rate of freezing promotes supercooling (Fig. 13.6B), and the formation of small ice crystals. At present, novel methods are being investigated to improve freezing. Investigators are examining the effect of cooling combined with high pressure treatment, ultrasound, or antifreeze protein to promote small ice crystals (Li and Sun 2002; Petzold and Aguilera 2009).

3.2.4 Freeze-Thaw Cycling

Repeated freezing and thawing will occur if freezer doors are opened too frequently. Comparatively large variations in the freezer temperature, due to a faulty thermostat, will also promote freeze-thaw cycling, and a gradual migration of moisture from small to large ice crystals. This Oswald ripening process leads to progressively larger diameter ice crystals during storage. Freeze-thaw cycling can be a problem for frozen fresh foods (Li et al. 2015) as well as frozen prepared foods (Supratim and Coupland 2006).

3.2.5 Purposeful Slow Freezing

Some foods, such as whole fish and fruit in barrels (used for manufacturing jams and jellies), are frozen in bulk by placing them in a cold room on racks (individually or in pans) or standing them on cold room floors. The temperature of such rooms may be as low as −10 to −30 °F (−23.3 to −34.4 °C) and some air circulation may be used within the storage room. In such instances, freezing proceeds at a slow rate. Fish frozen in this manner are immersed in cold water or sprayed with cold water (in pans) to form a glaze coating of ice) that helps to protect it against dehydration during frozen storage. The glaze is built up in layers and, during long storage periods, must be replaced as it is lost by sublimation.

3.2.6 Dehydrofreezing

Dehydrofreezing is an adjunct or variant of food freezing where products are subjected to a partial drying pretreatment before freezing (James et al. 2014). The aim of dehydrofreezing is to remove between 30–50% of moisture content of foods and then freeze. The process, as originally developed in 1946 by the USDA Western Regional Research Lab and patented, involved air-drying before freezing. Numerous variations on the basic dehydrofreezing technique have since been developed by altering the pre-dehydration step, e.g. by using osmotic dehydration, vacuum drying, combination dehydration using vacuum-microwave treatment, infra-red dehydration etc. Dehydrofreezing is suited particularly to suited to fruits and vegetables tissues which are prone to ice damage during conventional freezing (James et al. 2014).

Table 13.9 Claimed benefits of dehydrofreezing compared to conventional freezing

Processing benefits	Product benefits
Decreased weight: Volume ratio	Decreased drip loss
Decrease freezing time	Texture improvement
Lower energy requirements	Color retention
Increased freeze concentration	Nutrient retention
Increased glass transition temperature	Enzymic browning decrease

Source: Adapted from James et al. (2014)

The partial dehydration pretreatment before freezing is intended to reduce the degree of tissue damage arising from ice formation. Creating void spaces by partial dehydration should allow greater degrees of ice formation without damaging organelle and cells wall structures. The effect of the pre-concentration achieved by partial dehydration may explain the reductions in energy demand for freezing by about 25%; freezing time is also reduced. Finally, some products subjected to dehydrofreezing show decreased drip loss when thawed (James et al. 2014).Table 13.9 shows a listing of some possible benefits of dehydrofreezing.

Vacuum dehydrofreezing relies on the fact that the boiling point of water decreases with decreasing pressure. Accordingly, when a product is subjected to a vacuum at pressures of 4.6 mm Hg, cellular water "boils off" at a temperature of 32 °F (0 °C) creating spaces that may be filled with ice during subsequent freezing. Vacuum dehydrofreezing (Lamb 1965), has not been widely discussed in the general literature (James et al. 2014; Ramallo and Mascheroni 2010). Osmotic dehydration has also been applied frequently for dehydrofreezing. To achieve osmodehydrofreezing, a product is immersed in a hypertonic (50% sugar) solution for water removal prior to freezing at temperatures between −4 °F (−20 °C) and −40 °F (−40 °C) (Biswal et al. 1991; Dermesonlouoglou et al. 2007).

3.3 Freezing Equipment

The earliest recorded use of *artificial* freezing appear to in the mid- 1800s, when fish was frozen in pans surrounded by ice and salt. In the late 1800s, fish, meat and poultry were frozen by ammonia refrigeration equipment, with fish being the most important in terms of volume. The commercial freezing of fruits and vegetables started in the early 1900s, the former preceding the latter. Today, there are a number of freezing equipment that were developed for speed, quality, or for specific types of food. Food freezing technology is covered by a number of comprehensive volumes, handbooks, as well as textbooks (Barbosa-Cánovas et al. 2005; Desrosier and Tressler 2012; Evans 2008b). A selection of the freezing equipment currently available is described below (Table 13.10).

3.3.1 Air-Blast Freezing

Air-blast freezing is one of the most commonly used procedures for freezing foods. The foods are packaged and placed on racks, and the racks are wheeled into insulated tunnels (see Fig. 13.7) where air at −20 to −40 °F (−28.9 to −40 °C) is blown over the product at a speed of 500–1500 ft./min (152.4–457.2 m/min). When the temperature of the product reaches 0 °F (−17.8 °C) in all parts, the packages are put into cases and the cases are placed in storage at 0 °F (−17.8 °C) or below.

Air-blast freezing may also be applied to packaged products placed on a belt and carried through cold air tunnels (tunnel blast freezer). A modification of air-blast freezing is used to obtain free-flowing frozen products, such as peas. In this case, the product is frozen, prior to packaging, on a belt operating in a refrigerated, insulated tunnel. Air at −20 °F (−28.9 °C) or below is blown over the belt as it moves along. The frozen product empties from the belt into a hopper from which it is promptly removed, then packaged, cased, and stored. One drawback of the cold air-blast method is that moisture is lost to the cold air because the product is not packaged. Peas, for example, lose about 5% moisture when frozen by this method.

Table 13.10 Some commonly available food freezing equipment

Air-blast freezers
Belt freezers
Contact belt freezers
Contact freezers
Cryogenic freezers
Fluidized bed freezers
Immersion freezers
Indirect contact freezers
Liquid carbon dioxide freezers
Liquid nitrogen freezers
Plate freezers
Tunnel freezers

Source: Adapted from Barbosa-Cánovas et al. (2005)

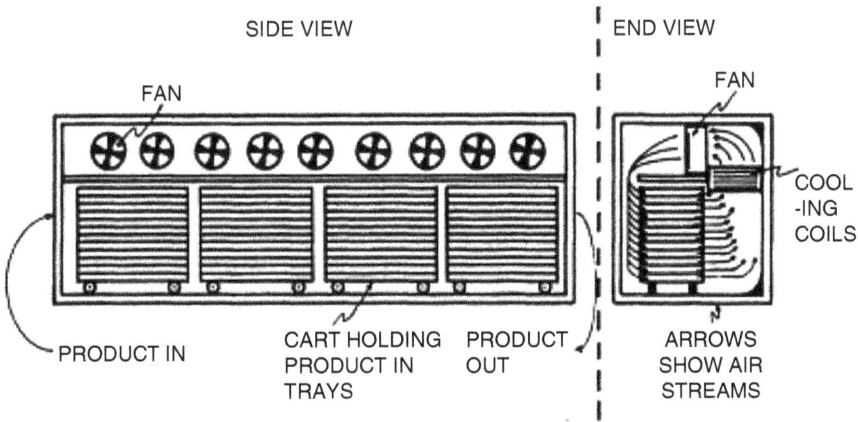

Fig. 13.7 Tunnel blast freezer

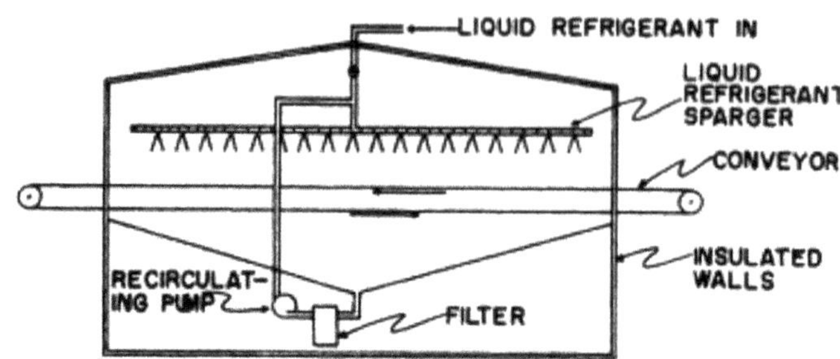

Fig. 13.8 Continuous liquid-refrigerant freezer

3.3.2 Fluidized-Bed Freezing

The fluidized-bed process can be used to freeze foods when cold air is used as the medium. The use of the word *fluidized* derives from the fact that the food particles in air-suspended motion undergo a flow from the starting edge of the bed or perforated plate to the opposite edge where the process ends, thus simulating fluidity. A major advantage of this process is that the food particles are individually quick-frozen (IQF). Even particles that normally tend to agglomerate may be IQF.

3.3.3 Plate Freezing

During plate-freezing layers of the packaged product are sandwiched between metal plates. The refrigerant is allowed to expand within the plates to provide temperatures of −28 °F (−33.3 °C) or below, and the plates are brought closer together mechanically so that full contact is made with the packaged product. In this manner, the temperature of all parts of the product is brought to 0 °F (−17.8 °C) or below within a period of 1.5–4 h (depending on the thickness of the product). The packages are then removed, put into cases, and stored. This method can be used for meat, fish, dairy and other products.

Continuous-operating plate freezers are used in commercial plate freezing. In one such system, the freezer is loaded at the front and unloaded at the rear after completion of the freezing cycle. This is done automatically and continually. In another continuous system, the packages are fed automatically on belts that place them in front of eight levels of refrigerated plates. The packages are forced into the spaces between the plates and the plates closed to provide contact. As freezing proceeds, the packages are advanced so that with each opening of the plates the packages are advanced by one row with a new set of packages entering the front row.

3.3.4 Liquid Nitrogen Freezers

Fluid refrigerants such as liquid nitrogen and dry ice (carbon dioxide) may be used for the quick freezing of foods. Liquid nitrogen has a temperature of −320 °F (−195.6 °C), and CO_2 is at −110.2 °F (−79 °C) in its solid state (Fig. 13.8).

With liquid nitrogen, the individual food portion is placed on a moving stainless steel mesh belt in an insulated tunnel where it is sprayed with liquid refrigerant (Fig. 13.8). Excess refrigerant is recovered, filtered, and recycled. The food leaves the freezer in the frozen state and is thereafter packaged, cased, and stored. This method, often referred to as the IQF (individually quick-frozen) method, provides very fast freezing and is used mostly for certain marine products, such as various forms of frozen shrimp. When using carbon dioxide, two general methods are employed. In one, powdered CO_2 is physically mixed with the food to quickly freeze it. In the other, liquid CO_2 under pressure is sprayed onto the food surface. As the pressure is released during the spraying process, the liquid CO_2 becomes dry ice at −110.2 °F (−79 °C) and the food is quickly frozen.

3.4 General Considerations of Food Freezing Preservation

There are three recognized methods of freezing foods: fast or quick freezing, sharp freezing, and slow freezing. There is no generally accepted definition for differentiating among the various freezing rates. It is obvious that with some methods, such as for food frozen in bulk, the product is frozen slowly.

It is generally agreed that the quality of foods that are frozen quickly is better than the quality of foods that are frozen slowly, and that the *lower* the freezing temperature to which the food is brought the better (Rahman and Velez-Ruiz 2007; James and James 2014). The reasons are that (1) rapid freezing results in the formation of many very small ice crystals that are evenly distributed, and this causes less damage to the tissues of the food; (2) soluble components move about within the food to a lesser degree when the product is quick-frozen because the time required for solidification of the food is shortened; (3) the rates of chemical and biochemical changes are reduced or prevented by decreasing the temperatures rapidly to a certain point. "Quick freezing" as we know it today was started in the US in the early 1920s.

Although a high rate of freezing is important for the quality of frozen foods, the temperature at which a frozen product is held after freezing is more important than the temperature to which it is brought during freezing. It is obvious that if a food is frozen to 0 °F (−17.8 °C), then stored at 10 °F (−12.2 °C), the same changes will take place and at the same rate as if the food were brought to only 10 °F (−12.2 °C), originally. In fact, additional damage is incurred when the product goes initially from 10 to 0 °F (−12.2 to −17.8 °C) and back to 10 °F (−12.2 °C), as any freezing cycle has an adverse effect on the quality of foods. Because frozen foods are stored for much longer periods than those required for freezing, storage changes are of much greater significance to the quality of the product than are the initial changes due to freezing.

The deterioration of many frozen foods will occur approximately twice as fast at 5 °F (−15 °C) as at 0 °F (−17.8 °C). Depending on the product, the Q10 value for frozen foods is 2–20 (Chap. 9, Sect. 3.3). Therefore, a 10- degree rise of temperature will produce a 2 to 20-fold increase in the rate of deterioration. A temperature rise from 4 °F (−15.6 °C) to 14 °F (−10 °C) increased the deterioration rates for pork and fatty fish by 8-9 fold. However, the deterioration rate for beef, fried poultry and ground beef increased by about 3-fold (Fu & Labuza 2012). Actually, freezer warehouses maintain a temperature of 0 °F (−17.8 °C), an FDA requirement for all foods shipped interstate and a requirement for all food-handling establishments, including restaurants. It should be remembered that the lower the temperature, the longer the shelf life and the better the quality.

3.5 Cold Supply Chain

Although changes often take place in frozen foods during storage after freezing, the major changes take place during the distribution. Frozen foods may be carried to retail outlets in trucks that are not refrigerated or in which the refrigeration is not adequate to hold the temperature at 0 °F (−17.8 °C). Frozen foods may be delivered to the retail outlet and left on the unloading platform at ambient temperatures for several hours, sometimes even subjected to the direct rays of the sun. At the retail outlet, products are sometimes placed out of refrigeration, either in an open top display case above the load line, or out of the case entirely. Open-shelf-type frozen food display cases now used in many retail stores are usually incapable of maintaining temperatures of 0 °F (−17.8 °C), especially around the products displayed at the front of the shelf, because refrigerated air is blown from the back of the shelf to the front, and cold air, being heavier than warm air, tends to flow out and down to the floor area in front of the case. Actually, in stores using this type of display case, the cold air can be felt as one walks through the aisle where such storage cases are located. Because high storage temperatures greatly accelerate deteriorative changes in frozen foods, loss of their quality occurs more often during distribution and display at retail stores than during freezing and subsequent storage in frozen food warehouses.

To maintain the quality of perishable goods and avoid temperature abuse requires a well-developed cold chain infrastructure (Fig. 13.9). A comprehensive analysis of cold-chain man-

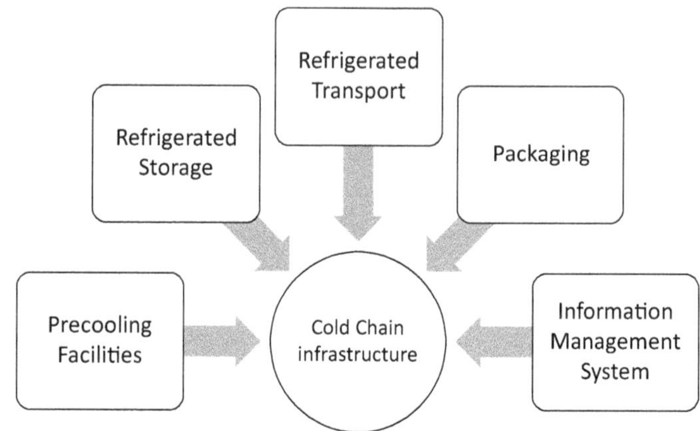

Fig. 13.9 Elements of cold chain infrastructure

agement is beyond the scope of this text (Fuller, 1998). The monograph edited by Da-Wen Sun devotes a whole section (6 Chapters) to the elements of cold chain, including freezing methods and equipment, cold store design, transportation, retail displays, household refrigeration, and finally monitoring and control of cold chains (Sun 2012). The monograph edited by Christopher Kennedy is also recommended reading on the topic of cold chain management (Kennedy 2000). A recent report predicted that the world "cold chain market" (refrigerated storage or transport) would be worth over US$233 billion by 2019 with annual growth rate of 15.6% for the period 2014–2015. Interestingly, it is worth highlighting the role of effective cold-supply chains in the globalization of food trade (Markets and Markets 2014).

3.6 Pre-processing Foods for Freezing

3.6.1 Blanching

Many foods require pretreatment before freezing. With a few exceptions, such as onions, vegetables must be blanched before they are frozen and stored. Vegetables are blanched by heating in steam or hot water (about 210 °F [98.9 °C]), while being carried along a perforated metal belt or screw conveyor, until the temperature of all parts reaches 185–200 °F (85–93.3 °C), then cooling with sprays of water or in water flumes. Vegetables are blanched prior to freezing in order to deactivate enzymes which are present naturally in living matter, and which if not inactivated would facilitate spoilage reactions within the food during storage. Reactions catalyzed by endogenous enzymes are known to produce off-flavors (hay like flavors) in the vegetable as well as changes in color and texture.

Blanching times vary with the type of food, the type of heat used (steam or hot water), and the bulk of material being heated. The heating must reach a high temperature in the product and remain at that temperature long enough to inactivate the particular enzymes that cause off-flavors. Typically, inactivation of the peroxidases and lipoxygenase can be used as indication that blanching is adequate. Blanching times used for frozen vegetables may be relatively short (about 1 min in water at 210–212 °F [98.9–100 °C] for peas) or relatively long (about 9 min in steam for corn on the cob). Besides destroying enzymes that would cause off-flavors and texture changes, blanching has a cleaning effect and destroys contaminating bacteria. Moreover, blanching fixes the color of green vegetables by removing air from beneath surface tissue and possibly affecting the chlorophyll.

3.6.2 Other Pre-processes for Freezing

Citrus juices are often concentrated to about one-fifth of their original volumes, and then diluted back to one-fourth of the original volumes with fresh juice prior to freezing. This dilution with fresh juice is done to add back flavor components because most of the flavor compounds are removed during vacuum concentration of the juice. When this product is to be reconstituted, it requires the addition of water equal to three times its volume.

Other solid natural products undergo simple pre-processes before freezing involving, sorting, cleaning, pealing, slicing, and removing pips or stones. Some foods may be treated with additives (e.g. ascorbic acid) to prevent oxidation during prolonged storage. Cooked as well as fresh foods (mainly fruits) may be stored in the frozen state (Barbosa-Cánovas et al. 2005).

3.7 Packaging Frozen Foods

Packaging is an important factor in the freezing of foods. It is not possible to maintain high humidity in frozen storage rooms because moisture present in the atmosphere of freezer storage rooms tends to condense and freeze on the expansion pipes or other apparatus used to cool the room, thereby reducing the cooling efficiency of the system. Packages used for frozen foods, therefore, must be reasonably moisture-vapor-proof. Otherwise, the product will dehydrate, causing undesirable changes in its appearance and accelerating loss in flavor and juiciness. Toughness and other manifestations of deterioration in quality will also be accelerated (Sun 2012; Erickson and Yen-Con 1997; Hui et al. 2004; Kennedy 2000).

Even when food is packaged in vaporproof material, however, some moisture can be lost from the food. This may happen when there is too much space inside the package not occupied by the food itself. This results in the formation of what is called cavity ice or ice within the package. In such instances, the product, which is warmer than the package, loses water to the cavity space by evaporation, as any water vapor in the cavity space is condensed on the colder package. Under these conditions, there is a transfer of water from the food to the inside of that part of the package surrounding the cavity. Cavity ice formation may cause the same undesirable changes to take place as moisture loss to the atmosphere outside the package, although it usually occurs in localized areas.

The frozen-food package should have sufficient mechanical strength; that is, it should have adequate burst strength and tear strength at low temperatures and high wet strength against water exposure. During freezing, the package will be subjected to stresses because of the expansion of the product. Because of this and possible cavity ice formation, the package should conform closely to the shape of the food. A suitable level of flexibility is, therefore, a desirable property of the packaging material.

Packages should be liquid-tight because some frozen foods (such as fruits packed in syrup) may have some free liquid that could leak out, and some are thawed in the package, which could also cause leakage.

Transfer of moisture-vapor through packaging may take place through pores or cracks in the container or by diffusion of moisture through the packaging material. The loss of moisture through seals, and package imperfections such as pores, may account for the greatest loss of moisture from frozen foods.

Tin or aluminum cans, waxed paper tubs or cylindrical containers, rectangular paper cartons treated with special waxes, plastic bags of polyvinylidene chloride or polyethylene, and aluminum foil wraps or aluminum dishes are all used for the packaging of frozen foods. Many paper containers are used in combination with cellophane, waxed paper liners, or overwraps. Various reduced oxygen packaging is recommened for storing frozen foods at 41 °F (5 °C) or lower (U.S. Food and Drug Administration 2013).

3.8 Frozen Food Quality

A number of changes occur during the frozen storage of foods, which affects the sensory, nutritional, storage life and safety (Erickson and Yen-Con 1997). These changes may be physical, chemical, enzymatic, and, in rare cases, microbial. Microbial changes occur when refrigeration is inadequate. The storage life of frozen foods, at temperatures that today are considered to be

Table 13.11 Freezer storage times at temperatures of 0 °F or below

Freezer storage time (weeks)	Product
4–8	Bacon
	Sausages
	Ham
	Hot dogs
	Luncheon meats
8–12	Casseroles
	Meat (cooked)
	Soups and stews
12	Gravy
	Meat
	Poultry
12–16	Frozen dinners
	Entres
	Meat uncooked (ground)
	Poultry (uncooked giblets)
16	Poultry (cooked)
16–48	Meat (uncooked roast)
	Meat (uncooked steaks and chops)
32–48	Wild game (uncooked)
36	Poultry (uncooked parts)
48	Egg whites and substitutes (unopened)
	Poultry (uncooked whole)

Note: Freezer storage times are for quality only. Frozen foods remain safe for periods of years
Source; USDA website (USDA Food Inspection Services 2010)

economically practicable, is not without limits. Some idea of the length of time that frozen foods may be expected to retain high quality may be obtained from Table 13.11.

The storage life of frozen foods refer to those periods at which a quality difference between the frozen and the fresh product can first be detected by an expert panel and do not refer to spoilage or rejection times (so-called use-by dates). Some of the characteristic causes of quality loss frozen foods are discussed in the following sections (Table 13.12).

3.8.1 Desiccation

Desiccation, freezer-burn or drying out is a kind of change that takes place in frozen foods under conditions of poor packaging or varying storage temperatures. In poultry, such loss of moisture occurs first around the area of the feather follicles, causing a speckled or "pock-marked" appearance. Protein and fat changes may also be accelerated when moisture is lost from the surface areas of frozen foods. The drying out occurs by the sublimation of ice from the product surface, which may increase for individually quick frozen (IQF) food (Schmidt & Lee 2009).

Table 13.12 Quality changes during frozen storage

Quality changes	Examples
Sensory	Desiccation (freezer burn), crystallization, colloidal instability, protein denaturation, atmospheric oxidation, non-enzymic browning, color changes, loss of volatiles
Nutritive	Oxidation, non-enzymic browning
Storage	Same as sensory
Safety	Thawing effects

3.8.2 Crystallization

Crystallization is a physical change that may occur in some types of frozen foods. Certain types of dairy products, such as ice cream and concentrated milks or creams, sometimes undergo this type of change, which is due to the crystallization of lactose or other sugars that do not readily dissolve on defrosting, causing an undesirable texture called sandiness. A similar change may occur in some types of sweetened citrus juices during long storage.

3.8.3 Loss of Volatiles

Loss of flavor components may occur in some frozen foods, such as fruit, during frozen storage because some volatile compounds boil or evaporate at temperatures lower than the temperature at which the foods are stored. This causes a loss of the typical flavor components of the foods.

3.8.4 Colloid Instability

The breaking of gels or emulsions during defrosting may take place with some foods. High-moisture fruits, such as tomatoes, the liquid of which is held by pectin gels, are subject to this type of change and have, therefore, not been frozen successfully. Foods packed with a white sauce or gravy are also subject to this type of physical change and it has been found that fluctuation of storage temperatures greatly accelerates the curdling and "weeping" of the gravy or white sauce.

3.8.5 Protein Denaturation

Protein denaturation is a general term for the physiochemical change occurring especially in flesh-type foods during frozen storage. It results in a toughening of the tissues and dryness or loss of succulence. Cold induced changes in proteins are partly chemical and physical in nature. Cold-denaturation of proteins is a physical process and the counterpart of the more familiar process termed high temperature denaturation. Loss of protein solubility can occur also when proteins undergo chemical modification by reacting with acids and aldehydes generated enzymatically. For example, enzymes may hydrolyze fatty acids from phospholipids in tissues or decompose trimethylamine oxide and produce formaldehyde. The free fatty acids or aldehydes then combine with protein chains, which causes them to aggregate resulting in a decline in the water-holding capacity and increased tendency for drip occurring during defrosting. Protein changes also produce firming or toughening of the tissues. Protein changes of this type occur especially in lean fish but also, to some extent, in poultry and meats. Grinding of tissues of this kind appears to accelerate protein changes during frozen storage (Huff-Lonergan and Lonergan 2005).

3.8.6 Oxidation of Food Components

The oxidation of some of the components of frozen foods is a cause of some changes that may take place during storage. Oxidation of ascorbic acid or vitamin C is a change of this kind. In fruits in which enzymes have not been inactivated by heating, oxidation may be accelerated by enzyme action, but it can also occur through non-enzymic or atmospheric oxidation. Aside from water-soluble nutrients, oxidation may also affect fat-soluble nutrients, and food colors (Nicolas and Potus 1994). Oxidative changes affecting susceptible nutrients will results in a loss of the nutritional quality of the food.

The fats of foods may oxidize during frozen storage, resulting in a change recognized as oxidative rancidity. The fats of meats may undergo this change but fish fats are especially subject to this kind of deterioration because the fatty acids in fish fats contain many adjacent carbons not fully saturated (combined) with hydrogen, i.e., polyunsaturated fats. They are, thus, very subject to reacting with oxygen, a process that eventually leads to the formation of compounds that cause the off-flavors recognized as rancid flavors. Fatty fish become rancid faster in frozen storage than do lean fish, but even lean fish are subject to rancidification during relatively short periods of frozen storage. As noted above, some colors from fruits and vegetables may also deteriorate through oxidation during frozen storage, resulting in a less desirable appearance.

3.8.7 Enzymatic Changes

Enzymes may cause various changes in foods during frozen storage. It is not possible to freeze and store completely uncooked lobster, because, during storage, enzymes react with the proteins, causing the meat to become soft and crumbly after cooking. On the other hand, if whole lobsters are cooked to inactivate the enzymes and then frozen and stored, the oil in the digestive gland (tomalley) oxidizes and becomes rancid, causing off-flavors that spread into the meat itself.

In fruits, enzymes, called polyphenolases, accelerate the oxidation of certain chemicals leading to the formation of brown- or black-colored compounds. This is the reaction that one sees when an apple or peach is cut and allowed to stand at room temperature for a short period. In preparing apples for freezing, therefore, peeled apple slices are treated with salt that liberates sulfurous acid (sulfur dioxide) and are held under refrigeration for sufficient time to allow this compound to diffuse into the tissues.

Sulfur dioxide inactivates or inhibits the enzymes that promote enzymatic browning. Also, in preparing sliced peaches for freezing, small amounts of ascorbic acid (vitamin C) are added to the syrup because this compound is a reducing agent (counter acts oxidation reactions).

3.9 Freezing and Food Safety

Recent advice from the USDA (USDA Food Inspection Services 2010) suggests that freezing food will preserve safety "indefinitely". Under ideal conditions, most microorganisms will not grow in frozen foods held at temperatures of 0 °F. At below freezing temperatures food borne microbes are "inactivated" or held in "suspended animation". In technical terms microorganism enter an extended lag-phase where there is no increase or decrease in numbers during storage (Marth 1998). The notion of indefinitely safe frozen foods seems to be supported by practical observations, but there are some important caveats.

The shelf life of frozen Nile Tilapia (*Oreochromis niloticus*) was declared arbitrarily as 9 months leading Ethiopian food scientists to examine changes in chemical and microbiological quality of this product (Emire and Gebremariam 2010). During storage for 90 days at 0 °F (−18 ± 2 °C) the microbiological load for Nile tilapia decreased threefold from 25.7 to 8 × 10^5 CFU/g and fecal coliforms decreased from 23 MPN (most probable number) to a negligible MPN value. Interestingly, freezing was accompanied also by linear increase of pH from pH 6.4 to about pH 6.65 whilst values for total volatile base nitrogen (a measure of spoilage) doubled. Frozen Nile tilapia was microbiologically acceptable after 90 days storage, but there were significant declines in protein and mineral content whilst fat content increased (Fig. 13.10).

Other studies showed that prolonged storage could lead to increasing growth of cold-adapted microorganisms in meat (Vieira et al. 2009) and meat products (Ambili et al. 2008) particularly when associated with unhygienic handling (Ambili et al. 2008).

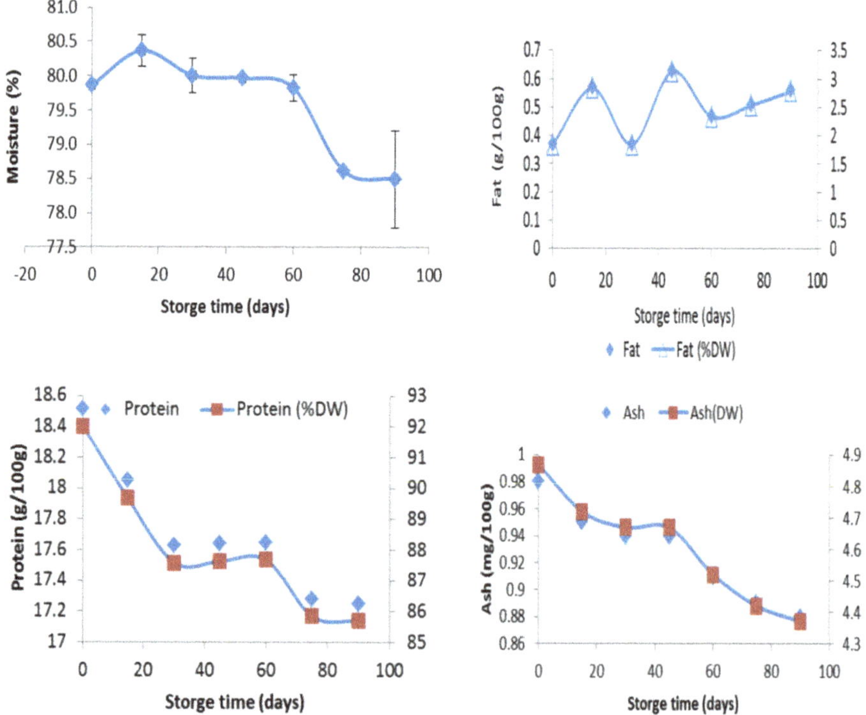

Fig. 13.10 Changes in proximate composition of the Nile tilapia during frozen storage for 3-months. Drawn using data from (Emire and Gebremariam 2010). Note that changes are also corrected for differences in moisture content

Whist freezing may lead to microbiologically safe products thawing foods will enable microbes to resume activity. There may be concerns also where there is freezer malfunction or where there is frequent opening-closing of a freezer leading to the formation of heat spots or where warm food is added to a freezer, or where a freezer is otherwise overloaded. Such instances of temperature abuse can render freezer content hazardous (Karthikeyan et al. 2015). In principle, foods may be maintained indefinitely and any limitation s in freezer storage are assumed to be due to loss in food quality and not food safety.

At temperatures above 15 °F (−9.4 °C) there may be enough free liquid in the food to allow the growth of some microorganisms, especially molds. A 1998 IFT position statement on the safety of refrigerated foods (as opposed to frozen) foods indicates that psychrotropic pathogens may grow at temperatures close to those for freezing water (Marth 1998). Currently, it appears that frozen foods have not often been involved in the transmission of foodborne disease, although, if grossly mishandled, such as by defrosting and then holding out of refrigeration for some time, they may constitute a public health hazard.

The safety of commercial frozen foods may be examined in a different light. Recent investigations have shown that frozen foods can be contaminated with pathogens to differing degrees, depending on the handling characteristics before freezing. For instance, imported frozen fish from different sources could demonstrate differing levels of microbiological quality (Elhadi et al. 2016; Kulawik et al. 2016).

4 Thawing or Defrosting

4.1 Potential Problems with Thawing

Thawing products for the food process plant, for foodservices applications or domestic/foodservice use can be challenging. Thawing is time-consuming and, in some cases, can lead to a loss of product quality (Javadian et al. 2013). It may also take longer to thaw food than to freeze under similar heat transfer conditions. The basic science behind thawing is similar to those for freezing (Quang Tuan 2014). There is a lot of interest in the effect of thawing on foods, especially meat (Leygonie et al. 2012; Li et al. 2015), and also fruits and vegetables (Ibanez et al. 1996).

Defrosting of bulk-packed foods, fruit in barrels, liquid egg frozen in 30-lb (13.6 kg) tins, may lead to a considerable loss of quality. When frozen, bulk-packed foods are allowed to defrost at room temperature, the outside layers of the food will have been held at room temperature for a long period of time before the inner layers become defrosted. This period of holding at room temperature may result in enzyme changes or in loss of quality caused by the growth of microorganisms.

Foods frozen in bulk, including fruit, large fish, or blocks of meat used for sausage products, may present a defrosting problem. The rate at which food defrosts is dependent on the temperature to which it is exposed and consequently, there may be a tendency to defrost the food at relatively warm temperatures. There are some well-recognized examples of bad practice in food thawing; e.g. thawing a bulk food on the porch or outdoors, thawing food on the tabletop counter, or using hot water; microwave heating to thaw and then not cooking immediately.. Some methods of alleviating the problems associated with the defrosting of bulk-frozen foods have been developed.

4.2 USDA Guidelines for Safe Thawing

USDA guidelines for safe thawing of foods is to avoid heating in hot water or leaving on the counter top for >2 h. Both hot water thawing and room temperature thawing risks keeping food in a temperature-danger zone where food pathogens can grow (USDA 2017). The National Restaurant Association and USDA all recognize three to four safe methods for thawing foods. Guidelines for thawing appear also alongside of rules for safe refrigerator practices in the food hygiene code (Sect. 2.6) (Table 13.13).

Table 13.13 Safe thawing conditions for frozen foods

Cold water thawing, at/below 70 °F (21.1 °C).
Cooking directly without thawing
Microwave thawing
Refrigerator thawing, at/below 40 °F (4.4 °C).

4.3 Refrigerator Defrosting and Cold Water Thawing

Refrigerator defrosting (holding at temperatures of 35–40 °F [1.7–4.4 °C]) is considered the best method of defrosting bulk-frozen foods when no fast method is available. This would apply to large whole fish, fruit in barrels, and apples in 30-lb (13.6 kg) containers, as bacterial or mold growth would be limited under these conditions. However, in industrial processing, where bulk-frozen products are thawed as an intermediate step in the manufacture of the company's line of products, the refrigeration space required may be so large as to discourage this practice. In defrosting eggs in 30-lb (13.6 kg) containers, it may be possible to use a machine that breaks up or grinds the product into a kind of slush that can be used in that form in preparing baked foods. Large blocks of meat used for manufacturing sausage products can be ground in the frozen state and used as such.

Defrosting under cold water is a viable proposition and quicker than refrigerator defrosting. The ideal approach recommended by the USDA is not to put the food product (usually meat) directly into contact with water, but rather to place in a watertight plastic container and then immerse in water. This approach avoids the product picking up water and cross-contamination. The water temperature is required to be below 40 °F (4.4 °C) during the thawing process and there should be frequent changes of water perhaps every 30 min (USDA 2017).

4.4 Defrosting by Cooking or Microwave Heating

Food can be thawed rapidly and with virtually no quality loss by the use of microwave energy. If microwave thawing is used, it is recommended that the food be cooked or processed immediately. Because the heat generated in foods by microwaves is quite rapid (about 10 times more rapid than by baking), when uneven heating in a frozen product does occur, the temperature differences within a food can become great. This, however, happens only under certain conditions, and it can be dealt with quite easily. For this condition, and when one wants to ensure uniform temperature control, one solution is to apply the microwave energy in intermittent bursts. By this technique, the absorbed thermal energy generated during a burst of microwaves is allowed to be distributed by conduction during the intervals between the bursts, thereby permitting the temperature of the food to increase more uniformly, albeit more slowly. Developments such as wave-guides and turntables have improved the distribution of microwave energy significantly (Kalla and Devaraju 2017).

The particular advantage of using microwave energy for thawing foods is that the process time is short and so deterioration by microorganisms is not a factor (Kalla and Devaraju 2017). The main difference between microwave thawing and thawing by conventional heating is that the former allows heating from the product interior thereby avoiding surface heating and opportunities for surface growth of microbes. The feasibility and benefits of microwave thawing of frozen meats and fish have been adequately demonstrated, especially for thawing frozen shrimp blocks.

4.5 Further Reading

A basic overview of pre-mechanized and conventional refrigeration technology is available from Wikipedia (2017). The topics of food chilling and freezing have been extensively documented in dedicated textbooks, and monographs (Desrosier and Tressler 2012; Erickson and Yen-Con 1997; Evans 2008a; Hui et al. 2004; Jeremiah 1996; Jul 1984; Kennedy 2000; Robinson 1985; Timm and Herrmann 1996; Wignall and Burt 1987). The FAO also produced a technical book, freely available, for businesses persons interested in starting a food freezing enterprise (Barbosa-Cánovas et al. 2005). Basic guidelines for home refrigeration and freezing have been produced by University extension agents based within most States (Garden-Robinson 2013; Schafer 2014).

References

Adams HS (1946) Refrigeration in a Food Control Program. Am J Public Health Nations Health 36(9):1007–1011. Retrieved from http://www.ncbi.nlm.nih.gov/pmc/articles/PMC1625930/

Ambili VS, Latha C, Nanu E, Prejit (2008) Effect of freezing on the quality of beef frankfurter with special emphasis on public health and food safety. J Vet Public Health 6(1):1–8

Anonymous (1912) The romance of refrigeration. J Public Health 2(11):840–848. Retrieved from http://www.ncbi.nlm.nih.gov/pmc/articles/PMC1089468/

Aworh OC (2015) Promoting food security and enhancing Nigeria's small farmers' income through value-added processing of lesser-known and under-utilized indigenous fruits and vegetables. Food Res Int 76(4):986–991. https://doi.org/10.1016/j.foodres.2015.06.003

Barbosa-Cánovas GV, Altunakar B, Mejía-Lorio DJ (2005) Freezing of fruits and vegetables: an agribusiness alternative for rural and semi-rural areas: FAO Agricultural Services Bulletin 158. Retrieved from http://www.fao.org/docrep/008/y5979e/y5979e00.htm#Contents

Barthel C, Götz T (2012) The overall worldwide saving potential from domestic refrigerators and freezers-With results detailed for 11 world regions. Retrieved from http://www.bigee.net/media/filer_public/2012/12/04/bigee_doc_2_refrigerators_freezers_worldwide_potential_20121130.pdf

Biswal RN, Bozorgmehr K, Tompkins FD, Liu X (1991) Osmotic concentration of green beans prior to freezing. J Food Sci 56(4):1008–1012. https://doi.org/10.1111/j.1365-2621.1991.tb14628.x

Bryce PH (1912a) The conservation of food products by refrigeration. Am J Public Health 2(5):325–328. Retrieved from http://www.ncbi.nlm.nih.gov/pubmed/18008662

Bryce PH (1912b) Physics of refrigeration. Am J Public Health 2(11):829–833. Retrieved from https://www.ncbi.nlm.nih.gov/pmc/articles/PMC1089463/

Carefully to Carry Committee (2006) Carriage instructions for refrigerated cargoes. From https://www.ukpandi.com/fileadmin/uploads/uk-pi/LP%20Documents/Carefully_to_Carry/Carriage%20instructions%20for%20refrigerated%20cargoes.pdf

Carpentier B, Lagendijk E, Chassaing D, Rosset P, Morelli E, Noel V (2012) Factors impacting microbial load of food refrigeration equipment. Food Control 25(1):254–259

Carriage instructions for refrigerated cargoes (2006). Retrieved from https://www.ukpandi.com/fileadmin/uploads/uk-pi/LP%20Documents/Carefully_to_Carry/Carriage%20instructions%20for%20refrigerated%20cargoes.pdf

Claussen IC (2011) Superchilling concepts enabling safe, high quality and long term storage of foods. Procedia Food Sci 1:1907–1909. https://doi.org/1910.1016/j.profoo.2011.1909.1280

Dermesonlouoglou EK, Giannakourou MC, Taoukis P (2007) Stability of dehydrofrozen tomatoes pretreated with alternative osmotic solutes. J Food Eng 78(1):272–280. https://doi.org/10.1016/j.joodeng.2005.09.026

Desrosier NW, Tressler DK (2012) Fundamentals of food freezing. Springer, Dordrecht. 630 pages

Elhadi N, Aljeldah M, Aljindan R (2016) Microbiological contamination of imported frozen fish marketed in Eastern Province of Saudi Arabia. Int Food Res J 23(6):2723–2731

Emire SA, Gebremariam MM (2010) Influence of frozen period on the proximate composition and microbiological quality of Nile tilapia fish (*Oreochromis niloticus*). J Food Process Preserv 34(4):743–757. https://doi.org/10.1111/j.1745-4549.2009.00392.x

Enochian RV, Woolrich RV (2012) The rise of frozen foods. In: Desrosier NW, Tressler DK (eds) Fundamentals of food freezing. Springer, Dordrecht, pp 1–27. 630 pages

Erickson MC, Yen-Con H (1997) Quality in frozen food, xxi + 484pp

Evans JA (2008a) Frozen food science and technology, x + 355pp

Evans JA (2008b) Frozen food science and technology. Retrieved from http://onlinelibrary.wiley.com/book/10.1002/9781444302325

Evans EW, Redmond EC (2015) Analysis of older adults' domestic kitchen storage practices in the United Kingdom: identification of risk factors associated with listeriosis. J Food Prot 78(4):738–745. https://doi.org/710.4315/0362-4028X.JFP-4314-4527

Evans EW, Redmond EC (2016) Time-temperature profiling of United Kingdom consumers' domestic refrigerators. J Food Prot 79(12):2119–2127

Food and Agriculture Organization of the United Nations (2008) An introduction to the basic concepts of food security. Retrieved from http://www.fao.org/docrep/013/al936e/al936e00.pdf

Fu B, Labuza TP (2012) Shelf life testing: procedures and prediction methods for frozen foods. In: Erickson MC, Hung YC (eds) Quality in frozen food. Springer, New York, pp 377–416. https://books.google.com.gh/books?id=6xfpBwAAQBAJ

Fuller RL (1998) A practical guide to the cold chain from factory to consumer. Retrieved from http://www.nutrifreeze.co.uk/Documents/THE%20COLD%20CHAIN.pdf

Garden-Robinson J (2013) Food freezing guide. Retrieved from https://www.ag.ndsu.edu/publications/food-nutrition/food-freezing-guide/fn-403-food-freezing-guide.pdf

Gast KLB (1991) Postharvest management of commercial horticultural crops: storage conditions fruits &

vegetables. Retrieved from http://www.ksre.ksu.edu/bookstore/pubs/mf978.pdf

Godwin SL, Fur-Chi C, Coppings RJ (2006) Correlation of visual perceptions of cleanliness and reported cleaning practices with measures of microbial contamination in home refrigerators. Food Prot Trends 26(7):474–480

Grace (2016) Food security & food access. Retrieved from http://www.sustainabletable.org/280/food-security-food-access

Gross KC, Wang CY, Saltveit M (2014) The commercial storage of fruits, vegetables, and florist and nursery stocks. USDA-ARS Agriculture Handbook Number 66 (HB66). Retrieved from http://ucanr.edu/datastoreFiles/234-2927.pdf

Heldman DR, Nesvadba P (2003) Food freezing history. Encyclopedia of agricultural, food, and biological engineering (Print), p 350. Retrieved from https://books.google.co.uk/books?id=fCRpUZzT2hMC

Henderson T (2016) Thermal physics – Lesson 1 – Heat and temperature: what is heat? Retrieved from http://www.physicsclassroom.com/class/thermalP/Lesson-1/What-is-Heat

Hueston WMA (2012) Overview of the global food system: changes over time/space and lessons for future food safety. In: Institute of Medicine (US). Improving food safety through a one-health approach: workshop summary. National Academies Press, Washington, DC. Retrieved from https://www.ncbi.nlm.nih.gov/books/NBK114491/

Huff-Lonergan E, Lonergan SM (2005) Mechanisms of water-holding capacity of meat: the role of postmortem biochemical and structural changes. Meat Sci 71(1):194–204

Hui YH, Cornillon P, Guerrero-Legaretta I, Lim MH, Murrell KD, Wai-Kit N (2004) Handbook of frozen foods, xi + 735pp

Ibanez E, Foin A, Cornillon P, Reid DS (1996) Study of different thawing methods for frozen fruits and vegetables. (1996 IFT annual meeting: book of abstracts): 1996. No Other Source Information Available (Conference proceedings)

Icelandic Fisheries (2017) Africa. Retrieved from http://www.fisheries.is/economy/markets/Africa/

IFT/FDA (2001) Evaluation & definition of potentially hazardous foods. A report of the Institute of Food Technologists for the Food and Drug Administration of the U.S. Department of Health and Human Services. Compr Rev Food Sci Food Saf 2(Supplement):108pp. https://www.fda.gov/files/food/published/Evaluation-and-Definition-of-Potentially-Hazardous-Foods.pdf

International Institute of Refrigeration (1996) Informatory note on refrigeration and food. The role of refrigeration in worldwide nutrition. Retrieved from http://www.iifiir.org/userfiles/file/publications/notes/NoteFood_01_EN.pdf

International Institute of Refrigeration (2009) The role of refrigeration on worldwide nutrition. 5th informatory note on refrigeration and food introduction. Retrieved from http://www.iifiir.org/userfiles/file/publications/notes/NoteFood_05_EN.pdf

International Institute of Refrigeration (2011) The cold chain, food security and economic development. Retrieved from http://www.iifiir.org/userfiles/file/publications/statements/Statement_2010_Cold_Chain.pdf

International Institute of Refrigeration (2015) The role of refrigeration in the global economy. 29th informatory note on refrigeration technologies retrieved from http://www.iifiir.org/userfiles/file/publications/notes/NoteTech_29_EN.pdf

James SJ, James C (2014) Chapter 5. Chilling and freezing of foods. In: Clark S, Jung S, Lamsal B (eds) Food processing: principles and applications. Wiley, 592 pages

James C, Purnell G, James SJ (2014) A critical review of dehydrofreezing of fruits and vegetables. Food Bioprocess Technol 7(5):1219–1234. https://doi.org/10.1007/s11947-014-1293-y

James C, Onarinde BA, James SJ (2017) The use and performance of household refrigerators: a review. Compr Rev Food Sci Food Saf 16(1):160–179

Javadian SR, Rezaei M, Soltani M, Kazemian M, Pourgholam R (2013) Effects of thawing methods on chemical, biochemical, and microbial quality of frozen whole rainbow trout (Oncorhynchus mykiss). J Aquat Food Prod Technol 22(2):168–177

Jeremiah LE (1996) Freezing effects on food quality, viii + 520pp

Jul M (1984) The quality of frozen foods, xiii + 292pp

Kaale LD, Eikevik TM (2014) The development of ice crystals in food products during the superchilling process and following storage, a review. Trends Food Sci Technol 39(2):91–103. https://doi.org/10.1016/j.tifs.2014.07.004

Kaale LD, Eikevik TM, Rustad T, Kolsaker K (2011) Superchilling of food: a review. J Food Eng 107(2):141–146. https://doi.org/10.1016/j.jfoodeng.2011.06.004

Kalla AM, Devaraju R (2017) Microwave energy and its application in food industry: a review. Asian J Dairy Food Res 36(1):34–47. Retrieved from http://www.arccjournals.com/uploads/articles/8DR1135.pdf

Karthikeyan JS, Kiran MD, Deepti S, Bruins R, Schaffner DW, Mukund VK (2015) Effect of temperature abuse on frozen army rations: Part 2: Predicting microbial spoilage. Food Res Int 76(Part 3):587–594. https://doi.org/10.1016/j.foodres.2015.07.012

Kennedy CJ (2000) Managing frozen foods. Woodhead Publishing, 304pp

Knútsson OH, Gestsson H (2001) The Icelandic fishing industry: a comprehensive overview until the end of 2001. Retrieved from http://hdl.handle.net/1946/1106

Kosa KM, Cates SC, Karns S, Godwin SL, Chambers D (2007) Consumer home refrigeration practices: results of a web-based survey. J Food Prot 70(7):1640–1649

Kraft AA (1992) Chapter 1. Health hazards vs. food spoilage. In: Psychotropic bacteria in foods disease and spoilage. Taylor & Francis, pp 2–26. https://books.google.com.gh/books?id=pczoPJMYNoQC

Kulawik P, Migdal W, Gambus F, Cieslik E, Ozogul F, Tkaczewska J, Szczurowska K, Walkowska I (2016)

Microbiological and chemical safety concerns regarding frozen fillets obtained from *Pangasius sutchi* and Nile tilapia exported to European countries. J Sci Food Agric 96(4):1373–1379. https://doi.org/10.1002/jsfa.7233

Lagendijk E, Assere A, Derens E, Carpentier B (2008) Domestic refrigeration practices with emphasis on hygiene: analysis of a survey and consumer recommendations. J Food Prot 71(9):1898–1904

Lamb FG (1965) US patent 3219463 A: process of dehydrofreezing foods. Retrieved from http://www.google.com/patents/US3219463

Le Meste M, Champion D, Roudaut G, Blond G, Simatos D (2002) Glass transition and food technology: a critical appraisal. J Food Sci 67(7):2444–2458

Leygonie C, Britz TJ, Hoffman LC (2012) Impact of freezing and thawing on the quality of meat: review. Meat Sci 91(2):93–98. https://doi.org/10.1016/j.meatsci.2012.01.013

Li B, Sun DW (2002) Novel methods for rapid freezing and thawing of foods – a review. J Food Eng 54(3):175–182. https://doi.org/10.1016/s0260-8774(01)00209-6

Li Y, Zhao J, Han Q, Kong B (2015) Influence of freeze-thaw cycles on the quality of meat: a review. Food Ind 8:243–248

MarketsandMarkets (2014) Cold chain market by type (refrigerated storage, refrigerated transport), product type (chilled, frozen), application (fruits & vegetables, bakery & confectionery, dairy & frozen desserts, meat, fish & seafood) & region – Global trends & forecast to 2019. Retrieved from http://www.marketsandmarkets.com/Market-Reports/cold-chains-frozen-food-market-811.html

Marklinder IM, Lindblad M, Eriksson M, Finnson AM, Lindqvist R (2004) Home storage temperatures and consumer handling of refrigerated foods in Sweden. J Food Prot 67(11):2570–2577

Marth E (1998) Extended shelf life refrigerated foods: microbiological quality and safety. Retrieved from http://www.ift.org/knowledge-center/read-ift-publications/science-reports/scientific-status-summaries/extended-shelflife-refrigerated-foods.aspx

McLintock AH (2015) Refrigeration. In: An encyclopedia of New Zealand 1966

Meat and Wool New Zealand (2015). SS Dunedin loading 1882.JPG. Retrieved from http://commons.wikimedia.org/wiki/File:SS_Dunedin_loading_1882.JPG

Medina I, Gallardo JM, Aubourg SP (2009) Quality preservation in chilled and frozen fish products by employment of slurry ice and natural antioxidants. Int J Food Sci Nutr 44(8):1467–1479. https://doi.org/1410.1111/j.1365-2621.2009.02016.x

Ministry of Culture & Heritage (2015) First shipment of frozen meat leaves NZ 15 February 1882. Retrieved from http://www.nzhistory.net.nz/first-shipment-of-frozen-meat-leaves-nz

Miyawaki O (2018) Water and freezing in food. Food Sci Technol Res 24(1):1–21. https://doi.org/10.3136/fstr.3124.3131

Nicolas J, Potus J (1994) Enzymatic oxidation phenomena and coupled oxidations. Effects of lipoxygenase in breadmaking and of polyphenol oxidase in fruit technology. Sci Aliment 14(5):627–642

Pease HD (1912) Hygienic results of refrigeration in the conservation of fish and mollusks. Am J Public Health 2(11):849–854

Peck M (2006) Clostridium botulinum and the safety of minimally heated, chilled foods: an emerging issue? J Appl Microbiol 101(3):556–570. https://sfamjournals.onlinelibrary.wiley.com/doi/full/510.1111/j.1365-2672.2006.02987.x

Peck M, Grinyer L, Goodburn K (2018) Guidelines for setting shelf life of chilled foods in relation to non-proteolytic clostridium botulinum. Quadram Institute, 16pp. https://quadram.ac.uk/wp-content/uploads/2018/2007/Non-proteolytic-Clostridium-botulinum-shelf-life-guidance-FINAL-2011st-Ed-2019-2017-2018.pdf

Pennington ME (1912) The hygienic and economic results of refrigeration in the conservation of poultry and eggs. Am J Public Health 2(11):840–848. Retrieved from http://www.ncbi.nlm.nih.gov/pubmed/18008742

Petzold G, Aguilera JM (2009) Ice Morphology: fundamentals and technological applications in foods. Food Biophysics 4(4):378–396

Quang Tuan P (2014) Food freezing and thawing calculations, xv + 153pp

Rahman MS, Velez-Ruiz JF (2007) 26. Food preservation by freezing. In: Handbook of Food Preservation. CRC Press, Boca Raton, pp 635–665. http://labgraos.cotm.br/manager/uploads/arquivo/capDOUBLEHYPHEN626-handbook-of-food-preservation-pdf-(profDOUBLEHYPHENmauricio-de-oliveira).pdf

Ramallo LA, Mascheroni RH (2010) Dehydrofreezing of pineapple. J Food Eng 99(3):269–275. https://doi.org/10.1016/j.jfoodeng.2010.02.026

Rees J (2013) Refrigeration nation: a history of ice, appliances, and enterprise in America. Retrieved from https://books.google.co.uk/books?id=JeoEAQAAQBAJ

Robinson RK (1985) Microbiology of frozen foods, x + 290pp

Rohit J, Banwet DK, Ravi S (2010) Consumer link in cold chain: Indian scenario. Food Control 21(8):1137–1142

Roos YH (2010) Glass transition temperature and its relevance in food processing. Annu Rev Food Sci Technol 1(1):469–496

Schafer W (2014) The science of Freezing foods. Retrieved from https://www.extension.umn.edu/food/food-safety/preserving/freezing/the-science-of-freezing-foods/

Schmidt SJ, Lee JW (2009) How does the freezer burn our food? J Food Sci Educ 8(2):45–52. https://ift.onlinelibrary.wiley.com/doi/full/10.1111/j.1541-4329.2009.00072.x

SICCODE.com (2017) NAICS code 311412 frozen specialty food manufacturing. Retrieved from http://siccode.com/en/naicscodes/311412/frozen-specialty-food-manufacturing

Stankiewicz BA, Poinar HN, Briggs DE, Evershed RP, Poinar GO (1998) Chemical preservation of plants and insects in natural resins. Proc R Soc Lond B Biol Sci 265(1397):641–647. Retrieved from https://www.ncbi.nlm.nih.gov/pmc/articles/PMC1689027/

Statistica (2015) Statistics and facts on the frozen foods market in the U.S. Retrieved from http://www.statista.com/topics/1339/frozen-foods-market/

Stonehouse GG, Evans JA (2015) The use of supercooling for fresh foods: a review. J Food Eng 148:74–79. https://doi.org/10.1016/j.jfoodeng.2014.08.007

Sun W (ed) (2012) Handbook of frozen food processing and packaging, 2nd edn. CRC Press, New York, 936pp

Supratim G, Coupland JN (2006) Factors affecting the freeze-thaw stability of emulsions. Food Hydrocoll 22(1):105–111

The Economist (2014) Fridge ownership: Cool developments – how chilled food is changing lives. Retrieved from http://www.economist.com/news/international/21603031-how-chilled-food-changing-lives-cool-developments

Timm F, Herrmann K (1996) Frozen foods, xiv + 345pp

Towns RE, Cullen RW, Memken JA, Nnakwe NE (2006) Food safety-related refrigeration and freezer practices and attitudes of consumers in Peoria and surrounding counties. J Food Prot 69(7):1640–1645

U.S. Food and Drug Administration (2013) Food code (2013). Retrieved from http://www.fda.gov/Food/GuidanceRegulation/RetailFoodProtection/FoodCode/ucm374275.htm

United Nations Economic and Social Council (2017) The role of science, technology and innovation in ensuring food security by 2030. Retrieved from http://unctad.org/meetings/en/SessionalDocuments/ecn162017d3_en.pdf

USDA (2013) Treatment manual. Retrieved from https://www.aphis.usda.gov/import_export/plants/manuals/ports/downloads/FAVIR/IR_labeling_packaging.pdf

USDA (2017) The Big Thaw – safe defrosting methods for consumers. Retrieved from https://www.fsis.usda.gov/wps/portal/fsis/topics/food-safety-education/get-answers/food-safety-fact-sheets/safe-food-handling/the-big-thaw-safe-defrosting-methods-for-consumers/CT_Index

USDA Food Inspection Services (2010) Freezing and food safety. Retrieved from https://www.fsis.usda.gov/wps/wcm/connect/cce745c9-0fc9-4ce6-a50c-84363e5b5a48/Freezing_and_Food_Safety.pdf?MOD=AJPERES

Vieira C, Diaz MT, Martinez B, Garcia-Cachan MD (2009) Effect of frozen storage conditions (temperature and length of storage) on microbiological and sensory quality of rustic crossbred beef at different states of ageing. Meat Sci 83(3):398–404. https://doi.org/10.1016/j.meatsci.2009.06.013

Weightman G (2012) The frozen water trade. Retrieved from https://books.google.co.uk/books?id=4SSPtu46zecC

Wignall J, Burt JR (1987) A study of the frozen food distribution system in the United Kingdom, vi + 117pp

Wikipedia (2017) Refrigeration. Retrieved from https://en.wikipedia.org/wiki/Refrigeration#Earliest_forms_of_cooling

Yeong CL (2005) NW salmon sent to China before reaching U.S. tables. Retrieved from http://www.seattletimes.com/business/nw-salmon-sent-to-china-before-reaching-us-tables/

14 Food Drying

1 Introduction

1.1 General Principles

Drying is an ancient method of food preservation whether this is drying cod in the artic or sun drying grains in Ancient China, Egypt or India.[1] Sun drying tomatoes in the Mediterranean and raisins in California remain popular. However, novel drying methods are being developed (Maisnam et al. 2017). The aim of this chapter is to introduce some key aspect of food drying and dehydration including, definitions, general principles, markets and the range of drying methods (Earle and Earle 1983a, b).

1.1.1 Definitions of Food Drying

Food drying and dehydration refer to the removal of water in order to create dried products or concentrates using controlled temperature conditions (Vega-Mercado et al. 2001). However, membrane separation or related physical methods of water removal are excluded from the definition (Fellows 2000). Virtually all major food groups can be dried, including processed foods and basic agricultural commodities (Table 14.1). An overview of different categories of dried foods is available from Wikipedia (Wikipedia 2020).[2]

1.1.2 Purpose and Advantages of Food Dehydration

Commercial food drying has three goals (i) extension of shelf life.[3] By reducing the amount of moisture present in a product, drying inhibits many forms of deterioration and so shelf life is extended. A longer shelf life allows greater time for marketing. Increases of stability also maintains the physical and nutritional functionality of food ingredients until such a time as these are delivered at the point of use (Fitzpatrick and Ahrné 2005); (ii) reductions of weight and increased portability. Fresh foods (in particular fruits and vegetables) contain up to 80% moisture (w/w) by weight. Removing moisture reduces weight rendering the product more portable or easy to transport. Exporting milk powder is preferable compared with fresh milk, because of the reduced weight. (iii) Creating convenience and instant foods. Instantized products including, beverages, soup mixes, desserts, and milk are ready-to-eat. However, products like custards and instant noodles may need moderate cooking.

[1] Dehydration, freezing and heating are examples natural phenomenon, which affected food items during the times of hunter gathers.

[2] The practical steps for drying fruit and vegetables can be found in many excellent guides produced by University extension agents, e.g. from UC Davis, and University of Georgia Cooperative Extension services.

[3] Shelf life and storage stability are used interchangeably in books and other reference sources

R. Owusu-Apenten, E. R. Vieira, *Elementary Food Science*, Food Science Text Series,
https://doi.org/10.1007/978-3-030-65433-7_14

Table 14.1 Examples of dehydrated foods

Food Groups	Processed dried foods	Applications
Beverages Cereals and grains Dairy products Fruits Legume, beans & pulses Meat (ham, pork) Seafood Seeds & nuts Vegetables	Beverages Dehydrated soup Dessert mixes Dried pasta Dried ready meals Dry mixes Egg powder Freeze dried foods Instant noodles Instant soup Plain noodles Potato powder Powders Rice Soups & Sauces	Animal feed Food for relief Food ingredients Instant foods Ready-to-eat meals Space foods Military rations

Adapted from various sources

A representative list of dried foods is shown in Table 14.1.

The listing is arranged according to major food groups, types of processed dried foods and some general areas of applications. Interestingly, many early "space foods" were dehydrated powders or concentrates and pastes. Military rations and foods for expeditions tend to be dried for reduced weight as well as the prolonged shelf life. Dehydrated foods find applications also in cases of, disaster and emergency relief where low costs, and convenient forms of transportation become paramount.

S*ome* of the advantages of food dehydration are shared by other forms of processing. Canned foods are shelf-stable, portable, instantized or ready-to eat. Some frozen foods exhibit convenience. However, one unique benefit of dehydrated foods is the *ease* and the comparatively low cost of storage. Compared with fresh, canned or frozen foods, dried foods exhibit lower density and weight which facilitates storage. Furthermore, dehydrated products can be stored using a moisture-proof container or polythene bag in the dark. In fact, any container that helps decrease exposure to moisture, oxygen and light can be used for dried food storage including cardboard. No additional effort or energy is required for storing dehydrated foods in contrast with food preservation by refrigeration. Drying is probably one of most economical preservation methods, and accessible for small scale producers in both developed and developing countries (Table 14.2).

Table 14.2 Segmentation of the global dried food market

Segmentation	Name
Drying technology	Freeze dried
	Spray dried
	Sun dried
	Vacuum dried
	Hot air dried
	Drum dried
Dried food products	Pastor and noodles
	Fruit and vegetables
	Dried soups and source
	Ready to eat foods, snacks
	Dessert and mixes
	Meat & meat products
	Seafood
	Dairy products
Regional players	North America
	Europe
	Asia Pacific
	Middle East
	Latin America
	Africa

Adapted from (Market Data Forecast 2019; MicroMarketMinitor 2017)

1.2 Global Markets for Dried Food

1.2.1 Market Segmentation

The dried processed food market can be segmented by (i) type of drying technology, (ii) type of dried product, and (iii) by different geographic regions (Market Data Forecast 2019). However, segmentation by economic metrics did not overlap fully with those applied in academic discussions.

R. L. Earle & Earle M. D. classified food-drying *processes* into three categories; (i) air and contact drying where heat is transferred to the product through air or by contact with a surface, (ii) vacuum drying, and (iii) freeze drying. Except for contact driers, vacuum driers or freeze driers most forms of drying are hot air or convective driers, e.g., tunnel drier, cabinet drier or spray driers (Table 14.2).

1.2.2 Value of the Dried Processed Food Market

The dried processed food market was valued at about US$286.7 billion (2019) and expected to reach US$ 378.8 billion by 2024. The main economic drivers behind the dried foods sector were are changes of population demographics and consumer expectations. A rise in two-worker families, increasing urbanization, rising incomes, need for leisure and rising demand for convenience foods and ready-to eat foods are all cited as reasons for the demand for dried foods. By contrast, loss of nutritional quality, texture loss and impaired sensory characteristics, remain a challenge for the dried foods industry as a whole (Market Data Forecast 2019).

The order of importance according to different drying methods was; freeze-dried foods - US$85.3 billion (Research and Markets 2020), spray dried foods – US$48 billion (Modor Intelligence 2019), and hot air dried foods – 19.3 US$ billon. There were no readily available figures for vacuum dried or sun dried foods. The popularity of different kinds of dehydrated food products varies with geographic region. The demand for pasta and noodles was highest in the Asia Pacific region. However, dried seeds and nuts were more important in North America and Europe, for snack foods and the breakfast cereal market. Beverages, instant coffee and chocolate were also more favored in North America and Western Europe (Market Data Forecast 2019). A great many dried foods, were predominant in lower income countries, including grains, cereals, spices and fish.

1.3 Water and the Shelf Life of Foods

1.3.1 Moisture and Food Deterioration

Dried foods have improved storage stability. The extended shelf life is due to the low moisture content of dried foods. A classic food stability – moisture graph (Fig. 14.1) shows the relations between water activity (A_W) and the rate of several deteriorative changes (y-axis) in foods.[4] Microbial activity, enzyme action, hydrolysis reactions and microbial activity, all increase with increasing moisture content with the exception of lipid oxidation (Barbosa-Cánovas et al. 2006).

Most food deterioration processes (Fig. 14.1) require moisture (Troller and Christensen 1978; Barbosa-Cánovas et al. 2006). The moisture in foods may be present as free water, loosely bound water (inside of pores and matrices, or chemically bound water. Preservation can be achieved by drying to achieve a low moisture content, and keeping the product within a low relative humidity (RH) atmosphere. The relations below show how A_W and RH are related.

$$A_W = \frac{\text{Vapor pressure}(\text{food})}{\text{Vapor pressure}(\text{pure water})} = \frac{P_s}{P_O} \quad (14.1)$$

$$\text{RH}(\%) = (\text{A}_\text{W})\text{x}100 \quad (14.2)$$

Briefly, A_W represents a measure of the ability of water to act as solvent. Addition of salt or sugar generates solvation interactions that restrict the mobility of water producing a drop of water vapor pressure (literally amount of vapor formed) over a solution. The ratio of water vapor pressure over a solution or food (P_S) compared with water vapor pressure of a pure solvent (P_O) equals A_W. Reductions of A_W occur when other solutes (sugar, amino acids) are dissolved in water (Anonymous 1979; Gailani and Fung 1986; Rockland 1987; Troller and Christensen 1978).

Deterioration reactions like non-enzymic browning, hydrolysis, microbial growth all decline at low A_W. The growth of most microbes fails at A_W below 0.6. However, the limit is A_W = 0.97 for Gram −ve bacteria, A_W = 0.90 for Gram +ve and A_W = 0.88 for yeast (Adams and Moss 2008, p.40). Enzymatic reactions in food appear to stop when A_W is below 0.73. Research showed a U-shaped relation between A_W and lipid oxidation. The concept of A_W explains how processes like salting, addition of honey, and freezing help

[4] Water activity is roughly related to the amount of "free" water rather than total water present in the product. A rigorous discussion of water activity is provided below.

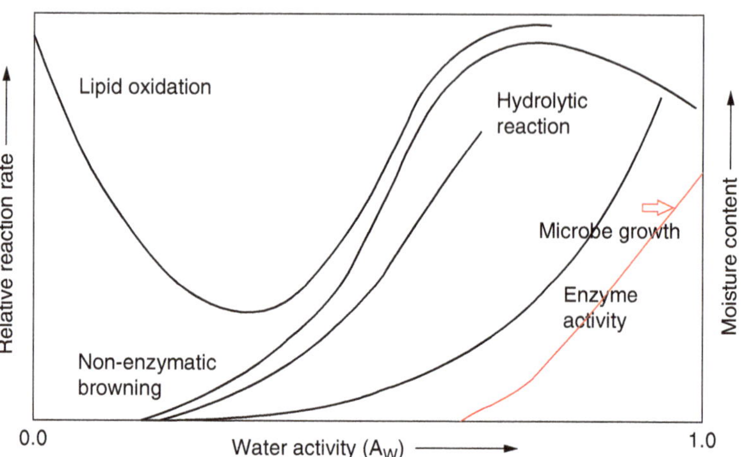

Fig. 14.1 A moisture – stability graph for foods. Adapted from (Troller and Christensen 1978)

to preserve foods (Troller and Christensen 1978; Barbosa-Cánovas et al. 2006; Troller 2012).

1.3.2 Food Polymer Model, Mobility and Stability

The concept of A_W assumes a moisture equilibrium condition exists between foods and surrounding atmosphere. A state of equilibrium infers that the rates of moisture ingress and exit (vaporization) between, a food sample, and the surrounding atmosphere are constant, but not necessarily equal. Under such circumstances, then A_W can be defined in a manner that is similar to an "equilibrium constant" from high-school chemistry (Eq. 14.1). Nevertheless, the presence of hysteresis in food-moisture adsorption curves indicates that those systems are not in equilibrium.[5]

An alternative theory, called the food-polymer molecular mobility model was developed to explain how drying improves food stability (Slade and Levine 1991, 1993). According to the food-polymer model, products exhibit two extreme levels of mobility and diffusion: foods exist as either an amorphous state or glassy state. The glassy state demonstrates extremely high levels of localized viscosity (like all glasses), leading to enhanced food stability and prolonged shelf life. In contrast, the amorphous state allows high internal mobility, diffusion, low stability and a low shelf life (Slade and Levine 1991, 1993; Le Meste et al. 2002; Khalloufi and Ratti 2003; Roos 2010). The theory states that interconversions between the glassy and amorphous state is controlled by moisture and temperature.

A food is converted from the glassy to the amorphous state (or the reverse) at the glass transition temperature (Tg). It was shown that Tg decreases exponentially with increasing moisture. For instance, Tg had negative values for high moisture foods, but the value increases above room temperature for low moisture foods. For any given fixed moisture level, dropping the temperature below Tg promotes the transition from an amorphous (low shelf life) state to the glassy state (enhanced shelf life). Drying increases Tg and the likelihood that food will exists as a glassy state at "room temperature". The food-polymer model brings together stabilization processes such as food freezing and dehydration.

Food dehydration issues were discussed mainly in terms of food-polymer approach over the past two decades (Slade & Levine 1991,

[5]The existence of hysteresis for food-moisture isotherms provides the visual indication that equilibrium conditions do not exist for foods-moisture interactions. Recall that a food-moisture isotherm can be produced by plotting moisture content (Y-axis) against values for A_W (X-axis). Isotherms produced by dehydrating foods are usually different from those produced by hydration of foods to the same A_W. So the direction of water addition to food samples, seems to matter contrary to what would be expected assuming that food-water interactions were equilibrium processes. Another important objection to the thermodynamic basis for water activity is that such relations were originally formulated for highly dilute (ideal) solutions – not what is found in foods.

1993; Le Meste et al. 2002; Khalloufi and Ratti 2003; Roos 2010). For example, changes to food structure, porosity and shrinking during drying are explained in terms of glass transition phenomenon (Anjali and Vir Singh 2015; Khalloufi and Ratti 2003; Rahman 2001). Lactose crystallization during spray drying (Debolina et al. 2014) and the quality of freeze-dried foods also seem to depend on the amorphous-glass stste transition phenomena (Khalloufi and Ratti 2003).

Research suggests that enzymic activity cannot occur or is severely depressed below Tg. Non-enzymic browning was also dependent on amorphous-glassy state transition (Kouassi and Roos 2000). Changes in the Tg with moisture may account for texture changes and loss of crispiness in foods with rising moisture content (Tolstoguzov 2000). Investigators have noted some limitations of the food-polymer science approach with indications that molecular mobility and diffusing could occur below the Tg contrary to predictions (Le Meste et al. 2002).

<u>Drying factors</u>

Rate ∝ $P_{air} - P_{surf}$

Rate = k ($P_{air} - P_{surf}$)

where P = vapor pressure

k = Constant

- **Air temperature**
- **Air speed**
- **Food surface area (A)**

Fig. 14.2 Basic relations describing drying rate for foods (Earle and Earle 1983b)

Table 14.3 Summary of factors affecting drying rate

Product characteristics	Drier characteristics
Shape (spheres, cylinders, slabs)	Drier configuration
Thickness	Air speed
Surface area	Operating temperature(s)
Porosity	Product residence time
Initial moisture	Relative humidity
Surface wax	Air volume flow
Varietal differences	Air flow pattern

Adapted from Mercer (2014)

1.4 Drying Rate

1.4.1 Factors Influencing the Rate of Drying

The rate of drying is dependent on the *difference* in the water vapor pressure or concentration at the food surface compared to the water vapor pressure in the air; the rate of drying is proportional to $P_{air} - P_{surf}$. As the difference in vapor pressure increases the rate of dehydration increases. The preceding relation holds true only if all "other" factors are held constant, e.g. air temperature, air speed and food surface area (Fellows 2016; Heldman and Hartel 1997; Maisnam et al. 2017; Mercer 2014). In terms of conventional design for food experiments, the aim is to keep most parameters constant, whilst varying one variable of interest (Fig. 14.2).[6]

The rate of food drying will increase along with the surface area e.g. by cutting or dicing solid foods. Liquids in the spray-drier are "atomized" into small droplets with a net increase of surface area. Raising the air speed facilitates moisture removal (Fellows 2016; Heldman and Hartel 1997). Factors that affect drying rate were classed as related to product characteristics or to drier/drying characteristics (Table 14.3).

1.4.2 Food Drying Curve

The progress of a food drying process can be presented in the form food-drying curve divided into three phases (Fig. 14.3): (i) the heating phase, (ii) a constant-rate zone and (iii) falling-rate zone. The heating phase allows for temperature rises when a product first enters a drier. Consequently, there is little or no perceptible change in moisture content (lbs. water/pound fiber) during the heating phase. Secondly, the heating phase is followed by the constant-rate phase. During the constant-rate phase of drying, water is lost from

[6]The need for constant conditions is one reason, why most research students maintain an almost paranoid need to avoid disturbances to their experimental "set up", including avoidance of noise/vibrations, draughts/ air movement/ temperature fluctuations/ in their lab.

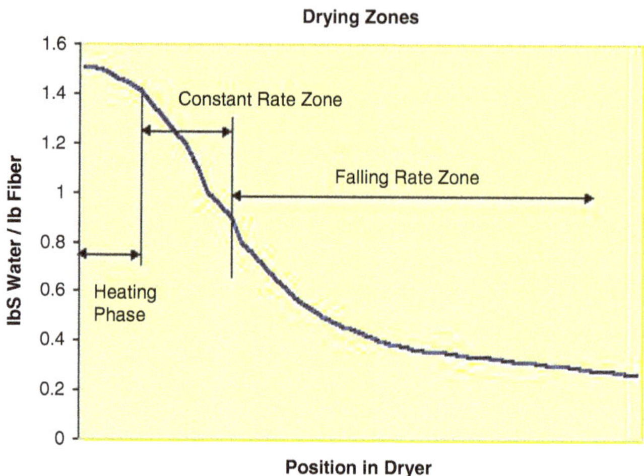

Fig. 14.3 A typical drying curve showing three phases of moisture loss from a food product. →Lag Phase →Constant Rate Zone →Falling Rate Zone

the product rapidly, at a constant rate.[7] The loss of moisture occurs from the surface of the product. Finally, the drying curve enters the falling-rate zone, where the rate of dehydration (slope of the graph) declines with time. This third and final stage of drying is explained by the receding waterfront within the product. Therefore, evaporating moisture travels longer distances from the interior pores of the product. Heating through the dried (air) layer is also less efficient. Prolonged drying can also lead to shrinking and a decrease of surface area. The formation of surface films may also impede water loss. Overall it is difficult to modify the later (falling) rate processes (Fellows 2016; Heldman and Hartel 1997).

1.4.3 Drying Kinetics and Food Quality

Drying rate is a fundamental characteristic for selecting different driers. Fast drying could save time and money (Leon et al. 2002). However, the effect of drying kinetics is difficult to study without confounding factors. Drying rate is affected by many variables (Table 14.3) which are potential confounding factors. Increased drying rate can be achieved by increasing the drying temperature – but this can affect nutrient or color adversely (Singh et al. 2008).

Table 14.4 Effect of drying conditions on garlic

Treatment	Moisture %	Drying time (days)
Oven drying	2.1	5
Sun drying	4.4	7
Shade	7.6	9

Adapted from Heikal et al. (1972c)

To illustrate how confounding factors can complicate experiments dealing with drying kinetics, consider one study which examined the effect of drying-rate on the quality of garlic. From the findings (Table 14.4) oven drying was more rapid compared to sun drying or shade drying, but color was best for the shade-dried garlic (Heikal et al. 1972a). Drying garlic in the shade was slow but notice that the differing drying times were achieved using *different temperatures*.

Another study showing why drying kinetics is difficult to study involved leafy vegetables dried by three methods: cabinet drying, solar drying and sun drying (Singh et al. 2006). In this case, the cabinet drier achieved faster drying compared to other methods. In addition, quality indices were improved with the cabinet drier including, a lower NEB index, final moisture, rehydration, and nutrient retention. However, the drying temperatures were unequal during this study (Singh et al. 2006). In fact most studies that compare the

[7]The rate of moisture loss can be seen from the slope of the

Table 14.5 Optimization of drying conditions for tomato quality

Factor studied	Range of values
Temperature levels °C	40–60
Air velocity (m/S)	1.0–2.0
Sample thickness (mm)	7–11
#Quality variables	Lycopene, NEB, vitamin C, brightness

Adapted from Abano et al. (2014)

effect of drying kinetics on product quality, also employ different driers and process temperatures (Di Scala and Crapliste 2008; Figiel 2009; Leeratanarak et al. 2006; Mandala et al. 2005; Methakhup et al. 2005). Non-thermal methods for increasing drying rate need to be examined, e.g. food particle size or airflow, can increase drying rate.

Multi-factorial experiments also provide another means for researchers to separate drying rate effects from coincidental changes of temperature, surface area, pressure etc. As an example, Abano and others used a "designed experiment" approach to examine variables affecting the quality of dried tomatoes (Abano et al. 2014). Overall, the optimum conditions for drying tomatoes involved rapid drying under mild conditions: low temperatures (44 °C), high airspeed (2 m/s) and low sample thickness (7.2 mm). The optimum drying time for tomatoes was 527 min. Therefore, a higher drying rate (achieved with higher air speed and a modest temperature) may be better for avoiding nutrient losses (Abano et al. 2014) (Table 14.5).

2 Pretreatment for Food Drying

2.1 Common Operations Prior to Drying

Some basic unit operations are needed before drying, e.g. the food product must be washed. Washing uses portable water, which is water suitable for human consumption. In addition, some fruits and vegetable can be peeled and/or destoned, to remove pips. Cutting or dicing will promote water loss, and heat transfer into the product. Finally, the product is usually spread on a drying platform before loading into a drier.

The presence of surface of a waxy cuticle will impede water loss. Therefore, several pretreatments are applied (Sections 2.2. +) that aim to improve water permeability and evaporation from plant cell walls. Other pretreatments can help reduce damage due to heating, oxygen or light exposure. Pretreatments applied to fruits and vegetables prior to drying were classified as chemical or physical (Deng et al. 2019; Oliveira et al. 2016; Wang et al. 2016).

2.2 Chemical Pretreatments

A variety of chemical pretreatments can improve the quality of dried foods (Lewicki 1998). Antioxidants help to control oxidation *as well as* non-enzymatic browning (NEB). Treating with sulfur dioxide (sulfuring) or sulfite prevents NEB. Sulfite restricts NEB by reacting with carbonyl (C=O) intermediates formed during browning (Wedzicha 1987).[8] Sulfite reacts also with quinones formed by enzymic oxidation catalyzed by polyphenol oxidase (PPO). Safety concerns indicate that sulfite use can cause irritations in some individuals and reduce levels of thiamine in foods. Therefore, the amount of sulfite used for pretreatment could be lowered by increasing drying temperature to accelerate the rate of drying and to reduce levels of NEB (Caboni et al. 2005).

Sorbic acid is an antimicrobial agent and antioxidant (Earle and Putt 1984). Pretreating apricots with sorbic acid (~500 mg/Kg) had multiple benefits for dried product depending on the final moisture content. The extent of NEB increased in low moisture dried apricot. For high-moisture dried apricot sorbic acid inhibited microbe

[8]The intersection of (non-enzyme) oxidation and NEB occurs at carbonyl compound (C=O) produced by these reactions. Sulfite and other nucleophilic compounds react with C=O intermediates from oxidation and NEB.

growth. Sorbic acid also protected beta-carotene from oxidation (Alagoz et al. 2015).

Grapes undergo chemical pretreatments in order to remove a naturally occurring waxy coating and promote water loss. In ancient times, a mixture of olive oil and wood ash was used for dewaxing grapes before drying. The older dewaxing agent was replaced by a mixture of 2.5% (w/w) potassium carbonate and 2% (v/v) ethyl oleate dissolved in cold water. The resulting emulsion was sprayed on grapes before tray drying (Christensen and Peacock 2000). Current pretreatments for grapes uses potassium carbonate, ethyl oleate and/or olive oil dissolved with water (Deng et al. 2019; Oliveira et al. 2016; Wang et al. 2016).

Grapes were also pretreated using dilute lye (0.2–0.5% w/v sodium hydroxide) solution at a temperature of 180 °F (82 °C) for about 12 s in preparation for tunnel drying (Christensen and Peacock 2000). After lye treatment, the grapes were rinsed with cold water and treated with sulfur dioxide to inhibit browning and so produce red or golden raisins (Christensen and Peacock 2000; Heikal et al. 1972b). The mode of action of alkaline pretreatments is uncertain. Potassium carbonate and fatty acid esters produce a synergistic affect to disrupt the structure of the plant waxy cuticle, and neutralize cell wall acidic groups (Christensen and Peacock 2000).

Dipping in ascorbic acid solution, lemon juice or honey improved the quality of dried fruits and vegetables (Deng et al. 2019; Lewicki 1998). The addition of sugars (Maltodextrins, glucose) before spray drying seemed to modify Tg and physical attributes of food powders (Anjali and Vir Singh 2015; Debolina et al. 2014)

2.3 Blanching

The browning enzyme PPO and most other enzymes found in fresh produce can be controlled by heat processing (Lopez et al. 1997).[9] Water blanching involves boiling a product at about 90–95 °C for between 5 and 10 min to inactivate enzymes before drying. Blanching helps deactivate pathogenic microbes making the food safer.

Some reports suggest that blanching can produce adverse effects on the quality of dried products. Lowering the temperature and time for the pretreatment can help reduce these bad effects (Aguilar et al. 2004). A study of various pretreatment methods on banana drying at 50 °C found that blanching (alongside of chilling, freezing, blanching with freezing) affected product flavor adversely (Dandamrongrak et al. 2003). Blanching potatoes led to adverse effects on quality with most effects occurring within the first 2 min of heat treatment (Mate et al. 1998). A loss of water-soluble vitamins by leaching may occur during blanching (Oliveira et al. 2016).

2.4 Edible Coatings

Edible coatings were applied to prevent oxygen exposure and to reduce the rate of oxidation. Fruits and vegetables treated with edible coatings (made from starch, modified starch, sodium alginate or protein) had improved retention of carotenoids, and colors. The edible coatings may hinder the uptake of oxygen and pro-oxidant minerals into the product. The use of edible coatings as pretreatment is relatively new and so more research will be needed before their widespread use in industry (Ahvenainen 1996; Oliveira et al. 2016).

2.5 Electrical Pretreatments

Ohmic blanching can inactivate endogenous enzymes in foods prior to drying. Solid or semisolid foods are placed between two electrodes and subjected to alternating currents of 20–40 Volt/cm at frequencies of up to 50 Hz. Temperatures of 80 °C were attained in order to

[9] The degree of enzyme retention during hot air drying has not been extensively studied. Freeze-drying retains virtually all-enzymatic activity though low moisture content immobilizes and temporarily deactivates enzymes. Enzymic activity in dehydrated organic solvent systems have been extensively studied,

inactivate plant peroxidases, used as indicator enzymes for blanching efficiency. Other electrical pretreatment methods e.g. pulsed electric field (PEF) processing help to increase water permeability of plant cell membranes and thereby improve dehydration. In addition, PEF pretreatment might be useful for controlling food enzymes and microbe activity (Ciurzynska et al. 2016; Evrendilek et al. 2012; Palmieri 2010; Terefe et al. 2015).

2.6 Enzymatic Pretreatment

Dried egg products (egg white, egg yolk, and whole egg products) are subject to browning and the development of off-flavors (Caboni et al. 2005). The deterioration involves small amount of glucose, which is naturally present in eggs, reacting with egg proteins. Therefore, egg products especially egg whites, may be treated with glucose oxidase and catalase. The glucose oxidase converts glucose to gluconic acid (which does not combine with amino groups) and hydrogen peroxide. The purpose of the catalase is to convert the peroxide to water and oxygen.

2.7 Osmotic Dehydration

Cut pieces of fruit or vegetables are immersed in a concentrated solution in order to achieve dewatering prior to drying (Oliveira et al. 2016; Raoultwack 1994). The movement of water from a low concentration food product to a highly concentrated solution across a semi-permeable cell-membrane is due to osmotic forces but some solute will also permeate into the food (Shi and Le Maguer 2002). Conditions used for osmotic dehydration include sucrose or sodium chloride (5% to 30% w/v) and soaking times between 2 and 7 h. With the right choice of conditions about 50% of the initial moisture content in the product may be removed, which could result in a significant decline in the subsequent energy needs for drying (Raoultwack 1994). As well as its use as a pretreatment, osmotic dehydration is employed as an independent dehydration unit operation whose efficiency is influenced by a host of variables (Ramya and Jain 2017; Suresh and Durvesh 2015).

Table 14.6 Effect of physical abrasion on the drying rate and quality of grape

Index	Comment
Drying rate	Increased
Drying time	Decreased by ~50%
Sugar & acids	No significant changes
Shrinkage	Reduced
Rehydration rate	Increased
Texture	Non change
Enzymic browning	Increased

Summarized from (Adiletta et al. 2015, 2016; Di Matteo et al. 2000)

2.8 Physical Abrasion

Though surface wax impedes water loss, the effect of physical abrasion on drying kinetics and quality was not tested until recently (Di Matteo et al. 2000). Overall, pretreating grapes by physical abrasion reduced drying times by about 50%. After abrasion, the grapes also shrunk less during drying and rehydrated more readily when wetted. The physical abrasion method might enable less reliance on chemical methods, and reduce chemical waste. However, physical abrasion was associated with higher levels of enzymatic browning (Adiletta et al. 2016; Adiletta et al. 2015) (Table 14.6).

3 Radiative Drying Methods

3.1 Classification of Drying Methods

A multitude of equipment are available for food drying (Arun 2017; Christopher 1997; Deng et al. 2019; Maisnam et al. 2017; Vega-Mercado et al. 2001). Some experts classify food driers according to, the type of energy source, operating temperature, pressure, mode of operation (batch or continuous), drying times, and form of material handling (Table 14.7). In addition, R. L. Earle & Earle M. D., classified food-drying *processes* as (i) air and contact drying, (ii) vacuum drying, or (iii) freeze drying (Table 14.8). Added to this there have been developed novel radiative driers. The choice of drier depends foremost on type of food commodity or ingredient to be dried, product quality required, and acceptable level of cost.

The main forms of food dehydration as of 1995 are reviewed by experts from the US Army Natick Research, Engineering and Development Centre (Cohen and Yang 1995). Others classified drying methods according to a semi-historic order of development into four generations (Vega-Mercado et al. 2001); (i) First generation cabinet driers for solid goods, (ii) Second generation. Spray drier and drum drier for food pastes and slurries. (iii) Third generation, freeze drier and osmotic drying, 4th generation driers – microwave driers, and radio-frequency driers. Each category of drier comes in a multitude of forms.

Table 14.7 Classification of food driers by different criteria

Classification	Comments
Heat transfer	Convective – Hot air driers Conduction – Direct contact, e.g. drum drier Radiative – Solar, microwave driers
Pressure	Ambient, vacuum driers
Temperature	Low temperature, high temperature
Mode of operation	Batch, continuous driers
Material handling	Fluidized bed, stationary, falling liquid
Drying times	Flash driers

Table 14.8 Some current drier equipment and drying techniques

Convectional	Novel
Belt driers	Osmotic dehydration
Bin driers	Microwave driers
Drum or roller driers	Dielectric driers
Fluidized-bed driers	Centrifugal fluidized bed
Freeze driers	Ultrasonic driers
Pneumatic driers	
Rotary driers	
Smoking	
Solar (open-air) driers	
Spray driers	
Tray driers	
Trough driers	
Tunnel driers	
Vacuum driers	

Source (Earle and Earle 1983a, b) and (Cohen and Yang 1995)

3.2 Sun Drying

The interest in sustainability has renewed attention on drying food and agricultural commodities using the sun. Advocates for solar energy point out the sun is a source of limitless energy, renewable, clean and free. The use of sunlight for food preservation scores highly in terms of the benign environmental impact. Sun drying remains important for considerable numbers of products ranging from cereals, cocoa to table grapes (Maisnam et al. 2017).

Sun drying is used widely in developing countries, which coincides with areas with relatively longer periods of high temperature, low humidity, and little or no rainfall.[10] Either direct sunlight or shade can be used for so-called atmospheric drying. Standard sun drying is now distinguished from solar drying also known as greenhouse drying, which is a form of sun drying that takes place in an enclosed space.

For successful sun drying, small fruits are prepared and spread on trays to dry in the sun for several days, then stacked to complete the drying cycle in shaded areas. Larger fruits, (apricots, peaches, and pears) can be halved and pitted, and apples are peeled, cored, and sliced prior to drying. Such fruits are "sulfured" to prevent enzymatic browning. Sun drying times vary between 4 and 25 days, depending on the size of the product, the type of pretreatment. During sun drying, we must prevent contamination from wind-blown dust. Moisture contained in the sun-dried products vary between 10% and 35%. After drying, some fruit may require moisture-vapour proof containers.

Sun dried products exposed to the elements must be pasteurized before storage. The dry heat pasteurization of fruits involves heating the sealed product at 160 °F (71.1 °C) for 30 minutes as first described in 1930s (Fellers

[10]The short-comings of sun drying can be overcome by solar drying.

1930). Alternatively, the dried product is exposed to low temperatures of 0 °F (−18 °C) for 48 hours (University of Georgia Extension 2020).

There is no truly comprehensive assessment dealing of the disadvantages and advantages of sun drying, that is from disinterested parties without an axe to grind. Heiket et al. reported that garlic could be dried using an oven, by indirect sun-drying (shade), or using direct sunshine with a drying time of 5, 7 and 9 days respectively and a final moisture content of 2.1%, 4% and 7.5%. The color of garlic was better following sun drying and was improved by pretreatment with sulfur dioxide (Heikal et al. 1972b). Joubert and others found that sundried rooibos tea had a darker color and a lower chlorophyll content compared with tea dried under controlled conditions (40, 50, 60 or 70 °C). However, the subjective quality of rooibos tea was not significantly affected by the drying method (Joubert and deVilliers 1997). Novel forms of sun drying are being developed in recent times (Maisnam et al. 2017). The three main designs for so-called solar dryers are (a) direct absorption type driers (b) convection type driers and (c) solar dryers combining both principles.

3.3 Other Radiative Drying Methods

Microwave appliances, which are common for cooking, can also be applied (with some modification) for food drying (Chandrasekaran et al. 2013; Murphy 1973; Wray and Hosahalli 2015). In fact, microwaves are part of the EMR and their use exemplifies the application of nonionizing radiation for dielectric heating for drying including infra-red radiation and radio waves (Hussain Riadh et al. 2015; Wang et al. 2011). Interestingly, dielectric driers are being used in combination with conventional driers, though these novel applications are in the development stage (Wray and Hosahalli 2015).

4 Convective - Hot Air Driers

4.1 Tunnel and Cabinet Hot Air Drying

For hot air drying, the product is placed on metal mesh belts in a tunnel, or in a cabinet on trays. Blowers circulate heated air and the air temperature, relative humidity, and air velocity are controlled. Hot-air driers are classified as parallel flow, counter flow, direct flow, or cross flow, depending on the direction in which the product moves in relation to the direction of the flow of the heated air. The hot air used in a tunnel drier or cabinet drier may or may not be recirculated. If air is recirculated, the relative humidity must be carefully regulated because during each passage over the food, the air takes up moisture, raising its relative humidity (Fig. 14.4).

The time for drying depends on the characteristics of the raw material (moisture content, composition, shape, and size), the temperature and humidity of the air in the drier, and the rate of air circulation in the drier. The drying profile is as described in Sect. 1.3.

Air velocities are usually regulated at about 1000 ft./min (5.08 m/s) at the start of drying cycle, but in the later stages, the air velocity is lowered to about 500 ft./min (2.5 m/s), as is adequate to remove moisture available at surfaces at this point of the drying cycle. High initial air speeds prevent adherence of the dried particles to drying trays and belts, thus facilitating unloading. Water-repellent plastics, such as polyethylene or Teflon, may be used for coating trays and belts to prevent sticking of the dried food.

For the drying of vegetables, the initial temperature of the air is typically 180 to 200 °F (82.2–93.3 °C). In the later stages of drying, the temperature is reduced to 130 to 160 °F (54.4–71.1°C) (Aguilera et al. 2003; Cohen and Yang 1995; Ratti 2001; Sagar and Kumar 2010; Vega-Mercado et al. 2001).

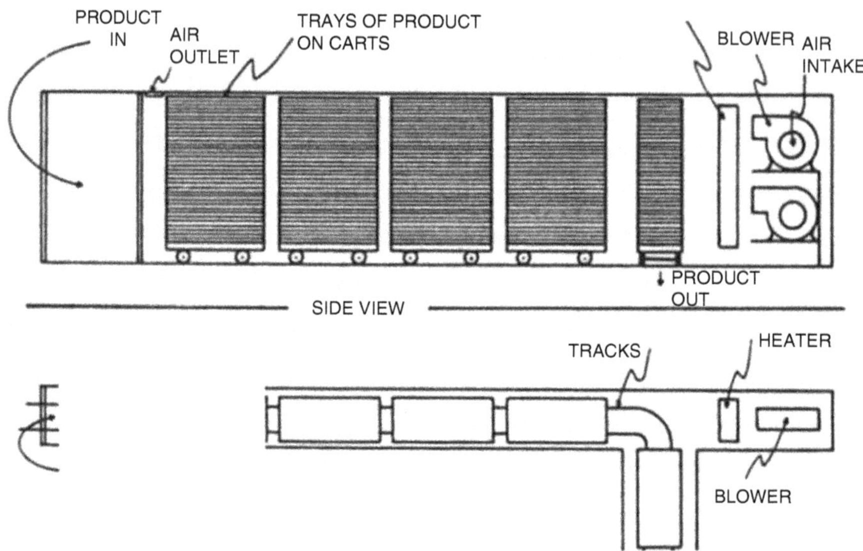

Fig. 14.4 Continuous tunnel drier

4.2 Fluidized-Bed Drying

In fluidized-bed drying (a special type of hot-air drying), the product is fed in at one end of the dryer to lie on a porous plate and is agitated and moved along toward the exit at the other end by hot air. The heated air that is blown upward through the product picks up moisture and exits through an outlet at the top of the drier (Beran 1980; Bhadania and Shah 1989; Cohen and Yang 1995; Santivarangkna et al. 2007).

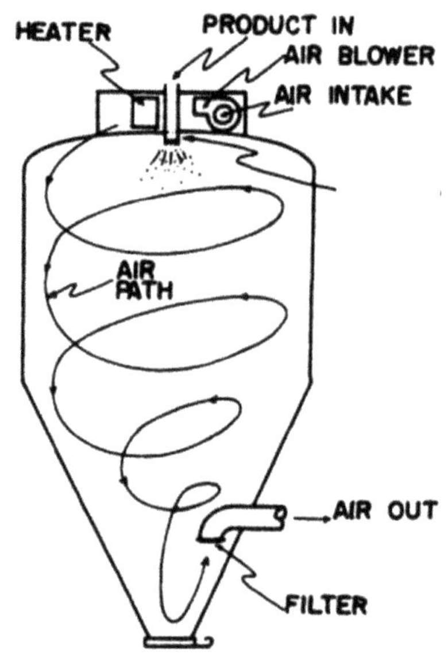

Fig. 14.5 Spray drier

4.3 Spray Drying

Milk, eggs, instant coffee, syrups, and other liquid foods or semiliquid products can be spray-dried (Anjali and Vir Singh 2015; McHugh 2018; Murugesan and Orsat 2012; Schuck et al. 2017) (Fig. 14.5). A new group of high-value, food commodities are also being spray dried for applications as dietary supplements and nutraceuticals, e.g. whey protein, egg white protein, and casein. Spray drying can be applied also for drying some heat sensitive pharmaceutical agents.

During spray drying a liquid, slurry or semisolid food material is first atomized, and then sprayed into the top of a chamber along with hot air, which is also blown in at the top. The cool, moist air exits near the bottom and the dried particles fall to the bottom and are collected by gravity flow, or with the aid of scrapers that may also be used to remove dried material from the walls or bottom. Cyclones (conical-shaped collectors) may be used to collect particles escaping with the exit air.

Particle size is important in spray drying, and the liquid is, therefore, dispersed into the drying chamber through an atomizer or pressure nozzle. This disperses the fluid by centrifugal force generated by a disc rotating at high speed. The atomizer produces droplets of a size that will be dried by the heated air through which they must fall on their way to the bottom of the drying chamber. The material may be pre-concentrated in some instances, or aids, such as gums, pectin, or milk solids, may be added prior to spray drying. In any case, it is necessary to lower the moisture content and control the temperature of the product to the point where particles will neither stick together nor stick to the wall of the spray drier.

In spray drying, the temperature of the food reaches a maximum of about 165 °F (73.9 °C) and remains there for a short period of time. The moisture content can be lowered to about 2.5% (w/w). The shearing action of the atomization step of spray drying may affect the functional characteristics of some proteins, such as those of egg products (whipping qualities, etc.).

Spray drying can produce fine powders that do not rehydrate readily; the dried powders neither wet nor disperse readily when reconstituted. The material tends to clump and form a mass through which water does not penetrate. To avoid this, a process known as agglomeration may be used. In this case, the dried material is slightly re-humidified; for instance, dried milk powder may be heated with steam under controlled conditions that surface-moisten the product. By other methods, the pre-concentrated product is injected into the drying chamber with a steam injector. The latter procedure causes the dried product to form in the shape of beads or bubbles, which break into: flakes that disperse readily in water (Islam Shishir and Wei 2017; Padma Ishwarya et al. 2015; Samaneh et al. 2015; Schuck et al. 2017). Rehydration of food powders is discussed further in Chap. 15.

5 Other Drying Methods

5.1 Conduction – Drum Drying

The drum drier is used for dehydrating milk, fruit, vegetable juices, purees, and cereals (Bonazzi et al. 1996; Holdsworth 1971). The product slurry is allowed to flow onto the surface of two heated stainless steel drums located side by side and rotating in opposite directions with little clearance between them. The product dries on the drums and is scraped off by stationary blades fixed along the surface of the drum (Fig. 14.6). Refrigeration may be used to lower the temperature of the dried product quickly.

Drum drying can also be carried out under vacuum, in which case the drying is accomplished at lower temperatures and the product is protected from oxidation. In ordinary drum drying, the process is controlled by varying the moisture content of raw material (pre-concentration), the temperature of drum surfaces, the space between the drums, the speed of rotation of the drums, and the amount of vacuum applied. However, drum drying is seen as a classical method suited for low-value products.

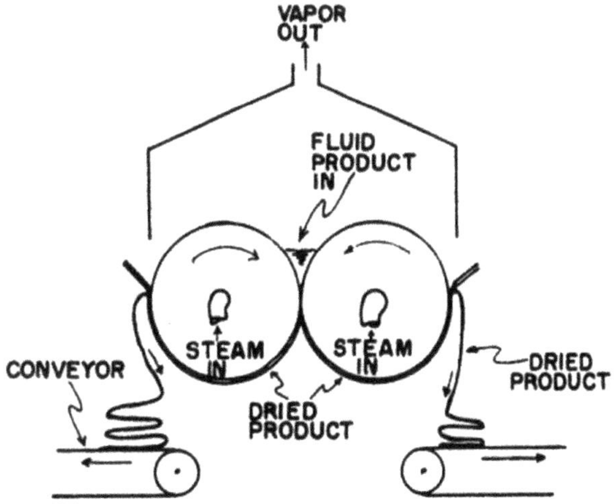

Fig. 14.6 Drum drier

5.2 Freeze Drying/Lyophilisation

Freeze drying is carried out by freezing the product and then subjecting it to a very high vacuum (Ciurzyńska and Lenart 2011; King 1971; Li et al. 2013; Oetjen 2000; Ratti 2001). The combination of low temperatures (−100 °C) and high vacuum (~0–2 mmHg) leads to a process of sublimation, meaning ice is transformed directly to vapor without passing through a liquid phase. The vapor from a food product is condensed and collected outside the freeze drier within a condensation chamber. The low temperature conditions used with a freeze-drier promotes the retention of organic volatiles and minimizes the loss of flavor and aroma.

Product changes during freeze-drying were described by Felix Franks one of the foremost experts on the behavior water in low temperature systems (Franks 1998). In the typical freezing scenario, some 99% of water from the product will become frozen as ice, thereby dehydrating the product. The 1% unfrozen liquid present within a frozen product experience a phenomenon of freeze concentration, which is the removal of water as solid ice, resulting in 20-fold or greater rise in the level of dissolved solutes and H^+ ions. One result of freeze concentration is that pH within the product changes, which could affect food components. Franks notes also that reaction rates, in freeze-concentrated system may be higher than anticipated by the Arrhenius equation, which predicts that reaction rates should decrease with decreasing temperature.[11] As the product temperature continues to fall, the freeze-concentrated liquid phase transforms into a glassy state at the glass transition temperature.

Subjecting the freeze-concentrated food to low pressure encourages ice sublimation, i.e. the direct transformation of solid ice to water vapor (Franks 1998). During sublimation drying, the ice front recedes into the product with time. This primary drying phase continues until all ice is evaporated from the product. During this phase, the temperature of the product may fall owing to energy for sublimation. At any rate, the temperature of the product is expected to remain constant until all the ice present in the food has sublimated (Oetjen et al. 2000).

In the secondary stage of freeze-drying, the adsorbed water must be removed from the product. Interestingly, the product temperature may be raised (provided it remains below the glass transition temperature) to ensure that residual moisture is removed (Oetjen et al. 2000). Research shows that keeping the product temperature below the glass transition temperature whilst freeze-drying is important for product quality (Franks 1998).

Freeze drying is arguably the most important method for sensitive products (Franks 1998). However, the manufacture of dehydrated probiotics and dairy ingredients are applying other methods besides freeze-drying due to economic considerations (Broeckx et al. 2016).

The quality of freeze-dried products is judged as significantly higher compared to most other dried products (Ratti 2001). Typically, a freeze-dried product has a fluffy porous structure with a high surface area. However, improper control of temperature in relation to the glass transition temperature can result product collapse, shrinking, reduced porosity and stickiness. High quality freeze-dried products maintain their original shape, and volume, whilst the density is reduced compared to the original food. The rehydration ability and color of freeze-dried products is better than for hot air dried products (Ratti 2001).

It may be necessary to blanch or to pretreat the foods prior to freezing them, or to treat them with sulfite to inactivate or inhibit enzymes. This is especially the case with vegetables (garlic and onions excepted) and mushrooms. Meats are sliced or cut into small enough portions to permit reasonably fast drying. Chicken is usually pre-cooked, boned, and diced prior to freezing and drying (Sect. 2) (Fig. 14.7).

[11] Food changes during freezing are described in Chap. 13.

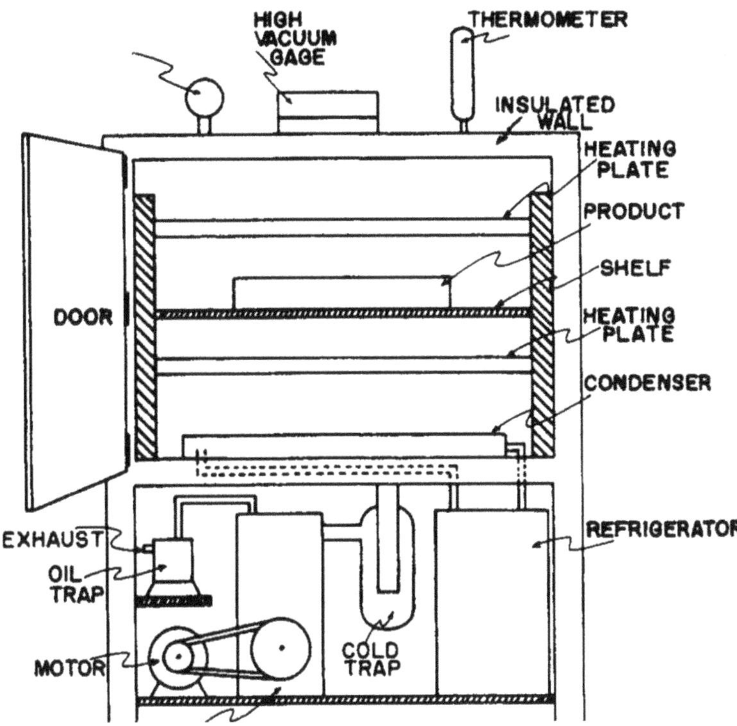

Fig. 14.7 Freeze drier

5.3 Cooking Incidental Food Dehydration

Food dehydration is incidental to a number of unit operations and cooking methods. A common but readily overlooked dehydration process, involves food freezing. During freezing, an increasing proportion of water within the food is transformed into ice. The ice crystals then separate as clusters leaving the rest of the food "freeze concentrated" and dehydrated. That proportion of unfrozen water within freeze concentrated food has reduced activity.

Cooking by dry heat (baking, roasting, frying etc.) leads to water loss. For example, as a food product is baked in an oven, water vaporizes above its boiling point turning from a liquid into a gaseous state. The initial loss of moisture occurs from pores located at the food surface, leaving a receding front of dryness. Prior to vaporization, a water-moistened surface would be limited to a maximum temperature of 100 °C regardless of the surrounding (oven) temperature. Once the food surface is denuded of liquid water, the local temperature rises rapidly towards the oven temperature, until the food material browns and then chars.

Deep frying food leads to dehydration similarly to that described for air drying (above). First, during frying the food is immersed in oil, which acts as the heat transfer medium. By contrast, air is the heat transfer medium during oven baking. The cooking temperature for a deep fryer of 350–375 °F (180–190 °C) is comparable to temperatures used for oven baking.[12] Deep frying leads to surface dehydration of the product, creating pores that become filled with oil mostly located in the crust and the percentage of oil in the product rises accordingly (Ziaiifar et al. 2008).

[12]Determining cooking temperatures is highly complicated, but an overriding factor is to cook in order to reach safe internal cooking temperature, irrespective of the cooking method. See Chap. 11.

References

Abano EE, Ma H, Qu W (2014) Optimization of drying conditions for quality dried tomato slices using response surface methodology. J Food Process Preserv 38(3):996–1009. https://doi.org/10.1111/jfpp.12056

Adams R, Moss O (2008) Food microbiology. RSC Publishing, 463 pp

Adiletta G, Senadeera W, Liguori L, Crescitelli A, Albanese D, Russo P (2015) The influence of abrasive pretreatment on hot air drying of grape. Food Nutr Sci 6:355–364

Adiletta G, Russo P, Senadeera W, Di Matteo M (2016) Drying characteristics and quality of grape under physical pretreatment. J Food Eng 172:9–18

Aguilar CN, Rodriguez-Herrera R, Montanez-Saenz JC, Reyes-Vega MD, Contreras-Esquivel JC (2004) Blanching at low temperatures: a thermal bioprocess applied to fruits and vegetables to improve textural quality. Food Sci Biotechnol 13(1):104–108

Aguilera JM, Chiralt A, Fito P (2003) Food dehydration and product structure. Trends Food Sci Technol 14(10):432–437. https://doi.org/10.1016/s0924-2244(03)00122-5

Ahvenainen R (1996) New approaches in improving the shelf life of minimally processed fruit and vegetables. Trends Food Sci Technol 7(6):179–187. https://doi.org/10.1016/0924-2244(96)10022-4

Alagoz S, Turkyilmaz I, Tagi S, Ozkan M (2015) Effects of different sorbic acid and moisture levels on chemical and microbial qualities of sun-dried apricots during storage. Food Chem 174:356–364. https://doi.org/10.1016/j.foodchem.2014.11.075

Anjali V, Vir Singh S (2015) Spray drying of fruit and vegetable juices-a review. Crit Rev Food Sci Nutr 55(5):701–719

Anonymous (1979) Water activity - an important parameter in the food industry. Food Trade Rev 49(8):451

Arun SM (2017) Handbook of industrial drying, 4th edn. CRC Press, Boca Raton, 4(2015)

Barbosa-Cánovas GV, Fontana AJ Jr, Schmidt SJ, Labuza TP (2006) Water activity in foods: fundamentals and applications. Wiley Online, Chichester, 435pp

Beran Z (1980) Fluidized-bed drying of baker's yeast. Kvasny Prumysl 26(5):109–114

Bonazzi C, Dumoulin E, Raoult-Wack AL, Berk Z, Bimbenet JJ, Courtois F et al (1996) Food drying and dewatering. Dry Technol 14(9):2135–2170

Broeckx G, Vandenheuvel D, Claes IJ, Lebeer S, Kiekens F (2016) Drying techniques of probiotic bacteria as an important step towards the development of novel pharmabiotics. Int J Pharm 505(1–2):303–318

Caboni MF, Boselli E, Messia MC, Velazco V, Fratianni A, Panfili G, Marconi E (2005) Effect of processing and storage on the chemical quality markers of spray-dried whole egg. Food Chem 92(2):293–303. https://doi.org/10.1016/j.foodchem.2004.07.025

Chandrasekaran S, Ramanathan S, Basak T (2013) Microwave food processing—a review. Food Res Int 52(1):243–261

Christensen PL, Peacock W (2000) The rasin drying process. Raisin production manual, 207–216. Retrieved from http://iv.ucdavis.edu/files/24413.pdf

Christopher GJB (1997) Industrial drying of foods. Retrieved from https://books.google.co.uk/books?id=P9kaMxIdm3UC

Ciurzyńska A, Lenart A (2011) Freeze-drying-application in food processing and biotechnology-a review. Polish J Food Nutr Sci 61(3):165–171

Ciurzynska A, Kowalska H, Czajkowska K, Lenart A (2016) Osmotic dehydration in production of sustainable and healthy food. Trends Food Sci Technol 50:186–192

Cohen JS, Yang TCS (1995) Progress in food dehydration. Trends Food Sci Technol 6(1):20–25. https://doi.org/10.1016/s0924-2244(00)88913-x

Dandamrongrak R, Mason R, Young G (2003) The effect of pretreatments on the drying rate and quality of dried bananas. Int J Food Sci Technol 38(8):877–882. https://doi.org/10.1046/j.0950-5423.2003.00753.x

Debolina D, Sormoli ME, Imtiaz Ul Islam M, Langrish TAG (2014) The effect of different plasticizers on lactose crystallization during spray drying. Dry Technol 31(15):1856–1862

Deng L-Z, Mujumdar AS, Zhang Q, Yang X-H, Wang J, Zheng Z-A, Gao Z-J, Xiao H-W (2019) Chemical and physical pretreatments of fruits and vegetables: effects on drying characteristics and quality attributes–a comprehensive review. Crit Rev Food Sci Nutr 59(9):1408–1432

Di Matteo M, Cinquanta L, Galiero G, Crescitelli S (2000) Effect of a novel physical pretreatment process on the drying kinetics of seedless grapes. J Food Eng 46(2):83–89

Di Scala K, Crapliste G (2008) Drying kinetics and quality changes during drying of red pepper. LWT-Food Sci Technol 41(5):789–795. https://doi.org/10.1016/j.lwt.2007.06.007

Earle RL, Earle M (1983a, May 2019) Unit operations in food processing: Web edition. Retrieved from https://nzifst.org.nz/resources/unitoperations/index.htm

Earle RL, Earle MD (1983b) Chapter 7: Drying. Unit operations in food processing: Web edition. Retrieved from https://nzifst.org.nz/resources/unitoperations/index.htm

Earle MD, Putt GJ (1984) Microbial spoilage and use of sorbates in bakery products. Food Technol N Z 19(11):25

Evrendilek GA, Baysal T, Icier F, Yildiz H, Demirdoven A, Bozkurt H (2012) Processing of fruits and fruit juices by novel electrotechnologies. Food Eng Rev 4(1):68–87. https://doi.org/10.1007/s12393-011-9045-5

Fellers CR (1930) Pasteurized dried fruits. Am J Pub Health 20(2):175–181. https://ajph.aphapublications.org/doi/pdf/110.2105/AJPH.2120.2102.2175

Fellows PJ (2000) Food processing technology: principles and practice, 2nd edn. Taylor & Francis, 608pp

Fellows PJ (2016) Food processing technology: principles and practice, 4th edn. Woodhead Publishing, 1152 pages

Figiel A (2009) Drying kinetics and quality of vacuum-microwave dehydrated garlic cloves and slices. J Food Eng 94(1):98–104. https://doi.org/10.1016/j.jfoodeng.2009.03.007

Fitzpatrick JJ, Ahrné L (2005) Food powder handling and processing: industry problems, knowledge barriers and research opportunities. Chem Eng Process Process Intensif 44(2):209–214

Franks F (1998) Freeze-drying of bioproducts: putting principles into practice. Eur J Pharm Biopharm 45(3):221–229

Gailani MB, Fung DYC (1986) Critical review of water activities and microbiology of drying of meats. CRC Crit Rev Food Sci Nutr 25(2):159–183

Heikal HA, El-Manawaty H, Kamel SI, Abo-Zeid M (1972a) Chemical and technological studies on the dehydration of Egyptian grapes. Agric Res Rev 50(4):171–183

Heikal HA, El-Sanafiri NY, Shooman MA (1972b) Some factors affecting the quality of dried mango sheets. Agric Res Rev 50(4):185–194

Heikal HA, Kamel SI, Awaad KE, Khalil NF (1972c) A study on the dehydration of garlic slices. Agric Res Rev 50(5):243–253

Heldman DR, Hartel RW (1997) Principles of food processing. Springer Science & Business Media, New York

Holdsworth SD (1971) Dehydration of food products. A review. J Food Technol 6(4):331–370

Hussain Riadh M, Binti Ahmad SA, Hamiruce Marhaban M, Soh AC (2015) Infrared heating in food drying an overview. Dry Technol 33(3):322–335

Islam Shishir MR, Wei C (2017) Trends of spray drying a critical review on drying of fruit and vegetable juices. Trends Food Sci Technol 65:49–67

Joubert E, deVilliers OT (1997) Effect of fermentation and drying conditions on the quality of rooibos tea. Int J Food Sci Technol 32(2):127–134. https://doi.org/10.1046/j.1365-2621.1997.00388.x

Khalloufi S, Ratti C (2003) Quality deterioration of freeze-dried foods as explained by their glass transition temperature and internal structure. J Food Sci 68(3):892–903. https://doi.org/10.1111/j.1365-2621.2003.tb08262.x

King CJ (1971) Freeze-drying of foods. Chemical Rubber Co, Cleveland, 86pp

Kouassi K, Roos YH (2000) Glass transition and water effects on sucrose inversion by invertase in a lactose-sucrose system. J Agric Food Chem *48*(6):2461–2466. https://doi.org/10.1021/jf9902999

Le Meste M, Champion D, Roudaut G, Blond G, Simatos D (2002) Glass transition and food technology: a critical appraisal. J Food Sci 67(7):2444–2458

Leeratanarak N, Devahastin S, Chiewchan N (2006) Drying kinetics and quality of potato chips undergoing different drying techniques. J Food Eng 77(3):635–643. https://doi.org/10.1016/j.jfoodeng.2005.07.022

Leon MA, Kumar S, Bhattacharya SC et al (2002) A comprehensive procedure for performance evaluation of solar food dryers. Renew Sust Energ Rev 6(4):367–393

Lewicki PP (1998) Effect of pre-drying treatment, drying and rehydration on plant tissue properties: a review. Int J Food Prop 1(1):1–22

Li H, Jiao X, Geng L et al (2013) The present status and prospect of research on rapid freeze-drying food in China. J Adv Chem 1(1). Retrieved from http://citeseerx.ist.psu.edu/viewdoc/download?doi=10.1.1.876.4136&rep=rep1&type=pdf

Lopez A, Pique MT, Ferran A, Romero A, Boatella J, Garcia J (1997) Influence of drying conditions on the hazelnut quality .2. Enzymatic activity. Dry Technol 15(3–4):979–988. https://doi.org/10.1080/07373939708917272

Maisnam D, Rasane P, Dey A, Kaur S, Sarma C (2017) Recent advances in conventional drying of foods. J Food Technol Preserv 1(1):25–34

Mandala IG, Anagnostaras EF, Oikonomou CK (2005) Influence of osmotic dehydration conditions on apple air-drying kinetics and their quality characteristics. J Food Eng 69(3):307–316. https://doi.org/10.1016/j.jfoodeng.2004.08.021

Market Data Forecast (2019) Dried processed food market. Retrieved from https://www.marketdataforecast.com/market-reports/dried-processed-food-market

Mate JI, Quartaert C, Meerdink G, van't Riet K (1998) Effect of blanching on structural quality of dried potato slices. J Agric Food Chem 46(2):676–681. https://doi.org/10.1021/jf970671p

McHugh T (2018) The significance of spray-drying. Food Technol Mag 74(4). Retrieved from https://www.ift.org/news-and-publications/food-technology-magazine/issues/2018/april/columns/processing-spray-drying-in-the-food-industry

Mercer DG (2014) An introduction to the dehydration and drying of fruits and vegetables. Part 1 food drying basics. Retrieved from https://www.uoguelph.ca/foodscience/sites/default/files/Drying-Part%201.pdf

Methakhup S, Chiewchan N, Devahastin S (2005) Effects of drying methods and conditions on drying kinetics and quality of Indian gooseberry flake. LWT-Food Sci Technol 38(6):579–587. https://doi.org/10.1016/j.lwt.2004.08.012

MicroMarketMinitor (2017) Global dried foods market research report Report Code: FO 1107. Retrieved from http://www.micromarketmonitor.com/market-report/dried-foods-reports-6776539440.html

Modor Intelligence (2019) Spray dried food market – growth, trends and forecasts (2020–2025). Retrieved from https://www.mordorintelligence.com/industry-reports/spray-dried-food-market

Murphy RP (1973) Microbiological contamination of dried vegetables. Process Biochem 8(10):17–19

Murugesan R, Orsat V (2012) Spray drying for the production of nutraceutical ingredients – a review. Food Bioprocess Technol 5(1):3–14

Oetjen G-W (2000) Freeze-drying. In: Wilson I (ed) Encyclopedia of separation science. Elsevier Science, London, pp 1023–1034

Oetjen G-W, Hasely P, Klutsch H, Leineweber M (2000) Method for controlling a freeze drying process. U.S. Patent 6,163,979, issued December 26, 2000.

Retrieved from https://patents.google.com/patent/US6163979A/en

Oliveira SM, Brandao TRS, Silva CLM (2016) Influence of drying processes and pretreatments on nutritional and bioactive characteristics of dried vegetables: a review. Food Eng Rev *8*(2):134–163. https://doi.org/10.1007/s12393-015-9124-0

Padma Ishwarya S, Anandharamakrishnan C, Stapley AGF (2015) Spray-freeze-drying: a novel process for the drying of foods and bioproducts. Trends Food Sci Technol 41(2):161–181

Palmieri L (2010) Technological innovation to improve food safety and quality. Prog Nutr 12(4):289–293

Rahman MS (2001) Toward prediction of porosity in foods during drying: a brief review. Dry Technol 19(1):1–13. https://doi.org/10.1081/drt-100001349

Ramya V, Jain NK (2017) A review on osmotic dehydration of fruits and vegetables: an integrated approach. J Food Process Eng 40(3):2017

Raoultwack AL (1994) Recent advances in the osmotic dehydration of foods. Trends Food Sci Technol 5(8):255–260. https://doi.org/10.1016/0924-2244(94)90018-3

Ratti C (2001) Hot air and freeze-drying of high-value foods: a review. J Food Eng 49(4):311–319

Research and Markets (2020) Freeze dried food market- growth, trends, and forecast (2020–2025). Retrieved from https://www.researchandmarkets.com/reports/4771892/freeze-dried-food-market-growth-trends-and

Rockland (1987) Water activity: theory and applications to food. Taylor & Francis. 424pp

Roos YH (2010) Glass transition temperature and its relevance in food processing. Annu Rev Food Sci Technol 1:469–496

Sagar VR, Kumar PS (2010) Recent advances in drying and dehydration of fruits and vegetables: a review. J Food Sci Technol 47(1):15–26

Samaneh K, Wan Daud WR, Nourouzi MM, Farideh N, Mostafa G (2015) Spray drying: an overview on wall deposition, process and modeling. J Food Eng 146:152–162

Santivarangkna C, Kulozik U, Foerst P (2007) Alternative drying processes for the industrial preservation of lactic acid starter cultures. Biotechnol Prog 23(2):302–315. https://doi.org/10.1021/bp060268f

Schuck P, Jeantet R, Bhesh B, Xiao Dong C, Perrone IT, Carvalho AFD et al (2017) Recent advances in spray drying relevant to the dairy industry: a comprehensive critical review. Dry Technol 34(15):1773–1790

Shi J, Le Maguer M (2002) Osmotic dehydration of foods: mass transfer and modeling aspects. Food Rev Intl 18(4):305–335. https://doi.org/10.1081/fri-120016208

Singh U, Sagar VR, Behera TK, Suresh KP (2006) Effect of drying conditions on the quality of dehydrated selected leafy vegetables. J Food Sci Technol-Mysore 43(6):579–582

Singh GD, Sharma R, Bawa A, Saxena D (2008) Drying and rehydration characteristics of water chestnut (Trapa natans) as a function of drying air temperature. J Food Eng 87(2):213–221

Slade L, Levine H (1991) Beyond water activity – recent advances based on an alternative approach to the assessment of food quality and safety. Crit Rev Food Sci Nutr 30(2–3):115–360

Slade L, Levine H (1993) Water relationships in starch transitions. Carbohydr Polym 21(2/3):105–131

Suresh C, Durvesh K (2015) Recent development in osmotic dehydration of fruit and vegetables: a review. Crit Rev Food Sci Nutr 55(4):552–561

Terefe NS, Buckow R, Versteeg C (2015) Quality-related enzymes in plant-based products: effects of novel food processing technologies part 2: pulsed electric field processing. Crit Rev Food Sci Nutr 55(1):1–15. https://doi.org/10.1080/10408398.2012.701253

Tolstoguzov VB (2000) The importance of glassy biopolymer components in food. Nahrung/Food 44(2):76–84. https://doi.org/10.1002/(sici)1521-3803(20000301)44:2<76::aid-food76>3.3.co;2-4

Troller J (2012) Water activity and food. Elsevier Science, 198pp

Troller J, Christensen J (1978) Water activity and food (1st edn). Retrieved from https://books.google.co.uk/books?id=14PXhbYMIVAC

University of Georgia Extension (2020) Preserving food. Drying fruits and vegetables. Retrieved from https://extension.illinois.edu/sites/default/files/drying_resources_from_outside_sources.pdf

Vega-Mercado H, Gongora-Nieto MM, Barbosa-Canovas GV (2001) Advances in dehydration of foods. J Food Eng 49(4):271–289. https://doi.org/10.1016/s0260-8774(00)00224-7

Wang Y, Li Y, Wang S, Zhang L, Gao M, Tang J (2011) Review of dielectric drying of foods and agricultural products. Int J Agric Biol Eng 4(1):1–19

Wang J, Mujumdar AS, Mu W, Feng J, Zhang X, Zhang Q, Fang XM, Gao ZJ, Xiao HW (2016) Grape drying: current status and future trends. In: Grape and wine biotechnology, pp 145–165. Retrieved from https://webcache.googleusercontent.com/search?q=cache:AEZbj25uHmsJ:https://cdn.intechopen.com/pdfs/52017.pdf+&cd=6&hl=en&ct=clnk&gl=us

Wedzicha BL (1987) Chemistry of Sulphur dioxide in vegetable dehydration. Int J Food Sci Technol 22(5):433–450

Wikipedia (2020) List of dried foods. Retrieved from https://en.wikipedia.org/wiki/List_of_dried_foods

Wray D, Hosahalli SR (2015) Novel concepts in microwave drying of foods. Dry Technol 33(7):769–783

Ziaiifar AM, Achir N, Courtois F, Trezzani I, Trystram G (2008) Review of mechanisms, conditions, and factors involved in the oil uptake phenomenon during the deep-fat frying process. Int J Food Sci Technol 43(8):1410–1423

15 Quality of Dried Foods

1 Physical Quality of Dried Foods

1.1 General Principles

Food quality refers to the degree to which a product meets consumer expectations, needs and wants (Chap. 5). Since different consumers have different needs and wants, it follows that quality is a subjective construct. Nevertheless, manufacturers have standards or brand attributes against which they evaluate their products (Botta 1995). Quality is affected by all unit operations involved in manufacturing (dried) processed foods. For instance, the choice of raw material, type of pre-processing, and choice of drying method will all affect the quality of a dried food product (Caric and Kalab 1987; Krokida and Maroulis 2000).

A loss of moisture, and changes of physical and chemical properties, translates into changes of quality for dried foods (Bonazzi and Dumoulin 2011). With the possible exception of some intermediate-moisture products (e.g. meat jerky), most dried foods are consumed after rehydration. Therefore, rehydration quality is another important consideration. This is why the quality of dried foods must consider the dehydrated as well as the rehydrated product. As we shall see, quality attributes (e.g. porosity) of dried foods is separate but often times linked with the quality of the rehydrated product. Interestingly, there is general recognition that rehydration will not restore the original product (Bonazzi and Dumoulin 2011). The rehydration degree is one of several quality attributes of dried foods (Krokida and Maroulis 2000).[1]

Food engineering theories indicate that drying involves two processes; (i) heat transfer to the product resulting in a rise of temperature, and (ii) moisture transfer from the product to the surrounding atmosphere via diffusion and evaporation.[2] High temperatures applied for some food dehydration methods will accelerate chemical and biochemical reactions. By contrast, dehydration per se, vitrifies and stabilizes the product. The receding waterfront allows the product temperature to rise to such an extent that browning or charring may occur. In general, the quality of dried foods are expected to rise to a maximum and then fall with a rising intensity of processing (Bonazzi and Dumoulin 2011).

The quality of dried foods was discussed by experts, emphasizing sensory, functional, nutritional, shelf life and the safety (SFNSS) characteristics (Anjali and Vir Singh 2015; Bonazzi and Dumoulin 2011; Deng et al. 2019; Devahastin and Niamnuy 2010; Nijhuis et al. 1998; Oikonomopoulou and Krokida 2013; Oliveira et al. 2016; Omolola et al. 2017; Warawaran and

[1] Rehydration of dried whole foods and food powders is discussed in Sect. 8 of this chapter.

[2] This generic account excludes the special case of freeze-dried foods.

R. Owusu-Apenten, E. R. Vieira, *Elementary Food Science*, Food Science Text Series,
https://doi.org/10.1007/978-3-030-65433-7_15

Table 15.1 Dried food quality from the consumer's point of view

Sensory quality	Color, brightness (dried/rehydrated)
Appearance	"Natural" looking
Shape or aspect	Big particles, powder
Color	Low browning, carotenoids, chlorophyll loss
Taste and aroma	Sweetness, sourness,(dried/rehydrated), aroma retention
Texture	Handedness, springiness, density, porous
Ease of use	Rehydration or dissolving rate Apparent density (variation for proportioning), Stability over time, Packaging enabling re-use of product No agglomeration or sticking
Special properties	No breakage during storage, short cooking time, composition, Reproducibility
Additives:	Origin and concentration
Nutritional value, composition	Calories, fat, protein, vitamins,
Microbiological safety	Low microbial count
Chemical safety	No nitrate or pesticide residuals
Biological safety	No allergens

Adapted from (Bonazzi and Dumoulin 2011; Krokida and Maroulis 2000)

Bronlund 2015). Many of the underlying physical or chemical changes for quality of dried foods was reviewed by Bonazzi and Dumoulin (2011), Oliveira et al. (2016), Mercer (2014a) and Krokida and Maroulis (2000). Nonetheless, scrutiny reveals insufficient replicated studies, and there is therefore scope for more research (Table 15.1).

1.2 Physical Structure and Quality

The structure of fresh foods is cellular, tissue, and organ based. A leafy vegetable can be modelled using a high-school biology description of plant leaf: varietal shape, waxy cuticle, characteristic disposition of cells, stomata, phloem etc. as vascular tissue. Animal derived foods are modelled based on the principle of "muscle structure as meat". In contrast, most prepared foods are colloidal structures. Typical food colloids are molecular ensembles which include, food dispersions, gels, foams, emulsions, glasses, or the amorphous solids, and or combinations of such elements (Aguilera et al. 2003; Joardder et al. 2017; Oikonomopoulou and Krokida 2013).

Food structure can be discussed using different orders of scale from atoms, molecules, to supermolecular assemblies, organelles, cells, tissues, organs to whole organisms. Other discussions have as their focus specific food commodities, e.g. starch granule structure, casein micelles, milk fat globule, fat crystal structure, the glassy state in the context of sugars and confectionary (Heertje 1993).

The physical quality of dried foods is determined by structural characteristics, e.g. porosity, shrinkage and collapse. In an ideal world, dried products will be light and fluffy. Other products, e.g. air dried fruits can be dense, rigid or almost stretchable in their texture. Parameters such as true density (ρ_T), apparent density (ρ_A), true volume (V_T), apparent volume (V_A), specific volume (ν) and porosity (ε) can be measured objectively (Bonazzi and Dumoulin 2011; Krokida and Maroulis 2000; Ramos et al. 2003). Table 15.2 shows a definition for each of these terms.

Recall that the falling-rate stage of drying indicates migration of moisture from the product interior to the surface. Subsequently, the evacuated spaces may shrink or else remain intact as air-filled cells. The porosity (ε) of a dried product is determined air-filled spaces, where $\varepsilon = 1 - (\rho_A/\rho_T)$. A product with few pores or void spaces will attain maximum density (Bonazzi and Dumoulin 2011; Ramos et al. 2003).

A light porous structure affects products texture and rehydration ability. Changes of porosity were a function of several variables; (i) type of drying method, (ii) choice of drying conditions, e.g. temperature and air speed for convective driers, (iii) final moisture content of the product, and (iv) type of food product. Values for porosity range from zero to nearly 1.0 (0–100%).[3] In general, porosity of products was highest for freeze-

[3]Literature values for porosity are generally reported from a range (0–1.0) where zero means no pores and 1.0 means a with maximum porosity. In the current text porosity will be referred to from 0–100%.

Table 15.2 Some physical attributes of dried foods

True density (ρ_T)	Mass divided by true volume
Apparent density (ρ_A)	Mass divided by apparent volume
True volume (V_T)	Volume excluding void spaces, and pores
Apparent volume (V_A)	Volume plus pores
Porosity (ε)	Fraction of empty volume, or void species
Specific volume(ν)	Inverse of density

Adapted from Krokida and Maroulis (2000)

Table 15.3 A summary of porosity values for selected foods

Product	Drying method/Conditions	Porosity (%)
Apple	Air drying (50 °C)	20
Apple	Vacuum dried (50 °C)	40
Apple	Air dried (60 °C)	60
Apple	Freeze dried	80
Banana	Air dried	20
Banana	Freeze dried	84–90
Blue berries	Air dried (50 °C)	20
Blue berries	Freeze dried	70
Cornmeal	Extrusion (140 °C)	95
Potato	Air dried (50 °C)	10
Potato	Air dried (50–60 °C), 1-2 m/s	10–20
Potato	Vacuum dried, (70 °C)	20
Potato	Freeze dried	70
Radish	Air dried (40 °C)	12
Raisins	Air dried (37–51 °C)	37–47
Seeds	Air dried	50
Sweet potatoes	Air dried (40 °C)	4

Final product moisture content is 10% unless stated otherwise. Selected values as compiled by (Krokida and Maroulis 2000)

dried products, followed by vacuum drying and air-drying (Table 15.3). The porosity of dried foods increased with declining moisture content. However, a product may undergo cracking, shrinkage, and structural collapse. So removing moisture may not result in a simple linear increase in porosity. The products specific volume (inverse of density) also decreases with increasing final moisture content.

The effect of drying methods and drying conditions on product porosity reflects inherent differences in temperature, pressure and or rates of heating used with different equipment. Interestingly, higher temperatures and/or rapid rates of heating appear necessary to maintain porosity. Drying the surface of a product rapidly can create a hardened shell, which prevents stickiness and structural collapse during the falling-rate phase of drying. In such cases, the outer surface of the product achieves a glass transition temperature above the drying temperature, and so remains stable (Bonazzi and Dumoulin 2011).

1.3 Texture of Dried Foods

Food texture refers to attributes perceived via the sense of touch and mouthfeel (Chen and Opara 2013). The loss of moisture when food is dried results in shrinkage, contraction and structural collapse. By contrast, increased porosity leads to declining hardness and brittleness. Dried products can also exhibit stickiness and caking (Warawaran and Bronlund 2015). Poor texture such as increased hardness can lead to failure to rehydrate (Oikonomopoulou and Krokida 2013; Oliveira et al. 2016; Omolola et al. 2017). Current research shows the properties of dried products are related to the glass transition temperature. Maintaining the drying temperature below the glass transition temperature may help to, maintain a glassy state, desirable physical characteristics, and avoid structure collapse, caking and stickiness (Khalloufi and Ratti 2003; Saito 1985).

Instrumental measurements for texture are discussed in Chap. 6 (Sect. 4.3). Descriptors for texture measurable using instruments include adhesion, chewiness, crispness, firmness, fracturability, gumminess, hardiness, springiness (Rahman and Al-Farsi 2005). Not all texture descriptors apply to the dry powders, but terms like gumminess, thickness, smoothness, may apply to the initial material and intermediate products during drying. In general, food products harden after dehydration but there are also examples of texture loss. For instance, products subjected to blanching, sulfite, acid or carbon dioxide and other forms of pretreatment (cf. Sect. 2) could experience a loss of texture compared with control samples dried without pretreatment (Deng et al. 2019). Texture loss may occur due to sub-optimal drying using different types of driers. The texture of bell pepper was increased by blanching at low temperatures in presence of added calcium (Dominguez et al. 1996).

2 Chemical Changes and Quality of Dried Foods

2.1 General Principles

The chemical changes observed with dried foods arise from two non-enzymatic reactions: autoxidation of lipids and the Maillard reaction (Marcus Karel 1984). Some of the products formed from chemical deterioration affect the color and flavor of foods. Other products from chemical deterioration, e.g. advanced glycosylation products (AGES) and advanced oxidation products (AOEPS) may have long-term adverse health effects. Chemical changes affecting nutritive components, e.g. vitamin C can occur also via oxidation, hydrolysis and nucleophilic processes.

A detailed account of color science is beyond the scope of the chapter. Briefly, color science is based on three fundamental principles; (i) the light spectrum, (ii) Color as visible light interacting with objects, (iii) physiology of the eye and processing of light sensory information in the brain (Anonymous 2004). The light spectrum consists of electromagnetic radiation (EMR) emitting from the sun. According to basic physics, light behaves like particles called photons. Particles travel in straight lines. However, some characteristics associated with light resemble wave phenomenon, e.g. two overlapping beams of light passed through a narrow slit will produce a zebra-like interference pattern, showing areas of increased and decreased intensity, which is what happens when two waves interact. Interestingly, also photons possess both electrical as well as magnetic fields oscillating at right angles to each other and the direction of travel.

EMR has wavelengths ranging from several meters (low energy EMR) to nanometers (high energy EMR). The long wavelength/low energy EMR consists of radio waves, microwaves, and infrared radiation. By contrast, high energy EMR includes ultraviolet rays, X-rays and gamma rays. About 1% of the EMR from the sun is visible by the human eye (University of Washington 2009b) (Fig. 15.1).

The interaction of visible light with a physical object produces color due the absorption, transmittance or reflection of EMR. Sir Isaac Newton showed that a prism could separate white light into seven rainbow colors: red, orange, yellow, green, blue, indigo and violet (R.O.Y.G.B.I.V).

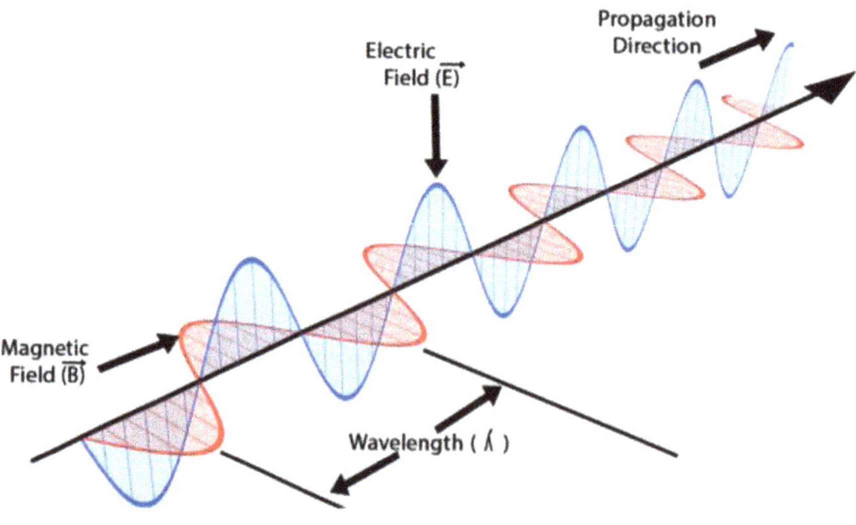

Fig. 15.1 A diagram of electromagnetic wave propagation (wiki commons)

Upon striking an object, a portion of (R.O.Y.G.B.I.V) light is *preferentially* absorbed, transmitted or reflected. For instance, absorption of red, orange and yellow light leaves an object appearing as blue, indigo or violet. By contrast, absorption of blue, indigo and or violet end of the spectrum leaves an object appearing as red, orange or yellow hue. The wavelength of light reaching the human eye changes accordingly.

Most materials are non-translucent with the exception perhaps of air, water and glass. According to the principles of color science, visible light interacts with dye molecules by energy transfer. Electrons in dye molecules are elevated from low to higher energy orbitals after energy is absorbed from EMR. The absorbed energy can be re-emitted as fluorescence or as heat. There are strict rules governing which molecules absorb energy from EMR, e.g. there must a good match between the energy absorbed and electronic structure and availability of higher energy orbitals for newly energized electrons. In short, there is a sound structural basis, why natural food colors and dyes appear as they do (University of Washington 2009a).

2.2 Color and Flavor of Dried Foods

The color of dried foods can be measured directly using a reflectance colorimeter which determines CIE (L, a, b) parameters related to lightness (ΔL), redness (Δa) or yellowness (Δb) of the product[4]. The color of dried foods is derived from color which is present before drying (see above) and pigments formed from the Maillard reaction. The extent of color change is dependent on, (i) the type of product, (ii) choice of drying equipment and (iii) drying conditions used. The literature contains trends on how color parameters change with different drying conditions (Krokida and Maroulis 2000).

[4]The color of liquid food extracts can be measured using absorption colorimeter, but poor extraction efficiency will lead to error. Instrumentation for objective color measurements are described in Chap. 6.

Table 15.4 Effect of drying conditions on the color of apple slices

	Color Parameter§		
Drying conditions	L	a	b
Apple slices			
Freeze dried	76.0	1.0	12.8
Air dried (1–21 °C)	45.0	14.0	4.5
Air dried (70 °C)	64.8	6.7	20.2

§Average of literature of range of values. L = lightness, a = redness/grey scale, b = yellow/blue scale. Adapted from (Krokida and Maroulis 2000)

Table 15.5 Effect of drying conditions on the color of banana slices

	Color Parameter§		
Drying conditions	L	a	b
Banana			
Freeze dried	53	3.6	14.0
Air dried (70 °C)	10	41.0	18.3
Air dried (90 °C)	−4.0	0.7	2.0

§Average of literature of range of values. L = lightness, a = redness/grey scale, b = yellow/blue scale. Adapted from (Krokida and Maroulis 2000)

The effect of drying temperature on product color is summarized in Tables 15.4, 15.5 and 15.6. Freeze-drying is usable as the reference method against which the effect of other drying methods can be gauged. Drying apples, led to decreased lightness, increased redness, and fall in yellowness (Table 15.4). Drying bananas, decreased lightness, increased redness turning to grey, and the retention of yellowness (Table 15.5). Finally, potatoes drying led to decreased lightness, increased redness turning to grey, and no changes to yellowness (Table 15.6).

Additionally, the degree of yellowness increases with falling process temperature, indicating that Maillard intermediates are more persistent. A fall in yellowness with rising temperature suggest that Maillard intermediates are converted to red/brown pigments. Overall, redness changes (Δa) increased with rising temperature (50-90 °C) whilst this parameter declines with decreasing air relative humidity. By comparison, yellowness (Δb) increased with decreasing drying temperature and rising air humidity (Krokida and Maroulis 2000).

The degree of redness of products (Δa) dried using different methods decreased in the order:

Table 15.6 Effect of drying conditions on the color of dried potatoes

	Color Parameter[§]		
Drying conditions	L	a	b
Potatoes			
Freeze dried	40	5.1	13
Air dried (70 °C)	45	30.0	16.0
Air dried (90 °C)	20	0.0	13.0

[§]Average of literature of range of values. L = lightness, a = redness/grey scale, b = yellow/blue scale. Adapted from (Krokida and Maroulis 2000)

Table 15.7 Five assorted classes of food taste compounds

Bitterness – plant phytochemicals (Drewnowski and Gomez-Carneros 2000)
Salty taste –sodium chloride
Sweet tastes – sugar, proteins, non-caloric sweeteners (Meyers and Brewer 2008)
Unami – glutamate, 5' ribonucleotides, (Yamaguchi and Ninomiya 2000)
Sour taste- organic acids, acidulants (Da Conceicao Neta et al. 2007)

air-dried > vacuum dried > microwave dried > freeze- dried = osmotic dried. The degree of yellowness of dried foods, (Δb) decreased according to the drying method; air-drying ≅vacuum dried > microwave dried >, freeze- dried > osmotic dried (Krokida and Maroulis 2000). Such data is informative, though with some obvious limitation s. For instance, current information is drawn from a limited set of products. Moreover, investigations are usually on a small scale or using pilot scale equipment. More research is needed to establish whether the preceding points have wider applications. Nevertheless, the knowledge of such trends are clearly important for food product development.

2.2.1 Flavor and Aroma of Dried Foods

Flavor arises from the combination of aroma and taste. Aroma or odor of foods originates from volatile compounds detected via the sense of smell (Takeoka et al. 2001). Depending on the vapor threshold, small compounds evaporate into the gas phase and can then be detected by the sense of smell. There is less information available on the effect of drying on food flavor and the trends are less clear by comparison with color

High temperature drying can lead to losses of flavor and aroma – but not as much as expected from the rise in temperature. For instance, hundreds of aroma compounds were formed when wheat bread baked products are cooked at high temperature, and partly because of interactions with the food matrix (Pico et al. 2015). Flavor retention in dried dairy products was affected by surface fat deposited on particles (Park and Drake 2014). Dried foods in the amorphous state retain more flavor compounds compared with the crystalline state (Bonazzi and Dumoulin 2011).

2.2.2 Taste of Dried Foods

Comparatively few studies report on the taste of dried foods presumably because this parameter requires human sensory analysis by trained panelist (Chap. 6). There are believed to be 5 main types of tastes. Recent advances in understanding of protein taste receptors as transmembrane G-proteins suggests that taste reception can be diverse. An important generalization appears to be that most taste molecules are relatively large, non-volatile, and at least partially water soluble compounds. Therefore, the expectation is that food taste is mostly intensified by drying as solutes become concentrated after the loss of moisture (Table 15.7).

2.3 Autoxidation and Browning Reactions

The development of off-flavor and bleaching of natural colors are important routes for quality loss in dried foods (Goldman et al. 1983; Karel 1986).

2.3.1 Oxidation

Both oxidative rancidity and bleaching arise from lipid autoxidation reactions that occur in foods rated high in unsaturated lipids. Fish, shrimp, crabmeat, lobster, seafood, microencapsulated fish oil are high in unsaturated fatty acids. The carotenoid pigment astaxanthin that provides the red color of salmon, shrimp and lobsters can undergo bleaching through autoxidation. Research suggests that products with high levels of lipid-soluble natural antioxidants tend to be more resistant to auto-oxidation. For instance, varieties of macadamia nuts from Hawaii with higher levels of total antioxidant activity measured as trolox equivalents were more stable as dried products (Wall 2010). Interestingly, the stability of macadamia nut oils from New Zealand were explained in terms of the high (80%) mono-unsaturated fatty acid content and less dependent on tocopherol content which is low (Kaijser et al. 2000).

Synthetic antioxidants can also be added to foods to inhibit oxidation. Antioxidants are chemicals that prevent oxidation of unsaturated carbon-carbon double bonds (C=C) in food compounds including, fats, tocopherols (Vitamin E) and carotenoids (Goldman et al. 1983). Small amounts of chemical antioxidants are required to provide a considerable protection against the oxidation of fats.

The packaging of dried products in such a manner as to exclude light is another procedure that helps to prevent fat oxidation because light energy accelerates fat oxidation and rancidification. Packaging under vacuum or in an inert atmosphere, such as nitrogen, and in such a manner as to exclude oxygen, may be used to protect dried foods against oxidation of fats (Pruthi 1978).

2.3.2 Maillard Reaction

Some color changes arising within dried products are the result of non-enzymatic browning (NEB) also known as the Maillard reaction. In general, NEB reactions are undesirable, as they lead to the formation of browning, loss of available lysine and the development of some off-flavors. NEB can also produce so-called advanced glycation products (AGES) which are suspected to be toxic (Aalaei et al. 2016; Aalaei et al. 2019; Aalaei et al. 2017; Oral et al. 2011). Ordinarily, NEB starts with the combination of certain carbonyl compounds (often sugars but also lipid oxidation intermediates) with proteins, leading eventually to the formation of brown- or black-colored compounds that not only cause off-flavors but νalso discolor food (Karel 1984). The best protection against such changes is to lower the moisture content to the point where the rate of NEB decreases. In some foods, this means lowering the moisture to less than 2%. Sulfite can be used to inhibit NEB. Browning may also change proteins, lowering their nutrient quality and affecting their rehydration properties.

2.4 Nutritional Changes

In principle, nutritional quality can be referenced to a wide range of components: carbohydrates, lipids, proteins, vitamins and nutrients (Sablani 2006).[5] The effect of drying on the nutritional quality of dried foods were thoroughly reviewed with a focus on the role of pretreatment methods and the choice of drying methods (Oliveira et al. 2016). There is general concern that dried foods may have lower levels of heat sensitive nutrients, e.g. vitamin C, vitamin A, thiamine, carotenoids as well as plant polyphenols (McSweeney and Seetharaman 2015; Omolola et al. 2017).

Of the most common pretreatment methods, blanching can produce large losses in water-soluble vitamins. Water blanching is more notorious compared with steam blanching, because the latter avoids leaching. A number of cases of enhanced nutrient (e.g. oil soluble vitamins) content were attributed to the inactivation of oxidative enzymes by blanching pretreatment before drying (Sect. 2, this chapter).

Drying methods, also affect nutritional quality. Approaches that involve high temperatures and long process times would be predicted to produce greater losses of nutrients. Interestingly, certain combinations of high temperature and short time drying methods could be beneficial for nutritional content, but it is difficult to produce generalizations. It seems likely that specific drying methods may be more suited for the retention of different nutrients; it is widely agreed that freeze-drying is the best method for retaining nutrients.

There is a growing interest in a wider group of phytochemicals e.g. polyphenols and how these respond to drying (McSweeney and Seetharaman 2015; Oliveira et al. 2016). Past research suggests that these diverse polyphenols (now classed as lipids) undergo oxidative degradation simi-

[5] Appendix 1 lists some common nutritional parameters.

larly to other unsaturated lipids (LIPIDMAPS 2014). Moreover, the determination of total phenols usually examines extractable fraction. Therefore, drying methods, which affect the release or irreversible binding of phenols to the insoluble food fraction, will register as increasing total phenols content. Interestingly, changes in the apparent total phenols content due to drying are important as the changes may well affect the bioavailability or degree of degree of absorption in the intestinal tract or function as dietary fiber.

Table 15.8 Factors influencing the rehydration of whole foods

Internal factors	External Factors
Density	Stirring or mixing degree
Physical State	Soluble matter
Porosity	Solvent composition
Product geometry	Solvent density
Tortuosity	Temperature
Water content	Viscosity
Molecular structure	Drier configuration

Source: Adapted from (Fernandes et al. 2011; Krokida and Maroulis 2000; Marabi and Saguy 2009)

3 Rehydration of Dried Foods

Products that rehydrate and dissolve readily are preferable to those that rehydrate poorly or form into lumps. Reconstitution or rehydration of dried foods is essential before consumption. Rehydration is defined as a process of moistening a dried product whether this is a whole food product such as fruit and vegetables (Lewicki 1998; Marabi and Saguy 2009), or food powders (Hogekamp and Schubert 2003). The rehydration of dried whole foods has different requirements compared to the rehydration of food powders. For instance, whole foods are rehydrated under quiescent conditions, without sample mixing or stirring in order to avoid them breaking up.

3.1 Whole Food Rehydration

The rehydration of whole foods including fruits and vegetables involves three stages; water absorption, swelling and leaching of components into solution (Fernandes et al. 2011; Krokida and Marinos-Kouris 2003; Lewicki 1998; Marabi and Saguy 2009). Water absorption occurs via capillary suction, osmosis and diffusion. The uptake and movement of liquid into surface pores of the food is generally rapid. There follows a slower diffusion phase into the product. Indeed, the rate and extent of rehydration of whole foods depends on a series of internal and external factors, cf. Table 15.8 (Marabi and Saguy 2009).

Geometric factors that increase a products surface area (small diameters, low volume: surface area) or which enhances moisture uptake (high porosity, low tortuosity, low initial water content) improve rehydration. Water uptake into regions possessing amorphous microstructure is thought to be higher than water ingress into glassy regions (Marabi and Saguy 2009). Water absorption is also inhibited by the presence of air bubbles within the food matrix (Lewicki 1998). Descriptions refer to role of capillary forces, and diffusion in water uptake. Moreover, it seems the degree of adhesion of water to food matrix and water-water adhesion forces play a role in food rehydration (Saguy et al. 2010). The rate of water uptake is usually rapid at first, and slows down during later stages. The loss of soluble matter (amino acids, vitamins, sugars, minerals) also occurs by diffusion (Maldonado 2015). Fernandes and co-workers reported that rehydration efficiency for many exotic tropical fruits varies with the drying method (Fernandes et al. 2011).

Research into rehydration of whole fruits and vegetables showed that[6]; the rate of rehydration could be described by 1st order kinetics, which means the uptake of water follow an exponential curve with a rapid rise in moisture content initially followed by a gradual decline in the rate of uptake. Secondly, the rate of rehydration increases with rising temperature from 68 to 176 °F (20 °C 80 °C). Thirdly, rehydration was generally incomplete, and so moisture content is not restored to the pre-drying levels. Apparently, food dehydration was irreversible indicating that

[6]Food samples were cut into 20 × 10 mm cylindrical shapes were air dried at temperature of 70 °C with air having relative humidity of 7%. Rehydration was with plain water at 40, 60 or 80 °C.

drying produces permanent changes to the product. Using a rehydrating temperature of 140 °F (60 °C), the following whole foods rehydrated slowly: mushrooms < bananas < green peppers <potatoes. In comparison, the following fruit and vegetables products underwent rapid rehydration: garlic < onion < apple < pumpkin < carrot < pepper < tomato. Interestingly, leaching of soluble matter into the water occurred above a temperature threshold of 140–158 °F (60–70 °C) (Krokida and Marinos-Kouris 2003).

The rehydration rate for edible dried mushrooms was slowest amongst whole foods. Blanching the mushrooms prior to drying affected the rehydration adversely. The two most common equation applied to rehydration research (Peleg equation and Weibull equation) described *Boletus edulis* mushrooms well.[7] Detailed investigations showed that rehydration increased with rising temperature 77, 104, 122 and 140 °F (25, 40, 50 & 60 °C) then declined at 158 °F (70 °C) owing to structural changes in the product. Mushroom samples attained 55–65% of the water content of fresh mushrooms (García-Pascual et al. 2005). Another study looking at the morel mushroom (Morchella esculenta) found similar results (García-Pascual et al. 2006).

Falade and Abbo (Falade and Abbo 2007) noted that date palm (popular in North Africa, Middle East and Nigeria) could be eaten fresh or allowed to ripen and dry on trees.[8] However, the introduction of mechanized drying could enable more stable products with enhanced storage. Their study using three varieties of date palm dried at 122–176 °F (50–80 °C) and rehydrated at 59–113 °F (15–45 °C) focused on rates, and physical parameters (Falade and Abbo 2007). The findings showed that the date palm had a drying time of 20 hours at 60 °C decreasing to about 12 hrs. If samples were dried at 70 & 80 °C. Rehydration of date palm was consistent with water diffusion into the product. The high activation energy (30–46.3 kJ/mol) showed date palm rehydration was sensitive to temperature.

Singh and coworkers (Singh et al. 2008) found that the best temperature for water-chestnut (singhara) rehydration varies with the initial drying temperature. For samples dried at temperatures between 122–194 °F (50–90 °C) the best product quality was achieved by drying at 158 (70 °C) and rehydrating at (40 °C) in order to preserve color. Intermediate temperatures produced the optimum color despite the fact that the rate of dehydration and rehydration both increased with temperature.

3.2 Food Powder Rehydration

Food powders and ingredients powders are nearly always reconstituted or rehydrated before or during use (Fitzpatrick and Ahrné 2005; Hogekamp and Schubert 2003; Ong 2019; Ortega-Rivas et al. 2010; Richard et al. 2013). In an ideal world, a powder will dissolve instantly and completely using cold or warm water with little in the way of stirring. Instead, powders have a tendency to, float when added to a solution, refuse to sink and to form lumps. According to food engineering principles, the efficiency of food powder rehydration is a function of three attributes; (i) wettability –or water absorption ability (ii) sinkability or ability to sink below water, and (iii) dissolvability or dispersibility, which refers to the ability to dissolve into solution (Gaiani et al. 2005).

Technical aspects powder rehydration have been reviewed (Ong 2019). Powder wetting is partly affected by water surface tension linked with particle floating and sinking. Capillary forces are thought to play a role in water uptake. Wetting is dictated partly by the water-sorption and diffusion into the food powders. Swelling and volume increases are related to the volume of water absorbed into a food. Formation of gels on powder surfaces may limit water uptake. Powder characteristics, such as density, surface roughness, presence of fat can all affect rehydration behavior. The process called granulation or agglomeration, whereby food powders are wetted and re-dried to form larger particles should increase powder wetting, but success seems to depend on the type of powder (Ji et al. 2016).

[7]The mushroom Boletus edulis (cep, porcino or California king belleta) is widely used in cooking as dried and rehydrated product. See Wikipedia a description of culinary uses (https://en.wikipedia.org/wiki/Boletus_edulis#cite_note-108)

[8]Date palm is also available in many western supermarkets

Some experts reported a need for more training in powder technology for food science students and engineers. Knowledge from industry and equipment suppliers need to be transferred systematically to end- users (Fitzpatrick and Ahrné 2005). Interestingly, rehydration seems to be even less well discussed as compared with other aspects of powder technology (Ortega-Rivas et al. 2010). Most of the research into food powders seem to deal with dairy foods, in particular milk powder and various derivations, e.g., milk protein concentrates, caseinate or whey protein concentrates (Fitzpatrick and Ahrné 2005; Haque et al. 2010, 2012; Heino et al. 2007; Ji et al. 2016; Le et al. 2011; Mao et al. 2012; Richard et al. 2013; Sikand et al. 2011). However, the overall principles apply to other industrial powder ingredients, including some pharmaceutical formulations. A couple of examples of power rehydration research will be cited in detail to as illustration of some of the features.

The efficiency of powder rehydration influenced by a wide range of factors including, composition and form of drying. Obviously, the conditions for rehydration (agitation speed, solution temperature) are paramount. For instance, powders could be rehydrated with warm water (45 °C) and varying degrees of stirring, or homogenizer setting – however, there are currently standardized conditions for testing (Table 15.9).

Table 15.9 Some conditions that affect food powder rehydration

Property	References
Composition	Heino et al. (2007)
Form of drying	Heino et al. (2007)
Heat load	Heino et al. (2007)
Ageing and storage	Celestino et al. (1997)
	Haque et al. (2010)
Mineral composition	Sikand et al. (2011)
Granulation, agglomeration	Gaiani et al. (2005)
Maillard reaction	Le et al. (2011)
Rehydration conditions ¥	Richard et al. (2013)
Agglomeration, granulation	Ji et al. (2016)

Source: compiled from various, ¥ = temperature, and stirring

4 Microbiology of Dried Foods

4.1 General Principles

General food microbiology is concerned with three themes: food safety, food spoilage, and the application of microorganisms for biotechnological purposes. The safety of dried foods can be compromised by the same factors observed for undried or rehydrated foods (See Chaps. 7, 8, 9, and 10). The range of hazards associated with dried foods is similar to hazards from moist food (Table 15.10). Good hygiene is necessary in order that rehydrated foods do not pose a risk to consumers. Acidic foods (fruits) can be sun dried at low temperature. Vegetables (low acidity) and meat drying requires more care, including meat smoking or use of salt and other preservatives to produce low or intermediate moisture meat products, like jerky(Holley 1985).

Food safety is improved by drying since microbes cannot grow when water activity is below 0.7. An equally interesting but less frequently discussed issue is how microorganisms become *inactivated* during the drying process. Intuitively, high-temperature drying methods (e.g. roller drying or spray drying) will inactivate pathogens more effectively than freeze-drying (Santivarangkna et al. 2008).

The inactivation of microorganism during drying involves various processes. For instance, freezing subjects cells to cryo-inactivation stresses. The rate of freezing seems to impact on the sizes of ice crystals formed which in turn affects the extent of cell damage and eventual viability when dried foods are reconstituted (Santivarangkna et al. 2008). Inactivation by high temperature drying,

Table 15.10 Typical hazards associated with dried foods

Biological hazards	Insects, rodents
Microbiological hazards	Growth of pathogens
Biochemical hazards	Enzymic, allergens, anti-nutrients
Chemical hazards	Mycotoxin, non-enzymic browning products, advanced lipid oxidation products, pesticide residues, food additives etc.

may involve heat damage to critical components in cells, including protein or DNA synthesis machinery (Santivarangkna et al. 2008).

Ironically, some classical drying methods have become quite refined in recent times in terms of their compatibility with micobes. Recently, spray drying has been proven useful for drying live cultures, depending on highly precise temperature-time profile control during the drying process (Peighambardoust et al. 2011; Santivarangkna et al. 2007).

4.2 Microbiological Safety of Dried Foods

One of the aims of food dehydration is to lower A_W in order to impede the growth of microorganisms (bacteria, molds, and yeasts). For a given moisture content, one food may support the growth of microorganisms while another will not. Whether microorganisms grow depends on the A_W not moisture content. Generally, molds will grow at lower water activities than bacteria, which has implications for mycotoxin risk (Sect. 4.4). For this reason, molds are more apt to grow in dried foods than are yeasts or bacteria.

In drying foods, A_W needs to decrease to below where microorganisms will grow and must be maintained at those values through packaging in order to exclude moisture. Low A_W is not achievable only by drying but also by addition of soluble components, such as sugar or salt. Thus, certain syrups and salted, partially dried foods (e.g., fish) are relatively stable as far as the growth of microorganisms is concerned, although there may be conditions in which they become subject to the growth of yeasts or molds because of their ability to grow at lower A_W values.

4.3 Drying Live Cultures

Dried starter cultures are alternatives to concentrated or frozen cultures for biotechnological applications (Tan et al. 2018). Some microorganism are also dried for applications as probiotics; probiotic microorganisms are living bacteria ingestion of which may be beneficial for health (Terpou et al. 2019). Starter cultures are used for manufacturing fermented food products whilst the probiotic microorganisms are used as functional food agents or dietary supplements. Techniques for drying food grade bacterial include convective air-drying, freeze-drying, spray drying, and vacuum drying (Huang et al. 2017; Santivarangkna et al. 2007). The various approaches induce stress on microorganisms, which can be partially mitigated by the addition of protective agents (Broeckx et al. 2016; Fenster et al. 2019). Avoding cell inactivation during drying ensures high degrees of viability during storage (Santivarangkna et al. 2008).

4.4 Mycotoxins in Dried Foods

Mycotoxins are toxic secondary metabolites produced by some fungi that grow fresh or dried plant foods. Eating food contaminated with mycotoxin is associated with increased risks of some forms of cancer, liver disease, kidney failure and also immune-suppression. Mycotoxins are associated with plant food commodities (cereals, grains, fruits, nuts) and some animal foods (milk, eggs and meat) owing to carryover from contaminated feed (Abbas and Shier 2009; Adeyeye 2016; Chelkowski 1991; Christensen 1982; Drusch and Ragab 2013; Karaca et al. 2010; Karlovsky et al. 2016; World Health Organization 2018).

Contamination by mycotoxins can affect international import/export trade in major agricultural commodities (World Health Organization 2018). Many importing countries have upper limits for mycotoxins, backed by surveillance systems. Economic losses due to mycotoxin contamination may be significant for producer countries.

There are hundreds if not thousands of different mycotoxins produced by diverse fungi. However, aflatoxin, ochratoxin, patulin and alternaria are some of the most common mycotoxins of concern within the human food chain. The main class of mycotoxins are cyclic peptides, with significant resistance to high temperature

Table 15.11 Processing methods to address food mycotoxins

Acidification
Alkaline treatment
Dehuling
Floatation
Heating
Milling
Oxidation
Sieving
Sorting
UV irradiation
Washing

From Karlovsky et al. (2016)

treatments. These mycotoxin can form all along the food chain under conditions of high humidity (A_W = 0.78–0.85).

Various methods for reducing the risk of mycotoxin intoxication were proposed (Table 15.11). Some preharvest strategies that reduce crop stress (with regard to water, nutrient, and pest) can also reduce the risk of mycotoxin formation. Dry weather harvesting, avoiding of physical damage, and storage in low relative humidity can also reduce the risk of mycotoxin formation. Also low temperatures can prevent fungal growth and mycotoxin formation. The removal of mycotoxins from food can be achieved by absorption as well as other common unit operations: cleaning, sorting to remove fruits infected with fungi (Karlovsky et al. 2016; Mousavi Khaneghah et al. 2018).

Owing to their stability, mycotoxins in fresh and rotting foods will carry over to dried products. Also dried and improperly stored foods posse a further mycotoxins risk, because fungi can grow under relatively moist conditions, during prolonged storage of "dried" foods. The topic of mycotoxins has been reviewed recently, and the reader is referred to these discussions for further information (Abbas and Shier 2009; Adeyeye 2016; Chelkowski 1991; Christensen 1982; Drusch and Ragab 2013; Karaca et al. 2010; Karlovsky et al. 2016; World Health Organization 2018).

5 Shelf Life of Dried Foods

5.1 General Principles

Product shelf-life is the time required for unacceptable loss of quality (best by date) or loss of safety (use by date). [9] Since quality is subjective, it follows that shelf life will be open to negotiation. [10] In addition, shelf life will be affected by storage conditions, and instrumental and non-instrumental methods used for monitoring quality (Villota et al. 1980).

To determine the shelf life of a dried food, batches of product are stored under well-controlled set of conditions of temperature, A_W or relative humidity, light exposure and the head-spacc gas composition (Caboni et al. 2005). Periodically, batches of product are removed and tested for quality. In most cases, quality decreases exponentially with time. Plotting an X-Y graph of log. (Quality) versus time, yields a straight-line. Storage-time is determined by the period required for Y% change in quality. The method of "stability testing", which was popularized by Ted Labuza is also used by the pharmaceutical industry (Labuza 1972). Stability testing requires that one controls experimental conditions to avoid unintended changes in storage conditions (Fig. 15.2).

5.2 Deterioration of Dried Foods

The loss of quality arises from food spoilage or deterioration. The common forms food deterioration are physical, chemical, biological or microbiological (Table 15.12). Physical and chemical changes are probably more important routes for quality loss in dried foods because other deterio-

[9] The quality of dried foods as influenced by physical changes (Sect. 1) and chemical changes (Sect. 2).

[10] The subjective nature of shelf life is a problem related to unnecessary food waste. It is widely recognised that significant amounts of food that reach their best by date should not be thrown away since they are safe to eat; Some donated foods may often fall into this category.

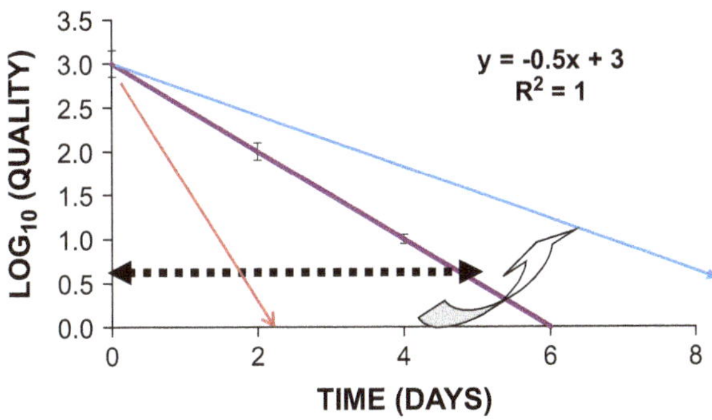

Fig. 15.2 Graph showing deterioration of food quality. Shelf life (dotted line) is time for unacceptable quality loss. Flipped arrows shows effect of decreasing temperature on deterioration rate (line slope)

Table 15.12 Major food deterioration reactions

Physical	Moisture movement, drying
	Toughening of texture (§)
	Moisture uptake, softening of texture, aggregation
	Sedimentation of dispersions (§)
	Creaming of dispersions(§)
Chemical	Oxidation, oxidative rancidity, loss of color
	Maillard reactions, discoloration, change in texture
Enzymatic	PPO, enzymic browning(§)
	LOX, oxidative rancidity(§)
	Lipase, lipolytic rancidity(§)
	Protease, fllavor and texture
Microbial	*Spoilage organisms* -quality deterioration
	pathogenic organisms, –food borne disease

Adapted from (Gould 2000). § Less relevant for dried products but may still occur during the drying process

rative processes (e.g. enzymatic and microbial) are attenuated under low moisture conditions.

Probably, the most important physical factors that determine the use by date or best-by date of dried foods is water/moisture as determined by A_W of the food and/or the relative humidity of gas in contact with the food. High temperature is another variable important for food stabilization. Storing dried foods, at high temperature will hasten deterioration. Oxygen exposure should be minimized in order to reduce the tendency for oxidation and bleaching which can produce off-flavors and loss of color and oil-soluble vitamins (Sect. 2.2).

5.3 Further Reading

5.3.1 Textbooks and Monographs

Food drying is an ancient and still very popular food technology. Not surprisingly, the amount of technical material available on this topic is vast. The preceding brief discussion will hopefully enable readers to seek out more in-depth reading available in the form of dedicated books and monographs (Arsdel et al. 1973a, b; Arun 2017; Gillies 1974; Greensmith 1971; Kudra and Mujumdar 2001; NIIR Board of Consultants and Engineers 2015). Some of the books provide in-depth focus on drying methods or specific food commodities; spray drying (Masters 1972, 1985), freeze drying (Goldblith et al. 1975; Gutcho 1977; Hua et al. 2010; King 1971; Oetjen 1999), vacuum drying (Reis 2017), drying for convenience foods (Noyes 1969), soups (Binsted and Devey 1970), fish (Burt 1988; Doe 1998) and also fruits and vegetables (Mercer 2014a, b; NIIR Board of Consultants and Engineers 2015; Torrey 1974)

5.3.2 General Reviews

Food dehydration has been reviewed by food engineers and technologist (Bourdoux et al. 2016; Cohen and Yang 1995; Raoultwack 1994; Ratti 2001; Salunkhe et al. 1973). Other papers describe dehydration methods for developing countries (Chua and Chou 2003), storage stability of dehydrated foods (Villota et al. 1980), food dehydration related to food structure (Aguilera et al. 2003), drying of fruit and vegetables (Sagar and Kumar 2010a, b), drying fruit juices (Anjali and Vir Singh 2015), and the effect of drying on food quality (Oliveira et al. 2016)

References

Aalaei K, Rayner M, Sjöholm I (2016) Storage stability of freeze-dried, spray-dried and drum-dried skim milk powders evaluated by available lysine. LWT 73:675–682. https://doi.org/10.1016/j.lwt.2016.07.011

Aalaei K, Sjöholm I, Rayner M, Tareke E (2017) The impact of different drying techniques and controlled storage on the development of advanced glycation end products in skim milk powders using isotope Dilution ESI-LC-MS/MS. Food Bioprocess Technol 10(9):1704–1714. https://doi.org/10.1007/s11947-017-1936-x

Aalaei K, Rayner M, Sjöholm I (2019) Chemical methods and techniques to monitor early Maillard reaction in milk products; a review. Crit Rev Food Sci Nutr 59(12):1829–1839

Abbas HK, Shier WT (2009) Mycotoxin contamination of agricultural products in the Southern United States and approaches to reducing it from pre-harvest to final food products. In: Mycotoxin prevention and control in agriculture, vol 1031. American Chemical Society, Washington, DC, pp 37–58

Adeyeye SA (2016) Fungal mycotoxins in foods: a review. Cogent Food Agric 2(1):1213127

Aguilera JM, Chiralt A, Fito P (2003) Food dehydration and product structure. Trends Food Sci Technol 14(10):432–437. https://doi.org/10.1016/s0924-2244(03)00122-5

Anjali V, Vir Singh S (2015) Spray drying of fruit and vegetable juices-a review. Crit Rev Food Sci Nutr 55(5):701–719

Anonymous (2004) Technical advisory service for images advice paper color theory: understanding and modelling color. Retrieved from http://dl3.takbook.com/doc/download-article-document-takbook-project/colour.20.pdf

Arsdel WBV, Copley MJ, Morgan AI Jr (1973a) Food dehydration. Vol. 1. Drying methods and phenomena, 2nd edn. AVI, Westport, p xi+347pp

Arsdel WBV, Copley MJ, Morgan AI Jr (1973b) Food dehydration. Vol. 2. Practices and applications, 2nd edn. AVI, Westport, p xi+529pp

Arun SM (2017) Handbook of industrial drying, 4th edn. CRC Press, Boca Raton, 4(2015)

Binsted R, Devey JD (1970) Soup manufacture. Canning, dehydration and quick-freezing, 3rd edn. Food Trade Press, London, p xi+260pp

Bonazzi C, Dumoulin E (2011) Quality changes in food materials as influenced by drying processes. In: Tsotsas E, Mujumdar AS (eds) Modern drying technology volume 3: product quality and formulation, 1st edn. KGaA/Wiley-VCH Verlag GmbH & Co., Weinheim, pp 1–20

Botta JR (1995) Evaluation of Seafood freshness quality. Retrieved from https://books.google.co.uk/books?id=HmHgGL4e2w0C

Bourdoux S, Li D, Rajkovic A, Devlieghere F, Uyttendaele M (2016) Performance of drying technologies to ensure microbial safety of dried fruits and vegetables. Compr Rev Food Sci Food Saf 15(6):1056–1066. https://doi.org/10.1111/1541-4337.12224

Broeckx G, Vandenheuvel D, Claes IJ, Lebeer S, Kiekens F (2016) Drying techniques of probiotic bacteria as an important step towards the development of novel pharmabiotics. Int J Pharm 505(1–2):303–318

Burt JR (1988) Fish smoking and drying. The effect of smoking and drying on the nutritional properties of fish. Elsevier Applied Science Publishers, New York, p xii + 166pp

Caboni MF, Boselli E, Messia MC, Velazco V, Fratianni A, Panfili G, Marconi E (2005) Effect of processing and storage on the chemical quality markers of spray-dried whole egg. Food Chem 92(2):293–303. https://doi.org/10.1016/j.foodchem.2004.07.025

Caric M, Kalab M (1987) Effects of drying techniques on milk powders quality and microstructure: a review. Food Microstruct 6(2):171–180

Celestino EL, Iyer M, Roginski H (1997) The effects of refrigerated storage of raw milk on the quality of whole milk powder stored for different periods. Int Dairy J 7(2–3):119–127. https://doi.org/10.1016/s0958-6946(96)00041-6

Chelkowski J (1991) Cereal grain. Mycotoxins, fungi and quality in drying and storage. Elsevier, Amsterdam, p xxii + 607pp

Chen L, Opara UL (2013) Texture measurement approaches in fresh and processed foods - a review. Food Res Int 51(2):823–835. https://doi.org/10.1016/j.foodres.2013.01.046

Christensen CM (1982) Storage of cereal grains and their products. American Association of Cereal Chemists, St. Paul

Chua KJ, Chou SK (2003) Low-cost drying methods for developing countries. Trends Food Sci Technol 14:519–528

Cohen JS, Yang TCS (1995) Progress in food dehydration. Trends Food Sci Technol 6(1):20–25. https://doi.org/10.1016/s0924-2244(00)88913-x

Da Conceicao Neta ER, Johanningsmeier SD, McFeeters RF (2007) The chemistry and physiology of sour taste—a review. J Food Sci 72(2):R33–R38

Deng L-Z, Mujumdar AS, Zhang Q, Yang X-H, Wang J, Zheng Z-A, Gao Z-J, Xiao H-W (2019) Chemical and physical pretreatments of fruits and vegetables: effects on drying characteristics and quality attributes–a comprehensive review. Crit Rev Food Sci Nutr 59(9):1408–1432

Devahastin S, Niamnuy C (2010) Modelling quality changes of fruits and vegetables during drying: a review. Int J Food Sci Technol 45(9):1755–1767. https://doi.org/10.1111/j.1365-2621.2010.02352.x

Doe PE (1998) Fish drying & smoking: production and quality. CRC Press, Boca Ratón, 250pp

Dominguez R, Quintero A, Bourne MC, Talamas R, Anzaldua-Morales A (1996) Texture of dehydrated bell peppers modified by low-temperature blanching and calcium addition. Int J Food Sci Technol 36(5):523–527

Drewnowski A, Gomez-Carneros C (2000) Bitter taste, phytonutrients, and the consumer: a review. Am J Clin Nutr 72(6):1424–1435

Drusch S, Ragab W (2013) Mycotoxins in fruits, fruit juices, and dried fruits. J Food Prot 66(8):1514–1527

Falade KO, Abbo ES (2007) Air-drying and rehydration characteristics of date palm (Phoenix dactylifera L.) fruits. J Food Eng 79(2):724–730

Fenster K, Freeburg B, Hollard C, Wong C, Rønhave Laursen R, Ouwehand AC (2019) The production and delivery of probiotics: a review of a practical approach. Microorganisms 7(3):83

Fernandes FAN, Rodrigues S, Law CL, Mujumdar AS (2011) Drying of exotic tropical fruits: a comprehensive review. Food Bioprocess Technol 4(2):163–185

Fitzpatrick JJ, Ahrné L (2005) Food powder handling and processing: industry problems, knowledge barriers and research opportunities. Chem Eng Process Process Intensif 44(2):209–214

Gaiani C, Banon S, Scher J, Schuck P, Hardy J (2005) Use of a turbidity sensor to characterize micellar casein powder rehydration: influence of some technological effects. J Dairy Sci 88(8):2700–2706

García-Pascual P, Sanjuán N, Bon J, Carreres JE, Mulet A (2005) Rehydration process of Boletus edulis mushroom: characteristics and modelling. J Sci Food Agric 85(8):1397–1404

García-Pascual P, Sanjuán N, Melis R, Mulet A (2006) Morchella esculenta (morel) rehydration process modelling. J Food Eng 72(4):346–353

Gillies MT (1974) Dehydration of natural and simulated dairy products, Food technology review, 15. Noyes Data Corporation

Goldblith SA, Rey LI, Rothmayr WW (1975) Freeze-drying and advanced food technology. Academic, London/New York/San Francisco, p xxxi + 730pp

Goldman M, Horev B, Saguy I (1983) Decolorization of β-carotene in model systems simulating dehydrated foods. Mechanism and kinetic principles. J Food Sci 48(3):751–754

Gould GW (2000) Preservation: past, present and future. Br Med Bull 56(1):84–96

Greensmith M (1971) Practical dehydration. vi+174pp. SBN, 900397(03)

Gutcho MH (1977) Freeze drying processes for the food industry, Food technology review, 14. Noyes Data Corporation

Haque E, Bhandari BR, Gidley MJ, Deeth HC, Moller SM, Whittaker AK (2010) Protein conformational modifications and kinetics of water-protein interactions in milk protein concentrate powder upon aging: effect on solubility. J Agric Food Chem 58(13):7748–7755. https://doi.org/10.1021/jf1007055

Haque E, Whittaker AK, Gidley MJ, Deeth HC, Fibrianto K, Bhandari BR (2012) Kinetics of enthalpy relaxation of milk protein concentrate powder upon ageing and its effect on solubility. Food Chem 134(3):1368–1373. https://doi.org/10.1016/j.foodchem.2012.03.034

Heertje I (1993) Structure and function of food products: a review. Food Struct 12(3 Article 7):343–364

Heino AT, Uusi-Rauva JO, Rantamaki PR, Tossavainen O (2007) Functional properties of native and cheese whey protein concentrate powders. Int J Dairy Technol 60(4):277–285. https://doi.org/10.1111/j.1471-0307.2007.00350.x

Hogekamp S, Schubert H (2003) Rehydration of food powders. Food Sci Technol Int 9(3):223–235. https://doi.org/10.1177/1082013203034938

Holley RA (1985) Beef jerky: viability of food-poisoning microorganisms on jerky during its manufacture and storage. J Food Prot 48(2):100–106

Hua TC, Liu BL, Zhang H (2010) Freeze-drying of pharmaceutical and food products. Woodhead, Oxford, 256pp

Huang S, Vignolles M-L, Chen XD, Le Loir Y, Jan G, Schuck P, Jeantet R (2017) Spray drying of probiotics and other food-grade bacteria: a review. Trends Food Sci Technol 63:1–17

Ji J, Fitzpatrick J, Cronin K, Maguire P, Zhang H, Miao S (2016) Rehydration behaviors of high protein dairy powders: the influence of agglomeration on wettability, dispersibility and solubility. Food Hydrocoll 58:194–203

Joardder MUH, Kumar C, Karim MA (2017) Food structure: its formation and relationships with other properties. Crit Rev Food Sci Nutr 57(6):1190–1205. https://doi.org/10.1080/10408398.2014.971354

Kaijser A, Dutta P, Savage G (2000) Oxidative stability and lipid composition of macadamia nuts grown in New Zealand. Food Chem 71(1):67–70

Karaca H, Velioglu YS, Nas S (2010) Mycotoxins: contamination of dried fruits and degradation by ozone. Toxin Rev 29(2):51–59

Karel M (1984) Chemical effects in food stored at room temperature. J Chem Educ 61:335–339. ACS Publications

Karel M (1986) Control of lipid oxidation in dried foods. In: MacCarthy D (ed) Concentration and drying of foods. Elsevier, London, pp 37–51

Karlovsky P, Suman M, Berthiller F, De Meester J, Eisenbrand G, Perrin I, Recker T (2016) Impact of food processing and detoxification treatments on mycotoxin contamination. Mycotoxin Res 32(4):179–205

Khalloufi S, Ratti C (2003) Quality deterioration of freeze-dried foods as explained by their glass transition temperature and internal structure. J Food Sci 68(3):892–903. https://doi.org/10.1111/j.1365-2621.2003.tb08262.x

King CJ (1971) Freeze-drying of foods. Chemical Rubber Co, Cleveland, 86pp

Krokida M, Marinos-Kouris D (2003) Rehydration kinetics of dehydrated products. J Food Eng 57(1):1–7

Krokida M, Maroulis Z (2000) Quality changes during drying of food materials. In: Mujumdar AS (ed) Drying technology in agriculture and food sciences. Science Publishers, pp 61–68

Kudra T, Mujumdar AS (2001) Advanced drying technologies. Marcel Dekker, New York, p vii + 459pp

Labuza TP (1972) Nutrient losses during drying and storage of dehydrated foods. CRC Crit Rev Food Technol 3(2):217–240

Le TT, Bhandari B, Holland JW, Deeth HC (2011) Maillard reaction and protein cross-linking in relation to the solubility of milk powders. J Agric Food Chem 59(23):12473–12479

Lewicki PP (1998) Effect of pre-drying treatment, drying and rehydration on plant tissue properties: a review. Int J Food Prop 1(1):1–22

LIPIDMAPS (2014) Tutorials and lectures on lipids. Retrieved from http://www.lipidmaps.org/resources/tutorials/lipid_tutorial.html#D

Maldonado S (2015) Drying and rehydration kinetics of mangoes: solute effects. Int J Innov Res Eng Manag (IJIREM) 2(6):71–84

Mao XY, Tong PS, Gualco S, Vink S (2012) Effect of NaCl addition during diafiltration on the solubility, hydrophobicity, and disulfide bonds of 80% milk protein concentrate powder. J Dairy Sci 95(7):3481–3488. https://doi.org/10.3168/jds.2011-4691

Marabi A, Saguy S (2009) Rehydration and reconstitution of food. In: Ratti C (ed) Advances in food dehydration. CRC Press, Boca Raton, pp 237–284

Masters K (1972) Spray drying, London, p xxxv+668pp

Masters K (1985) Spray drying handbook. Godwin, London, p 4

McSweeney M, Seetharaman K (2015) State of polyphenols in the drying process of fruits and vegetables. Crit Rev Food Sci Nutr 55(5):660–669

Mercer DG (2014a) An introduction to the dehydration and drying of fruits and vegetables. Part 1 food drying basics. Retrieved from https://www.uoguelph.ca/foodscience/sites/default/files/Drying-Part%201.pdf

Mercer DG (2014b) An introduction to the dehydration and drying of fruits and vegetables. Part 2. Drying of specific fruits and vegetables. Retrieved from https://www.uoguelph.ca/foodscience/sites/default/files/Drying-Part%201.pdf

Meyers B, Brewer M (2008) Sweet taste in man: a review. J Food Sci 73(6):R81–R90

Mousavi Khaneghah A, Fakhri Y, Sant'Ana AS (2018) Impact of unit operations during processing of cereal-based products on the levels of deoxynivalenol, total aflatoxin, ochratoxin A, and zearalenone: A systematic review and meta-analysis. Food Chem 268:611–624. [Abstract]. https://doi.org/10.1016/j.foodchem.2018.06.072

NIIR Board of Consultants & Engineers (ed) (2015) The complete technology book on processing, dehydration, canning, preservation of fruits & vegetables (3rd rev edn)

Nijhuis HH, Torringa HM, Muresan S, Yuksel D, Leguijt C, Kloek W (1998) Approaches to improving the quality of dried fruit and vegetables. Trends Food Sci Technol 9(1):13–20. https://doi.org/10.1016/s0924-2244(97)00007-1

Noyes R (1969) Dehydration processes for convenience foods. Noyes Development Corp, Park Ridge, 367pp

Oetjen GW (1999) Freeze-drying. ix + 276pp

Oikonomopoulou VP, Krokida MK (2013) Novel aspects of formation of food structure during drying. Dry Technol 31(9):990–1007

Oliveira SM, Brandao TRS, Silva CLM (2016) Influence of drying processes and pretreatments on nutritional and bioactive characteristics of dried vegetables: a review. Food Eng Rev 8(2):134–163. https://doi.org/10.1007/s12393-015-9124-0

Omolola AO, Jideani AIO, Kapila PF (2017) Quality properties of fruits as affected by drying operation. Crit Rev Food Sci Nutr 57(1):95–108

Ong XY (2019) Fundamental and applied aspects of powder dispersion in liquids. University of Surrey, PhD thesis

Oral RA, Dogan M, Sarioglu K (2011) Monitoring of the hydroxymethylfurfural amount as an indicator of Maillard products in sweet whey and skim milk powder stored under different conditions. Milchwissenschaft 66(4):398–401

Ortega-Rivas E, Juliano P, Yan H (2010) Food powders: physical properties, processing, and functionality. Springer US

Park CW, Drake MA (2014) The distribution of fat in dried dairy particles determines flavor release and flavor stability. J Food Sci 79(4):R452–R459

Peighambardoust SH, Tafti AG, Hesari J (2011) Application of spray drying for preservation of lactic acid starter cultures: a review. Trends Food Sci Technol 22(5):215–224. https://doi.org/10.1016/j.tifs.2011.01.009

Pico J, Bernal J, Gomez M (2015) Wheat bread aroma compounds in crumb and crust: a review. Food Res Int 75:200–215. https://doi.org/10.1016/j.foodres.2015.05.051

Pruthi JS (1978) Quality, control, packaging, and storage of dehydrated foods. Indian Food Packer 32(1):30–40

Rahman MS, Al-Farsi SA (2005) Instrumental texture profile analysis (TPA) of date flesh as a function of moisture content. J Food Eng 66(4):505–511. https://doi.org/10.1016/j.jfoodeng.2004.04.022

Ramos IN, Brandão TR, Silva CLM (2003) Structural changes during air-drying of fruits and vegetables. Food Sci Technol Int 9(3):201–206

Raoultwack AL (1994) Recent advances in the osmotic dehydration of foods. Trends Food Sci Technol 5(8):255–260. https://doi.org/10.1016/0924-2244(94)90018-3

Ratti C (2001) Hot air and freeze-drying of high-value foods: a review. J Food Eng 49(4):311–319

Reis FR (2017) Vacuum drying for extending food shelf life. xi + 72pp, 2014

Richard B, Le Page J-F, Schuck P, André C, Jeantet R, Delaplace G (2013) Towards a better control of dairy powder rehydration processes. Int Dairy J 31(1):18–28

Sablani SS (2006) Drying of fruits and vegetables: retention of nutritional/functional quality. Dry Technol 24(2):123–135

Sagar VR, Kumar PS (2010a) Recent advances in drying and dehydration of fruits and vegetables: a review. J Food Sci Technol 47(1):15–26

Sagar VR, Kumar SP (2010b) Recent advances in drying and dehydration of fruits and vegetables: a review. J Food Sci Technol Mysore 47(1):15–26

Saguy IS, Troygot O, Marabi A, Wallach R (2010) Rehydration modeling of food particulates utilizing principles of water transport in porous media. In: Food engineering interfaces. Springer, pp 535–552

Saito Z (1985) Particle structure in spray-dried whole milk and in instant skim milk powder as related to lactose crystallization. Food Microstruct 4(2):333–340

Salunkhe DK, Do JY, Bolin HR (1973) Developments in technology and nutritive value of dehydrated fruits, vegetables and their products. CRC Crit Rev Food Technol 4(2):153–192

Santivarangkna C, Kulozik U, Foerst P (2007) Alternative drying processes for the industrial preservation of lactic acid starter cultures. Biotechnol Prog 23(2):302–315. https://doi.org/10.1021/bp060268f

Santivarangkna C, Kulozik U, Foerst P (2008) Inactivation mechanisms of lactic acid starter cultures preserved by drying processes. J Appl Microbiol 105(1):1–13

Sikand V, Tong PS, Roy S, Rodriguez-Saona LE, Murray BA (2011) Solubility of commercial milk protein concentrates and milk protein isolates. J Dairy Sci 94(12):6194–6202

Singh GD, Sharma R, Bawa A, Saxena D (2008) Drying and rehydration characteristics of water chestnut (*Trapa natans*) as a function of drying air temperature. J Food Eng 87(2):213–221

Takeoka GR, Guentert M, Engel KH (2001) Aroma active compounds in foods: chemistry and sensory properties. American Chemical Society (ACS), Washington, DC, 289 pages

Tan D, Poh P, Chin S (2018) Microorganism preservation by convective air-drying—a review. Dry Technol 36(7):764–779

Terpou A, Papadaki A, Lappa IK, Kachrimanidou V, Bosnea LA, Kopsahelis N (2019) Probiotics in food systems: significance and emerging strategies towards improved viability and delivery of enhanced beneficial value. Nutrients 11(7). https://doi.org/10.3390/nu11071591

Torrey M (1974) Dehydration of fruits and vegetables, Food technology review 13. Noyes Data Corporation

University of Washington (2009a) Color. Retrieved from http://depts.washington.edu/cmditr/modules/lum/color.html

University of Washington (2009b) Electromagnetic radiation. Retrieved from http://depts.washington.edu/cmditr/modules/lum/electromagnetic_radiation.html

Villota R, Saguy I, Karel M (1980) Storage stability of dehydrated food. Evaluation of literature data. J Food Qual 3(3):123–212

Wall MM (2010) Functional lipid characteristics, oxidative stability, and antioxidant activity of macadamia nut (*Macadamia integrifolia*) cultivars. Food Chem 121(4):1103–1108

Warawaran R, Bronlund JE (2015) A review of drying processes in the production of pumpkin powder. Int J Food Eng 11(6):789–799

World Health Organization (2018) Mycotoxins. Retrieved from https://www.who.int/news-room/fact-sheets/detail/mycotoxins

Yamaguchi S, Ninomiya K (2000) Umami and food palatability. J Nutr 130(4):921S–926S

Part IV

Handling and Processing Food

Food Additives 16

1 Introduction

1.1 History of Chemical Additives

Substances added to foods in the United States (US) number about 10,787. About one-half of the compounds (5292 substances) are designated as food additives (Table 16.1). Another 4646 of the substances added to foods have Generally Recognized As Safe (GRAS) status; the significance of this is explained later in the chapter. Another 849 of the substances added legally to foods are classed as "other" referring to colors and compounds in common use prior to 1958 (Nicole 2013). About 3000 designated food ingredients are listed in a document titled "Everything added to food in the United States (EAFUS)" (U.S. Food and Drug Administration 2014a).

There was little control over chemical additives used as preservatives or colorings up until about 1906. In 1903, Dr. Harvey Washington Wiley, then the Chief of Bureau of Chemistry of the U.S. Department of Agriculture, established a "poison squad" that consisted of young men who consumed foods treated with known amounts of chemicals commonly used in foods. The goal of the project was to determine whether these compounds were deleterious to health. The result of the efforts of Dr. Wiley and the "squad" was the passage of the Food and Drug Act of 1906, which is also referred to as "The Pure Food Act." In September of 1958, the FD&C Act was amended to prohibit the use of food additives that had not been adequately tested to establish their safety (Lewis 2002).

1.2 Significance of Food Additives

The intentional use of food additives is unavoidable in the modern food industry though often attached with controversy (Fennema 1987). One reason for considering food additives as indispensable is that, they are used in order to achieve a variety of pre-defined technological goals; additives are intended to improve sensory, nutritional, safety and shelf life of products. Additives are also employed as processing aides, in order to assist with the manufacturing process. Frequently also, the distinction between food additive and ingredient is not so clear-cut. However, exposure to food additives is regulated in the same way as exposure to other environmental agents.

Each additive is approved for use only after prolonged evaluation for their safety and upper limits of intake by the FDA (Neltner et al. 2011). The scientific discipline concerned most with the evaluation of food additives, is toxicology. Some basic concepts must be remembered when dealing with food additives:

R. Owusu-Apenten, E. R. Vieira, *Elementary Food Science*, Food Science Text Series,
https://doi.org/10.1007/978-3-030-65433-7_16

Table 16.1 Food additives and other substances added to foods

1.	Food additives Direct food additives Indirect food additives (few additions after 1997) Substances covered by FCS notifications (began in 1997) FCS below threshold of regulation (began in 1995) Radiation sources
2.	GRAS substances Common food ingredients in use before 1958 Manufacturer self-determined Association expert panel-determined FDA-listed (ended in 1973) FDA-affirmed (began in 1973 and effectively replaced after 1997) Substances covered by FDA-reviewed GRAS notification (began in 1997)
3.	Prior-sanctioned substances (federally sanctioned before 1958)
4.	Color additives (began in 1960)
5.	Pesticide chemicals or residues (modified in 1996)
6.	Drugs in animal feed (modified in 1968)
7.	Dietary supplements (began in 1994)

Adapted from Neltner et al. (2011), FCS = Food contact substance

Table 16.2 Some food additives and packaging terms[a]

Food Additive. A food additive is defined in Section 201(s) of the FD&C Act as "any substance the intended use of which results or may reasonably be expected to result, directly or indirectly, in its becoming a component or otherwise affecting the characteristic of any food (including any substance intended for use in producing, manufacturing, packing, processing, preparing, treating, packaging, transporting, or holding food; and including any source of radiation intended for any such use); *if such substance is not GRAS or sanctioned prior to 1958 or otherwise excluded from the definition of food additives"*.
Food Contact Substance (FCS) – Section 409 of the FD&C Act defines an FCS as "any substance that is intended for use as a component of materials used in manufacturing, packing, packaging, transporting, or holding food if such use of the substance is not intended to have any technical effect in such food".
Colorant – "A colorant is a dye, pigment, or other substance that is used to impart color to or to alter the color of a food-contact material", but that does not migrate to food in amounts that will contribute to that food any color apparent to the naked eye. The term 'colorant' includes substances such as optical brighteners and fluorescent whiteners, which may not themselves be colored, but whose use is intended to affect the color of a food-contact material" (21 CFR 178.3297(a)).

From Food and Drug Administration (USDA) (2015). [a]Closely follows the regulatory text

- All foods are composed of chemical compounds. Any food component can be extracted and added to other foods in which case the extracted substance is classed as an additive.
- Any additive can be injurious to health but only at sufficiently high levels.
- Any additive can be safe to use at low levels
- Each additive must be assessed for toxicity using scientific procedures, regardless of how safe its proponents say it is and how toxic its opponents say it is.
- The use of radiation for preserving foods is declared an additive.

In general, additives cannot be added to foods, unless the FDA has approved their use. Moreover, additives are tested for toxicity in concentrations much greater than those allowed in foods. Most food additives are components obtained from natural foods and without such additives, the quality of many foods would be inferior to that to which we have become accustomed. The shelf life or availability of many foods would also be greatly limited. Food additives are classified by their function (see below) because they overlap each other in terms chemical structure.

1.3 Definitions of Food Additives

A food additive is defined in Section 201(s) of the United States FD&C Act (Table 16.2). Many nutrients, extracted food components as well as gasses are used as additives. However, many compounds added to foods, are excluded from regulatory oversight as food additives (David 1988).

Similarly, a food additive is defined by the WHO/ FAO as "*any substance not normally con-*

sumed as a food by itself and not normally used as a typical ingredient of the food, whether or not it has nutritive value, the intentional addition of which to food for a technological (including organoleptic) purpose in the manufacture, processing, preparation, treatment, packing, packaging, transport or holding of such food results, or may be reasonably expected to result (directly or indirectly), in it or its by-products becoming a component of or otherwise affecting the characteristics of such foods" (Codex Alimentarius Commission 1995). The codex definition implies that contaminants or and substances that improve nutritive characteristics should be considered additives – but this latter requirement is not always adopted by nation states – as shown by discussions of amino acids and other nutrients, below.

Food additives are defined in EU legislation in manner similar to the statements offered by codex. e.g. "*any substance not normally consumed as a food in itself and not normally used as a characteristic ingredient of food whether or not it has nutritive value, the intentional addition of which to food for a technological purpose ... results ... in it or its by-products becoming directly or indirectly a component of such foods*" (Europa 2014).

1.3.1 Categories of Food Additives

Several subcategories of food additives have been defined (Neltner et al. 2011). (a) direct food additive (from 21 CFR 173) refers to materials used as processing aides for their technical effect in the food but which are largely absent in the finished products. (b) Indirect food additive – a substances or substances that come into contact with foods as a result of packaging, or processing equipment, (c) Food Contact Substance (FCS) – materials that come into contact with foods, by virtue of their use for manufacturing packaging and not directly added to foods for some technical function. Some FCS occur in very low concentrations below that deemed to require regulation. (d) Radiation sources (Table 16.1).

1.3.2 Distinguishing Additives and Food Ingredients

The distinction between food additives and food ingredients is important. A food ingredient refers to "*substance, excluding a food additive, used in the manufacture or preparation of a food and present in the final product*" (Codex Alimentarius Commission 1981).

1.3.3 Other Exclusion from Additives

Some materials added to foods are not considered either food additives or ingredients (Table 16.1). One such group of "non-additives" are substances for which there is a prior history of use before 1958; these can be added to foods without regulatory oversight. Other substances not covered by food additives legislation are those with GRAS status (Burdock and Carabin 2004), color additives, dietary supplements, and drugs used for livestock treatments (Neltner et al. 2011).

1.4 Functional Classifications for Food Additives

There are currently 27 functional classes of food additives recognized as listed in Table 16.3 (Codex Alimentarius Commission 1989b). The

Table 16.3 Functional classification of food additives

CLASSES 1–14	CLASSES 15–27
1. Acidity regulator	15. Flour treatment agent
2. Anticaking agent	16. Foaming agent
3. Antifoaming agent	17. Gelling agent
4. Antioxidant	18. Glazing agent
5. Bleaching agent	19. Humectant
6. Bulking agent	20. Packaging gas
7. Carbonating agent	21. Preservative
8. Carrier	22. Propellant
9. Color	23. Raising agent
10. Color retention agent	24. Sequestrant
11. Emulsifier	25. Stabilizer
12. Emulsifying salt	26. Sweetener
13. Firming agent	27. Thickener
14. Flavor enhancer	

Adapted from Codex Alimentarius Commission (1989b)

different category names are also used in the US (U.S. Food and Drug Administration 2014b) and internationally.

The World Health Organization is also behind the International Numbering System (INS) for food additives (Codex Alimentarius Commission 1989a). Accordingly, each additive is assigned a numerical identifier, which is useful for labeling purposes. The INS identifier comprises three digits – sometimes followed with a letter to differentiate closely related additives. The various classes of food additives are divided according to the function of different substances, and not according to structure. For example, very many different, unrelated, chemical entities may function as food colors, bulking agents, flavor enhancer etc. A comprehensive treatise on food additives, edited by Smith & Hong-Shum, covers 14 of the 27 classes in detail (Smith and Hong-Shum 2011).

Some of the technological functions performed by food additives are summarized in Fig. 16.1. The sensory additives can affect appearance (colors, bleaching agents, glazing agents), texture (bulking agents, anticaking agents for food powders and ingredients, stabilizers for food dispersions), and taste (sweeteners, acidity regulators). The nutritive food additives group covers vitamins, amino acids and minerals. Food additives which extend shelf life decrease the rate of spoilage involving, chemical, physical, biological or microbiological changes.

2 Additives Legislation

2.1 The 1958 Food Additive Act

Food additives are regulated by, the 1958 Food Additive Amendment to the Food & Drugs and Cosmetic Act of 1938. The 1958 act for food additives brought into effect the process of pre-market approval. Safety assessment for food additives involves a process of risk assessment. In general terms, all chemicals introduced after 1958 are presumed to be potentially hazardous. It is necessary to identify the types of adverse effects associated with a compound, and to establish the dose-adverse response characteristics. Ultimately, experts will establish the no effect level (NOEL) concentration below which no harm will befall the consumer. Toxicity testing is performed using animals (typically rodents) as models. Values for the NOEL are then adjusted (e.g. NOEL values for rats are divided by 100) to allow for the fact that tests were done with animals, and to allow for inter-person variations in responses (Hanlon et al. 2016). Ultimately, such analysis produces a safe-use level, or allowed daily intake (ADI) for an additive, which then passes into law.

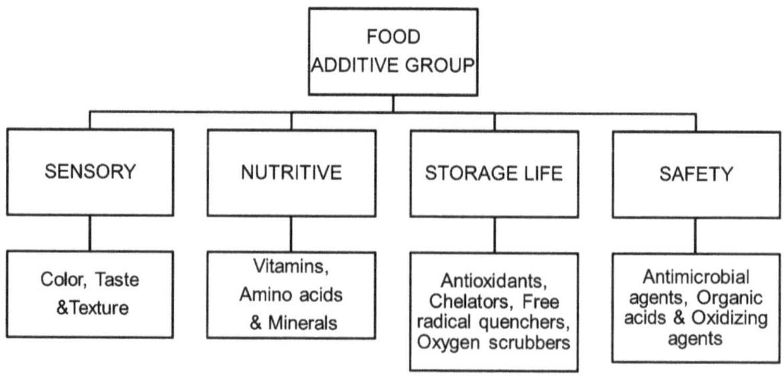

Fig. 16.1 Overview of food additives groupings (unofficial)

2.2 The Delaney Clause 1959

The Delaney Clause is named after US Representative James Delaney of New York, who introduced it into food legislation. The essence of the Delaney clause is that: "… *no additive shall be deemed to be safe if it is found to induce cancer when ingested by man or animal, or if it is found, after tests which are appropriate for the evaluation of the safety of food additives, to induce cancer in man or animal*" (Lester and Staples 1996). The Delaney amendment was considered controversial perhaps because it opted to dismiss concerns about dosage (Hoyle 1996; Lester 1996; Picut and Parker 1992). Critics of the Delaney clause noted that all chemicals are toxic at high dosages; readers can find more detailed arguments for and against the Delaney clause in the following texts (Picut and Parker 1992; Weisburger 1996).

2.3 Gererally Recognized as Safe Compounds

Compounds previously established by the FDA as GRAS can be added legally to foods without classing them as food additives. GRAS substances are outside of the food additive framework (Burdock and Carabin 2004). Substances are considered for GRAS if they were used historically, without harm to consumers. In addition, new substances can be evaluated as GRAS, after a petition is submitted to the FDA. Completion of a GRAS certification may require several years because the supporting evidence includes a 2-year feeding study using two species of animals. Such studies must reveal no long-term and short-term adverse effects (Neltner et al. 2013).

2.4 Approved Lists for Additives

Approved food additives have well defined chemical identities with standards for quality, and limits for use. Currently, approved listings of food additives are available online (Food and Drug Administration (USDA) 2013) (Food and Agriculture Organization of the United Nations 2013) (Codex Alimentarius Commission 2015). A core list of approved food additives in the US is available from the Codes Of Federal Register 21 CFR 172 (U.S. Government Publishing Office 2014) (U.S. Government Publishing Office 2012) European lists of approved food additives can be found from the Eur-lex website (Eur-Lex 2011). International listing of approved food additives and conditions for their use have been compiled by the JEFCA, short for Joint FAO/WHO Expert Committee on Food Additives (JEFCA 2015; Joint FAO/WHO Expert Committee on Food Additives 2001).

Some basic requirements apply for all approved food additives and GRAS substances: (i) Intentional additives must perform their intended function. (ii) Additives must not deceive the consumer or conceal faulty ingredients or defects in manufacturing practices, (iii) Additives should not reduce the food's nutritional value. (iv) An additive cannot be used to achieve an effect that could be gained by good manufacturing practices and (v) A method of analysis must exist to monitor the use of the additive in foods or its incidental occurrence in foods (such as migration from a packaging material).

2.5 International Regulations for Food Additives

International regulations for food additives can be traced to JEFCA which, is a committee formed from international experts, nominated by the WHO and FAO. JECFA evaluated the safety of chemical substances added to foods since the 1950's. Currently, JECFA publishes authoritative data on the toxicity of additives, which is the basis for international regulations. Specifically, JECFA also produces unique identification numbers for different compounds and their standards of purity required for use in foods. The CAC standard CXS_192e available online (Codex Alimentarious Commission 1995, 11,917), provides details of international food additive regulations. The 475pp document contains a core of 15 pages setting out clear definitions for food additives, foods for which food additives can and cannot be used, key principles of food additive use and justifications. For example, food additive may be justified on the basis that their use is advantageous and poses no risk for the consumer.

Manufacturing foods for people with special dietary needs provides another justifiable use of food additives. Additives are also justified if they aide the in the achievement of established end-goals in the food system: processing and manufacturing, packaging, storage and distribution. By contrast, the use of additives will not be permitted if alternative economically practical ways exist to achieve the intended end goal. In accordance with principles of GMP, the use level for food additives should be the lowest level required to achieve the stated goals (Codex Alimentarious Commission 1995).

2.6 Labelling Requirements for Food Additives

The labeling requirements for food additives, when sold to food manufactures and caterers are as described in the appropriate CAC standards (Codex Alimentarius Commission 1981); In fact, the labeling requirements for food additives, when sold as such, are similar to these for other general food labels (Codex Alimentarius Commission 1981, 1985).

3 Sensory Additives

This section provides an overview of some major food additive classes. The reader is referred to several sources for more information (Armstrong 2009; Jackson 2009; Rulis and Levitt 2009; Smith and Hong-Shum 2011).

3.1 Colorants

Color additives can be categorized into three major types: natural, nature-identical, or synthetic. Natural colorants include the yellow from the annatto seed; green from chlorophyll; orange color from carotene; brown from caramelized sugar; and red from beets, tomatoes, and the cochineal insect. Natural colors are obtained from animal, vegetable, or mineral sources.

The term "nature-identical" applies to synthetic counterparts of colors and pigments that are derived from natural sources. These include the pure carotenoids such as canthaxanthin (red), apo-carotenal (orange-red), and beta-carotene (yellow-orange).

The synthetic color additives are of two types described as "Food Drugs and Cosmetics (FD&C) dyes and lakes. Dyes are water soluble and are available in powders, granules, liquids, blends, and pastes. GMP suggest that they not be used in amounts exceeding 300 ppm. The lakes are water-insoluble FD&C certified dyes on a substratum of aluminum hydrate or aluminum hydroxide. The color lakes are useful in foods that have very little water e.g. coloring oils, icings, fondant coatings, cake and doughnut mixes, hard candy, and gum products. They do not solubilize, as do dyes but impart color by dispersion rather than as solutions. In 1960, the Color Additive Amendments separated "color additives" from "food additives." Colors (which include black, white, and intermediate grays) no longer were to be classified as food additives.

Some compounds are not color additives but are used to produce a white color. Thus, oxidizing agents including benzoyl peroxide, chlorine dioxide, nitrosyl chloride, and chlorine are used as bleaching agents, and to whiten wheat flour, which is pale yellow in color if untreated. Titanium dioxide, on the other hand, is considered a color additive and may be added to some foods, such as artificial cream or coffee whiteners to add a white color (U.S. Food and Drug Administration 2014b).

3.2 Flavorings and Flavor Enhancers

3.2.1 Flavors

Flavorings are compounds, natural or synthetic, that are added to foods to produce flavors or to modify existing flavors. The variety of natural and synthetic flavorings available to the modern food technologist is very large. Essential oils provide a major source of flavorings. (Jelen 2011). Because of the large production of orange juice, quantities of essential oil of orange are produced as byproducts so there is little need for the production of synthetic orange flavoring.

Fruit extracts have been used as flavorings, but these are relatively weak when compared to essential oils and oleoresins. Synthetic flavorings are usually less expensive and more plentiful than natural flavorings. On the other hand, natural flavorings are often more acceptable, but they are quite complex and difficult to reproduce synthetically. Many artificial flavors, such as amyl acetate (artificial banana flavor), benzaldehyde (artificial cherry flavor), and ethyl caproate (artificial pineapple flavor), are added to confectioneries, baked products, soft drinks, and ice cream. These flavorings are added in very small amounts, often 0.03% or less.

3.2.2 Flavor Enhancers

Flavor enhancers, intensify flavors already present in a food, where the desirable flavors are relatively weak. Monosodium glutamate (MSG) is one of the best-known flavor enhancers (Schiffman 1998). MSG occurs naturally in many foods and in a certain seaweed that was used for centuries as a flavor enhancer. In the last hundred years, isolated MSG is used. Other flavor enhancers that also improve flavors are extremely powerful, effective in parts per million and even per billion. These compounds have been identified as nucleotides, including disodium inosinate and disodium GMP. Interestingly, there is a tendency to reference MSG to a new taste called UMAMI (Beauchamp 2009)

Several theories attempt to explain how MSG and other flavor enhancers work. One theory is that they increase the sensitivity of the taste buds. A second suggests that an increase in salivation as a result of the flavor enhancers will increase flavor perception. A third theory of intensified flavor perception is based on the observation that flavor enhancers produce certain physical sensations in the mouth such as coolness and heat (Beauchamp 2009).

3.3 Sweeteners

Sweetening agents are added to a large number of foods and beverages. Table sugar (sucrose) is the most commonly used sweetener. Sweeteners include other sugars, as well as an abundance of natural and synthetic agents of varying sweetening power and caloric values. Many sweeternes are intened to function as sugar subtitutes or sugar replacements (Struck et al. 2014; Van der Sman and Renzetti 2019).

3.3.1 Fructose

Fructose (also known as levulose), is the sweetest and most soluble of the natural sweeteners used in foods (nearly twice as sweet as table sugar, sucrose). Fructose functions as a humectant in baked goods to retard dehydration. Solutions of fructose have a low viscosity that results in lower "body" feel than sucrose but have greater flexibility of use over a wide range of temperatures. Fructose also does not crystallize out of solution, whereas sucrose will. Fructose (fruit sugar) occurs in many fruits and berries; It also occurs in honey, corn syrup, cane sugar, and beet sugar. Since glucose cannot be metabolized by people with diabetes, fructose may be a better substitute for sucrose for diabetics but there are also concerns that this sugar may promote increased energy intake and weight gain (Bantle 2009; Choo et al. 2018).

3.3.2 Molasses

Molasses is a byproduct of sugar production. The use of molasses as a sweetener in human foods is largely in baked goods that include bread, cookies, and cakes. In addition to sweetening, molasses adds flavor and acts as a humectant. It is also used in baked beans and in the production of rum and molasses alcohol. (The greatest use of molasses, however, is in the production of animal feed). Molasses comprises about 60% sucrose. Molasses is being evaluated as a health promoting food ingredient (Filipčev et al. 2010).

3.3.3 Honey

Honey, a natural viscous syrup, comprises mainly invert sugar. It is produced from the nectar of flowers, which is mainly sucrose, by the action of an invertase enzyme that is secreted by the honey bee. Honey is used as a direct sweetener, as an additive in a number of products, including baked goods, as well as in other ways. It is relatively expensive, e.g. Manuka honey used for therapeutic applications.

3.3.4 Maple Sugar

Maple sugar is produced from the sap of the sugar maple tree. It is comprised mainly of sucrose and small amounts of other sugars, including invert sugar. Maple sugar is used in the manufacture of candies, fudge, baked goods, and toppings. It is among the most expensive of sweeteners.

3.3.5 Lactose

Lactose the sugar component of mammalian milk, is less sweet and less water-soluble than sucrose. Although most babies and young children are able to metabolize this sugar, some are unable to do so. The ability to metabolize the sugar appears to decrease with age. When a person is unable to metabolize lactose, the ingestion of milk may cause intestinal discomfort, cramps, and diarrhea. The major source of lactose is whey, a cheese byproduct. Because lactose is not as sweet as sucrose, larger amounts can be used in those foods in which the texture benefits from a high solids content.

3.3.6 Maltose

Maltose or malt sugar, is produced during the malting process in brewing (enzyme conversion of starch). It is converted to alcohol by the action of yeasts through an intermediate conversion to dextrose. This sugar is much less sweet than sucrose, and it is used mainly in the manufacture of baked goods and infant foods.

3.4 Nutritive and Non-nutritive Sweeteners

Many substances are classified as nonnutritive sweeteners. Although this classification might imply a lack of nutritional value, the implication is correct only in a relative sense. For example, the caloric value of a nonnutritive sweetener such as Aspartame is about 4 calories/g, which is the same as that for sugar. However, because it takes only 1 g of Aspartame to provide the same sweetness level as about 200 g of sugar (sucrose), it can be seen that the caloric contribution of Aspartame is only about 0.5% that of sucrose (Table 16.4). It is on this basis that a nonnutritive sweetener is classified as such (Butchko et al. 2002). According to the U.S. FDA, those tabletop sweeteners that contribute less than 2% of the caloric content of sucrose are referred as nonnutritive (U. S. Food and Drug Administration 2018).

Table 16.4 Some high-intensity tabletop sweeteners used in the US

Name[b]	Common names[b]	Sweetening effect[a]
Acylsulfame K	Sweet one	200
Advantame		20,000
Aspartame	Equal, Nutrasweet	200
Luo Han Guan extract	Nectresse, PureLo	100–250
Neotame	Neotame	7000–14,000
Saccharine	Sweet-N-Low	200–700
Steviol glycoside	Truvia	200–400
Sucralose	Splenda	600

Adapted from (U. S. Food and Drug Administration 2018), [a]Sweetening effect compared with an equal weight of sucrose. [b]Assume all names are trademarked registered trade names

3.4.1 Sugar Alcohols, Xylitol, Sorbitol and Mannitol

Xylitol, sorbitol and mannitol are the most common examples of sugar alcohols or polyols, suggested as low calorie sugar-replacers in a variety of products including baked products; sugar alcohols are used as sweeteners and bulking agent (Ghosh and Sudha 2012; Rice et al. 2020). The polyols have between 50–100% of the sweetening power of sucrose. Owing to their low metabolism, the caloric value ranges from 1.6–2.4 Cal/g (compared with 4 Cal/g for sucrose). Laterly polyols were used as sweetener in chewing gum, mainly because of their non-cariogenic property (they have not been found to cause tooth decay). Xylitol occurs naturally as a constituent of many fruits and vegetables, and is a normal intermediary product of carbohydrate metabolism in humans and in animals. Commercially, xylitol is produced by the hydrolysis of xylan (which is present in many plants) to xylose, which is then hydrogenated to produce xylitol.

Sorbitol occurs in red seaweed and in fruits (apples, cherries, peaches, pears, and prunes). It was first isolated from the sorb berries of the mountain ash, hence its name. Sorbitol is used as an additive for its humectant property and sweetening effect. Mannitol is a naturally occurring sweetener in many plants, algae, and mold. It occurs in the sap of the manna tree, an ash native to southern Italy, and can also be made by the reduction of the monosaccharides mannose or galactose.

Industrially, sugar alcohols can be produced by electrochemical reduction or catalytic hydrogenation methods. They are purified and crystallized. Xylitol imparts a sweet taste, which also appears to have a cooling effect.

3.4.2 High Intensity Sweeteners, Aspartame, Saccharine and Others

Aspartame is the common name for aspartylphenylalanine. It is a combination of the two amino acids. First produced in 1969, it is reputed to be about 200 times sweeter than sucrose. Unlike saccharin and cyclamate, Aspartame leaves no bitter aftertaste. (Butchko et al. 2002). Saccharin, the imide of o-benzosulfonic acid, is used as a sodium or calcium salt. It is about 300 times sweeter than sucrose (table sugar).

A new generation of high-intensity sweeteners are also approved for use in the US including products that are between 600-fold and 20,000–fold sweeter than sucrose (Table 16.4). Currently, Neotame™ and Advantame™ appear to be most sweetening of the commercial sweeteners. It was suggested that high-intensity sweetener may help reduce the intake of added sugar. A statement from the American Heart Association and also American Diabetic Association suggested that non-nutritive sweeteners might be useful for lowering health risks related with high intakes of added sugar, but that more studies were needed before more definite conclusions could be reached (Gardner et al. 2012)

3.5 Thickening Agents

3.5.1 Starches

Although starches differ from each other somewhat, depending on the plant from which they are extracted, they are sufficiently similar chemically to be often classified together. The properties of starch make it useful as a thickening agent. The major source of starch is corn, but some starch is also produced from sorghum, potatoes, and wheat (Eliasson 2004; BeMiller and Whistlerr 2009). Modified starches are used for thickener additives including those that are acid treated, heated, enzyme treated or chemically modified (Table 16.5).

3.5.2 Gums

Gums, a class of complex polysaccharides, are defined as materials that are dispersible in water and capable of making the water viscous. Many gums occur naturally in certain land and sea plants. Examples are gum Arabic and agar. Many gums, such as the cellulose derivatives, are modified or semisynthetic, and some gums, such as the vinyl polymers, are synthetic. Gums are used to stabilize ice cream and desserts, thicken certain beverages and preserves, stabilize foam in beer, emulsify salad dressings, and form protective coatings for meat, fish, and other products. A significant potential for the use of gums lies in the production of certain low-calorie foods. For example, the oil in salad dressing can be replaced with gums to result in a product with the normal appearance, texture, and taste but without the calories normally associated with the product.

3.5.3 Polyhydric Alcohols

In addition to their use as sweeteners, many polyhydric alcohols (sugar polyols, cf. Section 3.4) are used to improve texture and moisture retention because of their affinity for water. Many polyols occur in foods naturally, glycerine (glycerol) being the predominant one. However, only four of the many polyols are allowed as food additives. They are glycerin, sorbitol, mannitol (see above), and propylene glycol. All but the last have a moderately sweet taste although none is as sweet as sugar. Propylene glycol has a somewhat undesirable

Table 16.5 The major food additives used as thickeners or thickening agents

Name	Type	INS (¥)
Agar	Polysaccharide	406
Alginic acid	Polysaccharide	400
Alginate	Polysaccharide	402, 403, 404, 405
Arabinoglycan	Polysaccharide	409
Bakers yeast glycan	Polysaccharide	408
Beeswax	Lipid	901
Glycerol phosphate	Sugar	383
Candellilla wax	Lipid	802
Carob bean gum	Polysaccharide	410
Carrageenan	Polysaccharide	407
Cassia gum	Polysaccharide	427
CMC	Polysaccharide	468
Curdlan gum	Polysaccharide	427
Ethyl cellulose	Polysaccharide	462
Gelatin	Protein	428
Gellan gum	Polysaccharide	418
Guar gum	Polysaccharide	412
Gum Arabic	Polysaccharide	414
Gum ghatti	Polysaccharide	419
Hydroxypropyl cellulose	Polysaccharide	464 also 463 & 465
Cellulose (powdered)	Polysaccharide	460
Cellulose (microcrystalline)	Polysaccharide	460
Karaya gum	Polysaccharide	416
Konjac flour	Polysaccharide	425
Maltitols	Sugars	965
Pectins	Polysaccharide	440
Oat gums	Polysaccharide	411
Pullulan	Polysaccharide	1204
Soybean hemicellulose	Polysaccharide	426
Tannic acid	Polyphenol	181
Tara gum	Polysaccharide	417
Tragacanth gum	Polysaccharide	413
Xanthan gum	Polysaccharide	415
Xylitol	Sugar	967
Starch (acylated)	Polysaccharide	1422, 1414
Starch (acid treated)	Polysaccharide	1401
Starch (alkaline treated)	Polysaccharide	1402
Starch (roasted)	Polysaccharide	1400
Starch (bleached)	Polysaccharide	1403

Adapted from Codex Alimentarius Commission (1989a). (¥) Shows the international number system code for each additive

bitter taste, but is acceptable in small amounts. Polyols are used in the production of dietetic products including beverages, candy, gum, and ice cream to contribute to texture as well as to sweetness. They have less adverse effect on teeth than sugar, because they are not fermented as quickly as sugar and are usually washed away before being utilized by microorganisms.

3.6 Gelling Agents

Many polymer food additives function as gelling agents for controlling the water in foods. The compounds are usually large carbohydrate entities (polysaccharides) or proteins. Many of the gelling additives have additional functions at low concentrations where they form thickened solutions and function as thickeners. When applied to food surfaces, some gelling agents function as glazing agents to impart a shiny tint. The main classes of gelling food additives are shown in the table below (Table 16.6).

4 Nutrient Additives

The need for a balanced and ample nutrient intake by the human body is well known. Although nutrients are available in foods, losses of fractional amounts of some of them through processing and increasing frequency of improper dieting have led to the practice of adding minimum daily requirements or sizable fractions of minimum daily requirements of a number of nutrients to popular foods, such as breakfast cereals, baked goods, pasta products, and low-calorie breakfast drinks. Nutrient additives include mainly vitamins, proteins, and minerals.

4.1 Vitamins

Vitamin D is an exceptional example of the value of the food additive concept. The major source of vitamin D for humans is a precursor compound called 7-dehydrocholesterol which is produced in the liver. It circulates to an area just under the skin and is converted to previtamin D3 by the ultraviolet rays of sunlight. Previtamin D3 then goes through a number of steps and is converted to vitamin D3 and finally

Table 16.6 The major food additives used as gelling agents

Name	Type	INS[a]
Agar	Polysaccharide	406
Alginate	Polysaccharide	403
Alginic acid	Polysaccharide	400
Arabinogalactan	Polysaccharide	409
Carboxymethyl cellulose (CMC)	Polysaccharide	466
Carrageenan	Polysaccharide	407
Cassia gum	Polysaccharide	424
Curdlan	Polysaccharide	424
Euchema seaweed	Polysaccharide	407a
Gelatine	Protein	428
Konjac flour	Polysaccharide	425
Pectins	Polysaccharide	440
Tara gum	Polysaccharide	417
Yeast glycan	Polysaccharide	408

Adapted from Codex Alimentarius Commission (1989a). [a]Shows the international number system code for each additive

to active vitamin D. However, in many cases, exposure to the sun is sporadic and insufficient, especially in areas where there is normally insufficient sunshine or in cases where sunlight exposure is of insufficient duration. Thus, vitamin D is added to nearly all commercial milk in a ration of 10 micrograms (as cholecalciferol). This is equivalent to the old 400 IU per quart (0.95 liter). Vitamins A and C and some of the B vitamins are also added to some foods.

4.2 Proteins and Amino Acids

The addition of protein concentrate (produced from fish or soybeans) to components of the diet of inhabitants of underdeveloped countries has been used successfully to remedy the high incidence of protein malnutrition. The range of applications for protein isolates, ingredients, and supplements have increased greatly for wellbeing and health, e.g. to deal with protein needs for the elderly, protein and weight management, and protein uses for sports nutrition (Hoffman and Falvo 2004; Michelfelder 2009; Hertzler et al. 2020).

4.3 Minerals

Among minerals, iron has received major attention as a food additive, mainly because of its role in preventing certain anemias. The addition of potassium chloride to foods is increasing as a replacement for table salt.

5 Processing Aides

5.1 Acidulants

From the root word, *acid,* in acidulants, one can conclude that this class of compounds tends to lower the pH of any food into which the compounds are incorporated. Acidulants also enhance desirable flavors, and in many cases, such as in pickled products, are the major taste component. Vinegar (acetic acid) is added to relishes, chili sauce, ketchup, and condiments as a flavor component and to aid in the preservation of these products. Because the microbial spoilage of food is inhibited as the pH is lowered, acidulants are used for that purpose in many cases. Many acidulants occur naturally in foods (e.g., citric acid in citrus fruits, malic acid in apples, acetic acid in vinegars; all three are contained in figs). Tartaric acid is widely used to lend tartness and enhance flavor. Citric acid is widely used in carbonated soft drinks. Phosphoric acid is one of the very few inorganic acids used as an acidulants in foods. It is widely used, comprising 25% of all the acidulants in foods. Citric acid accounts for 60% of all acidulants used in foods.

In addition to their preservative and flavor enhancing effects, acidulants are used to improve gelling properties and texture. Acidulants are also used as cleaners of dairy equipment. Acidulants may be used in the manufacture of processed cheese and cheese spreads for the purpose of emulsification as well as to provide a desirable tartness.

Acid salts may be added to soft drinks to provide a buffering action (buffers tend to prevent changes in pH) which will prevent excess tartness. In some cases, acid salts are used to inhibit mold growth (e.g., calcium propionate added to bread).

As has been pointed out, all microorganisms have a pH at which they grow best (See Chap. 7, Sect. 3), and a range of pH above or below which they will not grow. Generally, it is not possible to preserve all foods by adding acid to the point where microorganisms will not grow. Most foods would be too acid to be palatable. The amount of acid may be enough to inhibit the growth of microorganisms if such treatment is combined with some other method of preservation. Certain dairy products, such as sour cream, and fermented vegetables, such as sauerkraut, are preserved with lactic acid produced by the growth of bacteria. Addition of the acid, along with holding at refrigerator temperatures above freezing, in combination will prevent growth of pathogenic and spoilage organisms. When sauerkraut is canned, it is given a heat process sufficient to destroy all spoilage and disease microorganisms.

Pickles are preserved by the addition of some salt, some acid, and a heat process sufficient to raise the temperature of all parts of the food to or near 212 °F (100 °C). Pickled herring are preserved by the addition of some salt, some acetic acid (vinegar), and holding at refrigerator temperatures above freezing. In this case, the nonacid part of the acetic acid molecule has an inhibiting effect on the growth of microorganisms.

5.2 Alkaline Compounds

Alkaline compounds are compounds that raise the pH. Alkaline compounds, such as sodium hydroxide or potassium hydroxide, may be used to neutralize excess acid that can develop in natural or cultured fermented foods. Thus, the acid in cream may be partially neutralized prior to churning in the manufacture of butter. If this were not done, the excess acid would result in the development of undesirable flavors. Sodium carbonate and sodium bicarbonate are used to refine rendered fats. Alkaline compounds are also added to chlorinated drinking water to adjust the pH to high enough levels to control the corrosive effects of chlorine on pipes and equipment. Sodium carbonate is also used in conjunction with other compounds to reduce the amount of hardness in drinking water. Sodium hydroxide is used to modify starches and in the production of caramel. Sodium bicarbonate is used as an ingredient of baking powder, which is used in baked products. (It is also a common household item used in a variety of cooking recipes.) Its action is described in the "Leavening Agents" section of this chapter. Alkaline compounds are used in the production of chocolate and to adjust the acidity level in grape juice and other fruit juices that are to be fermented in the production of wine.

Itis important to note that some alkaline compounds, such as sodium bicarbonate, are relatively mild and safe to use, while others, such as sodium hydroxide and potassium hydroxide, are relatively powerful reagents and should not be handled by inexperienced people.

5.3 Enzymes

Enzymes occur naturally in foods, and their presence may be either beneficial or detrimental, depending on the particular enzyme. When the presence of enzymes is undesirable, steps are taken to inactivate them. When their presence is desirable, either the enzymes or sources of the enzymes are intentionally added to foods.

The use of enzymes as food additives presents no problem from the standpoint of safety, because enzymes occur naturally, are nontoxic, and easily inactivated, when desired. Enzymes applications in the food industry are too diverse to cover in the limited space available. A few examples of enzyme uses in food manufacturing are given in this section.

For baking (bread, cakes and biscuits); we find microbe derived proteases (usually) being used to modify dough properties. In the dairy industry – plant derived proteases and rennet substitutes are used for cheese manufacturing. In the meat and protein industry – papain and other proteases are used for tenderizing meat and for protein recovery. The brewing and beverage industry employs pectinases to reduce haze. The fruit and vegetable processors use pectinases for juice clarification and tomato juice manufacture. The starch and sugar sector uses starch degrading enzymes and others in the production of high fructose corn syrup. Lipases are used in modifying and producing specialty fats for confectionary applications and for degumming (Miguel et al. 2013; Whitehurst and Van Oort 2009; Novozyme 2014). Some other enzymes uses are described below.

5.3.1 Invertase

Certain enzymes, such as invertase, split disaccharides, such as sucrose (table sugar), to lower sugars (glucose and levulose). Invertase has many applications, and is used, for example, to prevent crystallization of the sucrose that is used in large amounts in the production of liqueurs. Without invertase, the liqueurs would appear cloudy.

5.3.2 Pectinase

Pectinases are enzymes that split pectin, a polysaccharide that occurs naturally in plant tissues, especially those of fruit. Pectin holds dispersed particles in suspension, as in tomato juice. Because it is desirable to keep the thick suspension in tomato juice, pectinases that occur naturally in it are inactivated by heat. On the other hand, products such as apple juice are customarily clear, and this is accomplished by adding commercial pectinase to the product, which degrades the pectin in the apple juice, resulting in the settling out of the suspended particles, which are then separated from the clear juice. In the manufacture of clear jellies from fruits, it is first necessary to add pectinase to destroy the naturally occurring pectin in order to clarify the juice. This pectinase must now be inactivated by heat. Then more pectin must be added to the clarified juice to produce the thick consistency of jelly. If the pectinase is not inactivated after clarification, the enzyme would also break down the newly added pectin required to produce the thick consistency.

5.3.3 Cellulases

Cellulases are enzymes that can break down cellulose, said to be the most abundant form of carbohydrate in nature. Cellulose, the principal structural material in plants, is insoluble in water and is indigestible by humans and many animals. Ruminants are able to digest cellulose because of a cellulase (produced by microorganisms in the large stomach) contained in their gastric juice. Commercial applications of cellulases are not widespread at present. Cellulases are used for tenderizing fibrous vegetables and other indigestible plant material for the production of foods or animal feed.

5.3.4 Proteases

Proteases are enzymes that break down proteins, polypeptides, and peptides. Peptides are the structural units of which polypeptides consist, and polypeptides are larger structural units that make up the protein. There is a large number of specific proteases, and each attacks protein molecules at different sites, producing a variety of end products. Proteases are used to manufacture soy sauce from roasted soybeans, cheese from milk, and bread dough from flour. They are also used to tenderize meat and chill-proof beer, which, if untreated, develops an undesirable haze when chilled (see above).

5.3.5 Lipases

Lipases, the lipid (fat or oil) splitting enzymes, have commercial application (as mentioned above). Lipases prepared from microbial sources are used the production of certain cheeses and other dairy products, as well as lipase-treated butterfat used in the manufacture of candies, confections, and baked products. Lipases are also used to remove fat residuals from egg whites and in drain cleaner preparations.

5.3.6 Glucose Oxidase

Glucose oxidase is an enzyme that specifically catalyzes the oxidation of glucose to gluconic acid. This reaction is important in preventing nonenzymatic browning, because glucose is a reactant in the undesirable browning reaction. The most important application of this enzyme is in the treatment of egg products, especially egg whites, prior to drying. Eggs treated with this enzyme before they are dried do not undergo nonenzymatic browning during storage, because the sugar has been removed. In some cases, the enzyme is added to remove traces of oxygen to prevent oxidative degradation of quality. Examples of this type of application are mayonnaise and bottled and canned beverages (especially beer and citrus drinks).

5.3.7 Catalase

Catalases are used to break down hydrogen peroxide to water and oxygen. Therefore, catalases are used when the presence of hydrogen peroxide is undesirable or when hydrogen peroxide is used for specific purposes, such as in bleaching, but then must be removed from the system. Examples of the latter case are the uses of hydrogen peroxide for preserving milk in areas where heat pasteurization and refrigeration are unavailable and in the manufacture of cheese from unpasteurized milk.

5.4 Surface-Active Agents and Emulsifiers

Surface-active agents affect physical forces at interfaces of surfaces. Commonly called surfactants or emulsifies, they are present in all natural foods, because by their nature they play a role in the growth process of plants and animals. They are defined as organic compounds that affect surface activities of certain materials. They act as wetting agents, lubricants, dispersing agents, detergents, emulsifiers, solubilizes, and so forth. Emulsifiers, such as lecithin, mono- and diglycerides, and wetting agents, such as "tweens," may be added to bakery products (to improve volume and texture of the finished products and the working properties of the dough and to prevent staling of the crumb), cake mixes, ice cream, and frozen desserts (to improve whipping properties). Except for the tweens, the chemicals cited above are natural components of certain foods (Goecmen 1993; Schuster and Adams 1984; Stampfli and Nersten 1995).

5.5 Leavening Agents

Leavening agents are used to enhance the rising of dough in the manufacture of baked products. Inorganic salts, especially ammonium and phosphate salts, favor the growth of yeasts, which produce the carbon dioxide gas that causes dough to rise. Chemical reagents that react to form carbon dioxide are also used in baked goods. When sodium bicarbonate, ammonium carbonate, or ammonium bicarbonate is reacted with potassium acid tartrate, sodium aluminum tartrate, and sodium aluminum phosphate, or tartaric acid, carbon dioxide is produced. Baking powder is a common household leavining agent that contains a mixture of chemical compounds that react to form carbon dioxide, producing the leavining effect. Baking powder can be either single acting or double acting, giving the desired leavining effect in different products. The carbon dioxide gas is liberated when sodium bicarbonate (the base) reacts with potassium acid tartrate (the potassium salt of tartaric acid).

The double-acting baking powder has two acid-reacting ingredients (monocalcium phosphate monohydrate and sodium aluminum sulfate). The hydrated form of monocalcium phosphate reacts with sodium bicarbonate to release some carbon dioxide during mixing a batter or dough. The remaining sodium bicarbonate will react with sulfuric acid that is produced from the sodium aluminum sulfate; the major portion of carbon dioxide is released whilst the product is heated in the oven.

6 Chemical Preservatives

6.1 Antioxidants

Antioxidants are food additives used, since about 1947, to stabilize foods that by their composition would otherwise undergo significant loss in quality in the presence of oxygen (Finley and Given 1986). Oxidative quality changes in foods include: (i) the development of rancidity from the oxidation of unsaturated fats resulting in off-odors and off-flavors and (ii) discoloration from oxidation of pigments or other components of the food.

There is a large number of antioxidants, and although they may function in different ways, the purpose of each is to prevent, delay, or minimize the oxidation of the food to which they are added. One of the ways by which some antioxidants function involves their combination with oxygen. Others prevent oxygen from reacting with components of the food. When only a limited amount of oxygen is present, as in a hermetically sealed container, it is possible for some antioxidants to use up all of the available free oxygen, because they have a relatively great affinity for it. Some antioxidants lose their effectiveness when they combine with oxygen; therefore, there is no advantage to using this type of antioxidant unless the food is enclosed in a system from which oxygen or air can be excluded. With the use of antioxidants, it should be noted that other precautions are necessary to minimize oxidation, because heat, light, and metals are prooxidants, that is, their presence favors oxidative reactions.

6.1.1 Natural Antioxidants

Many of the antioxidants used commercially occur naturally (e.g., vitamin C, Vitamin E, citric acid, and certain phenolic compounds). However, the amines and the phenolic compounds can be toxic to humans in low concentrations; therefore, their use and that of synthetic antioxidants require strict regulation. It should be pointed out that the potency of the naturally occurring antioxidants is not as great as that of the commonly used synthetic antioxidants.

6.1.2 Synthetic Antioxidants

Fats and shortenings, especially those used in bakery goods and fried foods, are subject to oxidation and the development of rancidity after cooking. To prevent this, chemical antioxidants in concentrations up to 0.02% of the weight of the fat component may be added. The main synthetic antioxidants and most widely used are butylated hydroxyanisole (BHA), butylated hydroxytoluene (BHT), and propylgallate. These are generally used in formulations that contain combinations of two or all three of them, and often in combination with a fourth component, frequently citric acid. The main purpose of adding citric acid is that it serves as a chelator or sequestrant (a chelator ties up metals, thereby preventing metal catalysis of oxidative reactions).

The use of antioxidants is regulated by the FDA and is subject to other regulations, such as the Meat Inspection Act and the Poultry Inspection Act. Their use is limited so that the maximum amount that can be added is generally 0.02% of the fat content of the food; however, there are some exceptions to, and variations of, that rule.

6.1.3 Benzoates

The benzoates and para-benzoates have been used as preservatives mainly in fruit juices, syrups (especially chocolate syrup), candied fruit peel, pie fillings, pickled vegetables, relishes, horseradish, and some cheeses. The probable reason that the benzoates and the related para-benzoates have been allowed as additives to food is that benzoic acid is present in cranberries as a natural component in concentrations that are higher than 0.1%. The benzoates are most effective in acid foods in which the pH is as low as 4.0 or below where compound exists as benzoic acid. The para-benzoates are said to be more effective than the benzoates over a wider range of pH (Del Olmo et al. 2017).

6.1.4 Sequestrants

Because metals catalyze oxidative reactions, a sequestrant can be considered to have antioxidant properties. The role of sequestrants is to combine with metals, forming complexes with them and making them unavailable for other reactions. Sequestrants, like many other additives used for enhancing specific properties of foods, occur naturally in foods. Many sequestrants have other properties; for example, citric, malic, and tartaric acids are acidulants but they also have sequestering properties.

Thus, they stabilize foods against oxidative rancidity and oxidative discoloration. Sequestrant antioxidants protect vitamins, which are unstable when exposed to metal catalyzed oxidation. Sequestrants are used to stabilize the color of many canned products e.g. color and lipids in canned fish and shellfish, which have high concentrations of metal and poor color stability. Sequestrants are also used to stabilize the flavors and odors in dairy products and the color in meat products.

6.2 Antimicrobial Agents

The practice of preserving food by the addition of chemical is quite old, Many chemical substances used in the preservation of foods occur naturally. When they are used with the proper intent, they can be used to preserve foods that cannot be easily preserved by other means. To preserve food, it is necessary either to destroy all of the spoilage microorganisms that contaminate it or to create and maintain conditions that prevent the microbes from carrying out their ordinary life processes. There are additive free methods for food preservation including food drying, freezing and refrigeration.

6.2.1 Antibiotics

Conventional antibiotics can be used as feed additives, growth promoters or for treatment of animal diseases. The use of antibiotics in the food system has led to concerns with antimicrobial resistant strains (CDC 2020).

6.2.2 Fatty Acids

The salts of certain short chain fatty acids have an inhibitory effect on the growth of microorganisms. Thus, sodium diacetate (a mixture of sodium acetate and acetic acid) and sodium or calcium propionate are added to bread and other bakery products to prevent mold growth, as well as the development of a slimy condition known as "ropiness," which results from the growth of certain aerobic, spore-forming bacteria (see Chap. 3 for the definition of spore forming bacteria). Caprylic acid, CH3CH2-CH2-CH2-COOH, or its salts or the salts of other fatty acids may be used in cheese to prevent the growth of mold.

6.2.3 Sequestrant Antimicrobial Agents

The high mineral requirement by microorganism allows the use of sequestrants or chelating agents (chemicals that tie up metals and prevent the catalytic action of metals) to function as antimicrobial agents. The compound EDTA and citric acid from lemon and citrus fruit are good chelating agents, alongside of many plant-derived chemicals – polyphenols.

6.2.4 Sodium Chloride (Salt)

When sufficient salt is added to food, it makes water unavailable to microorganisms. Because microorganisms require water to survive, they cannot exist when their water requirement is diminished by the addition of salt. We can reduce the amount of water available to microorganisms by lowering the water activity (A_W). (For more on A_W see Chap. 7 -Sect 3 & Chap. 14 -Sect. 1.3) Some precautions must be observed in the salting preservation of flesh-type foods, such as fish or meats. The usual procedure is to hold products under refrigeration at 40–60 °F (4–15 °C) during salting until there has been an adequate "take-up" of salt throughout the food. If, such, precautions are not observed, the growth of spoilage or even disease-causing bacteria may occur in some parts of the food before enough salt has diffused into the product to inhibit growth.

6.2.5 Sulfur Dioxide

Sulfur dioxide or a source of this compound such as sodium bisulfite (NaHS03) may be added to foods to inhibit a narrow range of microorganisms. Therefore sulfur dioxide is usually applied together with another chemical inhibitor to prevent the growth of undesirable yeasts or bacteria in fruit juices, which are stored prior to fermentation, in the production of wine or vinegar. For many years, sulfiting agents have been classified as GRAS substances by the FDA for use as food preservatives when used in accordance with GMP (good manufacturing practice). However, the GRAS status does not apply for sulfur dioxide uses in three scenarios; (i) use with meat, (ii) use with fruits and vegetable products served or sold as raw to the consumer and (iii) foods that are sources of thiamine (vitamin B1). Sulfites may still be used in foods that have not been excluded by the FDA, their presence must be declared on the label when their concentrations exceed 10 ppm (parts per million). Research has shown that in concentrations of 10 ppm or less, these agents should not cause adverse reactions in humans (U. S. Food and Drug Administration 2019).

6.2.6 Sorbic Acid

Sorbic acid inhibits the growth of both yeasts and molds. This compound is most effective at pH 5.0 or below. Humans can metabolize this compound, as can fatty acids, and hence it is generally recognized as safe. Sorbic acid is used in certain bakery products (not yeast-leavened products, because it inhibits yeast growth), in cheeses, and in some fruit drinks, especially for the purpose of preventing molding. It is believed to inhibit the metabolic enzymes required by certain microorganisms for growth and multiplication.

6.2.7 Sodium Nitrite

Sodium nitrite is added to some food products to inhibit bacterial growth (especially *Cl. botuli-*

num) and to enhance color (U. S. Food and Drug Administration 2015). It is added to most cured meats, including hams, bacon, cooked sausage (such as frankfurters, bologna, salami), and to some kinds of corned beef. Nitrite provides for the red or pink color of the cured and cooked sausages and of the other cured products after cooking. The nitrite combines with the reddish pigment of meat, the myoglobin, and prevents its oxidation. If the meat were not treated with nitrite, it would discolor to a brown color during cooking or during storage. When red meat is heated, as in cooking, the color turns from red to gray or brown because of the conversion of myoglobin to the oxidized form, metmyoglobin. With intense heating or exposure to light and oxygen, even the nitrited myoglobin may be oxidized to metmyoglobin, with the result that the red or pink color is lost (Cammack et al. 1999; Alahakoon et al. 2015).

In addition to stabilizing the color of cured or cured and cooked meats, nitrite acts as a preservative to prevent the germination of spores and subsequent toxin production by *Cl. botulinum* (Chap. 17, Sect. 4)

It is uncertain whether or not nitrites should be allowed in food in any concentration. At present, the maximum usage levels are 10 ppm (10 mg/kg) nitrite for cured/smoked tuna fish and 200 ppm nitrite for cured/smoked salmon and smoked whitefish e.g. shads and chubs (21 CFR 172.170). However, sodium or potassium nitrite may not be used for fresh meat or fresh fish sold as such. The level of nitrite used for cured meat is 200ppm or lower with or without 500 ppm nitrate (U. S. Food and Drug Administration, 2015). National survey results showed the average level of nitrite were 4.54-ppm together with 37.07-ppm nitrate across the entire range of processed meats (Bryan et al. 2015).

Nitrite-cured products, especially those cooked at high temperatures, such as bacon, may develop nitrosamines, which are known to be extremely carcinogenic or cancer promoting. It is clear that the amount of nitrites used must be minimized but the level must remain high enough to protect against botulism and other microbiologically based foodborne diseases (Hord et al. 2009).

The levels of nitrite occurring naturally in some vegetables is surprisingly high (Santamaria 1997; Ferysiuk and Wójciak 2020). Therefore, some vegetable powders (e.g. spinach, kale, Chinese cabbage etc.) were used as sources of nitrite or nitrate for processed meat products labelled as, "No Nitrate or Nitrite Added" or "uncured" by manufacturers (Pennisi et al. 2020; Yong et al. 2021). Following objections from US consumer groups (Porter et al. 2020), these alternatively cured or "indirectly cured" food may not in the future qualify as additives-free or clean-label products (Devenyns 2020; McCarthy 2020).

6.2.8 Oxidizing Agents

Oxidizing agents, such as chlorine, iodine, and hydrogen peroxide, are not ordinarily used in food, but they are used to sanitize food-processing equipment and apparatus and even the walls and floors of areas where food is processed (Lopes 1986). Furthermore, some sanitizers are used for the disinfection of fresh produce surfaces, e.g. poultry, fruits and vegetables (Bolder 1997; Gómez-López et al. 2009; Prado-Silva et al. 2015). Thus, there is no doubt that small residuals, especially of chlorine or iodine, can get into food.

Hydrogen peroxide may be used to destroy the natural bacterial flora of milk, prior to inoculation with cultures of known bacterial species, for producing specific dairy products. In such cases, all of the residual hydrogen peroxide must be removed by treatment with the enzyme catalase. This treatment with catalase must be carried out prior to the inoculation of milk with cultures of desirable bacteria; otherwise the hydrogen peroxide will destroy the added bacterial culture, the growth of which is the objective of culturing milk.

Oxidizing agents are believed to inhibit and destroy the growth of microorganisms by destroying certain parts of the enzymes essential to the metabolic processes of these organisms.

6.3 Ionizing Radiation

The Food Additives Amendment of 1958 included ionizing radiation as a food additive. Ionization radiation is considered an additive because the irradiation may induce changes in food. In this case, the FDA tests foods that have been irradiated and approves irradiation sources and maximum dosages.

Irradiation does not leave a residue in food and it does not make it radioactive (Institute of Food Science and Technology (IFST) 2015). The levels of irradiation allowed in food processing do not induce measurable radioactivity. Any radioactivity found in irradiated foods has been shown to be "background radiation" or that which is already present naturally. Irradiation does cause small chemical changes in the food, as do other methods of food processing (Fig. 16.2).

Foods that have been irradiated must be labeled with the green international logo (see Fig. 16.2) to inform the consumer that the food has been processed by ionizing radiation. The words "Treated with Radiation" or "Treated by Irradiation" must also appear and must be in the same print style as the product name and be no smaller than one-third the size of the largest letter in the product name.

Long-term effects of consumption of irradiated products have been discussed. The safety of irradiated foods has been tested in feeding studies for over 40 years. The studies include both animal and human subjects. Chemistry studies, feeding studies, and mutagenicity and teratogenicity studies have not revealed any confirmable negative evidence as to the wholesomeness of foods preserved by ionization radiation. Nutrient retention of irradiated foods is comparable to that of heat-processed foods. Irradiated foods may be more susceptible to oxidation but this can be controlled by use of low temperatures and elimination of oxygen.

Fig. 16.2 International radiation logo (green in color)

Food irradiation remains a polarizing topic. However, there are over 50 countries where the use of ionizing radiation to sanitize and disinfect selected types of food is allowed (Institute of Food Science and Technology (IFST) (2015); Parnes and Lichtenstein 2004; Mahapatra et al. 2005; Maherani et al. 2016; Munir and Federighi 2020). The following articles may provide some insights about current debates concerning food irradiation (Louria 2000; Thayer 2000).

References

Alahakoon AU, Jayasena DD, Ramachandra S, Jo C (2015) Alternatives to nitrite in processed meat: up to date. Trends Food Sci Technol 45(1):37–49. https://www.sciencedirect.com/science/article/abs/pii/S0924224415001429

Armstrong DJ (2009) Food chemistry and U.S. food regulations. J Agric Food Chem 57(18):8180–8186

Bantle JP (2009) Dietary fructose and metabolic syndrome and diabetes. J Nutr 139(6):1263S–1268S

Beauchamp GK (2009) Sensory and receptor responses to umami: an overview of pioneering work. Am J Clin Nutr 90(3):723S–727S. https://doi.org/10.3945/ajcn.2009.27462E

BeMiller JN, Whistler RL (2009) Starch: chemistry and technology. Elsevier Science, 894pp. https://books.google.co.uk/books?id=Anbz_whRM2YC

Bolder NM (1997) Decontamination of meat and poultry carcasses. Trends Food Sci Technol 8(7):221–227

Bryan N, Osburn W, Keeton J, Hardin M (2015) A national survey of the nitrite/nitrate concentrations in cured meat products and non-meat foods available at retail, from https://www.nal.usda.gov/fsrio/research-projects/national-survey-nitritenitrate-concentrations-cured-meat-products-and-non-meat-foods-available

Burdock GA, Carabin IG (2004) Generally recognized as safe (GRAS): history and description. Toxicol Lett 150(1):3–18

Butchko HH, Stargel WW, Comer CP, Mayhew DA, Benninger C, Blackburn GL et al (2002) Aspartame: review of safety. Regul Toxicol Pharmacol 35(2):S1–S93. https://doi.org/10.1006/rtph.2002.1542

Cammack R, Joannou CL, Cui XY, Torres Martinez C, Maraj SR, Hughes MN (1999) Nitrite and nitrosyl compounds in food preservation. Biochim Biophys Acta 1411(2-3):475–488. https://pubmed.ncbi.nlm.nih.gov/10320676/

CDC (2020) Antibiotic/antimicrobial resistance (AR/AMR). Retrieved from https://www.cdc.gov/drugresistance/index.html

Choo VL, Viguiliouk E, Blanco Mejia S, Cozma AI, Khan TA, Ha V et al (2018) Food sources of fructose-containing sugars and glycaemic control: systematic review and meta-analysis of controlled intervention studies. BMJ 363:k4644. https://doi.org/10.1136/bmj.k4644

Codex Alimentarious Commission (1995) General standard for food additives. Codex stan 192–1995. Adopted in 1995. Revision 1997, 1999, 2001, 2003, 2004, 2005, 2006, 2007, 2008, 2009, 2010, 2011, 2012, 2013, 2014, 2015, 2016, 2017, 2018, 2019. Retrieved from http://www.fao.org/fao-who-codexalimentarius/sh-proxy/en/?lnk=1&url=https%253A%252F%252Fworkspace.fao.org%252Fsites%252Fcodex%252FStandards%252FCXS%2B192-1995%252FCXS_192e.pdf

Codex Alimentarius Commission (1981) Codex general standard for the labeling of food additives when sold as such: CODEX STAN 107–1981. Retrieved from http://www.fao.org/docrep/005/y2770e/y2770e03.htm#fn8

Codex Alimentarius Commission (1985) Codex general standard for the labeling of prepackaged foods: CODEX STAN 1–1985 (Rev. 1–1991)[1]. Retrieved from http://www.fao.org/docrep/005/y2770e/y2770e02.htm

Codex Alimentarius Commission (1989a, 2015) Class names and the international numbering system for food additives: CAC/GL 36-1989 Adopted in 1989. Revision: 2008. Amendment: 2015. Retrieved from http://www.fao.org/input/download/standards/13341/CXG_036e_2015.pdf

Codex Alimentarius Commission (1989b) Codex class names and the international numbering system for food additives CAC/GL 36-1989. Retrieved from http://www.codexalimentarius.org/input/download/standards/13341/CXG_036e.pdf

Codex Alimentarius Commission (2015) GSFA online: codex General Standard for Food Additives (GSFA) online database Retrieved from http://www.codexalimentarius.net/gsfaonline/index.html

David TJ (1988) Food additives. Arch Dis Child 63(6):582–583

Del Olmo A, Calzada J, Nunez M (2017) Benzoic acid and its derivatives as naturally occurring compounds in foods and as additives: uses, exposure, and controversy. Crit Rev Food Sci Nutr 57(14):3084–3103. https://doi.org/10.1080/10408398.2015.1087964

Devenyns J (2020) USDA plans to crack down on nitrate and nitrite labeling claims in meat. From https://www.fooddive.com/news/usda-plans-to-crack-down-on-nitrate-and-nitrite-labeling-claims-in-meat/592240

Eliasson AC (2004) Starch in food: structure, function and applications. Taylor Francis, 605pp

Eur-Lex (2011) Commission Regulation (EU) No 1130/2011 of 11 November 2011 amending Annex III to Regulation (EC) No 1333/2008 of the European Parliament and of the Council on food additives by establishing a Union list of food additives approved for use in food additives, food enzymes, food flavourings and nutrients Text with EEA relevance Retrieved from http://eur-lex.europa.eu/legal-content/EN/TXT/?qid=1410296955393&uri=CELEX:32011R1130

Europa (2014) Food additives and flavourings. Retrieved from http://ec.europa.eu/food/fs/sfp/flav_index_en.html

Fennema OR (1987) Food additives–an unending controversy. Am J Clin Nutr 46(1 Suppl):201–203. https://doi.org/10.1093/ajcn/46.1.201

Ferysiuk K, Wójciak KM (2020) Reduction of nitrite in meat products through the application of various plant-based ingredients. Antioxidants (Basel) 9(8):28pp. https://mdpi-res.com/d_attachment/antioxidants/antioxidants-09-00711/article_deploy/antioxidants-00709-00711-v00712.pdf

Filipčev B, Lević L, Bodroža-Solarov M, Mišljenović N, Koprivica G (2010) Quality characteristics and antioxidant properties of breads supplemented with sugar beet molasses-based ingredients. Int J Food Prop 13(5):1035–1053

Finley JW, Given P Jr (1986) Technological necessity of antioxidants in the food industry. Food Chem Toxicol 24(10–11):999–1006. https://doi.org/10.1016/0278-6915(86)90280-2

Food and Agriculture Organization of the United Nations (2013) Online edition: "combined compendium of food additive specifications". Retrieved from http://www.fao.org/food/food-safety-quality/scientific-advice/jecfa/jecfa-additives/en/

Food and Drug Administration (USDA) (2013) Everything added to food in the US (EAFUS). Retrieved from http://www.accessdata.fda.gov/scripts/fcn/fcnnavigation.cfm?rpt=eafuslisting&displayAll=true

Food and Drug Administration (USDA) (2015) Food ingredients and packaging terms. Retrieved from http://www.fda.gov/Food/IngredientsPackagingLabeling/Definitions/default.htm

Gardner C, Wylie-Rosett J, Gidding SS, Steffen LM, Johnson RK, Reader D, Lichtenstein AH (2012) Nonnutritive sweeteners: current use and health perspectives: a scientific statement from the American Heart Association and the American Diabetes Association. Circulation 126(4):509–519

Ghosh S, Sudha M (2012) A review on polyols: new frontiers for health-based bakery products. Int J Food Sci Nutr 63(3):372–379

Goecmen D (1993) The effect of flour and additives on bread quality and staling. Gida 18(5):325–331

Gómez-López VM, Rajkovic A, Ragaert P, Smigic N, Devlieghere F (2009) Chlorine dioxide for minimally processed produce preservation: a review. Trends Food Sci Technol 20(1):17–26

Hanlon P, Brorby GP, Krishan M (2016) A risk-based strategy for evaluating mitigation options for process-formed compounds in food: workshop proceedings. Int J Toxicol 35(3):358–370

Hertzler SR, Lieblein-Boff JC, Weiler M, Allgeier C (2020) Plant proteins: assessing their nutritional quality and effects on health and physical function. Nutrients 12(12).: 27pp. https://res.mdpi.com/d_attachment/nutrients/nutrients-12-03704/article_deploy/nutrients-03712-03704-v03703.pdf

Hoffman JR, Falvo MJ (2004) Protein - which is best? J Sports Sci Med 3(3):118–130. https://www.ncbi.nlm.nih.gov/pmc/articles/PMC3905294/pdf/jssm-3905203-3905118.pdf

Hord NG, Tang Y, Bryan NS (2009) Food sources of nitrates and nitrites: the physiologic context for potential health benefits. Am J Clin Nutr 90(1):1–10

Hoyle R (1996) Delaney clause knocked out by surprise compromise act. Nat Biotechnol 14(9):1056

Institute of Food Science and Technology (IFST) (2015) Food irradiation. Retrieved from https://www.ifst.org/resources/information-statements/food-irradiation

Jackson LS (2009) Chemical food safety issues in the United States: past, present, and future. J Agric Food Chem 57(18):8161–8170. https://doi.org/10.1021/jf900628u

JEFCA (2015) Online edition: "combined compendium of food additive specifications". Retrieved from http://www.fao.org/ag/agn/jecfa-additives/index.html

Jelen H (2011) Food flavors. Chemical, sensory and technological properties. CRC Press, 504pp

Joint FAO/WHO Expert Committee on Food Additives (2001) Gellan gum (WHO Food Additives Series 28). Summary of Evaluations Performed by the JEFAC for gellan gum. Retrieved from http://www.inchem.org/documents/jecfa/jecmono/v28je17.htm

Lester SU (1996) Delaney is still the best protection. Environ Health Perspect 104(4):352

Lester SU, Staples RE (1996) Delaney still the best protection comment: teratogens and the Delaney Clause. Environ Health Perspect 104(4):352

Lewis C (2002) The 'Poison Squad' and the advent of food and drug regulation. FDA Consum 36(6):12–15

Lopes JA (1986) Evaluation of dairy and food plant sanitizers against *Salmonella typhimurium* and *Listeria monocytogenes*. J Dairy Sci 69(11):2791–2796

Louria DB (2000) Counterpoint on food irradiation. Int J Infect Dis 4(2):67–69. https://www.ncbi.nlm.nih.gov/pubmed/10737841

Maherani B, Hossain F, Criado P, Ben-Fadhel Y, Salmieri S, Lacroix M (2016) World market development and consumer acceptance of irradiation technology. Foods 5(4):21pp. https://res.mdpi.com/d_attachment/foods/foods-05-00079/article_deploy/foods-00005-00079.pdf

McCarthy R (2020) FSIS addresses 'nitrates/nitrites' meat product labeling. Food Business News: 2pp. https://www.foodbusinessnews.net/articles/17544-fsis-addresses-nitratesnitrites-meat-product-labeling

Michelfelder AJ (2009) Soy: a complete source of protein. Am Fam Physician 79(1):43–47

Miguel A, Martins-Meyer TS, Figueiredo EVDC, Lobo BWP, Dellamora-Ortiz GM (2013) Enzymes in bakery: current and future trends. Food industry: IntechOpen, pp 278–321. https://www.intechopen.com/chapters/41661

Munir MT, Federighi M (2020) Control of foodborne biological hazards by ionizing radiations. Foods 9(7):23pp. https://res.mdpi.com/d_attachment/foods/foods-09-00878/article_deploy/foods-00809-00878.pdf

Novozyme (2014) Enzymes at work. Retrieved from http://www.novozymes.com/en/about-us/brochures/Documents/Enzymes_at_work.pdf

Neltner TG, Kulkarni NR, Alger HM, Maffini MV, Bongard ED, Fortin ND, Olson ED (2011) Navigating the US food additive regulatory program. Compr Rev Food Sci Food Saf 10(6):342–368. https://doi.org/10.1111/j.1541-4337.2011.00166.x

Neltner TG, Alger HM, O'Reilly JT, Krimsky S, Bero LA, Maffini MV (2013) Conflicts of interest in approvals of additives to food determined to be generally recognized as safe: out of balance. JAMA Intern Med 173(22):2032–2036

Nicole W (2013) Secret ingredients: who knows what's in your food? Environ Health Perspect 121(4): a126–a133

Parnes RB, Lichtenstein AH (2004) Food irradiation: a safe and useful technology. Nutr Clin Care 7(4):149–155

Pennisi L, Verrocchi E, Paludi D, Vergara A (2020) Effects of vegetable powders as nitrite alternative in Italian dry fermented sausage. Ital J Food Saf 9(2):8422. https://www.pagepressjournals.org/index.php/ijfs/article/download/8422/8907

Picut CA, Parker GA (1992) Interpreting the Delaney Clause in the 21st century. Toxicol Pathol 20(4):617–627, discussion 628–619

Porter ES, Castles C, Laudon KG, Fries AN (2020) United States: consumer groups file petition urging the USDA to change meat labeling requirements. Mondaq.com: 4pp. https://www.mondaq.com/unitedstates/consumer-law/879964/consumer-groups-file-petition-urging-the-usda-to-change-meat-labeling-requirements

Prado-Silva L, Cadavez V, Gonzales-Barron U, Rezende AC, Sant'Ana AS (2015) Meta-analysis of the effects of sanitizing treatments on Salmonella, Escherichia coli O157:H7, and Listeria monocytogenes inactivation in fresh produce. Appl Environ Microbiol 81(23):8008–8021. https://doi.org/10.1128/AEM.02216-15

Rice T, Zannini E, Arendt EK, Coffey A (2020) A review of polyols–biotechnological production, food applications, regulation, labeling and health effects. Crit Rev Food Sci Nutr 60(12):2034–2051

Rulis AM, Levitt JA (2009) FDA'S food ingredient approval process: safety assurance based on scientific assessment. Regul Toxicol Pharm 53(1):20–31

Santamaria P (1997) Occurrence of nitrate and nitrite in vegetables and total dietary intakes. Industrie Alimentari 36(364):1329–1334

Schiffman SS (1998) Sensory enhancement of foods for the elderly with monosodium glutamate and flavors. Food Rev Intl 14(2–3):321–333

Schuster G, Adams WF (1984) Emulsifiers as additives in bread and fine baked products. Adv Cereal Sci Technol 6:139–287

Smith J, Hong-Shum L (2011) Food additives data book. Wiley, 1128 pages

Stampfli L, Nersten B (1995) Emulsifiers in bread making. Food Chem 52(4):353–360

Struck S, Jaros D, Brennan CS, Rohm H (2014) Sugar replacement in sweetened bakery goods. Int J Food Sci Technol 49(9):1963–1976

Thayer DW (2000) Are there valid concerns about food irradiation? Int J Infect Dis 4(4):232–234. author reply 234-235 https://www.ncbi.nlm.nih.gov/pubmed/11231191

U. S. Food and Drug Administration (2015) 21.CFR Part 172.175 sodium nitrite. Retrieved from https://www.accessdata.fda.gov/scripts/cdrh/cfdocs/cfcfr/CFRSearch.cfm?fr=172.175

U. S. Food and Drug Administration (2018) Additional information about high-intensity sweeteners permitted for use in food in the United States. Retrieved from https://www.fda.gov/food/food-additives-petitions/additional-information-about-high-intensity-sweeteners-permitted-use-food-united-states

U. S. Food and Drug Administration (2019) CFR 21CFR182.3862. Substances Generally Recognized As Safe. Subpart D–Chemical Preservatives. Sec. 182.3862 Sulfur dioxide

U.S. Food and Drug Administration (2014a) Overview of food ingredients, additives & colors international food information council (IFIC) and U.S. Food and drug administration November 2004; revised April 2010. Retrieved from http://www.fda.gov/Food/IngredientsPackagingLabeling/FoodAdditivesIngredients/ucm094211.htm

U.S. Food and Drug Administration (2014b) Overview of food ingredients, additives & colors. Retrieved from http://www.fda.gov/Food/IngredientsPackagingLabeling/FoodAdditivesIngredients/ucm115326.htm

U.S. Government Publishing Office (2012) Code of Federal Regulations. Title 21–Food and Drugs; Chapter 1; Food and Drug Administration Department Of Health And Human Services, Subchapter B. Food for human consumption (continued). Part 172 Food additives permitted for direct addition to food for human consumption. Subpart I–Multipurpose additives Sec. 172.892 Food Starch-Modified. *21 CFR 172.892 – Modified Starch.* Retrieved from http://www.accessdata.fda.gov/scripts/cdrh/cfdocs/cfCFR/CFRSearch.cfm?fr=172.892

U.S. Government Publishing Office (2014) Electronic Code of Federal Regulations: Title 21: Food and Drugs Part 173—Secondary direct food additives permitted in food for human consumption. Retrieved from http://www.ecfr.gov/cgi-bin/retrieveECFR?gp=1&SID=761464347534256da7af5f3b0b73e700&h=L&mc=true&n=pt21.3.173&r=PART&ty=HTML

Van der Sman RGM, Renzetti S (2019) Understanding functionality of sucrose in biscuits for reformulation purposes. Crit Rev Food Sci Nutr 59(14):2225–2239

Weisburger JH (1996) Human protection against non-genotoxic carcinogens in the US without the Delaney Clause. Exp Toxicol Pathol 48(2–3):201–208

Whitehurst RJ, Van Oort M (2009) Enzymes in food technology. Wiley, 384pp. https://books.google.co.uk/books?id=GHQBMyTzJ-oC

Yong HI, Kim TK, Jang HW CHD, Jung S, Choi YS (2021) Clean label meat technology: pre-converted nitrite as a natural curing. Food Sci Anim Resour 41(2):173–184. https://www.ncbi.nlm.nih.gov/pmc/articles/PMC8115001/pdf/kosfa-8115041-8115002-8115173.pdf

17 Meat

1 Introduction

1.1 Defining Meat

United States food legislation (7CFR54.1) defines meat as "*the edible part of the muscle of an animal, which is skeletal, or which is found in the tongue, in the diaphragm, in the heart, or in the esophagus, and which is intended for human food, with or without the accompanying and overlying fat and the portions of bone, skin, sinew, nerve, and blood vessels which normally accompany the muscle tissue and which are not separated from it in the process of dressing. This term does not include the muscle found in the lips, snout, or ears*". The code of Federal regulations (40 CFR Part 432) also lists meat and poultry products for human consumption as "*cattle, hogs, sheep, chickens, turkeys, ducks and other fowl as well as sausages, luncheon meats and cured, smoked or canned or other prepared meat and poultry products from purchased carcasses and other materials*". Poultry is described as "*derived from the slaughter and processing of broilers, young chickens, mature chickens, hens, turkeys, capons, geese, and ducks; small game fowl such as quail or pheasants, and small game such as rabbits*" (Legal Information Institute (Cornel University Law School) 2015).

The WHO defined meat to be, "*all parts of an animal that are intended for, or have been judged as safe and suitable for, human consumption*" (Codex Alimentarius Commission 2005). The EU describes meat as: derived from domestic ungulates, poultry, lagomorphs and game and also includes products from blood (European Commission 2016);

1.2 The US Meat Industry at a Glance

The US meat industry (NAICS 3116 and 3111) contains about 5000 separate businesses and one-half million employees (Table 17.1). Currently, the US meat industry processes nearly 34 million head of cattle equal to about 10 million metric tons live weight (Table 17.2). The total sales ascribed to the meat industry amounted to approximately US$133 billion for meat and $52 billion for poultry packing and processing.

The range of meat products available for the consumer can be appreciated by noting the different standards of identity for the purpose of labelling (Table 17.3). Meat products are long established and so, new products are required to be labelled using standards of identity with new names being created only if new products cannot be mapped using existing names (Table 17.3).

R. Owusu-Apenten, E. R. Vieira, *Elementary Food Science*, Food Science Text Series,
https://doi.org/10.1007/978-3-030-65433-7_17

Table 17.1 The United States meat industry

NAICS	Code description	Employees (2012)	Number of Businesses
31161	Animal slaughtering & processing	488,848	3642
311611	Animal (except poultry) slaughtering	156,041	1494
311612	Meat processed from carcasses	101,442	1345
311613	Rendering and meat byproduct processing	8699	220
31111	Animal food manufacturing	50,200	1738
311111	Dog and cat food manufacturing	22,422	282
311119	Other animal food manufacturing	27,778	1456

United States Census (2014)

Table 17.2 Estimated processing capacity for the US meat industry

Meat source	Million Units	million metric ton
Beef & calves	33,722	8.1
Hog	11,316	7.25
Sheep & Lamb	2185	0.52
Chickens	86,000	10.0
Turkeys	250	0.707

Adapted from North American Meat Institute (NAMI) (2016)

Table 17.3 Standards of identity for US meat and meat products, 9 CFR319

US Standard of Identity	Example products
(B) Raw meat products	Chopped beef, ground beef, beef patties, hamburgher
(C) Cooked meats	Barbecued meats.
(D) Cured meats, unsmoked and smoked	Corn beef, cured beef country ham, bacon
(E) Fresh sausage	Beef, Pork, lamb, etc. Italian sausage
(F) Uncooked, smoked sausage	Smoked pork sausage
(G) Cooked sausage	Frankfurter, hotdog, wiener, bologna, knockwurst,
(I) Semi-dry fermented sausage	[Reserved]
(J) Dry fermented sausage	[Reserved]
(K) Luncheon meat, loaves and jellied products	Luncheon meat
(L) Meat specialties, puddings and nonspecific loaves	Scrapple
(M) Canned, frozen, or dehydrated meat	Chili con carne, hash, meat stews, spaghetti with meat balls
(N) Meat food entree products, pies, and turnovers	Meat pies
(O) Meat snacks, hors d'oeuvres, pizza, and specialty items [reserved]	
(P) Rendered animal fat or mixture	
(Q) Meat soups, soup mixes, broths, stocks, extracts	Meat extract. Stock extract
(S) Meat baby foods	[Reserved]
(T) Dietetic meat foods	[Reserved]
(U) Miscellaneous	Breaded products, liver meat products

Adapted from U. S. Government Publishing Service (2016)

1.3 Global Meat Production and Consumption

Current world meat production amounts to about 260 million Tons per year (Table 17.4). There was a steady increase in meat production from 2011–2015. The level of meat exports and imports are similar at 10% production (USDA Foreign Agricultural Services 2016). Not all meat producers are major consumers, but meat consumption is generally higher in countries with high production levels with a notable exception of India.

Per capita meat consumption was higher in the US (~92 kg per person per year or 250 g per day) compared to other countries (Fig 17.1). The top-10 meat consuming nations alongside of the US are Australia, Argentina, Israel, Brazil, Uruguay, New Zealand, Chile, and Canada (OECD 2016). Meat was least in demand in Bangladesh and India, which have significant vegetarian populations.

The meat preferences pattern within the US is for poultry (49.9%) > beef (26.0%) > pork (23.8%). Mutton or lamb is less popular (0.4%) and goat meat hardly figures in the US records (Table 17.5). The pattern of meat preferences is similar for the other top -10 nations for meat

Table 17.4 Global meat production and export data (million metric ton)

Production/Year	2011	2012	2013	2014	2015	2016 (§)
Beef & Veal	58.2	58.5	59.7	59.7	58.4	59.2
Pork	103.6	106.9	108.8	110.6	111.5	112.0
Broiler meat	81.2	83.3	84.5	86.5	87.9	89.3
Total	242.9	248.7	253.1	256.9	257.8	260.5
Exports/Year	2011	2012	2013	2014	2015	2016 (§)
Beef & Veal	8.1	8.1	9.1	10.0	9.6	9.9
Pork	7.0	7.3	7.0	6.9	7.1	7.3
Broiler meat	9.6	10.1	10.2	10.5	10.2	10.7
Total	24.6	25.5	26.4	27.3	27.0	27.9

Summarized from USDA Foreign Agricultural Services (2016). (§) Projected data

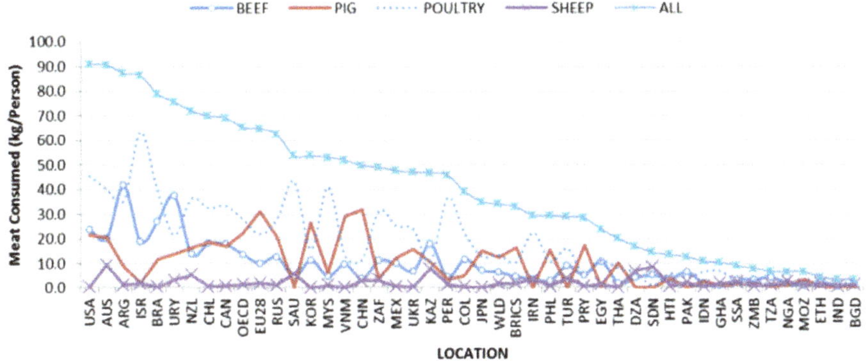

Fig. 17.1 World meat consumption in 2015. Drawing based on OECD data (OECD 2016). The per capita meat intake (kg-meat per person per year) from either poultry, pig, beef or sheep

consumption with the notable exception of the EU where pork is the predominant meat commodity (Table. 17.5).

1.4 Meat Preferences World-Wide

Meat preferences differ from one part of the world to another (Table 17.5). The low intake of pork in Israel and Muslim countries is explained readily based on religious strictures. There is a notable preference for chicken in Saudi Arabia, Peru, Malaysia, Iran, Israel, Ghana, South Africa, India, and Indonesia. The major pork eating regions are, China, Vietnam, Philippines – but Paraguay and Mozambique are other pork consumers. A definite preference for beef is found in Tanzania, Ethiopia, Pakistan and Uruguay (>50% of meat choice). Finally, sheep meat is most consumed in Sudan, Algeria, Nigeria, Bangladesh and Ethiopia.

Horses have not been an important source of meat in the USA, except during times of war when meat shortages have occurred, perhaps because of the high-regard and esteem associated with this species historically. Horsemeat is legitimate food in some parts of Europe whilst horse is still a companion animal. A market exists also for domestically produced game-type meats, e.g., alligator, buffalo, bear, elk, kangaroo, and rabbit. In many parts of the developing world, the list of meat animals may include "micro-livestock" including grass cutter, snails and guinea pigs; mention could be made also for "bush meat" which has significant economic importance though their importation into the US and Europe is currently illegal (Chaber and Cunningham 2016; Ordaz-Németh et al. 2017).

Except for tongues, head meats are generally used in the production of sausages. Tripe, prepared from the first and second stomachs of cat-

Table 17.5 Global meat consumption preferences by region

Region	Poultry'%	Region	Pork %	Location	Beef%	Region	Sheep%
SAU	80.7	CHN	64.0	TZA	63.6	SDN	58.5
PER	80.1	PRY	61.1	ETH	53.8	DZA	40.1
MYS	77.3	VNM	56.0	PAK	49.6	NGA	35.9
IRN	76.4	PHL	52.1	URY	49.5	BGD	35.1
ISR	73.5	THA	51.0	ARG	47.8	ETH	32.7
GHA	65.2	MOZ	49.7	EGY	45.2	SSA	22.9
ZAF	64.0	BRICS	49.6	ZMB	41.3	IND	18.9
IDN	58.9	KOR	49.4	KAZ	37.8	PAK	16.7
IND	57.2	EU28	47.8	SSA	35.0	KAZ	16.6
COL	56.8	JPN	42.8	SDN	35.0	GHA	16.2
TUR	55.7	WLD	36.9	BRA	34.2	TZA	16.0
MEX	53.1	OECD	34.2	TUR	30.8	TUR	13.4
UKR	51.1	UKR	33.6	COL	29.5	MOZ	12.8
NZL*	50.5	RUS	32.9	BGD	27.8	IRN	11.8
BRA	50.4	CHL	26.8	HTI	26.6	SAU	10.8
USA	49.9	HTI	26.6	USA	26.0	AUS	10.4
CAN	48.3	MEX	25.1	CHL	25.9	ZMB	7.5
EGY	46.8	CAN	24.8	CAN	25.6	NZL	7.4
CHL	46.4	USA	23.8	NGA	25.6	ZAF	6.3
RUS	44.8	AUS	22.9	DZA	23.3	EGY	6.1
AUS	44.3	NZL	22.7	ZAF	22.7	CHN	5.7
OECD	42.7	KAZ	22.4	AUS	22.4	BRICS	5.2
HTI	42.5	IDN	21.9	ISR	22.0	WLD	5.1
ARG	40.6	URY	18.0	KOR	21.3	HTI	4.3
THA	39.4	NGA	17.2	OECD	21.0	URY	4.1
WLD	39.1	ZMB	16.2	MEX	20.8	IDN	3.9
BGD	37.0	BRA	14.8	RUS	20.4	EU28	2.8
JPN	36.7	SSA	12.8	JPN	20.1	PER	2.8
DZA	36.6	COL	12.8	NZL	19.3	ISR	2.2
PHL	36.2	MYS	11.7	WLD	18.9	OECD	2.1
ZMB	35.0	ARG	10.2	VNM	18.6	RUS	2.0
EU28	33.8	GHA	9.5	PRY	18.3	MYS	1.9
PAK	33.6	PER	7.5	IND	17.9	PHL	1.8
BRICS	31.7	ZAF	7.0	EU28	15.6	PRY	1.8
SSA	29.2	IND	6.0	IDN	15.3	ARG	1.4
KOR	29.1	TZA	3.9	UKR	14.5	CAN	1.2
URY	28.4	ISR	2.3	MOZ	14.2	MEX	1.0
VNM	25.1	EGY	1.8	BRICS	13.5	COL	1.0
MOZ	23.4	SAU	0.5	IRN	11.8	UKR	0.8
KAZ	23.2	ETH	0.4	PHL	10.0	CHL	0.8
CHN	23.0	TUR	0.1	PER	9.6	BRA	0.5
NGA	21.3	DZA	0.0	THA	9.4	JPN	0.4
PRY	18.9	SDN	0.0	GHA	9.1	USA	0.4
TZA	16.5	BGD	0.0	MYS	9.0	VNM	0.3
ETH	13.2	IRN	0.0	SAU	8.0	KOR	0.3
SDN	6.5	PAK	0.0	CHN	7.3	THA	0.2

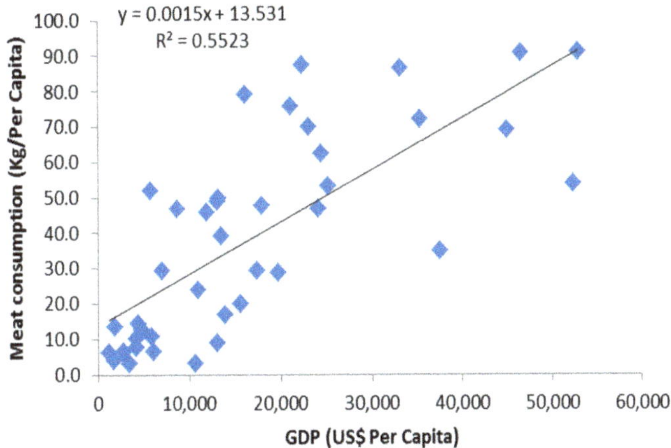

Fig. 17.2 The relations between income and meat consumption, (from authors analysis using data from (Sans and Combris 2015) and GDP data from Wikipedia

tle, is cooked for about 3 h and sold as is, or it may be cured. The sweetbreads (pancreas and thymus) are marketed either fresh or frozen. Both gelatin and animal glue are prepared from collagen derived mainly from the bones and the hide, but also from other parts, such as the sinews, ears, snouts, and trimmings. Oxtails, although quite tough, have a distinctly desirable flavor, and are used in making oxtail soup.

When livestock animals are grown for meat production, the males are castrated, which increases fat and the meat is succulent and tenderer. The strong odor associated with viable males also tends to disappear. The tenderness of beef is inversely related to the age of the animal, and marbling, an effect describing a uniform distribution of relatively high amounts of fat in the muscle. Beef and pork producers may change feed formulas to lower the fat levels in the beef and pork.

1.5 Meat Intake and Income

Income is a predictor for per capita meat consumption (Guenther et al. 2005; Henchion et al. 2014; Sans and Combris 2015). In general, meat consumption increases with rising income. Interestingly, meat protein choice also changes with income. Increasing household incomes lead to greater consumption of chicken whilst lower income household chose more processed pork products. Levels of education were also important with higher intakes of beef and pork trending with less formal education (Guenther et al. 2005).

International comparisons showed that income was directly correlated (R2 = 0.62) with meat intake based on the GDP data for 183 countries (Sans and Combris 2015). However, countries with similar GDP could still show large difference in meat intake, indicating that other variables also influence consumer behaviour including, cultural, geographic, historical, and religious factors (Sans and Combris 2015).

An additional test for the correlation between meat intake and GDP for a selection of countries showed (R2 = 0.55) from which it could be inferred that about 55% of differences in meat intake were accounted for by income.[1] Therefore, 46% of differences in meat intake between the different countries were not influenced by income. Meat intake was least affected by earnings when annual average household income rises above US$21k. With household earnings of US$21,000 to US$52,000 there was a 2% effect of income on meat intake (Fig. 17.2).

A possible reason suggested for the link between wealth and meat intake is that the costs and resources needed for rearing livestock (fuel, land, feed) is greater compared to the costs for producing arable food for human consumption. It takes 10 lb (4.5 kg) of plant food to produce about 1 lb (0.45 kg) of beef. However, closer

[1]This analysis was performed by authors using readily available data.

examination of the "wealth and meat" thesis shows inconsistencies because meat is affordable somewhat and not an luxury-item. There is some tendency for higher intake in those regions where meat is produced in comparative abundance (Argentina, Uruguay, and Brazil) which suggests availability is an important determinant for intake. Religion and culture also influence meat consumption. In some traditionally vegetarian communities such as Hindus, cost/income is unlikely to affect meat consumption (Falahi et al. 2012; Mathijs 2015; Sans and Combris 2015). There may be a trend for decreasing meat consumption in very high-income countries (Mathijs 2015; Sans and Combris 2015) but the evidence is not yet conclusive.

Table 17.6 Selected biological and chemical hazards for meat

Species	Biological hazards	Chemical hazards
Cattle	*Escherichia coli* (VTEC)[a], *Salmonella*	Dioxins, dioxin-like polychlorinated biphenyls (DL-PCBs)
Sheep & goats	*Escherichia coli* (VTEC)[a], *Toxoplasma*	Dioxins, other polychlorinated biphenyls (DL-PCBs)
Solipeds, horse	*Trichinella*	
Farmed deer	*Toxoplasma*	None
Farmed boar	*Salmonella, Toxoplasma*	None
Farmed game	None	None

[a]Verocytotoxin-producing *Escherichia coli*

1.6 Potential Safety Issues Related to Eating Meat

Table 17.7 Some concerns associated with meat

Concern	Comment
Polycyclic aromatic hydrocarbons (PAH)	For heated, fried or roasted meat
Nitrite/Nitrates & Salt	Curing agent
Foodborne disease	Fresh meat, undercooked meat
Heterocyclic aromatic amines (HAA)	Formed at high temperature
Cholesterol	Cardiovascular risk
Zoonotic diseases	Animal/human disease transfer
Saturated fats	Natural lipids
N-nitroso-compounds (NOC),	From nitrite curing agents

Animal derived foods contains many of the nutrients that are required by humans such as, essential amino acids, vitamins and minerals. Meat is particularly beneficial for the growth of infants and children (Krebs et al. 2011). Current dietary guidelines (2015–2020) shows that meat intake is above recommendations for teens and adult men. A decrease in meat intake and adoption of protein alternatives was recommended (U.S. Department of Health and Human Services and U.S. Department of Agriculture 2015).

A number of food safety concerns were associated with eating meat (Tables 17.6 and 17.7). First, there is a need to consider food safety concerns about the possible biological, microbiological, and chemical hazards associated with meat. Consumers remain uncertain also whether meat consumption is associated with specific health risks, such as cancer or cardiovascular disease, which are controversial topics. Meat consumption trends with increasing intakes of saturated fats, salt, and other food components thought to be associated with a variety of disease risks. Some controversies relating meat and health are briefly covered in this section.

The following topics are currently trending in the literature (i) low/higher intake groups for meat may experience different risk of developing colorectal cancer (Carr et al. 2016; Chan et al. 2011), (ii) Pork and poultry intake may lead to decreased risks of cancer, compared with eating red meat (Lippi et al. 2016). (iii) Meat intake does not strongly affect the risk of prostate cancer (Bylsma and Alexander 2015). (iv) Group 1 and 2 carcinogen status. The International Agency for Research on Cancer (IARC) designated processed meat as a Group 1 carcinogen, alongside of other substances "carcinogenic to humans"(World Health Organization 2015). According to IARC, eating red meat and processed meat produced an extra 34,000 deaths annually compared with 1000,000 extra deaths from tobacco smoking and 600,000 deaths from alcohol use (Bouvard et al. 2015). Other compounds or substances listed as Group-1 carcinogens include alcoholic beverages, areca nut, Chinese style salted fish, solar radiation and outdoor pollution. Currently IARC classified red meat (alone) as a Group 2A substances meaning "probably carcinogenic to humans" (Bouvard et al. 2015).

To explain the possible carcinogenic effects arising from processed meat the IARC pointed to the fact these foods contain significant levels of compounds previously found to be carcinogenic in animal testing; N-nitroso-compounds (NOC), polycyclic aromatic hydrocarbons (PAH), and heterocyclic aromatic amines (HAA). It is known that PAH and HAA levels in meat increase during high temperature cooking such as broiling (grilling), barbecues and frying – and less so during boiling (Sinha et al. 2005). N-nitroso-compounds can be produced from nitrite added to meat during curing. Vegetable derived N-nitroso-compounds also contributes to dietary intake (Joosen et al. 2009). Hemin entities derived from myoglobin, which contributes to the red color of meat, have been implicated in the carcinogenic load associated with meat.

Other reports indicated that processed and unprocessed meat were associated with increasing T2D risk (Fretts et al. 2015), inflammatory bowel disease (Ge et al. 2015), ischemic heart disease (Lippi et al. 2015), breast cancer (Guo et al. 2015), obesity (Rouhani et al. 2014), cardiovascular disease (Abete et al. 2014), hypertension, and increase the risk of all-cause mortality (Larsson and Orsini 2014).

Discussions surrounding possible risks from foods are interesting to the General public, but few are in the position evaluate the underlying facts directly. For instance, the IARC report for red meat – was based on 700 studies most of which were "observational" and were possibly affected by confounding effects. It is also difficult to establish a "cause and effect relationship" relationship based on observational studies.

More studies will be needed in order to ascertain the health benefits and other effects of meat consumption. Meanwhile a variety of advice has been proposed on the internet "just in case"; for instance it is been suggested that reducing fat intake associated with meat and using lean meat cuts, avoidance of extra oil when cooking may be sensible strategies to adopt when preparing meat (Santarelli et al. 2008).

2 General Meat Processing

2.1 Slaughter

The basic process for slaughtering livestock is similar the world over though differences occur in terms of equipment, and levels of hygiene and oversight (Heinz 2008). At the slaughterhouse or abattoir, animals are allowed to rest for a period just before slaughter. This is done in order that muscle starch (glycogen) is not all used up. When the animal is slaughtered with an adequate amount of glycogen in its body, the glycogen is converted under anaerobic conditions to lactic acid that produces a pH decline. The presence of this acid has a preservative effect on the meat as well as its color and texture. If the animal is not rested before slaughter and it uses up its glycogen, there will not be enough lactic acid formed to have the preservative effect, and thus the meat will spoil sooner.

Generally, animals must be rendered insensible to pain before slaughter by *stunning*. With cattle, this is done by hitting them on the head with a bolt from a specially devised stunning gun. The stunned animal is shackled at the hind legs, raised from the floor in an upside-down position, and then bled by severing its jugular vein. Most modern slaughter operations, employ a line-slaughter approach where carcass is hauled up off-of the floor for all subsequent stages, thereby ensuring hygienic conditions (Heinz 2008).

In compliance with religious laws, cattle for kosher meats are produced by "Shechita" or the Jewish method for slaughter. Therefore the throat is cut with a special knife, in one motion, severing the jugular vein, followed by bleeding. Slaughter is conducted by trained personnel, called *Shocet*. Certain parts of meat may be rejected as unacceptable or *treifa* or unfit to be

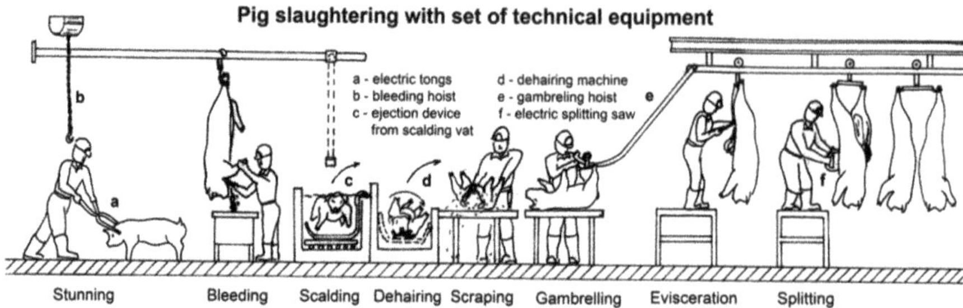

Fig. 17.3 General scheme for slaughtering animals

eaten. Halal (Islamic) requirements for slaughter require a sharp knife incision to the front of the throat severing the major blood vessels but leaving the spinal cord intact. The carcass is bled. Both Jewish and Islamic slaughter practices avoid pre-stunning, in part to ensure heart action and more thorough bleeding. The *Jhatka* method (Sikh requirement) for slaughter involves a single stroke or decapitation using a special sword. In all cases, religious slaughters involve specially trained practitioners using licensed premises where there is dispensation from stunning (Haluk Anil 2012) (Fig. 17.3).

After bleeding the carcass, the skin is cut ventrally (along the abdomen), pulled back from the belly portion, and the lower legs are removed. Hooks are inserted in the cords of the hind leg section, and the body is raised to the half-hoist position. Most of the skin is removed in this position, after which the carcass is raised to full hoist and skinning is completed. Also, in this position, the sternum (breastbone) is opened with a saw, the belly is split, the bung and esophagus are tied off, and the viscera are removed. The viscera are inspected by veterinarians, and the carcass is split into halves with a saw. The hide is inspected, salted, and sent to the tannery.

It is essential to avoid contamination of the carcass by, digester, fecal matter, hair and other sources of microbial contamination. Immediately after slaughter the carcasses may undergo postmortem inspection by qualified personnel before labeling and cold storage at 37–44 °F (3–7 °C); further details on slaughtering practices ca be found in later parts of the chapter.

The split carcass is washed, and, if it is not to be boned out soon after cooling, it is covered with a cloth to shape the surface fat. Sides handled in this manner are sent to a cooler where in 24–48 h it will be required to bring the temperature of all parts to about 35 °F (1.7 °C).

2.2 Cutting and Dressing

Dressing, including skinning, may be carried out completely while the animal is hanging from a rail. Mechanical instruments, such as hide pullers, may be used to increase the efficiency of rail dressing.

In the past, beef shipped out as sides or, more often, as quarters. Today, a considerable amount of beef is cut into various primal cuts, such as rounds, ribs, loins and packaged, under vacuum using a package impervious to moisture and oxygen. The packages are boxed and shipped to distributors and retailers. Because the meat has been boned and trimmed of excess fat, meat department personnel at the stores have little to do but cut the meat into individual portions, package the various cuts, and place them on display. Primal cuts packaged correctly and held at 32 and 40 °F (0 and 4.4 °C), have a storage life of at least 21 days.

2.3 Carcase Ageing

After slaughter, the carcass portions are aged in cold rooms at low temperatures just above freezing, usually 35 °F (1.7 °C). This process may take up to one month, depending on temperature, humidity, and other conditions, but ordinarily, meat is held for about 12 days prior to consumption. The aging is accelerated if the beef is stored at higher temperatures, but the higher temperatures permit the growth of bacteria and surface spoilage. However, because the spoilage is on the surface, this problem can be resolved by storing the meat at the higher temperatures in the presence of ultraviolet light, which has a bactericidal effect.

Sometimes a processor is willing to accept the microbial growth in order to hasten the tenderization. Because the bacteria and mold growth create a certain amount of spoilage on the surface of the meat, the processor must then trim the surface to remove the effects of microbial growth and spoilage. The main purpose of aging is, of course, to tenderize the meat. Meat cuts can be tenderized just prior to consumption by the addition of commercially prepared enzymes, which have a proteolytic effect; that is, they tend to break down the proteins. However, this is a slow process if the enzymes are added only to the surface, and it is believed that injection into the meat or into the bloodstream of the living animal just before slaughter is a more effective way of producing tender beef.

Unfrozen meat should be stored in a temperature range of 28–38 °F (−2.2 to 3.3 °C). The relative humidity should be in the range of 85–90%. The high humidity will prevent excessive drying and shrinkage. In addition, high humidity tends to preserve the white color of the fat in the meat. The storage room should have a reasonable circulation of air to ensure heat removal and more uniform distribution of the humidity. Meats stored for long periods are generally frozen and stored at temperatures in the range of −10 to 0 °F (−23.3 to −17.8 °C).

2.4 Meat Inspection

Because zoonotic diseases can be transmitted from animals to humans, all meats shipped interstate in the US are subject to inspection for wholesomeness under the authority of the Meat Inspection Act. "Wholesome" meat is safe to eat and is without adulteration. The Food Safety and Inspections Services (FSIS), examine meat for adulteration or misbranding in the broadest sense (Chap. 3) including inspections to check the health of livestock at the meat plant prior to slaughter. (USDA Food Safety and Inspection Service 2014). Table 17.8 summarizes some of the pre-slaughter and postmortem inspections undertaken by officials.

Table 17.8 Pre-slaughter and postmortem inspections for meat

Inspections	Goals
Pre slaughter inspection	Livestock health, check for humane slaughter methods (stunning)
Post mortem inspections	Check carcass markings, labeling, assure condemned meat is destroyed, inspect meat for export
Compliance	Certification and labeling
Compliance	Examining night-time operations
Compliance	Offering or receipts of gifts or bribes

Adapted from USDA Food Safety and Inspection Service (2014)

Inspectors from the USDA evaluate beef carcasses and specify various grades. These are "prime" (the best beef from steers and heifers), "choice" (the next best from heifers, steers, and young cows), and "select" (the third best). There are grades of lesser quality, but they are generally used for purposes other than as steaks or roasts.

In addition to the wholesomeness inspection, meat may be graded for quality. Although lean meat is more valuable than fat, the best cuts of beef have some fat, and the fat is well distributed, giving the marbling effect cited earlier. Inspectors from the USDA grade beef carcasses by removing a standard piece from between the 12th and 13th rib and examining this for the degree of marbling. For steers, heifers and cows the grading is: (1) prime, (2) choice, (3) select, (4) standard, (5) commercial, (6) utility, (7) cutter, and (8) canner. The grading for bullock carcasses is: (1) prime, (2) choice, (3) select, (4) standard, and (5) utility. Beef rated as "prime" has the highest extent of marbling and the highest quality, but in general is limited in amount and served by full-service restaurants. Meat of lower grades are of lesser quality, and are generally used for purposes other than as steaks or roasts. It should be noted that in federally regulated plants, the inspection for wholesomeness is mandatory. The grading of the meat is optional (USDA Food Safety and Inspection Service 2016; Griffin and Savell 2018).

2.5 Cooking Meat

Fresh meat such as steak may be subjected to a range of cooking methods, e.g. roasting, broiling, pan broiling, or cooking via impingement oven (American Meat Science Association 2015). Cooking produces desirable changes in sensory characteristics (flavor, texture, color etc.). Cooking also improves food safety. To inactivate any disease causing bacterial meat should be heated to prescribed minimum internal temperatures measured using a cooking thermometer (Table 17.9). Since these temperatures represent the minimum settings for safety, there is considerable flexibility for culinary practice. Meat can be cooked according to such designations as "rare, medium and well-done".

Formal regulations apply for cooking of meat and animal products. For instance, 9 CFR 318.17 describes requirements for heating in order to reduce salmonella or pathogen numbers by a factor about 3 million-fold (6.5 log10 reduction). Cooked beef, roast beef, and cooked corned beef products, should also avoid possible germination of *Clostridium botulinum* spores. The products must be processed under a HACCP food safety management system (U.S. Government Publishing Office 2012a). Regulation 9 CFR 318.23 prescribes a set of internal cooking temperatures from 151 °F (66 °C) to 157 °F (69.4 °C) with the corresponding holding times of 41 s and 10 s. Patty internal temperatures should be checked at rate of one per hour (U.S. Government Publishing Office 2012b).

Table 17.9 USDA safe minimum internal temperature recommendations for cooking meat USDA Food Safety and Inspection Service (2016)

Product	Minimum Internal Temperature
Meat[a]	145 °F (62.8 °C)
Ground meats	160 °F (71.1 °C)
Ham, fresh or smoked (uncooked)	145 °F (62.8 °C)
Fully Cooked Ham (reheat)[b]	140 °F (60 °C)
Fully Cooked Ham (reheat)	165 °F (73.9 °C).
Poultry[c]	165 °F (73.9 °C)
Eggs	160 °F (71.1 °C)
Fish & Shellfish	145 °F (62.8 °C)
Leftovers	165 °F (73.9 °C)
Casseroles	165 °F (73.9 °C)

[a]Beef, pork, veal & lamb, steaks, chops, roasts, – allow 3 min rest time

[b]for USDA packing plants, otherwise heat to higher temperature

[c]Breasts, whole bird, legs, thighs, and wings, ground poultry, and stuffing

As noted above, cooking meat produces a number of sensorial changes, e.g. taste, flavor, color, tenderness and juiciness. There is concerned also with shrinkage and cooking losses. The majority of the cooking changes observed with meat arise from the effect of heat on two muscle proteins – collagen and myosin. Collagen denatures at temperatures of about 50–60 °C, depending on the age of the livestock, leading to rising tenderness during cooking. By contrast heating myosin at >50 °C leads to shrinkage, hardening, and decreasing water holding.

Heating without water (e.g. dry heat) causes the connective tissue (collagen) to become tougher. The amount of connective tissue present in the meat depends on the location of the meat in the animal. The tenderloin, for example, has very little connective tissue, especially in the younger animals. Although boiling tends to make meat less tough, one of the disadvantages is that taste components are extracted from the meat, and, therefore, boiled meat is less tasty. Of course, the water in which the meat has been boiled acquires the meat flavor components and can be concentrated and sold as a meat extract, which can be used as a base for soups and stews. Also, little browning (which enhances meat flavor) occurs during boiling.

When a mass of comminuted (ground) meat is heated, proteins denature to form a network or gel which then holds, water, oil as well as small pieces of meat together in the form of a meat emulsion gel; this is found in products like sausages and meat loaves. Meat proteins undergo gel formation, like the process observed by heating egg-white (Cross et al. 1976; Modzelewska-Kapitula et al. 2012; Schmidt et al. 1970; Simmons et al. 1985).

3 Meat Products

3.1 Beef and Veal

The most important beef animal breeds in the United State are the Angus, Hereford, Galloway, and Shorthorn. Less important are the Santa Gertrudis; the Chambray, a cross between a Charolaise bull (French) and a Brahma cow; and the Brangus, a cross between an Angus bull and a Brahma cow. Much of the cattle population of the US is raised on the Western range, which includes the Great Plains, the Rocky Mountains, and the intermountain and Pacific Coast regions (Drouillard 2018).

The calves produced in these areas may be finished (final stages of feeding before slaughter) in feedlots elsewhere. These feeding areas include the best pulp feeding area of Colorado; the cottonseed cake section of Texas; the Corn Belt area (especially Iowa) and certain southern states, especially Florida; and California. Discarded (retired) dairy cows comprise a significant amount (approx. 10%) of the US beef supply (Moreira et al. 2021). Veal calves are produced mainly in dairying areas (especially Wisconsin) and slaughter calves (3–12 months old) are also produced in southwestern areas.

Farmers and ranchers may sell cattle to terminal markets (who sell to various buyers), local markets, auctioneers, or packers. Usually, cattle are shipped by truck to stockyards near or at the point of slaughter. Steers (males castrated prior to sexual maturity) and heifers (females beyond the veal and calf age that have never calved) make up the major portion of beef animals. Cows (females that have calved), bulls (mature males), and stags (castrated after reaching sexual maturity) make up the remainder of the beef animals (Drouillard 2018).

The term conformation is used to describe the structural characteristics of the beef animal. For the best conformation, the animal should be short and compact. A rather blocky, or large-bodied, short-legged appearance is good. Basically, the best set of structural characteristics or conformation is valuable to the retail butcher because apparently he is able to make more profitable cuts (chuck, round, and loin) from the animals having the top conformation, and because there is a relationship between good conformation and good quality (Fig. 17.4).

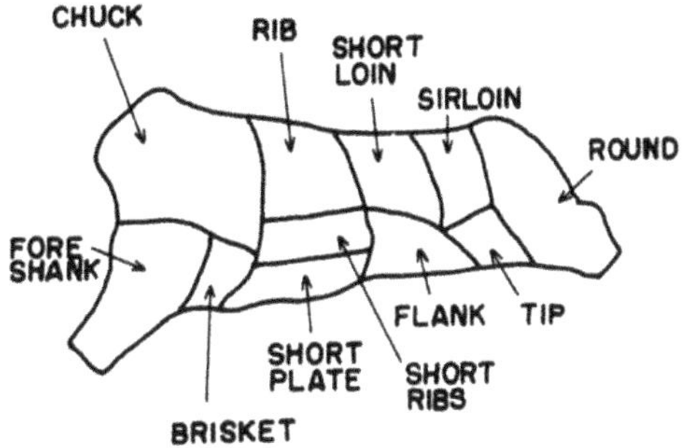

Fig. 17.4 Cattle carcass

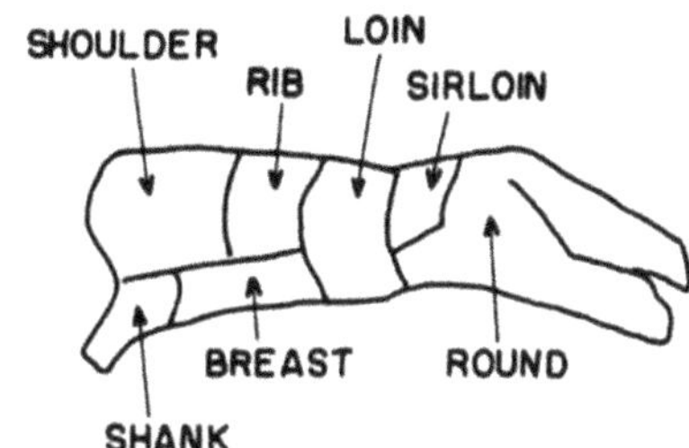

Fig. 17.5 Veal carcass

A veal carcass is shown in Fig. 17.5. Grades cannot be assigned until the animal has been slaughtered. A new technique, the use of ultrasound energy to show the structure of the meat prior to slaughter, has been developed and may influence grading techniques in the future.

3.2 Pork

There are many breeds of hogs that, in this country, have become pretty well mixed in recent years in order to produce meat-type animals (more lean and less fat) rather than lard-type animals. Part of the reason for getting away from the high-fat type of hog has been that vegetable oils have largely replaced lard, which previously was the chief fat used for cooking. Some breeds of hogs, such as the Poland-China, Berkshire, Chester White, and Hampshire, are not mixed.

The north-central states, especially Iowa, produce most of the hogs raised in the US. Hog production is said to be increasing in the southern states.

Hogs are usually fed soybean meal, meat byproducts, fish meal, or milk, or combinations of these as a source of protein. Carbohydrate foods consist mainly of corn feed, but other grains are used. Minerals may be added as supplements. Some vitamins, especially B12, may be added to hog feed, as they may serve as growth factors.

Slaughter hogs are classified as barrows (male castrated prior to sexual maturity), gilts (young females that have not produced young pigs), sows (mature females that have produced young or that have reached an advanced stage of pregnancy), stags (males castrated after reaching maturity), and boars (uncastrated males).

Hogs are slaughtered and processed as described in Sect. 2.

In general, hogs are cut up into ham, loins, shoulders, bellies, and back fat before shipment to distributors or to processing plants that manufacture various pork products. See Fig. 17.6 for an example of a hog carcass.

3.3 Lamb and Mutton

The bulk of the sheep produced in the US is said to be made up of crossbreeds developed for the production of both meat and wool. Pure breeds include Cheviots, Lincolns, Oxfords, Shropshires, Rambouillets, and Merinos. Sheep are raised on ranges in arid and semiarid regions and on farms. On farms, flocks of sheep are smaller than those raised on ranges. Where conditions allow, lambs may be fattened on the range, but during winter storms, feed must be supplemented with protein concentrates, corn, hay, barley, and oats.

Fattening of lambs may take place in the Corn Belt states, especially during the fall and winter months. This may be done in corn fields, stubble fields, or pastures. Some fattening occurs while the animals are confined to barns. Most sheep fattening is carried on from September through May. Sheep are classified as ewes (females), wethers (males castrated prior to reaching maturity), rams (uncastrated mature males), and lambs about 3–12 months in age that do not show their first permanent teeth. Lambs are identified by the break joint- the temporary cartilage in the leg bone just above the hoof. As the animal grows older, the break joint loses its redness and saw-toothed appearance, and the cartilage is replaced by bone. Spring lambs are those from new crops marketed before July 1 and weighing 70–90 lb. (31.8–40.9 kg). As with beef cattle and hogs, grading of slaughter lambs is based on conformation, finish, and quality.

Sheep are sold directly to the final buyer or indirectly through a commission agent. Because most sheep are produced west of the Mississippi and are consumed in the east, they must be transported for comparatively long distances for slaughter. Lambs and sheep are shipped both by truck and by rail, in double-decked cars.

After slaughter, sheep carcass is not split prior to shipment to the distributor or retailer. Sheep carcasses are classified into two categories: mutton and sheep (the latter is under 12–14 months of age). Lamb and mutton carcasses are graded as

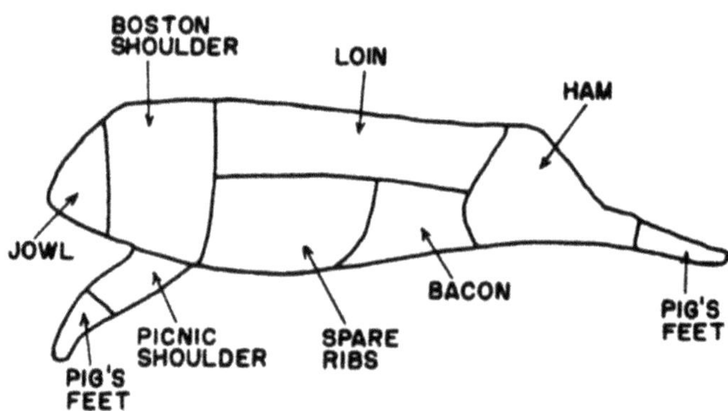

Fig. 17.6 Hog carcass

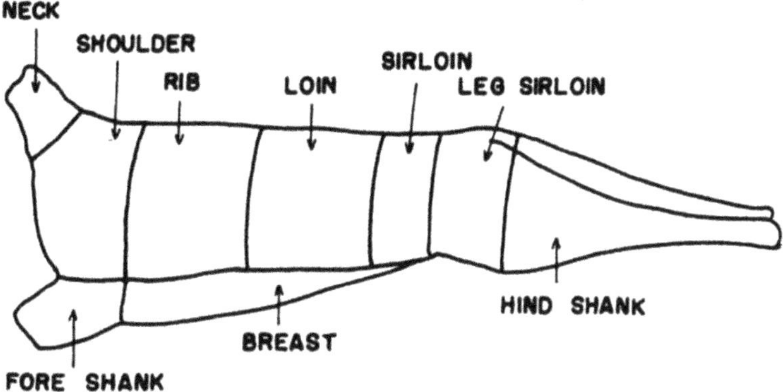

Fig. 17.7 Sheep carcass

prime, choice, good, utility, and cull. A sheep carcass is shown in Fig. 17.7.

3.4 Processed Meat Products

Meat attached to bone may be recovered for use in comminuted products, soups, and stock. Mechanically recovered meat is legally classed as meat. The Table below lists some of the other major types of processed meat products including, cured products, cooked sausages, fermented and air cured or smoked products. Table 17.10 lists some different meat products.

Table 17.10 Some common meat products

Whole and brine-injected, cured meat products
Bacon
Ham
Pastrami
Reformed meat products
Cooked sausages
Frankfurters (Austria, Germany, USA)
Chicken sausages
Luncheon meat (Australia)
Weisswurst (Bavaria)
Meatballs (Asia)
Fresh sausages
Nuremberg (Germany)
Cumberland sausages (UK)
Fermented salami
Salami (raw/cooked)
Spreadable raw fermented
Air cured meat products
Parma ham (Italy)
Serrano ham (Spain)
Beef jerky (USA)
Burgers, patties and crumbled products
Beef patties
Chicken patties
Chicken nuggets

Adapted from Feiner (2006)

4 Cured Meat Products

Most pork and some beef are cured or processed in some manner (Veal, lamb, and mutton are ordinarily not cured.) During the cutting up of these animals, there are always some portions trimmed off from the carcass. Much of the trimmings from both beef and pork carcasses are used for the production of fresh or cooked sausage. Fat that is trimmed off, and used for the manufacture of soap. Bones may be used for the manufacture of bone meal or gelatin. Pork skin is used almost entirely for the manufacture of gelatin.

Curing agents for meat include salt (sodium chloride), for manufacture of New England-style corned beef. However, more often, curing agents include salt, sodium nitrite, and table sugar (sucrose) or glucose (Table 17.11). Sometimes

Table 17.11 Composition of some curing solutions

Component (¥)	Ham	Bacon
Brine (% w/v)	55–80	65–75
Sugar (lbs)	20–50	20–100
Sodium Nitrite (lbs)	1.5	1–1.5
Water (Gallons)	100	100

(¥) weights added to 100 gallons or 378.5 l

reducing agents, such as vitamin C or the related salt of iso-ascorbic acid, sodium erythrobate, are used to facilitate the coloring effect of nitrate and nitrite by converting both to nitric oxide. In other instances spices, or materials such as monosodium glutamate, are used as flavor enhancers (Epley et al. 1992; Sindelar and Milkowski 2011; Sindelar 2012). The upper limit of nitrite addition for cured or smoked processed meat is currently 200 ppm (or lower) alongside a maximum of 500 ppm of nitrate (Chap. 16).

When freshly cut, the color of meat is purplish-red owing to the pigment myoglobin. After cutting the meat, the myoglobin soon becomes loosely bound to oxygen to form oxymyoglobin (bright red), and, if exposed to oxygen for some time, the bond becomes more permanent forming gray or brown metmyoglobin. If nitrite is allowed to react with myoglobin, oxymyoglobin, or even metmyoglobin, it forms nitrosomyoglobin which is bright red (Epley et al. 1992; Sindelar and Milkowski 2011; Sindelar 2012).

4.1 Ham

A curing mixture for ham is shown in Table 17.11 but small amounts of ascorbic acid or iso-ascorbic acid, or sodium erythrobate may also be added. Phosphate (for the retention of water) is allowed in cured meats in amounts not to exceed 0.5%. This pickle is pumped into the ham and they are placed in tierces (casks) and may or may not be covered with pickle. Curing in this manner may require 5–10 days at 40 °F (4.4 °C), depending on the size of the product.

4.2 Smoke Cured Ham

After curing, the ham are washed, hung to dry (with or without a stockinet cover), then smoked. In smoking, moistened hardwood sawdust is usually burned in a mechanically controlled smoke generator (located outside the chamber where the product is hung) for purposes of treating with smoke and heat. Oak and hickory sawdust or logs (logs may be used in a special smoke generator) are mostly used for the production of smoke.

The time and temperature of smoking must be sufficient to raise the temperature of all parts of the ham to at least 137 °F (58.3 °C). This is a USDA regulation that has been introduced to make certain that a roundworm, *Trichinella spiralis*, is destroyed. Cured ham to be boiled or canned may be smoked, but generally are not. Ham for canning is cured, skinned, boned, and then placed in the can with optional gelatin and various spices or flavoring. The cans are sealed under a slight vacuum and then heated in an agitated water bath at 165–180 °F (73.9–82.2 °C) until an internal temperature of 150–165 °F (65.6–73.9 °C) is reached. Neither boiled nor canned ham is commercially sterile and must be refrigerated to prevent spoilage. Since it is not sterile the canned ham must be labelled "keep refrigerated" to inform the consumer of the fact (Chap. 9, Sect. 3.4).

4.3 Dry-Cured Ham

Dry-cured ham is produced by repeated salting at low temperature, storing and allowing the product to equilibrate to higher ambient temperature. The dry-cured hams are produced and often named according to the country/region of origin. Space restrictions means that the varying types of dry-cured hams will not be discussed in detail, but the reader can find more information on these and country harms from the following sources (Ockerman et al. 2002; Rentfrow et al. 2012) (Table 17.12).

Table 17.12 Some dry-cured ham products

Dalmatian ham	Italy
D. O. Guijuelo hams	Spanish
French jambon	French
Gammon	English
Iberian ham	Iberian
Jinhua ham	Chinese
Mainz ham	German
Parma ham	Italian
Prosciutto ham	Italian
Spanish-style jamón	Spanish
Country ham	(US)

Adapted from Ockerman et al. (2002)

4.4 Bacon

The belly portion of the hog is stitch-pumped with the curing solution using machines that through needlelike projections inject the curing solution simultaneously and automatically into the meat in a uniform pattern. The amount of pickle used for bacon is 5–10% of the green (uncured) weight. The curring solution is as shown in Table 17.11. Curing times are short, and in some cases, the product is smoked almost immediately to an internal temperature of 120–125 °F (48.9–51.7 °C). If the temperature were raised to 137 °F (58.3 °C) in all parts, as required for ham, much fat would be rendered out and lost. Since bacon will be cooked prior to consumption it is deemed uneccesary to raise the temperature of all parts to 137 °F (58.3 °C) or higher.

Canadian bacon is prepared in a similar manner to that of regular bacon but using pork loins instead of bellies. Also, curing times are somewhat longer than for regular bacon, and during smoking, the internal temperature is raised to 137 °F (58.3 °C) or higher. Pork shoulders and shoulder picnics are cured and smoked much in the same manner as ham.

4.5 Corned Beef and Pastrami

Some beef products are cured, or cured and dried. Among these are corned beef, pastrami, and chipped or dried beef. Corned beef is usually produced from the brisket (lower forward portion of cattle near the foreleg). The fresh briskets are pumped with pickle and placed in cover pickle for 7–14 days at a temperature of 40 °F (4.4 °C) or below. In most cases, nitrite is used as an ingredient of the pickle, but in New England style corned beef, only salt is used. In producing pastrami, table salt and nitrite (in limited amounts as stated previously), together with spices, are used in curing solutions.

Chipped beef is produced from beef ham or other portions that make up certain parts of the round section of the hind quarter. Usually cutter and canner grade cattle are used for this product. The beef ham are placed in tierces and covered with a pickle of high salt content containing not more (usually less) than 2 lb (907.2 g) of nitrite per 100 gallons (3.78.5 l) of pickle. They are held in this manner for 50–65 days at a temperature below 40 °F (4.4 °C) (but above 34 °F [1.1 °C]). After curing, the product is hung in a room of low relative humidity and allowed to dry to the point where the salt content is 10–14%. It is then sliced into thin sheets and placed in glass jars that are capped under vacuum. This product is not heat-processed. The high salt content and the nitrite are sufficient to prevent the development of food spoilage or disease-causing bacteria. The vacuum packing is necessary to retain the pink color of chipped beef, and it prevents the development of mold growth.

5 Sausage Products

The meat ingredients used in a variety of sausage products may come, in part, from trimmings resulting from the cutting of beef quarters into rounds, loins, and other cuts and from the cutting of hogs into loins, ham, and shoulders. However, the chief source of meat for sausage comes from low-grade cattle and hogs. Bull meat is a prime ingredient of certain types of cooked sausage. Sausages are produced worldwide, but perhaps the most influential sausage recipes in the United States are of German, Italian or Polish descent.

Types of sausage vary, depending on the style of cutting or grinding of the meat, the size and shape of the product, the seasonings used, the degree of smoking, and the cooking procedure. The main types of sausages are the fresh sausages, cooked sausages and smoked / dried sausages. A few different types of sausages include kielbasa, knockwurst, cooked salami, and Polish sausage (USDA Food Safety Inspection Services, 2013; Govinfo.gov 2021).

5.1 Fresh Sausages

Fresh sausages is made from fresh or frozen meat (beef, pork, swine) usually without added byproducts. Fresh pork sausage is not cooked or smoked. The meat is ground and seasoned (only salt, sugar, sage, and pepper are added for seasoning, but ginger or fennel may also be used). Some ice, up to 3% by weight, may be added and no more than 50% fat as total ingredients. After mixing using a bowl-chopper the resulting paste is stuffed into natural casings made using sheep or pig small intestines. Cellulose casings may also be used for sausage.

Fresh sausages should be held at temperatures very near 32 °F (0 °C) after manufacture, because it is quite subject to bacterial spoilage, to oxidation of the fat (rancidification), and to loss of the typical pink color. Usually, when bacterial spoilage occurs, other types of spoilage do not.

5.2 Frankfurters or Hot Dogs

Frankfurters or hot dogs are the most popular type of cooked sausage prepared usually from pork, beef or poultry (equal to 15% or less of the total weight of ingredients, excluding water). The product is seasoned, cured, and usually smoked. The seasonings usually include sugar and spices (German Food Guide 2016).

Pepper, nutmeg or mace, mustard, garlic, and coriander are added to improve flavor. The curing salts usually contain sodium nitrite and sodium erythrobate (sodium iso-ascorbic acid) to prevent the growth of *Cl. botulinum* and to develop the color.

There have been studies suggesting that nitrites, in combination with protein and heat, produce carcinogenic nitrosamines. The USDA sets limits on the amount of nitrites, nitrates, or a combination of the two at 156 ppm (156 mg/kg of meat). Other research has noted that endogenous production of nitrites occurs naturally in the body accounting for ~50% of daily exposure. Vegetables are also a significant source of nitrite and related compounds. Finally, nitrite has been discovered to be important in lowering blood pressure (Hord et al. 2009).

Work is being done to develop curing solutions that do not contain nitrites but still inhibit *Cl. botulinum*. One such ingredient is sodium hypophosphite, which closely resembles nitrite in its preservative effect and is a GRAS substance (Rhodehamel et al. 1990). Alternatively, low-nitrite curing agents and nitrite-free meat curing agents that inhibit *C. botulinum* and also help to maintain meat color were formulated using mixtures of antioxidants, spices, antimicrobial agents and low amounts of nitrites (Eskandari et al. 2013). A further option (Chap. 16, Sect. 6.2.7) involves using celery and other natural sources of nitrite for meat curing (Pennisi et al. 2020; Yong et al. 2021).

Approved phosphates may be added for water retention, and products such as dry milk, cereal, and soy flour may be added for a total of 3.5% of the finished product. It may contain meat byproducts such as hearts, snouts, and tripe but must be labeled as "frankfurter with byproducts" or "frankfurter with variety meats." Water or ice, or both, may be added to facilitate chopping or to dissolve the curing ingredients. If smoking is done, it may be with hard wood smoke or with liquid smoke spray. When the process is complete, the frankfurters may not contain more than 30% fat and not more than a 40% combination of fat and water.

Frankfurters may also be made from a single species and may be labeled "beef frankfurter" or "turkey frankfurter." The poultry franks have grown in popularity because they are often formulated so that there is less fat in the finished product. Some of the beef producers are also making lower-fat varieties. These products are quickly becoming some of the most popular items on the meat processors' production lines. Whatever the case, all ingredients must appear on the label in decreasing order, by weight.

For the production of frankfurters the meat is passed through a meat grinder and then finely comminuted in a silent cutter (a large metal bowl in which knives rotate at right angles to the plane of the bowl, which also rotates). While the meat is in the silent cutter, ice is added to prevent the temperature of the meat from rising above 60 °F (15.6 °C). The finished product may contain up to 10% added water (federal regulations), the purpose of which is to facilitate cutting and to provide more succulence to the finished product. Curing agents, spices, and fillers (if used) are added while the meat is in the silent cutter. The nitrite is added separately from other additives, because if it is mixed with the other additives before it is added to the frankfurter mix, the preservative effect of the nitrite will be diminished.

When ascorbic acid or iso-ascorbic acid is used to assist in color development, it is added about 1 min before the end of chopping in the silent cutter. Colloid mills or similar equipment may be used instead of the silent cutter to comminute the meat. From the silent cutter, the meat emulsion is put into metal cars, which may be placed in a vacuum chamber to eliminate trapped air bubbles. The air may also be removed in a mixer under vacuum.

The meat emulsion consists of a continuous phase, which is the water, and a dispersed or discontinuous phase, which is the fat. The fat is dispersed more or less uniformly in the water. The agent that maintains this uniform dispersion of the fat (emulsifier) is actually proteins that has been solubilized, especially in the presence of salt. In preparing the emulsion, one must heed the temperature rise during preparation, and the rate at which the fat is added as well as the speed with which the emulsion is mixed. The temperature should not be too high. If it exceeds about 68 °F (20.0 °C), the emulsion may break down. The optimum rate at which the fat should be added depends a great deal on the rate at which the fat can be emulsified.

If the fat is added more slowly than the emulsification rate, then the mixing time will be unnecessarily extended, creating an increase in temperature and a decrease in the amount of fat that can be emulsified. If the rate at which the fat is added greatly exceeds the rate at which it can be emulsified, then it is quite possible that little or no emulsion will be formed. The viscosity of the mix, the size of the fat droplets, the type of fat, and a considerable number of other factors affect the emulsion characteristics. The meat emulsion is next placed in a stuffing machine where a powered piston extrudes the product into a stuffing horn and then into artificial (cellophane or other plastic) or natural (sheep or other animal intestine) casing. The stuffed casing comes out as a long cylinder, which is then made into link form, generally by machine, and the linked sausage is hung on racks and then smoked and cooked.

During smoking and heating, the frankfurters are hung in air-conditioned or natural draft smokehouses. For the first phase in the smokehouse, a temperature of 130–140 °F (54.4–60 °C) is used for 10–20 min. Smoke is then introduced, and the temperature is slowly raised to 165 °F (73.9 °C) and held until the internal temperature of the product reaches about 155–165 °F (68.3–73.9 °C). The internal temperature of frankfurters should never be less than 150 °F (65.6 °C), and preferably should be at 155 °F (68.2 °C) or slightly higher during cooking and smoking. After cooking, the frankfurters are hung in a cooler.

Frankfurters stuffed into natural casings are not skinned after manufacturing, but those stuffed into cellulose casings are removed from the casing before they are sold. The links of sausage are passed through a bath of warm water, then into the skinning machine, where the casing is slit with a blade, then rolled off or blown off. After skinning, the frankfurters are packaged and immediately returned to the cooler, which should be held at a temperature near 32 °F (0 °C), until shipped out.

Frankfurters, like other perishable foods, eventually spoil when held for long periods at temperatures near but above freezing. When spoiled, frankfurters either become slimy, as a result of the growth of bacteria or yeasts, or they turn green on parts of the surface. Greening is due to hydrogen peroxide produced by the growth of bacteria of the Lactobacteriaceae group. The hydrogen peroxide reacts with the nitrosohemo chrome (color complex of nitrite and myoglobin) and oxidizes it to form a green color. In continuous frankfurter manufacturing, automatic stuffers extrude the emulsion into molds, which pass through an automated tempering, smoking, cooking, and cooling unit. The product is packaged immediately after exit from the processing unit.

5.3 Bologna

Bologna is another type of smoked and cooked sausage. The ingredients and manufacturing procedures for manufacturing bologna are similar to those used in making frankfurters, but the size of the casing, hence the size of the bologna, is much larger in diameter, and the bologna link is proportionately longer than the frankfurter link. When not sufficiently cooked, bologna may develop green cores after processing.

There are many other types of meat sausage produced. Among the smoked, uncooked sausages are country-style pork sausage and country-style sausage that contains beef as well as pork. These products contain moderate amounts of curing agents. Cooked sausages are made both with and without curing agents added, and are stuffed into natural cellulose or saran casings. In the manufacture of Braunschweiger or liver sausage, pork liver and jowls, with or without beef, are used.

5.4 Dried Fermented Sausages

Curing salts are among the ingredients used in the production of fermented sausages, but nitrate sometimes replaces nitrite. During processing, bacteria remove oxygen from the nitrate, reducing it to nitrite. Added ascorbic acid can also reduce nitrate to nitrite and then nitric oxide, which combines with myoglobin to form the typical red or pink color of cured meat (Savic 1985).

After mixing with the curing agents, the meat for semidried sausage is held in pans for 48–72 h in a refrigerated room at 38–40 °F (3.3–4.4 °C). The meat is then remixed, stuffed into casings, and held at 50–60 °F (10–15.6 °C) for 12–48 h, then smoked. Smoking is accomplished in stages: first, at a temperature of 80–90 °F (26.7–32.2 °C) for 12–16 h, then, at 100 °F (37.8 °C) for 24 h, and finally, at 137 °F (58.3 °C) for 4–5 h.

During the processing of semidried sausage, conditions at first favor the growth of bacteria that reduce nitrate to nitrite. This is followed by fermentation involving lactic acid-producing bacteria. To ensure the type of fermentation desired, it is possible to obtain cultures of appropriate bacteria, which may then be added to the product. Typical semidried sausages are cervelat, Lebanon bologna, pork roll and Thüringen sausage.[2]

Dried sausages, such as Italian salami and pepperoni, are not heated to temperatures above 90 °F (32.2 °C), but they are stabilized against microbial spoilage because of their low pH (resulting from the action of lactic acid-producing bacteria), their low moisture content, and their high salt content. Spices and curing salts contribute to the preservation of the product.

5.5 Other Products

There are various loaf-type specialty products, such as chicken loaf, luncheon meat, meat loaf, and head cheese, and these may be processed in casings (similarly to sausages) or in metal containers. In the latter instance, they are usually placed in casings after cooking. This type of product may be cooked in hot water or baked in an oven. The products may be deep fat fried, after baking, in order to brown the surface.

References

Abete I, Romaguera D, Vieira AR, Lopez de Munain A, Norat T (2014) Association between total, processed, red and white meat consumption and all-cause, CVD and IHD mortality: a meta-analysis of cohort studies. Br J Nutr 112(5):762–775

American Meat Science Association (2015) Research guidelines for cookery, sensory evaluation, and instrumental tenderness measurements of meat. Retrieved from http://www.meatscience.org/docs/default-source/publications-resources/amsa-sensory-and-tenderness-evaluation-guidelines/research-guide/2015-amsa-sensory-guidelines-1-0.pdf?sfvrsn=6

Bouvard V, Loomis D, Guyton KZ, Grosse Y, El Ghissassi F, Benbrahim-Tallaa L, Guha N, Mattock H, Straif K (2015) Carcinogenicity of consumption of red and processed meat. Lancet 16(16):1599–1600

[2]Within the European Union, the designation Thüringen sausage has Protected Geographic Indicator (PGI) status which means that only products made in the German state of Thüringen can be sold with this label https://en.wikipedia.org/wiki/Thuringian_sausage

Bylsma LC, Alexander DD (2015) A review and meta-analysis of prospective studies of red and processed meat, meat cooking methods, heme iron, heterocyclic amines and prostate cancer. Nutr J 14:125

Carr PR, Walter V, Brenner H, Hoffmeister M (2016) Meat subtypes and their association with colorectal cancer: systematic review and meta-analysis. Int J Cancer 138(2):293–302

Chaber AL, Cunningham A (2016) Public health risks from illegally imported African bushmeat and smoked fish: public health risks from African bushmeat and smoked fish. EcoHealth 13(1):135–138. https://doi.org/10.1007/s10393-015-1065-9

Chan DS, Lau R, Aune D, Vieira R, Greenwood DC, Kampman E, Norat T (2011) Red and processed meat and colorectal cancer incidence: meta-analysis of prospective studies. PLoS One 6(6):e20456

Codex Alimentarius Commission (2005) Guide to the hygienic practice for meat CAC/RCP 58-2005. Retrieved from http://www.fao.org/fao-who-codexalimentarius/standards/list-standards/en/?no_cache=1

Cross HR, Stanfield MS, Koch EJ (1976) Beef palatability as affected by cooking rate and final internal temperature. J Anim Sci 43(1):114–121

Drouillard JS (2018) Current situation and future trends for beef production in the United States of America –a review. Asian Aust J Anim Sci 31(7):1007–1016. https://www.ncbi.nlm.nih.gov/pmc/articles/PMC6039332/

Epley RJ, Addis PB, Warthesen JJ (1992) Nitrite in meat. Retrieved from http://conservancy.umn.edu/bitstream/handle/11299/50792/00974.pdf?sequence=1

Eskandari MH, Hosseinpour S, Mesbahi G, Shekarforoush S (2013) New composite nitrite-free and low-nitrite meat-curing systems using natural colorants. Food Sci Nutr 1(5):392–401. https://www.ncbi.nlm.nih.gov/pmc/articles/PMC3967773/

European Commission (2016) Animal products-meat products. Retrieved from http://ec.europa.eu/food/animal/animalproducts/meatproducts/products_en.htm

Falahi E, Ebrahimzadeh F, Anbari K (2012) Determination of the causes of tendency toward red meat and meat products in the west of Iran. J Res Med Sci 17(4):373–377

Feiner G (2006) Meat products handbook: practical. Sci Technol. Retrieved from https://books.google.co.uk/books?id=LHTAAgAAQBAJ

Fretts AM, Follis JL, Nettleton JA, Lemaitre RN, Ngwa JS, Wojczynski MK et al (2015) Consumption of meat is associated with higher fasting glucose and insulin concentrations regardless of glucose and insulin genetic risk scores: a meta-analysis of 50,345 Caucasians. Am J Clin Nutr 102(5):1266–1278

Ge J, Han TJ, Liu J, Li JS, Zhang XH, Wang Y et al (2015) Meat intake and risk of inflammatory bowel disease: a meta-analysis. Turk J Gastroenterol 26(6):492–497

German Food Guide (2016) Wurst (Sausages & Cold Cuts). Retrieved from http://www.germanfoodguide.com/wurst.cfm

Govinfo.gov (2021). 9CFR319: title 9: animals and animal products part 319—definitions and standards of identity or composition. From https://www.ecfr.gov/cgi-bin/text-idx?SID=d5de369b547e044e0d77673aa9c22770&mc=true&node=pt9.2.319&rgn=div5#se9.2.319_1140

Griffin DB, Savell JW (2018) Understanding USDA beef quality grades. Texas A&M University AgriLife Extension. Retrieved from https://meat.tamu.edu/files/2018/10/Understanding-USDA-beef-quality-grades.pdf

Guenther PM, Jensen HH, Batres-Marquez SP, Chen CF (2005) Sociodemographic, knowledge, and attitudinal factors related to meat consumption in the US. J Am Diet Assoc 105(8):1266–1274

Guo J, Wei W, Zhan L (2015) Red and processed meat intake and risk of breast cancer: a meta-analysis of prospective studies. Breast Cancer Res Treat 151(1):191–198

Haluk Anil M (2012) Religious slaughter: a current controversial animal welfare issue. Anim Front 2(3):64–67. https://doi.org/10.2527/af.2012-0051. Retrieved from https://www.animalsciencepublications.org/publications/af/abstracts/2/3/64?search-result=1

Heinz G (2008) Abattoir development: options and designs for hygienic basic and medium-sized abattoirs. Retrieved from http://www.fao.org/docrep/010/ai410e/AI410E00.htm#Contents

Henchion M, McCarthy M, Resconi VC, Troy D (2014) Meat consumption: trends and quality matters. Meat Sci 98(3):561–568

Hord NG, Tang Y, Bryan NS (2009) Food sources of nitrates and nitrites: the physiologic context for potential health benefits. Am J Clin Nutr 90(1):1–10

Joosen AM, Kuhnle GG, Aspinall SM, Barrow TM, Lecommandeur E, Azqueta A et al (2009) Effect of processed and red meat on endogenous nitrosation and DNA damage. Carcinogenesis 30(8):1402–1407

Krebs NF, Mazariegos M, Tshefu A, Bose C, Sami N, Chomba E et al (2011) Meat consumption is associated with less stunting among toddlers in four diverse low-income settings. Food Nutr Bull 32(3):185–191

Larsson SC, Orsini N (2014) Red meat and processed meat consumption and all-cause mortality: a meta-analysis. Am J Epidemiol 179(3):282–289

Legal Information Institute (Cornel University Law School) (2015) 40 CFR Part 432 – Meat and poultry products point source category. Retrieved from https://www.law.cornell.edu/cfr/text/40/432.1

Lippi G, Mattiuzzi C, Sanchis-Gomar F (2015) Red meat consumption and ischemic heart disease. A systematic literature review. Meat Sci 108:32–36

Lippi G, Mattiuzzi C, Cervellin G (2016) Meat consumption and cancer risk: a critical review of published meta-analyses. Crit Rev Oncol Hematol 97:1–14

Mathijs E (2015) Exploring future patterns of meat consumption. Meat Sci 109:112–116

Modzelewska-Kapitula M, Dabrowska E, Jankowska B, Kwiatkowska A, Cierach M (2012) The effect of muscle, cooking method and final internal temperature on

quality parameters of beef roast. Meat Sci 91(2):195–202. https://doi.org/10.1016/j.meatsci.2012.01.021
Moreira L, Rosa GJM, Schaefer DM (2021) Beef production from cull dairy cows: a review from culling to consumption. J Anim Sci 99(7):skab192. https://doi.org/10.1093/jas/skab1192
North American Meat Institute (NAMI) (2016) The meat and poultry industry: basic statistics. Retrieved from http://www.themarketworks.org/stats
Ockerman HW, Basu L, Crespo FL, Sánchez FJC (2002) Comparison of European and American systems of production and consumption of dry-cured hams. National Pork Board. American Meat Science Association, Des Moines, pp 1–12
OECD(2016)Meatconsumption(indicator).Retrievedfrom https://data.oecd.org/agroutput/meat-consumption.htm
Ordaz-Németh I, Arandjelovic M, Boesch L, Gatiso T, Grimes T, Kuehl HS et al (2017) The socio-economic drivers of bushmeat consumption during the West African Ebola crisis. PLoS Negl Trop Dis 11(3):e0005450. https://doi.org/10.1371/journal.pntd.0005450
Pennisi L, Verrocchi E, Paludi D, Vergara A (2020) Effects of vegetable powders as nitrite alternative in Italian dry fermented sausage. Ital J Food Saf 9(2):8422. https://www.pagepressjournals.org/index.php/ijfs/article/download/8422/8907
Rentfrow G, Chaplin R, Suman S (2012) Technology of dry-cured ham production: science enhancing art. Anim Front 2(4):26–31
Rhodehamel E, Pierson MD, Leifer AM (1990) Hypophosphite: a review. J Food Prot 53(6):513–518. https://doi.org/10.4315/0362-4028X-4353.4316.4513
Rouhani MH, Salehi-Abargouei A, Surkan PJ, Azadbakht L (2014) Is there a relationship between red or processed meat intake and obesity? A systematic review and meta-analysis of observational studies. Obes Rev 15(9):740–748
Sans P, Combris P (2015) World meat consumption patterns: an overview of the last fifty years (1961–2011). Meat Sci 109:106–111
Santarelli RL, Pierre F, Corpet DE (2008) Processed meat and colorectal cancer: a review of epidemiologic and experimental evidence. Nutrition and Cancer-an International Journal 60(2):131–144. https://doi.org/10.1080/01635580701684872
Savic IV (1985) FAO animal production and health paper 52. Retrieved from, Small-scale sausage production http://www.fao.org/3/x6556e/X6556E00.htm#TOC
Schmidt JG, Kline EA, Parrish FC Jr (1970) Effect of carcass maturity and internal temperature on bovine longissimus attributes. J Anim Sci 31(5):861–865
Simmons SL, Carr TR, McKeith FK (1985) Effects of internal temperature and thickness on palatability of pork loin chops. J Food Sci 50(2):313–315
Sindelar JJ (2012) What about meat products without nitrates or nitrites? From http://fyi.uwex.edu/meats/files/2012/02/Nitrate-and-nitrite-in-cured-meat_10-18-2012.pdf
Sindelar JJ, Milkowski AL (2011) Sodium nitrite in processed meat and poultry meats: a review of curing and examining the risk/benefit of its use. Am Meat Sci Assoc White Paper Ser 3:1–14. https://meatscience.org/publications-resources/white-papers/docs/default-source/publications-resources/white-papers/2011-2011-amsa-nitrite-white-paper
Sinha R, Peters U, Cross AJ, Kulldorff M, Weissfeld JL, Pinsky PF, Hayes RB et al (2005) Meat, meat cooking methods and preservation, and risk for colorectal adenoma. Cancer Res 65(17):8034–8041
U. S. Government Publishing Service (2016) 9 CFR Part 319 – Definitions and standards of identity or composition. Subpart B – Raw meat products Section 319.15 – Miscellaneous beef products. Retrieved from https://www.gpo.gov/fdsys/granule/CFR-2012-title9-vol2/CFR-2012-title9-vol2-sec319-15
U.S. Department of Health and Human Services and U.S. Department of Agriculture (2015, December) 2015–2020 Dietary guidelines for Americans, 8th edn. Retrieved from http://health.gov/dietaryguidelines/2015/guidelines/
U.S. Government Publishing Office (2012a) 9 CFR 318.17 – Requirements for the production of cooked beef, roast beef, and cooked corned beef products. Retrieved from https://www.gpo.gov/fdsys/granule/CFR-2012-title9-vol2/CFR-2012-title9-vol2-sec318-17
U.S. Government Publishing Office (2012b) 9 CFR 318.23 – Heat-processing and stabilization requirements for uncured meat patties. Retrieved from https://www.gpo.gov/fdsys/pkg/CFR-2012-title9-vol2/xml/CFR-2012-title9-vol2-sec318-23.xml
United States Census (2014) 2011 county business patterns (NAICS). Retrieved from http://censtats.census.gov/cgi-bin/cbpnaic/cbpsect.pl or http://censtats.census.gov/cgi-bin/cbpnaic/cbpdetl.pl
USDA Food Inspection Services (2013) Sausages and food safety. Retrieved from https://www.fsis.usda.gov/food-safety/safe-food-handling-and-preparation/meat/sausages-and-food-safety
USDA Food Safety and Inspection Service (2014) Federal Meat Inspection Act. Retrieved from http://www.fsis.usda.gov/wps/portal/fsis/topics/rulemaking/federal-meat-inspection-act
USDA Food Safety and Inspection Service (2016, January 15, 2015) Safe minimum internal temperature chart. Retrieved from http://www.fsis.usda.gov/wps/portal/fsis/topics/food-safety-education/get-answers/food-safety-fact-sheets/safe-food-handling/safe-minimum-internal-temperature-chart/ct_index
USDA Foreign Agricultural Services (2016) Livestock and poultry: world markets and trade
USDS Food Safety and Inspection Service (2016) Inspection & grading of meat and poultry: what are the differences? Retrieved from http://www.fsis.usda.gov/wps/portal/fsis/topics/food-safety-education/get-answers/food-safety-fact-sheets/

production-and-inspection/inspection-and-grading-of-meat-and-poultry-what-are-the-differences_/inspection-and-grading-differences

World Health Organization (2015) Q&A on the carcinogenicity of the consumption of red meat and processed meat. Retrieved from http://www.who.int/features/qa/cancer-red-meat/en/

Yong HI, Kim TK, Jang HW CHD, Jung S, Choi YS (2021) Clean label meat technology: pre-converted nitrite as a natural curing. Food Sci Anim Resour 41(2):173–184. https://www.ncbi.nlm.nih.gov/pmc/articles/PMC8115001/pdf/kosfa-8115041-8115002-8115173.pdf

Dairy Products 18

1 Introduction

1.1 Definitions

The Code of Federal Regulations (CFR 21, part 131) describes milk as "*the lacteal secretions practically free of colostrum, obtained by complete milking one or more healthy cows* (U. S. Food and Drug Administration 2015). Milk may be obtained from hoofed other mammals in addition to cattle. Commercial milk animals include the bovine family [Bovidae] *water buffalo, sheep, goats, yaks*, the camel family [Camelidae], *llamas, alpacas, camels*; the deer family [Cervidae] wild *deer, reindeer, moose; and the horse family* [Equidae] – *horses, donkeys*. All commercial milk must be produced with the same hygienic standards expected for cow's milk (U.S. Department of Health and Human Services/Public Health Service/Food and Drug Administration 2014).

The average herd size for US dairy farms increased from ~61 cows to 144 cows over the period 1992 to 2012 (see Table 18.1). Meanwhile, the number of dairy farms with >999 cows increased from 564 to 1807 farms whilst those farms with <100 cows declined in number from 134,931 to less than 50,000. Profitability increased with larger herd sizes (USDA Economics Statistics and Market Information System 2010; MacDonald and Newton 2016).

1.2 Milk Standards

Milk and milk products have traditionally been priced on the valued component, the fat. Standards have been established for this component and other components at both the federal and state levels. Standards for milk and milk products are described in CFR 21, part 131 (U. S. Food and Drug Administration 2015). Standards for raw milk intended for pasteurization should have 3.25% milk fat and 8.25% total solids- not fat. CFR standards also cover vitamin addition, optional ingredients, approved methods for analysis, allowed nomenclature and product labeling (Milk Facts 2016b) (Table 18.2).

Sanitary standards for "Grade A" milk are described in the "The Grade A Pasteurized Milk Ordinance" (U.S. Department of Health and Human Services/Public Health Service/Food and Drug Administration 2014). The Grade A pasteurized milk ordinance is discussed again in Sect. 3.2.

The composition of cow's milk varies depending on diet and species but whole milk is adjusted to contain approximately, 3.3% fat, 3.3% protein, 4.8% carbohydrate (mainly lactose), 0.7% ash (minerals), and almost 88% water (Table 18.3). Milk also contains vitamins and other nutrients in small amounts, making it a most complete of foods since young mammals survive on it exclusively.

R. Owusu-Apenten, E. R. Vieira, *Elementary Food Science*, Food Science Text Series,
https://doi.org/10.1007/978-3-030-65433-7_18

The fat content of milk from Ayrshire and Brown Swiss, and especially from Guernsey and Jersey breeds, is slightly higher than that from Holstein cows, but the latter breed generally produces much more milk than the others do. Most milk is produced on farms that primarily raise dairy cattle.

2 Dairy Industry Oveview

2.1 Classification of the Dairy Industry

The dairy industry (NAICS 3115) comprises businesses that manufacture non-frozen milk products (NAICS 31151) or frozen dairy products (NAICS 31152). The former sector covers fluid milk, butter, cheeses and dehydrated products. The frozen dairy products are mainly ice cream and frozen yogurts (Table 18.4).

Table 18.1 Definition of a dairy farm

A dairy farm is any place or premises where one (1) or more lactating animals (cows, goats, sheep, water buffalo, or other hooved mammal) are kept for milking purposes, and from which a part or all of the milk or milk product(s) is provided, sold or offered for sale to a milk plant, receiving station or transfer station

Adapted from U.S. Department of Health and Human Services/Public Health Service/Food and Drug Administration (2014)

Table 18.2 Standards of identify for dairy products – Code of Federal Regulations

21CFR131	– Milk and cream
21CFR133	– Cheeses and related cheese products
21CFT135	– Frozen desserts

Reference: Adapted from Milk Facts (2016b)

2.2 United States Milk Production Statistics

The Holstein breed outnumbers all others used in the United States for the production of milk (Campbell and Marshall 2016). Jersey and Guernsey breeds tolerate hot weather better than Holsteins. Some Ayrshire, and Brown Swiss or Shorthorn breeds are used in colder areas. Current data (2014) shows the US accounts for 19% of the world total production for cow milk with 50% of American milk coming from just six States: California (20.5%), Wisconsin (13.7%), New York (6.69%), Idaho (6.67%) and Pennsylvania (5.25%). (Federal Milk Market Administrator 2015). The US produced 2.5 times more milk than any European Country, 4.5-fold more milk than New Zealand, and ~9-fold more milk compared to Australia (USDA Economic Research Services 2016; USDA Foreign Agricultural Services 2014) About 200 whey manufacturing establishments in the USA account for 25% of the global lactose production. Milk exports are predominantly as dried powder (40–50% US production) compared with 15% of fluid milk, 6% cheese, 3–8% butter, and 67–78% whey and lactose. (Progressive Dairy Man 2016; U.S. Dairy Export Council 2004).

2.3 Global Economic Significance of Milk

The global total amount of cow milk produced in 2016 was nearly 500 million metric tons (Table 18.5). The data confirms the US as the leading cow milk producing country. Considering all milk production from all livestock, shows the leading producer countries to be, India (18%),

Table 18.3 The composition of milk from a variety of sources

Source	Water	Protein	Casein	Whey P	Fat	Lactose	Ash
Buffalo	83.9	4.0	3.5	0.5	6.3	4.8	0.8
Cow	*86.9*	*3.5*	*2.8*	*0.7*	*3.5*	*4.8*	*0.7*
Goat	86.5	3.6	2.7	0.9	4.0	4.7	0.8
Sheep	82.0	4.6	3.9	0.7	7.2	4.8	0.85
Camel	86.5	3.6	2.7	0.9	4.0	5.0	0.5
Horse	88.8	2.5	1.3	1.2	1.9	6.2	0.5
Human	87.5	1.0	0.5	0.5	4.5	7.1	0.2

Table 18.4 Dairy product manufacturing industries (NAICS 3115)

	Dairy products manufacturing	Employees 2012	Number of businesses
31,151	Dairy product (except frozen)	115,728	1186
311,511	Fluid milk	54,833	420
311,512	Creamery butter	1849	38
311,513	Cheese	44,679	540
311,514	Dry, condensed, evaporated dairy product	14,367	188
31,152	Ice cream and frozen dessert	–	–
311,520	Ice cream and frozen dessert	20,933	401
Total		136,661	1587

Table 18.5 Global production and consumers of Cow's milk (1000 metric Ton)[a]

Area/region	Produced	%World	Cons	%Diff
EU	151,600	30.3	34,000	78
USA	96,343	19.3	26,521	72
India	68,000	13.6	62,750	8
China	38,000	7.6	15,570	59
Russia	30,085	6.0	9185	69
Brazil	27,100	5.4	10,095	63
New Zealand	21,150	4.2	497	98
Mexico	11,934	2.4	4183	65
Ukraine	10,100	2.0	5124	49
Argentina	10000.0	2.0	1800	82
Australia	9700	1.9	2700	72
Canada	8685	1.7	2945	66
Japan	7340	1.5	3900	47
Belarus	7175	1.4	1050	85
Korea South	2193	0.4	1529	30
Taiwan	380	0.1	381	0
Others	22	0.0	62	−182
World Total	499,807	100	182,292	64

[a]Data from 2016. Percentage difference (%Diff) between production & consumption (Cons) is potentially available for export. (Adapted from USDA Economic Research Services 2016; USDA Foreign Agricultural Services 2014)

USA (12%), Brazil (5%), China (5%), Germany (4%), France (3%), New Zealand (3%), turkey (2%), and Pakistan (2%).

About 10 species are currently farmed for milk production for human consumption. About 85% of the global milk supply is cow milk; the remainder is 11% buffalo milk, 2.5% goat milk, 1% sheep milk and 0.4% camel milk (FAO 2016). Exports of milk amounts to about 10% of global production. New Zealand is an exception where 95% of milk production is destined for export as ingredients and powders. Other major exporters for milk include the EU (28), US and Australia. India and some of the other large milk producer countries are not major importers and/ or exporters. China, the Russia Federation, Mexico, Indonesia and Algeria are some of the major importers (Table 18.6).

The per capita consumption of milk (2014) ranged from 100 liters per person per year (Finland, Ireland, Estonia, Australia) to 60–80 liters per person per year for some middle-consuming nations such as United States. Comparatively low per capita consumption of milk was observed for countries such as Iran, South Africa, Chile, Ukraine, Bulgaria, Egypt and China (Fig. 18.1).

2.4 Dairy Products

The US fluid milk sector (NAICS 311511), which comprises ~30% of the dairy industry by employment number, is divided into three sub-sectors. (i) bottled, vitaminized or flavored milk products. (ii) Manufactures of cream, sour cream and whipped toppings. (iii) Manufacturers of non-frozen yogurts (Table 18.4).

Manufacturers of creamery butter (NAICS 311512) are essentially dealing with various types of butter. Likewise, cheese manufacturing (NAICS 311513) includes real cheese, cheese spread, and a range of substitute cheese products. Whey by-product from cheese manufacturing is important economically. Globally, the main forms of dairy products from milk were fresh fluid milk (~45%), Cheese (25.4%), butter and ghee (23.2%), whole milk powder (3.7%) and skimmed dried milk (4.5%).

The total dairy products manufacture in the U.S (2014) exceeded 24 million metric tons derived from over 2500 manufacturing plants (Table 18.7). The largest percentage of products was ice cream and ice cream mixes (~50% total), followed by cheese and cottage cheese

Table 18.6 Global milk exports and imports (1000 metric tons)

Region/country	Export	%	Region/country	Import	%
New Zealand	18,375	27.1	China	13,345	20
EU	16,235	23.9	Russia Fed.	5158	8
USA	10,727	15.8	Mexico	2962	5
Australia	3277	4.8	Indonesia	2630	4
World	67,862		Algeria	2480	4
			World	65,133	

Reference: 2014 forecasts. (Adapted from Ragonnaud 2014)

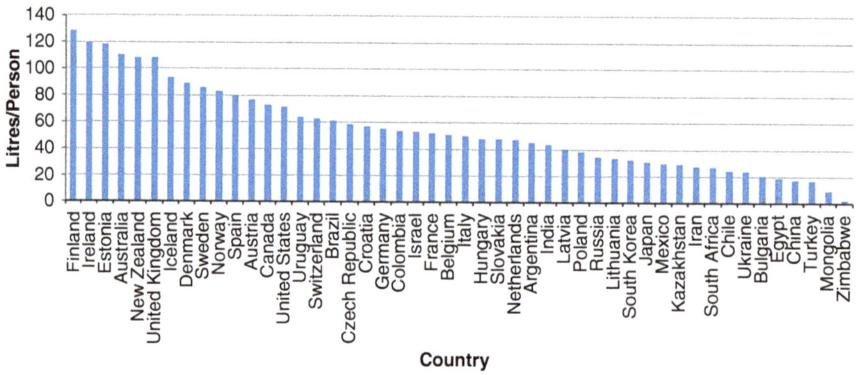

Fig. 18.1 Per capital consumption of milk in 52 countries. (Canadian Dairy Information Centre 2016b)

Table 18.7 Dairy products manufacture for the United States (2014)

Dairy product	No. plants[a]	Tonnes/1000	% total	Comment
Ice creams	581	9380.8	39.3	Regular, low fat, soft & hard
Cheese	536	5195.2	21.8	Excl. cottage cheese
Dry mix	484	2613.1	10.9	Regular, low fat, no fat ice cream
Yogurt	147	2158.1	9.0	Plain & flavored
Butter	81	842.3	3.5	
Condensed milk	62	828.8	3.5	Skim, whole, (un) sweetened,
NFDM	47	800.6	3.4	
Sour cream	107	591.9	2.5	
Cheese, cottage	56	477.8	2.0	Curd, creamed and low fat
Dry whey	29	388.4	1.6	For human use only
Frozen yogurt	139	251.4	1.1	Regular, low fat, no fat categories
Sherbet	161	171.9	0.7	
Dry mix (sh)	118	98.8	0.4	Sherbet dry mix
Buttermilk	25	50.4	0.2	Dry only
Whole dry milk	12	32.7	0.1	
Total	2585	23882.1	100.0	

Adapted from USDA Economics Statistics and Market Information System (2015)
[a]Manufacturing plants which produce more than one product may be counted twice-see Table (18.4)

(~24% total) and then yogurt (~9%). (USDA Economics Statistics and Market Information System 2015).

3 Fluid Milk

3.1 Milking

Milking on a commercial scale is mostly carried out using machines. (Food and Agriculture Organization of the United Nations 1998). At first, a small amount of milk is drawn and examined for impurities, after which the cups of the milking machine are applied to the clean, sanitary teats of the cow. After milking, the cups are immersed in a nonirritating bactericidal agent before they are applied to the teats of another cow. Milk from the cow passes through the milking machine from which it flows through a glass or stainless steel pipe to a bulk cooling refrigerated at 40 °F (4.4 °C). (American Cheese Society 2015).

3.2 Transportation of Fluid Milk

Milk is transported from the farm to the receiving station or fluid milk processing plant in a milk tank truck or pickup tanker. Pickup tankers are insulated, stainless steel tanks, usually having a holding capacity of more than 5000 gallons (18,925 liters) on a trailer handled by a motorized vehicle. There is no significant rise in the temperature of the milk during the period required for transportation and delivery to the processing plant. Grade "A" raw milk is expected not to exceed 10 °C and be cooled for <7 °C for on-farm storage (U.S. Department of Health and Human Services/Public Health Service/Food and Drug Administration 2014).

Upon receiving the milk, the product undergoes a series of tests for odor and flavor, overall volume, butter fat content and for the presence of antibiotics (Sect. 3.3). The milk is then pumped from the bulk tank into the tanker truck through a sanitized plastic hose after which the hose is capped. The final step is to prepare a weight ticket for the farmer and tabulate the weight, temperature, and other data of the product on a record sheet.

In accordance with the milk ordinance, pickup tankers, including auxiliary equipment such as hoses, must be cleaned and sanitized, as are dairy farm milking and milk-holding equipment, after delivery of the product to the processing plant (U.S. Department of Health and Human Services/ Public Health Service/Food and Drug Administration 2014).

3.3 Milk Quality Testing, Quality Assurance

3.3.1 Testing Farm Milk

When receiving the milk, the operator of the tanker tests the product, which has been stored in a bulk tank, for odor and flavor and, if not suitable, the milk is rejected. If acceptable, the volume of the product in the bulk tank is measured with a rod. It is then agitated, after which a sample is taken in a glass or plastic bottle from which the butterfat content will be determined, because farmers are paid on the basis of butterfat content.

A commingled sample is also taken to be tested for the presence of antibiotics using the *Bacillus stearothermophilus* disc assay or equivalent as required by the Pasteurized Milk Ordinance of the U.S. Quick tests approved by the FDA such as the Charm II Beta-lactam Tests (Charm Sciences Inc. of Malden, Massachusetts) also enable farmers, processors, and quality laboratories to examine milk for antibiotics quickly and accurately. Another tests-kit (SNAP Beta-lactam Test Kit, IDEXX Laboratories, Maine USA) was approved in 2003 for detection of penicillin G, amoxicillin, ampicillin, ceftiofur and cephapirin residue in co-mingled milk (U. S. Food and Drug Administration 2003). If there is no problem with the collected milk, then samples from the individual farmers need not be tested. If there is a problem, however, each individual farm sample must be tested, and the farmer responsible for the contaminated milk may be liable for ensuing losses. Because of such severe financial consequences, farmers are very careful not to

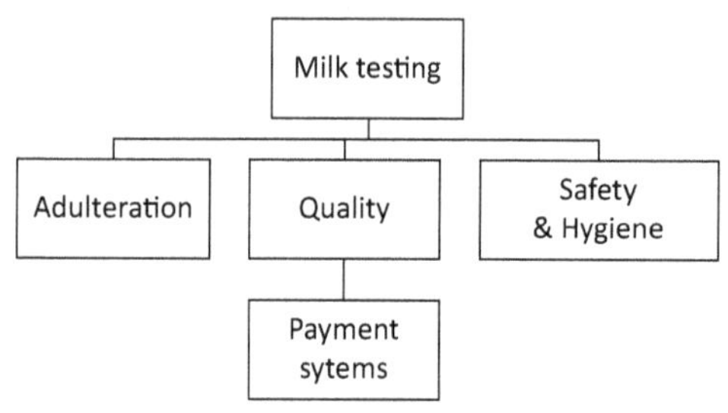

Fig. 18.2 Rational for milk quality testing

attempt to sell milk from cows that have been recently treated with antibiotics (Vieira 2013).

3.3.2 Laboratory Milk Tests

Milk testing is performed for a number of common reasons to do with assuring quality and determining payments to milk producers. Other quality issues include the potential to increase yield, detection of milk adulteration, as well as the evaluation of standard attributes. The following is a brief account of milk quality testing based on a FAO handbook intended for small milk producers (Draaiyer et al. 2009) (Fig. 18.2).

Sampling milk from the producer is one of the first steps before testing. Milk may be sampled periodically, randomly or samples may be collected and combined to provide a comingled batch for testing. Accurate and representative sampling is important. First, a specialized instrument is used for milk agitation, and a dipper is applied for sample removal and transfer to presterilized containers. In some cases, preservatives (potassium dichromate, bronopol, formaldehyde or hydrogen peroxide) may be added to the milk before transportation to the laboratory.

A range of techniques are frequently applied in the milk testing laboratories and briefly outlined below (Table 18.8).

Table 18.8 Milk quality testing methods

Tests	Comment
Quantity	Volume or weight of milk produced
Organoleptic testing	Appearance, Taste, odor, flavor
Composition	Water, protein, fat, lactose, minerals etc.
Physical / chemical	Density, Water addition, pH, Alcohol test
Hygiene	Total bacterial count
Hygiene	Somatic cell count for mastitis
Adulteration	Water, skim milk, fat removal
Drug residues	Antibiotics

Reference: Summarized from Draaiyer et al. (2009) and FAO (2016)

1. The quantity of milk can be measured by weighing or by volume measurements. For larger quantities, it is easier to weigh than to measure volume.
2. Organoleptic testing. The purpose of organoleptic testing is to determine appearance, taste, and odor, and flavor characteristics of milk; the appearance of milk should be whitish yellow. The taste should be light and slightly sweet. A sour taste is indicative of acidic milk, which suggests some degree of fermentation implying unacceptable contamination or age. The flavor of milk should not be rancid or bitter and nor should there by other taints (feed, malty, salty etc.).
3. The chemical or proximate composition of milk is clearly an important aspect of quality. Determination of fat content is one of the most important tests, frequently using the Gerber test or Babcock test. In both cases, milk is mixed with sulfuric acid and the volume of fat remaining undissolved is recovered by centrifugation and measured. Protein content of milk can be readily determined using formaldehyde titration.
4. The total solids (TS) content of milk is normally measured from density measurements using a hygrometer also called a lactometer in dairy chemistry. The hydrometer is very useful and versatile and able to be applied for lactose, TS and also solids not fat (SNF) content;

$$\mathrm{TS}(\%) = 0.25(\mathrm{L}) + 1.22\,\mathrm{fat}\% + 0.72$$

$$\mathrm{SNF} = \mathrm{TS}(\%) - \mathrm{Fat}\%$$

In the preceding equations, L is last two integers from the lactometer (hygrometer) reading, and fat% is fat content determined from chemical analysis.

Additional physical and chemical characteristics may be determined for milk such as, water, pH, and coagulation degree with heating or alcohol. The water content can be used to detect illegal addition by producers. Adulteration by water addition changes in the specific gravity of milk, which can be readily detected by the ubiquitous hygrometer. The readings are used in conjunction with fat% values to estimate water addition. Incidentally a range of adulterations go beyond water, and include re-addition of skimmed milk, addition of whey proteins or non-whey proteins, and or additions of milk from species other than cow.

Measurement of pH or titrable acidity is to check for milk fermentation. A couple of indirect measurements of acidity are also available which depend on a measurement of milk protein coagulation degree after the addition of alcohol or after a short degree of boiling. The following texts provide additional materials on milk testing (Al-Saqer et al. 1999; Rombaut et al. 2002; Tamime 2009).

4 Milk Hygine and Safety

4.1 The Grade "A" Milk Ordinance

The US standard for milk hygiene or "the Grade A Pasteurized Milk Ordinance" (PMO) provides detailed recommendations for milk production, handling, processing and retail (U.S. Department of Health and Human Services/Public Health Service/Food and Drug Administration 2014). Different States may adopt all or some parts of the PMO but the ordinance applies for all milk traded across State lines (Meyer 2015).[1] The Codex standards for milk and milk products considers "milk safety" as the responsibility of all stakeholders within the Agri-food system: producers, processors, distributers and retailers (Codex Alimentarius Commission 2004). Hygiene Issues surrounding the premises and fixtures (premises, utensils, personnel training etc.) for milk production are covered by the general requirements for Good Manufacturing Practices (GMP) (Table 18.9).

Table 18.9 Primary production factors and milk hygiene

Environmental hygiene Clean water, no cross contamination of milk
Hygienic production of milk Cf. GMP Prescriptions Regarding Premises, Good Animal Health, General Hygienic Practice Feeding, veterinary drugs, Hygienic milking
Handling, storage and transport of milk Milking equipment Storage equipment Premises for, and storage of, milk and milking-related equipment Collection, transport and delivery procedures and equipment

Adapted from Food and Agriculture Organization of the United Nations (1998)

The milk hygiene principles for smallholders (less than 10 milking cow per farmer or herd) involve broadly similar requirements as the GMP for commercial-scale units. However, it may be that small-scale milk producers experience other challenges, owing to the lower degree of mechanization, greater likelihood for hand milking, and use of small-scale milk collection units. Milk hygiene is attracting attention in the developing world, where the smallholder dairy farms are more common (Kurwijila 2006; Shija 2013).

4.2 Microorganisms in Milk

Though raw milk from healthy livestock is sterile, bulk tank milk is not free from microorganisms (Angulo et al. 2009; Vissers & Driehuis 2009; Verraes et al. 2014). Most of the microorganisms associated with raw milk arise by cross contamination from the farm environment, i.e. from animal feed, feces, bedding, or soil. Some bacteria derive also from the udder and teats particularly where these become infected, and swollen in the condition called mastitis (Table 18.10).

The raw milk microorganisms belong to one of the following classes, (i) normal, harmless or potentially beneficial, (ii) pathogens found also in other foods items (Chap. 8, Sect. 2), or (iii) zoonosis microorganisms that transmit animal

[1]At the time of writing the 2013 edition of the Grade A PMO (430 pages) is available online.

Table 18.10 Major bacteria from raw milk

Gram negative, psychrotroph
Pseudomonas species
Enterobacteria
Other
Gram positive spore forming
Bacillus species
Bacillus licheniformis
Bacillus cereus
Clostridium species
Lactic acid bacteria
Pathogenic bacteria
Campylobacter
E. coli
Listeria monocytogenes
Yersinia
Staphylococcus aureus

Adapted from Touch & Deeth (2009), Vissers & Driehuis (2009)

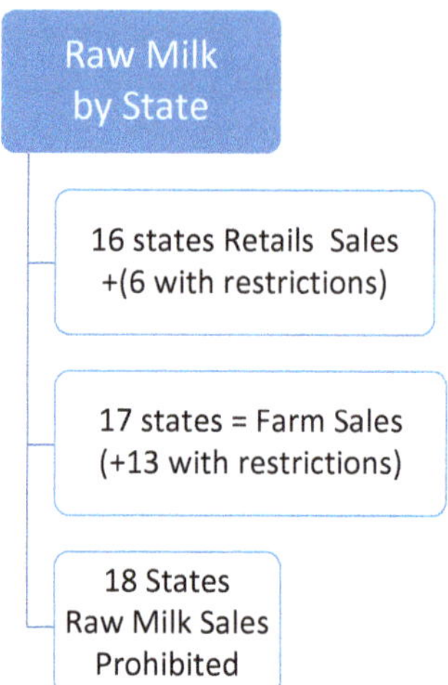

Fig. 18.3 Legislation surrounding the sales for raw milk. (Adapted from Marler Clark Associates (Clark 2016))

diseases to humans via raw milk. In addition to those listed in Table (18.10), the milk-associated pathogens can include, Salmonella spp, streptococcus species, *Clostridium botulinum*, *Mycobacterium bovis*, and *Toxoplasma gondii* (Vissers & Driehuis 2009, Quigley et al., 2013; Verraes et al., 2014; Verraes et al., 2015).

In the United States, certified milk is produced from herds that have been inspected, tested, and found to be free from disease. There are bacterial standards for certified milk that limit the number and types of bacteria that are allowed (The American Association Of Medical Milk Commissions Inc. 1999).

4.3 Raw Milk Controversy

It is claimed that raw milk contain higher levels of nutrients, which could be lost by heating (Anon 2015; Food Safety News 2016). Other reasons suggested for raw milk consumption include, a superior taste and flavor, and curative properties (Lejeune and Rajala-Schultz 2009, Lucey 2015, 6442; Oliver et al. 2009).

Currently, raw milk can be purchased within 29 States. Some 18 States prohibit the sale of raw milk for human consumption (Fig. 18.3). Cow-share schemes or direct sale methods are some of the ways for sourcing raw milk (Oliver et al. 2009). One popular cow-share scheme involve consumers buying shares in dairy farms, and then receiving "dividends" as co-owners, in the form of raw milk. Cow-share schemes and the range of intra-state legislations surrounding raw milk have been extensively discussed (Lejeune and Rajala-Schultz 2009; Lucey 2015; Oliver et al. 2009); the following documents reflect well the history and legal aspects of raw milk provision in the United States (Angulo et al. 2009; Meyer 2015).

The risks of food poisoning following the consumption of non-pasteurized milk was estimated to be 150-fold greater than the risks from pasteurized milk. Moreover, the severity of infections and the likelihood of hospitalizations were higher for consumers of non-pasteurized milk compared with pasteurized milk. Most certified milk is pasteurized. Between1993–2006 some 60% of food-borne disease outbreaks associated with dairy foods were linked with raw unpasteurized milk resulting in two deaths and more than 2000 hospitalizations (Langer et al. 2012). A further survey covering 2007–2013 (Table 18.11) found 81 outbreaks due to non-pasteurized milk, 979 cases of illness and 73 hospitalizations (Mungai et al. 2015).

Worldwide, the consumption of raw milk and milk products (cheeses and butter) seems rather common. Many people within the Mediterranean, East Asia, Sub-Sahara Africa, and the Middle East consume raw milk from goat, sheep, camel,

Table 18.11 Raw milk hazards – statistics, sources and myths

Outbreaks 2007–2012
Number of outbreaks 81
Number of cases 900+
Hospitalizations = 73, deaths = 0
Non-pasteurized milk products
Unpasteurized milk or cream
Soft cheeses (Brie and Camembert)
Mexican-style soft cheeses made from unpasteurized milk
Yogurt made from unpasteurized milk
Pudding made from unpasteurized milk
Ice cream or frozen yogurt made from unpasteurized milk
Raw milk (RM) myths
Pasteurized milk is less nutritious than RM
RM products (soft cheese, ice cream, and yogurt) are safe.
Raw milk labeled "organic." is safe
RM from non-symptomatic animals is safe
RM has curative properties

References: compiled from multiple sources. (Mungai et al. 2015)

water buffalo or donkey. Interestingly, researchers also found infected herds were common in some of the areas mentioned (Colavita et al. 2016; Verraes et al. 2014). Diseases likely to be transmitted to humans by bacteria in raw milk, include, Q fever (*Clostridium burnetii*), bovine tuberculosis (*Mycobacterium bovis*), caprine or goat tuberculosis (*Mycobacterium caprae*), brucellosis (Brucella species) and tick-borne encephalitis virus (Leedom et al. 2006; Verraes et al. 2014; Asante et al. 2019; Conte and Panebianco 2019).

Pasteurization was adopted in North America in the last century partly to deal with the high risks of bovine tuberculosis, typhoid or scarlet fever from drinking cow's milk. Initial activism for pasteurization came from doctors and pediatricians concerned with securing clean milk for feeding infants (Lederle 1907; Rickards 1909; Rotch 1907). By 1948 pasteurization was mandatory in many states. Sales of raw milk across State-lines were prohibited from 1987 (Centers for Disease Control and Prevention 2015; U. S. Food and Drug Administration 2016).

In summary, pasteurization of milk (and other liquid foods and beverages) improves safety as well as shelf life. In general, pasteurized milk is not sterile and must be refrigerated. However, the

Table 18.12 Measures for reducing bacteria in milk

Cleaning and Sanitation (teat, udder area, utensils, storage tanks)
Cold storage of milk (2 °C)
Carbon dioxide addition
Sterilization H202 /SCN (lacto-peroxidase
Lactoferrin addition
Lactic acid bacteria
Heat treatment, (Pasteurization, HSTS)

Reference: Adapted from Touch and Deeth (2009)

temperature /time controls are designed to reduce the number of microbial pathogens present by thousands to million-fold. Starting with good quality milks from a healthy herd (see below), and then pasteurizing this product hygienically helped to improve infant mortality and general public health (Ranking et al. 2017; Currier and Widness 2018). It is noteworthy that human donor milk is now being pasteurized to protect infants from transmissible diseases (Goya and Calvo 2018; Meeks et al. 2019).

4.4 Controlling Bacterial Contamination of Milk

Unless the udder is infected, it may not be an important source of contamination. Other sources of microbial contamination of milk are the body of the cow; milking machines and other equipment and utensils (Harding 1995; Vissers & Driehuis 2009). Some measures to limit the number of bacteria present in raw milk are summarized in Table 18.12 (Touch and Deeth 2009).

Large dairy farms often have a special wash pen for cows to be milked. Utensils, including the milking machine, should be cleaned and disinfected either with live steam or with a solution of chlorine (about 200 ppm of available chlorine). Bulk milk tanks may be cleaned with detergent and water then sanitized with chlorine solution. Outlet valves and the outside of the tank must be cleaned and sanitized manually. Cleaning in place (CIP) may be used to clean, sanitize, and rinse the milk pipeline, the teat cup assembly, and the bulk milk tank if a vacuum or pressure system is available. As noted above, milk producers and processors must comply with GMP requirements closely (Chaps. 7, 8 and 9).

5 Processing of Fluid Milk

Fluid milk and dairy products undergo a range of unit operations or processing e.g. heating, centrifugation, homogenization, dehydration concentration, membrane separation (ultrafiltration), fermentation and packaging. Processing transforms milk from a comparatively low cost and perishable commodity into higher value products. Elements of diary technology have been discussed in several important texts (Goff 2016; Robinson 1993a, b; Spreer 1998; Varnam and Sutherland 2012; Walstra et al. 2005). The following is a brief outline of the processing methods for fluid milk.

5.1 Whole Milk

Processing whole milk starts with clarification, which involves passing the fluid through a low speed centrifuge similar to a cream separator (Fig 18.4). The purpose of the clarifier is to separate out dirt and sediment that might be present. The milk is then pumped into a storage tank equipped with an agitator and standardized for butterfat content by adding enough cream or skim milk (milk from which cream has been removed) to provide the fat content required by State regulations. Some milk is fortified with vitamin D so that one serving of milk (8 oz. or 236 ml) will then contain 25% of the Daily Value (DV). Low fat and skim milk will usually be fortified with vitamin A also to give 10% of the DV per serving.

The next step in fluid milk processing is to pasteurize it. In the old "vat method," milk must be heated to 145 °F (62.8 °C) and held at this temperature for 30 min (Lewis 2012). Other temperature-time combinations (Table 18.13) are available for milk pasteurization in order to achieve similar effects (U.S. Department of Health and Human Services/Public Health Service/Food and Drug Administration 2014).

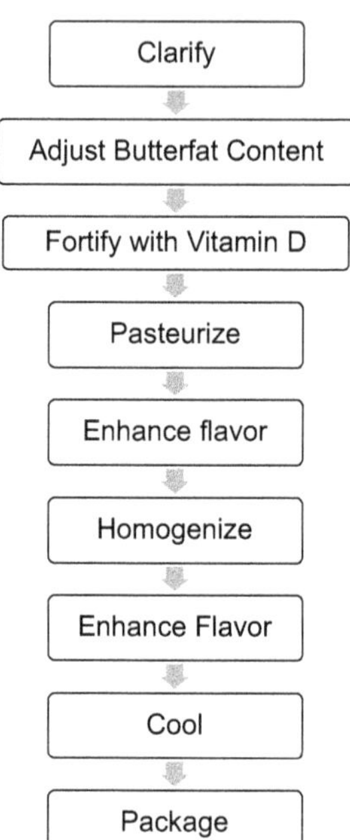

Fig. 18.4 A process chart for pasteurized whole milk

Table 18.13 Temperatures for food pasteurization

Temperature		Time
145 °F[a]	62.8 °C	30 min
161 °F[a, b]	71.7 °C	15 s
191 °F	88.3 °C	1 s
204 °F	95.6 °C	0.05 s
212 °F	100.0 °C	0.01 s
280 °F	137.8 °C	>2 s

Notes:[a]For products with >10% fat the process temperature is increased by 5 °F (~3 °C), [b]High Temperature Short time (HTST) pasteurization. (Adapted from U. S. Food and Drug Administration 2015)

Pasteurization temperatures ensure the destruction of the most heat-resistant pathogen in milk, *Coxiella burnetti*, the bacterium that causes Q fever (Cerf and Condron 2006). Milk may be pasteurized using an insulated vat, the plate heat exchanger (Fig. 18.5) or a tubular heat exchanger. Most commercial operations adopt one of several equivalent pasteurization temperatures (Table 18.13). Pasteurization is described further in Chap. 12 (Sect. 8.5).

In a process called deodorization, milk may be given a flavor treatment to provide a product that is uniform in odor and taste. Milk is instantly heated to about 195 °F (90.6 °C) with live steam

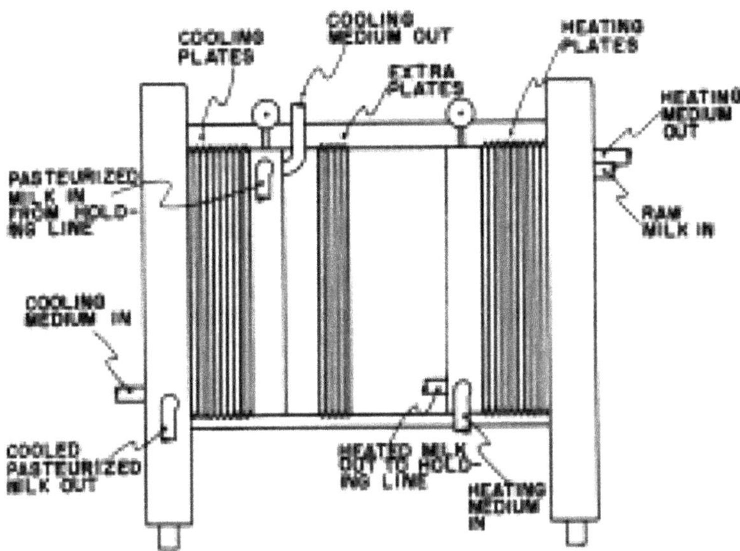

Fig. 18.5 A plate pasteurizer

(injected directly into the product) and then subjected to a vacuum of about 10 in. (25.4 cm) in one chamber and to a vacuum of about 22 in. (55.9 cm) in another chamber. The high-vacuum treatment serves to regulate flavor, to cool the milk to about 150 °F (65.6 °C), and to evaporate water that may have been added through the injection of steam. After the flavor treatment, and while the milk is still hot, it is usually homogenized by passing it through a small orifice that breaks up the fat globules to a small size, preventing the separation of cream from the milk. The milk is then cooled rapidly to about 35 °F (1.7 °C).

During HTST pasteurization, deodorization treatment and homogenization, milk is passed through heating and cooling cycles at such a rapid rate that at no time is it held for long periods at high temperatures (Kurtz et al. 1971).

Following the processing and cooling stages, milk is filled mechanically into containers made of waxed or plastic-coated cartons of volumes between 2 quarts (1.9 liters) and 1gallon (3.8 liters), and the containers are sealed. In this state, milk should be held as close to 32 °F (0 °C) as possible until consumed.

Sometimes, milk may be stored for hours or even days before processing. In these cases, large dairy plants employ a thermization step where milk is heated 145.4–149 °F (63–65 °C) for about 15 s and then chilled to 40 °F (4.4 °C) or below rapidly. The spore-forming spoilage organisms are not eliminated in this process but may revert to their vegetative state where, they are more vulnerable towards pasteurization eventually. Some countries prohibit double pasteurization so a phosphatase (enzyme) test must be done on the milk to ensure that it had not been entirely pasteurized before it goes through the second heat treatment and final pasteurization step.

Ultra-High-Temperature (UHT) milk is produced by heating at temperatures of 275–284 °F (135–140 °C) for a few seconds. The milk is then homogenized, cooled, and filled aseptically into sterile containers. The UHT plants can use two major types of heating to reach these high temperatures, indirect heating or direct steam heating (see Figs. 18.6 and 18.7). These systems are further described by the following references (Gillis et al. 1985; Hill 1988; Horner et al. 1980). The shelf life of UHT milk is between 40 and 45 days at 40 °F (4.4 °C) in situations where the shelf life using conventional HTST and packaging in nonsterile containers is often less than 20.

Alternative methods for pasteurizing milk and other liquid foods do not involve direct heating. Pulsed electrical fields (PEF), oscillating magnetic fields, high hydrostatic pressure, intense light pulses, and irradiation are being considered as methods for milk pasteurization. Whilst using short-duration, high-intensity PEF the temperature of milk does not exceed 131 °F (55 °C). Treatment of the liquid food takes place between two electrodes and lasts for less than 1 s (Fig. 18.8). Raw skim milk processed by PEF

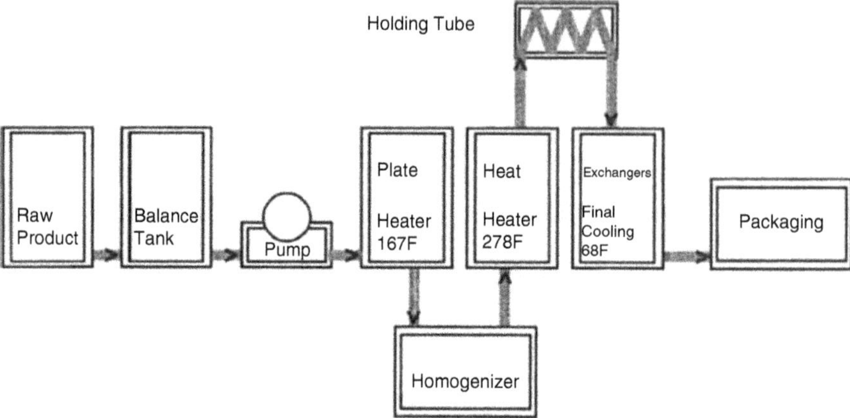

Fig. 18.6 UHT processing with indirect steam injection

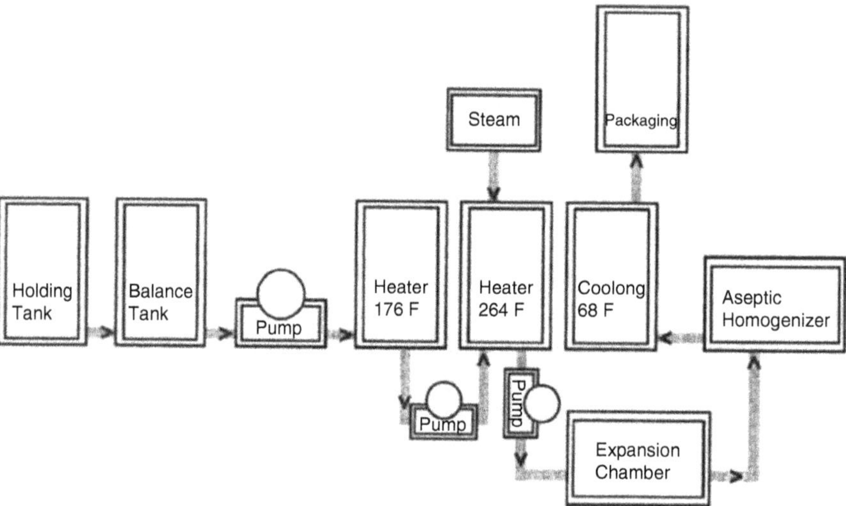

Fig. 18.7 UHT processing with direct steam heat

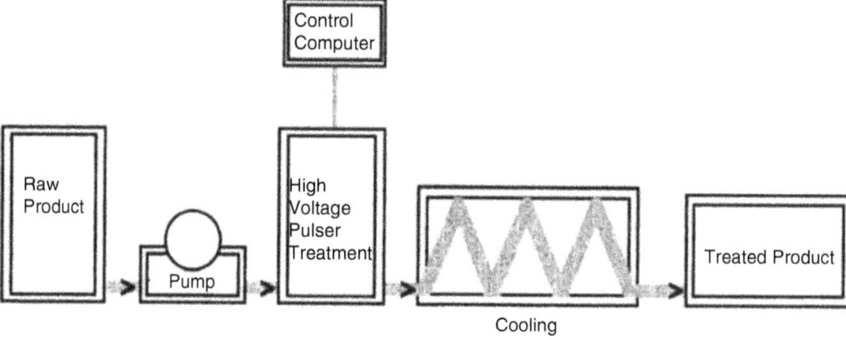

Fig. 18.8 Continuous PEF process for liquid foods

technology showed few or no changes in the physical and chemical properties and a sensory panel found no difference between heat-pasteurized milk and PEF milk (Morales-De la Pena and Martin-Belloso 2009; Mosqueda-Melgar et al. 2008; Otunola et al. 2008; Sepulveda et al. 2009). The storage life of milk processed by PEF treatment was 2 weeks at 40 °F (4.4 °C) and sometimes longer (Buckow et al. 2014).

5.2 Skim Milk

Skim milk (0.5% fat) and reduced-fat milk (0.2–.5% fat) are derived from whole milk which is passed through a centrifuge at high speeds, after the milk has been heated to 90–110 °F (32.2–43.3 °C), to remove the butterfat as cream. These products are usually fortified with vitamins A and D prior to pasteurization and cooling. In some cases, sodium caseinate (a derivative of casein, the main protein in milk) is also added.

The cream from the centrifuge may be separated as approximately 40% butterfat (heavy cream), 30% butterfat (all-purpose cream), or 20% butterfat (light cream). The creams higher in butterfat may be diluted with skim milk to provide the various fat densities or to produce a product known as half-and-half (about 10.5% butterfat). Cream tends to spoil more quickly than milk, and so it is usually given a more drastic heat treatment during pasteurization. For batch pasteurization cream is heated to 150–155 °F (65.6–68.3 °C) for 30 min prior to cooling. When the HTST method is used, cream is heated to 166–175 °F (74.4–79.4 °C) for 15 s prior to cooling. A product called table cream (light cream or half-and-half) is homogenized after pasteurization. After pasteurization, cream is cooled to 35 °F (1.7 °C) and containerized. Cream is held at 35–40 °F (1.7–4.4 °C) until consumed or subjected to additional processing.

5.3 Non-fat Dried Milk

Large quantities of skim and low-fat milk are dried by, a variety of methods, most commonly spray drying or drum drying (Chap. 14). Dried milk (usually the spray-dried type that contains about 5% moisture) may be re-humidified to slightly higher moisture content after drying. This treatment agglomerates the fine milk particles to form clumps of milk powder, which results in a product that dissolves or disperses in water much more readily than the finely powdered dried milk (Chap. 15, Sect. 3.2).

6 Other Dairy Products

6.1 Ice Cream

6.1.1 Economic Significance

The characteristics of good ice cream and allied products have been well documented (Goff and Hartel 2013). The estimated value for the global ice cream trade was US$74billion with the largest markets being, US >Italy > China, followed by Australia, Brazil, Russia and UK. The per capita consumption of ice cream was highest in New Zealand, Australia, USA and Finland. Some of the lowest consumers of ice cream occur in China, Brazil and Mexico (Clarke 2012; Goff and Hartel 2013).

6.1.2 Standards for Ice Cream

Standards for American ice cream and frozen desserts are described in 21CFR135.110 (Milk Facts 2016b; U. S. Food and Drug Administration 1978; US Government Publishing Office 1978). Ice cream is defined as, a product manufactured by freezing and aerating a pre-pasteurized mix of dairy ingredients. The major frozen desserts include, non-dairy ice cream made using milk protein and vegetable oil, gelato – Italian ice cream containing custard and eggnog, milk ice, sorbet – aerated dessert containing fruit juice and syrup, sherbet – like sorbet with some cream and or milk added (Clarke 2012) (Table 18.14).

Table 18.14 Ice cream and other frozen desserts (21 CFR135)

135.110 Ice cream and frozen custard.
135.115 Goat's milk ice cream.
135.130 Mellorine.
135.140 Sherbet.
135.160 Water ices

Reference: From (Milk Facts 2016b; U. S. Food and Drug Administration 1978; US Government Publishing Office 1978)

Table 18.15 Composition of ice cream (21 CFR135.110)

Composition	Quantity	Comments
Total solids	>1.6lbs/ gallon	>160 g/l equivalent or 16% (w/v)
Milk fat	10–16%	Cream
Milk Not fat, solids	9–12%	Casein, whey proteins and lactose
Sweeteners	12–16%	Mainly high-fructose corn syrup
Stabilizers & emulsifiers	0.2–0.5%	Polysaccharide thickening agents
Water	55–64%	From added Grade A milk
Density	>4.51lbs/ gallon	450 g/l liquid & 550 cc of Air (cf. over-ran, foaming)

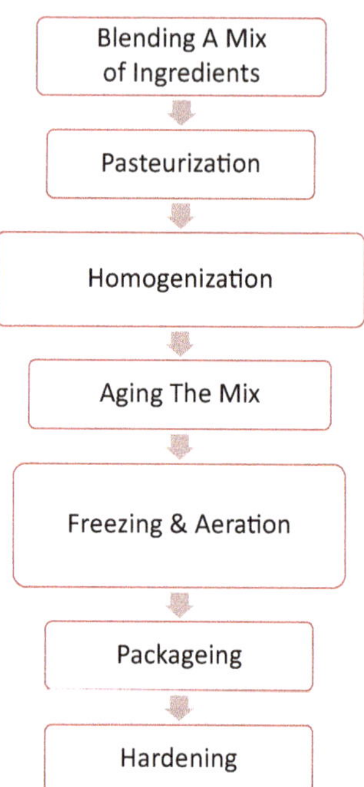

Fig. 18.9 A process diagram for ice cream manufacture

Standards for diary ice cream vary in different countries and regional blocks (Clarke 2012; Goff and Hartel 2013). American dairy ice cream (Table 18.15) shows a minimum of 10% (average 14%) milk fat, and a range of optional ingredients including cream; milk or skimmed milk; sugars (sucrose, dextrose, corn syrup or high-fructose corn syrup); stabilizers which may be gelatin, vegetable gums, or modified food starches; pasteurized eggs or egg whites; flavorings such as vanilla; fruit juices or extracts; cocoa; chocolate; candy; cookies; nuts; or just about anything the imagination and food science will allow. When nut is added this is not counted as part of the solids content, but rather as flavor.

The compositions of ice cream from New Zealand, Australia and Canada were similar to ice cream from the US, as regards milk fat content (10–16%), total solids content (16–20%) and density (450 g/l). In the UK ice cream tends to have a minimum of 5% milk fat, and 2.5% protein (Clarke 2012; Goff and Hartel 2013). French-style ice cream or custard ice creams have egg yolk added to improve texture. These must be pasteurized and use of homemade recipes that call for raw eggs is not recommended since salmonella might be present. The acceptable protein content of ice cream is not specified in many jurisdictions; indeed milk protein may be optional. Vegan ice cream does not containing animal (dairy) derived ingredients (Mullan 2016).

6.1.3 Ice Cream Manufacturing

The manufacturing of dairy ice cream involves mixing liquid dairy ingredients and liquid sweeteners separately (Fig. 18.9). These are then combined, blended together, and then the mixture is allowed to "soak" for a period of time (20 min) before it is pasteurized, homogenized, cooled with agitation to 36°F (2.2°C) and held in storage tanks.

Ice cream pre-mix is combined with flavors and then added to the ice cream freezer, which is a tube with walls refrigerated to 5–15 °F (−15 to −9.4 °C). The freezer is fitted with blades that rotate and scrape the mix from inner walls as it whips air into the mixture. The amount of air whipped into the ice cream mix is under government regulation because ice cream cannot contain less than 1.6 lb. of total solids per gallon and the total weight must be at least 4.5 lb. Air incorporated into the mix (Percentage overrun) can be calculated by the following formula:

$$Overran = 100 \times \left(\frac{Volume\ of\ ice\,cream - Volume\ of\ mix}{Volume\ of\ ice\,cream} \right)$$

6.1.4 Low-Fat Ice Cream

The nutritional profile for ice cream shows a high level of nutrients of concern: saturated fat, cholesterol and sugar (Appendix 1). However, the principles of food literacy (Chap. 6) imply that ice cream can be eaten *in moderation* as part of a balanced diet (Berkheiser 2019). Many recipes for ice cream are available that have reduced levels of fat. In addition to regular ice cream, the consumer can choose from low-fat ice cream or no-added sugar varieties. Compared to regular ice cream, low fat ice cream and no-added sugar varieties may have 2–3 fold lower Calories, total fat, cholesterol / or sugar. (Berkheiser 2019). Premium ice cream varieties have the highest total Calories, which implies that these expensive brands should be eaten with still greater moderation. The nutrient profile for some typical ice cream varieties can be found in Appendix 1.

Dairy soft-serves and ice milks use low amounts of butterfat and fat-free ingredients, skim milk, water, and thickeners such as modified food starches and other gums, e.g. locust bean, and carrageenan. These retain large amounts of water and entrap air when whipped into the product during the freezing process, thus giving the product a fatlike mouth feel with reduced calories and cholesterol. Ice cream substitutes are discussed further in Sect. 6.8.

6.1.5 Quality of Ice Cream

The quality of ice cream is discussed extensively in several dedicated texts and monographs on the subject (Arbuckle 2013; Clarke 2012; Goff & Hartel 2013; Marshall 2003; Stogo 1998). In the dairy quality control laboratory, many of the tests done on other dairy products are also done on ice cream. Tests for ice cream quality include the analysis of butterfat, protein, total solids, antibiotics, and a variety of microbial tests including standard plate counts and coliforms.

6.2 Yogurt

6.2.1 Cultured Milk Products

Yogurt belongs in a group of products made from cultured milk (e.g. cheese, buttermilk, sour cream, and yogurt). Cultured milk products are products formed from the controlled fermentations of milk. The reactions for milk fermentations transform lactose into lactic acid through the actions of lactobacilli.

Because of lactose fermentation, the acidity of milk drops to pH 4.6 where casein has no net electrical charge and therefore coagulates to form a curd. The sharp "prickly" taste of cheeses and yogurt is from lactic acid produced by bacterial fermentation. Codex defines fermented milks products as "*...products obtained by fermentation of milk, by the action of suitable microorganisms and resulting in reduction of pH with or without coagulation,,..*[where] *the starter microorganisms shall be viable, active and abundant in the product to the date of minimum durability*". (Codex Alimentarius Commission 2003).

Currently, the best-known fermented milk products in North America include products such as acidophilus milk, cultured buttermilk, sour cream, kefir, koumiss, filmjoik, viili and crème fraiche (MilkIngredients.Ca 2011a). An international listing of some fermented milk products is shown in Table (18.16) but this is a small fraction of the variety of products documented in the "Encyclopedia of fermented fresh milk products" (Kurmann et al. 1992).

The science behind fermented milk products (in particular yogurt) is well documented and the reader should refer to the following texts for more in-depth information (Tamime and Robinson 1999a, b; Chandan et al. 2008; Tamime 2008).

6.2.2 Economic Significance of Yogurt

Yogurt is growing in popularity in the United States but the largest producers and consumers are found in Europe. For 2015, the world production total for yogurt was just over 40 million metric tons representing a 34% increase compared to 2010 (Fig. 18.10). For the period 2010–2015, the value for yogurt trade increased from US$31 billion to US$71 billion per year. The yogurt producing regions include Asia, Middle East + Africa, Western Europe and Eastern Europe. The countries having the highest yogurt manufacturing volumes were, China >Iran >Turkey, Russia >, USA > Germany. Some commercial producers of yogurt are listed in Table 18.17 (Preben Mikkelsen (PM) Food and Dairy Consulting 2013).

6.2.3 Standards for Yogurt

The US Code of Federal Regulations (CFR21 CFR.131.200) describes yogurt as, "*the food produced by culturing one or more of optional dairy ingredients .... with a characterizing bacterial culture that contains the lactic acid-producing bacteria culture that contains Lactobacillus bulgaricus and Streptococcus thermophilus*". The ingredients for yogurt are listed as "*Cream, milk, partially skimmed milk, or skim milk, used alone or in combination*". The overall standards for yogurt composition are flexible and allow for a wide range of optional ingredients and additives (Table 18.18).

Sometimes there are requirements for specific numbers of viable microorganisms within yogurt. For instance, Canadian yogurt should contain ten million viable microorganisms per gram (Milk Ingredients.Ca, 2016). The requirement for live organisms is not applicable where yogurt has been purposely heat treated after fermentation in order to prolong the storage life.

Yogurt is used in developed countries as a dessert, between-meal snack, complete lunch, and diet food. People eat yogurt as part of a balanced

Table 18.16 Selected fermented milk products from around the world

Product	Origins
Airag	Mongolia
Amasi	Southern Africa
Cheese	Worldwide
Chhu	India, Nepal, China
Chhurpi	India, Nepal, China
Dahi	India, Nepal, China
Dadih, Dadiha	Indonesia, Sumatra
Kefir	Russia
Koumiss (Kumys)	Russia, Mongolia
Laban rayeb	Egypt
Leben	Africa
Misti	India
Nunu (nono)	West Africa
Philu	India
Shrikhand	India
Somar	India
Viili	Finland
Yogurt	Worldwide

Adapted from Tamang et al. (2016)

Table 18.17 Global yogurt production companies (1000 metric tons)

Company	Production
Danone	6000
GM/Yoplait	1200
Nestle	900
Yakult	850
Lactalis/Parmalat	840
Friesland Campina	700
WBD-Pepsico	680
Muller	580
Wahaha	550
Fronterra	460

From (Preben Mikkelsen (PM) Food and Dairy Consulting 2013)

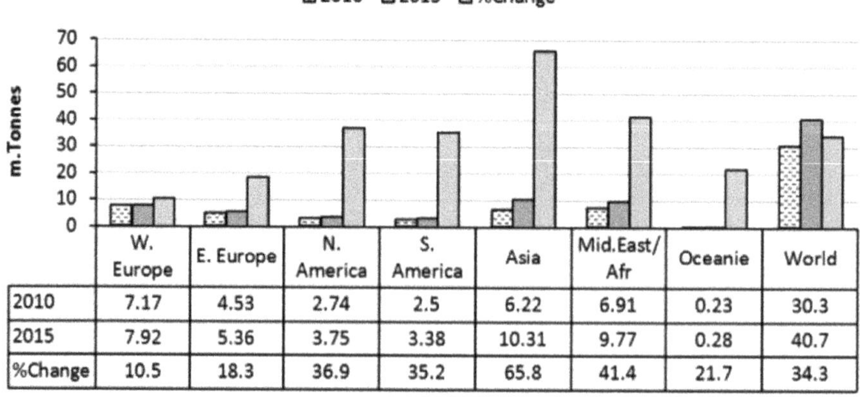

	W. Europe	E. Europe	N. America	S. America	Asia	Mid.East/ Afr	Oceanie	World
2010	7.17	4.53	2.74	2.5	6.22	6.91	0.23	30.3
2015	7.92	5.36	3.75	3.38	10.31	9.77	0.28	40.7
%Change	10.5	18.3	36.9	35.2	65.8	41.4	21.7	34.3

Fig. 18.10 World yogurt production by region (million metric tons). (Drawn using data from Preben Mikkelsen (PM) Food and Dairy Consulting 2013)

Table 18.18 Composition of yogurt (21 CFR131.200)

Composition	Quantity	Comments
Optional ingredients	Single or combinations	Cream, milk, partially skimmed milk, or skim milk
Milk fat	>3.2%	Roese-Gottlieb Method
Not fat, solids (NFS)	>8.25%	Proteins (caseins and whey proteins) and lactose
Titrable acidity	>0.9%	Lactic acid
Nutritive carbohydrate sweeteners	Unspecified	Sugar, brown sugar; molasses (not blackstrap); high fructose corn syrup; fructose; fructose syrup; maltose; maltose syrup, dried maltose syrup; malt extract, malt syrup, dried malt honey; maple sugar; or other sweeteners except table syrup
Flavoring ingredients	Unspecified	Nature unspecified
Stabilizers & emulsifiers	0.2–0.5%	Mainly polysaccharide thickening agents
Vitamins (optional)	Vit. A & D	4000 Units Vitamin A, 2000 Units Vitamin D per quart

Reference: Milk Facts (2016b)

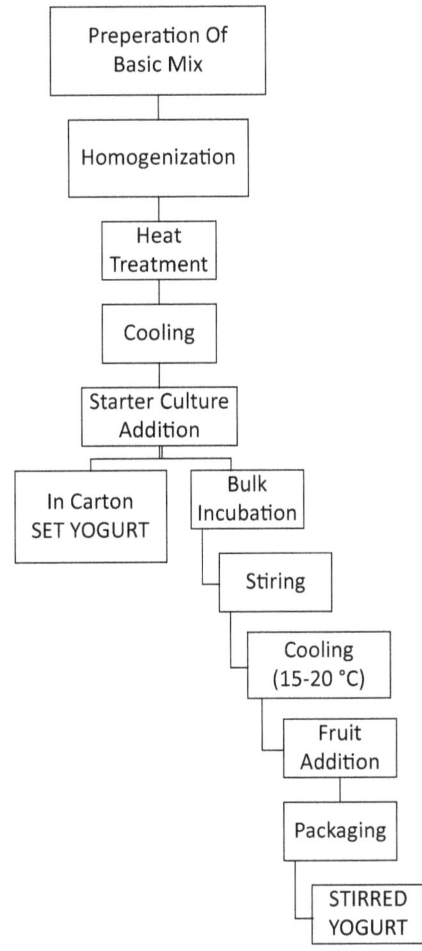

Fig. 18.11 Process flow chart for yogurt manufacturing. (Adapted from Robinson and Tamine 1993)

diet because they like it and feel that it favorably affects their health. Flavored yogurts made from low-fat or skim milk are very popular and unflavored yogurts are eaten plain or are used as substitutes for some of the more fatty dairy products such as sour cream or some cheeses.

Yogurt may have a consistency of a smooth viscous liquid, a soft curd, or a solid frozen dessert. The soft curd, one of the most popular products, is easily spooned out of its container without free whey. Acceptable colors and stabilizers may be added to improve appearance and texture but unlike in cultured buttermilk, no salt is added. Sugar may also be added to neutralize the sour taste of acid and many yogurts have fruit added. The amount of fat in yogurt varies depending on the milk used in the process. Most yogurts in the United States are made with skim or low-fat milk and have a fat content of up to about 1.7%. Some yogurt is still made with whole milk and will contain about 3.25% fat.

6.2.4 Yogurt Manufacturing

Modern cultured yogurt manufacturing is associated with high degrees of mechanization, but its center point is a unique bacterial fermentation. The basic steps for yogurt manufacture involve; preparation of milk mix, homogenization, heat treatment, addition of starter culture or fermentation microorganisms and then packaging (Fig. 18.11).

Preparation of Basic Mix

The only dairy product needed for yogurt production is milk but skim milk, condensed milk, nonfat dry milk solids, modified starches, alginates, and gums may be added to produce low-fat or nonfat products (Aziznia et al. 2009; Peker and Arslan 2013). These mixes must be blended, homogenized, and pasteurized. The initial stage for yogurt manufacture involves a range of separate unit operations.

Milk for yogurt manufacture undergoes standardization for protein and fat content. With inline milk standardization there is continuous sampling, and testing for protein, fat, and lactose to enable efficient process control. Upon completion, of this stage the milk has undergone a process of standardization in relation to fats, protein and carbohydrate content. The mixture is then homogenized to achieve efficient mixing and to reduce milk fat globule size.

Heat Treatment

The aims of heat treatment are several fold; (i) destruction of microorganisms present in their vegetative state, (ii) denaturation of whey proteins to promote their interactions with casein and (iii) changes and redistribution of minerals in milk. The pasteurization temperatures are higher than those employed for the destruction of thermophiles. Vat pasteurization can be achieved at 190.4 °F (88 °C) for 30 min or HTST at 203 °F (95 °C) for 5–10 min. The mixture is then cooled to an optimum incubation temperature of 113 °F (45 °C) needed for the growth of thermophilic microorganisms (Rasic and Kurmann 1978; Robinson and Tamine 1993).

Starter Culture, Yoghurt Fermentation

Yogurts are produced using mixed cultures of two microorganisms, *Lactobacillus delbrueckii* subsp. *bulgaricus* and *Streptococcus thermophilus* (Matalon and Sandine 1986; Tamine and Robinson 1976, 2007a). The microorganisms may be present naturally in milk but quality starter cultures produced from commercial stocks are added to milk previously heat-treated to inactivate any indigenous microbes.

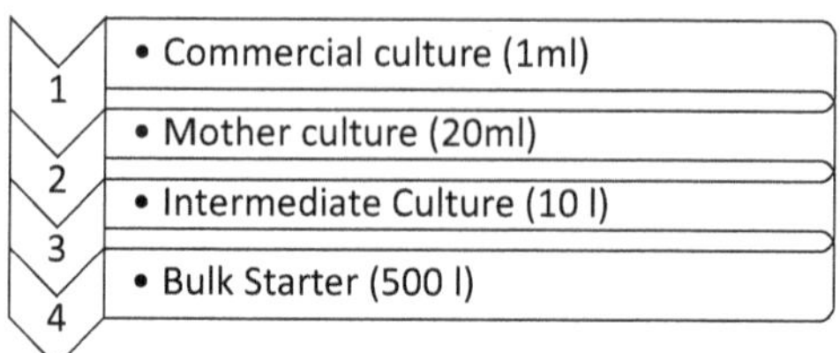

Fig. 18.12 Maintenance and propagation of yogurt starter culture

The job of maintaining and propagating starter cultures, in sufficient quantities for large scale manufacture is the responsibility of the food microbiologists (Matalon and Sandine 1986; Tamine and Robinson 1976, 2007a). Starting with a 1 ml batch of commercial culture, these will be grown up to 500 ml in order to produce sufficient quantity to transform a 25,000 liter vat of yoghurt, at an addition rate of 2% (see below). For long term storage and or transportation, starter cultures are lyophilized to powdered form and transported dry (Chap. 15, Sect. 4.3) (Fig. 18.12).

The two bacteria are kept as a mixed culture usually. The starter culture is added as a suspension comprising 2–5% of the total mix. In the case of non-stirred yogurt (Fig. 18.11) the inoculated milk is filled into containers that may contain fruit and is incubated until a pH of 4.4 (0.9–1.2% titrable acid as lactic) is attained (Table 18.19).

As noted earlier, the aim of yogurt fermentation is to lower the pH sufficiently to promote casein coagulation but a great many other changes occur which affect product quality (Matalon and Sandine 1986). The rate of acidification is important and a slow rate of acidification produces better texture. Alongside of the pH decrease, starter cultures produce changes in lipids, protein and carbohydrates thereby changing flavor, texture and other attributes.

The main flavor compounds in yoghurt are believed to be acetaldehyde, diacetyl, acetic acids (Fig. 18.12) produced from glucose metabolism (Hefa 2010; Routray and Mishra 2011).

Microorganisms forming a starter culture work cooperatively (Crow and Curry 2002). The faster growing *S. thermophilus* produces lactic acid which lowers the medium pH and encourages *L.*

Table 18.19 Some bacteria used as starter culture for milk fermentation

Species	Temp °C	Lactic acid %	Prot[a]	Used in
Streptococcus thermophilus	40–45	0.7–0.8	Yes	Acidified milk,
Lactococcus lactis[a]	25–30	0.5–0.7	Yes	Acidified milk
Lactococcus cremoris	25–30	0.5–0.7	Yes	Acidified milk
Lactococcus diacetylactis	25–30	0.3–0.6	Yes	Acidified milk,
Leuc cremoris	25–30	0.2–0.4	Yes	Acidified milk
Lactobacillus acidophilus	37	0.6–0.9	–	Acidified milk
Lactobacillus helveticus	40–45	2.0–2.7	Yes	Acidified milk
Lactobacillus bulgaricus	40–45	1.5–2.0	Yes	Acidified milk
Bifidobacterium	37	0.4–0.9	–	Acidified milk

[a]Protease producing strains, During the 1980's the taxonomy of lactic acid bacterial underwent revision and streptococcus strains important for food fermentations were reclassified as Lactococcus (Stiles and Holzapfel 1997)

Table 18.20 Listing of some different types of yogurts*

Basic types	Lifestyle/demographic products
Greek style yogurt	Yogurt for toddlers
Kefir	Yogurt for kids
Frozen yogurt	Yogurt for men
Drinkable yogurt	100 calorie yogurt
Savory yogurt	Yogurt for men
Breakfast cereal yogurt	Functional yogurt
Set yogurt	High protein yogurt
Fruit yogurt	Probiotic yogurt
Low fat, not fat, regular	Tube yogurt
Swiss type	Organic yogurt
Yogurt cheese	Lactose free yogurt
	Enriched yogurt

*From various internet sources

bulgaricus growth. *L. bulgaricus* is more proteolytic and helps liberates amino acids and peptides that stimulate the growth of *S. thermophilus*. As the acidity decreases to about pH 5.0 due to the action of *S. thermophilus*, the production of acetaldehyde by *L. bulgaricus* is inhibited and eventually the medium pH falls to pH 4.0–pH 4.4. Milk used for yogurt must be penicillin-free, for *S. thermophilus* is sensitive to it.

Typically, yogurt is cooled and held at about 39.2 °F (4 °C). Sundae style just described above requires mixing by the consumer. Yogurt may also be made in a "Swiss" style where fruit puree, sugar, and stabilizer are blended with fresh fermented yogurt at 60.8 °F (16 °C), packaged and cooled.

Some varieties of yogurt available are listed in Table 18.20. The basic yogurt can be modified. For example, fruit preserves can be added at about a 15% level but have been added up to 27%. Stabilizers are not usually required for plain whole-milk yogurt but flavored, low fat, or nonfat varieties may require the addition of gelatin, agar, modified food starches, or gums at levels of about 0.4–0.5%. The stabilizers can be added prior to pasteurization but care must be taken that excess amounts are not used or a sticky, gummy, hard mouth feel and free whey may result.

Frozen flavored yogurts are made from a mix that is prepared and pasteurized much like ice cream mixes. Frozen yogurt is then frozen in much the same manner as ice cream with various levels of overrun. Liquid yogurts are made from whole and skim milk, and low-lactose yogurts that have been produced with milk treated with a lactase enzyme prior to fermentation, are examples of other yogurt-based products.

6.2.5 Yogurt Quality

Virtually all stages of yogurt manufacture may affect product quality, including the nature of ingredients-milk, cream and proteins added manufacturing / processing factors, packaging and storage. As noted earlier food quality encompasses a great number of dimensions (Chap. 6) but this discussion will be limited to nutritional and sensory quality.

Quality problems that arise with yogurt include separation of whey or syneresis, which can be overcome by higher temperature treatment prior to inoculation, in order to facilitate whey protein-casein interactions, (Kroger 1976). The texture of yoghurt is also improved by increasing the degree of homogenization.

Obviously, the inoculation and fermentation stage can also affect product quality. For instance, too much acid production, caused by too rapid a growth of *L. bulgaricus*, is undesirable. Normal yogurt has 0.9–1.5% acid but rapid growth of starter culture can lead to 2–3% acidity. Careful maintenance of fermentation temperatures and the use of pure cultures help to solve this problem. Premature curdling of milk may occur with yogurt that has had fruit added, especially Swiss style yogurt. When the fruit rests on the bottom, this rarely occurs. Another problem occurs owing to the presence of thermophilic bacteria or yeast in the milk. Off-flavors, such as bitterness caused by the bacteria, and alcohol and gas produced by the yeast will result from low quality or improperly handled milk. Use of high-quality milk and proper quality control will solve these problems (Byland et al. 2003).

6.2.6 Nutritional Quality of Yogurt

Nutritionally, yogurt is an excellent food, giving all of the protein benefits from milk but especially in the case of low-and nonfat yogurts, with less fat and cholesterol. Yogurt is frequently discussed also in terms of probiotic qualities derived from the presence of live microorganisms. Flavored low-fat yogurts may have more calories than expected because of added fruits and sugar (Table 18.21).

Lactobacillus acidophilus and *Bifidobacterium* strains may be added to yogurt because of the believe that they survive transit through the GI tract and function as probiotic organism (Guarner et al. 2005; Morelli 2014; Sarkar 2008). For instance, health claims indicating yogurt could help deal with lactose digestion were approved in Europe (EFSA Panel on Dietetic Products 2010). However, claims that yogurt could reduce the risks of acute diarrhea for patients taking antibiotics were not approved (Table 18.22).

There are also speculations currently about other health-benefits of yogurt e.g. related to cardiovascular disease, type 2 diabetes, weight-loss, cardiovascular disease, gastroenteritis, and hypertension (Astrup 2014; Chen et al. 2014; Marette and Picard-Deland 2014). Yet many yogurt health claims have yet to be evaluated and (dis)approved by food legislators before these

Table 18.21 Nutritional information for one cup (227 gm) of various yogurts

	Calories	Fat (g)	Sat. Fat (g)	Calories from fat
Nonfat plain yogurt	126	<1	0.3	3
Low-fat plain yogurt	143	4	2.3	36
Whole milk plain yogurt	139	7	4.8	63
Nonfat flavored yogurt (fruit added)	204	<1	0.3	3
Low-fat flavored yogurt (fruit added)	232	2	1.6	18
Nonfat flavored yogurt sweetened with Aspartame (fruit added)	100	<1	<1	0

Note: These values vary with different brands. Read the labels for accurate amounts

can be used for labeling purposes. Meanwhile unsubstantiated claims risks court action.[2] Therefore, more research into the health effects of yogurt and other dairy products is needed (World Health Organization 2009).

6.3 Cheese

Cheese as a fermented milk product. The basic idea behind cheese manufacturing is to concentrate milk solids (lipid, protein, and minerals) by coagulation to form a semi-solid material (Fox and McSweeney 2004). Cheese has 50% less moisture and roughly six to 7-fold increased level of protein and lipids compared to fluid milk (Table 18.23). The recovery of milk solids is quite efficient during cheese production. Using 100 g of milk will produce 10.4 g of cheese (~5.4 g dry matter).

[2]An example court action associated with the advertising claims for a popular brand of probiotic yogurt is described in Wikipedia https://en.wikipedia.org/wiki/Activia#Class_action_in_2008.E2.80.932009 (Accessed Aug, 2016)

Table 18.22 Nutrition and health claims related to some yogurt components

Reference	Authorized
2010;8(10):1763	"Live yoghurt cultures in yoghurt improve digestion of lactose in yoghurt in individuals with lactose mal-digestion (EFSA Panel on Dietetic Products 2010)
	Non-Authorized
ACTIMEL® Lactobacillus casei DN-114 001 plus yoghurt	Fermented milk with probiotic Lactobacillus casei DN-114 001 and yoghurt decreases Clostridium difficile toxins in the gut (of susceptible ageing people). Presence of Clostridium difficile toxins is associated with the incidence of acute diarrhea
2011;9(4):2029 Lactobacillus rhamnosus LB21 NCIMB 40564	Helps to strengthen and maintain balance in the gut flora
2009;7(9):1247 Yoghurt with probiotic bacteria Bifidobacterium animalis ssp. lactis & Lactobacillus acidophilus	Contains millions of bacteria, maintain the balance of natural flora in your body, can aid digestion and general well-being. Helps to maintain harmony in your digestive system
Dairy products (milk, cheese and yoghurt)	Three portions of dairy food every day, as part of a balanced diet, may help promote a healthy body weight during childhood and adolescence

Reference: Adapted from (Europa 2016) with wording edited lightly to retain original form

Table 18.23 Comparing cheese and milk composition (%Fresh weight basis)

	Water	Protein	Fat	Lactose	Ash
Cow milk	87.5	3.5	3.9	4.8	5.4
Cheese	41.0	25	31	2.0	4.5
% Change	−53	614	695	−58	−17

6.3.1 Nutritional characteristics of cheese

Cheese is considered a good source of, calcium, protein, and vitamin D (Walther et al. 2008). For instance, one study found that cheese was the #2,

Table 18.24 Effect of cheese on the nutrient intake of US children (1–18 years)*

Nutrient	% Intake	Rank	#Food Groups
Total Fat	9.3	1	18
Saturated Fat	16.3	1	18
Calcium	19.4	2	11
Sodium	8.1	3	15
Protein	9.7	4	15
Cholesterol	9.6	5	11
Energy	4.7	7	21
Vit. D	2.3	8	8
Potassium	2.0	14	15

*Summarized from, Keast et al. (2013)

#4 and #8 source of calcium, dietary protein and vitamin D for US children between the ages of 2–18 years old as shown in Table 18.24 (Keast et al. 2013). Milk was the #1 source of calcium, protein and Vitamin D. The contribution of cheese to dietary protein intake was just below poultry and meat ranked #2 and #3, respectively. Eggs and fish ranked far below cheese as a source of dietary protein at #14 and #15, respectively (Keast et al. 2013). Cheese contributed negligible dietary carbohydrates, total sugar or added sugar. The dietary fiber contribution from cheese was negligible also (Keast et al. 2013). Nutritional surveys of the type described above (Keast et al. 2013) are important because they reflect (i) food composition and (ii) habitual intake. For instance, it may be inferred that cheese is eaten in appreciable amounts by children in the US and that this product provides them with high amounts of specific nutrients some of which are desirable (Keast et al. 2013).

Cheese has been found to influence the diet of adults from the Republic of Ireland, leading to significant changes in 10 or more nutrients, e.g. total energy, protein, fat, calcium, vitamins, sodium, total fat, saturated fat, monounsaturated fat etc. People consuming high levels of cheese tended to be males and younger compared to non-consumers (average age 39.5 vs 51 years). The body mass index or BMI for consumers and non-consumers were not different (Feeney et al. 2016). More generally, cheese is considered a potential source of functional food components

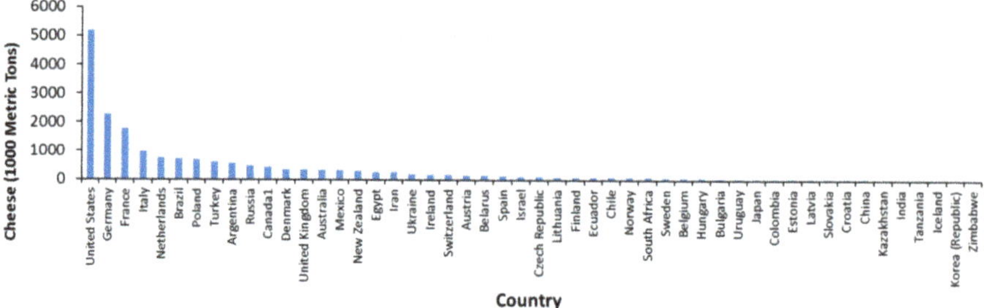

Fig. 18.13 Cheese production in selected countries ('000 metric tons). (Data from for 2016 (Canadian Dairy Information Centre 2016c))

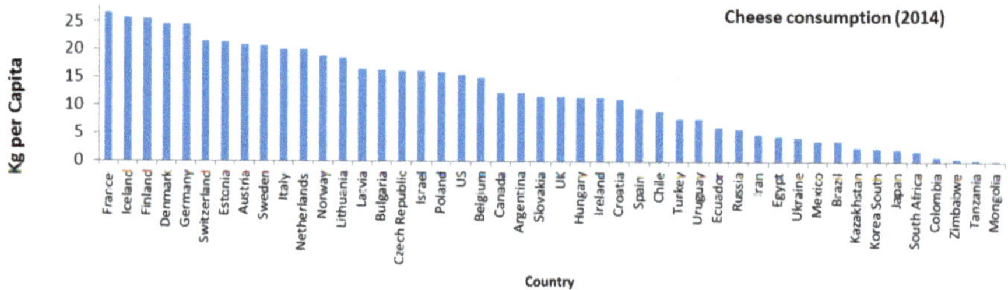

Fig. 18.14 Per capita cheese consumption (Kg/person). (Data from 2016 (Canadian Dairy Information Centre 2016a))

including, medium chain fatty acids and bioactive peptides (Haelein 2004; Walther et al. 2008; Feeney et al. 2021).

A number of so-called nutrients of concern are associated with cheese (Table 18.24). The US children's survey (Keast et al. 2013) showed cheese was ranked #1 as source of total fat and #1 for saturated fat. Dietary sodium from cheese (rank #3) fell behind of table salt (rank #1) and yeast bread/rolls (rank #2). Cholesterol intake from cheese, ranked 5th out of 11 food groups examined. Although saturated fat is a risk factor for coronary heart disease, dairy fat sources were not associated with heart disease risk. One possible explanation maybe that dairy fat may contain saturated fat alongside of counterbalancing (health promoting) dairy components (Huth and Park 2012; O'Sullivan et al. 2013).

6.3.2 Economic Significance of Cheese

Cheese is consumed the world over but the vast majority of the global production of cheese takes place in the US (22%) and the European Union (66%). The value of the global cheese market is estimated at US$79billion per annum (2012) projected to reach US$105 billion by the year 2019 with an annual growth rate of 4.5% (Transparency Market Research 2014). The cheese industry is described as mature, sophisticated and demanding (Law and Tamime 2011). The US is ranked #1 for individual countries, with an annual cheese production of 5.7 million metric tons or about 11.4 billion lbs. (Canadian Dairy Information Centre 2016c) (Fig. 18.13).[3]

Over the period 2008–2012, the 27% of global cheese production, which was exported, was valued at US$27billion (Vlahović et al. 2014). About 80% of the international trade in cheese involved European nations with EU(27) accounting for 76% of world exports (Table 18.25). Most of the US cheese production was consumed domestically. France, Iceland, Finland and Denmark are the biggest consumers for cheese per capita (Fig. 18.14).

[3]The US cheese production matches the net output from the next four top-ranked nations (France, Germany, Italy and Netherlands) combined.

Table 18.25 Top global exporters & importers for cheese (2008–2012)

Export country	% World	Import	% World
Germany	18.6	Germany	12.1
France	11.8	Italy	9.1
The Netherlands	11.6	Gt. Britain	8.2
New Zealand	5.2	Russia	5.7
Italy	5.0	The Netherlands	4.2

Reference: Adapted from Vlahović et al. (2014)

Table 18.26 Some well known cheeses globally

Type	Origins
Brie, Camembert	France
Edam, Gouda, *Limburger*	Netherlands
Cheddar, Cheshire	England
Emmentaler, Gruyere	Switzerland
Parmesan	Italy
Brick, Coldy	US

6.3.3 Standards for Cheese

There are between 400–1200 different types of cheeses globally (Beattie 2008; Helweg 2010). About 1750 cheeses were categorized according to the type of milk (by species), the texture, color, country and whether they are vegetarian (Cheese.com 2016). More generally, cheeses are classified based on, (i) the use of raw or pasteurized milk, (ii) level of moisture, fat content and overall texture – e.g. soft, semi-hard or hard cheese, (iii) type of coagulation – acid, rennet or acid and heat, (iv) extent of ripening – fresh versus ripened, (v) international standards or style of manufacture (American Cheese Society 2015) (Table 18.26).

The identity of some European cheeses are protected by their geographic origins in accordance with the protected designation of origin (PDO), protected geographical indication (PGI), or traditional specialties guaranteed (TSG) schemes. Within the European market, only cheeses produced in specific regions can be named using those designations (Vlahović et al. 2014).

The most popular types of cheese in the US are the Italian-style cheeses (42% production) such as mozzarella cheese used as pizza toping. The American type cheeses (Cheddar, Colby, Monterey, and Jack) are also popular and together make up 40% of the total production (Table 18.27). It is feasible also, that some-minor volume cheeses could have passionate followers and be suited for highly specific culinary applications. The largest quantities of cheeses are produced in the states of Wisconsin, California, Idaho, New Mexico and New York (Table 18.27).

Table 18.27 Different types of cheeses common in the United States

Cheeses	2016 Annual (000 metric tons)	%
American types	2314	40
Cheddar[a]	1707	29
Other Am. types[b]	608	10
Blue and Gorgonzola	47	1
Brick and Muenster	84	1
Cream and Neufchatel	406	7
Feta	58	1
Gouda	15	0
Hispanic	124	2
Italian types	2506	43
Mozzarella	1967	34
Parmesan	176	3
Provolone	188	3
Ricotta	106	2
Romano	32	1
Other Italian types	37	1
Swiss	152	3
All other types	83	1
Total cheese	5790	100

[a]American type cheese = Cheddar, Colby, Monterey, and Jack

Extracted from (USDA Economics Statistics and Market Information System 2016). [a]Annual forecast

6.3.4 Cheese Manufacturing

The critical steps in cheese making involve, coagulation of milk protein using bacterial fermentation and rennet, draining of the curd, pressing, combination of cheeses (process cheese), aging with further fermentation by bacteria or molds, packaging and storage (American Cheese Society 2015; Hill 2016a). Some cheeses may be manufactured using raw milk, or milk that has been heat-treated at temperatures below those for pasteurization (Fig. 18.15).

Starter cultures are used for most cheeses in a manner similar to yogurt (Sect. 6.2). The cultures may be one strain or a mixed culture of bacteria

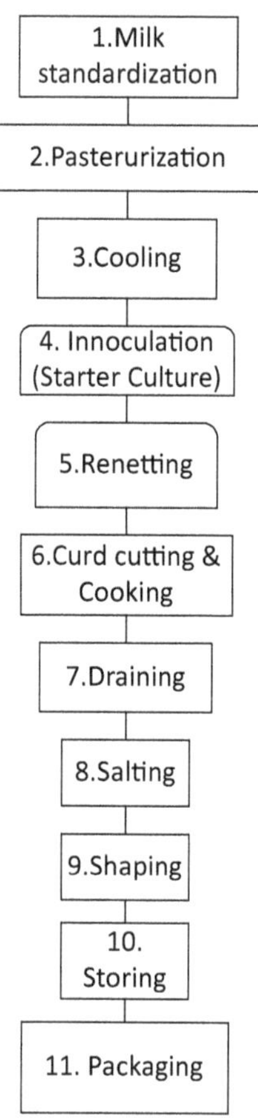

Fig. 18.15 A process chart for cheese manufacturing. (Adapted from Milk Facts 2016a)

that work in a symbiotic manner. Cheese starter cultures may be lyophilized (freeze-dried) with milk components, or frozen with liquid nitrogen at −320.8 °F (−196 °C) and held in this state. Both methods yield viable cultures but the dry method requires no special containers or storage facilities other than a freezer or refrigerator. Most cultures are lactobacillus (L) or streptococcus (S) strains (Table 18.28).

Some cheesemakers have successfully held lactic starters for several years but most usually start a new one from a dry, lyophilized culture every 3 or 4 weeks. A continual sub-culturing of microbial starters is necessary to ensure consistency of quality. This rotation must be adhered to or problems could develop.

Setting the milk means preparing warm milk with a starter culture or rennet extract to form a smooth curd block. The curd may be either an acid casein curd set with a starter or a sweeter, calcium para-casein curd set with rennet and starter. The acid curd takes from 5 to 16 h whereas the sweeter curd can be completed in 15–30 min. With acid curds casein micelles coagulate when the acidity of milk reaches about pH 4.6 which the pH where they have not net charge. With renneting curd, kappa-casein undergoes enzymatic hydrolysis, leading to a loss of negative charge so that the remaining casein particles coagulate.

Table 18.28 Some bacterial starter cultures used for cheese and other cultured products (§)

Culture	Temp. °C	Functions	Product use
Propionibacterium freudenreichii subsp. shermanii	15–40	Flavor, CO_2 eye formation	Swiss cheese, Emmental cheese
Lactobacillus delbrueckii subsp. bulgaricus	40–45	Acid and flavor	Buttermilk
Lactobacillus delbrueckii subsp. lactis	40–45	Acid, proteolytic	Parmesan, Emmental
Lactobacillus helveticus	40–50	Acid, flavor, Probiotic	Swiss cheese, Emmental, Cheddar, Parmesan
Streptococcus thermophilus	40–45	Acid	Emmental, Cheddar, Italian cheese,
Streptococcus diacetilactis, now Lactococcus lactis	20–35	Acid and flavor	Sour cream, ripened cheese, butter, buttermilk, starter cultures, cheese All types of cheese

(§) (From, Matalon & Sandine 1986; Tamine & Robinson, 1976; Tamime & Robinson, 2007a; Tamine 2008).

Cutting the curd must be done carefully and will increase the curd surface area many times. This leads to effective whey expulsion and permits the equal-sized smaller curds to be cooked throughout. Larger cubes give a higher-moisture cheese. Cooking causes the curds to contract and also drives off the free whey. Cooking also influences curd texture, gains time for lactic acid development, and arrests it. It also suppresses the growth of spoilage organisms and influences the final moisture content.

Draining simply separates the whey from the curd by passing it through a straining device. In industry, this is accomplished by lining the exit gate of the vat with a sieve or having a sieve lay the entire length of the vat. Dipping is usually done in smaller operations by scooping curds into perforated molds or colanders or by inserting a coarse cloth into the kettle and bagging all the curds. Regardless of the method, his step accomplishes separation of whey and concentration and coalescing of curds, and pro vides additional time for lactic acid development.

A process of knitting and transformation of curds includes the process of "cheddaring" in cheddar cheese where slabs of the curd are piled one on top of another and acid is allowed to develop to about 0.5% and more moisture is expelled. Other knitting processes include preliminary packing and pressing of brick and blue cheese curd, and the pulling and processing of acid-ripened provolone and mozzarella cheeses. These processes result in the characteristic texture of the various cheeses.

Salting can be done by spreading salt over the curds or by dipping the pressed cheese into brine solutions. Salting improves flavor, texture, and appearance while slowing or stopping the lactic acid fermentation when it has reached its optimum level. Salting also depresses the growth of spoilage organisms and reduces moisture, thus controlling it in the final product.

Pressing removes moisture and gives the cheese its shape. Care must be taken not to press with too much force in some cheeses such as Roquefort or blue cheese. If they are pressed too compactly, no air will be able to seep into the cheese and carbon dioxide will not be able to escape easily. Both of these are necessary for the successful growth of the molds needed to develop the characteristic flavors, aromas, and textures in the finished cheese.

The special applications cover a wide range of processes, from creaming cottage cheese to the addition of special microorganisms to ripen cheeses such as brick or Lim burger.

Process cheeses are very popular in the United States and are manufactured by combining different types of cheeses as ingredients. Generally, not fully aged (green) cheese is ground together with some aged cheese and water, emulsifiers (i.e., disodium phosphate), salt, and powdered skim milk are added. The ingredients are heated, mixed, and extruded into molds.

Cheese making is an art as well as an application of advanced sciences. The success in this industry depends on a logical sequence of fixed steps where the emphasis is varied and the incorporation of special applications gives cheeses their various similarities and differences (Davies and Law 1984; Fox et al. 2004; Fox and McSweeney 2004; Hill 2016a).

6.3.5 Cheese Quality

Many bacteria including *S. lactis, L. bulgaricus, and S. diacetilactis* can perform the milk fermentation reaction. In some cheeses such as Swiss and Gruyere, the propionic acid fermentation leads to the typical flavor and eyes. A distinct flavor may arise when a fine balance between diacetyl, propionic acid, acetic acid, and other related compounds exists. In some fermentations, large amounts of $C0_2$ and hydrogen gas are produced and the cheese body is distorted. Cheeses affected include Swiss and Gruyere and large blowholes occur several months after the cheese is placed in curing rooms.

Another fermentation that causes spoilage and deterioration of the food is caused by the coliform group of bacteria. *Escherichia coli* and *Aerobacter aerogenes* are included in this coliform group and they are usually present in milk

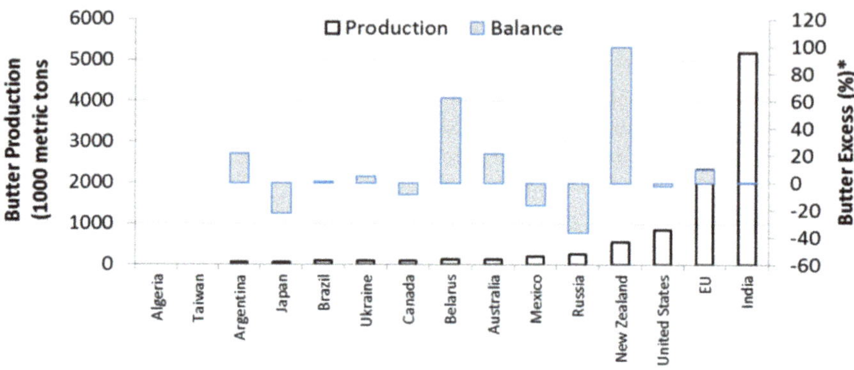

Fig. 18.16 Butter production in selected countries and excess (%) compared to consumption (2016)

of poor quality. Their fermentation produces large amounts of carbon dioxide, resulting in slits in the cheese curd and obnoxious and unclean flavors in the cheese. Rapid production of lactic acid by starter culture bacteria will prevent their growth.

6.4 Whey

The fluid byproduct of cheese making is whey that comprises 5% lactose, 2% other milk components, and 93% water. Some whey is used for the production of ricotta cheese. Currently, cheese whey is mined as a source of ingredients, mainly lactose and whey protein. Whey protein is used as protein supplements in drinks, baked goods, ice cream, sherbet, candy, fudge, and other confections (Diaz et al. 2009).

6.5 Butter

Butter is made from cream having a butterfat content of 25–40%, into which the bacterium *Streptococcus diacetyllactis* has been added. Bacterial fermentation produces diacetyl, the major flavor component in butter. Butter contains 20% water in the form water-in-oil emulsion. Some bacteria may grow in this water and produce lipase, which will de-esterify some fatty acids and cause rancidity, giving off-flavors and odors. Salt of about 1–2.5% by weight is added to butter to add flavor and inhibits the growth of the rancidity-causing bacteria. Sweet butter, which contains no salt, is more perishable than salted butter (Fig. 18.16).

The global production of butter (2016) amounts to approximately ten million metric tons. The leading butter producers are, India > EU > USA > New Zealand > Russia. As most butter is destined for domestic consumption, the nominal "buffer excess" or difference between production and consumption is 0% for most countries. There can be shortfalls in butter supply for some countries which is met through imports, notably Russia > Japan > Mexico > Canada. Countries that produce butter in excess of domestic consumption include, New Zealand > Belarus > Argentina > Australia (USDA Foreign Agricultural Services 2016a).

6.6 Buttermilk

Buttermilk is the liquid produced when 40% cream is churned leading to the separation of milk lipids as butter. Buttermilk contains whey protein, casein, milk-fat globule membrane (MFGM) fraction and phospholipids (Sodini et al. 2006; MilkIngredients.Ca, 2011b). Two other buttermilk variants are available from churning, i.e. whey milkfat (milk fat recovered from whey) or cultured cream used for European-style butter (Sodini et al. 2006). The higher fat content for buttermilk compared to whole milk increases the tendency towards oxidative rancidity and the need for preservation by spray drying to form buttermilk powder. The average composition of buttermilk powder was reported as, moisture (3%), protein (34%), fat (6–20%) and ash (8%) (MilkIngredients.Ca, 2011b); the phospholipid content was 1.2–1.8 (g/100g) dry weight basis for buttermilk powder (Sodini et al. 2006). Buttermilk is a functional ingredient for baked products, ice cream, dry mixes, breading, batter, puddings and beverages (MilkIngredients.Ca, 2011b).

Cultured buttermilk is distinct from sweet buttermilk obtained by churning butter directly. Some commercial production of cultured buttermilk starts with skim milk, which is then subjected to pasteurization, fermentation, cooling, flavoring and packaging (Gettys and Davidson 1985). Cultured buttermilk is also produced from cream. Various starter cultures were reported for buttermilk fermentation including, e.g. *Streptococcus lactis*, *Streptococcus cremoris* – renamed *Lactococcus lactis*, *Lactococcus cremoris* respectively. The flavor of buttermilk is due to a combination of lactic acid, amino acids, diacetyl, acetaldehyde formed by fermentation. Section 6.2.1 and Chap. 7, Sect. 5.3 contain additional information on cultured milk products.

6.7 Sour Cream

Sour cream as the product formed by fermenting pasteurized cream. The production of sour cream is similar to buttermilk. It is prepared from light cream (16–20% butterfat) to which 8–9% nonfat milk solids and sometimes stabilizers (e.g., gelatin, gums, or modified starches) are added. After pasteurizing and homogenizing, the same bacteria used for buttermilk are added for the conversion of sugars to acid. (Lactococcus lactis, Table 18.28). The product is packaged and cooled and should be held at 35–40 °F (1.7–4.4 °C) until consumed. USDA standards require that full-fat sour cream has not less than 18% milk fat, whilst "reduced-fat" and "lite" sour cream have 25% and 50% less milk fat compared to the full-fat product (United States Department of Agriculture 2000). Consumer preference for sour cream was linked with products having high viscosity, dull appearance and cooked/ milky flavor (Shepard et al. 2013).

6.8 Dairy Product Substitutes

Dairy product substitutes have been developed for economic, religious, and health considerations (Table 18.29). Dairy-free options may be chosen by consumers affected with health issues such as lactose intolerance, lactose sensitivity or milk allergy. Vegetarian and vegans may also choose dairy free products, in order to void cow's milk. A Jewish dietary law (kashrut) which requires the separation of meat from milk may also lead to the adoption of non-dairy drinks after meat-based meal (Feder 2016; Schaeffer 2016).

Historically, one of the earliest non-dairy analogues and certainly one of the most controversial was margarine, which led to legislations to protect butter producers in the US as well as France (Brown 2011). These times, dairy product substitutes are received with less hostility. Non-dairy creamers or coffee whiteners are widely known. Tofu-based frozen desserts, and fluid milk-like products and cheese substitutes (milk fat is replaced with vegetable fat and vegetable protein) continue to be developed for direct

Table 18.29 Selected dairy product substitutes or analogues

Analogue	Comments	Examples
Plant Milk	Plant based W/O emulsions, suspensions	Soy Milk, Almond Milk, Rice Milk, Oat Milk, Almond Milk, Corn Milk, Potato Milk, Hazelnut Milk,
Cream	DehydratedW/O emulsions	Coffee Whitener, Non-dairy Creamer, toppings
Cheese	Soy protein systems	Tofu based cheese analogues
Sour cream	Soy based	fermented crème analogue
Ice cream	Plant based	Rice dream
Butter	W/O emulsion	Margarine

Reference: Adapted from Feder (2016) and Schaeffer (2016)

consumption, and as primary ingredients for catering applications.

The plant-based beverages (soymilk, almond milk, rice milk) were worth US$8 billion in 2014 and projected to reach US$19.5 billion in 2020 (MarketsandMarkets 2016). Soymilk, almond milk and rice milk were the most important product types with sales of soymilk expected to reach US$13.6 billion in 2020. The largest regional market share for plant-based beverages was Asia-Pacific (45%) followed by North America, and Europe. As noted above a high incidence of lactose intolerance in certain consumers groups (e.g. from Asia-Pacific, African, and the Mediterranean) is one possible reason for switching from milk to non-milk based substitutes (Simons 2016).

6.9 Further Reading

Dairy product manufacturing has been covered by highly regarded textbooks and monographs (Robinson 1993b; Tamime and Robinson 1999, b; Varnam and Sutherland 2012; Walstra et al. 2005). An interesting source book titled, the Tetra Pak "Dairy Process Handbook" is worth viewing in particularly because many dairy machinery and process equipment discussed in this chapter are illustrated there (Byland et al. 2003). The American Cheese Society's "Best practices guide for cheese makers" contains a great deal of advice on milking and handling for the small-scale producer (American Cheese Society 2015).

Appendix 1: The nutritional characteristics of vanilla ice cream varieties (§)

Nutrient composition (per 100 g sample)	(1) Vanilla	(2) Fat free	(3) Rich/premium	(4) Light	(5) Light (no sugar)	(6) Light, soft serve
Water (g)	61.0	64.4	57.2	59.8	65.5	69.6
Energy (kcal)	207.0	138.0	249.0	180.0	169.0	126.0
Protein (g)	3.5	4.5	3.6	4.8	4.0	4.9
Total lipids (g)	11.0	0.0	16.2	4.8	7.5	2.6
Saturated fatty acids (g)	6.8	0.0	10.3	2.9	4.1	1.6
Cholesterol (mg)	44	24	92	27	27	12
Ash (g)	0.9	1.1	0.8	1.1	1.2	1.1
Carbohydrate	23.6	30.1	22.3	29.5	21.4	21.8
Sugars (g)	21.2	6.3	20.6	22.1	6.5	18.7
Fiber (g)	0.7	1.0	0.0	0.3	0.0	0.0
Calcium (mg)	128	149	117	161	136	157
Phosphorous (mg)	105	150	105	103	75	121
Potassium (mg)	199	302	157	208	196	221
Sodium (mg)	80	97	61	74	96	70

(§)Notes: Table design based on Berkheiser (2019). Data is for 6 out of 21 legacy files for vanilla ice cream extracted from FoodData Central (USDA Agricultural Research Service 2021)

Sample database numbers and descriptions: (1) NDB 19095 Ice creams, vanilla; (2) NDB 19867 Ice creams, vanilla, fat free; (3) NDB 19089 Ice creams, vanilla, rich; (4) NDB 19088 Ice creams, vanilla, light; (5) NDB 19260 Ice creams, vanilla, light, no sugar added, and (6) NDB 19096 Ice creams, vanilla, light, soft-serve

References

Al-Saqer JM, Sidhu JS, Al-Hooti SN, Al-Mazeedi HM (1999) Physico-chemical and sensory quality of milk being produced in the state of Kuwait. Adv Food Sci 21(1/2):1–9

American Cheese Society (2015) Best practices guide for cheesemakers. Retrieved from http://www.cheesesociety.org/wp-content/uploads/2015/07/Best-Practices-Guide-for-Cheesemakers-Chapters-1-5.pdf

Angulo FJ, LeJeune JT, Rajala-Schultz PJ (2009) Unpasteurized milk: a continued public health threat. Clin Infect Dis 48(1):93–100. https://doi.org/10.1086/595007

Anon (2015) Raw milk revolution: it's tastier – but is unpasteurized dairy safe? Retrieved from http://www.telegraph.co.uk/food-and-drink/news/is-raw-unpasteurised-milk-safe-does-it-taste-better/

Arbuckle WS (2013) Ice cream (4th edn). Springer, New York, 483pp. https://books.google.com.gh/books?id=_pAkBAAAQBAJ

Asante J, Noreddin A, El Zowalaty ME (2019) Systematic review of important bacterial zoonosis in Africa in the last decade in light of the 'one health' concept. Pathogens 8(2, 50):29pp. https://doi.org/10.3390/pathogens8020050

Astrup A (2014) Yogurt and dairy product consumption to prevent cardiometabolic diseases: epidemiologic and experimental studies. Am J Clin Nutr 99(5):1235

Aziznia S, Khosrowshahi A, Madadlou A, Rahimi J, Abbasi H (2009) Texture of nonfat yoghurt as influenced by whey protein concentrate and Gum Tragacanth as fat replacers. Int J Dairy Technol 62(3):405–410. https://doi.org/10.1111/j.1471-0307.2009.00507.x

Beattie SJ (2008) Dairy: cheese. In: Smith JS, Hui YH (eds) Food processing principles and applications. Wiley, pp 273–289

Berkheiser K (2019) Is ice cream good for you? Nutrition facts and more. Retrieved from https://www.healthline.com/nutrition/ice-cream

Brown MA (2011) The debacle of margarine's utility. Retrieved from https://mises.org/library/debacle-margarines-utility

Buckow R, Chandry PS, Ng SY, McAuley CM, Swanson BG (2014) Opportunities and challenges in pulsed electric field processing of dairy products. Int Dairy J 34(2):199–212

Byland G, Malmgren B, Holanowski A, Hellman M, Mattsson G, Svensson B, Pålsson H, Lauritzen K, Vilsgaard T, Verweij E, Bronsveld E (2003) Dairy processing handbook. Retrieved from www.dairyprocessinghandbook.com

Campbell JR, Marshall RT (2016) Dairy production and processing: the science of milk and milk products. Waveland Press

Canadian Dairy Information Centre (2016a) Global cheese consumption (Kg per capita). Retrieved from http://www.dairyinfo.gc.ca/index_e.php?s1=dff-fcil&s2=cons&s3=consglo&s4=tc-ft

Canadian Dairy Information Centre (2016b) Global milk consumption (liters per capita). Retrieved from http://www.dairyinfo.gc.ca/index_e.php?s1=dff-fcil&s2=cons&s3=consglo&s4=tm-lt

Canadian Dairy Information Centre (2016c) Production of Cheese in selected countries ('000 metric tonnes). Retrieved from http://www.dairyinfo.gc.ca/index_e.php?s1=dff-fcil&s2=proc-trans&s3=psdp-pvpl&s4=cp-pf&page=prodcglo

Centers for Disease Control and Prevention (2015) Food safety and raw milk. Retrieved from http://www.cdc.gov/foodsafety/rawmilk/raw-milk-index.html

Cerf O, Condron R (2006) Coxiella burnetii and milk pasteurization: an early application of the precautionary principle? Epidemiol Infect 134:946–951. Retrieved from http://www.ncbi.nlm.nih.gov/pmc/articles/PMC2870484/

Chandan RC, White CH, Kilara A, Hui YH (ed) (2008) Manufacturing yogurt and fermented milks. Blackwell Publishing Asia, Carlton, 364pp. https://books.google.com.gh/books?id=_2QagyHQHboC

Chandry PS, Ng SY, McAuley CM, Swanson BG (2014) Opportunities and challenges in pulsed electric field processing of dairy products. Int Dairy J 34(2):199–212

Cheese.com (2016) Find over 1750 specialty cheeses from 74 countries in the world's greatest cheese resource. Retrieved from http://www.cheese.com/alphabetical/?per_page=20

Chen M, Sun Q, Giovannucci E, Mozaffarian D, Manson JE, Willett WC, Hu FB (2014) Dairy consumption and risk of type 2 diabetes: three cohorts of US adults and an updated meta-analysis. BMC Med 12:215

Clark M (2016) Real raw milk facts: state by state. Retrieved from http://www.realrawmilkfacts.com/raw-milk-regulations

Clarke C (2012) The science of ice cream. (2nd edn). Royal Society of Chemistry, Cambridge, 213pp. https://books.google.com.gh/books?id=Zd10DZiL2LAC

Codex Alimentarius Commission (2003, 2008) Codex standard for fermented milks. Retrieved from http://www.fao.org/input/download/standards/400/CXS_243e.pdf

Codex Alimentarius Commission (2004, 2009) Code of hygienic practice for milk and milk products: CAC/RCP 57–2004. Retrieved from http://www.fao.org/fao-who-codexalimentarius/standards/list-of-standards/en/

Colavita G, Amadoro C, Rossi F, Fantuz F, Salimei E (2016) Hygienic characteristics and microbiological hazard identification in horse and donkey raw milk. Vet Ital 52(1):21–29. https://doi.org/10.12834/VetIt.180.545.1

Conte F, Panebianco A (2019) Potential hazards associated with raw donkey milk consumption: a review. Int J Food Sci 2019:5782974. https://doi.org/10.1155/5782974/2019

Crow V, Curry B (2002) Lactobacillus spp: lactobacillus delbrueckii group. Encyclopedia of dairy sciences (Roginski H, ed). Elsevier,

Oxford, pp 1494–1497. https://doi.org/10.1016/B1490-1412-227235-227238/200243-227231
Davies FL, Law BA (1984) Advances in the microbiology and biochemistry of cheese and fermented milk. Sole distributor in the USA and Canada, Elsevier Science, 260pp
Diaz O, Pereira CD, Cobos A (2009) Application of whey protein concentrates and isolates in the food industry. Alimentaria 400:108–115
Draaiyer J, Dugdill B, Bennett A, Mounsey J (2009) Milk testing and payment systems resource book. A practical guide to assist milk producer groups. Retrieved from http://www.fao.org/docrep/012/i0980e/i0980e00.htm
EFSA Panel on Dietetic Products NAAN (2010) Scientific Opinion on the substantiation of health claims related to live yoghurt cultures and improved lactose digestion (ID 1143, 2976) pursuant to Article 13(1) of Regulation (EC) No 1924/2006. EFSA J 8(10):1763 [18 pp]. Retrieved from http://www.efsa.europa.eu/en/efsajournal/pub/1763
Europa (2016) EU Register on nutrition and health claims. Retrieved from http://ec.europa.eu/nuhclaims/?event=search&CFID=2245312&CFTOKEN=62fa5706da6100c4-BD8F1AE7-9B58-B8CF-D5CCABFED7E8C28C&jsessionid=93121fbb83870ed1931e25658f463b940227TR
FAO (2016) Dairy production and products: quality and testing. Retrieved from http://www.fao.org/agriculture/dairy-gateway/milk-and-milk-products/quality-and-testing/en/#.V_KiMk3rtLM
Feder D (2016) Dairy analogs. Retrieved from http://www.preparedfoods.com/articles/114081-dairy-analogs
Federal Milk Market Administrator (2015) 2013 Milk production
Feeney EL, Nugent AP, McNulty B, Walton J, Flynn A, Gibney ER (2016) An overview of the contribution of dairy and cheese intakes to nutrient intakes in the Irish diet: results from the national adult nutrition survey. Brit J Nutr 115(4):709–717. https://doi.org/10.1017/S000711451500495X
Fleming DW, Cochi SL, MacDonald KL, Brondum J, Hayes PS, Plikaytis BD et al (1985) Pasteurized milk as a vehicle of Infection in an outbreak of listeriosis. N Engl J Med 312(7):404–407. Retrieved from http://www.nejm.org/doi/full/10.1056/NEJM198502143120704
Food and Agriculture Organization of the United Nations (1998) Milk hygiene on the dairy farm: A practical guide for milk producers: FAO Animal Production and Health Paper 78. Retrieved from http://www.fao.org/docrep/004/T0218E/T0218E00.htm#TOC
Food Safety News (2016) Raw milk. Retrieved from http://www.foodsafetynews.com/tag/raw-milk/#.V5x3AdwZH8o
Fox PF, McSweeney PLH, Cogan TM, Guinee TP (2004) Cheese: chemistry, physics and microbiology, vol 2
Fox PF, McSweeney PLF (2004) Cheese: An overview. In: Fox PF, McSweeney PLF, Cogan TM, Guinee TP (eds) Cheese: chemistry, physics and microbiology. Third edition. Volume 1. General aspects. Academic, pp 1–18
Gaya A, Calvo J (2018) Improving pasteurization to preserve the biological components of donated human milk. Front Pediatr 6:288
Gettys SC, Davidson PM (1985) A comparison of buttermilks made using fermentation direct acidification and a combination of both methods. J Dairy Sci 68(3):620–625. https://www.sciencedirect.com/science/article/pii/S0022030285808675
Gillis WT, Cartledge MF, Rodriguez IR, Suarez EJ (1985) Effect of raw milk quality on ultra-high temperature processed milk. J Dairy Sci 68(11):2875–2879
Goff DH (2016) The dairy science and technology eBook. Retrieved from https://www.uoguelph.ca/foodscience/book-page/dairy-science-and-technology-ebook
Goff HD, Hartel RW (2013) Ice cream, 7th edn. Springer US, New York
Guarner F, Perdigon G, Corthier G, Salminen S, Koletzko B, Morelli L (2005) Should yoghurt cultures be considered probiotic? Br J Nutr 93(6):783–786
Haenlein G (2004) Goat milk in human nutrition. Small Rumin Res 51(2):155–163. https://www.amalattea.com/documenti/158-latte-di-capra-e-nutrizione-umana.pdf
Harding F (1995) Milk quality. Springer
Hefa C (2010) Volatile flavor compounds in yogurt: a review. Crit Rev Food Sci Nutr 50(10):938–950. https://doi.org/10.1080/10408390903044081
Helweg R (2010) The complete guide to making cheese, butter, and yogurt at home: everything you need to know explained simply. Atlantic Publishing Group
Hill AR (1988) Quality of ultra-high-temperature processed milk. Food Technol 42(9):92–97
Hill AR (2016a) Cheese making technology eBook Retrieved from https://www.uoguelph.ca/foodscience/book-page/cheese-making-technology-ebook
Hill AR, Ferrer MA (2021) Table 3.2: typical composition (% by weight) of some cheese varieties. Cheese Making Technology eBook Retrieved from https://books.lib.uoguelph.ca/cheesemakingtechnologyebook/chapter/other-technological-criteria/
Horner SA, Wallen SE, Caporaso F (1980) Sensory aspects of UHT milk combined with whole pasteurized milk. J Food Prot 43(1):54–57
Huth PJ, Park KM (2012) Influence of dairy product and milk fat consumption on cardiovascular disease risk: a review of the evidence. Adv Nutr 3(3):266–285. https://doi.org/10.3945/an.3112.002030
International Dairy Food Association (2019) Pasteurization. Retrieved from https://www.idfa.org/news-views/media-kits/milk/pasteurization
Keast DR, Fulgoni VL, Nicklas TA, O'Neil CE (2013) Food sources of energy and nutrients among children in the United States: national health and nutrition examination survey 2003–2006. Nutrients 5(1):283–301. https://doi.org/10.3390/nu5010283
Kroger M (1976) Quality of yogurt. J Dairy Sci 59(2):344–350

Kurmann JA, Rasic JL, Kroger M (1992) Encyclopedia of fermented fresh milk products: an international inventory of fermented milk, cream, buttermilk, whey, and related products. Springer, US, 368pp. https://books.google.com.gh/books?id=ucPf5kCNGjMC

Kurtz FE, Tamsma A, Pallansch MJ (1971) Organoleptic properties of foam spray-dried products made from deodorized milk fat and skim-milk. J Dairy Sci 54(2):173–177

Kurwijila LR (2006) Volume 1: hygienic milk handling, processing and marketing: reference guide for training and certification of small-scale milk traders in Eastern Africa. Retrieved from https://cgspace.cgiar.org/bitstream/handle/10568/1697/TrainerGuideVol--1_C.pdf

Langer AJ, Ayers T, Grass J, Lynch M, Angulo FJ, Mahon BE (2012) Nonpasteurized dairy products, disease outbreaks, and state laws-United States, 1993–2006. Emerg Infect Dis 18(3):385–391. https://doi.org/10.3201/eid1803.111370

Law BA, Tamime AY (2011) Technology of Cheese making. Wiley

Lederle EJ (1907) Pasteurization of milk. Am J Public Hyg 17(2):164–181. Retrieved from http://www.ncbi.nlm.nih.gov/pubmed/19599214

Leedom JM (2006) Milk of nonhuman origin and infectious diseases in humans. Clin Infect Dis 43(5):610–615. https://doi.org/10.1086/507035

Lejeune JT, Rajala-Schultz PJ (2009) Food safety: unpasteurized milk: a continued public health threat. Clin Infect Dis 48(1):93–100

Lewis MJ (2012) Heat treatment of milk. In: Robinson RK (ed) Modern dairy technology Vol 1. Advances in milk processing, vol 1, 2nd edn. Springer, pp 1–60

Lucey JA (2015) Raw milk consumption: Risks and benefits. Nutr Today 50(4):189–193

MacDonald JM, Newton D (2016) Milk production continues shifting to large-scale farms. Retrieved from http://www.ers.usda.gov/amber-waves/2014-december/milk-production-continues-shifting-to-large-scale-farms.aspx#.V7N37036tLM

Marette A, Picard-Deland E (2014) Yogurt consumption and impact on health: focus on children and cardio-metabolic risk. Am J Clin Nutr 99(5):2014

MarketsandMarkets (2016) Dairy alternatives market by type (soy milk, almond milk, rice milk, others), formulation (plain sweetened, plain unsweetened, flavored sweetened, flavored unsweetened), application (food & beverages) & region – Global forecast to 2020. Retrieved from http://www.marketsandmarkets.com/Market-Reports/dairy-alternative-plant-milk-beverages-market-677.html

Marshall Revise Text, Goff HD, Hartel RW (2003) Ice cream. Springer, New York, 371pp. https://books.google.com.gh/books?id=DN9Ju6oiSWkC

Matalon ME, Sandine WE (1986) Lactobacillus bulgaricus, Streptococcus thermophilus and yogurt: a review. Cult Dairy Prod J 21(4):6–12

Meeks M, Franks A, McGregor H, Lamb R, Webb G (2019) Supporting mothers, protecting babies for long-term health: establishing a pasteurized human milk bank. NZ Med J 132(1505):83–91. https://www.nzma.org.nz/journal-articles/supporting-mothers-protecting-babies-for-long-term-health-establishing-a-pasteurised-human-milk-bank

Meyer AL (2015) Don't Have a Cow, Man! Recognizing herd share agreements for raw milk. Case Western Reserve University Health Matrix: Journal of Law-Medicine 25. Retrieved from http://scholarlycommons.law.case.edu/healthmatrix/vol25/iss1/13

Milk Facts (2016a) Cheese production. Retrieved from http://www.milkfacts.info/Milk%20Processing/Cheese%20Production.htm

Milk Facts (2016b) Standards of identity for dairy products. Retrieved from http://www.milkfacts.info/Milk%20Processing/Standards%20of%20Identity.htm

MilkIngredients.Ca (2011a) Fermented milk products. Retrieved from http://www.milkingredients.ca/index-eng.php?id=180

MilkIngredients.Ca (2011b) Buttermilk powder. Retrieved from https://application.cdc-ccl.gc.ca/ddpp-ppdpl/PearsieCopy/milking/index-eng.php?id=172

Morales-De la Pena M, Martin-Belloso O (2009) High-pulsed electric fields: extending shelf life. Food Beverage International 8(6):38–39

Morelli L (2014) Yogurt, living cultures, and gut health. Am J Clin Nutr 99(5 Suppl):1248S–1250S

Mosqueda-Melgar J, Elez-Martineza P, Raybaudi-Massilia RM, Martin-Belloso O (2008) Effects of pulsed electric fields on pathogenic microorganisms of major concern in fluid foods: a review. Crit Rev Food Sci Nutr 48(8):747–759

Mullan WMA (2016) Compositional standards for ice cream. [On-line] From: https://www.dairyscience.info/index.php/202-uncategorised-sp-269/296-standards.html

Mungai EA, Behravesh CB, Gould LH (2015) Increased outbreaks associated with nonpasteurized milk, United States, 2007-2012. Emerg Infect Dis 21(1):119–122. https://doi.org/10.3201/eid2101.140447

Oliver SP, Boor KJ, Murphy SC, Murinda SE (2009) Food safety hazards associated with consumption of raw milk. Foodborne Pathog Dis 6(7):793–806. https://doi.org/10.1089/fpd.2009.0302

O'Sullivan TA, Hafekost K, Mitrou F, Lawrence D (2013) Food sources of saturated fat and the association with mortality: a meta-analysis. Am J Public Health 103(9):e31–e42. https://www.ncbi.nlm.nih.gov/pmc/articles/PMC3966685/

Otunola A, El-Hag A, Jayaram S, Anderson WA (2008) Effectiveness of pulsed electric fields in controlling microbial growth in milk. Int J Food Eng 4(7). https://doi.org/10.2202/1556-3758.1494

Peker H, Arslan S (2013) Effects of addition of locust bean gum on sensory, chemical, and physical properties of low-fat yoghurt. J Food Agric Environ 11(2):274–277

Preben Mikkelsen (PM) Food & Dairy Consulting (2013) World yoghurt market report. Retrieved from http://pmfood.dk/upl/9747/DINMarch5backpage.pdf

Progressive Dairy Man (2016) U.S. dairy exports end five-year growth streak. Retrieved from http://www.progressivedairy.com/news/industry-news/u-s-dairy-exports-end-five-year-growth-streak

Quigley L, Sullivan O, Stanton C, Beresford TP, Ross RP, Fitzgerald GF, Cotter PD (2013) The complex microbiota of raw milk. FEMS Microbiol Rev 37(5):664–698. https://doi.org/10.1111/1574-6976.12030

Ragonnaud G (2014) Trends in EU-third countries trade of milk and dairy products. Retrieved from http://www.europarl.europa.eu/RegData/etudes/STUD/2015/529076/IPOL_STU(2014)529076_EN.pdf

Rasic JL, Kurmann JA (ed) (1978) Fermented fresh milk products, vol. I. Yoghurt. Scientific grounds, technology, manufacture and preparations. Copenhagen, Technical Dairy Publishing House, 466pp. https://books.google.com.gh/books?id=9xi_mQEACAAJ

Rickards BR (1909) The commercial pasteurization of milk. Am J Public Hyg 19(3):507–513

Robinson RK (ed) (1993a) Modern dairy technology, volume 1: advances in milk processing. Springer, New York, 485pp. https://www.springer.com/gp/book/9780834213579

Robinson RK (ed) (1993b) Modern dairy technology, volume 2: advances in milk products. Springer, New York, 516pp. https://books.google.com.gh/books?id=hRBu1w_drVoC

Robinson RK, Tamine AY (1993) Manufacture of yoghurt and other fermented milk products. In: Robinson RK (ed) Modern dairy technology, Volume 2: advances in milk products, vol 2. Springer, pp 1–48

Rombaut R, Dewettinck K, Mangelaere G d, Vooren L v, Huyghebaert A (2002) Raw milk microbial quality and production scale of Belgian dairy farms. Milchwissenschaft 57(11/12):625–628

Rotch TM (1907) Pasteurization of milk for public sale. Am J Public Hyg 17(2):181–211. Retrieved from http://www.ncbi.nlm.nih.gov/pubmed/19599385

Routray W, Mishra HN (2011) Scientific and technical aspects of yogurt aroma and taste: a review. Compr Rev Food Sci Food Saf 10(4):208–220

Sarkar S (2008) Effect of probiotics on biotechnological characteristics of yoghurt: a review. Br Food J 110(7):717–740

Schaeffer J (2016) Dairy food substitutes — The sky's the limit. Today's Dietitian 14(8):38. Retrieved from http://www.todaysdietitian.com/newarchives/080112p38.shtml

Sepulveda DR, Gongora-Nieto MM, Guerrero JA, Barbosa-Canovas GV (2009) Shelf life of whole milk processed by pulsed electric fields in combination with PEF-generated heat. LWT Food Sci Technol 42(3):735–739. https://doi.org/10.1016/j.lwt.2008.10.005

Shepard L, Miracle RE, Leksrisompong P, Drake MA (2013) Relating sensory and chemical properties of sour cream to consumer acceptance. J Dairy Sci 96(9):5435–5454. https://www.journalofdairyscience.org/article/S0022-0302(5413)00488-00488/pdf

Shija F (2013) Assessment of milk handling practices and bacterial contaminations along the dairy value chain in Lushoto and Handeni districts, Tanzania. MSc thesis, Sokoine University of Agriculture. Morogoro, Tanzania: Sokoine University of Agriculture. Retrieved from https://cgspace.cgiar.org/handle/10568/73335

Simons M (2016) Explosive growth in dairy alternatives market expected through 2020, study finds Retrieved from https://www.plantbasedfoods.org/explosive-growth-dairy-alternatives-market-expected-2020-study-finds/

Sodini I, Morin P, Olabi A, Jimenez-Flores R (2006) Compositional and functional properties of buttermilk: a comparison between sweet, sour, and whey buttermilk. J Dairy Sci 89(2):525–536

Stiles ME, Holzapfel WH (1997) Lactic acid bacteria of foods and their current taxonomy. Int J Food Microbiol 36(1):1–29

Stogo M (1998) Ice cream and frozen desserts: a commercial guide to production and marketing. Wiley, New York, 560pp

Tamang JP, Watanabe K, Holzapfel WH (2016) Review: diversity of microorganisms in global fermented foods and beverages. Front Microbiol 7:377

Tamime AY (2008). Fermented milks. Blackwell Science Ltd, Oxford, 208pp. https://books.google.com.gh/books?id=xKAu9IYnK2wC

Tamime AY (2009) Milk processing and quality management. Wiley

Tamime, A. Y., & Robinson, R. K. (1999). Yoghurt: science and technology

Tamime AY, Robinson RK (2007a) Chapter 8: Preservation and production of starter cultures Tamime and Robinson's yoghurt: science and technology. (3rd edn, Tamime AY, Robinson RK, eds). Woodhead Publishing Ltd, Cambridge, pp 486–514

Tamime AY, Robinson RK (2007b). Tamime and Robinson's yoghurt: science and technology. 3. Cambridge, Woodhead Publishing Ltd, 791pp. https://books.google.com.gh/books?id=K1GkAgAAQBAJ

Tamine AY, Robinson RK (1976) Recent developments in the production and preservation of starter cultures for yogurt. Dairy Ind Int 41(11):408–411

The American Association of Medical Milk Commissions Inc (1999) Methods and standards for the production of certified milk Retrieved from http://web.ncf.ca/fk980/nm/aammc.htm

Touch V, Deeth HC (2009) Microbiology of raw and market Milks. In: Tamime AY (ed) Milk processing and quality management. Blackwell Publishers, Chichester, pp 48–100

Transparency Market Research (2014) Global cheese market – Industry analysis, size, share, growth, trends and forecast, 2013–2019. Retrieved from http://www.transparencymarketresearch.com/Global-cheese-market.html

United States Department of Agriculture (2000) USDA specifications for sour cream and acidified sour cream. Retrieved Aug 27, 2021, from https://www.ams.usda.gov/sites/default/files/media/sourcream.pdf

U. S. Food and Drug Administration (1978, 2015) 21CFR135.110 ice cream and frozen custard. Retrieved from http://www.accessdata.fda.gov/scripts/cdrh/cfdocs/cfcfr/cfrsearch.cfm?fr=135.110

U. S. Food and Drug Administration (2003) M-I-03-8: IDEXX Laboratories Inc. New Snap Beta-lactam Test. Retrieved from http://www.fda.gov/Food/GuidanceRegulation/GuidanceDocumentsRegulatoryInformation/Milk/ucm079111.htm

U. S. Food and Drug Administration (2015) CFR – Code of Federal Regulations Title 21: Part 131 Milk and Cream. Retrieved from https://www.accessdata.fda.gov/scripts/cdrh/cfdocs/cfcfr/CFRSearch.cfm?CFRPart=131

U. S. Food and Drug Administration (2016) The dangers of raw milk: unpasteurized milk can pose a serious health risk. Retrieved from http://www.fda.gov/Food/ResourcesForYou/Consumers/ucm079516.htm

U.S. Dairy Export Council (2004) Reference manual for U.S. Whey and lactose products. Retrieved from http://www.thinkusadairy.org/resources-and-insights/resources-and-insights/product-resources/reference-manual-for-us-whey-and-lactose-products.#sthash.JUaP9VgE.dpuf

U.S. Department of Health and Human Services/Public Health Service/ Food and Drug Administration (2014) Grade "A" Pasteurized Milk Ordinance: 2013 Revision. Retrieved from https://agriculture.az.gov/2013-pasteurized-milk-ordinance

US Government Publishing Office (1978) 21 CFR 135 – Frozen desserts. Retrieved from https://www.gpo.gov/fdsys/granule/CFR-2011-title21-vol2/CFR-2011-title21-vol2-part135

USDA Agricultural Research Service (2021) Food data central search results. Vanilla ice creams. Retrieved Aug 25, 2025, from https://fdc.nal.usda.gov/fdc-app.html#/food-details/167575/nutrients

USDA Economic Research Services (2016) Milk production and inventories continue shifting to larger herds. Retrieved from http://www.ers.usda.gov/data-products/chart-gallery/detail.aspx?chartId=56840

USDA Economics Statistics and Market Information System (2010) Overview of the U. S. Dairy Industry (Released September 22, 2010, by the National Agricultural Statistics Service (NASS), Agricultural Statistics Board, United States Department of Agriculture (USDA). Retrieved from http://usda.mannlib.cornell.edu/usda/current/USDairyIndus/USDairyIndus-09-22-2010.pdf

USDA Economics Statistics and Market Information System (2015) Dairy products annual summary. Retrieved from http://usda.mannlib.cornell.edu/MannUsda/viewDocumentInfo.do?documentID=1054

USDA Economics Statistics and Market Information System (2016) Dairy products. Retrieved from http://usda.mannlib.cornell.edu/MannUsda/viewDocumentInfo.do?documentID=1052

USDA Foreign Agricultural Services (2014) Cow's milk production and consumption: summary for selected countries Retrieved from http://apps.fas.usda.gov/psdonline/psdHome.aspx

USDA Foreign Agricultural Services (2016) Butter production and consumption: summary for selected countries Retrieved from http://apps.fas.usda.gov/psdonline/psdHome.aspx

Varnam AH, Sutherland JP (2012) Milk and milk products: technology, chemistry and microbiology. Springer

Verraes C, Claeys W, Cardoen S, Daube G, De Zutter L, Imberechts H, Dierick K, Herman L (2014) A review of the microbiological hazards of raw milk from animal species other than cows. Int Dairy J 39(1):121–130

Verraes C, Vlaemynck G, Van Weyenberg S, De Zutter L, Daube G, Sindic M, Uyttendaele M, Herman L (2015) A review of the microbiological hazards of dairy products made from raw milk. Int Dairy J 50:32–44. https://orbi.uliege.be/bitstream/2268/185614/185611/185611-s185612.185610-S095869461500120X-main.pdf

Vissers MMM, Driehuis F (2009) On-farm hygienic milk production. Milk processing and quality management (Tamime AY, ed). Wiley, Chichester, pp 1–22. https://books.google.co.uk/books?id=xBaz25ow23jZcC

Vlahović B, Popović-Vranješ A, Mugoša I (2014) International cheese market – current state and perspective. Eco Insights Trends Challenges III (LXVI) No. 1/2014:35–43. Retrieved from http://www.upg-bulletin-se.ro/archive/2014-1/4.Vlahovic_PopovicVranjes_Mugosa.pdf

Vieira ER (2013) Elementary food science, 4th edn. Springer, 423 pp

Walstra P, Wouters JTM, Geurts TJ (2005) Dairy science and technology, 2nd edn. CRC Press

Walther B, Schmid A, Sieber R, Wehrmüller K (2008) Cheese in nutrition and health. Dairy Sci Technol 88(4–5):389–405. https://doi.org/10.1051/dst:2008012

World Health Organization (2009) Europe puts health claims to the test. Bull World Health Organ 87(9):645–732. Retrieved from http://www.who.int/bulletin/volumes/87/9/09-020909/en/

19 Poultry and Eggs

1 Introduction

1.1 The Nature of Poultry

Poultry is described in the United States(US) Code of Federal Regulations as, "*meat derived from the slaughter and processing of broilers, young chickens, mature chickens, hens, turkeys, capons, geese, and ducks; rabbits, small game fowl such as quail or pheasants*". Some descriptions for poultry includes ratites (e.g. ostriches) and young pigeons (squabs) bred for the table (U.S. Government Publishing Office 2014). Poultry refers to farmed birds within the European Union. Poultry products refer to various poultry meat and dishes sold in the supermarkets and which have US standards for identify (Sect. 3.3) (Table 19.1).

1.2 The US Poultry Industry at Glance

The US poultry processing industry (NAICS 311615) is the largest of its kind globally. Poultry is the most popular single source of animal protein in the US; Americans eat greater quantities of chicken compared to either beef or pork. The poultry industry is part of the animal slaughtering & processing industry (NAICS 31161).

The US poultry industry (production and trade) is currently valued at about US$48 billion covering chicken, turnkey, and eggs. The total weight of poultry amounts to 40.1 billion pounds (broiler meat) of which, about 82% is used for domestic consumption and 18% is exported. Only Brazil exports more poultry globally compared with the US (discussed further in Sect. 1.3) (Table 19.2).

The value of the poultry sector grew by 19.5% over the past 10 years (see Tables 19.3 and 19.4) in-line with the ~19.8% rate of inflation over the same period (USDA Economic Research Services 2016; USDA National Agricultural Statistics Service 2016). In practical terms, the affordability of poultry has remained unchanging over the recent past. A summary of some key economic data, including the average wholesale and retail price for America poultry is shown in Tables 19.3 and 19.4.

Poultry is popular in the US as indicated by an annual consumption average of ~63 pounds for chicken, 13–14 pounds for turkey, and 250 eggs per person (North American Meat Institute (NAMI) 2016). OECD estimates indicates an American poultry consumption rate of 100 pounds per person per year, which ranks second behind of Israel and just above Saudi Arabia and Malaysia (OECD 2016). Values for US poultry consumption are compared with data for other big consumer nations in Table 19.5.

R. Owusu-Apenten, E. R. Vieira, *Elementary Food Science*, Food Science Text Series,
https://doi.org/10.1007/978-3-030-65433-7_19

Table 19.1 Definitions for poultry

Poultry/ EU	Farmed birds, birds not considered as domestic but farmed as domestic animals, with the exception of ratites.
USA	Any domesticated bird (chickens, turkeys, ducks, geese, guineas, ratites, or squabs - young pigeons from one to about 30 days of age), whether live, or dead.

Table 19.2 NAICS 311615 poultry processing

311 – Food manufacturing
3116 – Animal slaughtering and processing
31161 – Animal slaughtering and processing
311611 – Animal (except Poultry) Slaughtering
311612 – Meat Processed from Carcasses
311613 – Rendering and Meat Byproduct Processing
311615 – Poultry Processing[a]
Canning poultry (except baby and pet food)
Chickens, processing, fresh, frozen, canned, or cooked (except baby and pet food)
Chickens, slaughtering and dressing
Dressing small game
Ducks, processing, fresh, frozen, canned, or cooked
Ducks, slaughtering and dressing
Geese, processing, fresh, frozen, canned, or cooked
Geese, slaughtering and dressing
Hams, poultry, manufacturing
Hot dogs, poultry, manufacturing
Luncheon meat, poultry, manufacturing
Meat canning, poultry (except baby and pet food), manufacturing
Meat products (e.g., hot dogs, luncheon meats, sausages) made from a combination of poultry and other meats
Poultry (e.g., canned, cooked, fresh, frozen) processing
Poultry canning (except baby, pet food)
Poultry slaughtering, dressing, and packing
Processed poultry manufacturing
Rabbits processing (i.e., canned, cooked, fresh, frozen)
Rabbits slaughtering and dressing
Small game, processing, fresh, frozen, canned or cooked
Small game, slaughtering, dressing and packing
Turkeys, processing, fresh, frozen, canned, or cooked
Turkeys, slaughtering and dressing

[a]Subcategories of poultry processing industry are also listed. Adapted from (NAICS Association 2020)

1.3 World Poultry Trade

The total amount of poultry produced globally was estimated at 199.3 billion pounds and valued at about US$140 billion (USDA Foreign Agricultural Services 2016).[1]The top-four poultry producing regions (US> Brazil, China> EU) account for 60% of the world production (Table 19.6). Some of the big producers of poultry are also big consumers.

Currently, 12.6% of the global poultry production is traded internationally. The major poultry exporting regions are Brazil > US> EU > Thailand with 54%, 37%. 15.5% and 8.6% export market share, respectively. Brazil and the US accounted for 91% of global poultry export market (Fig. 19.1). The dollar-value for poultry trade was approximately US$19 billion assuming a wholesale price of about $0.5 per pound for frozen broiler meat (OECD 2016).

The major importers for poultry are, Japan > Mexico > Saudi Arabia >EU (Fig. 19.2). When considered on per population basis, the Gulf Corporation Councils states (Oman, Qatar, UAE, Kuwait, Bahrain and Saudi Arabia) are significant importers accounting for 15% of world imports. Many countries are believed to import poultry not only for consumption domestically but also for re-export (USDA Foreign Agricultural Services 2016) (Figs. 19.1 and 19.2).

1.4 Poultry Preferences

1.4.1 Consumption patterns for poultry

Preferences for different types of meat were discussed in Chap. 17. The high preferences for poultry (Saudi Arabia, Peru, Iran, Israel) may be related to the ease/ of raising poultry under adverse weather conditions compared with other livestock, imports degree, and extent of re-exports. There is clearly a high preference for poultry in countries such as Saudi Arabia, Peru, Malaysia and Israel where >70% of meat intake is in the form of poultry (Table 19.7). However, the big consumers for poultry are, Israel (140 lb./person) > USA (100 lb./ person) > Saudi Arabia (96 lb. / person) followed by Malaysia and Australia (91 lb. / person).

[1] Approximate value of world broiler production (US$140 billion) assuming a whole sale price of US$0.70 per pound of broiler meat.

Table 19.3 Poultry and egg industry overview

Year	Value (US$billion)	Farm Receipts (US$billion)	Prod. (Billion lb.)	Exports billion lb.	Export value (US$billion)	Wholesale Price (US$/lb.)	Retail Price (US$/lb.)
2006	41	17.7	35.5	5.2	1.8	0.560	1.571
2007	42	21.5	36.2	5.9	2.7	0.763	1.651
2008	44	23.2	36.9	7.0	3.5	0.711	1.746
2009	44	21.8	35.5	6.8	3.3	0.755	1.780
2010	45		36.9	6.8	3.1	0.743	1.753

Adapted from (USDA Economic Research Services 2016)

Table 19.4 Estimated value of production for poultry – United States, 2014 & 2015

Year	Broilers	Turkeys	Chickens	Eggs	Total
	1000$	1000$	1000$	1000$	1000$
2014	32,728,234	5,304,501	96,151	10,257,972	48,386,858
2015	28,709,834	5,707,671	100,899	13,499,904	48,018,508

Data not including states with less than 500,000 broilers, Adapted from (USDA National Agricultural Statistics Service 2016)

Table 19.5 US Poultry consumption compared to rest of the worldwide (Pounds/person/year)

RNK	Country	Poultry
1.	ISR	140
2.	USA	100
3.	SAU	96
4.	MYS	91
5.	AUS	89
6.	BRA	88
7.	PER	81
8.	NZL	80
9.	ARG	78
10.	CAN	74

Adapted from (OECD 2016)

Table 19.6 World broiler meat production and consumption (billion Pounds)[a]

Country	Produced 2016	Country	Consumed 2016
US	40.3	US	33.9
Brazil	30.0	China	28.0
China	28.0	EU	23.3
EU	24.4	Brazil	20.9
India	9.3	India	9.2
Russia	8.3	Mexico	9.0
Mexico	7.2	Russia	8.5
Argentina	4.6	Japan	5.2
Turkey	4.2	Argentina	4.3
Thailand	3.9	South Africa	4.0
Indonesia	3.6	Indonesia	3.6
Others	33.6	Others	43.2
Total	199.3	Total	193.2

[a]Adapted from (USDA Foreign Agricultural Services 2016) data converted from 1000-metric tons

In the US, most poultry used for food consists of chicken and turkey. Ducks and geese are relatively insignificant as food sources. The rise in poultry consumption in the US is mainly due to its low production costs and its dietary health benefits. The low production costs result from the high feed-conversion ratio for poultry. For every 1.8 lb. of food, the bird will develop 1 lb. of meat. This is a higher ratio than can be obtained raising beef, pork, or lamb (Bell and Weaver 2013; Parkhurst and Mountney 2012).

1.4.2 Nutritional considerations for poultry meat

The edible portion of chicken is about 54% of the carcass. This includes meat (about 39%), skin, and parts of the viscera that are also edible (about 15%). The nutritional composition for the meat from a roasted chicken is roughly 30% protein, 3% fat (2% saturated fat), and about 124 mg of cholesterol per one cup serving (140 g [4.9 oz.]); there is about 1% ash (minerals), and the remaining 64% is mostly water (Table 19.8).

Increasing public awareness of the excellent nutritional properties of poultry may be another reason for the high preference. Chicken and turkey are both relatively good sources of iron and low in fat (Table 19.8).

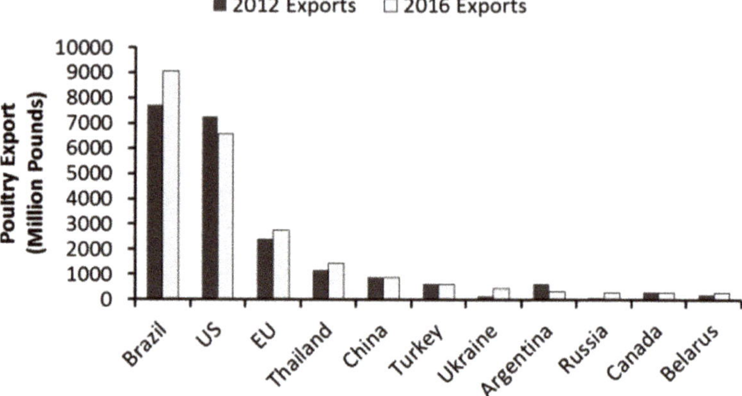

Fig. 19.1 World broiler meat exports by country

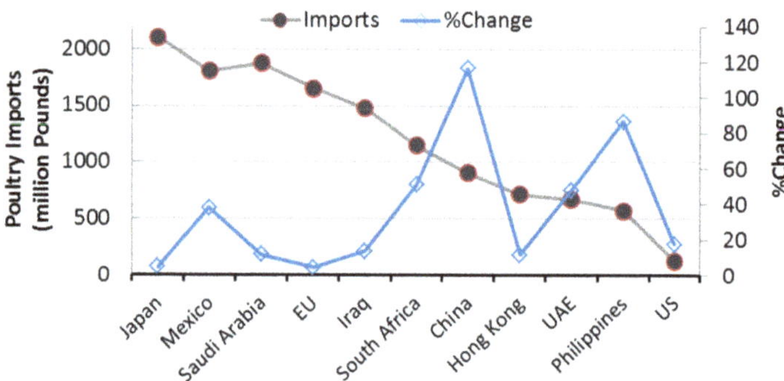

Fig. 19.2 World broiler meat imports by country for 2016, and percentage change since 2012

The different cooking methods for poultry had an effect on its nutritional characteristics. For example, broiled (grilled) chicken breast and boiled chicken breast meat had similar values for total fat, saturated fat and cholesterol. In contrast, frying led to a 2-fold increase in the total fat content for chicken breast meat compared with values obtained with boiling or broiling. Skinless chicken breast meat had only a-third of the total fat or saturated fat content compared with chicken breast meat with skin. The cholesterol value for chicken breast meat was not affected so much by the cooking method (Scherr and Ribeiro 2013). Oven cooking did not affect the total fat content of chicken, but removing the skin (before or after cooking) reduced the total fat and saturated fatty acid content by 2-3 fold (Prusa and Lonergan 1987). For both chicken and mutton, grilling (broiling) had a greater effect compared to roasting on the total fat content, perhaps due to the higher cooking losses with the former (Alina et al. 2012).

The total fat content for another type of poultry (quail) meat was 1.3% (uncooked, control), 1.5% (microwave cooking), 1.8% (fried with limited oil), 2.9% (steamed), 4.3% (roasted) and 5.9% (fried with excess oil). The saturated fatty acid content for quail decreased for most forms of cooking with the exception of steaming (Mohamadi et al. 2020). Overall, the effect of different cooking methods on fat content persist even after considering food moisture changes during cooking (Hewavitharana et al. 2020; Biandolino et al. 2021). Clearly, it would be useful to know more about how different cooking methods affect the composition of foods. Cooking methods are described in Chap. 12, Sect. 8.4).

There is currently, conflicting research on whether the amount of cholesterol ingested is as significant as the amount of saturated fat in the development of heart disease. Because of this, coupled with new data on cholesterol levels in the average egg (see end of this chapter), consumption of eggs may actually increase in future years.

The low cost of poultry (Table 19.3) and, especially, the public's perception of its health benefits, accounts for the increased use of poultry in the production of a hamburger like product and in the production of all-poultry frankfurters. Poultry is also allowed in the production of conventional frankfurters in amounts of up to 15% and poultry is being used in the production of a bologna-type and other cold-cut products. In all cases where poultry is used, either wholly or partially, the fat and cholesterol contents of these products are usually lower than in their beef counterparts. Check the nutrition label to be sure.

Table 19.7 Preferences for poultry meat worldwide

RNK	Region	Poultry ('%)[a]
1	SAU	80.7
2	PER	80.1
3	MYS	77.3
4	IRN	76.4
5	ISR	73.5
6	GHA	65.2
7	ZAF	64.0
8	IDN	58.9
9	IND	57.2
10	COL	56.8
11	TUR	55.7
12	MEX	53.1
13	UKR	51.1
14	NZL	50.5
15	BRA	50.4
16	USA	49.9
17	THA	39.4
18	WLD	39.1
19	BGD	37.0
20	JPN	36.7
21	DZA	36.6
22	PHL	36.2
23	ZMB	35.0
24	EU28	33.8
25	PAK	33.6
26	BRICS	31.7
27	SSA	29.2

[a]Protein (%) intake that is poultry

1.5 Small-Scale Family Poultry Production

Small-scale poultry production can make significant contributions to poverty alleviation for many rural poor (Sonaiya and Swan 2004). Family-poultry defined as "*small-scale poultry keeping by households using family labor*" is important for diversifying income, providing fertilizer and improving general food security. The FAO in particular has over many years supported developments of training schemes and information intended to support small-scale poultry producers (Sonaiya and Swan 2004). In addition, mention must be made of the numbers of poultry fanciers, hobbyist, homesteaders for whom breeding and keeping poultry is an important leisure activity as well as providing food.

1.6 Hygiene considerations for Poultry

Foodborne hazards from poultry are probably no more or less serious compared to other farm animals. However, the popularity of poultry items in the diet could mean that the frequency of exposure

Table 19.8 Nutritional characteristics of different types of meat[a]

	Energy (kcal)	Fat (g)	Protein (g)	Iron (mg)	Vit. B12 (mcg)	Niacin (mg)
Beef[b]	187.1	6.6	29.4	2.6	2.6	6.0
Pork	174.1	6.2	27.4	0.8	0.6	8.0
Lamb	205.9	9.4	27.8	2.2	2.8	7.1
Chicken	164.7	3.5	30.6	1.1	0.4	13.8
Turkey	135.3	0.7	30.6	1.5	0.4	7.5
Tofu	132.9	8.0	14.1	2.6	0.0	0.9
Broccoli	26.5	0.0	1.8	0.9	0.0	0.6
Peanuts	291.8	24.7	12.4	1.8	0.0	7.1
Black Beans	88.2	0.5	5.3	0.9	0.0	1.2

Adapted from (North American Meat Institute (NAMI) 2016)
[a]Value for 3 oz. adjusted to 100 g compared with major plant protein sources
[b]Average value for lean beef cuts

to foodborne hazards would be greater. Salmonella and campylobacter infections from poultry have been widely discussed by experts, as affecting both carcasses and eggs.

Chicken is the most consumed single species in the US and in many parts of the world. Chapters 16 and 17 cover food microorganisms and some hygiene issues important for the safe handling and preparation foods. There have been recent USDA programs aiming to educate the general consumer on the safe handling and cooking of chicken (USDA Food Safety and Inspection Service 2014).

Poultry production is an intricate business, and consequently safety and hygiene discussions assume a "farm to fork", or primary production-to-consumer approach. The avoidance of food-borne microbes requires action, by hatcheries, feed mills, farmers, processing and packaging, distributers, retailers and the consumer.

Chicken should be produced under hygienic conditions to minimize microbial contamination; indeed many producers adhere to written codes of practice that sets out the requirements for optimum animal welfare, requirements for staff training, building specifications in-line with GMP (National Farm Animal Care Council (Canada) 2016).

USDA inspectors examine poultry arriving at processing plants. Raw chicken presents a risk of pathogen exposure and must be handled carefully and cooked properly according to food hygiene guidelines. Some of the "farm to table" issues important for the safe handling and preparation of chicken are listed in (Table 19.9). A detailed examination of the principles needed to reduce the risk of salmonella and campylobacter infections from poultry is available from Codex. (Codex Alimentarius Commission 2011).

Factors that influence the risk of food-borne illness from poultry range from (i) Freezing and storage – birds are designated as fresh or frozen should be stored at designated temperatures. Frozen birds should be thawed in a refrigerator or under enclosed conditions which avoids cross contamination or can be cooked from frozen, e.g. during roasting, (ii) handling and cooking – birds are required to be cooked to a fixed minimum internal temperature, cf. Chap 17 Sect 2.5, (iii) Concerns about antibiotics. Whereas some antibiotics may be used in the diet, regulations require that birds are allowed a "washout" time before slaughter so that antibiotic levels are typically trace or undetectable. The principles for safe handling and thawing frozen foods are discussed further in Chap. 13 (Sect. 4). Safe cooking temperatures for different kinds of meat are described in Chap. 9 (Sect. 2.5) and Chap. 7 (Sect. 2.5).

Table 19.9 Food safety and hygiene issues for poultry

Chicken Grading
Cooking times
Dating of Chicken Products
Foodborne Organisms
Fresh or Frozen
Home storage time/temperatures
Hormones & Antibiotics
Inspection (process plant)
Marinating
Partial cooking
Rinsing or Soaking Chicken
Storage time
Stuffed Chicken
Thawing conditions
4C (Clean, Cross contamination, Cook, Chill)

Adapted from (USDA Food Safety and Inspection Service 2014)

2 Poultry

2.1 Chicken

For meat production, almost all flocks are started from 1-day-old chicks. The chicks are sexed at the hatchery and divided into males and females since these grow at different rates. Normally chickens are delivered in special trays and transferred to deep liter housing. Commercial growers commonly raise at least four flocks of broilers per year (Mead 2004; Parkhurst and Mountney 2012; Richardson and Mead 1999).

In raising birds for meat, the floor and walls of the chicken houses (brooders) must be cleaned and disinfected. Fresh litter (usually shavings) is put down on the floors. The day-old chicks, which may have been de-beaked to prevent self-harm are introduced. Lights are kept on continually, and the temperature of the brooders is brought to 95 °F (35 °C) in cold weather or 90 °F (32.2 °C) in hot weather. The temperature of the brooders is lowered 5 °F (2.8 °C) weekly until 75 °F (23.9 °C) is reached, and it is held there until the birds are well feathered (Mead 2004).

Troughs or hanging feeders may be used to hold food for meat-type birds and hoppers for grit and calcium supplement may be used. Feeders, which must be adequate in size and number, may be filled automatically. Water troughs or hanging waterers may be used with 16 ft. (4.88 m) of watering space per 200 birds when the temperature is at 75 °F (23.9 °C). When the temperature goes above 80 °F (26.7 °C), 20 ft. (6.1 m) per 200 birds is needed. Feeders and waterers should be kept clean, and waterers should refill automatically.

Food for meat-type birds, a complex mixture, is obtained from feed companies that are expert in formulating such rations.

Trace amounts of antibiotics may be added to feeds to prevent diseases. Among the above ingredients, corn and soybean meals constitute the major portion of the starter feeds. Finishing feeds, which may also include dried whey and steamed bone meal, are used and the combination of technologies from breeders, animal nutritionists, and feed manufacturers has made it possible to raise a 2 to 3 lb. broiler in as little as 6 weeks. Other larger birds such as capons and roasters take longer (Mead 2004).

For birds classed as organic poultry, there should be little or no antibiotics in the feed, as these may be replaced with other agents to improve growth (Griggs and Jacob 2005).

Predators, such as rats, must be kept out of poultry houses. Floors, litter, walls, roosts, and even the birds themselves may have to be treated with malathion insecticide to get rid of mites, lice, or ticks.

2.2 Turkey

The turkey is the second most important poultry in the US. There is a movement to develop strains that produce more meat, especially breast meat. This development has been limited mostly to white and bronze-feathered birds.

Turkey raising requires the same basic conditions as does chicken raising. For birds that are 1–3 weeks old, 1 ft2 (0.09 m2) of floor space is required. Birds that are 4–8 weeks of age require 1.5 ft2 (0.14 m2) per bird, and birds that are 8–15 weeks old require 2 ft2 (0.18 m2) per bird. Turkeys may be vaccinated to prevent erysipelas, salmonellosis, and other diseases.

In growing turkeys, a serious disease called blackhead has been known to break out. It was eventually learned that the disease was caused by a bacterial infection carried by chickens, and it was soon realized that if turkey flocks were going to be kept healthy, they would have to be kept away from chickens or from places where chickens had been, since the bacteria were found in heavy concentrations in chicken droppings and in areas that had been contaminated by chicken droppings.

Turkeys require a higher protein and vitamin content in their food than chickens, although, as with chickens, the protein content of the food is gradually reduced as the birds mature. Turkeys may be marketed as broilers 12–15 weeks old or as mature roasting birds 20–26 weeks of age.

2.3 Ducks and Geese

Ducks and geese are both raised for human consumption, although the amount of the consumption of these species is insignificant in comparison to that of chickens and turkeys. Ducks and geese, for human consumption, have been developed from wild species. Their production and processing are similar to those of chickens.

3 Poultry Processing

3.1 Chicken Slaughter and Handling

Birds grown for meat are the same age, and once marketing size is reached, they are placed in cages and removed from the growing house. They are slaughtered and processed elsewhere. Prior to installation of new flocks, the house is

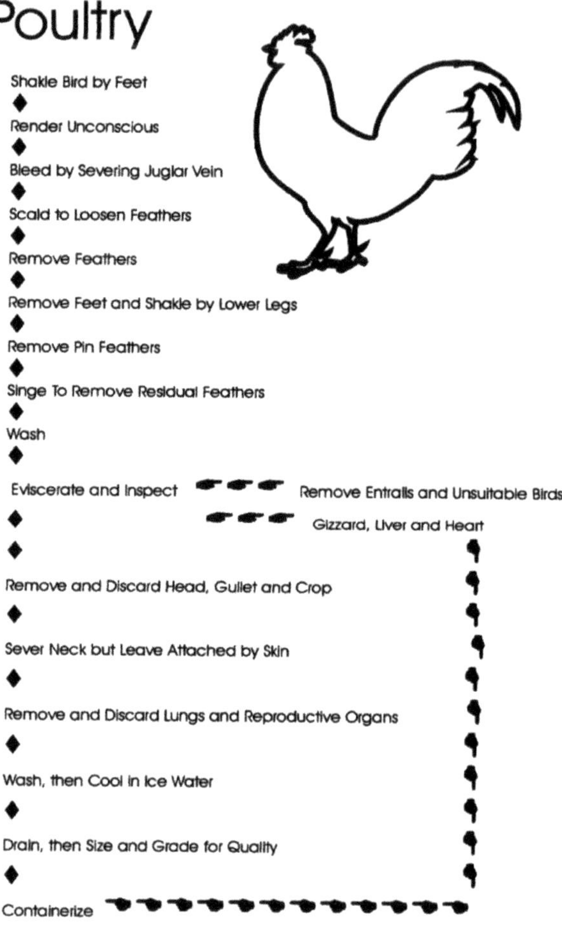

Fig. 19.3 Chicken processing

cleaned and disinfected. In order for a growing operation to be economical, the number of birds in roaster and capon flocks must be no less than 2000, and in broiler flocks, no less than 6000.

Some poultry processing plants are small, but the trend is to process poultry in plants capable of handling at least 10,000 birds per hour. These plants are usually divided into at least two rooms that separate bleeding, scalding, and defeathering from the eviscerating and chilling operations (Mead 2004) (see Fig. 19.3)

Chickens are not fed for about 12 h before they are to be slaughtered in order that their crops will be empty. This is important because it makes the operation much cleaner. The birds, shackled by their feet, are carried in the upside down position by conveyors from one operation to another. After shackling, they are slaughtered by slitting one or both of the jugular veins in the neck. An electrified knife or a stationary electric stunner may be used to render them unconscious prior to bleeding. Rendering birds unconscious prevents broken wings and bruising resulting from the flopping around during bleeding. After their jugular veins are severed, the birds are allowed to bleed from one to several minutes.

Still attached to the conveyor, to facilitate removal of feathers, they next pass through the scalding tank containing water at 135–40 °F (57.2–60 °C) for larger birds or 122–128 °F (50–53.3 °C) for broilers. Immersion times vary with the size of birds, but several minutes in the scalding tank are required even for broilers. If scalding time or temperature is too high, the skin may be damaged. Automatic picking machines are used to remove the feathers from poultry. In some systems, flexible rubber fingers beat the birds as they pass through the machine. In other systems, they are dropped into baskets where feathers are removed by flexible rubber fingers rotating on a central shaft.

After the birds are defeathered, their feet are cut off, and they are rehung on the moving shackles by the lower legs. They then pass along a line where workers remove pinfeathers by hand with the aid of a knife. Then they pass through a gas flame for singeing residual pinfeathers. They are washed externally by water sprays as they pass along the conveyor.

Evisceration is usually carried out as the birds pass along the conveyor. The oil gland may be cut out before or after evisceration. A circular cut is made around the vent and the intestine is then pulled out a few inches. Another cut is then made through the abdominal wall from the vent toward the breastbone for broilers. For larger specimens a horizontal cut is made. The gizzard, liver, heart, and intestines are pulled out and allowed to hang so that they may be examined by a government inspector (a veterinarian) for signs of disease. Diseased birds are removed and destroyed. Gizzards that pass inspection are opened, emptied, peeled, and washed, and then packaged together with the livers and hearts, either for insertion within the bird or for holding separately. Intestines are not used.

The heads, gullets, and crops are removed and discarded, and the neck may be cut off and allowed to hang by the skin. Suction tubes are used to remove lungs and traces of reproductive organs or these are scraped out by hand. Both the interior and exterior of the bird are thoroughly washed. The neck and neck skin are placed in the body cavity. The birds are next cooled, either in an air-agitated ice-water slush in tanks, in continuous ice-water chillers, or in moving refrigerated air. During chilling, the temperature of the birds, which may be 80–95 °F (26.7–35 °C), is lowered to 35 °F (1.7 °C), and they lose some residual blood and pick up a few percent of moisture from the chilled water. After chilling, they are drained, sized according to weight, and graded for quality (Fletcher 2002).

Grading chicken is based on its conformation, fleshing, covering with fat, and the presence of pinfeathers, torn skin, bruises, and so on. Chickens are classified, according to age and condition, as broilers or fryers, roasters, capons (male castrated prior to maturity), stags (young uncastrated males), hens, stewing chickens or fowl (hens older than 20 weeks), and cocks or old roosters. Poultry may be also classified for quality as grade A, B, or C or No. 1, No. 2, or No. 3. A sample grade shields is shown in Fig. (19.4).

Regulations introduced from 1997, required that birds designated as fresh should never have been cooled to below 26 °F the freezing point for poultry meat. Therefore, birds frozen and thawed could not qualify as “fresh” though such products would feel soft to the touch. On the other hand, the temperature of frozen poultry should be kept at 0.0 °F (Anonymous 1995a, b; Cunnungham 1975).

3.2 Turkey Slaughter and Processing

Turkeys are slaughtered, de-feathered, eviscerated, and further processed in much the same manner as chickens, except for some differences. Longer bleeding times are required for turkeys. They must also be scalded for longer times and at higher temperatures to facilitate the removal of feathers. With turkeys, the tendons in the legs are pulled out after defeathering and removing the feet. Further processing is essentially the same as for chickens. During cooling, turkeys pick up 4.5–8% water, the smaller birds absorbing the highest proportional amounts. Grading of turkeys is based on the same characteristics as those for chickens, and they may be graded for quality as A, B, or C.

3.3 Poultry Products

After the slaughtering operation, chicken is graded and packed in wooded or waxed-fiberboard boxes as ready to cook (Table 19.10). Packaged fresh chicken is surrounded with crushed ice and held at temperatures below 40 °F (4.4 °C). When the product is held at 28 °F (−2.2 °C), a significant extension in the shelf life may be realized though technically such product is technically fresh.

Fig. 19.4 Inspection and grade shields for poultry. Adapted from USDA Agricultural Marketing Services (2021), and North Dakota Meat and Poultry Inspection Program (2018)

Table 19.10 Standards of identify of some poultry products

381.157 Canned boned poultry
381.158 Poultry dinners (frozen) and pies.
381.159 Poultry rolls.
381.160 (Kind) burgers; (Kind) patties.
381.161 "(Kind) A La Kiev."
381.162 "(Kind) steak or fillet."
381.163 "(Kind) baked" or "(Kind) roasted."
381.164 "(Kind) barbecued."
381.165 "(Kind) barbecued with moist heat."
381.166 Breaded products.
381.167 Other poultry dishes and specialty items.
381.169 Ready-to-cook poultry

From Code of Federal Regulations **(CFR 9. Section 381.156)**

Some chicken products are battered and breaded, deep fat fried, and frozen. The high-quality storage life of uncooked chicken at 0 °F (−17.8 °C) can be more than a year if adequately packaged, while the high-quality storage life of fried chicken is less than 3 months at this temperature (Table 19.10).

Some chicken is also precooked, deboned, and heat-processed in glass jars or cans. Some deboned poultry is used as an ingredient of canned soups. Chicken may also be precooked, deboned, cut into cubes or small portions, frozen, and then freeze-dried. This product is used as an ingredient of dried soups. As mentioned, chicken is quite popular in the production of frankfurters.

A larger percentage of turkeys is frozen and sold to the consumer in the frozen state than is the case with chickens. However, a proportionally smaller amount of turkey meat than chicken meat is canned or dried. Some turkey is frozen, to be used later as an ingredient of pies or dinners. While the meat appears to be quite stable at 0 °F (−17.8 °C) as an ingredient of pies, it is somewhat less stable than chicken meat in frozen storage. As with chicken, turkey frankfurters are gaining in popularity and many varieties have total fat contents of less than 10% (the USDA standards allow up to 30%).

3.4 Further Reading

Poultry science is a highly active area as discussed in the following monographs (Davies and Board 1998; Mead 2004; Richardson and Mead 1999). Material on poultry and egg production have been published (Mead 2004; Parkhurst and Mountney 2012; Richardson and Mead 1999). A number of reviews have also appeared dealing with specialist topics, including factors, which affect the color and general quality of poultry meat (Barbut 1998; Berri 2000; Froning 1995; Solomon et al. 1998), microbiological safety of poultry (Bautista et al. 1997; Dincer and Baysal 2004; Keener et al. 2004; Kelly et al. 2003; Waldroup 1996). Some modifications of poultry diets are being considered also in order to improve nutritional characteristics of meat and eggs (Abril and Barclay 1998; Jensen et al. 1998; Williams 1997) and the effects of modified atmosphere packaging of poultry(Narasimha Rao and Sachindra 2002).

4 Eggs

4.1 Shell Egg Characteristics

Egg is defined as shell egg from avian species (e.g. hens, ducks, turkey quail, goose, guinea fowls etc.) but excluding egg from alligator, turtles and other reptiles (Food Code. p7; U.S. Food and Drug Administration 2017). Except for some variations such as differing eggshell color, the chemical composition of shell egg falls within narrow limits. Of the total weight for a typical egg, approximately 11% is shell, 58% is egg white, and 31% is yolk. However, the detailed chemical composition of hen egg (e.g. pH, % water, protein, fat, etc.) can vary with the age of the hen, egg size, and storage time between laying and processing (USDA Agricultural Research Service 1969, p2).

The shell, largely calcium carbonate, has an outer coating (the cuticle) that protects the pores of the main part of the shell, as long as it remains intact. Inside the shell, there are two membranes, the one next to the shell being thicker and tougher than the one covering the contents of the egg (Stadelman et al. 2013; Belyavin 2016).

4.1.1 Whole egg composition

The whole egg has about 12% protein, 10% fat, 1% carbohydrate, 1% ash, 75% to 76% water, and about 213 mg of cholesterol. The white of the egg is about 12% protein, less than 1% ash, about 88% water, and contains virtually no fat or cholesterol. Small amounts of sugar, carbon dioxide, and other constituents are present. The white is in two distinct parts: the thick, jellylike part that surrounds the yolk, and a less viscous "thin white" that spreads out when the egg is broken out of its shell. However, the thick white will also spread if it is cut, as its inner part is of thin consistency.

The yolk is about 17% protein, 49% water, 30% fat, and 1% ash. All of the 213 mg of cholesterol in the whole egg are contained in the yolk. Small quantities of numerous other constituents, including vitamins, are also present. The yolk is surrounded by the vitelline membrane, which when ruptured or cut, causes the yolk to spread when the egg is broken out of the shell. The yolk is much more complex chemically than the white, and accounts for the major nutritional composition of the egg.

The egg contains a small air pocket that develops after the egg is laid when it cools from the body temperature of the hen, contracting the contents and pulling in the inner membrane. Because the eggshell is actually porous, the space formed by the contraction of the contents is soon filled by air that can be drawn in through the shell.

4.1.2 Egg hygiene and handling

The eggshell presents a barrier to the entrance of dirt. However, there are pores in the shell large enough to allow the entrance of bacteria and microscopic molds. The number of pores that are found in the shell vary in the range 100–200 per cm^2. When the egg is just laid, a thin layer of protein, the cuticle, seals the pores of the shell.

The logical reason for porosity of the shell is to allow for the flow of gases into and out from the developing embryo in case the egg has been fertilized. The inner membranes tend to prevent the entrance of microorganisms. In egg white, there is an enzyme (lysozyme) that tends to lyse or disintegrate some bacteria. There is also a substance in raw eggs, avidin that ties up biotin, a required factor for the growth of some microorganisms. Finally, there is a material in fresh white that binds with iron, making it unavailable to several species of Pseudomonas bacteria that are responsible for more than 80% of the egg spoilage.

The cuticle may be lost not only by washing; droppings can also dissolve it. In any case, about 3 weeks after the egg is laid, the cuticle becomes brittle and particles chip off. Some bacteria are not affected by lysozyme and require little or no biotin. As they are held in storage, enzymes within the eggs cause chemical changes that deteriorate the iron-binding properties of the egg white. The defense mechanism against spoilage is, therefore, eventually lost, so that if microorganisms penetrate the shell, spoilage will occur.

Egg should be purchased already refrigerated, transferred home quickly and refrigerated for storage. Safe handling of eggs requires the same care and procedures used for meat; eggs should be cooked until there are no runny parts and the yolk is firm. Foods containing raw eggs (mayonnaise, some ice cream, tiramisu) should be avoided with the exception of those using pasteurized shell egg (USDA Food inspection Services 2019).

4.1.3 Washing and sanitizing shell egg

Shell egg may contain *salmonella* outside and within the egg including the yolk (Braden 2006; Galiş et al. 2013; Whiley and Ross 2015). USDA regulations require that shell egg (for retail or processing to liquid, frozen or dried egg) must be washed, and sanitized under strict conditions. Safe cleaning procedures for shell egg are described by regulations (9CFR590.515) that cover, the types of washing equipment, wash-water temperature (above 90 °F (32 °C)), choice of cleaning compounds, and arrangements for wash-water (portable water only, with changes of wash-water every 4hrs) and water disposal. Soaking of shell-eggs must be avoided (Wu 2014). The Environmental Protection Agency has a list of sanitizers approved for shell egg decontamination. After cleaning and sanitizing, shell egg must be stored at 40 °F (4.4. °C).

Dirty eggs can be cleaned by buffing or washing, but unless this is done under rigidly controlled conditions, the interior may become contaminated with bacteria during cleaning (Galiş et al. 2013; Whiley and Ross 2015). *Consumers are advised not to wash shell egg* at home (USDA Food inspection Services 2019).

Approaches for eggshell decontamination vary internationally. Egg washing and cold storage is a requirement in North America (US and Canada), Australia and Japan. However, the EU has concern that bacteria may penetrate inside of the eggs once the cuticle is removed by washing and egg washing is forbidden in Europe. Refrigerated storage is also not a requirement for egg producers, though cold storage slows the growth of salmonella inside of eggs. A possible reason for the policy differences between the EU and North America on egg decontamination is based on the fact that Europe uses vaccination to control salmonella in chicken flocks (Galiş et al. 2013; Whiley and Ross 2015).

Newly cleaned shell egg can be coated with a light spraying of mineral oil. Coating shell egg is optional and practiced by about 10% of US egg suppliers. Spraying with mineral oil can increase the storage life of shell egg (Waimaleongora et al. 2009; Nongtaodum et al. 2013). Different coating materials (mineral oil, chitosan, protein, edible oils, silicone etc.) were shown to hinder the loss of carbon dioxide from shell eggs, which curtails the pH rise that takes place otherwise. Weight loss due to dehydration declines after coating the shell. Shell egg coatings also helped to maintain grade quality (Waimaleongora et al. 2009; Nongtaodum et al. 2013).

Prior to shipment, fresh eggs are packed in clean, odorless containers (usually holding 30 dozen) at 45 °F (7.2 °C). Individual eggs are packed with their large ends up apparently to prevent the air pocket from migrating upward into the yolk, which may promote spoilage. The temperature requirement for transporting and receiving shell egg are the same as for storage. Egg packaging must be labeled "keep refrigerated" (USDA Food inspection Services 2019).

4.1.4 Egg quality and grading

High quality eggs are clean and not cracked. Indeed, damaged eggs must be withdrawn from sale (Food code, p. 413; U.S. Food and Drug Administration 2017). The consumer should avoid purchasing cracked eggs owing to the heightened risk of contamination by salmonella. As an option, packaging for good quality shell egg may also carry the USDA inspection shield that confirms wholesomeness.

The USDA recognizes six size-categories, according to the package weight per dozen eggs. The egg sizes are, Jumbo (30 oz. or 851 g per dozen), Extra-large (27 oz. or 766 g per dozen), Large (24 oz. or 680 g per dozen), Medium (21 oz. or 595 g per dozen, small (18 oz. 510 g per dozen), and peewee (15 oz. or 425 g per dozen).

Eggs are also graded by the USDA, according to their interior quality as "AA", "A" or "B" (see Fig. 19.5). The grade "AA" or "A" rating indicates that the broken-out egg will cover a "small" or "moderate" area when spread on a flat surface. Furthermore, the white will be reasonably thick and stand high, and the yolk will be firm and high. With grade "B" rating, the broken-out egg covers a wide area with only a small amount of white that can be considered thick, and the yolk is somewhat flattened. Eggs normally sold in shops, are grade "AA" or "A" standard. Grade "B" shell egg is processed to provide frozen liquid eggs or dried egg powder (USDA Food inspection Services 2019).

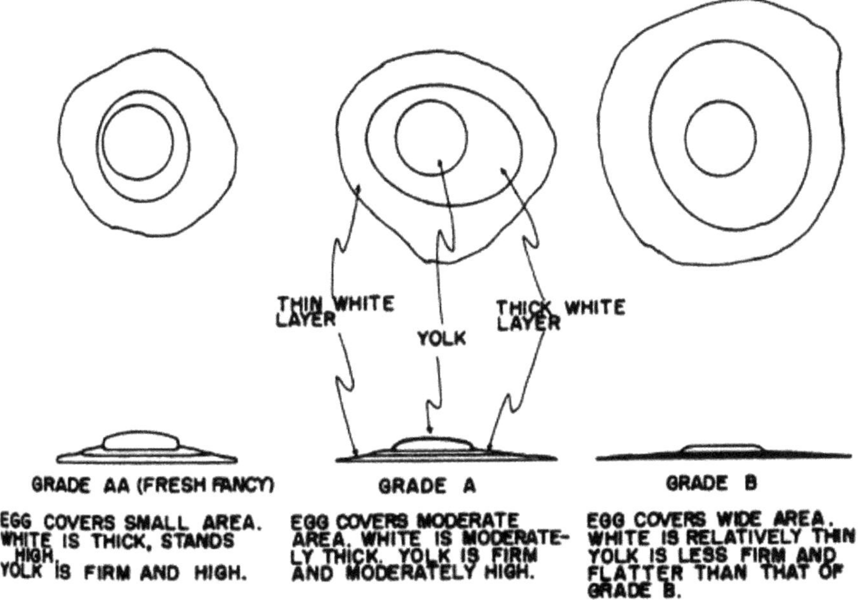

Fig. 19.5 U.S. grades for eggs (broken out)

4.2 Processing of Eggs

4.2.1 Breaking and separation of shell egg

Eggs broken out of their shells, as whole eggs, egg white, egg yolk or blends, are called egg-products (discussed above). Virtually all egg products used in bakeries and the foodservices are preserved by freezing or by drying after separating the yolk from the white (Wu 2014).

Machines break the egg out into cups, which may or may not also separate whites from yolks, depending on whether the process is to produce whites, yolks, or whole egg magma (mixed white and yolks). While the separation of the egg is such that the white can be separated so that it is free of yolk, the yolk cannot be separated so that it is entirely free of the white. In fact, standards or definitions concerning yolk allow as much as 20% white. As the eggs in cups are carried along by conveyor, they are examined by inspectors. If a particular cup contains a bad specimen, the cup with the egg is removed and replaced with a clean, chemically sanitized cup. The egg from the removed cup is then discarded and the cup is washed and sanitized by rinsing in a chemical solution (usually 50 ppm or more of chlorine). As the conveyor moves along, the contents of each cup are emptied into one container when eggs are to be kept whole, or the whites and yolks are emptied into separate containers when separation is desired.

4.2.2 Pasteurization of egg products

Eggs products, including whole egg magma, yolks, and whites produced in the US must be pasteurized in order to destroy Salmonella bacteria, because the disease salmonellosis has been traced to contaminated eggs (Braden 2006; Galiş et al. 2013; Whiley and Ross 2015; Jarvis et al. 2016). There are five methods approved for the pasteurization of liquid egg white on an industrial scale described in the “Egg pasteurization manual” (USDA Agricultural Research Service 1969).

After screening to remove chalaza (albuminous thread) and shell fragments, liquid egg white can be pasteurized using a plate-type heat exchanger. The pasteurization process involves heating to between 125 °F and 140 °F (51.7 °C and 60°C) for 1.75 to 4 min. The temperature-time required for pasteurization depends on the sample pH and if food additives are present (Chmielewski et al. 2013). Cooling may be accomplished using tanks provided with cooling coils and paddles that agitate the product to facilitate cooling, or in thin-film heat exchangers (USDA Agricultural Research Service 1969). The conditions approved for liquid egg white pasteurization are summarized in Appendix 1.

Satisfactory temperature-time combinations for liquid egg white pasteurization were developed only from 1964 onwards. Firstly, the pH for liquid egg white was adjusted from pH 9 to pH 7.0 to increase the stability of egg protein sufficiently to withstand 140 °F (60°C) pasteurization without denaturing and fouling the processing equipment. Alternatively, hydrogen peroxide was added to liquid egg white, in order to decrease the temperature for salmonella inactivation (USDA Agricultural Research Service 1969). The two so-called heat-plus-peroxide pasteurization processes used in industry were re-examined recently using a bench-top (small-scale) apparatus. The study confirmed that hydrogen peroxide is essential for salmonella inactivation (7-8 log cycle reduction) at 132.4 °F (55.8 °C) and a holding time of 2 min to 3.5 min at pH 8.6-pH 9.0. (Muriana 1997; Robertson and Muriana 2004).

Pasteurization of egg yolk, whole eggs, or blends containing liquid egg white and yolk require higher temperatures than those used for liquid egg white, e.g. at 140 - 146 °F (60-63.3 °C) with a holding time of 3.5 to 6.2 min. The relatively severe conditions for pasteurizing whole eggs is because the heat resistance of salmonella increases in the presence yolk, salt and sugar (USDA Agricultural Research Service 1969, p11). Interestingly, the stability of salmonella species is higher for eggs compared to most other food vectors (Doyle and Mazzotta 2000).

4.2.3 In-shell egg pasteurization

Pasteurization of intact egg is quite an innovation. The pasteurized shell egg provides a safe alternative to pasteurized liquid egg white or yolks for the foodservice industry. The process for in-shell pasteurization was developed independently by several groups between 1994 and 1996 and patented finally in 1999 (Stadelman et al. 1996; Vandepopuliere and Cotterill 1999). Nowadays, pasteurization of shell egg is done on a commercial scale by several companies under license from the University of Missouri (Mermelstein 2001). Intact shell egg must be pasteurized by heating at or below 57 °C for prescribed times, sufficient to inactivate salmonella, whilst retaining the egg white and yolk in a liquid state.

Pasteurized shell egg requires refrigeration but only to maintain quality. There is no legal requirement for refrigerating pasteurized shell egg. Restaurants can use pasteurized shell egg in ways that are not possible for normal eggs, e.g. preparation of fried eggs "sunny-side up" with reduced risk to the consumer. Another use for pasteurized shell egg is in preparing mayonnaise and other undercooked egg dishes (National Pasteurized Eggs Inc. 2014). A recent study showed that up to 80% of restaurants surveyed used unpasteurized eggs, and sometimes pooled eggs, which can spread infection (CDC Environmental Health Services (EHS) 2019). Australian researchers conducted in-shell pasteurization by heating with a sous-vide cooker set at 57 °C for 9 minutes without adversely affecting the properties of mayonnaise. The method was successful for eggs that are not internally contaminated with salmonella (Keerthirathne et al. 2020).

4.2.4 Frozen egg products

After cooling to 40 °F (4.4 °C), the pasteurized eggs may be placed in metal cans holding about 30 lb. (13.6 kg) of product. The filled cans are then placed in a cold room at 0 to −20 °F (−17.8 to −28.9 °C) until the product is frozen, after which it will be held at 0 °F (−17.8 °C) or lower until shipped out to the distributor or to the point of utilization. Frozen whole egg magma and frozen yolks are subject to deterioration during frozen storage. Gelation of yolk lipoprotein can give rise to a gummy mass at low temperature. In order to prevent this, 5–7% salt, glycerin, or 5–10% sugar may be added. Citric acid may also be added to help extend shelf life (USDA Agricultural Research Service 1969).

Restaurant and other foodservice recipes that require raw eggs or undercooked eggs must use pasteurized egg products or else use pasteurized shell egg (Sect 4.2.3.). Dishes such as mayonnaise, meringue, egg nog, home-made ice cream, and some egg-fortified beverages must use pasteurized egg and egg products (Food Code. p74; U.S. Food and Drug Administration 2017).

4.2.5 Spray dried egg products

Whites, yolks, or whole egg magma may be spray-dried by being forced through a nozzle (to form droplets) into a chamber of heated air where most of the moisture is removed from the droplets to the heated air, which is vented to the outside. The dried product falls to the bottom of the drier and is collected. Spray-dried eggs have a moisture content of about 2-4% (Belyavin 2016).

Nonenzymatic browning and lipid oxidation are well-known reactions that lead to the spoilage of dried foods (Karel 1984; Villota et al. 1980; Caboni et al. 2005). Non-enzymic browning involves glucose and is prevented by its removal. Dried egg albumin imported from China prior to the 1940s had superior color, due to a (secrete) de-sugaring process involving natural fermentation by mixed bacteria. De-sugaring egg white is performed currently using lactic acid bacteria or yeast (Sebring 2013; Delves-Broughton 2014; Wu 2014).

An enzymatic method for de-sugaring egg white was developed from research during the second world war (Sebring 2013; Delves-Broughton 2014; Wu 2014).Treating liquid egg white with glucose oxidase and catalase, transforms glucose to gluconic acid and hydrogen peroxide, and the hydrogen peroxide is decomposed to water and oxygen (Chap 14, Sect 2.6). As glucose oxidase is highly selective for glucose, this enzyme does not promote lipid oxidation. The storage life of spray dried egg white changes in the order; glucose-free egg white powder > acidified egg white powder >> untreated egg white powder. Egg white powder is more stable also if stored under an atmosphere of nitrogen and carbon dioxide, compared with nitrogen, or normal air. Lowering the moisture content from 4% to 2% and /or dropping the storage temperature from 100 °F (37.7 °C) to between 40 and 70°F (4.4 to 21.1 °C) also improved the storage life of egg powder. Whole-egg powder and dried egg yolk are comparatively less stable compared with egg white powder probably due to the formers higher lipid content (Villota et al. 1980).

The heating encountered during the pasteurization treatment and the physical forces encountered during spray drying have some effect on the functional characteristics (whipping quality, etc.) of egg products, especially those made from whites. Therefore, in some countries where pasteurization is not mandatory, egg white is allowed to undergo a natural fermentation and is then dried in cabinets on trays. In such instances, to eliminate disease-causing bacteria that may be present (because the product was not pasteurized prior to drying); the dried product is held at 130 °F (54.4 °C) or at higher temperatures for several days (Appendix 1). This dry-heat pasteurization process is designed to destroy salmonella during the production of pan-dried egg white powder flakes (McBee and Cotterill 1971, Northolt et al. 1978; Bergquist 1995; Baron et al. 2003). In addition, a form of dry heat treatment can also restore the functions of egg white powder partially lost during spray drying (Talansier et al. 2009; Lechevalier et al. 2017b). Recent investigations are showing more clearly how drying and storage conditions affect dried egg powders (Caboni et al. 2005; Lechevalier et al. 2007; Koç et al. 2011). Spray drying under moderate conditions can retain or improve the functional properties of egg white powder compared with fresh eggs (Ayadi et al. 2008). The characteristics of dried egg products have been reviewed (Wu 2014; Bergquist 1995).

4.3 Egg Consumption

The annual rate for egg consumption for the United States fell from 300 eggs per person (pp) to 235 eggs (pp) between 1971 and 1989 (Putnam 1990; Putnam and Allshouse 1999). Allowing for year-to-year fluctuations, egg consumption stood at about 248 eggs (pp) for the years 2000 and 2009. There was a rise in egg consumption from 2015/16 reaching 290 eggs (pp) for 2019. The former drop in US egg consumption coincided with consumer concerns over egg cholesterol and salmonella (Stadelman 1999; Schmit and Kaiser 1998). Salmonella is discussed in Sect. 4.2.2. The main sources of dietary cholesterol for American children were ranked, eggs > poultry > beef > milk> cheese (Keast et al. 2013). However, the relations between dietary and blood cholesterol levels is proving controversial (see below).

Health warnings by the American Heart Association (AHA) dating from 1968 stated that eggs are high in cholesterol. Therefore, people should consume no more than three eggs per week (McNamara 2000). In the late 1980s, there were claims by some egg producers that their eggs had less cholesterol than "the average" egg and it was even advertised on packaging. The truth was that older methods used to measure cholesterol were not as accurate as current ones. Studies with high-pressure liquid chromatography (HPLC), which had only then become available widely, showed eggs contained 195 (±17) mg cholesterol rather than 274 mg of cholesterol per egg accepted as USDA data. Canadian researchers demonstrated clearly that the average cholesterol value per egg was dependent on the method of analysis: 243 (±11.6) mg using an older colorimetric method, 205 (±13.9) mg using an enzyme-based assay, 195(±10.2) mg using gas chromatography and 195 (±10.5) mg using HPLC (Jiang et al. 1991). Current results show the typical class-A eggs (typical weight 50.5g) have a cholesterol value of 202 mg per egg (range 187-227 mg) (USDA Agricultural Research Service 2019). The 20% drop in egg cholesterol values after 1989 was due to improvements of scientific methodology rather than changes affecting the eggs themselves.

There were no changes to the AHA limit for egg intake from 3.0 eggs per week, as expected based on a 20% drop the amount of cholesterol value per egg. Meanwhile the egg industry funded more research into the relations between egg consumption and human health. From the year 2000 onwards, the AHA dropped specific recommendations about the numbers of eggs, which it would be safe to consume. However, the recommendation to limit dietary cholesterol to less than 300 mg per day remained for the AHA (Krauss et al. 2000) but was not supported by the US Federal authorities. There was no specific cholesterol limit in the 2015-2020 dietary guidelines for Americans (McNamara 2015).

Egg is acknowledged as a source of cholesterol. However, a positive association between dietary cholesterol and blood cholesterol is now considered doubtful by experts. There is also a focus on the LDL/ HDL cholesterol ratio as a measure of cardiovascular disease risk. The view is that eggs are a useful source of protein and vitamins for most healthy individuals (Harvard TH Chan School of Public Health 2021). The so-called "rehabilitation of eggs" (McNamara 2015) has been arduous and complex. For more insight about eggs, dietary cholesterol and health, the reader may refer to the following literature (Watson 2002; Kritchevsky 2004; Gray and Griffin 2009; McNamara 2015; Harvard TH Chan School of Public Health 2021).

4.4 Egg Substitutes

If it is desired to eliminate cholesterol, egg substitutes may be purchased. These products are marketed under names such as "scramblers®" and "egg beaters®." They do not have to be labelled as "imitation "eggs because they have inovatice names and because they have the same nutritional value as the product they copy (Chap. 4; Sect. 2.2.). They contain mostly egg whites. The other ingredients vary but both contain corn or soybean oil, emulsifiers (mono- and diglycerides), stabilizers (vegetable gums or modified starches), and a number of nutrients (vitamins and minerals) that are present in eggs. These products can be used as egg substitutes in most recipes and can be scrambled or used in omelets.

Appendix1: Current approved pasteurization conditions liquid egg white

1. Pasteurization at pH 7; adjust liquid egg white to pH 7 using lactic acid (with 0.013% aluminum sulfate). Heat liquid egg white (pH 7.0) to 140 °F (60 °C) for 1.75 min.
2. Heat-plus-peroxide (Armour & co. ltd); pre-heat liquid egg white (pH 9) to 125 °F (51.7 °C) and hold for 1.5 min. Add hydrogen peroxide (0.05% final concentration) and hold for 2 min, cool, and add catalase.
3. Heat-plus-peroxide (standard brand); add hydrogen peroxide to liquid egg white at pH 9. Heat to 125 °F (51.7 °C) and hold for 3.5 min. Cool the liquid egg white and add catalase.
4. Vacuum-heat pasteurization process; vacuum treat liquid egg white and heat to 134 °F (56.7 °C), hold for 3.5 min and cool.
5.Dry heat pasteurization; pack egg white solids or flakes (approx. 6% moisture) into a heating room and hold at 125–130 °F (51.7–54.4 °C) for 7–10 days

Adapted from, Egg pasteurization manual (ARS 74–48) (USDA Agricultural Research Service 1969)

References

Abril R, Barclay W (1998) Production of docosahexaenoic acid-enriched poultry eggs and meat using an algae-based feed ingredient. World Rev Nutri Diet 83:77–88

Anonymous (1995a) 'Frozen' poultry may be labelled 'fresh' but California can still ban word in ads. World Food Regul Rev 4(9)

Anonymous (1995b) USDA proposes new definition for 'fresh' poultry products. World Food Regul Rev: 4(10):21–22

Ayadi MA, Khemakhem M, Belgith H, Attia H (2008) Effect of moderate spray drying conditions on functionality of dried egg white and whole egg. J Food Sci 73(6):E281–E287

Barbut S (1998) Estimating the magnitude of the PSE problem in poultry. J Muscle Foods 9(1):35–49. https://doi.org/10.1111/j.1745-4573.1998.tb00642.x

Baron F, Nau F, Guerin-Dubiard C, Gonnet F, Dubois JJ, Gautier M (2003) Effect of dry heating on the microbiological quality, functional properties, and natural bacteriostatic ability of egg white after reconstitution. J Food Prot 66(5):825–832. https://doi.org/10.4315/0362-4028X-4366.4315.4825

Bautista DA, Sprung DW, Barbut S, Griffiths MW (1997) A sampling regime based on an ATP bioluminescence assay to assess the quality of poultry carcasses at critical points during processing. Food Res Int 30(10):803–809. https://doi.org/10.1016/S0963-9969(98)00049-0

Belyavin CG (2016) Eggs: use in the food industry. Encyclopedia of food and health (Caballero B, Finglas PM, Toldrá F, eds). Academic Press, Oxford, pp 476–479. https://doi.org/410.1016/B1978-1010-1012-384947-384942.300245-384942

Bell DD, Weaver WD (2013) Commercial chicken meat and egg production, 5th edn. Springer, Boston

Bergquist DH. (1995). Chapter 14. Egg dehydration. Egg science and technology (4th edn, Stadelman WJ, Cottrell OJ). The Hayworth Press, Binghamton, pp 335–376. https://books.google.com.gh/books?id=dwZDDwAAQBAJ

Berri C (2000) Variability of sensory and processing qualities of poultry meat. World's Poult Sci J 56(3):209–224. https://doi.org/10.1079/WPS20000016

Biandolino FPI, Denti G, Nardo VD, Prato E (2021) Effect of different cooking methods on lipid content and fatty acid profiles of mytilus galloprovincialis. Foods 10(2):416. https://doi.org/410.3390/foods10020416

Braden CR (2006) Salmonella enterica serotype Enteritidis and eggs: a national epidemic in the United States. Clin Infect Dis 43(4):512–517. https://doi.org/510.1086/505973

Caboni MF, Boselli E, Messia MC, Velazco V, Fratianni A, Panfili G, Marconi E (2005) Effect of processing and storage on the chemical quality markers of spray-dried whole egg. Food Chem 92(2):293–303

Chmielewski RA, Beck JR, Swayne DE (2013) Evaluation of the U.S. Department of Agriculture's egg pasteurization processes on the inactivation of high-pathogenicity avian influenza virus and velogenic Newcastle disease virus in processed egg products. J Food Prot 76(4):640–645

CDC Environmental Health Services (EHS) (2019). How restaurants prepare eggs. From https://www.cdc.gov/nceh/ehs/ehsnet/plain_language/how-restaurants-prepare-eggs.htm

Codex Alimentarius Commission (2011) Guidelines for the control of campylobacter and salmonella in chicken meat, CAC/GL 78–2011. Retrieved from http://www.fao.org/input/download/standards/11780/CXG_078e.pdf

Cunnungham FE (1975) Acceptability and use of frozen poultry – a review. World's Poult Sci J 31(2):136–148. https://doi.org/10.1079/WPS19750011

Davies A, Board R (1998) The microbiology of meat and poultry

Delves-Broughton J (2014) Eggs. Microbiology of egg products. Encyclopedia of food microbiology (2nd edn, Batt CA, Tortorello ML). Academic Press, Oxford, pp 617–621. https://doi.org/610.1016/B1978-1010-1012-384730-384730.300090-384732

Dincer AH, Baysal T (2004) Decontamination techniques of pathogen bacteria in meat and poultry. Crit Rev Microbiol 30(3):197–204. https://doi.org/10.1080/10408410490468803

Doyle ME, Mazzotta AS (2000) Review of studies on the thermal resistance of salmonellae. J Food Prot 63(6):779–795. https://doi.org/710.4315/0362-4028X-4363.4316.4779

Fletcher DL (2002) Poultry meat quality. World's Poult Sci J 58(2):131–145. https://doi.org/10.1079/WPS20020013

Froning GW (1995) Color of poultry meat. Poult and Avian Biol Rev 6(2):83–93

Galiş AM, Marcq C, Marlier D, Portetelle D, Van I, Beckers Y, Théwis A (2013) Control of Salmonella contamination of shell eggs—preharvest and postharvest methods: a review. Compr Rev Food Sci Food Saf 12(2):155–182. https://doi.org/110.1111/1541-4337.12007

Gray J, Griffin B (2009) Eggs and dietary cholesterol - dispelling the myth. Nutr Bull 34(1):66–70. https://doi.org/10.1111/j.1467-3010.2008.01735.x

Griggs JP, Jacob JP (2005) Alternatives to antibiotics for organic poultry production. J Appl Poult Res 14(4):750–756

Harvard TH Chan School of Public Health (2021). The nutrition source: eggs. From https://www.hsph.harvard.edu/nutritionsource/food-features/eggs/

Hewavitharana G, Perera DN, Navaratne SB, Wickramasinghe I (2020) Extraction methods of fat from food samples and preparation of fatty acid methyl esters for gas chromatography: A review. Arab J Chem 13(8):6865–6875

Jarvis NA, O'Bryan CA, Dawoud TM, Park SH, Kwon YM, Crandall PG, Ricke SC (2016) An overview of Salmonella thermal destruction during food processing and preparation. Food Control 68:280–290

Jiang Z, Fenton M, SIM JS (1991) Comparison of four different methods for egg cholesterol determination. Poult Sci 70(4): 1015-1019. https://doi.org/1010.3382/ps.0701015.

Jensen C, Lauridsen C, Bertelsen G (1998) Dietary Vitamin E: quality and storage stability of pork and poultry. Trends Food Sci Technol 9(2):62–72. https://doi.org/10.1016/S0924-2244(98)00004-1

Karel M (1984) Chemical effects in food stored at room temperature. J Chem Ed 61(4):335–339. https://pubs.acs.org/doi/abs/310.1021/ed1061p1335

Keast DR, Fulgoni VL, Nicklas TA, O'Neil CE (2013) Food sources of energy and nutrients among children in the United States: national health and nutrition examination survey 2003–2006. Nutrients 5(1):283–301. https://doi.org/210.3390/nu5010283

Keener KM, Bashor MP, Curtis PA, Sheldon BW, Kathariou S (2004) Comprehensive review of Campylobacter and poultry processing. Compr Rev Food Sci Food Saf 3(2):105–116. https://doi.org/10.1111/j.1541-4337.2004.tb00060.x

Keerthirathne TP, Ross K, Fallowfield H, Whiley H (2020) A successful technique for the surface decontamination of salmonella enterica serovar typhimurium externally contaminated whole shell eggs using common commercial kitchen equipment. Foodborne Pathog Dis 17(6):404–410. https://www.liebertpub.com/doi/pdfplus/10.1089/fpd.2019.2734

Kelly LA, Hartnett E, Gettinby G, Fazil A, Snary E, Wooldridge M (2003) Microbiological safety of poultry meat: risk assessment as a way forward. World's Poult Sci J 59(4):495–508. https://doi.org/10.1079/WPS20030031

Koç M, Koç B, Susyal G, Yilmazer MS, Ertekin FK, Bağdatlıoğlu N (2011) Functional and physicochemical properties of whole egg powder: effect of spray drying conditions. J Food Sci Technol 48(2):141–149. https://www.ncbi.nlm.nih.gov/pmc/articles/PMC3551060/

Krauss RM, Eckel RH, Howard B, Appel LJ, Daniels SR, Deckelbaum RJ, Erdman JW, Kris-Etherton P, Goldberg IJ, Kotchen TA, Lichtenstein AH, Mitch WE, Mullis R, Robinson K, Wylie-Roset J, St JS, Suttie J, Tribble DL, Bazzarre TL (2000) AHA dietary guidelines: revision 2000: a statement for healthcare professionals from the Nutrition Committee of the American Heart Association. Circulation 102(18):2284–2299. https://www.ahajournals.org/doi/pdf/2210.1161/2201.cir.2102.2218.2284

Kritchevsky SB (2004) A review of scientific research and recommendations regarding eggs. J Am Coll Nutr 23(sup6):596S–600S

Lechevalier V, Jeantet R, Arhaliass A, Legrand J, Nau F (2007) Egg white drying: influence of industrial processing steps on protein structure and functionalities. J Food Eng 83(3):404–413. https://www.academia.edu/download/45495792/Egg_white_drying_Influence_of_industrial20160509-45422272-45495795y45495790oo45495799.pdf

Lechevalier V, Guérin-Dubiard C, Anton M, Beaumal V, Briand ED, Gillard A, Le Gouar Y, Musikaphun N, Tanguy G, Pasco M, Dupont D, Nau F (2017) Pasteurization of liquid whole egg: optimal heat treatments in relation to its functional, nutritional and allergenic properties. J Food Eng 195:137–149. https://hal.archives-ouvertes.fr/hal-01454657/document

Lechevalier V, Guérin-Dubiard C, Anton M, Beaumal V, Briand ED, Gillard A, Le Gouar Y, Musikaphun N, Pasco M, Dupont D (2017b) Effect of dry heat treatment of egg white powder on its functional, nutritional and allergenic properties. J Food Eng 195:40–51

McBee L, Cotterill O (1971) High temperature storage of spray-dried egg white: 3. Thermal resistance of *Salmonella oranienburg*. Poult Sci 50(2):452–458

McNamara DJ (2000) The impact of egg limitations on coronary heart disease risk: do the numbers add up? J Am Coll Nutr 19(sup5):540S–548S

McNamara DJ (2015) The fifty year rehabilitation of the egg. Nutrients 7(10):8716–8722. https://doi.org/8710.3390/nu7105429

Mead GC (2004) Poultry meat processing and quality. Woodhead Publishing, 400 pp

Mermelstein NH (2001) Pasteurization of shell eggs. Food Technol 55(12):7pp. https://www.ift.org/news-and-publications/food-technology-magazine/issues/2001/december/columns/products-and-technologies_processing

Mohamadi S, Fallah AA, Dehkordi SH, Fizi A (2020) Effect of different cooking methods on nutritional quality, nutrients retention, and lipid oxidation of quail meat. J Nutr Fast Health 8(4 (Spe)):238–247. https://jnfh.mums.ac.ir/article_16500.html

Muriana P (1997) Effect of pH and hydrogen peroxide on heat inactivation of salmonella and listeria in egg white. Food Microbiol 14(1):11–19

NAICS Association (2020). 311615 – poultry processing. Retrieved from https://www.naics.com/naics-code-description/?code=311615

Narasimha Rao D, Sachindra NM (2002) Modified atmosphere and vacuum packaging of meat and poultry products. Food Rev Int 18(4):263–293. https://doi.org/10.1081/FRI-120016206

National Farm Animal Care Council (Canada) (2016) Code of practice: chickens, turkeys and breeders. Retrieved from http://www.nfacc.ca/codes-of-practice/chickens-turkeys-and-breeders

National Pasteurized Eggs Inc. (2014) Safest choice™ pasteurized shell eggs. From https://ulfweb.com/assets/PDF/nutritional/215625.pdf

Nongtaodum S, Jangchud A, Jangchud K, Dhamvithee P, No HK, Prinyawiwatkul W (2013) Oil coating affects internal quality and sensory acceptance of selected attributes of raw eggs during storage. J Food Sci 78(2):S329–S335

North American Meat Institute (NAMI) (2016) The meat and poultry industry: basic statistics. Retrieved from http://www.themarketworks.org/stats

North Dakota Meat and Poultry Inspection Program (2018) North Dakota poultry slaughter, processing, and sales guidelines for small-scale producers. From https://www.nd.gov/ndda/sites/default/files/resource/Poultry%20Brochure%2010_18.pdf

Northolt M, Wiegersma N, Van Schothorst M (1978) Pasteurization of dried egg white by high temperature storage. Int J Food Sci Technol 13(1):25–30

OECD (2016) Meat consumption (indicator). Retrieved from https://data.oecd.org/agroutput/meat-consumption.htm

Parkhurst C, Mountney GJ (2012) Poultry meat and egg production. Springer, Dordrecht

Prusa KJ, Lonergan MM (1987) Cholesterol content of broiler breast fillets heated with and without the skin in convection and conventional ovens. Poult Sci 66(6):990–994

Putnam JJ (1990) Food consumption. Food Rev Nat Food Rev 13(1482-2018-4364):1–9

Putnam JJ, Allshouse JE (1999) Food consumption. From https://www.ers.usda.gov/webdocs/publications/47097/14802_sb965f_1_.pdf?v=0

Richardson RI, Mead GC (1999) Poultry meat science

Robertson, WR., Muriana PM. (2004). Reduction of salmonella by two commercial egg white pasteurization methods. J Food Prot 67(6): 1177–83. https://doi.org/10.4315/0362-028X-67.6.1177

Scherr C, Ribeiro JP (2013) The influence of food preparation methods on atherosclerosis prevention. Rev Assoc Med Bras 59(2):148–154. https://doi.org/110.1016/S2255-4823(1013)70448-X

Schmit TM, Kaiser HM (1998) Egg advertising, dietary cholesterol concerns, and US consumer demand. Agric Resour Econ Rev 27(1):43–52

Sebring M (2013) Desugaring of egg products. Egg science and technology (Stadelman WJ, D Newkirk, L Newby, eds). Routledge, New York, pp 323–334. https://books.google.com.gh/books?id=dwZDDwAAQBAJ

Solomon MB, Laack RLJMV, Eastridge JS (1998) Biophysical basis of pale, soft, exudative (PSE) pork and poultry muscle: a review. J Muscle Foods 9(1):1–11. https://doi.org/10.1111/j.1745-4573.1998.tb00639.x

Sonaiya EB, Swan SEJ (2004) Small-scale poultry production: technical guide. Retrieved from http://www.fao.org/docrep/008/y5169e/y5169e0d.htm

Stadelman WJ, Singh RK, Muriana PM, Hou H (1996) Pasteurization of eggs in the shell. Poult Sci 75(9):1122–1125. https://citeseerx.ist.psu.edu/viewdoc/download?doi=1110.1121.1121.1943.1438&rep=rep1121&type=pdf

Stadelman W (1999) The incredibly functional egg. Poult Sci 78(6):807–811

Stadelman WJ, Newkirk D, Newby L (2013) Egg science and technology. Routledge, New York, 608 pp. https://books.google.com.gh/books?id=dwZDDwAAQBAJ

Talansier E, Loisel C, Dellavalle D, Desrumaux A, Lechevalier V, Legrand J (2009) Optimization of dry heat treatment of egg white in relation to foam and interfacial properties. Lwt-Food Sci Technol 42(2):496–503

U.S. Food and Drug Administration (2017) Food code (2017). Retrieved from https://www.fda.gov/media/110822/download

U.S. Government Publishing Office (2014) 9 CFR 381 – poultry products inspection regulations. Retrieved from https://www.gpo.gov/fdsys/granule/CFR-2011-title9-vol2/CFR-2011-title9-vol2-part381

USDA Agricultural Marketing Services (2021) Poultry grading shields. Retrieved August 30, 2021, from https://www.ams.usda.gov/grades-standards/poultry/grade-shields

USDA Agricultural Research Service (1969) Egg pasteurization manual (ARS 74-48). Albany, CA 94710, ARS/USDA, 47pp. https://naldc.nal.usda.gov/download/CAIN709025458/PDF

USDA Economic Research Services (2016) Poultry and eggs: statistics and information. Retrieved from https://www.ers.usda.gov/topics/animal-products/poultry-eggs/readings/

USDA Agricultural Research Service (2019) Food data central search results: eggs, Grade A, Large, egg whole. Retrieved September 9, 2021, from https://fdc.nal.usda.gov/fdc-app.html#/food-details/748967/nutrients

USDA Food inspection Services (2019) Shell eggs from farm to table. From https://www.fsis.usda.gov/food-safety/safe-food-handling-and-preparation/eggs/shell-eggs-farm-table

USDA Food Safety and Inspection Service (2014) Chicken from farm to table. Retrieved from https://www.fsis.usda.gov/wps/wcm/connect/ad74bb8d-1dab-49c1-b05e-390a74ba7471/Chicken_from_Farm_to_Table.pdf?MOD=AJPERES

USDA Foreign Agricultural Services (2016) Livestock and poultry: world markets and trade

USDA National Agricultural Library (2021) Eggs. From https://www.nal.usda.gov/fnic/eggs

USDA National Agricultural Statistics Service (2016) Poultry – production and value 2015 summary april 2016. Retrieved from usda.mannlib.cornell.edu/usda/current/PoulProdVa/PoulProdVa-04-28-2016.pdf

Vandepopuliere JM, Cotterill OJ (1999) US patent US6004603A. Method of controlling Salmonella in shell eggs. University of Missouri System at Columbia, Columbia, 10pp. https://patents.google.com/patent/US6004603A/en

Villota R, Saguy I, Karel M (1980) Storage stability of dehydrated food. Evaluation of literature data. J Food Qual 3(3):123–212. https://doi.org/110.1111/j.1745-4557.1980.tb00699.x

Waimaleongora EKP, Garcia KM, No HK, Prinyawiwatkul W, Ingram DR (2009) Selected quality and shelf life of eggs coated with mineral oil with different viscosities. J Food Sci 74(9):S423–S429

Waldroup AL (1996) Contamination of raw poultry with pathogens. World's Poult Sci J 52(1):7–25. https://doi.org/10.1079/WPS19960002

Watson RR (2002) Eggs and health promotion. Iowa State Press, Ames, p 218pp

Whiley H, Ross K (2015) Salmonella and eggs: from production to plate. Int J Environ Res Public Health 12(3):2543–2556. https://doi.org/2510.3390/ijerph120302543

Williams PEV (1997) Poultry production and science: future directions in nutrition. World's Poult Sci J 53(1):33–48. https://doi.org/10.1079/WPS19970004

Wu J (2014) Eggs and egg products processing. Food processing: principles and applications (2nd edn, Clark S, S Jung, B Lamsal, eds). Wiley, Chichester, pp 437–455

20 Fish and Shellfish

1 Introduction

1.1 Definitions for Fish

At least 1500 species of fish and shellfish are used as food within the United States (US). The FDA defines fish as *"fresh or saltwater finfish, crustaceans, other forms of aquatic animal life (including, but not limited to, alligator, frog, aquatic turtle, jellyfish, sea cucumber, and sea urchin) other than birds or mammals, and all mollusks, where such animal life is intended for human consumption*". Basically, fish is aquatic animal life intended for human consumption other than birds, and mammals (Food and drug Administration 2001).[1] Moreover, "fishery product" was also defined as "any edible human food derived in whole or in part from fish, including fish that has been processed in any manner." Seafood was taken to refer to "fish" and "fishery products" derived from the sea, as outlined above (Federal Register 1995).

The Codex Committee on Fish and Fishery Products (CCFFP) define fish as "*Any of the cold-blooded (ectothermic) aquatic vertebrates. Amphibians and aquatic reptiles are not included*". Codex rules allow for differences in what is considered "fish" in different jurisdictions. Therefore, alligator and frog qualifies as "fish" in the US but might not be so in other countries.

Awareness of the biological classification of aquatic species, including some that are used as foods, is also useful (Table 20.1). Seafood include fish captured from fisheries and those produced from aquaculture, using marine or inland waters. However, "seafood" presumably does not include fresh water fish.

1.2 Naming Fish for Retail: The Seafood List

The US Federal authorities have provided guidance on naming of fish unambiguously for interstate commerce (Food and drug Administration 2016b). The Seafood list (fully online) is a searchable database of over 1800 species of fish. All commercially important fish varieties are assigned four different kinds of names; (i) the acceptable market name – also considered a statement of identity, (ii) common name, (iii) the scientific – Latinized name and (iv) the vernacular name (Food and Drug Administration 2016a).

The accepted market name should be used for labeling. Indeed, seven common names for fish are mandatory for labeling purposes (Table 20.2). For the vast majority of seafood, manufacturers and producers should inspect the seafood lists in

[1]The FDA describe mollusks to include terrestrial gastropods such as, *Achatina fulica* (the African land snail).

R. Owusu-Apenten, E. R. Vieira, *Elementary Food Science*, Food Science Text Series,
https://doi.org/10.1007/978-3-030-65433-7_20

Table 20.1 Biological classification of aquatic animals

Phylum	Subphylum	Example[a]
Porifera		Sponges
Ctenophore		Comb jellies
Cnidaria		Jelly fish
Arthropoda	Crustacean	Hermit crab, shrimps
Mollusca	Mollusks	Snails, clams, squid
Echinoderms		Starfish, sea cucumber
Chordata	Vertebrates	
	Pisces	Sharks, rays, fish
	Mammals	Seals, whales, dolphins
	Birds	Puffins,
	Reptilians	Turtles, alligators,
	Amphibia	Frogs

[a]Not all are used as foods

Table 20.2 Some legally recognized common names for fish and shellfish[a]

Bonito (21 CFR 102.47),
Canned oysters (21 CFR 161.145),
Canned Pacific salmon (21 CFR 161.170),
Canned tuna (21 CFR 161.190)
Catfish (Federal Food, Drug, and Cosmetic Act (FD&C Act); Sec. 403(t) (21U.S.C. 343(t))
Crabmeat (21 CFR 102.50),
Greenland turbot (21 CFR 102.57),
Pacific whiting (21 CFR 102.46)

[a]*CFR* Code of federal regulations citation

order to determine the most appropriate labeling. Where an acceptable market name is not available, then the common name may be used in a manner that does not mislead (Food and Drug Administration 2016a) (Tables 20.2 and 20.3).

A recent search of the Seafood list using the term "herring" showed 30 different types of fish species that may be labelled as "herring" or else are commonly called herring (Table 20.3)[2]. Interestingly the Seafood data includes basic names and (in some cases) molecular data (e.g. DNA sequences), pictures of the different fishes, and external links to taxonomic information.

[2]A searchable database of acceptable market names for different species of fish that may be important for interstate commerce is available from the Seafood list produced by the FDA and updated every 6 months. Currently the Seafood list contains 1800 species of fish (as defined in Sect. 1.1.) which are either vertebrates, invertebrates and molluscs.

Table 20.3 The Seafood lists showing acceptable market names for herrings

Acceptable market name	Common name	Scientific name
Alewife or river herring	Alewife	*Alosa pseudoharengus*
Cisco or Tullibee	Lake herring	*Coregonus artedi*
Dorab	Dorab wolf-herring	*Chirocentrus dorab*
Herring	Round herring	*Etrumeus teres*
Herring	Flatiron herring	*Harengula thrissina*
Herring	African Ilisha	*Ilisha africana*
Herring	Pugnose Ilisha	*Ilisha elongata*
Herring	Panamanian Ilisha	*Ilisha fuerthii*
Herring	Bigeye Ilisha	*Ilisha megaloptera*
Herring	Indian Ilisha	*Ilisha melastoma*
Herring	Javan Ilisha	*Ilisha pristigastroides*
Herring	Tardoore	*Opisthopterus tardoore*
Herring	Indian Pellona	*Pellona ditchela*
Herring or river herring	Blueback herring	*Alosa aestivalis*
Herring or river herring	Skipjack herring	*Alosa chrysochloris*
Sea herring or Sild	Atlantic herring	*Clupea harengus*
Herring or sea herring or Sild	Pacific herring	*Clupea pallasii*
Herring, thread	Deepbody thread herring	*Opisthonema libertate*
Herring, thread	Atlantic thread herring	*Opisthonema oglinum*
Herrring	Araucanian herring	*Clupea bentincki*
Kahawai	Australian ruff	*Arripis georgianus*
Ladyfish	Ladyfish	*Elops hawaiensis*
Ladyfish	Ladyfish	*Elops saurus*
Mooneye	Mooneye	*Hiodon tergisus*
Mullet	Yelloweye mullet	*Aldrichetta forsteri*
Sardine	Perforated-scale sardine	*Sardinella albella*
Shad	Hickory Shad	*Alosa mediocris*
Shad, gizzard	Western Australian gizzard Shad	*Nematalosa vlaminghi*
Smelt	Great silver smelt	*Argentina silus*
Sprat	Sprat	*Sprattus* spp.

1.3 The United States (US) Seafood Industry at Glance

The seafood industry covers all five Agri-food business activities (Table 20.4): (i) aquaculture or finfish and shellfish farming (NAICS 1125), (ii) finfish and shellfish fishing from capture fisheries (NAICS 1141), (iii) seafood processing and manufacturing (NAICS 3117), (iv) wholesalers (NAICS 4244) and (v) retail of fish & seafood or "fish markets" (NAICS 4452).

Primary production by the US seafood sector (NAICS 1125 & NAICS 1141) amounted to 4.3 million tonnes (US$5.0 billion) in 2012. The top States for commercial fishing by volume were, Alaska > Louisiana > Virginia > Washington > California. However, the value of fish landed were ranked, Alaska> Massachusetts > Maine> Louisiana > Washington. The volume of US seafood export amounted to 3.7 million tonnes (US$ 5.0 billion) with imports of 4.7 million tonnes valued at US$15 billion. Aquaculture accounted for 17% of the seafood production by value whilst 83% was by wild-capture. (National Oceanic and Atmospheric Administration 2012). There are about 170, 000 jobs on 8623 vessels involved in US seafood fishing (OECD 2021).

Table 20.4 NAICS for the fish and aquatic foods

NAICS	Description[a]
112511	Fish farms, finfish
112512	Fish farms, shellfish
114111	Finfish fishing (anchovy, blue fish, cod, croaker, dolphin, eel, flounder, grouper, herring, salmon, seabass, shark, sword fish, whiting)
114112	Shellfish fishing (clam, crab, crayfish, lobster, mussel, octopus, oyster, scallop, sea urchin, shrimp, squid)
114119	Frog fishing, terrapin, turtle,
311710	Fish freezing (e.g., blocks, fillets, ready-to-serve products),
311710	Fish manufacturing, fish meal processing, fish and marine oil processing,
311710	Fish, curing, drying, pickling, salting, and smoking
311710	Seafood manufacturing, chowders, soups, dinners,
424420	Fish, packaged frozen, merchant wholesalers
424460	Merchant wholesalers, fresh, frozen, cured,
424490	Canned foods (e.g., fish, meat, seafood, soups) merchant wholesalers
445220	Fish markets

[a]NAICS = North America Industry Classification Scheme. Abbrided from NAICS Association (https://www.naics.com/search/)

The value for seafood exports (Table 20.5) was estimated at US$5.3 billion in 2014 (USDA Foreign Agricultural Services, 2015). By 2018, seafood production, export and imports were US$ 7.1 billion, US$5.2 billion and US$23.4 billion, respectively (Food and Agriculture Organization 2021; OECD 2021). The US has had a negative trade balance in seafood recently (National Oceanic and Atmospheric Administration 2016a). The major destinations for US seafood exports were the EU (23%), China (22%), Canada (17%), Japan (14%), South Korea (7%) and others (17%). Trade with China increased by 370% over a 5-year period (2009-2014), compared with an overall export growth of 57%. The changes were due to export volume rises, as prices increased only by 4%. Fish and Frozen fish export exceeded shellfish exports (67 vs 33%) but the former grew less strongly (USDA Foreign Agricultural Services, 2015; National Oceanic and Atmospheric Administration 2016a).

The seafood processing, and packaging industry (NAICS 3117) comprised 600 businesses, 32,000 employees and receipts amount-

Table 20.5 US seafood and shellfish receipts (2015)[a]

NAICS	Description	Shipments ($1000)	% Value
311710	Seafood product preparation and packaging	11,463,051	100.0
3117101	Prepared fresh fish and other fresh seafood	2,688,241	23.5
3117102	Prepared frozen fish	3,581,322	31.2
3117103	Prepared frozen shellfish	2,423,672	21.1
3117104	Other prepared fresh and frozen seafood	1,135,300	9.9
3117105	Seafood canning	1,182,317	10.3

[a]Value including domestic consumption. Adapted from (US Census Bureau 2017)

Table 20.6 United States Fish and Seafood Exports (2014)

	FY 2014 (US$million)	5-year increase (%)
Fish (fresh/frozen)	3233	39
Shellfish	1658	65
Other fish products	429	16
Aquaculture exports[a]	1219	
Total exports	5320	44%

Adapted from (USDA Foreign Agricultural Services 2015)
[a]Aquaculture products across all fish categories

ing to US$11.5 billion for 2015 (US Census Bureau 2017, p.10). About 92 businesses (2100 employees) were involved in seafood canning (NAICS 311711), whilst 552 businesses (29,158 employees) dealt with fresh and frozen seafood (cf. Chapter 2; Sect. 2.5). Approximately, 24%, 31% and 10% of earnings were for fresh, frozen or canned finfish respectively (Table 20.5).

1.4 Global Seafood Industry

The world production of seafood was 167 million tonnes (US$300 billion) live-weight in 2014. Remarkably, the value of fish produced by aquaculture (73.8 million tonnes; US$160.2 billion) exceeded the value seafood captured from the wild, for the first time (Food and Agriculture Organization of the United Nations 2018).

Most of the global seafood yield (63%) is consumed domestically whilst 37% (US$148 billion) is used for export. The main importers for seafood products were the US, Japan, China and Spain. From 2009 to 2014, fish consumption per capita increased from 16 kg to 20kg per person (Food and Agriculture Organization of the United Nations 2018). The "GLOBEFISH" website contains updated information on the status of the global fishing industry (Food and Agriculture Organization 2021).

1.5 Hygiene and Safety of Fish Products

1.5.1 Fish Handling

The general principles of food hygiene apply to fish (Borgstrom 2012). Some issues that are important for safety handling of fish in a domestic of foodservice setting are highlighted below. International guidelines for health and safety related to fish and fishery products are also briefly outlined. The fish flesh is readily digested, and is subjected to highly active bacterial enzymes. Therefore, fish can deteriorate rapidly and cannot be held at temperatures above freezing for long periods.

A simple principle that applies to all fresh food, and especially to fish, is the 3/H rule: Rule 1. Handle the product under strict sanitary conditions (to keep the microbial contamination at a minimum). Rule 2. Handle the product at a cool temperature (microbes multiply rapidly and spoilage reactions proceed rapidly at warm temperatures but both proceed slowly at cool temperatures). Rule 3. Handle the product quickly (fish deteriorate as a function of time as well as temperature).

To explain the importance of temperature, fresh-caught fish will generally last about 12 days if held in ice (temperature at about 32 °F or 0 °C) whereas they will last only about 4 days at 46 °F (7.8 °C), a temperature commonly found in domestic refrigerators. There are at least three reasons why fish spoil so rapidly at refrigerator temperatures: first, because they are readily digestible; second, because muscle glycogen is nearly depleted during harvesting, leaving little to be converted to lactic acid, which would act as a preservative; finally, because the bacteria found on fish are psychrophiles (they can grow well at low temperatures), and their enzymes are functional at low temperatures. Even among psychrophiles, there is a range of optimum growth temperatures for individual species, and it is known that some of the psychrophilic bacteria found naturally on fish grow at such low temperatures that they are not reliably detected by standard bacteriological plating techniques that incubate at elevated (warm) temperatures (Huss 1994a; Ghaly et al. 2010).

1.5.2 Safety Codes for Fish and Fishery Products

International guidelines exist to deal with health and safety aspects of fish and fishery products – covering production, storage, handling, transportation, import / export, and processing. The guidelines produced by the Codex Committee on Fish and Fishery Products (CCFFP) address safe production of fish and fishery products in order to facilitate trade and avoid international disputes. The CCFFP require that various stakeholders employ sound food safety management systems, based on two approaches: (a) the so-called prerequite safety program similar to GMP that addresses (i) fishing and harvest vessels, (ii) facilities and processing plants, (iii) Utensils and fittings, (iv) An ongoing hygiene program, (iv) Personnel hygiene and (v) staff straining. The GMP pre-requisite program is expected to be followed with a formal HACCP program that focusses on processing methods relevant for the specific food manufacturer (Codex Committee on Fish and Fishery Products (CCFFP) 2003).

1.5.3 Common Hazards Associated with Eating Fish

Fish are in large part carnivorous. Those predatory fish higher up in the food chain may accumulate significant levels of environmental pollutants (heavy metals, organo-mercuric pesticides can be a concern). Another particular hazard frequently discussed in relation to fish are biotoxins produced by algal blooms.

Disease outbreaks associated with seafood were grouped into several categories (Lipp & Rose 1997; Todd 1997). Firstly, norovirus is a frequent cause of seafood-linked outbreaks. The second group of outbreaks are due to toxic algal blooms (tiny microalgae that produce toxins, which accumulate in marine lifeforms). Algal toxins are linked with a cluster of life threatening conditions, e.g. amnesic, diarrheic, neurotoxic, and paralytic shellfish poisoning (Munday and Reeve 2013). Thirdly, scromboid poisoning is due histamine produced from the amino acid histidine during spoilage. Scomboid poisoning is associated often with tuna, skipjack and related fishes (Attaran and Probst 2002; Lavon et al. 2008). Finally, many salt-water bacteria pathogens are associated with seafood (e.g. vibrio and Clostridium spp) either directly or due to environmental contamination (Huss 1994b) as discussed in Chap. 2.

2 Commercial Species of Fish

2.1 Fish Capture

Seafood taken directly from the wild (wild-capture or capture fisheries) levelled out at 91.2 million tonnes annually for a decade (2009-2019). As the total fish production volume increased from 145.9 million tonnes (2009) to 177.8 million tonnes (2019), the proportion from wild-capture as part of the total declined from 60% (2009) to 51% (Food and Agriculture Organization of the United Nations 2018). Nevertheless, capture fisheries (worth US$140.0 billion) remain an important industry.

Of the variety of methods employed for commercial fishing, line fishing is one of the simplest (National Oceanic And Atmospheric Administration 2021; Australian Fisheries Management 2021). Examples of hand lines and long lines are shown in Fig. 20.1. Fish such as halibut, cod, and haddock are caught using these methods. Troll lines are used to catch certain species of salmon and other fish found near the ocean surface (Fig. 20.2).

Nets are used to catch popular species such as cod, haddock, flounder, and other bottom fish. (Fig. 20.3). Gill nets are used to catch salmon and shad, and sometimes herring, mackerel, cod, and haddock. Otter trawls are used to catch the bottom fish. A purse seines net is used to wall off large areas of the sea, to catch pelagic fish that swim together in large groups such as menhaden, tuna, salmon, herring, and mackerel. The relative ease of capture of pelagic fish leads to their being used not only for human food but also for fish feed in the aquaculture industry (Alder et al.

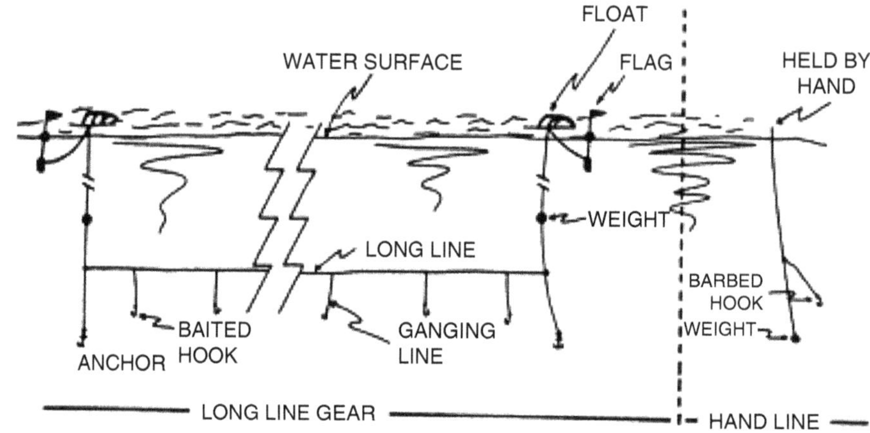

Fig. 20.1 Line fishing

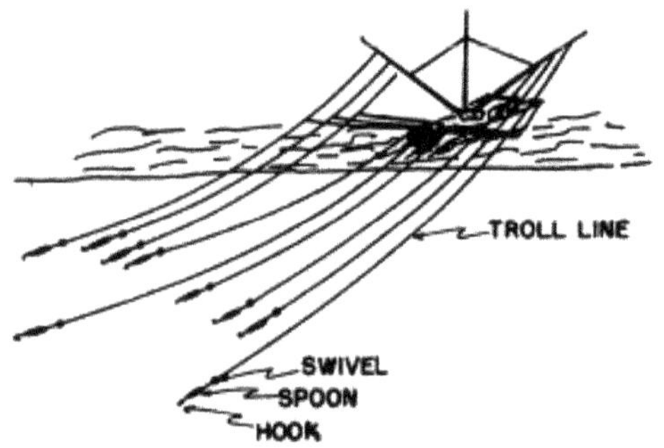

Fig. 20.2 Troll line fishing

2008). Crustaceans such as crabs and lobsters and some freshwater fish are caught in pots. Examples of these are shown in Fig. 20.4. Shellfish such as scallops, clams, and oysters are harvested by the use of dredges, tongs, rakes, or forks (Fig. 20.5). After fin fish are caught, ice is added to the catch to reduce deterioration. Different commercial fishing gear have their own advantages and disadvantages (National Oceanic And Atmospheric Administration 2021; Australian Fisheries Management 2021).

2.2 Aquaculture, Fish Farming

The value of seafood produced by aquaculture (US$160.2 billion) exceeded the value of seafood extracted from the wild from 2014 onwards (Food and Agriculture Organization of the United Nations 2018). Over the period 2017-2019, fish production by aquaculture increased by +3.9% compared to a 3.4% decline in fish production by wild-capture (Food and Agriculture Organization 2021). The growth of fish farming was faster than anticipated 20 years ago (Tidwell & Allan 2001).

Aquaculture covers more than 500 species including, finfishes (362 species), mollusks (104 species), crustaceans (62 species), aquatic invertebrates (9 species) and aquatic plants (37species). The rise in fish farming is partly because of the high feed / conversion ratio for fish where every 1 lb. (454 g) of feed yields about 0.67 lb. (300 g) of live weight. Fish also require less space than other livestock (Tidwell & Allan, 2001; FAO 2016).

Food scientists are getting involved in aquaculture in terms of adding quality attributes to the finished products (Fig. 20.6). Both color and flavor can be altered by additives to the fish feed. The Atlantic salmon aquaculture industry has found that the addition of carotenoid (astaxanthin) to the fish feed imparts a pinkish appearance to the fish flesh and this seems to be a positive quality factor to consumers. Taste can also be improved by the addition of bromophenols to the feed of aquaculture freshwater fish. This adds shrimp like flavor that is preferred by the consumer (Boyle et al. 1993; Anderson 2001).

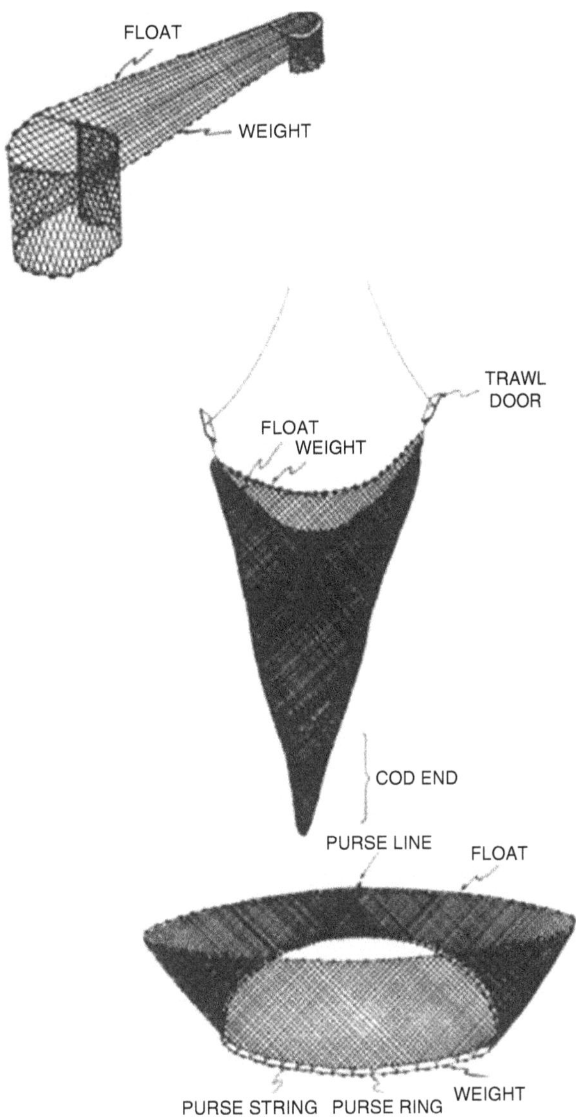

Fig. 20.3 Shows (top to bottom) the drift gill net, otter trawl net, and purse seine net

2.3 The Herring Family (Clupeidae)

2.3.1 Sea Herring

Sea herring (see Fig. 20.7) are found in ocean waters from Alaska to the State of Washington on the West Coast, and from Labrador to Cape Hatteras on the East Coast. Most sea herring are caught with purse seines, but some are caught with pound traps or weirs (similar to pound traps but constructed with poles and brush). Gill nets are sometimes used to catch these fish.

The larger fish may be exported, as there is a foreign market for this species where it is used for human food. The smaller herring, which are canned as sardines, must be held in the purse

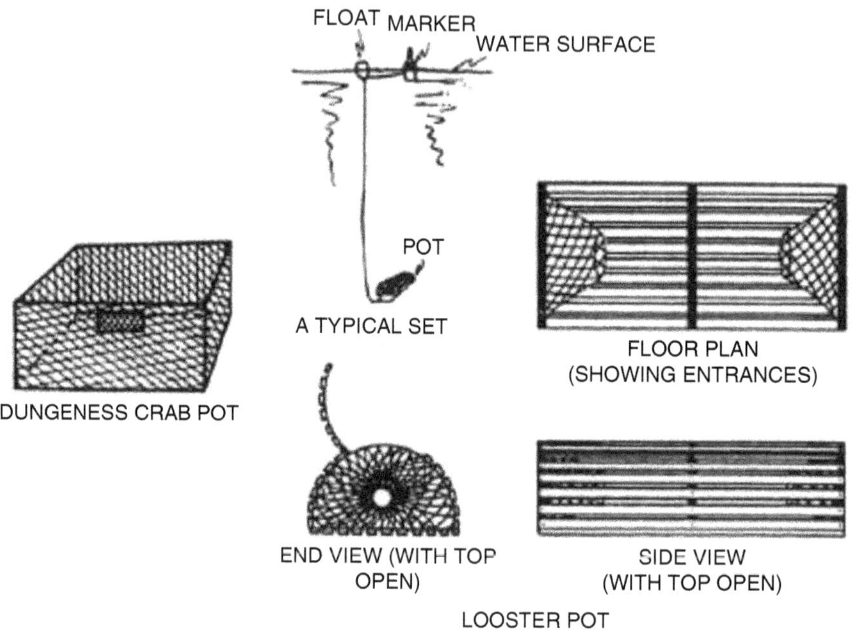

Fig. 20.4 Entrapment devices

Fig. 20.5 Shellfish harvesting devices

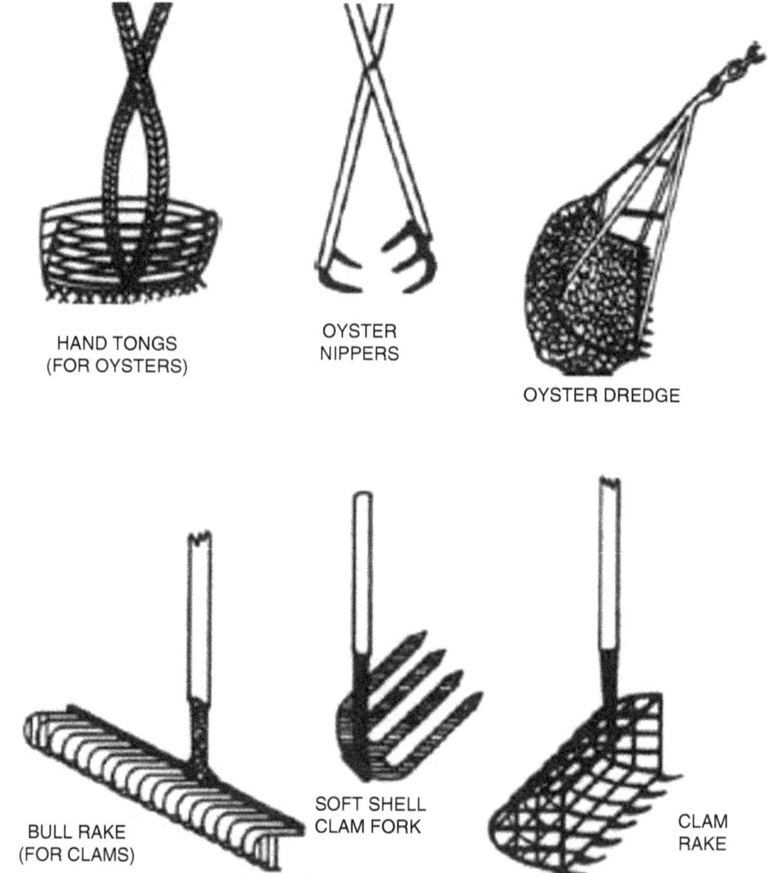

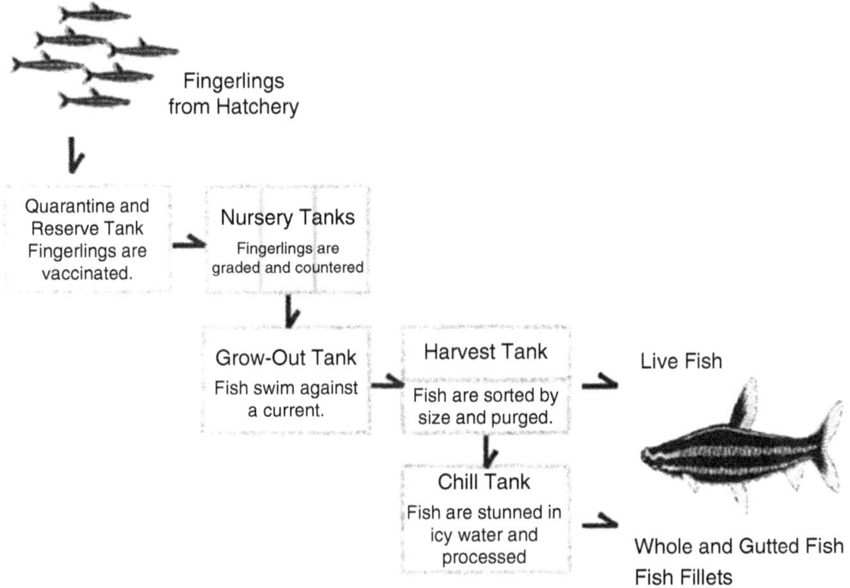

Fig. 20.6 Aquaculture of striped bass

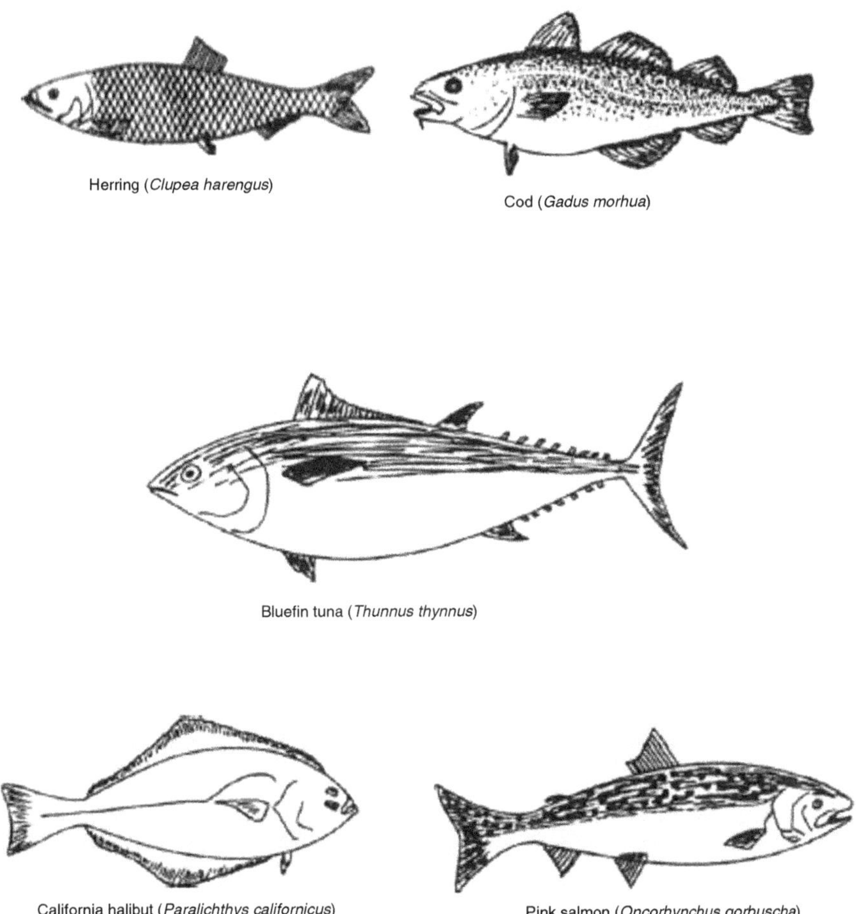

Fig. 20.7 Some important finfish species

seine alive (sometimes for longer than 24 h) until the stomach is free from feed. This is done to prevent enzyme action, which would reduce the quality of the fish. Fish from the nets are loaded on the boat with a large vacuum pump and are salted. More salt is added at the processing plant after the fish are beheaded and eviscerated. The small herring are then preheated, placed in small rectangular cans, and subjected to steam for 18–20 min. The liquid in the cans is drained and replaced with vegetable oil, tomato sauce, or mustard sauce and the cans are sealed, heat-processed and cooled. The resulting product is shelf-stable.

Herring is also processed "whole" but in larger oval cans, usually in tomato sauce. Some large herring are cut transversely into bite-sized pieces and packed in small cans in tomato sauce. This product is also shelf-stable. Some herring are pickled (salt, vinegar, sugar, and spices) and packed in glass jars. Sour cream, onions, or other flavoring ingredients may be added before sealing the jars. Other pickled products include rollmops (pickled herring strips wrapped around a piece of pickle or onion) and kippered herring (salted and lightly smoked). These products must be refrigerated. Some herring are highly salted and heavily smoked, and this product is shelf-stable.

In Alaska, and to some extent in eastern US, the roe (eggs) is taken from herring approaching the spawning stage, salted, packaged in containers, and sold at high prices to certain Asian countries.

Some herring is converted into fishmeal, which is used as a protein supplement for cattle and poultry; the availability of fishmeal is an important consideration for aquaculture (Olsen and Hasan 2012). When used for this purpose, some fish are brought to port in unrefrigerated vessels. At the processing plant, the fish is first cooked in live steam in a continuous cooker, and then pressed in a screw-type continuous press. The press cake is dried with hot gases from oil-produced flames in a rotary drier, to a moisture content of 5–8%. The liquid from the pressed product is not discarded. It is first centrifuged to remove oil, which is collected and sold for industrial uses. The remaining liquid (stick water), which contains proteins, peptides, and amino acids, is then vacuum concentrated to a solids content of 50% and acidified to prevent spoilage. This product may be sold as a protein supplement or it may be added back to the press cake before the latter is dried (Chadd et al. 2002).

2.3.2 Shad

Shad are anadromous fish (ascend rivers to spawn) that spend the greater part of their lives in ocean waters as far as 50 miles (80.5 km) from shore. Shad are used almost entirely as the fresh product, with only small amounts being frozen. They contain many small bones, but can be filleted to eliminate most of them from the flesh. The roe (unfertilized eggs), which prior to spawning is held together by a thin membrane, is highly prized. It is sold fresh, or packaged in moisture vapor proof material and frozen to be sold to the restaurant trade and, in this state, may be stored at 0 °F (−17.8 °C) for 6–8 months. Longer storage under these conditions usually results in a rancid product resulting from oxidation of the fat contained therein.

2.3.3 Menhaden

Four species of menhaden, sometimes called "pogy," "bunker," or "mossbunker," are found in the Western Atlantic. They range from Nova Scotia to Brazil. Menhaden are not used for human food. They are processed to produce fishmeal and oil in the manner described for herring. In the US, larger quantities of menhaden are caught (several hundred thousand metric tons) than that of any other fish or shellfish.

2.3.4 The Anchovy

The anchovy (family Engraulidae) is a small herring-like fish found off the coasts of California and Mexico (California Department of Fish and Game 2006). It is caught in purse seines, and at one time was used as line bait for tuna fishing and later mostly in the production of fishmeal and fish oil. However, a growing amount is used for human consumption in appetizers, garnishes, sauces, relishes, and especially in toppings for pizzas. It is packed in 2-oz (56-g) tins in oil for domestic use and in large tins for industrial use.

2.3.5 Other Clupeidae

Pilchards are members of the herring family, and at one time were plentiful off the coast of California, where they were caught with purse seines. They were used for canning as sardines and to produce fishmeal and oil.

2.4 The Cod Family (Gadidae)

2.4.1 The Cod

Cod (see Fig. 20.7) are found on both sides of the Atlantic and are most plentiful around Norway, Iceland, Newfoundland, Nova Scotia, and on Georges Bank off Cape Cod. In the Western Atlantic, they range from Greenland to North Carolina. Mollusks (clams, oysters, scallops, etc.) are said to make up an important part of the diet of the cod, but cod also eat small fish.

By far the largest quantities of cod are caught with otter trawls in waters ranging from 300 to 1500 ft. (91–457 m) in depth. Small quantities of this species are caught with long lines, hand lines, or gill nets. When caught with otter trawls, the fish are gutted and washed on the deck of the boat. During summer months, the gills must be removed, but the head is left intact. The fish are stored in boxes or pens, in either case layered with ice.

At the processing plant, the fish are washed and filleted, skinless. The fillets are then candled (observed over a bright light) to locate and remove parasites such as worms. Fillets to be sold as fresh are precooled, placed in metal tins of 10, 20, or 30 lb. (4.5, 9.1, or 13.6 kg), and the tins are refrigerated (mechanically or in ice) until they reach their destination. Some fillets are precooled, packed in small trays, overwrapped with a transparent plastic, and shipped to their destination in insulated containers.

Fillets that are to be frozen for the retail trade are packed in 1-lb (454-g) waxed cartons with or without being first wrapped in moisture-vapor-proof plastic. Some cod fillets are frozen, usually in a plate freezer into 16-lb (7.26-kg) blocks and then cut into fish sticks or fish portions that are then breaded in a batter and deep fried. Some battered and breaded product is not fried but frozen raw. This product is used by fast food restaurants and other food service facilities where it is deep-fried to order.

Cod may be salted. To produce salted cod, the fish are beheaded and split longitudinally; the backbone and the abdominal cavity lining are removed, and the fish is washed and layer-salted in closed casks (brine salting).

Fish cakes are prepared from salt cod by first cooking and freshening the fish to remove most of the salt, then mixing it with mashed cooked potatoes and small portions of oil, onions, and pepper. A proportion of about 40% shredded cooked fish and 60% of cooked potatoes is used. This product may be canned without forming or it may be formed into small cakes, deep fat fried to brown the surface, and frozen, or sold in the refrigerated state.

2.4.2 Haddock

The haddock is the second most important member of the cod family. The haddock is found on both sides of the Atlantic from Norway to New Jersey but are most plentiful in waters off Nova Scotia and Cape Cod (Georges Bank). In recent years, the stocks of haddock have been greatly depleted because of overfishing. Haddock are processed to produce fillets, fish blocks, and fish sticks in the same manner as that described for cod. Haddock is not salted and dried although some are lightly salted and lightly smoked, without heat, to produce a product called "finnan haddie."

2.4.3 Pollock

Pollock are found on both sides of the Atlantic from Norway to the Chesapeake Bay but are most plentiful in waters off Nova Scotia, Cape Cod (Georges Bank), and in the Gulf of Maine. Pollock are caught in waters at levels between the surface and a depth of 450 ft. (137 m). They are caught, handled aboard the boat, and processed in much the same manner as that described for cod. Small quantities of pollock are salted and dried.

2.4.4 Hake

There are several species of hake, the most important of which is the silver hake or whiting.

The whiting is most abundant in waters off Nova Scotia. It is caught and handled in much the same manner as are cod. Some whiting are headed, gutted, washed, and frozen in blocks without further cutting, for utilization as food.

2.4.5 Small Fish Species

Small fish, species not especially prized as food (red hake, etc.), and fish frames (the portion remaining after the fillet has been cut away) may be passed through mechanical meat/bone separators (machines that separate the flesh portion from bones and skin). This provides a significant yield of edible, ground fish flesh (resembling hamburger in texture) that may be used to produce frozen fish blocks to be further processed into products such as fish portions and fish sticks.

Handled in conventional fashion, such products are not stable in frozen storage because the fat oxidizes and becomes rancid, and the tissues get tough at a faster rate than that of the corresponding fillets held under the same conditions. The faster oxidation of fats is due to the exposure to oxygen because of the great increase in surface area. The increased rate of toughening may be due to a wider distribution of the enzyme that decomposes trimethylamine oxide to form dimethyl amine and formaldehyde (Sotelo et al. 1995). The latter compound is known to denature proteins. The spoilage reactions may be slowed considerably by storing at lower temperatures, for example, −20 °F (−28.9 °C). Rancidity can be prevented altogether by protecting the product with a wrapper of gas-impermeable plastic film, for example, polyester, polyvinyl chloride, nylon-II or aluminum laminate.

2.5 The Mackerel Family (Scombridae)

The mackerel family include various tuna, the Atlantic mackerel, the jack mackerel, and the Spanish mackerel. Tuna are torpedo-shaped fish tapering to a pointed nose and a slender caudal peduncle (the portion near the tail). Globally the more important species of tuna include the skipjack tuna, yellow fin tuna and big eye tuna that make up, 50.7%, 31% and 10% of catches, respectively (Miyake et al. 2010). The blue fin can attain a weight of 1000 lb. (454 kg), but average size of the tuna caught is about 30 lb. (13.6 kg). Mackerel is much smaller than tuna, but is similarly shaped.

2.5.1 Tuna

Yellowfin tuna is found on the West Coast from southern California to southern Chile. Bluefin tuna range from Nova Scotia to Brazil on the East Coast and from southern California to northern Mexico on the West Coast. The skipjack is found in the Pacific Ocean from southern California to Central and South America. The albacore ranges from Puget Sound (State of Washington) to lower California. The yellowtail is found in Pacific waters from southern California to the coast of Mexico (California Department of Fish and Game 2006). Several other species of tuna are found elsewhere, especially in the Eastern Pacific Ocean.

Tuna are caught mainly with purse seines, which, owing to the size of the fish, are made with heavy twine. Most tuna fishing boats make trips lasting for several months, and for this reason, the fish are frozen aboard the boat. In freezing, the whole fish are cooled in a large well with refrigerated seawater at 28 °F (−2.2 °C). Later the RSW is replaced with refrigerated brine at 10 °F (−12.2 °C). The fish are kept frozen by circulating mechanically refrigerated air.

Traditional fresh tuna steaks (sashimi) and dried skipjack sticks (katsuobushi), originally sold at Japanese auctions, were followed by the canning of tuna in the 1960s; interestingly the trade for sashimi is now a global phenomenon with supply chains shifting to involve supermarkets (Miyake et al. 2010).

Canned tuna are produced at fish processing plants, where the fish is first thawed with running water, eviscerated and washed (Fig. 20.8). They are then cooked in steam, under pressure, and then cooled. When cooled, the heads and

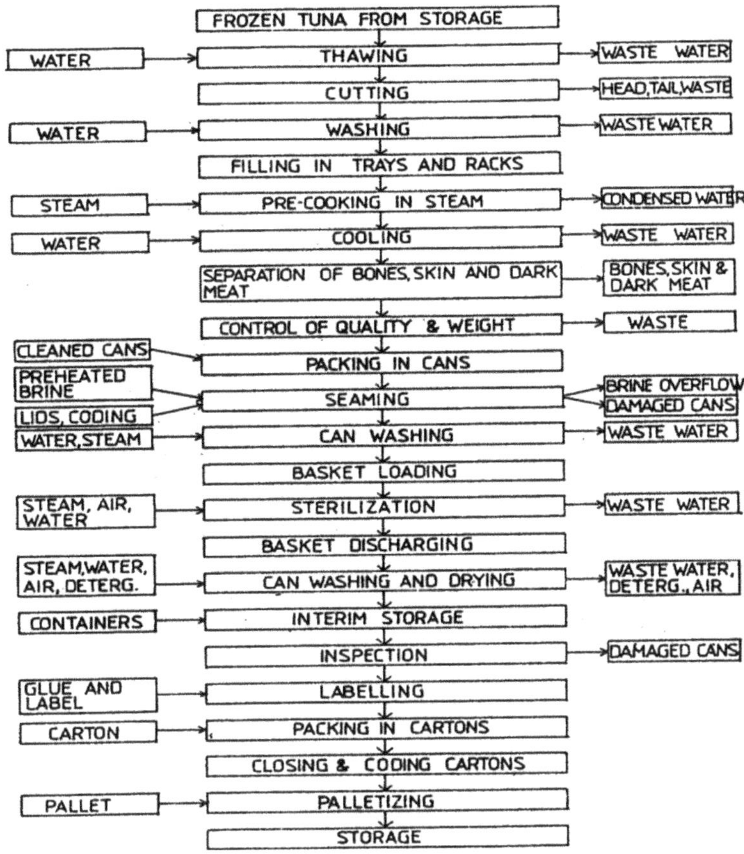

Fig. 20.8 Flow chart for processing tuna. Source: Food and Agriculture organization of the United Nations (1985), reproduced with permission

skins are removed. The fish are then cut longitudinally, after which all bones are removed. If all white meat varieties are being produced, the dark meat is also removed. The white meat is then shaped, mechanically, into a cylinder, fed to cans, and cut to length. To pack chunk-style tuna, pieces are filled into cans with an adjustable filler. Vegetable oil or a broth containing hydrolyzed vegetable protein in water is metered into the cans, which are then heated in steam, sealed, heat-processed, cooled, labeled, and stored (Dewberry 1969).

2.5.2 Mackerel

Atlantic mackerel are found from the Gulf of St. Lawrence to Cape Hatteras in America, and from Norway to Spain in the Eastern Atlantic. Spanish mackerel range from Maine to Brazil in the Western Atlantic but are mostly caught in waters off the Carolinas and southward. The jack mackerel ranges from British Columbia to Mexico in the Pacific Ocean.

Mackerel may be taken in pound traps or with gill nets, but by far the greatest quantities are taken with purse seines. If the boat is to remain out of port after the fish are caught, they are held in ice in the round, un-eviscerated state. Atlantic and Spanish mackerel are sold to retailers as the fresh product, either as fillets or as the round uncut fish. Some are frozen and sprayed with water for purposes of glazing to prevent dehydration until defrosted for sale to restaurants or retail outlets.

Jack mackerel are canned in 1-lb (454-g) tall containers. During the first pass on the conveyor belt the fish passes under circular knives. This cuts off the heads and tails whilst cutting the fish to can-size lengths. The entrails are then removed, after which the fish are washed and flumed to a container feeding the packing table, where they are filled into cans by hand. The open cans are then heated in a steam box to raise the product temperature to 145 °F (62.8 °C), after which they are inverted to drain off liquid formed during heating. Oil, brine, tomato sauce, or mustard sauce is then added to cover the fish, and the cans are then sealed and heat-processed.

2.6 The Salmon Family (Salmonidae)

2.6.1 Pacific Salmon

A number of commercially important species of the salmon family are found throughout the world. Some of the most important species of salmon are listed in Table (20.7). There are five types of Pacific salmon important in the US: pink, chum, Coho or silver, the sockeye or red salmon and Chinook or king salmon (Fig. 20.7). Chinook salmon is the largest salmon with a high oil content. Chum salmon, which has lowest oil content, is least desirable. Silver salmon has similar flavor to Chinook. The sockeye or red salmon is highly regarded for both its flavor and texture (Jensen 1976, p.10). The enormous salmon cannery businesses on the Pacific Northwest reached their peak in the 1900s (Newell 1989).

Table 20.7 Seafood listing for some commercially important salmon

Accepted market name	Vernacular name	Scientific name
Salmon, Pink, Humpback	Pink Salmon	*Oncorhynchus gorbuscha*
Salmon, Danube salmon	Danube salmon	*Hucho hucho*
Salmon, Chum, Keta	Chum Salmon	*Oncorhynchus keta*
Salmon, Coho, Silver Medium Red Salmon	Coho Salmon	*Oncorhynchus kisutch*
Salmon, Cherry salmon	Cherry Salmon	*Oncorhynchus masou*
Trout, Rainbow or Steelhead	Rainbow trout	*Oncorhynchus mykiss*
Salmon, Sockeye, Red Blueback	Sockeye Salmon	*Oncorhynchus nerka*
Salmon, Chinook, King	Chinook salmon	*Oncorhynchus tshawytscha*
Salmon, Atlantic salmon	Atlantic salmon	*Salmo salar*
Trout	Brown trout	*Salmo trutta*

Adapted from the seafood list of acceptable market names. (Food and drug Administration 2016b)

2.6.2 Atlantic Salmon

Wild populations of the North American Atlantic salmon (*Salmo salar*) have been in decline since the late nineteenth century. The decline of the Atlantic salmon may be partly due to overfishing, habitat loss and interruptions in the migratory routes. Catches of Atlantic salmon rose to their peak in the 1970s and diminished exponentially to present very low levels in about 2010. Currently, the Atlantic salmon is designated as" extinct or endangered" in many areas of West Coast. Indeed commercial fishing for Atlantic salmon is prohibited in US federal waters (NASCO 2016; National Oceanic and Atmospheric Administration 2021).

Commercial farming of Atlantic salmon is credited to systems developed in Norway from the 1960s. From 1980 to 2010, the global production increased from 5000 metric tons to 2.5 million metric tons or ~ 4% of global aquaculture production. About 50% of farmed Atlantic salmon is used in world trade. The main producer nations are Norway, the USA, Scotland, Chile, Australia and New Zealand (FAO Fisheries and Aquaculture Department 2016) (Fig. 20.9).

Salmon farming "has long" been controversial due to concerns with the possible negative impact on the environment, inadvertent interactions between farmed and wild fisheries, welfare of farmed salmons and the need to promote responsible aquaculture practices (FAO Fisheries and Aquaculture Department 2016). The FDA approved a fast-growing genetically modified Atlantic salmon as safe to eat as a non-genetically modified salmon. Manufacturers may adopt voluntary labeling to distinguish their products as "not bioengineered" (Food and Drug Administration, 2015). The status of US salmon fisheries has been discussed recently (Fay et al. 2006).

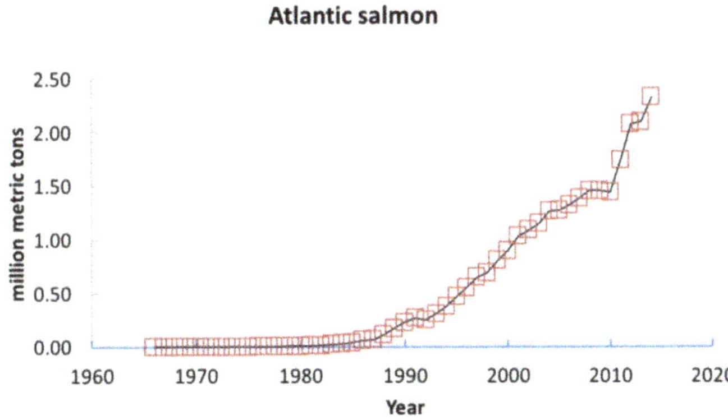

Fig. 20.9 World production of Atlantic salmon by aquaculture. (Drawn using data from (FAO Fisheries and Aquaculture Department 2016)

2.6.3 Salmon Processing

The salmon processing industry located at the Pacific Northwest was impacted greatly by the introduction of a powered iron-butchering machine for salmon invented by Edmond A. Smith in 1901/ 2 (Jones 2006). Salmon could be prepared more quickly for canning using the iron butcher, as compared to manual butchering. The iron butcher removed the head, tail, fins, and viscera from fish after which the salmon is trimmed by hand to remove extraneous material and then washed. The fish next passes under rotary blades on a slotted conveyor, where they are cut into can-sized lengths. The cut salmon pieces then pass on to a volumetric filling machine where salt is added to the can (about 1.25% by weight), and the cans are filled with fish. After filling, the covers are clinched on the cans, which are then sealed under vacuum, or they may be sealed without first clinching, a steam jet being used to remove air from the headspace in the can. After sealing, the cans are washed, heat-processed to provide commercial sterility, and then cooled in the retort. The cans are then labeled, if not lithographed, packed in cases, and stored in a warehouse until shipped out. Heat-processing times and temperatures are applied according to the weight of the product in the can.

Some coho and spring salmon are sold fresh as steaks; some are frozen, in gas-impermeable containers. Spring salmon are sometimes preserved by salting in casks, and held at 35 to 40 °F (1.7 to 4.4 °C) for 30 days. This product is usually shipped to processors who smoke the fish. In smoking, the salted fish is first soaked in water to remove salt, then smoked at temperatures below 90 °F (32.2 °C) or hot smoked at a temperature of about 175 °F (79.4 °C).

2.7 The Flatfish Family (Pleuronectidae)

Many species of flatfish are utilized as food. On the East Coast of the US, the halibut, the turbot, the sand dab, the fluke, the yellowtail flounder, the black back flounder, the lemon sole, the plaice, and other species are edible. On the West Coast, the halibut, petrale sole, English sole, rex flounder, arrowtooth flounder or turbot, Dover sole, starry flounder, rock sole, and other types are caught as edible fish. All the above are flounder; none are true sole. In shape, flounder are flat, comparatively thin fish.

Flatfish vary in size. Although halibut can grow to a very large size, the average weight of those caught today is about 40 lb. (18.2 kg); the average size of turbot is about 7 lb. (3.2 kg); plaice are about 10 lb. (4.54 kg); other species are smaller.

Small flounder are caught with otter trawls. Aboard fishing boats, they are held in pens or boxes in ice, as are cod, but these fish are not eviscerated prior to icing. On the West Coast, halibut are caught with long lines. The fish are eviscerated, the gills are removed, and they are placed in hold pens in ice much in the same manner as described for cod, but in this case, the "poke" (belly cavity) is also filled with ice. Small flounders, to be sold either fresh or frozen, are handled, packed, and distributed similarly to cod fillets.

Halibut are handled in both the fresh and frozen state. As the fresh product, the fish are beheaded, washed, and packed in ice in boxes. Then they are shipped from the West Coast to the Midwest or the East Coast under refrigeration. If the fish are small, they may be sold by distributors to retailers as received. If the fish are large, they may be sold to retailers as portions. Frozen halibut are decapitated, washed, and placed on racks in freezer rooms at 0 °F (−17.8 °C) or below until shipped to distributors in the frozen state. Small halibut, after freezing, may be sawed into steaks, trimmed, and packaged in moisture-vapor-proof plastic film as 12, 14, or 16-oz (340-, 397-, or 454-g) portions.

2.8 Other Fish

Many species of fish are not discussed to any extent in this chapter, for example, bluefish, butterfish, croaker, red and black drum, eels, groupers, mullet, ocean perch (fairly important fish of small size, caught with otter trawls handled aboard boats in the iced un-eviscerated state, and processed as fresh or frozen fillets), pompano, rockfish, sablefish, sea trout, red and other snappers, spot, striped bass, swordfish and other marine species, as well as freshwater fish, such as buffalo fish, carp, catfish, chubs, cisco, trout, and whitefish.

Currently, there is a considerable fish farming industry in the US in which catfish and trout are grown in freshwater ponds. In many countries, carp and tilapia are grown in freshwater ponds and harvested as food for humans (Sect. 2.2).

3 Shellfish

3.1 Bivalve Mollusks (Class Pelyopoda)

There are other mollusks, besides bivalves, that are used as food for humans; squid is among them. In the US, the mollusks used chiefly for food are oysters, a number of clam varieties, and scallops (Cheng and Capps Jr. 1988). As noted above the market for shellfish products is about half of the value for fin fish (Sect. 1.3).

Oysters and other bivalves may be eaten raw or without cooking. Therefore, great care must be taken to make sure that bivalve growing areas are not polluted with even traces of human excrement. Control of bivalve harvesting areas is supervised by a division of the Food and Drug Administration but must be effected by state authorities. Another safety issue, arises because clams and mussels that feed mainly on algae may at times become toxic to humans (shellfish poisoning). This happens when bivalves feed on certain algae (dinoflagellates) containing substances that are toxic to humans but not to mollusks. Public health officials periodically test bivalves for toxin and close the shellfish beds when there is danger of shellfish poisoning outbreaks (Lipp & Rose 1997; Sobel and Painter 2005; James et al. 2010).

3.1.1 Oysters

There are five species of oysters in the US, three on the East Coast, and two on the West Coast, one of which was introduced from Japan (Lavoie 2005; Vilanova 2014). Oysters are harvested with rakes, tongs, dredges, or with water-jet vacuum dredges (Fig. 20.5). The boat used to harvest bivalves should be outfitted with a chemical toilet so that oyster-growing areas will not be polluted with human discharges.

At the processing plant, oysters in the shell are washed with cold seawater (with or without sanitizer), packed in sacks or barrels, and shipped to restaurants at temperatures between 32 °F and 40 °F (0 °C and 4.4 °C). Most oysters are shucked (meats removed from the shell) by hand with the aid of a knife. The meat is washed with fresh potable water, graded for size, and packed in jars or metal containers and shipped to market in crushed ice at a temperature about 33 to 34 °F (0.6–1.1 °C). Some shucked oysters are frozen and stored at 0 °F (−17.8 °C) or below until shipped to market. Finally, oysters is also breaded, packed in waxed paperboard cartons holding 10–14 oz. (284–397 g) of product, then frozen. Oysters may be eaten raw from the half shell or in stews (lightly heated in milk with some butter), or breaded and deep-fat fried. In addition, there is available canned smoked oyster, canned boiled oyster, dried oyster, salt-fermented oyster and oyster juice (Park et al. 1988).

3.1.2 The Hard-Shell Clam

The hard-shell clam is similar to the oyster in its internal structure. The shell is rounded, symmetrical, and relatively smooth on the outside, coming to a gradual peak near the hinge. The shell is quite hard and thick.

Hard-shell clams may be harvested by hand (feeling for them with the hands or feet and removing them by hand). They may also be removed from shallow water with clam rakes. The largest quantities of this clam are harvested with scratch rakes, with tongs similar to those used to remove oysters, or with dredges. Dredges are used in comparatively deep water and may be of the basket or water-jet type. Aboard boats, hard-shell clams should be handled in the manner described for oysters.

In preparing them for market, hard-shell clams are washed with seawater, graded for size, and cooled. Some clams may be taken from semi polluted waters, provided they are depurated.

In the shell, hard-shell clams are marketed according to size. The different sizes are "chowders" (large size), which are used to prepare chowders, fritters, or stuffed clams; "cherrystones" (medium size), which are used for baking; and "littlenecks" (small size), which are used as steamed clams or for eating raw on the half shell. Hard-shell clams are neither canned nor frozen in significant quantities.

3.1.3 The Soft-Shell Clam

The soft-shell clam is found in the Western Atlantic as far north as the Arctic regions and as far south as Virginia, being most plentiful off the coasts of New England, New Jersey, and Virginia. In New England, soft-shell clams are harvested when the tide is low by digging into the mud with the short-handled clam hoe and removing them by hand. In the Chesapeake Bay area, clams are harvested from boats using water-jet dredges and an escalator.

They are placed in bags or baskets and brought to the processing plant, where they are washed with seawater and sorted according to size. Specimens 3 in. (7.6 cm) in length or smaller are usually cooked by steaming. The larger-sized clams are removed from the shell by hand and placed in metal containers, after which the containers are refrigerated by being surrounded with crushed ice. In this form, they are shipped to restaurants to be served as a breaded, deep-fat-fried product. Soft shell clams may be removed from restricted areas and depurated.

3.1.4 Surf Clams

The surf clam or "skimmer" is large, reaching a length of 8 in. (20.3 cm). It is found just below the surface of sandy bottoms in waters 30–100 ft. (9–30 m) deep off Atlantic Coast states from Massachusetts to Virginia. Most of the harvesting of this species is done off New Jersey with water-jet dredges having V-shaped scoops. Aboard the boats, the clams are placed in baskets or jute bags and brought to the processing plant without refrigeration.

Surf clams are used primarily for canning. The viscera are not utilized as food. At the canning plant, the clams are washed, then steamed lightly to cook the meat partially and so the shell will open. The meat is then removed by hand, the nectar (liquid left in the shell) being saved. The lower part of the neck (syphon), the mantle, the adductor muscle, and the foot (muscular portion that allows the clam to anchor itself in the mud) are then removed with scissors and diced into pieces about 3/8 in. (1 cm) wide. The diced portions are then filled into cans together with some hot nectar and salt, after which the cans are sealed and heat-processed.

3.1.5 Other Clams

Other species of clams used as human food include the butter clam and pismo clam harvested off the West Coast and the ocean quahog harvested off the East Coast. The latter is not as large, nor as much in demand, as the surf clam, but is used to help fill the demand for surf clam, which exceeds its supply.

Mussels that resemble soft-shell clams, except for the color of the shell and other minor differences, are used to some extent to help fill the demand for soft-shell clams, of which there is an insufficient supply.

3.1.6 Scallops

There are several types of scallops, of which the sea scallop and the bay scallop are best known. The internal anatomy of the scallop is similar to that of the oyster, but the adductor muscle, the only part of the scallop that is eaten, is much larger than those of oysters are and clams are. Once beyond the larval stage, the scallop may attach itself temporarily to some object, but the adult scallop is quite mobile. By closing the opened shell with its adductor muscle, thus forcing water through two holes in the top shell, the scallop becomes jet propelled. These bivalves cannot be held out of water in the live state, as can clams and oysters, since the water drains from the shell, which cannot be tightly closed.

The bay scallop is generally circular in shape with a grooved upper and lower shell and a rectangular projection at the back near the ligament (the bay scallop is the logo that can be seen in any Shell Gasoline sign). The bay scallop reaches several inches (1 in. = 2.54 cm) in diameter, and the adductor muscle may be as large as 1 in. (2.54 cm) in diameter. Bay scallops are harvested with basket rakes in shallow water and with dredges in deeper water.

The sea scallop is much larger than the bay type. It may reach a size of 8 in. (20.3 cm) in diameter, and the adductor muscle may be as large as 3 in. (7.6 cm) or more in diameter. Unlike the bay scallop, the shell of the sea scallop is not grooved. Sea scallops are found in ocean waters 60 ft. (18.3 m) or more in depth. While this bivalve ranges from Labrador to New Jersey, it is most plentiful on Georges Bank off Cape Cod. Sea scallops are harvested with dredges. Aboard the boat, the "eyes" or adductor muscles are removed from the bivalves with the aid of a knife, placed in muslin bags, and iced and brought this way to port. The remaining portions are discarded at sea. Sea scallops are sold in the fresh or frozen form. If frozen for purposes of selling after defrosting, they are placed in freezer rooms in muslin bags and held until shipped to retailers. Some sea scallops are breaded and may be deep-fat-fried prior to packaging and freezing. At 0 °F (−17.8 °C), the storage life of scallops is longer than 1 year. Other species of scallops include the calico scallop, found off the coast of Florida, and bay-type scallops, found off the coasts of Alaska and Australia and in the Irish sea.

3.2 Crustaceans(Class Decapoda)

Several types of crustaceans are used as food for humans, most of which are prized as delicacies. Included among these are shrimp (several species); lobsters (American, European, and Norwegian species); crabs (several species); and crayfish (several marine species and the freshwater species). Although the shells of Crustacea vary in color, they all turn pink when cooked.

Crustaceans have a hardened external skeleton made up of a calcified polymer of glucosamine (a six-carbon sugar containing an amine [NH2] group) called chitin.

The external anatomy of crustaceans consists of the mouth parts, the eyes, the antennae (varying greatly in size), the body or cephalothorax to which five pairs of legs are attached, and the abdomen or tail consisting of a number of jointed segments adjoined to the body. In some crustaceans, the first pair of legs is chelate or enlarged and developed into biting and crushing appendages called "claws." The end section of the tail has several parts, including the fan-shaped telson. In some species, the tail may be contracted or flexed to provide for movement in the water. On the underside of the tail, there are a number of attachments called pleopods or "swimmerets," which species are the main appendages providing for movement in the water. Some crustaceans will shed an injured claw, then generate and grow a new one.

Crustaceans grow by shedding the old shell (molting) to become soft-shelled for a short period (the new larger shell soon becomes hard) and filling up the new larger shell, which allows more room for growth; molting occurs most frequently in the early years of growth. Mating takes place when the female is in the soft shell stage. The fertilized eggs are attached to the swimmerets, are eventually hatched, and after several larvae stages the small crustaceans sink to the bottom and assume the general habits of the adult.

3.2.1 Lobsters

Lobsters have either a well-developed first pair of walking legs or biting claws. The European lobster is found around certain parts of the British Isles and mainland Europe. The Norwegian lobster is found mainly around the coast of Norway and the west coast of Sweden. The American lobster ranges from Labrador to the coast of North Carolina in an area that extends seaward for a distance of 50 miles (80 km). However, there are deep-sea lobsters found more than 200 miles (322 km) from the coast. The depth of water where lobsters are taken is usually 30–150 ft. (9–46 m), but deep sea lobsters live at a depth up to 1200 ft. (366 m). The American lobster is most abundant off the coasts of Maine and the maritime provinces of Canada. The average American lobster caught measures 9–10 in. (23–25 cm), weighs 1to 2 lb. (454–908 g), and is 4–7 years old. However, deep-sea lobsters are larger, and specimens weighing more than 40 lb. (18 kg) have been caught that are believed to be more than 50 years of age. The food of lobsters include fish, clams, and other mollusks.

Lobsters are caught in pots (Fig. 20.5) and are held in the live state aboard boats without refrigeration, because they are brought to port shortly after harvesting. Lobsters may be held in the live state, out of water at low temperatures above freezing, for more than 1 week, if given sufficient air space, because they are able to obtain oxygen from what dissolves in the water on their gills (the gills must be kept moist). They may also be held in the live state for a month or more in ocean pounds, which allow the free flow of water, or in tanks in which seawater is filtered, aerated, and circulated. When lobsters are held in tanks, the biting claws may be immobilized by the insertion of a wooden plug into the flesh above the thumb or by an elastic band encircling the thumb and claw.

Lobsters are sold mostly to restaurants or to the consumer, in the live state. Lobsters should be cooked from the live state or killed and cooked immediately. The reason for this is that lobsters have a very active proteolytic enzyme system that soon digests part of the tissue of the dead lobster, partially liquefying the meat or causing it to become soft and crumbly (a condition known as "short meated"). Some cooked lobster meat is sold as a canned or frozen product but does not make up a significant part of the catch. The storage life of cooked frozen lobster meat at 0 °F (−17.8 °C) is at least 8 months. Whole lobsters in the raw, cooked, or partially cooked state cannot be frozen successfully, because when frozen in the cooked state, the tomalley (liver) becomes rancid and affects the flavor of the meat, and, when frozen raw or in a partially cooked state, the flesh undergoes proteolysis.

3.2.2 Shrimp

There are numerous species of shrimp used as food for humans. The edible types vary in size from very small, about 2 in. (approximately 5 cm), to more than 10 in. (25 cm). The larger shrimp are called prawns. The overall size of Gulf shrimp as caught is 7–8 in. (17–20 cm). Most shrimp caught by U.S. fishermen are taken from the Gulf of Mexico, and these consist of three main types: white, brown, and pink. Some shrimp are taken from Atlantic waters off the Carolinas, Georgia, and Florida, and some are taken off Alaska, Maine, and Massachusetts. Shrimp are imported from Mexico, India, Panama, Venezuela, Brazil, Guiana, Ecuador, Nicaragua, Colombia, El Salvador, Honduras, Thailand, Surinam, Malaysia, and other countries.

Shrimp are caught with otter trawls that are somewhat modified from those used to catch cod and haddock. In some instances, two boats may be used to tow the trawl attached to an outrigger. Aboard the boat, all but the tails of the shrimps are discarded. The tails are washed and stored in boxes or pens in ice.

At the processing plant, the shrimps may be peeled and deveined (the intestinal tract is removed), and they are washed. In some cases, shrimps are dipped in a solution of sodium tripolyphosphate to prevent softening of the texture and loss of water during storage. However, if excess amounts of this salt are absorbed, the cooked product will appear and have the texture of raw shrimp.

Shrimps may be frozen in a plate freezer where boxes of the product are frozen into shrimp blocks or the shrimps may be frozen individually using a liquid freezant (e.g., liquid nitrogen). Prior to freezing, some shrimps are butterflied (split longitudinally), and some may be cooked. Some frozen shrimp may be tempered (partially thawed) in order to coat them with a breading, and then refrozen.

Raw shrimp in the shell, when protected against dehydration, have a high-quality storage life of at least 2 years at 0 °F (−17.8 °C) or below. Cooked shrimp, especially those cooked in hot oil, have a storage life of 3–5 months at 0 °F (−17.8 °C). Uncooked shrimp, prepared and frozen in the butterfly form, are subject to storage changes. There is so much air space in the package that dehydration occurs through a continuing two-step process: (1) moisture from the product vaporizes and fills the voids and (2) moisture from the voids condenses on the inner surface of those parts of the package that are adjacent to the voids.

Frozen shrimp imported from other countries must be defrosted before they can be processed. This may be done by tempering the product at about 40 °F (4.4 °C) for 24 h, then completing defrosting by holding the unpackaged shrimp in running water. A more sanitary defrosting method-employing microwave heating is also used.

Considerable quantities of shrimp are canned. For this purpose, they may be delivered to the cannery with the heads on. The shrimp are first washed and separated from the ice. The tails are then removed from the heads. The shell is removed and the vein taken. Individual specimens are then inspected, and broken and decomposed shrimp are discarded and may be used for pet foods or fertilizers. The product is then blanched or heated in boiling saturated salt solution (25%) for a period of 45 s to 3 min. After blanching, the shrimp are graded for size and filled by machine into one of several can sizes. Hot, dilute salt solution is added to the product in the cans, and the cans are sealed immediately. Heat-processing to provide commercial sterility is carried out at 250 °F (121.1 °C) for various times depending on the size of the container.

In the Pacific Northwest and Alaska, very small shrimp may be canned without deveining. Shorter blanching times are used for this product and small amounts of citric acid are added to the brine used to cover the shrimp. The brine is added cold, and the cans are vacuum-sealed prior to the heat processing.

3.2.3 Blue Crabs

Crabs have the general anatomy of other crustaceans, but thc body is oval-shaped or disc-like instead of cylindrical, as in lobsters, shrimp, and crawfish. Also, the abdomen (tail) is comparatively small, flattened, and permanently flexed under the body. Several species of crab are used as food by humans.

The blue crab is found from Nova Scotia to Mexico, including the Gulf of Mexico, and is especially abundant in the Chesapeake Bay region. It is commercially important only south of New Jersey. The semi oval body of the blue crab has spiked peaks near the back end. The first walking legs are well developed into biting claws and the last pair of legs (called back fins) is flattened and used to propel the crab in the water. When fully grown, these crustaceans measure 7 in. (15 cm) or more across the body. Crabs are caught with crab pots or traps or with trot lines.

Blue crabs are brought to the processing plant in the live state. They are then cooked in boiling seawater, or in steam. After cooking, the back shell, viscera, claws, and legs are removed; then the meat may be removed from the shell with the aid of a small, sharp knife. In hand picking, the body meat adjoining the back fin is separated from the finer body meat, as it is considered to be of better quality and of higher value. Certain machines are now available for removing crab meat from the shell. This may be done by impact, or a roller process may be applied to the cooked or partially cooked "debacked" body, legs, and claws. Somewhat better yields of meat are obtained by machine picking and much less labor is required to do the job, but machine picking does not provide for the separation of back fin lump meat from the other body meat, unless it is done by hand before the crabs are machine-processed.

After the meat has been picked, it is packed in metal cans that are then closed and heated in boiling water until an internal temperature (at the center) of 185 °F (85 °C) is reached. This temperature is maintained for 1 min. The product is then cooled and held at 33–38°F (0.6–3.3 °C) prior to distribution.

Blue crab meat is neither frozen nor heated-processed to the point of commercial sterility, because either treatment results in a product of poor quality. If blue crabs, when caught, are nearing the molting stage, they may be held in seawater pounds, until they shed their shells. They are then soft-shell crabs, and they are sold in the live state at a premium price, as soft-shell crabs are considered to be a delicacy. High mortality rates are usually encountered during the holding of crabs for molting.

3.2.4 Dungeness Crab

The Dungeness crab is found from the Alaskan peninsula to southern California but is most abundant in the area between San Francisco and southeast Alaska (California Department of Fish and Game 2000). It may attain a size of 9 in. (23 cm) across the back. It has well developed biting claws. Dungeness crabs are caught in water 12–120 ft. (3.7–36.6 m) deep. The circular pot is used in deep water, the rectangular pot in shallow water. Ring nets may also be used. Dungeness crabs are brought to port in the live state aboard the boat, held in wells of seawater, and may thereafter be held in tanks of seawater until sold to restaurants in the live state.

In processing, to obtain meat from Dungeness crabs, the back shell of the live crab is first removed, the viscera and gills are then torn away, and the body is broken in half with the legs attached. The sections are then cooked in boiling seawater for 10–12 min; then the meat is removed by hand by shaking or by affecting against a metal container or it may be removed by running the body and legs between mechanized rollers. Pieces of shell may then be separated by floating the meat in a salt solution of the appropriate specific gravity for the meat to float and the shell to sink. Fresh meat is packed in cans, the cans are sealed, and the product is held at 32–40 °F (0–4.4 °C) for purposes of distribution.

Some whole or eviscerated Dungeness crab is frozen in brine at 5 to 0 °F (−15 to −17.8°C), packaged or glazed, and stored at 0 °F (−17.8°C) for distribution in the frozen state. A larger amount of this type of crab meat is packed in hermetically sealed cans and frozen in moving air at 0 to −10 °F (−17.8 to −23 °C). This type of crabmeat is not especially stable in frozen storage but may be held for as long as 6 months at −10 °F (−23 °C) with good results.

Dungeness crab meat is also canned and heat-processed. The meat is packed in cans holding 6.5 oz. (185 g) of product. A weak solution of salt and citric acid (pH 6.6–6.8) is then added, covering the meat to prevent discoloration. The cans are then sealed and heat-processed at 240 °F (115.6 °C) for 60 min, then cooled in the retort. The quality of this product is inferior to that of the fresh meat.

3.2.5 King Crab

This species is not a true crab but is similar to crabs in structure and habits. It is much larger than other crabs, attaining a spread of about 5 ft. (1.5 m) and a weight of about 24 lb. (10.9 kg).

King crabs are caught off central Alaska to the Aleutian Islands and off the islands of northern Japan. They are harvested with large rectangular pots. Aboard boats, the crabs are held in the live state in wells of circulating seawater.

King crab meat is either canned or frozen. In canning, the whole crab is cooked in boiling water, after which the meat is squeezed out between rubber rollers. The meat is then washed, packed in cans in a weak brine, and the cans are sealed, heat-processed, cooled, and stored.

The meat may be frozen in large blocks for the restaurant trade. The legs and claws are also frozen for retail outlets and restaurants. If properly packaged, and frozen to a temperature of 0 °F (−17.8 °C) or below and held at this temperature, King crab meat has a high-quality storage life of at least 12 months. Lower storage temperatures provide for an even longer storage life.

3.2.6 Snow or Tanner Crab

This species is relatively large, reaching a size of 5–6 in. (12.7–15.2 cm) across the back and 2.5 ft. (76.2 cm) between the tips of the outstretched legs. The snow crab is taken in deep water off central and eastern Alaska and in the Bering Sea and some is taken off Nova Scotia and Newfoundland. Itis caught in large, baited pots, as is the King crab.

Snow crabs are handled and processed in much the same manner as are King crabs, but most of the meat is canned and heat-processed. The meat of the snow crab is inferior to that of the King crab.

3.2.7 Red Crab

The red crab is found from Nova Scotia to South America but is taken almost entirely in deep waters off southern New England. Red crabmeat is removed mechanically from the debacked, partially cooked specimens with machinery employing the roller process. Red crabmeat is sold mostly as the fresh cooked refrigerated product but some is sold as frozen.

3.2.8 Jonah Crab

The Jonah crab is found in waters from Nova Scotia to North Carolina and is caught in lobster pots. The meat of this crab is difficult to remove from the shell, and the product is sold mostly as cooked refrigerated or frozen whole crabs or claws.

3.2.9 Marine Crayfish

The crayfish or spiny lobster has become a popular food in the US. There are a number of different species of marine crayfish ranging from Florida and the Gulf of Mexico to Central and South America. They are also found off Australia, New Zealand, South Africa, and other countries. These species have the general anatomies of lobsters, but the first pair of walking legs is not developed into biting claws.

Because only the tail portion is eaten, this is removed from the live specimen, packaged, with shell on, in moisture-vapor-proof material, and frozen, for sale to restaurants or for the retail trade.

3.2.10 Freshwater Crayfish

Freshwater crayfish are grown in ponds. Although they have the general anatomy of the true lobster, with well-developed biting claws, they are much smaller, the maximum weight being about 8 oz. (227 g). There is at present a small industry in which the small crayfish are placed in rice fields after the rice has been harvested. Here, they eat the rice roots and also serve to fertilize the fields. By planting time, the fields can be drained and the crayfish harvested. Generally, these specimens are handled in the fresh, refrigerated state and are processed only by cooking.

4 Surimi

Japanese fisherman found nearly 900 years ago that if they minced and washed the fish and mixed it with salt and spices, ground it to a paste, then cooked it by steaming or broiling, it would last much longer than fresh fish (Nicklason 1993). The product is called surimi. Surimi is not itself a foodstuff but is an intermediate raw material from which the traditional Japanese kneaded foods called "kamaboko" are made (Park 2013).

The word "surimi" literally means "minced meat." Surimi is more than minced meat, as it has gel-forming capacity and has long-term stability in frozen storage. It gains these qualities from the addition of sugars as cryoprotectants. When fish muscle is separated from bones, skin, and entrails, and then comminuted, it is called minced meat (Park 2013).

Minced fish becomes raw or unfrozen surimi after it has been washed to remove fat and water-soluble constituents. The raw surimi is quite bland in flavor and the washing removes water-soluble components, thus isolating the meat's myofibrillar protein. This protein is water insoluble and gives the surimi the excellent gel-forming capacity.

When raw surimi is mixed with anti-denaturants and frozen, the product is called frozen surimi. Sugar compounds such as sucrose and sorbitol are often added as anti-denaturants or cryoprotectants. The production of frozen surimi is shown in Fig. 20.10. Surimi-based products have other ingredients added to give the product desired flavor, color, and texture. An example of the formulation of a crab leg-type product is shown in Figs. 20.11 and 20.12.

Surimi has been made from Alaska Pollock but other fish such as sardine and mackerel have been used and some very high quality surimi and surimi-based products have been produced. Some examples of different surimi-based products are shown in Figure 20.12.

5 Further Reading

About 75% of the earth's surface is covered with water, which provides a large and apparently limitless supply of seafood, except that it is not inexhaustible. Amongst the many interesting topics not covered in this short discussion, the sustainability of future fish stocks is amongst the most important (Jennings et al. 2016). The production of fish protein concentrates from fish waste and underutilized parts has also received considerable attention in the past (Ariyawansa 2000; Kristinsson and Rasco 2000). The potential for various sea foods to contain toxins such as heavy metals and organophosphorus compounds is one reason why expectant mothers are advised to reduce fish consumption (Pompa et al. 2003). Another interesting aspect worth further reading is in the area of fish and seafood allergy (Moonesinghe et al. 2016; Pedrosa et al. 2015).

Some of the main types of shellfish used for human food and the manner of their production, preparation and processing were described. However, the focus was on mollusks (oysters, clams and scallops) and crustaceans (lobsters, shrimps, crabs and crayfish) which are most important in the US. Other shellfish groups that are also deserving of further discussion include the, gastropods (e.g. snails and whelk) and cephalopods e.g. squid, octopus and cuttlefishes (Duncan 2003; Granata et al. 2012; Vilanova 2014).

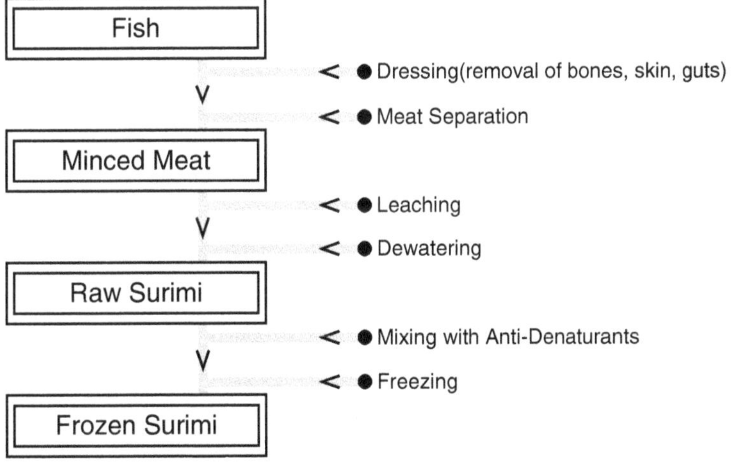

Fig. 20.10 Production of surimi

Fig. 20.11 Manufacturing of a crab-leg type product

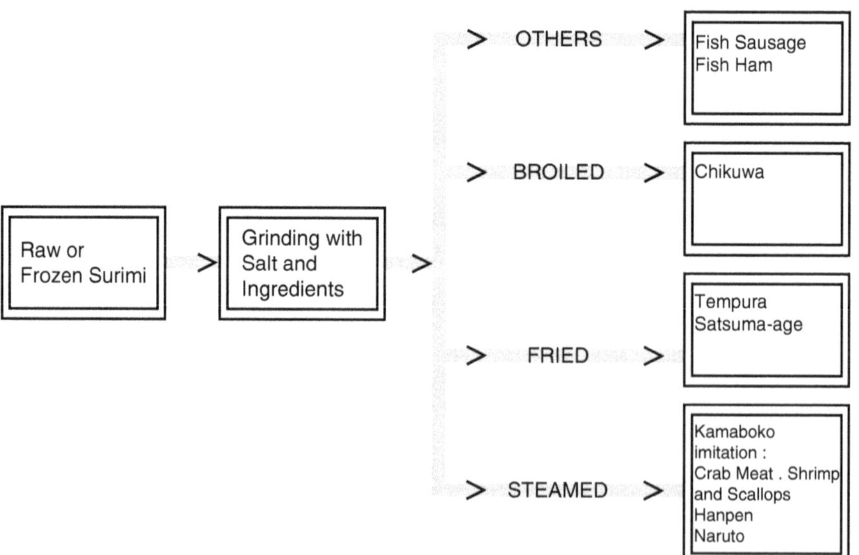

Fig. 20.12 Different surimi-based products

References

Alder J, Campbell B, Karpouzi V, Kaschner K, Pauly D (2008) Forage fish: from ecosystems to markets. Annu Rev Environ Resour 33

Allshouse J, Buzby JC, Harvey D, Zorn D (2016) International trade and seafood safety. Retrieved from www.ers.usda.gov/webdocs/publications/aer828/15641_aer828i_1_pdf

Anderson S (2001) Salmon color and the consumer. IIFET 2000 Proceedings 3pp. From https://ir.library.oregonstate.edu/downloads/h702q7167

Ariyawansa S (2000) The evaluation of functional properties of fishmeal. United Nations University, Fisheries Training Programme, Project Final, Sri Lanka, p 125

Attaran RR, Probst F (2002) Histamine fish poisoning: a common but frequently misdiagnosed condition. Emerg Med J 19(5):474–475

Australian Fisheries Management (2021) Methods and gear. Retrieved from https://www.afma.gov.au/fishing-gears

Borgstrom G (2012) Fish as food V1: production, biochemistry, and microbiology. Elsevier, 742 p

Boyle J, Lindsay RC, Stuiber DA (1993) Contributions of bromophenols to marine-associated flavors of fish and seafoods. J Aquat Food Prod Technol 1(3-4):43–63

California Department of Fish and Game (2000) Review of some California fisheries for 1999: market squid, Dungeness crab, sea urchin, prawn, abalone, ground fish, swordfish and shark, ocean salmon, nearshore finfish, Pacific sardine, Pacific herring, Pacific mackerel, reduction, white seabass, and recreational. California Cooperative Oceanic Fisheries Investigations, Report, 41, 8–25.

California Department of Fish and Game (2006) Review of some California fisheries for 2005: coastal pelagic finfish, market squid, Dungeness crab, sea urchin, abalone, Kellet's whelk, groundfish, highly migratory species, ocean salmon, nearshore live-fish, Pacific herring, and white seabass. California Cooperative Oceanic Fisheries Investigations, Report, 47, 9–29.

Chadd SA, Davies WP, Koivisto JM (2002) Practical production of protein for food animals – Stephen A Chadd, W. Paul Davies and Jason M Koivisto. Protein sources for the animal feed industry. Retrieved from http://www.fao.org/docrep/007/y5019e/y5019e00.htm#Contents

Cheng HT, Capps O Jr (1988) Demand analysis of fresh and frozen finfish and shellfish in the United States. Am J Agric Econ 70(3):533–542

Codex Committee on Fish and Fishery Products (CCFFP) (2003) Code of practice for fish and fishery products (CAC/RCP 52-2003). Retrieved from http://www.fao.org/fao-who-codexalimentarius/standards/list-of-standards/en/

Dewberry EB (1969) Tuna canning in the United States. Food Trade Rev 39(11):37–42

Duncan PF (2003) Shellfish: commercially important mollusks. Encyclopedia of food sciences and nutrition (2nd edn, Caballero B, ed). Academic Press, Oxford, pp 5222–5228. https://doi.org/5210.1016/B5220-5212-227055-X/201079-227058

FAO (2016) The state of world fisheries and aquaculture 2016.Contributing to food security and nutrition for all. 200 pp. Retrieved from http://www.fao.org/3/a--i5555e.pdf

FAO Fisheries and Aquaculture Department (2016) Cultured aquatic species information programme *Salmo salar* (Linnaeus, 1758). Retrieved from http://www.fao.org/fishery/culturedspecies/Salmo_salar/en

Fay C, Bartron M, Craig SD, Hecht A, Pruden J, Saunders R, et al (2006) Status review for anadromous Atlantic salmon (Salmo salar) in the United States. Report to the National Marine Fisheries Service and U.S. Fish and Wildlife Service. Retrieved from http://www.nmfs.noaa.gov/pr/species/statusreviews.htm

Federal Register (1995) 60 FR 65096 – procedures for the safe and sanitary processing and importing of fish and fishery products. Retrieved from http://www.gpo.gov/fdsys/granule/FR-1995-12-18/95-30332/content-detail.html

Food and Agriculture Organization of the United Nations (1985) FAO fisheries circular no. 784. Planning and engineering data 2. Fish Canning. (1). Canning principles. Retrieved from http://www.fao.org/docrep/003/r6918e/R6918E00.htm#Contents

Food and Drug Administration (2001) Seafood HACCP. Retrieved from http://www.fda.gov/Food/GuidanceRegulation/HACCP/ucm2006764.htm

Food and Drug Administration (2015, 03/20/2015) An overview of Atlantic salmon, its natural history, aquaculture, and genetic engineering. Retrieved from http://www.fda.gov/AdvisoryCommittees/CommitteesMeetingMaterials/VeterinaryMedicineAdvisoryCommittee/ucm222635.htm

Food and Agriculture Organization (2016a) GLOBEFISH – analysis and information on world fish trade. Retrieved from http://www.fao.org/in-action/globefish/fishery-information/world-fish-market/en/

Food and Agriculture Organization (2016b) GLOBEFISH – analysis and information on world fish trade. Retrieved from, World tuna trade http://www.fao.org/in-action/globefish/fishery-information/resource-detail/en/c/880744/

Food and Agriculture Organization of the United Nations (2018) The state of world fisheries and aquaculture 2016. Contributing to food security and nutrition for all. FAO, Rome, 204pp. http://www.fao.org/3/i5555e/i5555e.pdf

Food and Agriculture Organization (2021a) GLOBEFISH – information and analysis on world fish trade. Retrieved September 15, 2021, from http://www.fao.org/in-action/globefish/fishery-information/world-fish-market/en/

Food and Agriculture Organization (2021b) GLOBEFISH – information and analysis on world fish trade. Retrieved September 15, 2021, from http://

www.fao.org/in-action/globefish/fishery-information/world-fish-market/en/

Food and Drug Administration (2016a) Guidance for industry: the seafood list – FDA's guide to acceptable market names for seafood sold in interstate commerce. Retrieved from http://www.fda.gov/Food/GuidanceRegulation/GuidanceDocumentsRegulatoryInformation/ucm113260.htm

Food and Drug Administration (2016b) The seafood list. Retrieved from http://www.accessdata.fda.gov/scripts/fdcc/?set=seafoodlist

Ghaly AE, Dave D, Budge S, Brooks MS (2010) Fish spoilage mechanisms and preservation techniques. Am J Appl Sci 7(7):859–877. https://thescipub.com/pdf/ajassp.2010.2859.2877.pdf

Gram L (2009) Microbiological spoilage of fish and seafood products. Springer, Compendium of the microbiological spoilage of foods and beverages, pp 87–119

Granata LA, Flick GJ, Martin RE (eds) (2012) The seafood industry: species, products, processing, and safety, 2nd edn. John Wiley & Sons, Chichester. 488pp, https://books.google.com.gh/books?id=dynWPBe1LF0C

Huss HH (1994a) 3. Quality aspects associated with seafood. Assurance of seafood quality. FAO fisheries technical paper. No. 334. Rome Food and Agricultural Organization, 21pp. http://www.fao.org/3/T1768E/T1768E03.htm#ch3.3

Huss HH (1994b) Assurance of seafood quality. FAO Fisheries Technical Paper. No. 334. From http://www.fao.org/docrep/003/T1768E/t1768e00.HTM

Huss HH, Ababouch L, Gram L (2003) Assessment and management of seafood safety and quality FAO fisheries technical paper. No. 444. FAO, Rome, 230p. Retrieved from http://www.fao.org/docrep/006/y4743e/y4743e02.htm

James K, Carey B, O'halloran J, Škrabáková Z (2010) Shellfish toxicity: human health implications of marine algal toxins. Epidemiol Infect 138(7):927–940. https://doi.org/910.1017/S0950268810000853

Jennings S, Stentiford GD, Leocadio AM, Jeffery KR, Metcalfe JD, Katsiadaki I et al (2016) Aquatic food security: insights into challenges and solutions from an analysis of interactions between fisheries, aquaculture, food safety, human health, fish and human welfare, economy and environment. Fish Fish 17(4):893–938

Jensen WS (1976) The salmon processing industry. Part 1: the institutional framework and its evolution. Agricultural Experiment Station Circular of Information no. 654, Oregon State University Sea Grant College Program Publication no. ORESU-T-76-003, 37pp. https://ir.library.oregonstate.edu/downloads/dz010q030j

Jones PBC (2006) Revolution on a dare: Edmond a smith and his famous fish butchering machine. COLUMBIA The Magazine of Northwest History, Fall 2006, vol 20, No. 3. From https://www.washingtonhistory.org/wp-content/uploads/2020/04/fall-2006-jones.pdf

Kristinsson HG, Rasco BA (2000) Fish protein hydrolysates: production, biochemical, and functional properties. Crit Rev Food Sci Nutri 40(1):43–81

Lavoie RE (2005) Oyster culture in North America: history, present and future. The 1st international Oyster symposium proceedings. Oyster Res Inst New 17:14–21. https://worldoyster.org/wp/wp-content/uploads/2019/2004/news_2017e.pdf

Lavon O, Lurie Y, Bentur Y (2008) Scomboid fish poisoning in Israel, 2005–2007. Isr Med Assoc J 10(11):789–792

Lipp EK, Rose JB (1997) The role of seafood in foodborne diseases in the United States of America. Revue Scientifique Et Technique De L Office International Des Epizooties 16(2):620–640

Miyake MP, Guillotreau P, Sun C-H, Ishimura G (2010) Recent developments in the tuna industry: stocks, fisheries, management, processing, trade and markets. Food and Agriculture Organization of the United Nations, Rome, 125pp

Moonesinghe H, Mackenzie H, Venter C, Kilburn S, Turner P, Weir K, Dean T (2016) Prevalence of fish and shellfish allergy: a systematic review. Ann Allergy Asthma Immunol 117(3):264–272. e264

Munday R, Reeve J (2013) Risk assessment of shellfish toxins. Toxins (Basel) 5(11):2109–2137. https://res.mdpi.com/toxins/toxins-2105-02109/article_deploy/toxins-02105-02109.pdf?filename=&attachment=02101

NASCO (2016) North Atlantic Salmon Conservation Organization. Retrieved from http://www.nasco.int/index.html

National Oceanic and Atmospheric Administration (2012) Fisheries of the United States. Retrieved from https://www.st.nmfs.noaa.gov/Assets/commercial/fus/fus12/FUS_2012_factsheet.pdf

National Oceanic and Atmospheric Administration (2016a) Importing and exporting seafood commodities. From http://www.nmfs.noaa.gov/ia/our_work/importing_exporting_seafood.html

National Oceanic and Atmospheric Administration (2016b) The state of world fisheries and aquaculture 2010. From www.nmfs.noaa.gov/aquaculture/docs/aquaculture_docs/2010_fao_state_of_world_fish_aq.pdf

National Oceanic and Atmospheric Administration (2021a) Atlantic salmon (Protected). Retrieved September 2021, from https://www.fisheries.noaa.gov/species/atlantic-salmon-protected

National Oceanic and Atmospheric Administration (2021b) Bycatch: fishing gear and risks to protected species. From https://www.fisheries.noaa.gov/national/bycatch/fishing-gear-and-risks-protected-species

Newell DE (1989) Development of the pacific salmon-canning industry: a grown man's game, McGill-Queen's University Press, 303pp. https://books.google.com.gh/books?id=QrP9wXOnvPQC

Nicklason PM (1993) Applied technology in the high seas fishing industry. J Aquat Food Prod Technol 2(2):113–123. https://doi.org/10.1300/J030v02n02_07

OECD (2021) Fisheries and aquaculture in United States January 2021. From https://www.oecd.org/agriculture/topics/fisheries-and-aquaculture/documents/report_cn_fish_usa.pdf

Olsen RL, Hasan MR (2012) A limited supply of fishmeal: Impact on future increases in global aquaculture production. Trends Food Sci Technol *27*(2):120–128

Park JW (2013) Surimi and Surimi seafood, 3rd edn. Taylor & Francis, 666p

Park B, Park MS, Kim BY, Hur SB, Kim SJ (1988) Processing. Culture of the pacific oyster (*Crassostrea gigas*) in the Republic of Korea, Food and Agricultural Organization. http://www.fao.org/3/ab706e/AB706E00.htm#TOC

Pedrosa M, Boyano-Martínez T, García-Ara C, Quirce S (2015) Shellfish allergy: a comprehensive review. Clin Rev Allergy Immunol 49(2):203–216

Pompa G, Caloni F, Fracchiolla M (2003) Dioxin and PCB contamination of fish and shellfish: assessment of human exposure. Review of the international situation. Vet Res Commun 27(1):159–167

Sobel J, Painter J (2005) Illnesses caused by marine toxins. Clin Infect Dis 41(9):1290–1296. https://academic.oup.com/cid/article/1241/1299/1290/278063?login=true

Sotelo CG, Pineiro C, Perezmartin RI (1995) Denaturation of fish proteins during frozen storage – role of formaldehyde. Z Lebensm Unters Forch 200(1):14–23

Stern AH (2007) Public health guidance on cardiovascular benefits and risks related to fish consumption. Environ Health 6:31

Tidwell JH, Allan GL (2001) Fish as food: aquaculture's contribution. EMBO Rep 2(11):958–963

Todd EC (1997) Seafood-associated diseases and control in Canada. Rev Sci Tech 16(2):661–672

US Census Bureau (2017) Manufacturing and international trade report: 2015 and 2014(1), 94pp. https://www.census.gov/foreign-trade/Press-Release/MITR/2015/2015%2020Manufacturing%2020and%2020International%2020Trade%2020Report.pdf

USDA Foreign Agricultural Services (2015) U.S. Fish and seafood exports reach record levels. Retrieved from http://www.fas.usda.gov/data/us-fish-and-seafood-exports-reach-record-levels

Vilanova XM (2014) Species of meat animals. Shellfish. Encyclopedia of meat sciences (2nd edn, Dikeman M, Devine C, eds). Academic Press, Oxford, pp 380–387. https://doi.org/310.1016/B1978-1010-1012-384731-384737.300082-384739

Cereal Grains

21

1 Introduction

1.1 Definitions of Cereals and Cereal Products

Of all the plants humans have depended on for food those that produce the cereal grains are by far the most important, as they have been since earliest recorded time. The OECD differentiates the cereals, *wheat (Triticum* sp.*) and rice (Oryza* spp.*)*, from *coarse grains* which are used by member countries as livestock feed; the coarse grains are barley, oats, sorghum, triticale, and rye (OECD 2013).[1] Corn is considered a coarse grain in most OECD countries with the exception of Mexico.[2] The main cereal grains produced globally are, corn > wheat> rice > barley > sorghum > oats > rye (Fig. 21.1).

The Grain Inspection, Packers and Stockyards Administration (GIPSA), which is the US organization for grading supplies for import and export, describes cereals as "*the grass family cultivated for their starchy seeds*". GIPSA considered *coarse grain* or *feed grain* as referring to cereals used foremost as livestock feed (corn, barley, oats, and sorghum) and *food grain* to mean wheat and rice (Womach 2005). GIPSA noted also that "grain" may refer to both cereals and oilseeds, e.g. flaxseed, canola, soybean, and sunflower seeds (USDA Federal Grain Inspective Service 2016).

Corn and sorghum are dual purpose being widely used for both livestock and human consumption. Oat was well known as horse feed but is now an important human functional food (Tosh and Chu 2015). Several coarse grains are used for human food within the developing world, e.g. peal millet and finger millet or ragi (Table 21.1). A trade in "old rice" is emerging, as this commodity is proving suitable as livestock feed after mixing with distiller's dried grains solubles in parts of Asia (USDA Foreign Agricultural Services 2016).

About 15–17 distinct types of cereals are listed by the FAO (Table 21.1). The FAO defines cereals as seeds derived from the grasses family that are harvested as dry grain but considers sweet corn also as a cereal. The designation "mixed grain" refers to a mixture of different cereal species, which are sown and harvested together, for example wheat and rye (Food and Agriculture Organization of the United Nations (FAO) 1994). Those grasses cultivated for use as forage or cereals grown only for industrial use were excluded by the FAO from consideration as cereal grains.

Cereal products refer to primary ingredients and also palatable, convenient foods and

[1]OECD = Organization for Economic Cooperation and Development.

[2]An agreed definition of cereal commodities is for the purpose of quality assurance, and for collecting statistical data about international trade, remaining stocks and balances at any specific time. Most authorities consider cereal grains to be the seeds of cultivated grasses.

R. Owusu-Apenten, E. R. Vieira, *Elementary Food Science*, Food Science Text Series,
https://doi.org/10.1007/978-3-030-65433-7_21

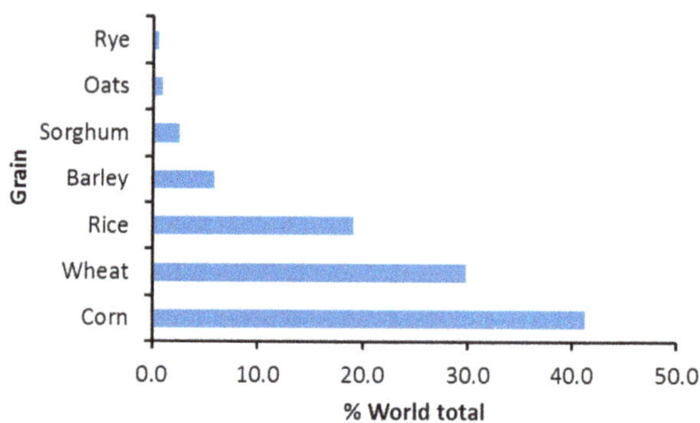

Fig. 21.1 Major cereal grains of the world

Table 21.1 Main types of cereals recognized by the FAO

Name	Synonyms, varieties, comments
Wheat	Triticum spp., common (*T. aestivum*), durum (T. durum), spelt (T. spelta)
Rice	*Oryza* spp., mainly *oryza sativa*
Barley	Hordeum spp., two-row barley (H. disticum), six-row barley (H. hexasticum), four-row barley (H. vulgare)
Maize,	*Zea mays*, corn, Indian corn, mealies
Popcorn	*Zea mays var. everta*
Rye	*Secale cereale*
Oats	*Avena* spp.,mainly *Avena sativa;* breakfast cereal and horse feedstuff
Millets	Barnyard or Japanese millet (*Echinocloa frumentacea*), ragi, finger or African millet (*Eleusine coracana*), teff (*Eragrostis abyssinica*), common, golden or proso millet (*Panicum miliaceum*), koda or ditch millet (*Paspalum scrobiculatum*), pearl or cattail millet (*Pennisetum glaucum*), foxtail millet (*Setaria italica*)
Sorghum,	Sorghum spp., *Sorghum bicolor* (L.) Moench, Guinea corn (*S. guineense*), milo, feterita, kaffir corn (*S. vulgare*), durra, jowar, kaoliang (*S. dura*)
Buckwheat	*Fagopyrum esculentum* (Polygonaceae)
Quinoa	*Chenopodium quinoa*(Chenopodiaceae)
Fonio	*Digitaria* spp., fonio or findi *(D. exilis),* black fonio, or hungry rice *(D. iburua)*
	Triticale
Canary seed	*Phalaris canariensis,* bird feed
Mixed grain	Mixed grain
Cereals Nes	Canagua or Coaihua, *(Chenopodium pallidicaule),* quihuicha or Inca wheat *(Amaranthus caudatus),* Job's tears *(Coix lacryma-jobi),* wild rice *(Zizania aquatica)*

Adapted from (Food and Agriculture Organization of the United Nations (FAO) 1994), taxonomic details are available from Uniprot. Website. (http://www.uniprot.org/taxonomy/4565)

beverages (Table 21.2). Cereals are also eaten directly (e.g. sweetcorn, popcorn or bulgur) or milled into flour for baking products such as, bread, pastry, biscuits cookies, and pizza. Dough can also be processed into non-baked products principally, macaroni and of pasta. Cereal flour serves as a basis for the extraction of protein (gluten) and starch. Finally, cereal germ is used for the extraction of oils (Wrigley et al. 2010).

1.2 Whole Grain and Health

According to the commonly accepted definition, whole grain (wholegrain) and whole grain products, "*are foods which contain all the essential parts and naturally-occurring nutrients of the entire grain seed in their original proportions*" (Korczak et al. 2016; Whole Grain Council 2016). Amaranth, buckwheat and quinoa are considered as honorary whole grains (pseudo-grains)

Table 21.2 Main classes of cereal products

Commodity	Comments
Beverages	Fermented drinks of varying alcohol content
Bran	Sharps and residue from the milling, sifting or other working of the grain. Contains a little flour.
Bread	Baked product of flour, ordinary, unleavened, crackers, rusks, etc.
Bulgur	Whole-wheat grains that are boiled dried and cracked, popular in the, near east.
Flour	Include meal, groats and pellets; strong flours (hard wheat) for bread, durum wheat for pasta, weaker flours for cakes, pastries, biscuits and noodles.
Germ of wheat	Seed embryo, whole or rolled germ for oil extraction, flaked or ground for dietetic preparations, feed supplements and in pharmaceutical preparations.
Gluten	A protein by-product of the industrial production of starch, for enriching protein in flours for bread, pasta, etc.
Macaroni	Pasta made from semolina, or flour, uncooked, stuffed or otherwise prepared.
Pastry	All baked products except bread. May contain milk, eggs, sugar, honey, starch, fats, fruit, seeds, etc.
Ready to eat breakfast cereals	A variety of flaked, puffed, dehydrated, cereal products generally eaten with milk or as cereal bars
Starch	White, odorless powder, insoluble in cold water, it forms a paste in hot water; produced from low-quality flour.
Whole grain	Any cereal grain product containing the original proportions of bran, germ and endosperm[a]

Modified from (Food and Agriculture Organization of the United Nations (FAO) 1994)
[a]*Defined by whole grain council (Whole Grain Council 2016)

because of their long history of use as cereals, if these contain the natural proportions of endosperm, germ and bran. Legumes and oilseeds are not considered whole grains from a nutritional viewpoint (Whole Grain Council 2016) though these seeds are considered grains from a trading perspective (Sect. 1.1).

In July 1999 the FDA approved the health claim statement – "*Diets rich in whole grain foods and other plant foods and low in total fat, saturated fat, and cholesterol, may help reduce the risk of heart disease and certain cancers*" in response to a petition submitted by General Mills Inc., under the Food and Drug Administration Modernization Act of 1997 (FDAMA) which allows label claims to fast-tracked based on authoritative statements from recognized federal bodies (Food and Drug Administration 1999; Marquart et al. 2003). In order to use the health claim, foods should contain 55% or more whole grain in relation to the reference amount customarily consumed (RACC). Both the 2010 (7th) and 2015–2020 (8th) editions of the Dietary Guidelines for Americans contain advice to "consume at least one half of grains as whole grain" (U.S. Department of Health and Human Services and U.S. Department of Agriculture 2015). Moreover, federally supported school lunch programs have requirements or guidelines for inclusion of whole grains as part of healthy diets (USDA Food and Nutrition Service 2016).

With refined grain or refined grain products, cereals are milled in a manner that removes some of the bran or germ (Fig. 21.2). The bran is removed because it is indigestible by humans and because of its adverse effect on the appearance and on some functional properties of flour and the germ because of its high oil content, which may subsequently become rancid. The germ is used to produce oil (e.g. corn oil). The bran goes to feed livestock. However, with new dietary guidelines people will be consuming more whole grain foods.

1.3 The US Cereal Foods Industry at Glance

The United States is a major producer of grains, currently ranked first for corn, fifth for wheat, and 17th for rice globally. US exports for corn and wheat are #1 and #2 globally. The vast pro-

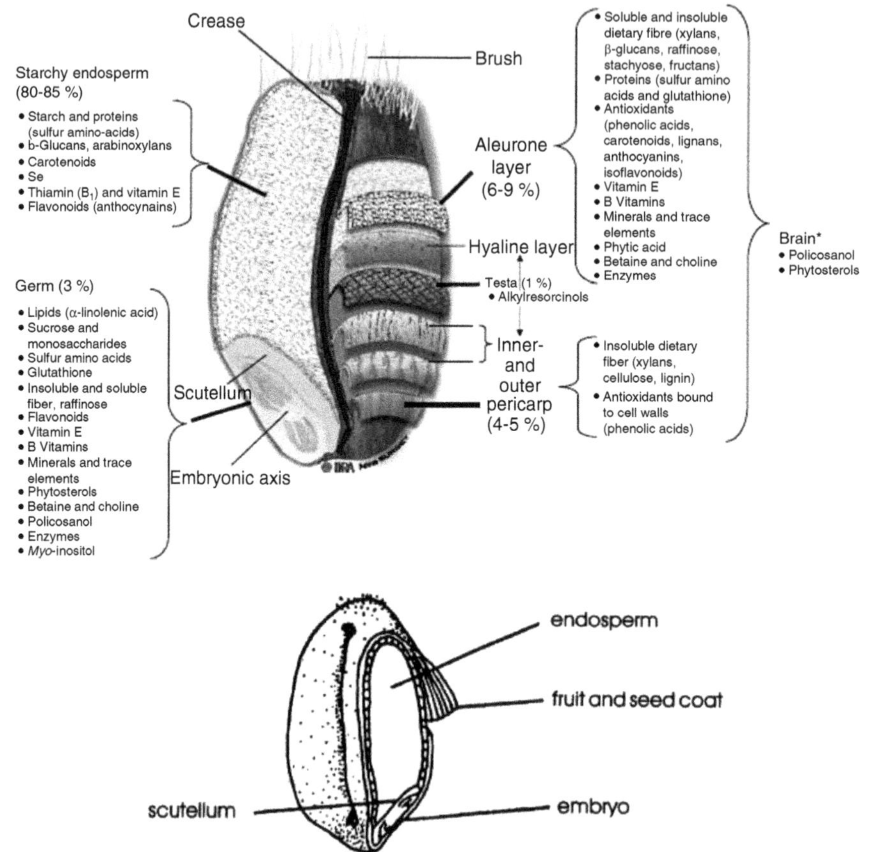

Fig. 21.2 General structure of a cereal grain kernel

Table 21.3 United States production, consumption and trade in major grains (1000 metric tons)

Grain	Production	Consumption	Export	Import
Corn	386,700	314,593	56,500	1250
Wheat	62,959	35,163	26,000	3400
Sorghum	11,740	5462	6300	25
Rice	7500	4220	3500	775
Barley	4339	4638	200	400
Oats	940	676	30	1500
Rye	342	554	5	200

Adapted from (USDA Foreign Agricultural Services 2016)

portion of US grain production is consumed domestically. The export/ import balances show a net excess in most cereals with the exception of barley, oats and rye (Table 21.3).

The grain and cereals industry is diverse covering, grain milling to flour, malt manufacturing, the manufacture of starches and oils, wet milling for starch and also gluten production. There are linkages to industries such as brewing, and the sweeteners industry (Table 21.4).

The first ready-to-eat cereals were produced just before the turn of the twentieth century with flaked and puffed cereals following within a decade (Fast and Caldwell 2000). Ready-to-eat cereals made from the endosperm of wheat, corn, rice or oats are convenient, nutritious and come in a large variety of forms textures and tastes. The processing of cereals into breakfast items was started in the United States. The most popular of the breakfast cereals are formed or puffed and oven-baked.

Table 21.4 Classification of some industries linked with cereals

NAICS	Description
311	Food manufacturing
3112	*Grain and oilseed milling*
31121	Flour milling and malt manufacturing
311211	Flour milling
311212	Rice milling
311213	Malt manufacturing
31122	Starch and vegetable fats and oils
311221	Wet corn milling
311225	Fats and oils refining and blending
31123	Breakfast cereal manufacturing

Adapted from (United States Census 2014)

Table 21.5 2016 World grain production (million metric tons)

Grain	Production	Export	*Export (%)*[a]
Corn	1039.7	141.6	*13.6*
Wheat	751.0	175.0	*23.3*
Rice	481.5	46.6	*9.7*
Barley	146.6	30.0	*20.4*
Sorghum	63.7	8.0	*12.5*
Millett[b]	29.9	–	–
Oats	22.8	2.1	*9.2*
Rye	12.9	0.4	*2.9*
Total	2518.2		

[a]The % global production used for trade/export
[b]Production estimate for 1990 (Food and Agriculture Origination of the United Nations 1995)

1.4 Global Cereal Foods Industry

The world production of cereal grains for the year 2016 amounted to 2.5 billion metric tons (Table 21.5) with the principal types being corn (40%), wheat (~30%) and rice (19%). Some 90% of the global cereal production was derived from three or four species depending on geographic factors. For instance, the EU cereals production involved wheat (47%), barley (21.5%) and maize (18.8%) with rice production at ~1% (Schoenlechner and Berghofer 2006). Cereal grains form the staple food of many people from developing countries providing them with 75% and above of their total caloric intake and 67% of their total protein intake. Grains are eaten in many forms, as a paste or frequently milled and further processed into various products. Cereal grains are also used as feed for livestock that provide us with meat, eggs, milk, butter, cheese and a host of other foods.

1.5 Quality of Cereals and Products

The principles of food safety and quality assurance apply to cereals and cereal products (Vasconcellos 2003). Cereal quality assurance covers harvest and raw materials handling, post-harvest handling, processing, storage and distribution (Fellows et al. 1995). Conditions for harvesting should ensure proper maturation, correct moisture content and avoidance of cracking.

1.5.1 Commercial Quality

Harvested grains may be dried to correct the moisture content before storing in order to avoid inadvertent growth of fungi and the risk of aflatoxin formation.

Quality assurance during the processing of cereals poses a range of challenges at each unit operation (Fig. 21.3). Briefly, grain must be received and graded. Before processing, cereals may be moistened, dried or blended in a process of conditioning immediately before milling, followed by sieving (if required) and then packaging (Fellows et al. 1995).

Cereal grains intended for interstate or international trade is graded, by GIPSA on a fees basis. The GIPSA employs pre-established standards for designating cereals, as Grade 1 (best quality), Grade 2, Grade 3, Grade 4 and Grade 5 (low quality) samples according to attributes listed in Table 21.6. Quality standards have been developed also for individual cereal grains (USDA Federal Grain Inspective Service 2016).

The performance of cereal ingredients is another aspect of quality; here a range of testing methods can be used to assess technological features such as the milling behavior, baking quality and brewing performances (Serna-Saldivar 2012). The microbiological safety of cereals can be a concern, where storage occurs under high moisture conditions that may allow the development fungal aflatoxins (Gazioglu and Kolak 2015). There can be safety concerns also related the detection of allergenic proteins and or trace amounts of gluten arising

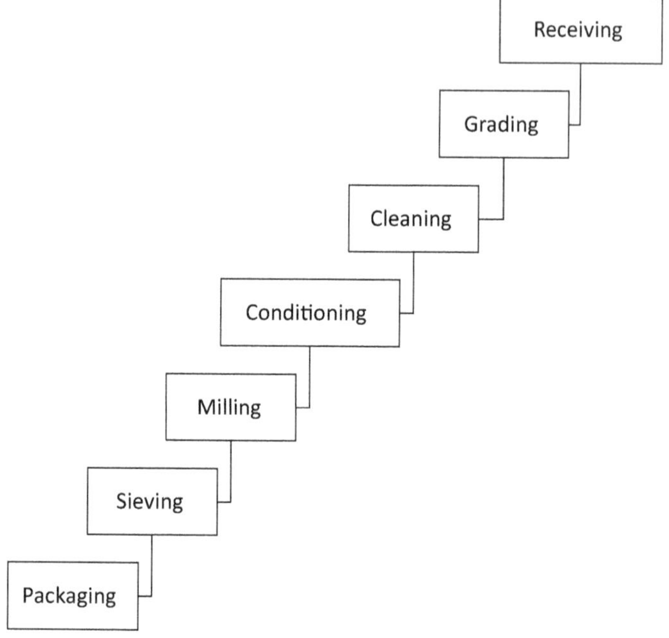

Fig. 21.3 A flow diagram showing unit operations for cereal processing

Table 21.6 Quality attributes for cereal grains

Test weight per bushel[a]
Floatation index
Hardness
Breakage[a]
Cracking and fissures
Damage kernels[a]
Viable germs
Damaged kernels -heat and other[a, b]
Foreign material[a, b]
Shrunken and broken kernels[a, b]
Wheat of other classesa, b
Animal filth[a,c]
Presence of other grains[a,c]
Glass[a,c]
Stones[a,c]
Unknown foreign substances[a,c]
Insect-damaged kernels[a,c]

Adapted from (Serna-Saldivar 2012)
Cereal grading attributes (a), indicating defects (b), or foreign material (c)

from cross contamination during processing (Flodrova et al. 2015).

1.5.2 Nutritional Quality

The nutritional quality of cereals is determined from their proximate composition, e.g., percent fat, protein, moisture, minerals, fiber, and carbohydrate content by difference (Table 21.7). Information on vitamins content is also important for the nutritional declaration on food packaging (Chap. 4, Sect. 5.4).

An estimated 3 million people in the US are diagnosed as celiac disease sufferers, which mean they exhibit gluten sensitivity. Celiac disease results in the production of antibodies, in response to peptide fragments from wheat gluten, which damages the gut lining thereby impairing nutrient absorption and allowing excessive absorption of other components that

Table 21.7 Nutritional profile of some cereals

Description	Water(g)Per 100 g	Energy(kcal)	Fiber, total dietary(g)	Protein(g)	Total lipid (g)	Ash(g)	Carbohydrate, (g)
Amaranth grain, uncooked	11.3	371	6.7	13.6	7.0	2.9	65.3
Barley flour or meal	12.1	345	10.1	10.5	1.6	1.3	74.5
Buckwheat	9.8	343	10.0	13.3	3.4	2.1	71.5
Corn flour, whole-grain, white	10.9	361	7.3	6.9	3.9	1.5	76.9
Corn flour, whole-grain, yellow	10.9	361	7.3	6.9	3.9	1.5	76.9
Corn flour, yellow, degermed,	9.8	375	1.9	5.6	1.4	0.5	82.8
Cornmeal, whole-grain, yellow	10.3	362	7.3	8.1	3.6	1.1	76.9
Couscous, dry	8.6	376	5.0	12.8	0.6	0.6	77.4
Millet, raw	8.7	378	8.5	11.0	4.2	3.3	72.9
Oats	8.2	389	10.6	16.9	6.9	1.7	66.3
Quinoa, uncooked	13.3	368	7.0	14.1	6.1	2.4	64.2
Rice, brown, long-grain, raw	11.8	367	3.6	7.5	3.2	1.2	76.3
Rice, white, glutinous,	10.5	370	2.8	6.8	0.6	0.5	81.7
Rye flour, dark	10.8	325	23.8	15.9	2.2	2.5	68.6
Rye flour, light	11.4	357	8.0	9.8	1.3	0.8	76.7
Rye flour, medium	11.0	349	11.8	10.9	1.5	1.2	75.4
Rye grain	10.6	338	15.1	10.3	1.6	1.6	75.9
Sorghum grain	12.4	329	6.7	10.6	3.5	1.4	72.1
Spelt, uncooked	11.0	338	10.7	14.6	2.4	1.8	70.2
Teff, uncooked	8.8	367	8.0	13.3	2.4	2.4	73.1
Triticale flour, whole-grain	10.0	338	14.6	13.2	1.8	1.9	73.1
Wheat, hard red spring	12.8	329	12.2	15.4	1.9	1.9	68.0
Wheat, hard red winter	13.1	327	12.2	12.6	1.5	1.6	71.2
Wheat, hard white	9.6	342	12.2	11.3	1.7	1.5	75.9
Wheat, soft red winter	12.2	331	12.5	10.4	1.6	1.7	74.2
Wheat, soft white	10.4	340	12.7	10.7	2.0	1.5	75.4
Averages	10.8	352.9	9.6	11.3	2.8	1.6	73.6

Source: Compiled from the USDA nutrient database. (USDA National Agricultural Library 2020)

would normally be filtered out. The management of celiac disease requires avoidance of gluten containing foods. Guidelines for labeling gluten-free foods have been produced by the FDA. In essence, products can be developed wherein other low-gluten cereals replace wheat. (Gallagher 2009; Pawlowska et al. 2014; Riemsdijk et al. 2011; U.S. Food and Drug Administration 2014).

2 Major Cereal Grains

2.1 Wheat

Whole wheat consisting of about 13% protein can contribute considerably to the diet. The flour made from the whole wheat is higher in biological value than white flour. However, comparison of the nutrient composition of flours is

Table 21.8 Nutrient composition of white flour and enriched white flour

Nutrient[a]	Per 100 g	Plain flour NDBNo. 20481	Enriched NDBNo. 20581
Water	g	11.9	11.9
Energy	kcal	364.0	364.0
Protein	g	10.3	10.3
Total lipid (fat)	g	1.0	1.0
Carbohydrate, by difference	g	76.3	76.3
Fiber, total dietary	g	2.7	2.7
Sugars, total	g	0.3	0.3
Calcium, Ca	mg	15.0	15.0
Iron, Fe*	mg	1.17	4.64
Magnesium, Mg	mg	22.0	22.0
Phosphorus, P	mg	108.0	108.0
Potassium, K	mg	107.0	107.0
Sodium, Na	mg	2.00	2.00
Zinc, Zn	mg	0.70	0.70
Thiamin*	mg	0.12	0.79
Riboflavin*	mg	0.04	0.49
Niacin*	mg	1.25	5.90
Vitamin B-6	mg	0.04	0.04
Folate, DFE*	µg	26.00	291.0
Vitamin A, IU*	IU	0.00	2.00
Vitamin E* (alpha-tocopherol)	mg	0.06	0.23
Vitamin K (phylloquinone)	µg	0.30	0.30
Fatty acids, total saturated	g	0.16	0.16
Fatty acids, MUFA	g	0.09	0.09
Fatty acids, PUFA	g	0.41	0.41

[a]Values per 100 g sample. Flour samples are wheat flour, white, all-purpose, unenriched (NDBNo. 20481) and wheat flour, white, all-purpose, enriched, unbleached (NDBo.20581). NDBNo = Nutrient database number, unique identifier for products (*) shows enriched nutrients

complicated by the large number of varieties now available, and by frequent enrichment. Table 21.8 shows a list of the nutrient content of plain white flour with a similar sample, enriched for iron and the B-group vitamins.

Wheat is the most popular cereal grain for the production of bread, cakes and pastries. In addition, the unique properties of the wheat protein gluten can produce bread doughs of the strength and elasticity required to produce low-density bread and pastries of desirable texture and flavor.

There are many varieties of wheat classified as hard red winter wheats, hard spring wheats, soft red winter wheats, white wheats, and durum wheats. Winter wheats are planted in the fall and harvested in the late spring or early summer. Spring wheats are planted in the spring and harvested in the late summer. Hard wheats are higher in protein content and produce more elastic doughs than soft wheats. Therefore, hard wheat is used for breads and soft wheats are used for cakes. Durum wheat is used most for alimentary pastes (spaghetti macaroni etc.) and for the thickening of canned soups.

The hardness of wheat varieties is basically related to kernel texture. More energy is required to mill hard kernels as compared with soft wheat kernels. In addition, wheat hardness is also associated with, moisture content, kernel color, vitreous appearance, average (flour) particle size after milling, protein content and baking performance (Yamazaki and Donelson 1983; Wu et al. 1990; Wu and Stringfellow 1992). Recent research showed a possible molecular distinction between soft and hard wheat, related to the presence of specific starch-granule associated proteins – reviewed by Pasha, Anjum and Morris (Pasha et al. 2010).

Wheat is harvested by combines that cut the stalk remove and collect the seed and either return the straw to the soil to be plowed under with the stubble and thus provide humus or compress and bale it for future uses such as for litter or ensilage.

Wheat may be bagged in jute sacks and stored in warehouses or it may be stored in bulk in elevators. The latter method provides the best protection against rodent and insect infestation. The moisture content of bulk-stored wheat should not be higher than 14.5% and that of sack-stored wheat not higher than 16%. Otherwise, microorganisms may grow and cause heating and spoilage. When it is necessary to lower the moisture content of wheat it may be dried in bins by blowing hot air (not higher in temperature than 175 °F (79.4 °C) across the bins (Table 21.9).

The unit operations for cereal processing (Fig. 21.3) are discussed in Chap. 22. Processing of wheat to wheat flour is now briefly described. To prepare wheat for milling, it is first blown into hopper scales that record the quantity of uncleaned wheat. Some of the coarser impurities are removed by this process. The grain then passes over a series of coarse and fine sieves that further remove contaminating materials including chaff, straw and stones.

Once this initial cleaning stage is completed, the wheat is passed through a magnetic separator to remove any metals present. A further cleaning process may involve dry scouring to remove adhering dirt. The wheat is then washed in water a process that both removes dirt and adds 2% to 3% water to the grain. The added water is necessary to provide desirable conditions for milling. A stone trap is included in the washer. Excess water is removed by centrifugation. A second wet cleaning stage with a light brushing action is ordinarily used followed by aspiration (blowing air through the grain) which is the final cleaning operation. The grain is then carried into a bin from which it is fed to the milling operation. This bin is located on the top floor of the flourmill the grain having been elevated to this position during the various cleaning operations.

Milling may be carried out by passing the grain through a series of corrugated rolls rotating toward each other, which remove chunks of the endosperm from the bran. After each passage through the break rolls the material is sifted through cloth or wire sieves and separated according to particle size. The streams of different-sized flour particles are finally blended to provide the different grades of flour. The more finely ground flour is nearer to white but less nutritious due to the more effective removal of bran and germ from finely ground flour (see Fig. 21.4). Impact milling is now used in some operations. With this method the seed is broken open by impact in a machine called an Entoleter. Disinfection of insect pests (eggs and adults stages) from grains and flours could be achieved also using the Entoleter (Vincent et al. 2003). Flour particles of different sizes are separated by air classification or by centrifugation.

Table 21.9 Varieties of wheat and their uses

Description	Product (protein content %)
Soft wheat	Cakes (7–8%)
	Cookies (7.5–9.5)
	Crackers (7–11)
Hard winter	All-purpose flour (8.5–12.5)
	Noodles (9.5–10.5)
	White bread (11.5–14)
Durum wheat	Pasta (12.5–15)
	Heath bread (13.5–15.5)
Hard spring	Whole bread (13.5–15.5)

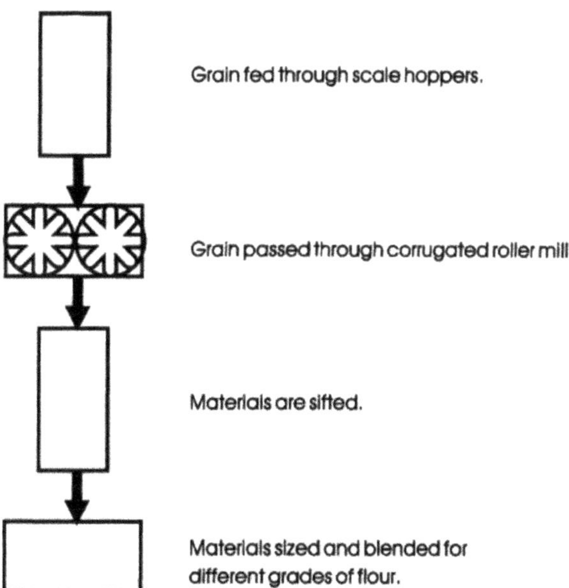

Fig. 21.4 General procedure for milling of grain

High-protein flour is desirable for some types of baked products, flour of moderate protein content for others. Flour particles can be separated by air classification using a turbomilll into various particle sizes and protein contents, and then blended to provide whatever protein or starch content is required by the baker or other users of flour. In the United States, wheat flour may be enriched with the minerals iron (as a salt such as ferrous sulfate) and calcium (in the form of salts) and by addition of the vitamins thiamin, riboflavin, folic acid and niacin (Table 21.8).

Wheat flour is used to prepare leavened products such as bread, cakes, pastries and doughnuts and unleavened products such as alimentary pastes (macaroni spaghetti noodles etc.). Cake mixes are also prepared with flour and flour is used for thickening canned and homemade stews, soups, gravies and white sauce.

Various breakfast cereal products are made from wheat (Fast and Caldwell 2000) by cooking, rolling and flaking. Alternatively, cereal products may be made using an extruder; extrusion produces "puffed" cereals, which give a different appearance and texture. An overview of production technology for breakfast cereals can be found in several texts (Fast and Caldwell 2000; Guine and Correia 2013).

2.2 Corn

Corn is the principal grain of the world in terms of total quantities produced (Table 21.3) (USDA Economic Research Services 2015). Sweet corn is produced as a vegetable and eaten fresh, canned or frozen. Popcorn is also used as a food. Corn has multiple other uses and is processed into a variety of end products including starches, sweeteners, industrial alcohol and fuel ethanol (Hull 2011; USDA Economic Research Services 2015; Wrigley et al. 2010).

Different varieties of corn are usually classified as starchy or waxy depending on the proportions of the two starches (amylose and amylopectin) present (Table 21.10). The development of hybrid strains has improved the yields of field corn. Corn is deficient in the amino acid lysine but a variety of high-lysine corn has been developed that may eventually have a great impact on human nutrition in some parts of the world.

Table 21.10 Types of corn

Dent Corn (*Zea mays indenata*)	Hard corn used for livestock
Flint corn (Zea mays indurata)	Indian corn, used for livestock
Flour corn (*Zea mays amylacea*)	Soft floury starch, baked products
Sweet corn (Zea saccharata)	Corn on the cob, 2.5-fold sugar
Popcorn (Zea mays everta)	
Special-purpose corns	
Waxy corn:	Has 80 + % amylopectin
High-amylose corn	General corns with >60+ amylose
High-lysine corn	

Adapted from (Brown et al. 1985)

Ears of field corn are harvested by a machine that strips the matured ears from the stalks. If harvested in wet weather corn may have to be dried before it is stored. Usually it is allowed to dry on the stalk in the field before harvesting and is stored in small roofed bins or silos with metal or wire mesh walls. Much of the corn storage is done on the farm as most of the corn crop is used as feed for livestock. Stalks and leaves may be harvested chopped and placed in piles or in silos to form ensilage for animal feed. Stalks and leaves may also be chopped and returned to the soil for humus.

Corn to be milled for flour (corn meal) is cleaned then moistened to a water content of 21%. The germ is removed mechanically. The endosperm is dried to a moisture content of 15% and passed through crushing rolls and sifted to remove the bran. With the use of sieves, milled corn is separated into grits (largest-sized particles), meals, and flours (smallest-sized particles).

2.3 Oats

Oats, which is a popular present-day breakfast cereal, was once regarded as useful for feeding livestock. Oats can grow in colder and wetter climates than can wheat. Oats are harvested much in the same manner as is wheat. The moisture content at the time of harvest should not be higher than 13% (Webster and Wood 2016).

Milling of oat kernels requires that they first be washed and cleaned and then dried in a rotary kiln or pan drier to a moisture content of about 12%. They are then hulled by the seeds being thrown from a rotating disc against a rubber ring that splits off the hull leaving the groat mostly intact. After the hulls are removed by passing the product through sieves the oats are steam heated and passed between rollers to produce rolled oats or they are cut into pieces about one-third of the original size and then steamed and rolled to produce quick-cooking rolled oats. Small amounts of oatmeal may be produced by grinding the steamed oats. Steaming facilitates cooking and inactivates enzymes that if not inactivated may cause bitter flavors to develop. Oat flour may be produced for use as an ingredient of bread or a thickener for soups. If made from unheated oats lipolytic enzymes may remain and there may be a problem with the development of rancidity (Webster and Wood 2016).

Oat bran has become a very popular item and much of this is due to the research that links consuming water-soluble fiber from oats with low blood cholesterol (Andersson and Hellstrand 2012; Kapica 2001) and effects on blood glucose (Malkki and Virtanen 2001; Tosh and Chu 2015). Many articles show that eating oat bran can lower blood cholesterol and produce benefits for cardiovascular health (Maier et al. 2000). Successful health claims for oats were approved by the FDA which could be quite profitable for those manufacturers able to bring such products to market (Fitzsimmons 2014) Currently, oat beta-glucans have been designated GRAS status for use as an ingredient in many prepared foods (U. S. Food and Drug Administration 2012). The reader may refer to the following texts for more details (Webster 2011; Webster and Wood 2016).

2.4 Barley

Barley is strongly associated with the brewing industry, and some uses as livestock feed. There is rising interest in barley also, much like oats, due to an ability to lower blood cholesterol and sugar levels (Abumweis et al. 2016; Paras and Kotari 2017). In general, barley is not widely used for human food with only about 2% being employed directly.

Malt is partially sprouted barley that provides a source of starch degrading enzymes and flavors for brewing beers, whiskey and other beverage(Bringhurst 2015; Carvalho et al. 2016; Skadhauge 2006).To produce malt, the grain is soaked in water for several days or until the moisture content reaches approximately 50%. Its then removed from the steep tank and placed in containers where air at 65–70 °F (18.3–21.1 °C) can be drawn through it over a period of approximately 1 week. This allows the barley to germinate or sprout. The sprouted barley is then kiln dried over a period of 24 h. Drying is begun at a low temperature that is gradually raised as drying proceeds. The purpose of malting barley is to produce enzymes that will hydrolyze starch to maltose a sugar that can be utilized by yeasts to produce ethyl alcohol and carbon dioxide. Malt is used in the brewing industry for converting the starches present in barley, rye, rice, corn or other grains to maltose which can be utilized by yeasts (Bringhurst 2015; Carvalho et al. 2016; Skadhauge 2006).

Producing beer starts with the process called mashing, where malt is suspended in warm water. The malt enzymes that are reactivated digest starch to form sugars, glucose and maltose. The resulting sweet liquid produced by this enzymatic action is called wort. With beer and related beverages, hops are added here for flavor. Once brewing takes place, the mixture is heated to elevated temperatures so the enzymes are now permanently inactivated (Fig. 21.5).

The hops are now strained out and the bittersweet liquid remaining (hopped wort) can now be fermented by yeast to produce ethyl alcohol and carbon dioxide. Beer is then filtered through diatomaceous earth filters that remove almost all of the yeast. It is then clarified and bottled or packaged in kegs. Draft beer must be refrigerated whereas bottled beer is pasteurized in the bottle and then cooled. Some beer is "cold pasteurized" a process in which it is sent through fine filters to remove all of the yeast before it is bottled. Beer may be carbonated naturally in a secondary fermentation before packaging. The science of brewing is

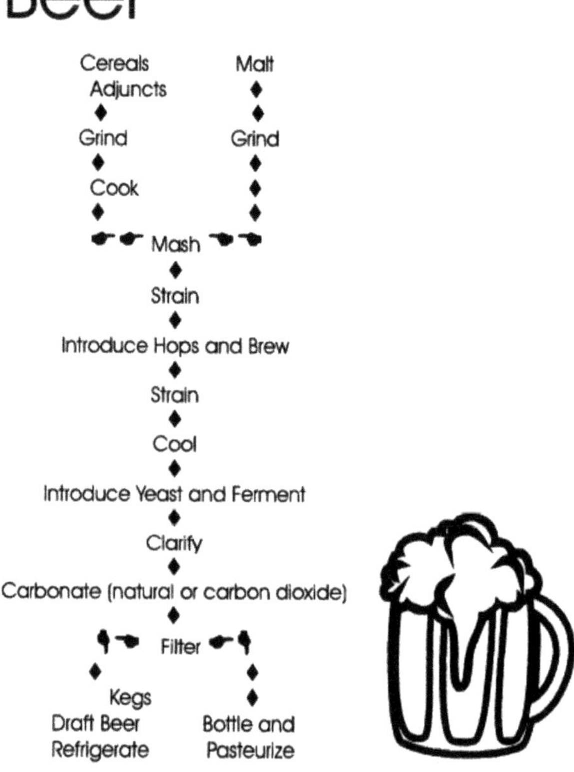

Fig. 21.5 Method of domestic beer production in the United States

discussed further in the following major articles (Foulds 2015; Hornsey 2013; Stewart et al. 2016) and textbooks on the topic (Boulton 2013; Gales 2007; Hornsey 2013).

Malt may also be used in bread baking for much the same reason as in beer production although in this case the purpose is to have the yeast produce carbon dioxide for leavening (raising the dough) the alcohol produced being largely dissipated during the heating involved in baking. The following publications provide in-depth discussions of barley and its uses in food processing (Shewry and Ullrich 2014; Slafer et al. 2002).

2.5 Rice

Rice is an ancient cereal having been domesticated up to 12,000 years ago and now grown in over 100 countries (Smith and Dilday 2002). Rice is a major staple crop for countries in South Asia, South East Asia, Latin America and Sub-Sahara Africa. Nearly 50% of the world's population depends on rice and for many such groups, the growing; harvesting, processing and cooking rice is considered an intimate part of their culture (Bhattacharya 2012). Currently the world production of rice is estimated at about 488.5 million metric tons, with the major producers being India (148.5 million metric tons) > China (106.5 million metric tons) > Indonesia (36.6. million metric tons) > Bangladesh (34.5 million metric ton) > Vietnam (27.8 million metric ton) > Thailand (18.6 million metric ton).

About 8% of the available global supply of rice is involved in international trade with the main exporter countries being; India (10 million metric tons) > Thailand > Vietnam > Pakistan > US. Rice is imported widely across of many countries including China (5 million metric tons) as the leading rice destination. There follows a second tier of rice importers who each purchase between 1 and 2 million metric tons each of rice per annum (Nigeria > EU > Saudi Arabia >Philippines >Indonesia >Cote d'Ivoire > Iran >Iraq. A third tier of rice importers each purchased 500–900 thousand metric tons annually.

Rice is consumed after boiling or steaming, and is also processed into many other products, including, rice puddings, rice breakfast cereals and infant foods. Rice is also the basis for several beverages including sake, and rice-milk substitutes for dairy milk (Houston 1972; Juliano 1985). Rice bran is used for the extraction of oil (rice bran oil) with ascribed health benefits due in part to the content of gamma-oryzanol (Lerma-Garcia et al. 2009; Sah and Srivastava 1985).

Rice has been discussed in relation to several health issues as briefly outlined here. Rice flour may is used for the manufacture of gluten-free products for those allergic to wheat flour. More speculatively, it was suggested that consumption of germinated brown rice could be important owing to its lower glycemic index and higher content of bioactive agents (Bhauso Patil and Khalid Khan 2011; Zeng et al. 2012). The possibility of arsenic accumulation for rice irrigated with poor quality water is being discussed in terms of risks to adults (Gousul Azam et al. 2016) and levels of heavy metals in rice-based infant foods (Anonymous 2016).

3 Other Cereal Grains

3.1 Sorghum (Milo)

Sorghum (milo) is the third most important cereal grain in the US (Arendt and Zannini 2013) and the fifth most important grain worldwide (Smith and Frederiksen 2000; USDA Foreign Agricultural Services 2016). Some 90% of land area devoted to sorghum is in the developing countries, which account for about 60% of world total sorghum. Comparing data from 1996–2016 shows, the total sorghum production (64.2 million metric tons) remained constant over the past 30 years though there were changes in the major producer countries. Sorghum is suited to arid conditions and contributes to the food security of many low income communities in the third world (Food and Agriculture Organization of the United Nations 1995). Sorghum (and millet, see below) is a staple food in many countries bordering the Sahel in Africa, where this cereal is used for making a thin porridge (koko) eaten along with fried bean curd (Kosi) or stiff porridges eaten with a spicy source (Kordylas 1991).

Currently 12.4% of grain sorghum is traded (USDA Foreign Agricultural Services 2016) and most of this is used as livestock feed. Some commercial sorghum supplies are also used in the production of malt for brewing up to 20 different traditional non-alcoholic and alcoholic beverages (Table 21.11) ranging from kaffir beer in South Africa to the bitter rich and fruity flavored pito in West Africa (Solange et al. 2014). Sorghum grain has a significant use also in brewing modern beer,

Table 21.11 Some traditional beverages produced using sorghum

Local name	Product type	Geographic origin
Amgba	Alcoholic beverage	Cameroon
Bushera	Non-alcoholic beverage	Uganda
Bili-bili	Alcoholic beverage	Tchad
Burukutu	Alcoholic beverage	Northern Nigeria
Dolo	Alcoholic beverage	Burkina Faso
Gowé	Non-alcoholic beverage	Benin
Ikigage	Alcoholic beverage	Rwanda
Kaffir beer	Alcoholic beverage	South Africa
Kunun-zaki	Non-alcoholic beverage	Northern Nigeria
Oti-oka	Alcoholic beverage	South-Western Nigeria
Pito	Alcoholic beverage	Nigeria, Ghana
Chakpalo	Alcoholic beverage	Benin, Togo
Tchapalo	Alcoholic beverage	Côte d'Ivoire
Tchoukoutou	Alcoholic beverage	Benin, Togo

Adapted from (Solange et al. 2014)

developed by several companies in Nigeria, South Africa and the US (Taylor et al. 2013). More recently also there has been rising interest in the potential health benefits believed to accrue from dietary sorghum owing to its content of resistant starch, and phytochemicals (Cardoso et al. 2017; Taylor and Duodu 2015).

3.2 Buckwheat

Buckwheat is not a true cereal grain. All the cereal grains belong to the botanical family Gramineae whereas buckwheat belongs to the family Polygonaceae. However, from a use standpoint it is considered a cereal food. Although it is a minor crop in the United States only the Soviet Union and France produce more buckwheat than the United States. It is grown mainly in New York, Pennsylvania, Michigan, Maine, and Ohio. Of the few varieties used, the Silverhull is used mainly for producing flour because of the higher yield of the endosperm. Buckwheat is dried to about 12% moisture cleaned, graded by size, and milled similarly to wheat. Most of the flour is used for making pancakes.

3.3 Millet

Until now, the literature on millet has been scarce and this grain is definitely less well documented compared to other cereals (National Research Council 1996). Millet is receiving attention as a drought-tolerant grain suitable in an era of climate change (Jukanti et al. 2016). There are thought to be five major types of millets the most widely grown being Pearl millet (*Pennisetum glaucum*) (Table 21.12).

Table 21.12 Commonly grown types of millet

Common name	Scientific name
Japanese barnyard millet	*Echinochloa crusgalli* var. *Frumentacea*
Finger millet	*Eleusine coracona*
Proso millet, common millet	*Panicum miliaceum*
Little millet	*Panicum sumatrense*,
Kodo millet	*Paspalum scorbiculatum.*
Pearl millet	*Pennisetum glaucum*
Foxtail millet	*Setaria italica*

The world production of millet is estimated at 26.6–29.8 million metric tons making this the sixth most important grain after sorghum. Nearly 90% of the global supply of millet is grown in the developing countries where virtually all (but 5%) is used for human food. The major producers are India >China> Nigeria = USSR >Niger > Mali. Like sorghum millet is seen as vital for food security in producing countries most of which are within semi-arid zones (Fig. 21.6).

Traditional methods of processing millet for subsistence uses are considered inefficient for commercial applications (Kajuna 2001). Traditional processes involve pounding with wooden pestle and mortar. Millet flour can be rolled and used for couscous. Much like sorghum, soaked and fermented millet is ground to form a meal which is the basis for producing the spicy, Hausa koko eaten for breakfast across much of the Sahel spanning East to West Africa. Traditional stable items are produced from millet also in India, ranging from, fluid porridges to more stiff-doughs or "tuo" usually eaten with a sauce. Modern processing for millet gains is possible using abrasive de-hullers and milling using modern equipment has become common for commercial operations. Millet flour can be used for baking leavened bread when mixed with wheat flour or for baking flat bread. Addition of millet flour as a meat extender has been described (Suman and Sharma 2015). Again as with sorghum, millet is used for manufacturing a range of traditional beverages (Amadou et al. 2011) but brewing modern beers is considered still at an experimental stage (Taylor et al. 2006).

3.4 Triticale

Triticale a hybrid of wheat and rye first produced in the late 1800s combines the high total protein content of wheat with the high lysine content of rye (Arendt and Zannini 2013; Kulp and Ponte 2000; Wrigley and Batey 2010). It is also more

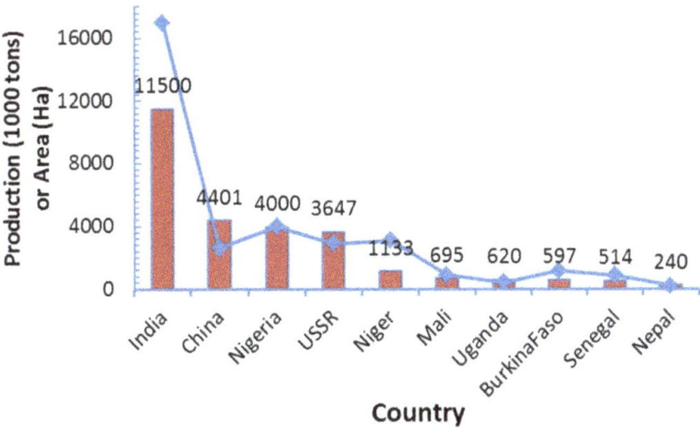

Fig. 21.6 World production for millet (1990)

adaptable to unfavorable growth conditions and seems to resist wheat rust (a disease caused by molds). The improvement of this hybrid is continuing and may lead to more beneficial genetic changes. This cereal is now being grown in more than 52 different countries.

3.5 Further Reading

The following texts deal with general cereal science and practical methods for quality determination (Serna-Saldivar 2012; Wrigley and Batey 2010) and general methods for proximate analysis (Nielsen 2010; Owusu-Apenten 2002). The following texts describe various aspects and general processing such as milling and also the major grains in turn (Arendt and Zannini 2013; Declour and Hoseney 2010; Guine and Correia 2013; Kulp and Ponte 2000; Owens 2001; Serna-Saldivar 2012).

References

Abumweis S, Thandapilly SJ, Storsley J, Ames N (2016) Effect of barley beta-glucan on postprandial glycemic response in the healthy human population: a meta-analysis of randomized controlled trials. J Funct Foods 27:329–342. https://doi.org/10.1016/j.jff.2016.08.057

Amadou I, Gbadamosi OS, Guo-Wei L (2011) Millet-based traditional processed foods and beverages – a review. Cereal Foods World 56(3):115–121

Andersson KE, Hellstrand P (2012) Dietary oats and modulation of atherogenic pathways. Mol Nutr Food Res 56(7):1003–1013. https://doi.org/10.1002/mnfr.201100706

Anonymous (2016) FDA proposes limit for inorganic arsenic in infant rice cereal. World Food Regul Rev 25(11):29–30

Arendt EK, Zannini E (2013) Cereal grains for the food and beverage industries. Woodhead Publishers, Cambridge, 512 p

Bhattacharya KR (2012) Rice quality. A guide to rice properties and analysis. Elsevier, Burlington, 608 p

Bhauso Patil S, Khalid Khan M (2011) Germinated brown rice as a value added rice product: a review. J Food Sci Technol 48(6):661–667. https://doi.org/10.1007/s13197-011-0232-4

Boulton C (2013) Encyclopedia of brewing. Wiley, Chichester, 720p

Bringhurst TA (2015) 125th Anniversary review: barley research in relation to Scotch whisky production: a journey to new frontiers. J Inst Brew 121(1):1–18

Brown WL, Zuber MS, Darrah LL, Glover DV (1985) National corn handbook: origin, adaptation, and types of corn. Retrieved from http://corn.agronomy.wisc.edu/Management/pdfs/NCH10.pdf.

Cardoso LdM, Pinheiro SS, Martino HSD, Pinheiro-Sant'Ana HM (2017) Sorghum (Sorghum bicolor L.): nutrients, bioactive compounds, and potential impact on human health. Crit Rev Food Sci Nutr 57(2):372–390. https://doi.org/10.1080/10408398.2014.887057

Carvalho DO, Goncalves LM, Guido LF (2016) Overall antioxidant properties of malt and how they are influenced by the individual constituents of barley and the malting process. Compr Rev Food Sci Food Saf 15(5):927–943. https://doi.org/10.1111/1541-4337.12218

Declour JA, Hoseney RC (2010) Principles of cereal science and technology, 3rd edn. AACC International, 270 p

Fast RB, Caldwell EF (2000) Breakfast cereals and how they are made, 2nd edn. American Association of Cereal Chemists, 562p

Fellows P, Axtell B, Dillon M (1995) Quality assurance for small-scale rural food industries. Retrieved from http://www.fao.org/docrep/v5380e/V5380E00.htm#Contents

Fitzsimmons R (2014) Oh, what those oats can do. Quaker oats, the food and drug administration, and the market value of scientific evidence 1984 to 2010. Compr Rev Food Sci Food Saf 11(1):56–99. https://doi.org/10.1111/j.1541-4337.2011.00170.x

Flodrova D, Benkovska D, Lastovickova M (2015) Study of quantitative changes of cereal allergenic proteins after food processing. J Sci Food Agric 95(5):983–990

Food and Agriculture Organization of the United Nations (FAO) (1994) Definition and classification of commodities (draft). Cereals and cereal products. Retrieved from http://www.fao.org/es/faodef/fdef01e.htm

Food and Agriculture Organization of the United Nations (1995) Sorghum and millets in human nutrition. Food and Agriculture Organization, Rome, 184pp. http://www.fao.org/3/t0818e/T0818E00.htm#Contents

Food and Drug Administration (1999) Whole-grain foods authoritative statement claim notification. Docket 99P–2209. Retrieved from http://www.fda.gov/ohrms/dockets/dailys/01/sep01/091901/let0003.pdf

Foulds M (2015) The ABC of brewing. Food Rev 42(5):28–30

Gales PW (2007) Brewing chemistry and technology in the Americas. American Society of Brewing Chemists, 248p

Gallagher E (2009) Gluten-free food science and technology. Wiley, New York, 256p

Gazioglu I, Kolak U (2015) Method validation for the quantitative analysis of aflatoxins (B1, B2, G1, and G2) and ochratoxin A in processed cereal-based foods by HPLC with fluorescence detection. J AOAC Int 98(4):939–945. https://doi.org/10.5740/jaoacint.14-211

Gousul Azam SMG, Tushar CS, Sabrina N (2016) Factors affecting the soil arsenic bioavailability, accumulation in rice and risk to human health: a review. Toxicol Mech Methods 26(8):565–579. https://doi.org/10.1080/15376516.2016.1230165

Guine RDPF, Correia PMDR (2013) Engineering aspects of cereal and cereal-based products. CRC Press, Boca Raton, 367p

Hornsey I (2013) Brewing. 2nd edn. Royal Society of Chemistry, Cambridge, 332p

Houston DF (1972) Rice chemistry and technology

Hull P (2011) Glucose syrups: technology and applications. Wiley, New York, 392pp

Jukanti AK, Laxmipathi Gowda CL, Rai KN, Manga VK, Bhatt RK (2016) Crops that feed the world 11. Pearl Millet (Pennisetum glaucum L.): an important source of food security, nutrition and health in the arid and semi-arid tropics. Food Sec 8(2):307–329. https://doi.org/10.1007/s12571-016-0557-y

Juliano BO (1985) Rice: chemistry and technology. American Association of Cereal Chemists, St. Paul, 774p

Kajuna STAR (2001, 04/05/2001) Millet: Post-harvest operations. Retrieved from http://www.fao.org/3/a--av009e.pdf.

Kapica C (2001) Oats – nature's functional food. Nutr Today 36(2):56–60. https://doi.org/10.1097/00017285-200103000-00004

Korczak R, Marquart L, Slavin JL, Ringling K, Chu Y, O'Shea M, Harriman C, Toups K, de Vries J, Jacques P, Klurfeld DM, Camire ME, Unnevehr L (2016) Thinking critically about whole-grain definitions: summary report of an interdisciplinary roundtable discussion at the 2015 Whole Grains Summit. Am J Clin Nutr 104(6):1508–1514

Kordylas JM (1991) Processing and preservation of tropical and subtropical foods. ELBS with Macmillan, 414p

Kulp K, Ponte JG Jr. (2000) Handbook of cereal science and technology. CRC Press, Boca Raton, 808p

Lerma-Garcia MJ, Herrero-Martinez JM, Simo-Alfonso EF, Mendonca CRB, Ramis-Ramos G (2009) Composition, industrial processing and applications of rice bran gamma-oryzanol. Food Chem 115(2):389–404. https://doi.org/10.1016/j.foodchem.2009.01.063

Maier SM, Turner ND, Lupton JR (2000) Serum lipids in hypercholesterolemia men and women consuming oat bran and amaranth products. Cereal Chem 77(3):297–302. https://doi.org/10.1094/cchem.2000.77.3.297

Malkki Y, Virtanen E (2001) Gastrointestinal effects of oat bran and oat gum - A review. Lebensmittel-Wissenschaft Und-Technologie-Food Sci Technol 34(6):337–347. https://doi.org/10.1006/fstl.2001.0795

Marquart L, Wiemer K, Jones J, Jacob B (2003) Whole grain health claims in the USA and other efforts to increase whole-grain consumption. Proc Nutr Soc 62(1):151–159. https://doi.org/10.1079/pns2003242

National Research Council (1996) Chap. 4 Millet. In: Lost crops of Africa. National Academy Press, Washington DC, pp 77–92

Nielsen S (2010) Food analysis laboratory manual. Springer, 177p

OECD (2013) Glossary of statistical terms. Coarse grains. Retrieved from https://stats.oecd.org/glossary/detail.asp?ID=369

Owens G (2001). Cereals processing technology. Woodhouse Publishers, Cambridge, 256p

Owusu-Apenten R (2002) Food protein analysis: quantitative effects on processing. CRC Press, Boca Raton, 488p

Paras S, Kotari SI (2017) Barley: impact of processing on physicochemical and thermal properties-a review. Food Rev Int 33(4):359–381. https://doi.org/10.1080/87559129.2016.1175009

Pasha I, Anjum FM, Morris CF (2010) Grain hardness: a major determinant of wheat quality. Food Sci Technol Int 16(6):511–522. https://wwql.wsu.edu/wp-content/uploads/2014/2012/Pasha-et-al.2010.Food-Sci-Tech-Int.2016.2511.pdf

Pawlowska P, Diowksz A, Kordialik-Bogacka E (2014) State-of-the-art incorporation of oats into a gluten-free diet. Food Rev Int 28(3):330–342. https://doi.org/10.1080/87559129.2012.660715

Riemsdijk LE, Goot AJ, Hamer RJ, Boom RM (2011) Preparation of gluten-free bread using a meso-structured whey protein particle system. J Cereal Sci 53(3):355–361. https://doi.org/10.1016/j.jcs.2011.02.006

Sah A, Srivastava UK (1985) Rice bran oil industry

Schoenlechner R, Berghofer E (2006) Alternative cereals for food processing. In: Proceedings of the third International Congress Flour – bread '05 and fifth Croatian Congress of Cereal Technologists, pp 134–139

Serna-Saldivar SO (2012). Cereal grains: laboratory reference and procedures manual. Taylor & Francis, Boca Raton, 394p

Shewry PR, Ullrich SE (2014) Barley. Chemistry and technology, 2nd edn. AACC International, St. Paul, 322p

Skadhauge B (2006) Malting barley update for 2006. Scand Brew Rev 63(6):19–20

Slafer GA, Molina-Cano JL, Savin R, Araus JL, Romagosa I (2002) Barley science: recent advances from molecular biology to agronomy of yield and quality. CRC Press, 665p

Smith CW, Dilday RH (2002) Rice: origin, history, technology, and production. Wiley, Hoboken, 656p

Smith CW, Frederiksen RA (2000) Sorghum: origin, history, technology, and production. Wiley, New York, 824 pages. https://books.google.com.gh/books?id=b7vxU44v794C

Solange AKA, Konan G, Fokou GF, Koffi Marcellin DJE, Bonfou B (2014) Review on African traditional cereal beverages. www.usa-journals.com. Am J Res Commun 2(5):103–153. Retrieved from http://www.usa-journals.com/wp-content/uploads/2014/04/Solange_Vol25.pdf.

Stewart GG, Maskell DL, Speers A (2016) Brewing fundamentals – fermentation. Tech Q Master Brew Assoc Am 53(1). https://doi.org/10.1094/TQ-53-1-0302-01

Suman T, Sharma BD (2015) Scope of millet grains as an extender in meat products. Crit Rev Food Sci Nutr 55(6):735–739

Taylor JRN, Duodu KG (2015) Effects of processing sorghum and millets on their phenolic phytochemicals and the implications of this to the health-enhancing properties of sorghum and millet food and beverage products. J Sci Food Agric 95(2):225–237

Taylor JRN, Schober TJ, Bean SR (2006) Novel food and non-food uses for sorghum and millets. J Cereal Sci 44(3):252–271. https://doi.org/10.1016/j.jcs.2006.06.009

Taylor JRN, Dlamini BC, Kruger J (2013) 125th Anniversary Review. The science of the tropical cereals sorghum, maize and rice in relation to lager beer brewing. J Inst Brew 119(1–2):1–14

Tosh SM, Chu Y (2015) Systematic review of the effect of processing of whole-grain oat cereals on glycemic response. Br J Nutr 114(8):1256–1262. https://doi.org/10.1017/s0007114515002895

U. S. Food and Drug Administration (2012) GRN No. 437: Beta-glucans from oat bran. Retrieved from http://www.accessdata.fda.gov/scripts/fdcc/index.cfm?set=GRASNotices&id=437

U.S. Department of Health and Human Services and U.S. Department of Agriculture (2015) 2015–2020 Dietary guidelines for Americans, 8th edn. December 2015. Retrieved from http://health.gov/dietaryguidelines/2015/guidelines/.

U.S. Food and Drug Administration (2014) Gluten-free labeling of foods. Retrieved from http://www.fda.gov/Food/GuidanceRegulation/GuidanceDocumentsRegulatoryInformation/Allergens/ucm362510.htm

United States Census (2014) 2011 County Business Patterns (NAICS) Retrieved from http://censtats.census.gov/cgi-bin/cbpnaic/cbpsect.pl or http://censtats.census.gov/cgi-bin/cbpnaic/cbpdetl.pl

USDA Economic Research Services (2015) Corn. Retrieved from http://www.ers.usda.gov/topics/crops/corn/.aspx

USDA Federal Grain Inspective Service (2016) Official U.S. Standards; US standards for grains. Retrieved from https://www.gipsa.usda.gov/fgis/usstandards.aspx

USDA Food and Nutrition Service (2016) Whole grain resource for the national school lunch and school breakfast programs. Retrieved from https://www.fns.usda.gov/tn/whole-grain-resource

USDA Foreign Agricultural Services (2016) Grain: World markets and trade. Retrieved from https://www.fas.usda.gov/data/grain-world-markets-and-trade

USDA National Agricultural Library (2020) USDA nutrient data laboratory. Retrieved from https://www.nal.usda.gov/fnic/usda-nutrient-data-laboratory

Vasconcellos JA (2003) Quality assurance for the food industry: a practical approach. CRC Press, New York, 448p

Vincent C, Hallman G, Panneton B, Fleurat-Lessard F (2003) Management of agricultural insects with physical control methods. Annu Rev Entomol 48(1):261–281. https://doi.org/210.1146/annurev.ento.1148.091801.112639

Webster FH (2011) Oats: chemistry and technology. Academic Press, 376p

Webster FH, Wood PJ (2016) Oats chemistry and technology, 2nd edn. Academic Press, 379p

Whole Grain Council (2016) Definition of a whole grain. Retrieved from http://wholegrainscouncil.org/definition-whole-grain

Womach J (2005) CRS report for Congress. agriculture: a glossary of terms, programs, and laws, 2005 edn. Retrieved from http://digital.library.unt.edu/ark:/67531/metacrs7246/

Wrigley CW, Batey IL (2010) Cereal grains. Assessing and managing quality. Woodhead Publishing, Cambridge, 552p

Wrigley C, Batey I, Miskelly D (2010) Industrial processing of grains into co-products of protein, starch, oils and fiber. In: Wrigley C, Batey I, Miskelly D (eds) Cereal grains: assessing and managing quality. Woodhead Publishing, Cambridge, pp 635–652

Wu YV, Stringfellow AC (1992) Air classification of flours from wheats with varying hardness: protein shifts. Cereal Chem 69(2):188–191. https://pubag.nal.usda.gov/download/24463/PDF

Wu YV, Stringfellow AC, Bietz JA (1990) Relation of wheat hardness to air-classification yields and flour particle size distribution. Cereal Chem 67(5):421–427

Yamazaki W, Donelson J (1983) Kernel hardness of some US wheats. Cereal Chem 60(5):344–350. https://www.cerealsgrains.org/publications/cc/backissues/1983/Documents/Chem1960_1344.pdf

Zeng YW, Pu XY, Du J, Yang SM, Yang T, Jia P (2012) Use of functional foods for diabetes prevention in China. Afr J Pharm Pharmacol 6(35):2570–2579. https://doi.org/10.5897/ajpp12.119

Bakery Products 22

1 Introduction

1.1 The United States (US) Baked Products Industry

The officially named bakeries and tortilla industry (NAICS 3118) is divided into three categories (Fig. 22.1); they are producers of (i) bread and bakery products (NAICS 31181), (ii) cookies, crackers and pastries (NAICS 31182) and (iii) tortilla (NAICS 31183). The three sectors differ considerably in size. The bread and bakery sector is largest by far and provides nearly four fold more jobs as compared with jobs based on the manufacture of cookies, crackers and pasta. The bread and bakeries sector provides 13-fold more jobs as compared with the tortilla sector (Table 22.1).

Official census data shows the United States (US) bread and bakery sector (NAICS 31181) can be further divided into, the retail (in-shop) bakeries (NAICS, 311811), the commercial bakeries (NAICS 311812) and a separate grouping of 231 enterprises (NAICS 311813) covering, frozen cakes, pies, and pastries (Table 22.1). The cookies, cracker and pasta sector (NAICS, 31182) makes not only cookies and cracker (NAICS 311821) but also flour mixes and dough (NAICS 311822) or dry pasta (NAICS 311823). The number of jobs and enterprises for each of these sectors is summarized in Table 22.1 (United States Census 2014).

A report from the American Bakers Association showed the bakery industry had annual earnings of about US$100 billion (2010) and is associated currently with 630,000 personnel (Table 22.1). Some 50% of jobs and earnings from the bakeries industry is concentrated inside of 7–8 US states; California > Illinois > Pennsylvania > Texas > New York > Ohio > Georgia > New Jersey > Michigan (Table 22.2 and Fig. 22.2).

Other estimates show the total number of jobs amounted to about 300,000 (Table 22.1) but as noted elsewhere food businesses financial data can vary depending on the metrics and research methods applied. Some reporting data sometimes also include, bakery ingredients and allied industries, including, enzymes, emulsifiers, fibers, and colors etc.

All told, the economic data for the bakery industry was somewhat fragmentary or proprietary and not freely accessible. Nevertheless, the indirect impact of the bakery industry is thought to be three fold greater than the direct impact. Two sources of data were found to be particularly comprehensive (Agriculture and Agri-Food Canada 2017; American Bakers Association 2017).

The main types of bakery products from the US can be placed in four categories; (i) bread, bread rolls and buns, (ii) morning goods, e.g., sweetened rolls, egg rolls, muffins, scones and croissants, (iii) cakes and pastry, and (iv) cookies

R. Owusu-Apenten, E. R. Vieira, *Elementary Food Science*, Food Science Text Series,
https://doi.org/10.1007/978-3-030-65433-7_22

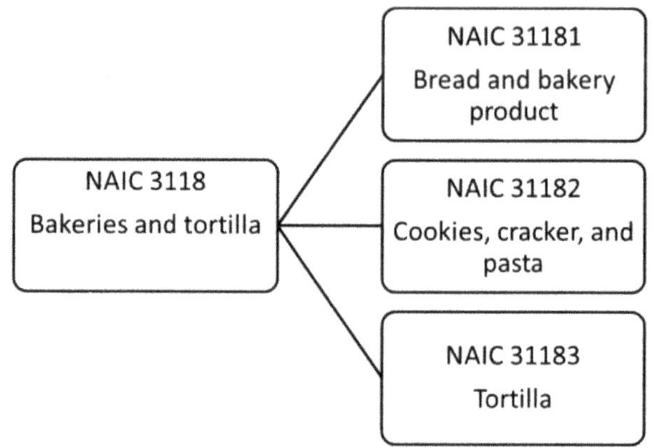

Fig. 22.1 Major sectors of the bakeries and tortilla industry

Table 22.1 The United States bakery products sector

NAICS code	Code description	Employees (2012)	No. Enterprises
3118	Bakeries and tortilla	280,317	10,520
31181	Bread and bakery product	208,824	9336
311811	Retail bakeries	55,166	6400
311812	Commercial bakeries	134,389	2705
311813	Frozen cakes, pies, and pastries	19,269	231
31182	Cookies, cracker, and pasta	55,043	822
311821	Cookie and cracker	31,519	380
311822	Flour mixes and dough	19,206	293
311823	Dry pasta	4318	149
31183	Tortilla	16,450	362

Data from March 2012 (United States Census 2014)

Table 22.2 Major concentrations of bakery industries by State

State	US$ Million	Value (%)	Total (%)
California	14,000	13.7	13.7
Illinois	7950	7.8	21.4
Pennsylvania	6660	6.5	27.9
Texas	6162	6.0	33.9
New York	5745	5.6	39.5
Ohio	5645	5.5	45.1
Georgia	4691	4.6	49.6
New Jersey	4083	4.0	53.6
Michigan	3656	3.6	57.2
All -US	102,453	100	100

*Net Value, Adapted from (American Bakers Association 2017)

and crackers. Bakery products may be wrapped or unwrapped and in the case of bread, either sliced or unsliced (Agriculture and Agri-Food Canada 2017; Frost and Sullivan 2008).[1] The range of baked products and recipes available for baseline formulations is virtually endless (Table 22.3).

Artisanal producers account for 40% of bakery products in the US compared with 60% from commercial bakeries. Other classifications for bakery products cuts across product types e.g., terms such as, organic, functional, high-fiber, and traditional. Bakery products of varying international styles and recipes are also available to the US consumer including, flatbreads, quick breads, and sourdough.

The pattern for baked product consumption in the US (Fig. 22.3) shows one specific group of products (bread, buns and rolls) account for roughly 50% of consumption compared with 20% for cookies and non-savory biscuits. Health and wellness is an important consideration for bakery products in particular as affects baked in the oven breakfast cereals (Agriculture and Agri-Food Canada 2017). The brief listing of some popular baked goods from America can be found in (Table 22.4); in all cases relevant recipes have been produced by the Wheat Foods Council,

[1] For other than North American Readers, the term cookies is equivalent to "biscuit" from UK and Canadian English. In this text, biscuit refers to the "US" biscuit-type product sometimes known as baking powder biscuit, which resembles the "scone" from UK.

The American Baker's Association
601 Pennsylvania Ave., NW, Suite 230
Washington, D.C. 20004
(202) 789-0300
info@americanbakers.org

Fig. 22.2 American Baker's Association details

Table 22.3 A common classification of major US bakery products

Product	Examples	Market Value (%)
Bread	White bread, Whole meal bread, Brown bread, Specialty bread artisan bread	50%
Morning goods	Buns, Crumpets, Rolls, Muffins, Scones, Bagels, Croissants, Tea Cakes, Pan Cakes, Hot Cross buns	20%
Cakes and Pastry	Soft cakes, pastries	10%
Cookies	Sweetened, sweetened or crackers	20%
	Other classification terms	
Baked Goods	Bread, pastries and cakes.	
Biscuits	Sweet biscuits, savory biscuits and crackers.	
Breakfast Cereals	Ready-to-eat and hot cereals	
Frozen Bakery	Semi-baked products, items sold frozen (excluding deserts)	
Frozen Desserts	Frozen forms of cakes, pies/tarts, chocolate cake, strawberry cake, black forest cake, lemon tarts and others, (Excludes ice cream and related)	

Adapted from (Frost and Sullivan 2008) and (Agriculture and Agri-Food Canada 2017)

which is a non-profit organization that promotes wheat based foods including, baked goods (Wheat Food Council 2016).

1.2 The European Union Bakery Industry

Bread and allied products are appreciated universally. Therefore, the bakery industry from European Union is briefly outlined based on two reliable sources of information. First, we consider data compiled as a part of an EU funded project to develop low energy ovens (LEO) for the bakery industry (LEO Project 2015). Secondly, EU bakery industry data were compiled also by Agri-food Canada (Agriculture and Agri-Food Canada 2016).

The LEO report considered a variety of metrics for EU bakery industry between 2009 and 2012 noting the effects of a major financial crisis of 2008 (LEO Project 2015). Currently, there are three main types of bakery producers; (i) commercial/industrial production (61%), (ii) artisanal or in-shop bakeries (35%) and artisanal/in-restaurant producers (5%). During 2012, the bakery industry had revenues equal to US$105.8 billion (€98.5 billion) which is 20% of the total GVA for EU food industry; another estimator put the EU bakery industry earnings at equivalent of US$171 billion (€124,800 million) in 2010 (Reduello 2011) suggesting it is the largest contributor to the world bakery industry total value of about US$340billion. It is agreed the bakery industry is ranked #1 in terms of total number of EU food enterprises and total jobs (LEO Project 2015).

The structure of the EU bakery industry is not uniform across different member states. Nearly 50% of EU bakery enterprises occur in France and Italy. For countries like Greece, France, Portugal, and Italy the artisanal element accounted for 50% or more of bakery production. By contrast, Netherlands, UK, Latvia, Bulgaria, Sweden, Ireland and Estonia had 80% or more of bakery products from commercial producers (LEO Project 2015). Across the EU, the main bakery product is bread (70%), compared with 10% cookies, 10% pastries and (8%) Viennoiserie products like croissant. The preference for cookies was highly variable, accounting for 20% by volume of bakery products in the UK, but only 5% in Germany (Tables 22.5 and 22.6).

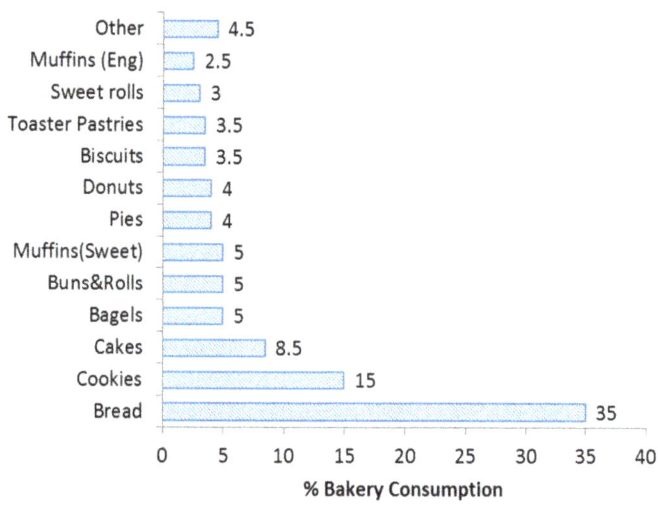

Fig. 22.3 Consumption patterns for different bakery products. (Redrawn from Agriculture and Agri-Food Canada 2017)

1.3 Semi-cooked Frozen Dough and Dry Mixes

Most bakery products are sold in a ready-to-serve form, but a significant quantity of products are also retailed as partially baked products and require a final baking step to serving (NAICS 311822; Table 22.1). For instance, some in-store bakeries and domestic producers will bake using par-baked products. Growing amounts of baked goods are also handled and sold in the frozen form, especially because the quality of baked goods is exceedingly well preserved by freezing. Some canned cookies and biscuit mixes are refrigerated until baked.

The dry ingredients used for some baked goods, such as cakes, are premixed industrially and sold as prepared mixes, and although the user must add the fluid ingredients and bake the product, it is still a convenient system for the consumer. With self-rising flours, chemical leavening agents are added directly to the flour before sale. Some of these mixes are formulated for baking in a microwave oven. These have the advantage of producing cakes in significantly shorter times than are required in conventional baking ovens.

2 Bread, Rolls, and Buns

2.1 Standards of Identity

Bakery products are either leavened or unleavened. The leavened products are those which are manufactured using carbon dioxide, produced by the growth of yeasts (i.e., breads, rolls, etc.) In addition, chemical leavening agents such as baking powder can be used for quick breads, cakes, doughnuts, and biscuits. Some bakery products can be produced by physical leavening or incorporation of air (e.g., batter-whipped breads, angel food cake) (Table 22.7).

Unleavened baked products, include a large number of flatbreads and ethnic-specialty products, that are increasingly available to US consumers, and also some crackers and pie crusts (Table 22.8).

Code of Federal Regulations 21CR136.110 describe standards of identity for several bakery products. Bread, buns, rolls, white bread, white buns, white rolls are similar products but differentiated according to their weight and dimensions (U. S. Food and Drug Administration 2016). Breads, rolls and buns must be prepared using yeast as a leavening agent, and a farina-

Table 22.4 Some popular US baked goods

Almond breakfast bread
Apricot streusel scones
Autumn apple pie & double pastry crust
Baked whole-wheat tortilla chips
Banana orange muffins
Basic focaccia
Basil and sun-dried tomato pizza crust
Breakfast biscuit
Bugsy bread
Caramel apple Bundt cake
Carrot cranberry bread
Chocolate chip biscotti
Cracked wheat bread
Cream DeMint chocolate chip bread
Double chocolate biscotti
Dried cranberry oat muffin
Easy sweet focaccia
French cookies
Fruited hearth bread
Grand old glory bread
Herb focaccia
Honey almond biscotti
Honey wheat bread
Ice cream muffins
Italian cheese bread
Lavash
Native American tortillas in a bag
Panettone
Pear and bacon waffle bake
Swiss cheese breakfast biscuits
Whole wheat banana bread
Whole wheat bread
Whole wheat bread bowls
Whole wheat chapattis
Whole wheat chocolate sheet cake
Whole wheat cinnamon rolls
Whole wheat cinnamon-sugar cookies
Whole wheat dream shoe cookies
Whole wheat flour tortillas

Adapted from, Wheat Food Council. See reference for baseline recipes (Wheat Food Council 2016)

ceous agent - flour from wheat or non-wheat sources. Bread is synonymous with white bread or brown bread since there are distinct standards of identity for whole wheat breads (Table 22.9).

The CFR allows much flexibility in terms of the range optional ingredients for bread, rolls and buns (Table 22.10). Not surprisingly, there are hundreds of different types of leavened bread. As noted previously, bread is the most important of the bakery products and accounts for 50% (USA) to 70% (EU) sales for bakery products.

2.2 Enriched Bread Rolls and Buns

Standards for bread and flour products shipped for interstate commerce and labeled "enriched," were developed by the federal government in 1952. These standards require the enrichment or fortification of bread with thiamin (vitamin B1), riboflavin (vitamin B2), niacin, and iron. Calcium and vitamin D may also be added as enrichment agents (U. S. Food and Drug Administration 2017).

Nutrients may be added to bread to restore any which is lost during processing. Alternatively, bread may be enriched using vitamins or minerals even where these were not present originally. In the past, bread as well as other staple foods were fortified if there was significant likelihood of a nutrient deficiency in the general population (Scrimshaw 1998).

Enriched bread, bread rolls and buns must conform to the standards of identity for ordinary bread and have added minerals and vitamins (Preedy et al. 2011; Rosell et al. 2015). A key requirement for enriched bread and buns is that any added nutrients should be stable during the shelf life of the product (Table 22.11).

Table 22.5 Characteristics of the EU bakeries industry

Index	Value	Comment
Earnings	US$105.8billion	20% Food Industry GVA
Jobs	1.36 million	36% All Food Jobs
Enterprises	154,803	54.% Food Companies
SME	99.7%	9 Employees/Enterprise

LEO Project (2015)

Table 22.6 Bakery product categories in the European Union

Product	Million Tons	Percent
Bread (+Cookies)	31.7	79
Pastries	4.0	10
Viennoiserie	3.21	8
Savory pastries	1.2	3

Adapted from (LEO Project 2015)

Table 22.7 Some leavened breads by country/region

Name	Origins
Bread roll	Europe
Baguette	France
Bagel	Poland
Canadian White	Canada
Carrot bread	All
Challah	Poland / Russia
Chickpea bread	Albania, Turkey
Filone	Italy
Ka'ak	Near East
Lahoh	Djibouti, Somalia, Yemen
Marraqueta	Chile
Miche	France
Michetta	Italy
Panbrioche	Italy
Pane di Altamura	Italy
Pane ticinese	Switzerland
Focaccia (Pan focaccia)	Italy
Peg bread	Jamaica and West Indies
Pistolet	Belgium
Potbrood	South Africa
Pretzel, soft	Italy, Germany, USA
Proja	Serbia
Pumpernickel	Germany
Quick bread	North America
Ryaninjun	New England, USA
Rye bread	Europe
Salt-rising bread	United States
Sgabeo	Italy
Taftan	Iran
Tsoureki	Greece
Vánočka	Czech Republic, Slovakia
Vienna bread	Austria (Vienna)
White bread	All
Whole wheat bread	All
Zopf	Switzerland, Germany
Zwieback	Germany

By the late 1990s, there was considerable evidence that daily intakes of folate (Vitamin B12) in the American population might be suboptimal putting some sectors of the population at risk of cardiovascular disease. Moreover, folate was proven to be vital for expectant mothers as this nutrient could decrease the risk of neural tube defects and spinabifida in babies. From June 1997 the USDA-FDA regulated that certain cereal foods including, some flours, cornmeal,

Table 22.8 Unleavened flatbreads by country/region

Name (¥)	Origin
Aish Merahrah	Egypt
Balep korkun	Tibet (Central)
Bammy [Cassava flatbread]	Jamaica
Bannock	Canada, Scotland, US
Barbari bread, Shirmal	Iran, Afghanistan
Bazlama	Turkey
Bing bread	China
Bolani, Perakai, Poraki	Afghanistan
Chapatti [Chapati]	India
Crisp bread	Scandinavia
Crumpet	United Kingdom
Farl [soda bread, potato bread]	Ireland & Scotland
Laobing [Green onion pancake]	China
Hallulla	Chile
Hardebrood	Germany
Himbasha, Injera	Eritrea, Ethiopia
Hubuz	Arab world, North African
Johnnycake or Hoecake	North America
Khanom bueang	Thailand, Cambodia
Lavash	Transcaucasia, Armenia
Lefse	Norway
Malooga	Yemen
Markook	Levant
Matzo	Jewish
Naan	All
Ngome	Mali
Pancake	All
Pane carasau	Sardinia
Papadum, Parotta Barotta, Kulcha, Roti	India, Pakistan & Punjab
Piadina, Pizza	Italy
Pita	Greece, Turkey
Puran poli, Obbatu, Bakshalu	India
Qistibi	Tatarstan, Bashkortostan
Tortilla	Mexico
Tunnbröd	Sweden
Yufka	Turkey

(¥) list of individual products [ ] shows synonyms

Table 22.9 Standards of identify for some bakery products

CFR 21.	Product
§ 136.110	Bread, rolls, and buns.
§ 136.115	Enriched bread, rolls, and buns.
§ 136.130	Milk bread, rolls, and buns.
§ 136.160	Raisin bread, rolls, and buns.
§ 136.180	Whole wheat bread, rolls, and buns.

From (U. S. Food and Drug Administration 2017)

Table 22.10 Standards of identify and ingredients for bread, rolls and buns (21CR136.110)

A, Yeast leavening
B, Farinaceous agent
C, Optional ingredients
Flour, brominated flour or phosphate flour
Water
Yeast - unspecified
Salt
Shortening
Lecithin, corn or soy oil, mono- or diglycerides
Milk or dairy products
Egg product
Nutritive carbohydrate sweeteners
Enzymes
Lactic acid bacteria
Non-wheat flour components, meals, grits, etc. (<3.5%)
Yeast nutrients and calcium phosphate
Strengtheners and, improvers (listed)
Bromate, iodate salts (0.0075%)
Azodicarbonamide (0.0045%)
Other Strengtheners (0.5%)
Spices, spice oils or extracts
Colours, prohibited unless as part of "spice"
Other ingredients, allowed if does not adversely affect product
Names as per statutory list
Egg bread, egg rolls etc., must have 2.56% (w/w) whole egg

rice, pasta and bread must be enriched with folic acid (McBride and Tucker 1998). Following the FDA regulation manufacturers increased folate levels in their formulations by two to three fold with the highest levels being found not in bakery products but rather oven baked breakfast cereals (Cho et al. 2002). By most accounts the fortification of cereal and bakery products led to increases in folate intake above those predicted (Choumenkovitch et al. 2002).

Table 22.11 Nutrients allowed to be added to bread

Nutrient	Amount (mg/ pound)
Thiamin,	1.8
Riboflavin,	1.1
Niacin,	15
Folic Acid,	0.43
Iron.	12.5

From (U. S. Food and Drug Administration 2017)

2.3 Milk Bread, Rolls, and Buns

Milk-bread, milk rolls or milk buns are products where milk is used as the dough-moistening ingredient. Alternatively, other dairy ingredients may be used so long as these are not added to an extent where milk solids do not exceed 8.2% w/w of flour and where the non-fat milk solids comprises 60–70% of total milk solids.

2.4 General Ingredients

Bread is the oldest and most important baked product (Bakerpedia 2021). It has been made from many grains, including wheat, com, rye, rice, barley, oats, and even buckwheat. The growth of its popularity has been due to a number of factors, but an important one is that grains of one type or another have been grown in nearly all the inhabited parts of the world. An example of the composition of the ingredients of a loaf of white bread is approximately 57% flour, 36% water, 1.6% sugar, 1.6% fat or shortening, 1% milk powder, 1% salt, 0.8% yeast, 0.8% malt, and 0.2% mineral salts. The following sections provide an overview of the main ingredients for bread and key manufacturing processes. Readers wishing for more information should refer to many dedicated publications on this topic (Al-Dmoor 2012; Arpita and Datta 2008; Cauvain 2013; Hamelman 2004; Jenkins 1975; Pomeranz 1980; Smith 1995).

2.4.1 Flour

The flour used for making bread is usually of the hard wheat type, which is higher in protein than that of soft wheat types. The reason for this is that in yeast leavened products, the gluten (protein) in the unbaked loaf must be sufficient in quantity and of adequate elasticity to form a stretched mass that will entrap bubbles of carbon dioxide. This increases the volume and forms the structure of the loaf. It also allows the retention of the structure until sufficient heating has occurred. When, as a result of heating that coagulates the gluten, a more rigid structure has been formed, the structure of the loaf of bread is fixed (Courtin et al. 2014; Gan et al. 1995; Stojceska and Butler 2012).

Strengtheners, conditioning agents and bleaching agents are usually added to darker flour. The bleaching agents have a discoloration role leading to whiter flour. The strengthening and conditioning agents are diverse materials which either reduce (add hydrogen) or oxidize sulfur residues in flour proteins thereby affecting its workability and rising characteristics. A great deal of scientific research has been done to understand the effect of ascorbic acid, L-cysteine, potassium bromate and other conditioners in baking technology (Barrett et al. 1971; Elkhalifa et al. 2007; Every et al. 1999; Selomulyo and Zhou 2007; Wood 1980).

Flour millers are able to supply bakers with flour, which has essentially the same protein content from delivery to delivery, as specified by the baker, by blending a variety of flours that have been classified according to protein composition. Also, quality control laboratories test for protein content and quality. There are other tests that determine the characteristics of different flours by measuring the physical properties of the dough made from them (Stojceska and Butler 2012).

2.4.2 Water

Water is a chief ingredient of the dough in baking. The amount of water added is such that the finished loaf does not contain more than 38% water. If the water available to a bakery is hard (contains minerals), the amount of yeast food (mineral salts) to be added may be modified. Also, during the mixing of the dough, a certain amount of heat is generated because of friction encountered during the forcing of the mixing bars through the dough, and from the motor that runs the mixing apparatus, as well. If the temperature rises above a certain point (82 to 85 °F [27.8 to 29.4 °C], the yeast may be destroyed and the gluten and starch of the dough adversely affected. Therefore, the water must be cooled prior to adding it to the ingredients in the mixer, or part of the water must be added as ice, which melts during the mixing and controls the temperature.

2.4.3 Sugar

Sugar (cane or beet sugar) in small amounts is used in baking bread, to serve as a source of readily utilizable carbohydrate for the yeast, providing for a suitable fermentation that produces carbon dioxide required to raise the dough.

2.4.4 Shortening

Some fat or shortening is added to bread mixtures. Ordinarily, this is a solid fat (or hydrogenated vegetable oil). This facilitates mixing, tenderizes the crumb of the loaf, and prevents staling of the bread. Today, as a rule, small amounts of monoglycerides are also added, as they are more active anti-staling agents than fats. Monoglycerides consists of a glycerol molecule that has had a fatty acid combine with only one of its three alcohol groups; the other two of its alcohol groups remain unchanged. Milk powder may be added to bread dough, because it has a desirable effect on the texture of the crumb (inside the loaf) of the finished bread.

2.4.5 Yeast

The yeast used in bread baking is *Saccharomyces cerevisiae*. This may be used as a dried material or as a moist compressed cake containing 70% water (the latter type must be held under refrigeration prior to use). Either type of yeast must be suspended in warm water prior to being added to the material in the mixer, in order to obtain an even distribution throughout the dough. A freeze-dried, vacuum-packed yeast is also available. This product has a greater number of viable yeast cells; therefore, theoretically, it is required in lesser amounts. It does not need to be suspended in water; therefore, it may be added with the dry ingredients. Because it is freeze-dried and vacuum-packed, it requires no refrigeration, and theoretically, has a longer shelf life.

Small amounts of salt (sodium chloride) are used in making bread, because it is desirable for the flavor of the finished loaf and trace amounts may be utilized by the yeast during growth. During mixing, proofing, and the early part of baking, the yeasts grow and produce ethyl alcohol and carbon dioxide. The latter, a gas, causes the dough to rise, and provides for the volume of the loaf. Leavening action may result to a degree from the vaporization of water in the dough when the temperature of the mass is raised sufficiently in the oven. The increase in loaf volume is also believed to be affected in the presence of shortening, which can entrap air during mixing and release it when the air expands during baking. The ethyl alcohol is largely dissipated during baking, although residual amounts of alcohol, esters, and other components may remain and contribute to the flavor of the loaf.

2.4.6 Malt

When malt is used in bread making, it is usually of the diastatic type (contains active enzymes that will convert starch to maltose or glucose). As proofing continues, the small amount of sugar present in flour, and the sugar that has been added to the dough mixture, may be mostly used up by the yeast. Therefore, to continue the yeast growth, the action of the malt enzymes on starch can provide a source of sugars during the latter stages of proofing and the early stages of baking. Because a small amount of sugar is essential to the browning of the crust of the loaf of bread, the malt is also important in that it provides the sugar in the development of crust color.

2.4.7 Other Ingredients

The mineral salts added to dough mixtures are called yeast nutrients. Yeasts require small amounts of nitrogen-containing and phosphorus-containing salts for growth and production of carbon dioxide. For this reason, small quantities of ammonium salts and phosphates are added to the ingredients of the dough. In addition to the above components, many special types of bread are produced that contain one or more additional ingredients (i.e., butter, extra milk powder, buttermilk or dried buttermilk solids, dried vegetable powders, honey, etc.).

3 Bread Manufacture, Dough Handling

There are two general methods of handling dough in baking bread: the straight dough method and the sponge method (NIIR Board of Consultants and Engineers 2014).

3.1 Straight Dough Method

This batch method for mixing dough would be familiar to anyone used to baking in a domestic setting. In the straight dough method (see Fig. 22.4), all the ingredients are added to one

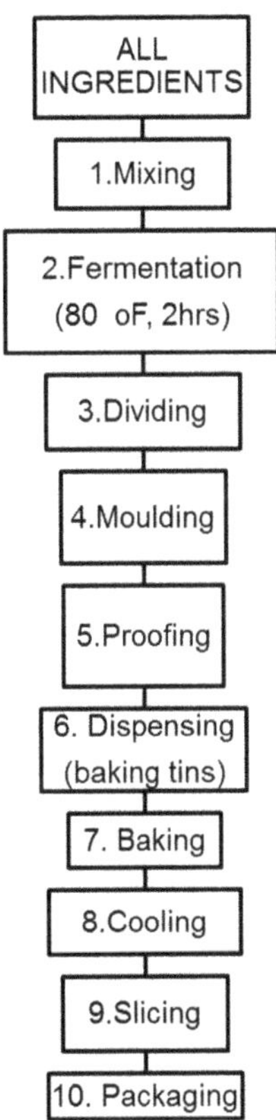

Fig. 22.4 Manufacture of bread by the straight dough method

container and then subjected to kneading manually or using mixer for the larger bakeries. When using a mixer, the flour and other ingredients are mixed first at low speed (about 35 rpm), then at high speed (about 70 rpm). The mixed dough is then placed in large metal containers or troughs and held in an insulated room at about 80 °F (26.7 °C) and in an atmosphere of high humidity to allow fermentation. During fermentation, the mass of dough is kneaded several times to allow the escape of some carbon dioxide, which is produced continually during fermentation. In addition, working of the dough in this manner assists in stretching and conditioning the gluten, which is the important ingredient responsible for the formation and retention of the structure of the loaf.

3.2 Sponge Dough Method

In the sponge dough method, 50–75% of the flour, enough water for a moderately stiff dough, all the yeast, the malt, and the yeast foods are added to the mixer and combined. This sponge is fermented for 3–4 h, then returned to the mixer and combined with the remainder of the flour and water, the shortening, the sugar, the milk powder, and the salt. The sponge method of baking bread produces a crumb of finer texture and with smaller gas holes than that obtained when the straight dough method of bread baking is used.

After fermentation, or after fermentation and mixing, the dough is divided into pieces that will eventually make up the finished loaves. This is done by a machine that measures the dough by volume and cuts off pieces of the desired size. When cut, the dough has an irregular shape with cut ends through which the leavening gas can escape. It is immediately dusted with flour and rounded. This is done by machine. Rounding dries the surface with flour and closes up the cut ends, thus preventing the escape of gas.

The rounded dough is then carried on a belt to a proofer where it is transferred to another belt and held at 80 °F (26.7 °C) and 76% relative humidity for a period during which the dough relaxes and increases in volume as more carbon dioxide is produced. The pieces of dough are then molded and shaped by machine. In this process, the floured dough is rolled out into a sheet, curled into a loose cylinder, again rolled, and the ends sealed. In some operations, two cylinders of dough are twisted together in the molding operation. The cylinders of dough then fall into pans conveyed to the oven for baking. During the first stages in the oven, the dough continues to ferment and increase in volume.

As the dough passes through the oven, the temperature is increased, further expanding the dough (increase in volume of the gas is due to the rising temperature), and eventually the gluten is set by the heat, the starch first gelatinized and then set by the heat, and some water and ethanol are evaporated. Eventually, the outer layers of the dough become browned to form the crust. Browning is probably due to both the reaction between proteins and sugars, and caramelization of sugars. After baking, the loaves of bread are cooled as they are carried through air-conditioned tunnels. The loaves are cut into slices by machine, and the sliced loaves are packaged automatically by bread-wrapping machines.

Bread and other baked goods are subject to spoilage by molds. Therefore, small amounts of mold inhibitors, such as sodium or calcium propionate, are used and allowed at levels of 0.32 parts per 100 parts of flour (0.32%) in white bread and 0.38% in whole wheat products. Because water and other ingredients are used in bread, the actual concentration of these inhibitors in the finished loaf is much lower than in the flour. Mold inhibitors not only delay the growth of molds in bread, but they also inhibit the growth of certain bacteria that produce a slime in the crumb of the loaf, a condition known as "ropiness."

3.3 Batter Whipped Process or Continuous Mix Baking

The batter whipped process, also known as the continuous mix baking, is a method that speeds up the baking process by reducing the length of time for fermentation. The method invented by

Dr. J.C. Baker in 1953 (Bobrow-Strain 2012) also called "No-time baking" employs a separate fermentation step. A liquid mixture of water, sugar, yeast, milk powder, salt, and yeast food; small amounts of flour; and some vitamins is fermented from 2 to 3 h and then cooled. This pre-fermented mixture is then be added to liquid shortening and flour, and agitated at high speed in a developer, which incorporates air. The dough is then extruded directly into baking pans which, after a short proofing period at 80 °F (26.7 °C), are conveyed directly to the baking oven.

Automation in bread manufacture has led to higher production volumes, lowered production costs, shorter production time, and better control over the properties of the finished product, making it easy to produce large volumes of uniform bread that can meet whatever specifications are desired. Bread made by the continuous-mix baking process is finer in texture, but less flavor is developed in processes of this type possibly because of the absence of bulk-fermentation step (Morad and D'Appolonia 1980).

4 Other Popular Cereal Products

4.1 The Cakes and Cookies

Leavening is produced for cakes and cookies by chemical means, instead of by yeast fermentation. The chemical source of carbon dioxide is sodium bicarbonate. This produces carbon dioxide when brought into contact with acids, such as potassium hydrogen tartrate (cream of tartar), sodium hydrogen pyrophosphate, calcium hydrogen phosphate, or sodium aluminum sulfate (alum). It is necessary to produce small bubbles of gas at a constant rate consistent with the period involved in mixing and baking to temperatures that set the structure of the cake. Cookies are formed with dies or by extrusion, and, in many instances; the process is quite complex, as the dough may have to be extruded around a central component of fig paste or other types of filling.

Many prepared mixes are also produced, some of which may be used by bakers, although many are used by the home baker. Soft wheat flour or flour of low or moderate protein content is used for cakes. In premixes, the flour, egg powder, shortening, fruit or flavoring components, and leavening agents are combined in the dry state, although, when mixed, the shortening is in the melted or liquid state, and emulsifiers, such as monoglycerides, which improve air incorporation during mixing, may be used. Much of the successful preparation of cake mixes depends on the kind and quantity of chemical leavening used. Angel cakes contain egg white and are leavened by the incorporation of air into the mixture. During baking, the air in the cake expands and acts as the leavening agent.

4.2 Doughnuts

The ingredients for doughnuts are similar to those for cake, especially pound cake, except that some doughnuts are leavened with the use of yeast. After mixing, the dough is extruded, cut into doughnut form, and cooked in hot oil (370 to 380 °F [187.8 to 193.3 °C]). Fat absorption is reported to be about 15%. Fat absorption may be higher when processing parameters (e.g., temperature of the dough) are not controlled, resulting in greasy doughnuts. When fat absorption is insufficient, the keeping quality of the doughnuts is diminished. When doughnuts are to be sugared, the temperature and relative humidity must be controlled (70 to 75 °F [21.1to 23.9 °C] and 85%, respectively) for optimum sugar pickup. Generally, doughnuts are prepared from materials already premixed elsewhere, and seldom do the manufacturers of doughnuts mix their own formula.

4.3 Crackers

Crackers are unleavened or only slightly leavened. The flour used for these products consists of wheat flour, although some rye flour is often used. For producing crackers, flour, liquid shortening, salt, and small amounts of a chemical leavening agent, with or without sugar, and sometimes a flavoring agent, such as onion powder, are mixed into a dough. The dough is then extruded into the desired shape and baked without proofing. Milk powder, whey powder, and emulsifiers, such as monoglycerides, may be used with some combination of the ingredients listed.

4.4 Pie Crusts

Pie crusts, which are also unleavened bakery products, can be made from all-purpose flour, but the highest quality pie crusts are obtained from unbleached soft pastry flour of low-protein content. When the protein content is too high, the desirable flakiness characteristic is minimized, and, to compensate, a larger amount of shortening must be used.

The shortening (see Chap. 24) should be a solid or hydrogenated one (e.g., lard or hydrogenated vegetable fat), and it should be medium firm. Milk powder or fluid milk may be used in small quantities to enhance color. Eggs have the same effect. Salt and sugar may be added. The composition of a high-quality pie crust is approximately 47.5% flour, 1.0% salt, 3.3% nonfat milk powder, 2.0% glucose, 32.4% shortening, and 13.8% water. To ensure a high-quality crust, all ingredients should be mixed at a temperature of 60 to 65 °F (15.6 to 18.3 °C), and the ingredients combined with a minimum of mixing and handling.

5 Bakery Products and Health

5.1 Low- and No-Fat Bakery Items

Food technologists and bakers have worked together to develop bakery products that have attributes such as mouthfeel, texture, and other physical and sensory qualities similar to many of the popular items that use fats and oils in their formulations. It is quite easy to find fat-free cookies, breads, cakes, and muffins in virtually any supermarket. The fat-free commercial muffin mixes are the most popular items on many bakery supply house product lines.

Gums, fruit and cereal fibers, fruit purees, sugars, polyhydric alcohols, and modified food starches are among the ingredients that have been used in a wide variety of combinations as substitutes for the fats and oils. Emulsifiers such as lecithin are also added to some formulas to improve texture and volume of the finished product. Lecithin is well known as a constituent of eggs but it also is a byproduct of soybean oil refining. It is a lipid (a phospholipid) and must be labeled as a fat. In some formulas, it may add 0.5 g of fat per serving to the end product. This product cannot contain the fat-free claim but could still contain a low fat claim (less than 3 g per serving).

5.2 Other Health Issues for Bakery Products

There is ongoing interest in a raft of health effects associated with the consumption of bakery products and other cereal based foods. For instance, increasing dietary fiber levels in bread seems to be associated with a lower tendency to raise blood sugar, i.e. lowering of the glycemic response (Burton et al. 2011; Fardet et al. 2006) (Scazzina et al. 2013).

The issue of gluten sensitivity remains ever present, but there is optimism here too. There is an ever-increasing range of non-wheat cereal products with nearly similar sensory properties as their wheaten counterparts. Gluten free products are now fairly commonplace in most medium to large supermarket chains (Anton and Artfield 2008; Capriles and Areas 2014). In the previous chapter we discussed the seemingly inverse relations between the consumption of whole grain bread and certain and chronic conditions though more work is needed to understand how or which grain components produce the health benefits. (Gil et al. 2011).

Resent concerns with decreasing sodium intake is leading to attempts produce low salt bakery products notably bread (Belz et al. 2012; Quilez and Salas-Salvado 2012). Finally, attempts are underway to develop functional bread. As well as high-fiber products, it seems likely that products fortified with enhanced levels of phytochemicals and antioxidants will continue to receive attention (Somayeh et al. 2014).

6 Further Reading

Bread making is a mature technology and relatively well documented. Readers wishing for more extensive information beyond of the material presented here could refer to one of the many excellent monographs for additional reading (Cauvain 2013; Coates and Stanford 1980; Hamelman 2004; Hanneman 1980; Miller 1981; Sluimer 2005). The following special reviews are also interesting dealing, with frozen doughs (Ertop et al. 2012; Giannou et al. 2003; C. M. Rosell and Gomez 2007; Selomulyo and Zhou 2007; Yadav et al. 2009) (Yadav et al. 2009), bread improvers and conditioners (Elkhalifa et al. 2007; Selomulyo and Zhou 2007; Stampfli and Nersten 1995; Wood 1980). There is considerable interest in oriental food and in particular the concept of steamed bread (Chao et al. 2015; Fan 2014; Jing-sheng et al. 2015) which is becoming increasingly popular and available as part of Western cuisine.

References

Agriculture and Agri-Food Canada (2016) Pathfinder – bakery in the European Union. Retrieved from http://www.agr.gc.ca/eng/industry-markets-and-trade/statistics-and-market-information/agriculture-and-food-market-information-by-region/europe/market-intelligence/pathfinder-bakery-in-the-european-union/?id=1462975152278

Agriculture and Agri-Food Canada (2017) Bakery products in the United States. Retrieved from http://www.agr.gc.ca/eng/industry-markets-and-trade/statistics-and-market-information/agriculture-and-food-market-information-by-region/united-states-and-mexico/market-intelligence/bakery-products-in-the-united-states/?id=1410083148478

Al-Dmoor HM (2012) Flat bread: ingredients and fortification. Qual Assur Saf Crops Food 4(1):2–8. https://doi.org/10.1111/j.1757-837X.2011.00121.x

American Bakers Association (2017) Baking industry economic impact study. Retrieved from https://www.americanbakers.org/industry-data/

Anton AA, Artfield SD (2008) Hydrocolloids in gluten-free breads: a review. Int J Food Sci Nutr 59(1):11–23

Arpita M, Datta AK (2008) Bread baking – a review. J Food Eng 86(4):465–474. https://doi.org/10.1016/j.jfoodeng.2007.11.014

Bakerpedia (2021) Bread. Retrieved from https://bakerpedia.com/specialties/bread/

Barrett S, Croft AG, Hartley AW (1971) Determination of cysteine hydrochloride, ascorbic acid and potassium bromate in admixture in bread improver formulations. J Sci Food Agric 22(4):173–175

Belz MC, Ryan LA, Arendt EK (2012) The impact of salt reduction in bread: a review. Crit Rev Food Sci Nutr 52(6):514–524. https://doi.org/10.1080/10408398.2010.502265

Bobrow-Strain A (2012) White bread: a social history of the store-bought loaf. Beacon Press. 252 pages

Burton PM, Monro JA, Alvarez L, Gallagher E (2011) Glycemic impact and health: new horizons in white bread formulations. Crit Rev Food Sci Nutr 51(10):965–982. https://doi.org/10.1080/10408398.2010.491584

Capriles VD, Areas JAG (2014) Novel approaches in gluten-free breadmaking: interface between food science, nutrition, and health. Compr Rev Food Sci Food Saf 13(5):871–890

Cauvain SP (2013) Breadmaking. Improving quality, 2nd edn

Chao T, Ling-yu Q, Wei-zhe S, Hao-chen D, Bao-guo S, Xiu-ting L (2015) The progress of traditional starter cultures applied to Chinese steamed bread. Food Res Dev 36(11):1–5. https://doi.org/10.3969/j.issn.1005-6521.2015.11.001

Cho S, Johnson G, Song WO (2002) Folate content of foods: comparison between databases compiled before and after new FDA fortification requirements. J Food Compos Anal 15(3):293–307. https://doi.org/10.1006/jfca.2002.1067

Choumenkovitch SF, Selhub J, Wilson PWF, Rader JI, Rosenberg IH, Jacques PF (2002) Folic acid intake from fortification in United States exceeds predictions. J Nutr 132(9):2792–2798

Coates H, Stanford JR (1980) Growing wheat and making bread on a small scale

Courtin CM, Rezaei MN, Jayaram VB (2014) Evaluating the impact of yeast fermentation on bread dough matrix rheology. Cereal Foods World 59(6S)

Elkhalifa AEO, Mohammed AM, Mustafa MA, El Tinay AH (2007) Use of guar gum and gum Arabic as bread improvers for the production of bakery products from sorghum flour. Food Sci Technol Res 13(4):327–331. https://doi.org/10.3136/fstr.13.327

Ertop MH, Etgu H, Hayta M (2012) The impacts of formulation components on improvement of frozen bread dough and final product quality. Gida 37(1):55–62

Every D, Simmons L, Sutton KH, Ross M (1999) Studies on the mechanism of the ascorbic acid improver effect on bread using flour fractionation and reconstitution methods. J Cereal Sci 30(2):147–158. https://doi.org/10.1006/jcrs.1999.0248

Fan Z (2014) Influence of ingredients and chemical components on the quality of Chinese steamed bread. Food Chem 163:154–162

Fardet A, Leenhardt F, Lioger D, Scalbert A, Remesy C (2006) Parameters controlling the glycaemic response to breads. Nutr Res Rev 19(1):18–25. https://doi.org/10.1079/NRR2006118

Frost & Sullivan (2008) Strategic assessment of the U.S. Bakery Industry N341-88. Retrieved from http://s3.amazonaws.com/zanran_storage/hdcGlobal.com/ContentPages/105980746.pdf

Gan Z, Ellis PR, Schofield JD (1995) Gas cell stabilisation and gas retention in wheat bread dough. J Cereal Sci 21(3):215–230. https://doi.org/10.1006/jcrs.1995.0025

Giannou V, Kessoglou V, Tzia C (2003) Quality and safety characteristics of bread made from frozen dough. Trends Food Sci Technol 14(3):99–108. https://doi.org/10.1016/S0924-2244(02)00278-9

Gil A, Ortega RM, Maldonado J (2011) Wholegrain cereals and bread: a duet of the Mediterranean diet for the prevention of chronic diseases. Public Health Nutr 14(12A):2316–2322. https://doi.org/10.1017/S1368980011002576

Hamelman J (2004) Bread. A baker's book of techniques and recipes

Hanneman LJ (1980) Bakery: bread & fermented goods

Jenkins SM (1975) Bakery technology. Book 1. Bread

Jing-sheng L, Xia-dong L, Hai-xia X, Xiao-qing R (2015) Study on the steamed bread of grains in China. Food Res Dev 36(23):180–183. https://doi.org/10.3969/j.issn.1005-6521.2015.23.047

LEO Project (2015) Bakery and bake-off market study. Retrieved from http://leo-fp7.eu/index.php/en/publications-en

McBride J, Tucker KL (1998) Foods to be fortified with folic acid. USDA-ARS. Retrieved from https://agresearchmag.ars.usda.gov/1997/jun/folate/

Miller BS (1981) Variety breads in the United States. 158 pp

Morad MM, D'Appolonia BL (1980) Effect of surfactants and baking procedure on total water-solubles and soluble starch in bread crumb. Cereal Chem 57(2):141–144

NIIR Board of Consultants & Engineers (2014) The complete technology book on bakery products (Baking science with formulation & production), 3rd edn. NIIR Project Consultancy Services. 672 pages

Pomeranz Y (1980) Molecular approach to breadmaking: an update and new perspectives. Bak Dig 54(1):20–27

Preedy V, Watson R, Vinood P (2011) Flour and breads and their fortification in health and disease prevention. Academic Press. 542 pages

Quilez J, Salas-Salvado J (2012) Salt in bread in Europe: potential benefits of reduction. Nutr Rev 70(11):666–678

Reduello F (2011) European baking market: in recovery mode. BakingBusiness.com. Retrieved from http://www.bakingbusiness.com/news/news-home/features/2011/2/european-baking-market.aspx?cck=1

Rosell CM, Gomez M (2007) Frozen dough and partially baked bread: an update. Food Rev Intl 23(3):303–319. https://doi.org/10.1080/87559120701418368

Rosell CM, Bajerska J, Sheikha AFE (2015) Bread and its fortification: nutrition and health benefits. CRC Press. 417 Pages

Scazzina F, Siebenhandl-Ehn S, Pellegrini N (2013) The effect of dietary fibre on reducing the glycaemic index of bread. Br J Nutr 109(7):1163–1174

Scrimshaw NS (1998) Food Nutr Bull 19(2). Retrieved from http://archive.unu.edu/unupress/food/V192e/begin.htm#Contents

Selomulyo VO, Zhou W (2007) Frozen bread dough: effects of freezing storage and dough improvers. J Cereal Sci 45(1):1–17. https://doi.org/10.1016/j.jcs.2006.10.003

Sluimer P (2005) Principles of breadmaking: functionality of raw materials and process steps

Smith M (1995) Cheaper, better bread for the people of SA. Food Rev 22(6):37

Somayeh R, Taghi Gharibzahedi SM, Hadi Razavi S, Mahdi Jafari S (2014) Recent developments on new formulations based on nutrient-dense ingredients for the production of healthy-functional bread: a review. J Food Sci Technol 51(11):2896–2906

Stampfli L, Nersten B (1995) Emulsifiers in bread making. Food Chem 52(4):353–360. https://doi.org/10.1016/0308-8146(95)93281-U

Stojceska V, Butler F (2012) Investigation of reported correlation coefficients between rheological properties of the wheat bread doughs and baking performance of the corresponding wheat flours. Trends Food Sci Technol 24(1):13–18. https://doi.org/10.1016/j.tifs.2011.09.005

U. S. Food and Drug Administration (2016) CFR - Code of federal regulations Title 21. Food and drugs, Chapter I. Food and drug administration, Department of Health And Human Services, Subchapter B-Food for human consumption, Part 136 -Bakery products. Subpart B-Requirements for specific standardized bakery products. Sec. 136.110 Bread, rolls, and buns. Retrieved from http://www.accessdata.fda.gov/scripts/cdrh/cfdocs/cfcfr/CFRSearch.cfm?fr=136.110

U. S. Food and Drug Administration (2017) 21 CFR Part 136 -Bakery products. Retrieved from http://www.accessdata.fda.gov/scripts/cdrh/cfdocs/cfcfr/CFRSearch.cfm?CFRPart=136

United States Census (2014) 2011 county business patterns (NAICS). Retrieved from http://censtats.census.gov/cgi-bin/cbpnaic/cbpsect.plor http://censtats.census.gov/cgi-bin/cbpnaic/cbpdetl.pl

Wheat Food Council (2016) Baked goods. Retrieved from http://www.wheatfoods.org/recipes/Baked%20Goods?page=1

Wood PS (1980) Recent advances in bread improvers. Process Biochem 15(6):12

Yadav DN, Patki PE, Sharma GK, Bawa AS (2009) Role of ingredients and processing variables on the quality retention in frozen bread doughs: a review. J Food Sci Technol Mysore 46(1):12–20

23 Fruits and Vegetables

1 Introduction

1.1 Significance of Fruits and Vegetables

Dietary guidelines for Americans shows a need for eating more fruit and vegetables with 50–70% of US adults now consuming less than the recommended daily intake (U.S. Department of Health and Human Services and U.S. Department of Agriculture 2015). On a worldwide basis, intakes of fresh fruit and vegetables are 50% below values recommended by the World Health Organization (Mason-D'Croz et al. 2019). Fruit and vegetable purchases decreased for lower income groups, and the elderly (Bihan et al. 2010; Nicklett and Kadell 2013; Plessz and Gojard 2013).

Vegetables are plant foods that include various edible parts such as leaves, shoot, roots, tubers, flowers, and stems. The popular definition of fruit applies to naturally sweet plant foods normally used for desserts. Therefore, tomatoes and olives are frequently treated as vegetables. Berries belong to a class of fruit that are usually small and delicate. However, melons as a class are usually large, often with a tough, and sometimes thick, outer skin. In consequence, watermelons, squash and zucchinis are listed alongside of other vegetables - though they are technically fruit as indicated by the complement of seeds.

Consuming more fruits, vegetables and nuts is associated with decreasing risk of several chronic conditions, e.g. diabetes, high blood pressure, stroke and cancer (Kader et al. 2014; Liu 2014; Lock et al. 2005). Fruit and vegetable intake is thought to help with weight loss if there is a net reduction in energy intake. Fruits and vegetables being bulky foods tend to increase the sense of satiety (Kaiser 2014). A low fruit and vegetables intake is ranked amongst the top-10 risk factors for mortality. The FAO indicated 5.2 million extra deaths annually were associated with low intakes of fruits and vegetables. (Food and Agriculture Organization of the United Nations 2017). Fruit and vegetables contribute to food security (Aworh 2015). The extent of processing may moderate the health benefits from fruits. Though fresh fruit had nil or positive effects on health, canned or sugared fruit juice had adverse effects on disease risk (Fardet et al. 2019).

Fruits and vegetables provide a good source of vital nutrients e.g. folate, vitamin C, dietary fiber and minerals like potassium and magnesium (U.S. Department of Agriculture and U.S. Department of Health and Human Services 2010). Both fresh and processed fruits and vegetables can provide sources of important nutrients, which is something to bear in mind where the availability of fresh produce is limited (Blanck et al. 2008; Dwyer et al. 2012). Consuming some fruit and vegetables may even improve appearance

R. Owusu-Apenten, E. R. Vieira, *Elementary Food Science*, Food Science Text Series,
https://doi.org/10.1007/978-3-030-65433-7_23

because of the increasing accumulation of carotenoids in the skin (Whitehead et al. 2012a, 2012b).

1.2 Fruit and Vegetable Production

1.2.1 Production of Fruit and Nuts

About 50% each of the 2 million farming enterprises in the US are concerned with crop production and animal production, respectively. However, just about 5% of farming enterprises are concerned with producing fruits, tree nuts, and vegetables (Table 23.1). The United States census data showed there were 93 thousand farming enterprises for fruits and tree nuts (NAICS 1113), with 10% of such farms concerned with oranges and other citrus fruits. Fruit is grown on approximately 3.5 million acres compared with 2.5 million harvested acres for tree nuts (USDA 2014). Table 23.1 list the main faming enterprises for fruits, nuts, and vegetables. The value or total cash receipts for the fruit and nut crop harvest (2012) was estimated at US$25.3 billion, compared to US$7.0 billion for tree nuts (USDA 2014).

The top-10 ten ranked fruits produced in the US by total acreages harvested, were grapes > apples > peaches > cherries (sweet) > plums& prunes >avocados >pears > olives > cherries (tart) > pomegranates. From 2007 to 2012 the acreages for passion fruit, pomegranates, mangoes, olive and coffee increased - though such products started from a low baseline. By contrast, nectarines, figs, prunes, peaches and papaya declined substantively in the same interval of time (Table 23.2).

1.2.2 Vegetable and Pulses Production

Vegetables and melon production is classified as NAICS 1112. The current US annual production of vegetables (and pulses) is estimated at about 140 billion pounds or 63.5 million metric tons (Table 23.2). From Fig. 23.1, the annual production amounted to roughly 40–45 billion pounds each of fresh vegetables, processed vegetables and nearly 50 billion pounds of potatoes. The farm-gate value for vegetables and pulses was estimated at US$20 billion (Figs. 23.1 and 23.2).

According to production volume, the most important fresh vegetables in the United States

Table 23.1 United States crop production classified by NAICS code (2012)

Sector description (NAICS)	Number of Farms	Farms (acres)	Harvested Acres	Crop Value (1000$)
Total	2,109,303	914,527,657	314,964,600	212,397,074
Crop production (111)	1,054,987	454,064,220	262,297,534	200,168,398
Oilseed and grain farming (1111)	369,332	289,765,872	213,227,680	120,552,373
Vegetable and melon farming (11121)	43,021	8,206,468	5,333,376	16,624,257
Potato farming (111211)	2529	2,630,696	1,999,092	3,995,512
Other vegetable & melon farming (111219)	40,492	5,575,772	3,334,284	12,628,746
Fruit and tree nut farming (1113)	93,020	13,162,702	5,641,474	25,346,460
Orange groves (11131)	6141	1,268,406	672,041	1,955,556
Citrus (except orange) groves (11132)	2839	357,519	175,912	632,859
Non-citrus fruit and tree nut farming (11133)	84,040	11,536,777	4,793,521	22,758,045
Apple orchards (111331)	8921	2,133,711	418,403	2,750,702
Grape vineyards (111332)	17,580	2,683,393	1,205,918	6,816,072
Strawberry farming (111333)	1545	109,972	67,848	1,860,428
Berry (except strawberry) farming (111334)	9928	1,041,594	240,168	1,339,096
Tree nut farming (111335)	23,414	3,841,338	2,101,087	6,942,604
Fruit and tree nut combination (111336)	979	247,633	112,660	456,220
Other non-citrus fruit farming (111339)	21,673	1,479,136	647,437	2,592,922
Greenhouse, nursery, and floriculture production (1114)	52,777	4,502,514	1,483,765	14,694,443

Source: Adapted from (USDA 2014). NAICS = North American Industry Classification system

Table 23.2 Additional highlights of the US Vegetable and pulses industry

Acres planted	7.412 million
Total production	140 billion pounds (65 million metric tons)
Farm gate value	US$20 billion
% Export share	~12%
Domestic use	120 billion pounds
Per capita consumption	1.7 cups / day (219g per day)

*See Appendices for details

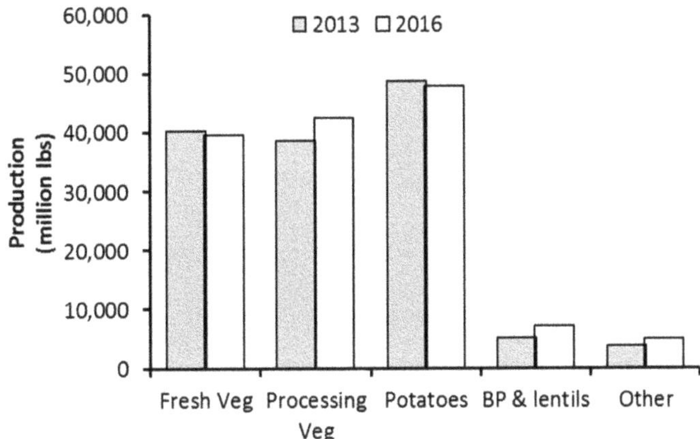

Fig. 23.1 United States annual production of vegetables and pulses by weight (million pounds). Categories are fresh vegetables (fresh veg), processing vegetables (processing Veg), potatoes, and dried beans, peas and lentils (BP & lentils)/Data does not include fruits. (Drawn using data from USDA Economic Research Services 2016a)

are onions, lettuce, watermelons and tomatoes. However, based on US$-value the most important vegetables are, lettuces and tomatoes (USDA National Agricultural Statistics Services 2016). A large part of the US vegetable industry is concerned with domestic consumption with 12% of the volume for export.

1.2.3 Melon Production

Melon is sometimes classed alongside of vegetables in the US (NAICS 1112) whilst FAO/WHO lists this product alongside of fruits. Since melons are a leading commodity classifying this as a fruit or vegetable affects the statistics coming out different jurisdictions. The current US production of melons (including, watermelon, cantaloupe, honey dew melons and related) is derived from 181 thousand acres of cultivation, leading to 2.6–3.0 million metric tons of yield, valued at about US$825 million (USDA National Agricultural Statistics Services, 2016). The US is ranked #5 or 6 in terms of producers for melons after China > Iran > Tukey > Brazil and Egypt (Table 23.3).

1.3 Postharvest Handing

Improper storage of vegetables after harvesting is detrimental to their quality. They undergo microbial spoilage, they lose water, in many cases they lose sugar and other nutrients (e.g., vitamins), and they give up considerable energy in the form of heat (value reported is more than 100,000 Btus per ton per day (27,784 Cal (kg)/MT/day). Of course, the heat produced by vegetables hastens their own deterioration by accelerating microbial spoilage (Barr and Poole 1941). Enzymatic deterioration, especially at sites where bruises occur, is also accelerated by higher temperatures (again within limits).

Advice on the storage of commercial fruits and vegetables is available from the USDA Agricultural Research Services in the form of the

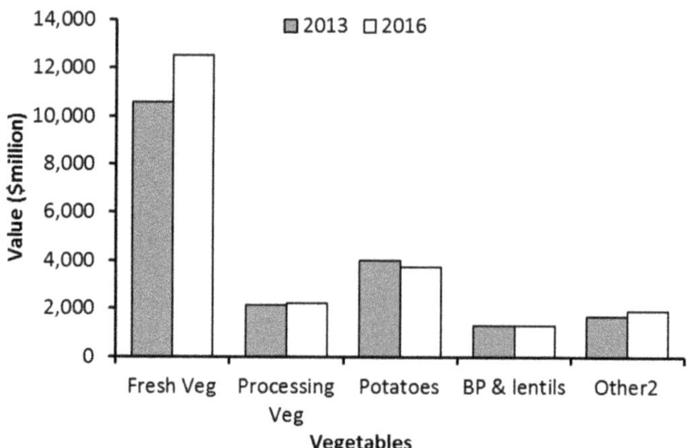

Fig. 23.2 United States annual production of vegetables and pulses by value (US$ million). Categories are fresh vegetables (fresh veg), processing vegetables (processing Veg), potatoes, and dried beans, peas and lentils (BP & lentils)/Data does not include fruits. (Drawn using data from USDA Economic Research Services 2016a)

Table 23.3 Watermelon production in leading countries and the world, 2008–2013 (million mTons)

Country	2008	2009	2010	2011	2012	2013
China	69.3	71.4	75.2	75.9	77.9	80.4
Iran	2.8	3.4	3.8	4.2	4.2	4.4
Turkey	4.4	4.2	4.1	4.3	4.5	4.3
Brazil	2.2	2.3	2.3	2.4	2.3	2.4
Egypt	1.6	1.8	1.8	1.7	2.1	2.1
United States	2.0	1.9	2.1	1.9	1.9	2.0
Others	22.0	23.7	22.5	23.6	24.9	25.3
World	104.3	108.7	111.7	113.9	117.8	120.8

Adapted from Ref. (USDA National Agricultural Statistics Services 2016); Data includes fresh and processed, converted from million CWT (1 million CWT = 50,000 US (metric) tons.

Agricultural Handbook 66 (AH66). This enormous resource (700 pages +) lists storage conditions for 138 fruit and vegetable of significance in the US (Gross et al. 2014). Temperature considerations are highly important in the postharvest period. So produce should be harvested early in the morning to avoid heating from solar rays. It is beneficial to subject fresh produce to a pre-cooling process, using a variety of large scale methods including, forced air cooling, hydro-cooling, packaging icing or refrigerated room cooling (Gross et al. 2014).

Fresh vegetables require washing before holding at low temperature and relative humidity. These conditions may be held during distribution, storage, and display on the retail level. If fresh vegetables are packaged for retail or wholesale distribution, it is important allow them to "breathe" and avoid built-up condensation in the package. Fruits and vegetables may be held in a temperature-controlled warehouse until it is distributed. Fresh fruit or vegetables may be held under controlled atmosphere storage conditions with CO_2 enriched air to slow deterioration (Gross et al. 2014). Vegetables are sold at retail as fresh produce, as heat-processed and canned, as frozen, and occasionally as dried products. The processing phase for fruits and vegetables are discussed in the next section.

1.4 Fruits and Vegetable Processing Industries

There are three main types of fruit and vegetables processing businesses (U.S. Census Bureau 2016). There are industries involved with post-harvest activities (NAICS 115114) for fresh produce e.g., sorting, grading, cleaning, irradiation, waxing, packing, and packing of farm dried fruit.[1] The second group includes industries

[1]The list of unit operations is not intended to suggest any composite process.

involved with processing fruit and vegetables (NAICS 3114) to produce frozen fruit, fruit juices, and vegetables (NAICS 311411),[2] canning brining/or pickling (NAICS 311421), or dehydration (NAICS 311423). Finally, the third group business are involved in marketing either wholesale, retail or via specialty stores (NAICS 44523) selling fresh fruits, berries, and vegetable. The outlets for the sale of fruits, vegetables and nuts are supermarkets, shops, and roadside stands (Statistics Canada 2017).

US census data showed that the fruit and vegetable processing industries support about 4458 enterprises and 175,495 jobs (Table 23.4) not including farming or agricultural production (cf. Sect. 1.2). Finally, a comparatively, smaller sector (NAICS 311340), manufactures candied and glazed fruits and vegetables, and peels. The major subdivisions of the fruit and vegetables industries are summarized in Fig. 23.4.

Table 23.4 US fruits and vegetable processing industries (NAICS 3114)

NAICS	Description	Employees (March 2012)	Number of Businesses
3114	*Fruit & vegetable preserving, specialty food*	155,163	1670
311411	Frozen food	85,424	659
311411	Frozen fruit, juice, and vegetable	29,754	225
311412	Frozen specialty food	55,670	434
31142	Canning, pickling, and drying	69,739	1011
311421	Fruit and vegetable canning	42,336	701
311422	Specialty canning	13,926	116
311423	Dried and dehydrated food	13,477	194
4452	Specialty stores		
44,523	Fruit & Vegetable Markets*	20,332	2788

* Retails fresh fruits and vegetables via stores or roadside stands (Statistics Canada 2017)

Some of the official data for NAICS 3114 were compiled using a variety of metrics. Some trade data refer to "fruit and nuts" or "vegetables and pulses" as the basis for measurement. Other trade data combine fruit with vegetables without mentioning nuts or pulses. Data for fresh and processed goods were mostly treated separately. However, some trade data (reviewed in Sect. 1.4) refer simply to horticultural products.

The fresh and processed vegetable sectors are discussed separately in part because these businesses have different quality specifications for fresh produce. Fruits and vegetables intended for processing are usually grown, and harvested under different conditions from those for fresh market produce. The varieties of vegetables selected for processing (freezing, canning or dehydration) have to be able to withstand industrial processing (USDA Economic Research Services 2016a).

1.5 United States Export and Import of Fruits and Vegetables

For 2015/16 the value for United States horticultural products exported annually (US$34 billion) exceeded any other single group of agricultural commodities (Table 23.5). The export value for fruit and vegetables (US$14.5billion) was divided nearly equally between processed fruits and vegetables (US$ 7.6 billion) and fresh produce (US$ 7.2 billion). Import and exports trade for fresh and processed fruits and vegetables (Tables 23.5 and 23.6) is substantial with a slightly larger balance in favor of exports (USDA Economics Statistics and Market Information System 2016).

The US trade totals for fruit and vegetables (US$ 15billion) should be viewed against a background of agricultural exports valued at US$118 billion and imports of US$125billion. As noted above, there was a positive trade surplus of US$7-9billion per year from the trade in agricultural commodities. The major export destina-

[2]NAICS 311411 Covers Frozen fruit, fruit juices, frozen vegetables, manufacturing drinks.

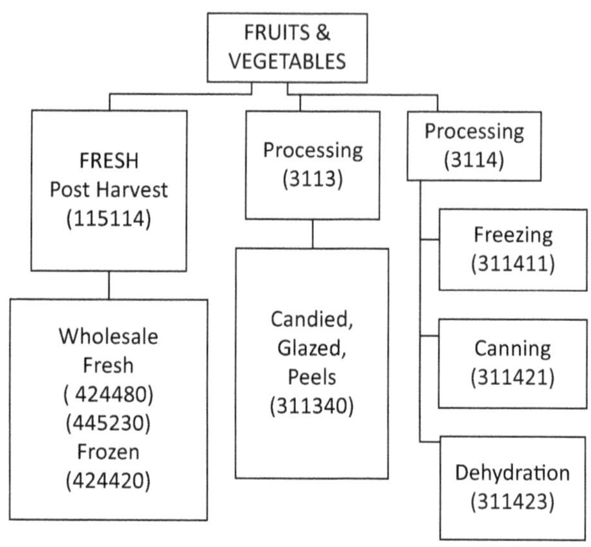

Fig. 23.3 A summary of the fruit and vegetable sector, not including agricultural production. NAICS identification codes are in brackets. (Drawn from U.S. Census Bureau 2016)

Table 23.5 Agricultural exports by commodity (US, 2015–2016)

US Exports	2015 US$ billion	2016 (Feb) US$ billion
Grains and feeds (A)	31.60	27.2
Oilseeds and products (B)	31.70	25.4
Livestock, poultry, and dairy (C) (D)	29.29	25.7
Tobacco unmanufactured	1.25	1.0
Horticultural products (E)	34.11	34.7
Fruits and vegetables fresh	7.22	7.1
Fruits and vegetables, processed	7.38	7.6
Tree nuts, whole and processed	8.91	9.0
Sugar and tropical products (F)	6.06	6.1
Total	139.74	125.0

Notes: (A) corn gluten feed and meal and processed grain products, corn, barley, sorghum, oats, and rye but excludes wheat flour, (B) excludes corn gluten feed and meal, (C) Chilled, frozen, and processed meats, (D) Includes broiler meat if federally inspected, (E) Includes fresh and process fruits and vegetables and juices, (F) Includes coffee and cocoa products, tea, and spices. Adapted from (USDA Economics Statistics and Market Information System 2016)

Table 23.6 Agricultural imports by commodity (US, 2015–2016)

US Imports	2015 US$ billion	2016 US$ billion
Grains and feed	10.86	11.0
Oilseeds and products	8.75	9.0
Livestock, dairy, & poultry	19.45	15.8
Horticulture products	49.7	53.9
Fruits, fresh	10.2	11.3
Fruits processed	5.1	5.6
Fruit juices	1.8	2.0
Nuts, whole and processed	2.6	3.1
Vegetables, fresh	*6.9*	*7.0*
Vegetables, processed	*4.4*	*4.6*
Sugar & tropical products	23.5	26.9
Cocoa and products	4.7	5.3
Coffee and products	6.4	7.3
Total agricultural imports	114.0	118.5

Rounded to 2DP by authors, Data adapted from (USDA Economics Statistics and Market Information System 2016)

tions for US agricultural products include Asia (43%), North America (28%), EU-28 (8.8%), South East Asia (7.8%) and Middle East (6%) – Table 23.7 lists the major regions for US agricultural exports.

1.6 World Production of Fruits and Vegetables

Fruits and vegetables are important for food security globally (Dauthy 1995; Diop and Jaffee 2004). For 2012, the world production of fruits

Table 23.7 U.S. Agricultural exports: Value by region, 2015 Fiscal year

Region/Country	US$ billion	% Total
Asia	60.8	43.5
East Asia	*47.9*	*34.3*
Japan	11.7	8.4
China	22.5	16.1
Hong Kong	3.9	2.8
Taiwan	3.3	2.4
South Korea	6.4	4.6
Southeast Asia	*10.7*	*7.6*
Indonesia	2.4	1.7
Philippines	2.4	1.7
Malaysia	0.9	0.6
Thailand	1.7	1.2
Vietnam	2.4	1.7
South Asia	*2.2*	*1.5*
India	1.1	0.8
Western Hemisphere	*53.7*	*38.4*
North America	*39.3*	*28.2*
Canada	21.3	15.3
Mexico	18.0	12.9
Caribbean	3.4	2.4
Dominican Rep	1.2	0.8
Central America	*3.8*	*2.7*
South America	7.1	5.1
Brazil	0.8	0.6
Colombia	2.6	1.9
Peru	1.2	0.9
Venezuela	0.9	0.6
Europe/Eurasia	13.6	9.7
European Union-28	12.3	8.8
Other Europe	0.7	0.5
FSU-12	0.6	0.5
Russia	0.4	0.3
Middle East	*1.4*	*5.9*
Turkey	1.6	1.2
Saudi	0.3	1.3
Africa	*3.7*	*2.7*
North	0.4	1.8
Egypt	1.0	0.7
Sub-Saharan	0.4	1.9
Nigeria	0.7	0.5
Oceania	*2.1*	*1.5*
Total	139.7	100.0

Ref. Adapted from (USDA Economics Statistics and Market Information System 2016)

was about 730 million metric tons whilst vegetables production amounted to 875 million metric tons (Tables 23.8, 23.9 and 23.10) (European Fresh Produce Association 2015). Five regions or countries accounted for over 50% of the global fruit production: China > India> EU-28 > Brazil > USA > Mexico. The global production of vegetables is skewed towards a limited number of producers. China accounts for >50% of the world production of vegetables, followed by India (11.2% world total) and the EU-28 (6.3%) and USA (2.8%). The world ranking for vegetable production is thus, China > India > EU-28 > USA > Turkey (Table 23.8).

Some six fruits account for 70% total production, i.e., watermelons, bananas, apples, guava and mangoes, and plantains. Arguably, the least widely grown fruits and those with the most limited geographic distribution include raspberries, currents, gooseberries, figs, kiwifruit, cherries, apricots and dates. A listing of some of the fruits produced globally is summarized below (Table 23.9).

A similar listing of vegetables shows tomatoes are the most popular type of vegetable globally, followed by cucumbers, and aubergines (Table 23.10). Interestingly, the world trade in fruit and vegetables is not very well documented. One of the interesting generalizations is that much of the world production is for domestic consumption. Major producers are not necessarily exporters (Diop and Jaffee 2004).

2 Fruits

2.1 Postharvest Fruit Maturation and Ripening

Ripening is desirable as it leads to improvements in quality but the same process decreases storage life. The maturation and ripening behavior of fruits, determines how such produce are harvested and handled as foods (Lelievre et al. 1997; Payasi and Sanwal 2010; Sam et al. 2014). Fruits can be categorized into two groups in terms of their postharvest characteristics; Group 1 non-climacteric fruits "ripen on the vine". Group 1 fruit do not ripen, or ripen very slowly once picked. Producers need to harvest Group 1 fruits at maturity as they begin ripening. By comparison, Group 2 climacteric fruits continue to ripen

Table 23.8 World fruit and vegetable production 2013 (million metric Tons)

Country	Fruit mTons (10^6)	%Total	Country	Vegetables mTons (10^6)	%Total
China	224.8	30.9	China	463.9	53.0
India	79.9	11.0	India	98.0	11.2
EU-28	46.0	6.3	EU-28	54.7	6.3
Brazil	35.1	4.8	USA	24.2	2.8
USA	22.1	3.0	Turkey	19.3	2.2
Mexico	17.4	2.4	Iran	15.1	1.7
Turkey	17.2	2.4	Egypt	14.0	1.6
Indonesia	16.1	2.2	Viet-Nam	13.4	1.5
Philippines	15.9	2.2	Russia	11.7	1.3
Iran	15.2	2.1	Italy	11.2	1.3
Spain	13.3	1.8	Nigeria	9.6	0.1
Other	266.3	36.6	Other	196.4	22.4
Total	726.9	100.0	Total	875.2	100.0

*Converted from tones to metric (US) tons. Adapted from Freshfel (European Fresh Produce Association 2015)

even after harvesting (Table 23.11). Therefore, climacteric fruit can be sold and ripened at home.

The ripening of both climacteric and non-climacteric fruit is usually characterized by a range of changes. Degradation of the green pigment chlorophyll resulting in the unmasking of carotenoid yellow pigment in the fruit, loss of the hardened texture due to the breakdown of plant cell walls, increased build-up of sugars, a decline in acids, and a marked in increased in aroma compounds occurs during ripening. Ripening is also associated with increased susceptibility to infection and physical damage (Table 23.11).

Some of the key differences in the biology of climacteric and non-climacteric fruit have been widely discussed (Alexander and Grierson 2002; Lelievre et al. 1997; Payasi and Sanwal 2010; Pech et al. 1994; Sam et al. 2014). In climacteric fruit, ripening coincides with a rise of the plant hormone ethylene, and an increase in the rate of respiration. By contrast, non-climacteric fruit do not show the ethylene increases and there is a gradual ripening transition whilst still attached to the parent plant. (Figure 23.4) (Alexander and Grierson 2002). Since, the appearance and concentrations of ethylene gas in plant cells controls ripening processes, then technologies for delaying the formation of ethylene or removal of this agent provides opportunities for managing fruit ripening. Historically, the application of ethylene in horticultural practices involved the use of smoke houses to hasten ripening (smoke and vehicle exhaust contain ethylene). In modern times, there are genetic methods to silence the genes controlling ethylene metabolism in plants and to produce slowly ripening varieties (Reid 2015).

2.1.1 Fruit Harvesting

Fruit may be picked green and allowed to ripen during distribution. Fruit is considered ripe when it reaches the optimum succulence, texture, and aroma and a balance occurs in the levels of sugars and acidity. Fruit that goes past its optimum quality leads to over-ripeness and senescence, followed by a physical breakdown. In this stage, there is a loss of texture or firmness. The levels of sugars, acids, and aroma elements generally all decline in senescent fruit. Some fruit, such as the banana, has an early senescence, and deterioration, once begun, is rapid. On the other hand, some fruit, such as apples, resists the onset of senescence as well as its progression.

The onset of senescence can be controlled by keeping the fruit at the lowest temperature that it can tolerate and by increasing the amount of atmospheric carbon dioxide to a controlled level. Modified atmosphere packaging and controlled atmosphere storage, have found many applications in the storage and general shipping of fruit and vegetables.

Table 23.9 Word production totals for some common fruits (2013)

Fruits	mTons/10^6	% Total
Watermelons	107.7	14.9
Bananas	104.9	14.5
Apples	71.0	9.8
Guava & mangoes	64.2	8.9
Oranges	55.0	7.6
Plantains	37.4	4.0
Other melons	29.1	3.7
Tangerines, mandarins, clementine's	26.8	3.5
Grapes	25.4	3.4
Pears and quinces	24.9	3.4
Pineapples	24.5	3.0
Peaches and nectarines	21.4	1.9
Lemons and Lime	14.0	1.7
Papayas	12.3	1.6
Citrus fruit	11.2	1.5
Plums and sloes	10.7	1.1
Grapefruit and pomelos	7.7	1.1
Strawberries	7.6	1.0
Dates	7.1	0.6
Avocados	4.7	0.6
Apricots	4.1	0.5
Cherries	3.5	0.4
Kiwifruit	3.2	0.3
Cranberries and blueberries	1.9	0.2
Figs	1.1	0.1
Currants, gooseberries	0.9	0.1
Raspberries + berries	0.5	5.6
Other	40.5	100.0
Total	723.4	

Adapted from (European Fresh Produce Association 2015)

Table 23.10 World production of vegetables (million metric tons, 2013)

Vegetables	mTons/10^6	% Total
Tomatoes	143.7	16.4
Cabbages	70.4	8.0
Cucumbers, gherkins	70.4	8.0
Aubergines, Eggplant	48.9	5.6
Carrots, turnips	36.8	4.2
Chilies, peppers	30.7	3.5
Pumpkins, Squash	24.4	2.8
Garlic	23.9	2.7
Lettuce, Chicory	23.5	2.7
Spinach	23.1	2.6
Beans	22.9	2.6
Cauliflowers, broccoli	22.1	2.5
Peas	17.2	2.0
Asparagus	7.9	0.9
Mushrooms, truffles	7.5	0.9
Onions	5.5	0.6
Leeks, and related	2.1	0.2
Artichokes	1.8	0.2
Leguminous vegetables	1.6	0.2
Total	875.2	

Adapted from (European Fresh Produce Association 2015)

2.1.2 Preservation of Fruit

Fruits are preserved by canning when heat-processed to attain commercial sterility. Most fruit because of their acidity (pH usually below 4.6) can be sterilized by heating the containers in boiling water to the point where all parts of the product reach a temperature of 180 to 200 °F (82.2 to 93.3 °C). Processing of fruit is thus less stringent compared to vegetables and high pH foods which require high temperatures from 240 °F to 250 °F (115.6 °C to 121.1 °C) for significant lengths of time to attain commercial sterility. The principles behind food canning are discussed in Chap. 12.

Fruit can be preserved by freezing at 0 °F (−17.8 °C) or below and thereafter held at 0 °F (−17.8 °C) or below. Ordinarily, fruit is packed in retail-sized containers and frozen in one of three ways: (1) the containers are placed on trays, the trays are placed on racks and frozen as the racks are moved through a tunnel in which blasts of cold air are circulated; (2) the cartons are placed on chain mesh belts that move slowly through a tunnel in which blasts of cold air are circulated; or (3) the cartons are placed on trays, the filled trays are placed between refrigerated metal plates, and the product is frozen with the containers in contact with the cold plates.

2.2 Grapes and Raisins

Grapes are grown mainly in California, New York, Pennsylvania, Michigan, and Ohio. Washington, Missouri, and Arkansas also produce some grapes. Grapes can be consumed as fresh grape juice, vinegar, jams, jellies, or as the fresh product. Grapes are planted as vines or cuttings from older plants. (Rosenstock 2007).

Table 23.11 A listing of non-climacteric and climacteric fruits

Group 1 or Non-climacteric Fruit	Group 2 or Climacteric fruits
Blackberry	Apple
Cherry	Apricot
Grape	Avocado
Grapefruit	Banana
Lemon	Cherimoya
Lime	Guava
Longan	Kiwifruit
Loquat	Mango
Lychee	Nectarine
Mandarin	Papaya
Muskmelons	Passion fruit
Orange	Peach
Pepper(bell)	Pear
Pineapple	Pepper (chili)
Pomegranate	Persimmon
Prickly pear	Plum
Rambutan	Quince
Raspberry	Sapodilla
Strawberry	Sapote
Tamarillo	Tomato
Watermelon	

2.2.1 Table Grapes

Table grapes are harvested in a mature ready-to eat state as indicated by coloration and a high soluble sugars content of 14–17.5% (Christensen et al. 1995a). The picked grapes may be packed in the vineyard or later in packing premises but the former is preferred in order to maintain freshness. Picking occurs simultaneously with grading to exclude defective or substandard grapes (Table 23.12).

Fresh grapes are packed in wooden crates, which should be lined to avoid bruising the contents. Cooling freshly harvested grapes is highly important for their preservation. To avoid general overheating, it helps to carry out the harvest during the morning. The newly gathered grapes should also be stored in the shade away from sunlight and then precooled to about 40 °F (4.4 °C) before transportation. Grapes that are to be stored for future shipment should be packed in crates precooled to 36 to 40 °F (2.2 to 4.4 °C), placed in refrigerated storage (29 to 32 °F [−1.67 to 0 °C]), fumigated with sulfur dioxide, and held in this manner until shipped (Rosenstock 2007). Additional periodic fumigation with sulfur dioxide may be required to prevent spoilage by molds. Under these conditions, grapes have a storage life of 1–7 months depending mainly on the variety.

2.2.2 Raisins

Raisins are produced from grapes by sun drying and artificial drying (Savoy et al. 1983). In sun drying, the grapes are picked as bunches and placed in a single layer on wooden trays between the rows of grape vines. The trays are tilted to face the sun. After they are partially dried, the grape bunches are turned and allowed to dry to the point where no juice can be pressed out. The trays are then stacked, and the air-drying is continued in the shade until a moisture content of about 17% is reached. After drying, the raisins are placed in sweatboxes to equilibrate, or to even out, the moisture that is present, and then they are packed in retail-size containers or in larger containers to be sold to the bakery trade. In artificial drying, grapes are first dipped in 0.25–1% lye (sodium hydroxide) and treated with sulfur dioxide.

Of the grape varieties introduced in the late nineteenth century, Thompson grapes are probably the most popular for raisins production (Table 23.13). More recently, the Fiesta variety is apparently also gaining in popularity. The USDA has an on-going program to review grape varieties from round the world for possible adoption in the United States. Desirable quality attributes include early ripening, to enable timely harvesting. Seedless varieties are a priority. Christensen from the University of California has written comprehensively on the topic of raisin grape varieties (Christensen 2000; Christensen et al. 1995a, 1995b).

2.3 Apples

The apple is the 2nd-most important fruit crop in the US (Sect 1.2.1) and is grown commercially in 35 States (USDA 2014, p. 498). Based on 2007 census results, the foremost production of apples occurs in the States of Washington (174,152 acres), New York (47,148 acres), Michigan

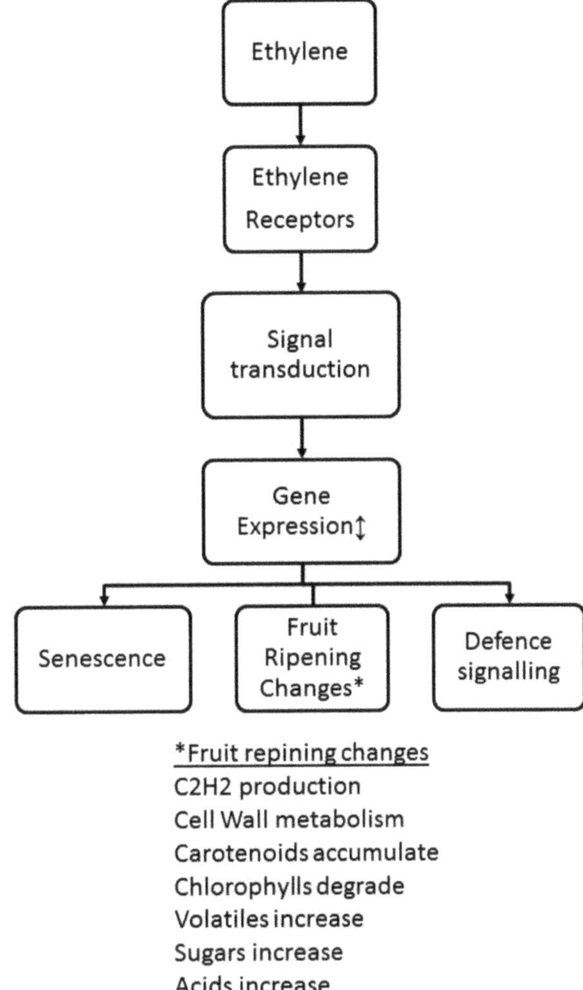

Fig. 23.4 The role of ethylene on ripening related changes in plants. (Adapted from Alexander and Grierson 2002)

(43,240 acres), Pennsylvania (21,556 acres), California (18,205 acres), and Virginia (11,929 acres). Cultivation of apples was at borderline (or below) commercial levels in the following States: Nebraska (298 acres), Arkansas (296 acres), Rhode Island (230 acres), Nevada (214 acres), South Dakota (197 acres), Delaware (163 acres), Florida (160 acres), Mississippi (111 acres), Wyoming (41 acres), Louisiana (28 acres), North Dakota (21 acres). Alaska (17 acres) and Hawaii (4 acres). There are hundreds of varieties of trees produced from seedlings that were grown in nurseries, or from grafts on existing apple trees. The fruit is developed on spurs formed by branchlets of three or more years of growth, the tree yielding fruit for many years thereafter. Fertilization of the soil and periodic pruning and thinning of apple trees are considered essential for good apple crops.

Apples are used as the fresh fruit, and apples not suitable for fresh fruit are used for the production of juice, cider, sauce, vinegar, jam, jelly, pie filling, and as an ingredient in a variety of baked goods. Pectin is extracted from the peels and cores.

For the production of apple slices, apples are size-graded, peeled, cored, sliced, and immersed in a 3% salt solution. Just before filling into cans, the slices are rinsed to remove salt. They are then filled into cans with 40% sugar solution. The cans are then heat-exhausted, sealed, and heat-processed.

Table 23.12 Defects for Table grapes

Unit Operation	Sorting/ rejection
Picking	Blackened
Sorting	Defective berries
Packaging	Clusters filled with shot berries
Cooling	Clusters too small
	Cracked
	Crushed
	Decayed
	Dried
	Inaccessible compact clusters
	Inadequately colored clusters
	Irregular-shaped
	Mildewed
	Scarred
	Shrunken
	Straggly clusters (+exposed stems)
	Sunburned
	Undersized

Source: Adapted from (Rosenstock 2007)

Table 23.13 Some common varieties of raisin grapes grown in the US

Grape variety	Date of Introduction
Sultana	1800 (mid)
Muscat of Alexandria	1851
Zante Currant (Black Corinth)	1854
Thompsons	1872
Black Monukka	1900s
'Fiesta	1965
Dovine	1995
Merbein Seedless	new

Source: Adapted from (Christensen 2000)

For the production of frozen apple slices, slices are immersed in brine, subjected to vacuum (to remove air), immersed in brine, washed, and packed with sugar in a ratio of 4 fruit to 1sugar. The product is then frozen. A bisulfite dip can be included to prevent nonenzymatic browning.

For drying, the peeled, cored, and sliced apples are first treated in a weak solution of citric acid and a bisulfite dip, which inhibits enzyme browning. Refrigeration for 24 h allows the sulfur dioxide to penetrate the apple slices. The apple slices are then dried using a loft-type kiln to a moisture content of 10%, before packaging in moisture proof containers. Dried apples are used in the bakery trade.

To make applesauce, peeled and cored apples are sliced, cooked in steam, pulped, and heated to 190 °F (87.8 °C) with added sugar. The sauce is filled into glass jars that are then sealed and heated to 200 °F (93.3 °C).

2.4 Bananas

The chief production of banana fruit occurs in Hawaii, Mexico and Central American countries, in Cuba, Jamaica, Haiti, the Dominican Republic, Honduras, Colombia, and Brazil. Bananas are also grown in some Asian and Middle Eastern countries. Green or cooking banana or plantain is indigenous and eaten as a vegetable in many countries within Africa and the Caribbean. The world production of banana fruit is foremost of all the fruits, greater than grapes, oranges, or apples (Aurore et al. 2009).

Banana fruit is handled mainly as fresh and shipped from the growing area while still green. At one time, bananas were shipped as bunches, on the stem, but today, they are handled mostly as hands or groups of single bananas, cut from the stem and packed in plastic-lined boxes. They may be treated with fumigants prior to boxing and should be precooled to 57 to 62 °F (13.9 to 16.7 °C). Bananas are subject to a chilling injury if held below 55 °F (12.8 °C); low temperatures kill certain surface cells and prevent normal ripening. Nor should bananas be held for extended periods at temperatures above 70 °F (21.1 °C); they must, therefore, be shipped under controlled temperatures. Bananas are ripened at 58 to 64 °F (14.4 to 17.8 °C), but the process can be accelerated by subjecting the fruit to ethylene gas.

Some bananas are peeled, pureed, and canned for the bakery and soda fountain trade (Singh et al. 2018). In such cases, the pureed banana pulp is quickly heated to about 280 °F (137.8 °C) for a few seconds, cooled and then filled into sterile, large sized cans, and sealed under aseptic conditions. All parts of the process are done with equipment and containers being pre-sterilized. Essentially no bananas are frozen but some are dried and sold as snack items and drinks, e.g. banana chips (Singh et al. 2018). Roasted, boiled or fried red plantain (cooking banana) is popular restaurant food (Newilah et al. 2005).

2.5 Oranges

The orange is utilized as a food more so than any other citrus fruit. The trees are sent out from nursery stock and must be protected from freezing weather. As with other citrus trees, some pruning has to be done each year. According to the 2012 census data (USDA 2014, p. 508), the five States which produce oranges or a commercial scale are Florida (465,001 acres), California (193, 087 acres), Texas (7,831 acres), Arizona (3,304 acres), Louisiana (777 acres) followed by Hawaii (377 acres), Alabama (9 acres). In the United States, about three-quarters of all oranges are used for frozen concentrated orange juice (FCOJ) production whilst a smaller but growing proportion goes to produce "not-from concentrate" (NFC) juice which, matches closely the attributes of freshly squeezed orange juice but is conveniently packaged in aseptic cartons or bottles (Johnson 2001).

Oranges should be picked when the ratio of total soluble solids (TSS) to titrable acidity (TA) indicates maturation (Lacey 2020). The requirements in Florida is for a TSS value of 8% to 11% brix and for the TSS/TA ratio to reach, 10.5:1 to 9.0:1 (Florida Department of State 2013). In some cases, oranges may be dyed by immersing in a solution of certified food dye at 120 °F (48.9 °C) for about 3 minutes prior to waxing, polishing, and cooling, because the color of the skin is often green when the fruit is picked. Some oranges are cooled to 32 to 40 °F (0 to 4.4 °C) and others to 40 to 44 °F (4.4 to 6.7 °C), depending on variety. They should be held at these temperatures until sold to the consumer. Under these conditions, they have storage life of 1to 3 months, depending on variety.

2.6 Other Fruits

Other fruits include blackberries, blueberries, cranberries, raspberries, strawberries, apricots, cherries, grapefruit, lemons, melons, peaches, pears, pineapples, and plums. Blackberries, raspberries, and strawberries are very popular fresh and in jellies and jams. They may also be found in fruit juice blends or packaged and frozen with or without sugar added. Apricots and plums are used fresh but many are also dried and some frozen and cherries are marketed fresh but are also quite popular as canned pie fillings. Peaches and pears are usually sold either fresh or are canned in light or heavy sugar syrups. Grapefruit, pineapples, and lemons are popular for juice production or used fresh. Pears can be processed much like apples and are used in ciders and juices, and may be canned.

All of the fruits can be canned, dried, and frozen using methods described in earlier chapters and adapted for the individual fruit. For more information on the most important methods for processing fruits the reader is referred to several dedicated book on this subject (Cano et al. 2014; Hui 2006; Pyke 1981; Salunkhe and Kadam 1995; Teskey and Shoemaker 1978).

2.7 Jelly and Jam

Many fruits are used for producing jelly and jam, with jelly making up the larger part. In jelly making, sugar and about 0.25–0.3% of dispersed pectin are mixed with the clarified fruit juice, and the mixture is concentrated in open kettles to a soluble solids content of about 65%. Citric acid solution is added to adjust the pH from 3.0 to 3.2. The product is then poured into glass jars, vacuum capped, and sprayed with hot water to bring the temperature of all parts to about 160 °F (71.1 °C), after which the product is cooled (Fig. 23.5). A similar process is applied to make jam, using crushed fruit in place of juice (USDA 1965).

3 Vegetables

Some fresh vegetables may be cut or shredded and packaged as pre-prepared salads or "slaws" (cabbage or broccoli). These are classified as potentially hazardous foods and must be pre-

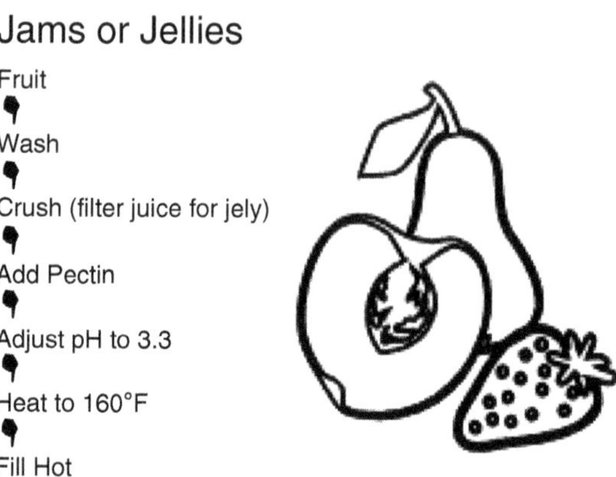

Fig. 23.5 General method for production of jams and jellies

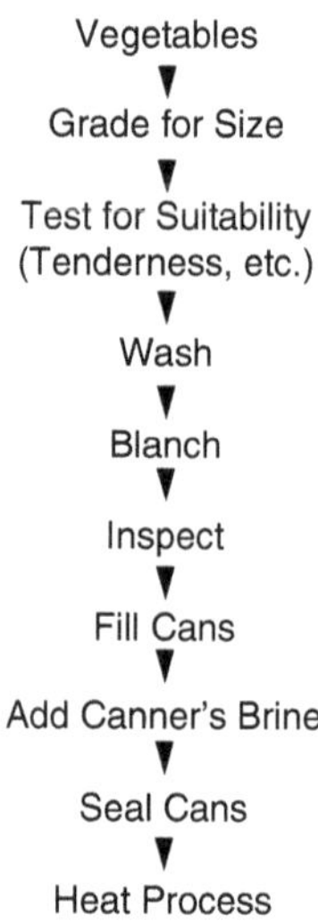

Fig. 23.6 Production of canned vegetables

pared, packaged, and stored under strict sanitary conditions (see above). Partial vacuums or other modified atmosphere packaging techniques are sometimes used here but extreme care must be taken to eliminate the possibility of *Clostridium botulinum* growth.

All vegetables sealed in cans or glass jars must be heat-processed (Fig. 23.6), usually at 240 or 250 °F (115.6 or 121.1 °C) and sometimes at higher temperatures to make them commercially sterile. Commercially sterile means that all disease-causing bacteria have been killed, and all bacteria and bacterial spores, which might grow out and cause spoilage under the conditions in which the product will be handled after processing, have been destroyed (Chap. 12).

The time for which a heat preserved vegetable must be processed to attain commercial sterility depends on the specific vegetable, the size of the container, the temperature at which the product is processed, the type of container (glass, metal, or plastic), whether or not the product is agitated during heating, whether the vegetable product heats by convection or conduction, and other factors.

As described with fruits (Sect. 2.1.2), freezing vegetables and vegetable products will require a temperature of 0 °F (−17.8 °C) or below in all parts, and holding under such conditions until the product is sold to the consumer. Vegetables are packaged and frozen on moving chain mesh belts in cold-air tunnels. Packages or vegetables are sometimes placed on trays, the trays placed on racks, and the product frozen as the racks are moved through cold-air tunnels. In other instances, vegetables are packaged, the packages placed on trays, and the trays placed between refrigerated metal plates for purposes of freezing the product. Some vegetables are loose-frozen on chain mesh belts in cold-air tunnels prior to packaging.

The following sections cover harvesting, processing, and general handling of the more important vegetables used in the United States (Gross et al. 2014).

3.1 Tomatoes

United States is ranked #2 in terms of global producers of tomatoes with a value (US$1.24 billion per year) and #5 in terms of overall volume (1.37 billion metric tons) compared with other vegetables. Though onions are produced in larger quantities compared to tomatoes, the value of the latter is grater (USDA Economic Research Services 2016b). The consumption of tomatoes seems likely to increase owing to the possible association of lycopene and health (Shi and Le Maguer 2000). Tomatoes are handled as the fresh product, as well as being processed as, canned whole tomatoes, tomato juice, tomato puree, tomato paste, ketchup, soup, and sauces (spaghetti, pizza, and chili sauce). Tomatoes are frozen only as an ingredient of prepared products such as pizza or cooked lasagna (Feld 1972; Thakur et al. 1996).

At the processing plant, tomatoes are washed and sorted. Green specimens are separated from the firm ripe type, and diseased specimens are discarded. Mature partially ripened tomatoes are packaged in cardboard, cellophane-topped cartons. They should be cooled to 55 to 60 °F (12.8 to 15.6 °C) and held at this temperature until sold to the consumer. In this condition, they have a storage life of 2–3 weeks.

Tomatoes to be canned should be in the firm ripe condition. They are washed and inspected, with green and rotten specimens culled out and mold and rot trimmed out from infected but otherwise good specimens. After trimming, the tomatoes are cored, that is, the pistil and stamen section is cut out. Tomatoes to be canned as whole are scalded with sprays of hot water and cooled with sprays of cold water, which facilitates peeling off the skin. Some varieties have a skin that does not require peeling. Ripened tomato has increased pH and for this reason, the addition of lemon juice is often recommended in home canning procedures.

A variety of tomato products can be found in the consumer basket. Tomato puree and tomato paste are prepared by vacuum concentrating tomato juice. By comparison, tomato ketchup or catsup is made by concentrating tomato juice, and then adding sugar and salt and then vinegar (10% acetic acid) extract of spices (headless cloves, black pepper, red pepper, cinnamon, mace, onions, garlic, etc.). Chili sauce is made from vacuum-concentrated, finely chopped, peeled, cored tomatoes to which a vinegar extract of spices and red peppers has been added, together with onions and garlic. This product is bottled and handled much in the same manner as ketchup.

Tomato soup, Italian tomato sauce, with or without meat, and other tomato products are canned and bottled to some extent. These products may require various heat treatments to attain commercial sterility, depending on the pH (acidity) and added ingredients of the finished product. Some tomato producers have incorporated genetic engineering into their products as to delay softening or rotting as quickly as familiar varieties (Thakur et al. 1996).

3.2 Cabbage

Most cabbage is handled in the fresh state or is made into sauerkraut. After harvest, cabbage should be held at 32 to 35 °F (0 to 1.7 °C) until sold to the consumer. Under these conditions, shelf life may be 3–4 months.

Cabbage to be used for sauerkraut may be held out of refrigeration until the outer leaves wilt. The cabbages are cored and the outer coarse and green leaves are removed. The trimmed and cored heads are now washed with sprays of water and then sliced into strips that have a width of about 1/32 in. (0.08 cm). The cabbage is then placed in large vats and salted at the rate of about 2.25% of the weight of the cabbage in layers to ensure even distribution. In this condition, the cabbage is then allowed to ferment for 40–90 days to allow lactic acid to be produced at the rate of at least 1.5%. Leuconostoc and lactobacillus are the main types of bacteria involved with sauerkraut fermentation (Fig. 23.7) (Battcock and Azam-Ali 1998).

Most sauerkraut is canned in either metal or glass containers. It is heated before being added to the container and then heat-processed to an internal temperature of about 200 to 210 °F (93.3 to 98.9 °C). It is then cooled and can be stored at room temperature like other commercially sterile products.

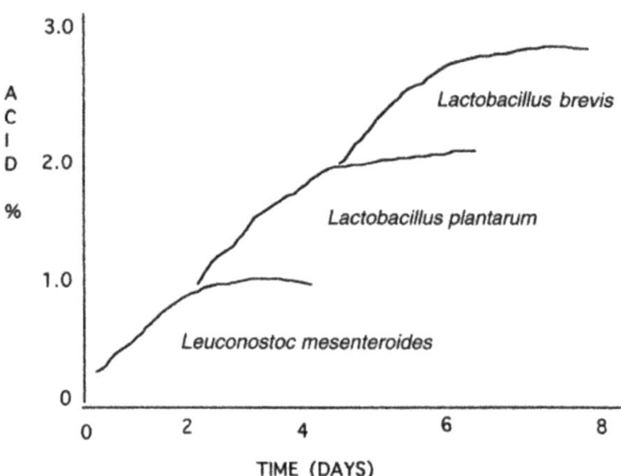

Fig. 23.7 Bacterial growth in sauerkraut production

3.3 Cucumbers

Cucumbers can be used as the fresh produce, or to make pickles and relishes. A period of 55–60 days is required from planting to harvesting. Cucumbers, grown for the fresh market, are of a variety that produces relatively large specimens. Varieties grown for processing include relatively large cucumbers used for the production of relish and medium- and small-sized specimens for pickling.

At the processing plant, cucumbers to be used as fresh produce are washed, and may be covered with a thin layer of wax and polished by machine. They should be promptly cooled to 45 to 50 °F (7.2 to 10 °C) and held at this temperature until sold to the consumer. In this condition, they have a high-quality storage life of 10–14 days. At temperatures lower than 45 °F (7.2 °C), they may develop pitting or dark-colored, watery areas.

3.4 Green and Wax Beans

Green and wax beans are planted from seed in the spring or when temperatures will allow. A period of 50–70 days is required from planting to harvesting. On harvesting, beans should be cooled to 40 °F (4.4 °C) and held at that temperature until sold as the fresh product, or until processed.

At the processing plant, beans are sorted for size, washed, and their ends are snipped off mechanically. They are then cut transversely or longitudinally (French-cut) and blanched. Some beans are frozen individually and packed in plastic pouches alone or as a component of mixed vegetables, or they may be first packaged and then frozen in a blast freezer. Some beans are canned in brine and heat-processed.

3.5 Broccoli

Broccoli requires 60–70 days from planting before it can be harvested, after which the broccoli must be cooled to, and held at, 35 to 40 °F (1.7 to 4.4 °C) until sold as fresh or processed. Cooling can be done by subjecting the product to cold-air blasts or by hydro-cooling. Most broccoli is sold as fresh, but some is frozen. Broccoli to be sold fresh is inspected to remove blossomed and insect-infested heads, washed, then bound at the stalk in small retail-sized bunches. The frozen product is prepared by splitting the heads to a smaller diameter and conveying to a station where they pass under a rotary cutter that trims the stalks to a length of 5 in. (12.7 cm). After washing, the trimmed stalks are blanched in free-flowing steam for 2.5–7 min (usually 3–5 min). The blanched product is cooled in water, and then packaged and frozen. Current US production of broccoli is valued at over US$993 million.

3.6 Lettuce

Lettuce is used entirely as a fresh product, because if heated or frozen, the leaves wilt. This is unacceptable, because lettuce is expected to be crisp and firm when eaten. There are four main varieties of lettuce in the US, the crisphead or iceberg lettuce, Romaine or cos lettuce, leaf lettuce and the butterhead or Boston lettuce (Saltveit 2014). Iceberg lettuce, which is the main lettuce for commercial use, requires a period of 75 to 80 days from planting to maturity. After harvesting, the lettuce should be conveyed quickly to pre-cooling and processing plant. At the processing plant, lettuces are trimmed, washed, and placed in crates; the crates are placed in refrigerated freight cars or trucks. The cars or trucks are then moved into large metal chambers. The chambers are closed and subjected to vacuum, which accelerates the evaporation of moisture from the lettuce, lowering the temperature of the product to about 33 °F (0.56 °C) through evaporative cooling. During shipment and until sold to the consumer, lettuce should be held at 32 to 34 °F (0 to 1.1 °C) to avoid freezing. At such temperatures lettuce has a storage life of 2–3 weeks.

3.7 Olives

Large and unblemished olives are used for canning as black or green olives. Small and blemished fruit are used for oil or for chopped olives. Olives must be treated with lye solution (sodium hydroxide) to destroy or decrease the content of a bitter substance, oleuropein. In the production of black olives, the sized, washed, and inspected fruit is placed in tanks and covered with a 1–5% lye (sodium hydroxide) solution at 60 °F (15.6 °C) for 4–8 h, the mixture being stirred occasionally. The lye solution is then drained off, and the fruit is exposed to air for 3–6 h with occasional stirring, to oxidize and set the color (to black). The fruit is then covered with water, agitated by compressed air, and held for 3–4 days. It is next drained and again treated with a 0.5–0.75% lye solution for 3–5 h, then drained and washed once more. Black olives may or may not be pitted prior to canning.

Green olives are placed in a brine (2.5–5% sodium chloride) solution and the salt increased daily until it reaches 7.5–10%. The fruit is held in this manner for 30–45 days, during which time it undergoes fermentation, owing to the growth of lactic acid bacteria. After washing without agitation in a tank of water for several days, the fruit is treated with lye solution, as in the case of black olives, and then rewashed. The fruit is inspected, graded for size, pitted or not pitted, and canned in the same manner as black olives. Olives in large containers (No. 10 cans) must be exhausted (heated) to bring the temperature in all parts to 180 °F (82.8 °C) prior to sealing the cans, then heat-processed at high temperatures.

3.8 Potatoes

Different varieties of white potatoes are used mainly as a fresh, dried, frozen, or canned. Other varieties are best suited to use as boiled or mashed potatoes. Others are processed into snack items such as chips. At the processing plant, potatoes to be sold as fresh are passed through a dry reel to remove soil, washed, and packed in paper or plastic bags of 5- or 10-lb (2.37- or 4.5-kg) capacity.

Frozen white potatoes are processed as French fries, baked stuffed, hash-browned, or as some other prepared product. To produce French fries, first the potatoes are washed, peeled in lye, and washed in a neutralizing weak acid solution. Alternatively, potatoes can be heated in steam at 80 psi (5.6 kg/cm^2) for about 10 seconds and subjected to water sprays to remove the peel. To prevent discoloration, they are then inspected to remove eyes, treated in a weak solution of citric acid and sodium bisulfite. The potatoes are then cut into strips. Deep fat frying is done for only sufficient time to give the French fried potatoes a light brown color. After cooling, the potatoes are frozen.

Baked frozen potatoes are baked "whole" with the skin on. They are then cooled and cut in half lengthwise. The inner material is then mashed and mixed with cooked onions, margarine, and various flavoring materials. The half shells are then filled with the prepared mashed product, placed in cartons, and frozen.

Hash-browned potatoes are prepared in the manner of French fried potatoes except that they are cut to a smaller size and cooked in oil for somewhat longer periods to provide a darker color. Some small-sized potatoes are canned. Potatoes preserved by drying are mostly used as mashed potatoes by rehydrating them just prior to use.

Potatoes may contain toxic glycoalkaloid compounds (Fig. 23.8). One of these is solanine, which appears as a green discoloration that may also involve the tissue below the skin. In such cases, the skin and tissue so affected should be removed. With the popularization of potato skins as entrees, the possible presence of these compounds and their side effects (headache, nausea, and diarrhea) should be considered. A cases study of solanine poisoning - relates that eating jacketed potatoes was more risky compared to boiled potatoes since water tended to disperse the alkaloid (Anonymous 1960).

4 Minimally Processed Fruits and Vegetables

4.1 Fresh-Cut Products

The rising popularity of minimally processed fruits and vegetables (MPFV) and organic certified products has been impressive in recent years (Askar 1998; Bourn and Prescott 2002; Hoover 1997; Martinez et al. 1997; Simons and Sanguansri 1997). Minimally processed products are defined by the FDA as food products which are produced without a lethal heat processing step, in order to retain an essentially fresh-like quality (U. S. Food

HO
HO
OH
OH
D-Glc
HO
OH
O
D-Gal
O
L-Rha
OH
OH
HO
CH3
H
H
N
H
H
H
H

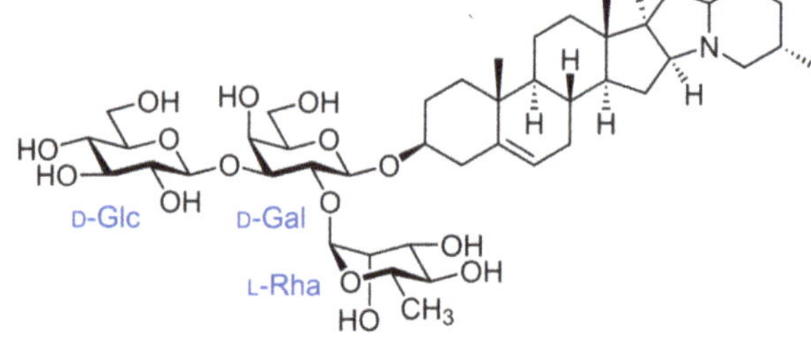

Fig. 23.8 The structure of solanine a toxic alkaloid found in potatoes and other members of the Solanaceae, (or nightshades) family including tomatoes and eggplant

Sorting & Grading
Washing
Pealing Or Equivalent
Cutting/ Slicing Etc.
Washing
Dewatering Or Optional Dehydration
Packaging

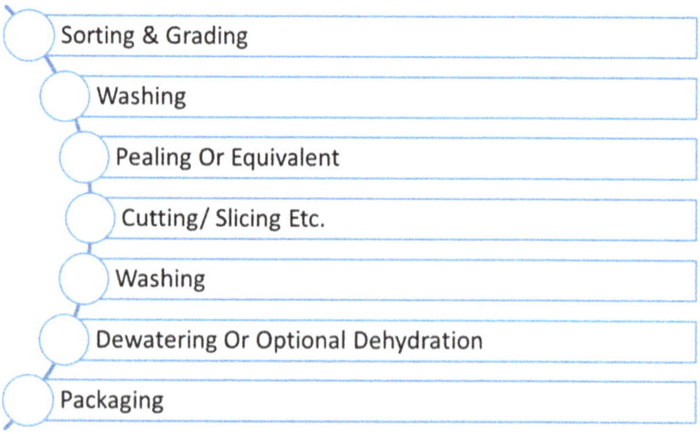

Fig. 23.9 Main unit operations for minimal processed fruits and vegetables

and Drug Administration 1998, 2008). Some commonly applied unit processes for minimally processed foods include, sorting, washing and cutting ending with a variety of packaging options (Fig. 23.9); most of the unit operations are also employed for conventional fruit and vegetable processing, but there is no blanching (Siddiqui and Rahman 2014).

The many benefits associated with minimally processed foods include, enhanced convenience, shorter preparation time, better food safety, and increased shelf life. Food packaging provides a means of protecting the product from air exposure, dehydration, and other changes that limits shelf life (Chap. 6). Packaging also helps convey information for the consumer. Such considerations lead to "added value", and help to account for the marked growth and popularity of minimally processed fruits and vegetables (Gill and Allende 2012; Pasha et al. 2014; Siddiqui and Rahman 2014).

4.2 Microbial Safety of Fruit and Vegetables

More than 70 outbreaks of food borne disease were reported for fresh fruits and vegetables in the United States from 1996 to 2006. Leafy vegetables were associated with 17% of food borne diseases compared with 16% for fruit and nuts and 21% for poultry (Centers for Disease Control and Prevention 2009). Surveillance work showed ~47% of food borne diseases incidences were attributable to vine-stalk vegetables along with fruits and nuts (Centers for Disease Control and Prevention 2011).

To retain the fresh-like character valued by consumers, only non-thermal methods for preservation should be applied to minimally processed food. The FDA issued guidelines on how to ensure food safety for minimally processed fresh-cut fruit and vegetables in 1998 and 2008 (U. S. Food and Drug Administration, 1998, 2008).

Manufacturers of fresh-cut produce are recommended to apply current Good Management Practices (cGMP) covering their premises, manufacturing equipment and fixtures, staff hygiene and training. As of 2008, the FDA did not consider Hazard Analysis Critical Control Point (HACCP) as mandatory for fresh-cut fruit and vegetable producers (U. S. Food and Drug Administration 2008). However, the principle of HACCP was also promoted to reduce major risks of food borne disease pro-actively even if HACCP is not mandatory for processing fresh cut produce. The microbiological food safety of fruits and vegetables has been discussed by a various experts (Rajwar et al. 2016). Food safety and sanitation are discussed further in Chaps. 17 and 18.

Fresh-cut produce may pose risk to the consumer because of a range of factors. Cutting fruit and vegetables disrupts the natural protective outer coverings, which prevent the ingress of microorganisms inside of intact fruits and vegetables. Cutting also releases juices that serve as nutrients for microbial growth (Koukkidis et al. 2016). The procedures for handling freshly cut products, can increase mixing and the chances of contamination by extraneous microorganisms. There is also the possibility of temperature variations as freshly cut and cooled produce is conveyed through the retail chain to the consumer. Finally, fresh fruits are not usually subjected to heat processing in order to retain a fresh-like quality wanted by consumers. There is no lethal heat process step for killing microorganisms, unlike processes such as pasteurization, blanching and canning.

There are a number of possible sources of contamination for fresh fruits and vegetables, along the Agri-food chain. First, all contact surfaces involved in harvesting, transportation, storage and postharvest handling, could be sources of contamination. Table 23.14 lists some of the bacterial pathogens identified from contaminated fruit and vegetables (Wadamori et al. 2016). Water for washing, cutting machinery, and packaging lines and any insanitary conditions encountered during processing may also pose some risks of microbial contamination.

Safe food handling reduces the risks of food borne diseases associated with fresh fruit and vegetables. Avoiding fecal contamination is pos-

Table 23.14 Listing of pathogens associated with freshly cut fruits and vegetables

Bacteria	Effects and symptoms	Sources of infection
Cyclospora	Diarrhea, appetite loss, weight loss, abdominal bloating, cramping, fever, nausea, vomiting, fatigue	Raspberries, lettuce, basil
E. coli O157:H7	Non-bloody diarrhea, hemorrhagic colitis, hemolytic uremic syndrome (HUS), thrombotic thrombocytopenic purpura (TTP)	Apple juice, coleslaw, water
Hepatitis A virus	Fever, malaise, nausea, anorexia, and abdominal discomfort	Salads, water
L. Monocytogenes	Listeriosis	Raw vegetables
Noroviruses	Gastroenteritis	Salad ingredients
Salmonella	Nausea, vomiting, abdominal cramps, fever, mild diarrhea, and headache	Fresh produce, unpasteurized fruit juice
Shigella spp	Abdominal pain, cramps, Diarrhea, fever, vomiting, bloody stools	Shredded lettuce, potato salad, green onions, parsley,

Adapted from (U. S. Food and Drug Administration 1998, 2008)

sible by avoiding or washing produce that has been in contact with compost; good practices include receiving, inspection and sorting foods, washing with good quality water, cutting and handling using hygienically sound utensils, cooling processed fruits and vegetables, and or subjecting such products to modified atmosphere packaging (MAP) to control bacterial growth (U. S. Food and Drug Administration 1998, 2008).

The challenges faced by producers of freshly cut fruits and vegetables and some modern methods for processing to ensure safety have been discussed (Rajwar et al. 2016). The following texts provide additional information on fruits and vegetables as minimally processed products (Cano et al. 2014; Gomez et al. 2011; Muhammad et al. 2012; Rodrigues and Fernandes 2013; Sinha et al. 2011).

Table 23.15 Some methods for the preservation of minimal processed products

Active packaging
Antimicrobial agents
Antioxidants
Chlorine dioxide
Edible films, edible coatings
High pressure processing
Ionizing radiation
Microwave heating*
Modified Atmosphere Packaging (MAP)
Natural preservatives
Ohmic heating*
Photosensitization
Pulsed electric fields
Pulsed light
Refrigeration
Sanitizers (food grade)
Vacuum packaging

From (Ahvenainen 1996; Gomez-Lopez et al. 2007; Gómez-López et al. 2009; Hoover 1997; King Jr. and Bolin 1989; Pasha et al. 2014; Tapia de Daza et al. 1996). *Heating is not permitted for commercially available minimally processed fruits and vegetables in the US

4.3 Deterioration and Preservation of Minimally Processed Products

Minimally processed fruit and vegetables deteriorate by similar biological, chemical or mechanical processes that apply to other foods. Metabolic changes continue with fresh-like products because they are still living. Plant cells undergo oxidative metabolism of sugars to from carbon dioxide, water and heat. In a low oxygen atmosphere, sugars are fermented to alcohol. (Pasha et al. 2014). Simple enzymatic changes may also take place in fresh produce leading to alterations in sensory properties. For instance, the action of polyphenol oxidases (PPO) leads to enzymic browning reactions. Pectinases are thought to produce changes in texture. The action of lipoxygenases produces a greeny flavor in some plant foods etc. Non-enzymic reactions for deterioration involve the two major pathways: Maillard reactions or lipid oxidation but the former is not important in non-heated products (Dauthy 1995).

The microbiological spoilage of fresh-cut produce is paramount, because these are not subjected to lethal heat treatments. To ensure product longevity the number of food spoilage bacteria may be reduced, by washing and sanitizer use (Pinela & Ferreira 2017). Other possible modes of preservation for minimally processed fruits and vegetables are shown in Table 23.15. Firstly, minimally processed foods undergo washing and cleaning with portable water and food-grade sanitizers (chlorine, hydrogen peroxide, ozone treatment) in order to reduce bacterial numbers. Once rinsed, dewatered and packaged the product are refrigerated-but not frozen.

The range of non-thermal methods of chemical perseveration include antioxidants (sulphite, ascorbic acid, metal chelators) and antimicrobial agents (Sorbic acid, lactic acid, and sulphite). The use of coatings for prolonging shelf life is now well established. Other newer and more speculative methods for preserving minimally processed products are emerging and undergoing active research such the use of UV light and other forms of irradiation (Table 23.15).

Space limitations restrict the range of commodities discussed in the chapter. Other vegetables that are popular include asparagus, lima beans, beets, Brussel sprouts, carrots, cauliflower, celery, sweet com, mushrooms, onions, green peas, sweet potatoes, spinach, squash, and turnips. All are marketed fresh and some are frozen, canned, or dried by methods described earlier.

4.4 General Conclusions and Further Reading

The literature on fruit and vegetable is very large and diverse with some very comprehensive texts available (Bates et al. 2001; Barrett et al. 2004; Cano et al. 2014; Muhammad et al. 2012; Rodrigues and Fernandes 2013; Sinha et al. 2011; Woodroof and Luh 2012). For in-depth discussions of fruit and vegetable processing, the reader is referred several excellent texts dedicated to the topic (Cano et al. 2014; Myskin and Ivanov 1986; Rodrigues and Fernandes 2013; Verma and Joshi 2000; Woodroof and Luh 2012). Some novel methods of processing and preservation are discussed by the experts including, processing of fruit juices using novel electrochemical technology (Evrendilek et al. 2012), and ultra-high pressure processing fruit juices (Wu et al. 2011). The Food and Agriculture organization, publishes a library of resources on food and vegetables, including discussions covering the use intermediate technology for small scale processing (Dauthy 1995; Fellows 1997), fermentation of fruit and vegetable products (Battcock and Azam-Ali 1998).

References

Ahvenainen R (1996) New approaches in improving the shelf life of minimally processed fruit and vegetables. Trends Food Sci Technol 7(6):179–187. https://doi.org/10.1016/0924-2244(96)10022-4

Alexander L, Grierson D (2002) Ethylene biosynthesis and action in tomato: a model for climacteric fruit ripening. J Exp Bot 53(377):2039–2055

Anonymous (1960) Solanine poisoning from potatoes. Br Med J 1:1264 [English]. Retrieved from http://www.accessdata.fda.gov/scripts/plantox/detail.cfm?id=1364

Askar A (1998) Minimally processed tropical fruits. Fruit Process 8(8):339–343

Aurore G, Parfait B, Fahrasmane L (2009) Bananas, raw materials for making processed food products. Trends Food Sci Technol 20(2):78–91

Aworh OC (2015) Promoting food security and enhancing Nigeria's small farmers' income through value-added processing of lesser-known and under-utilized indigenous fruits and vegetables. Food Res Int 76(4):986–991. https://doi.org/10.1016/j.foodres.2015.06.003

Barr T, Poole WD (1941) Precooling of strawberries. LSU Agricultural Experiment Station Reports. 679., Louisiana State University, 25pp. http://digitalcommons.lsu.edu/agexp/679

Barrett DM, Somogyi L, Ramaswamy HS (2004) Processing fruits: science and technology, 2nd edn. CRC Press. 864 pages

Bates RP, Morris JR, Crandall PG (2001) FAO Agricultural Services Bulleting. Principles and practices of small – and medium – scale fruit juice processing. Retrieved from http://www.fao.org/docrep/005/y2515e/y2515e00.htm#toc

Battcock M, Azam-Ali S (1998) Chapter 5. Bacterial fermentations. FAO Agricultural Services Bulletin No.134. Fermented fruits and vegetables: a global perspective. Retrieved from http://www.fao.org/3/x0560e/x0560e10.htm#5.6.2

Bihan H, Castetbon K, Mejean C, Peneau S, Pelabon L, Jellouli F et al (2010) Sociodemographic factors and attitudes toward food affordability and health are associated with fruit and vegetable consumption in a low-income French population. J Nutr 140(4):823–830

Blanck HM, Gillespie C, Kimmons JE, Seymour JD, Serdula MK (2008) Trends in fruit and vegetable consumption among U.S. men and women, 1994–2005. Prev Chronic Dis 5(2):A35

Bourn D, Prescott J (2002) A comparison of the nutritional value, sensory qualities, and food safety of organically and conventionally produced foods. Crit Rev Food Sci Nutr 42(1):1–34. https://doi.org/10.1080/10408690290825439

Cano MP, Barta J, Sidhu JS, Wu JSB, Sinha NK (2014) Handbook of fruits and fruit processing, 2nd edn. Wiley. 694p

Centers for Disease Control and Prevention (2009) Surveillance for foodborne disease outbreaks – United States, 2006. Morb Mortal Wkly Rep 58(22):609–615

Centers for Disease Control and Prevention (2011) Food safety epidemiology capacity in state health departments – United States, 2010. MMWR Morb Mortal Wkly Rep 60(50):1701–1704

Christensen LP (2000) Raisin grape varieties. Raisin production manual. University of California, Agricultural and Natural Resources Publication, 3393, pp. 38–47. Retrieved from http://www.ucanr.org/sites/intvit/files/24430.pdf

Christensen LP, Bianchi ML, Miller MW, Kasimatis AN, Lynn CD (1995a) The effects of harvest date on Thompson Seedless grapes and raisins. II. Relationships of fruit quality factors. Am J Enol Vitic 46(4):493–498

Christensen LP, Bianchi ML, Lynn CD, Kasimatis AN, Miller MW (1995b) The effects of harvest date on Thompson Seedless grapes and raisins. I. Fruit composition, characteristics, and yield. Am J Enol Vitic 46(1):10–16

Dauthy ME (1995) Fruit and vegetable processing. FAO Agricultural Services Bulletin No. 119 Retrieved from http://www.fao.org/docrep/v5030e/v5030e00.HTM

Diop N, Jaffee SM (2004) Fruits and vegetables: global trade and competition in fresh and processed product markets. In: Aksoy MA, Beghin JC (eds) Global agricultural trade and developing countries. World Bank Publications, Washington, DC, pp 237–257

Dwyer JT, Fulgoni VL 3rd, Clemens RA, Schmidt DB, Freedman MR (2012) Is "processed" a four-letter word? The role of processed foods in achieving dietary guidelines and nutrient recommendations. Adv Nutr 3(4):536–548

European Fresh Produce Association (2015) World production of fruits and vegetables

Evrendilek GA, Baysal T, Icier F, Yildiz H, Demirdoven A, Bozkurt H (2012) Processing of fruits and fruit juices by novel electrotechnologies. Food Eng Rev 4(1):68–87. https://doi.org/10.1007/s12393-011-9045-5

Fardet A, Richonnet C, Mazur A (2019) Association between consumption of fruit or processed fruit and chronic diseases and their risk factors: a systematic review of meta-analyses. Nutr Rev 77(6):376–387. https://doi.org/10.1093/nutrit/nuz004

Feld HN (1972) Tomatoes, tomato paste, canning and their international movements. Food Trade Rev 42(5):16–19

Fellows P (1997) Guidelines for small-scale fruit and vegetable processors. Fao Agricultural Services Bulletin No 127. Retrieved from http://www.fao.org/docrep/w6864e/w6864e00.htm#Contents

Florida Department of State (2013) Florida Administrative Code and Florida Administrative Register. Rule Chapter: 20–54. Fresh oranges–minimum ratios of solids to acid. Retrieved from https://www.flrules.org/gateway/ChapterHome.asp?Chapter=20-54

Food and Agriculture Organization of the United Nations (2017) Fruit and vegetables for health initiative. Retrieved from http://www.fao.org/publications/card/en/c/93711531-5331-4d35-9839-d6a78b2417ba/

Gill MI, Allende A (2012) Chapter 6. Minimal processing In: Gomez-Lopez M (ed) Decontamination of fresh and minimally processed produce. Wiley, pp 105–120

Gomez PL, Welti-Chanes J, Alzamora SM (2011) Hurdle technology in fruit processing. Annu Rev Food Sci Technol 2:447–465

Gomez-Lopez VM, Ragaert P, Debevere J, Devlieghere F (2007) Pulsed light for food decontamination: a review. Trends Food Sci Technol 18(9):464–473

Gómez-López VM, Rajkovic A, Ragaert P, Smigic N, Devlieghere F (2009) Chlorine dioxide for minimally processed produce preservation: a review. Trends Food Sci Technol 20(1):17–26

Gross KC, Wang CY, Saltveit M (eds) (2014) The commercial storage of fruits, vegetables, and florist and nursery stocks. USDA-ARS Agriculture Handbook Number 66 (HB66). U.S. Department of Agriculture, Agricultural Research Service, Washington, DC, 792pp. http://ucanr.edu/datastoreFiles/234-2927.pdf

Hoover DG (1997) Minimally processed fruits and vegetables: reducing microbial load by nonthermal physical treatments. Food Technol 51(6):66–69

Hui YH (2006) Handbook of fruits and fruit processing. Blackwell Publishing. 688 pages

Johnson TM (2001) Citrus juice production and fresh market extension technologies. China/FAO Citrus symposium 14–17 May 2001, Beijing, People's Republic of China. Retrieved from http://www.fao.org/docrep/003/X6732E/x6732e08.htm#9

Kader AA, Perkins-Veazie P. Lester GE (2014) Nutritional quality and its importance to human health. The commercial storage of fruits, vegetables, and florist and nursery stocks. USDA-ARS Agriculture Handbook Number 66 (HB66). Retrieved from http://www.ba.ars.usda.gov/hb66/title.html

Kaiser KA, Brown AW, Bohan Brown MM, Shikany JM, Mattes RD, Allison DB (2014) Increased fruit and vegetable intake has no discernible effect on weight loss: a systematic review and meta-analysis. Am J Clin Sci 100(2):567–576

King AD Jr, Bolin HR (1989) Physiological and microbiological storage stability of minimally processed fruits and vegetables. Food Technol 43(2):132–135

Koukkidis G, Haigh R, Allcock N, Jordan S, Freestone P (2016) Salad leaf juices enhance salmonella growth, colonization of fresh produce, and virulence. Appl Environ Microbiol 83(1)

Lacey K (2020) Measuring internal maturity of citrus. Retrieved from https://www.agric.wa.gov.au/citrus/measuring-internal-maturity-citrus?page=0%2C1

Lelievre JM, Latche A, Jones B, Bouzayen M, Pech JC (1997) Ethylene and fruit ripening. Physiol Plant 101(4):727–739

Liu RH (2014) Health-promoting components of fruits and vegetables in the diet. Adv Nutr 4(3):384S–392S

Lock K, Pomerleau J, Causer L, Altmann DR, McKee M (2005) The global burden of disease attributable to low consumption of fruit and vegetables: implications for the global strategy on diet. Bull World Health Organ 83(2):100–108

Martinez A, Fernandez PS, Rodrigo E, Rodrigo MC (1997) Minimal thermal processes for fruit and vegetables. Eur Food Drink Rev 39:41–42

Mason-D'Croz D, Bogard JR, Sulser TB, Cenacchi N, Dunston S, Herrero M, Wiebe K (2019) Gaps between fruit and vegetable production, demand, and recommended consumption at global and national levels: an integrated modelling study. Lancet Planet Health 3(7):e318–e329. https://doi.org/10.1016/s2542-5196(19)30095-6

Muhammad S, Jasim A, Lobo MG, Ozadali F (2012) Tropical and subtropical fruits: postharvest physiology, processing and packaging. Wiley, 648 pages

Myskin MM, Ivanov SV (1986) Technology of fruit, berries and vegetable processing

Newilah GN, Tchango JT, Fokou É, Etoa F-X (2005) Processing and food uses of bananas and plantains in Cameroon. Fruits 60(4):245–253

Nicklett EJ, Kadell AR (2013) Fruit and vegetable intake among older adults: a scoping review. Maturitas 75(4):305–312

Pasha I, Saeed F, Sultan MT, Khan MR, Rohi M (2014) Recent developments in minimal processing: a tool to retain nutritional quality of food. Crit Rev Food Sci Nutr 54(3):340–351. https://doi.org/10.1080/10408398.2011.585254

Payasi A, Sanwal GG (2010) Ripening of climacteric fruits and their control. J Food Biochem 34(4):679–710

Pech JC, Balague C, Latche A, Bouzayen M (1994) Postharvest physiology of climacteric fruits: recent developments in the biosynthesis and action of ethylene. Sci Aliment 14(1):3–15

Pinela J, Ferreira IC (2017) Nonthermal physical technologies to decontaminate and extend the shelf life of fruits and vegetables: trends aiming at quality and safety. Crit Rev Food Sci Nutr 57(10):2095–2111. https://doi.org/10.1080/10408398.2015.1046547

Plessz M, Gojard S (2013) Do processed vegetables reduce the socio-economic differences in vegetable purchases? A study in France. Eur J Pub Health 23(5):747–752

Pyke M (1981) Food science and technology

Rajwar A, Srivastava P, Sahgal M (2016) Microbiology of fresh produce: route of contamination, detection methods, and remedy. Crit Rev Food Sci Nutr 56(14):2383–2390

Reid M (2015) Biology of ethylene production and action in fruits. Retrieved from http://ucanr.edu/datastoreFiles/234-2811.pdf

Rodrigues S, Fernandes FAN (2013) Advances in fruit processing technologies. CRC Press. 472 pages

Rosenstock T (2007) Postharvest technology and methods for grapes and raisins: a manual. International Programs Office, College of Agricultural and Environmental Sciences, University of California, Davis. Retrieved from http://afghanag.ucdavis.edu/a_horticulture/fruits-trees/grapes/manuals/Posdtharvest_manual_grapes.pdf

Saltveit ME (2014) Lettuce. The commercial storage of fruits, vegetables, and florist and nursery stocks. USDA-ARS Agriculture Handbook Number 66 (HB66) (Gross KC, Wang CY, Saltveit M, eds). U.S. Department of Agriculture, Agricultural Research Service, Washington, DC, pp 386–389. http://ucanr.edu/datastoreFiles/234-2927.pdf

Salunkhe DK, Kadam SS (1995) Handbook of fruit science and technology: production, composition, storage, and processing. Taylor & Francis. 632 pages

Sam C, Figueroa CR, Helen N (2014) "Movers and shakers" in the regulation of fruit ripening: a cross-dissection of climacteric versus non-climacteric fruit. J Exp Bot 65(17):4705–4722

Savoy CF, Morris JR, Petrucci VE (1983) Processing muscadine grapes into raisins. Proc Florida State Horticult Soc 96:355–357

Shi J, Le Maguer M (2000) Lycopene in tomatoes: chemical and physical properties affected by food processing. Crit Rev Food Sci Nutr 40(1):1–42. https://doi.org/10.1080/10408690091189275

Siddiqui MW, Rahman MS (2014) Minimally processed foods: technologies for safety, quality, and convenience. Springer. 306 pages

Simons LK, Sanguansri P (1997) Advances in the washing of minimally processed vegetables. Food Australia 49(2):75–80

Singh R, Kaushik R, Gosewade S (2018) Bananas as underutilized fruit having huge potential as raw materials for food and non-food processing industries: a brief review. Pharma Innov J 7(6):574–580

Sinha NK, Hui YH, Evranuz EO, Muhammad S, Jasim A (2011) Handbook of vegetables and vegetable processing. Wiley. 772 pages

Statistics Canada (2017) North American Industry Classification System (NAICS) 2007: classification 44523 fruit and vegetable markets. Retrieved from http://stds.statcan.gc.ca/naics-scian/2007/cs-rc-eng.asp?criteria=44523

Tapia de Daza MS, Alzamora SM, Chanes JW (1996) Combination of preservation factors applied to minimal processing of foods. Crit Rev Food Sci Nutr 36(6):629–659

Teskey BJE, Shoemaker JS (1978) Tree fruit production. AVI Publisher Co.. 336 pages

Thakur BR, Singh RK, Nelson PE (1996) Quality attributes of processed tomato products: A review. Food Rev Intl 12(3):375–401

U. S. Food and Drug Administration (1998) Guidance for industry: guide to minimize microbial food safety hazards of fresh-cut fruits and vegetables. Retrieved from http://www.fda.gov/Food/GuidanceRegulation/GuidanceDocumentsRegulatoryInformation/ucm064574.htm

U. S. Food and Drug Administration (2008) Guidance for industry: guide to minimize microbial food safety hazards of fresh-cut fruits and vegetables -February 2008. Retrieved from https://www.fda.gov/regulatory-information/search-fda-guidance-documents/guidance-industry-guide-minimize-microbial-food-safety-hazards-fresh-cut-fruits-and-vegetables#ch2

U.S. Census Bureau (2016, Dec) North American industry classification system. Retrieved from http://www.census.gov/eos/www/naics/

U.S. Department of Agriculture and U.S. Department of Health and Human Services (2010) Dietary guidelines for Americans, 2010. 7th Edition. Retrieved from http://www.cnpp.usda.gov/DietaryGuidelines

U.S. Department of Health and Human Services and U.S. Department of Agriculture (2015) 2015 – 2020 dietary guidelines for Americans. 8th Edition. December 2015. Retrieved from http://health.gov/dietaryguidelines/2015/guidelines/

USDA (1965) How to make jellies, jams and preserves at home. Home and gardens bulletin no. 56. Retrieved Sept, 2020, from https://naldc.nal.usda.gov/download/CAT40000392/PDF.

USDA. (2014) 2012 census of agriculture: U.S. summary and state reports. Retrieved from https://www.agcensus.usda.gov/Publications/2012/

USDA Economic Research Services (2016a, Nov 06) Vegetables & pulses – market outlook. Retrieved from https://www.ers.usda.gov/topics/crops/vegetables-pulses/

USDA Economic Research Services (2016b, Nov 06) Vegetables & pulses – tomatoes. Retrieved from https://www.ers.usda.gov/topics/crops/vegetables-pulses/tomatoes.aspx

USDA Economics Statistics and Market Information System (2016) U.S. agricultural trade update: economic research service. Retrieved from http://usda.mannlib.cornell.edu/MannUsda/viewDocumentInfo.do?documentID=1412

USDA National Agricultural Statistics Services (2016) Vegetables 2015 summary: February 2016. Retrieved from http://usda.mannlib.cornell.edu/usda/current/VegeSumm/VegeSumm-02-04-2016.pdf

Verma LR, Joshi VK (2000) Postharvest technology of fruits and vegetables: handling, processing, fermentation and waste management. Indus Publishing. 1222 pages

Wadamori Y, Gooneratne R, Hussain MA (2016) Outbreaks and factors influencing microbiological contamination of fresh produce. J Sci Food Agric 97:1396–1403. https://doi.org/10.1002/jsfa.8125. [Epub ahead of print]

Whitehead RD, Coetzee V, Ozakinci G, Perrett DI (2012a) Cross-cultural effects of fruit and vegetable consumption on skin color. Am J Public Health 102(2):212–213

Whitehead RD, Ozakinci G, Stephen ID, Perrett DI (2012b) Appealing to vanity: could potential appearance improvement motivate fruit and vegetable consumption? Am J Public Health 102(2):207–211

Woodroof JG, Luh BS (2012) Commercial fruit processing. Springer. 678 pages

Wu H, Zhang Y-C, Han Q-H, Zhao Y-B (2011) Research development of ultra-high pressure processing on fruit juice. Packag Food Mach 29(1):42–45

Sugar and Confectionery Products 24

1 Introduction

Sugar is the common name for sucrose, which is extracted and refined from sugar cane and sugar beet. There are many substances classified as sugars, and when these are referred to, they are always used with a qualifier such as in milk sugar (lactose), corn sugar (dextrose), and malt sugar (maltose). When the word sugar is used without a qualifier, it refers to the common sweetener (sucrose) or table sugar. Additional sources of sucrose include maple trees and some fruits. The aims of this chapter is to introduce the sugar and confectionery industry (Edwards 2003; Wolf 2016).

1.1 The Sugar and Confectionery Industry

The sugar and confectionery manufacturing industry (NAICS 3113) consists of companies that produce sugar from cane or sugar beet. For the Unites States (US), the sector covers non-chocolate confectionery, chocolate confectionery from cocoa beans and confectionery using purchased chocolate; the corn syrup industry (NAICS 311930) is also relevant (Table 24.1). The United Nations industry code ISIC 1073 applies to the manufacture of cocoa, chocolate and sugar confectionery. The range of products include, cocoa butter, cocoa fat, cocoa oil, chocolate and chocolate confectionery, sugar confectionery: caramels, cachous, nougats, fondant, white chocolate, chewing gum, preserving in sugar of fruit, nuts, fruit peels and other parts of plants, confectionery lozenges and pastilles.

The synthetic sweeteners industry is classed within basic organic chemical manufacturing (NAICS 325199) and manufacturing of so-called "miscellaneous chemical product and preparations (NAICS 325998) which are fascinating areas though outside of the scope of the current discussion.

1.2 Commercial Sugar Supplies

Currently, 80% of the world's sugar is obtained from sugar cane and 20% from sugar beet and other sources (Fig. 24.1). In 2014, the world total production for cane sugar was 175 million metric tons. The major producers of sugar include Brazil (33.8 million tons), India (27.2 million metric tons), EU (16.3 million metric tons), China (13.3 million metric tons), Thailand (10.2 million metric tons), US (7.6 million metric tons), Mexico (6.5 million metric tons) and Pakistan (4.7 million metric tons) (Table 24.4). The market for sugar was reported to be worth about US$40–70 billion dollars (USDA Foreign Agricultural Services 2014).

R. Owusu-Apenten, E. R. Vieira, *Elementary Food Science*, Food Science Text Series,
https://doi.org/10.1007/978-3-030-65433-7_24

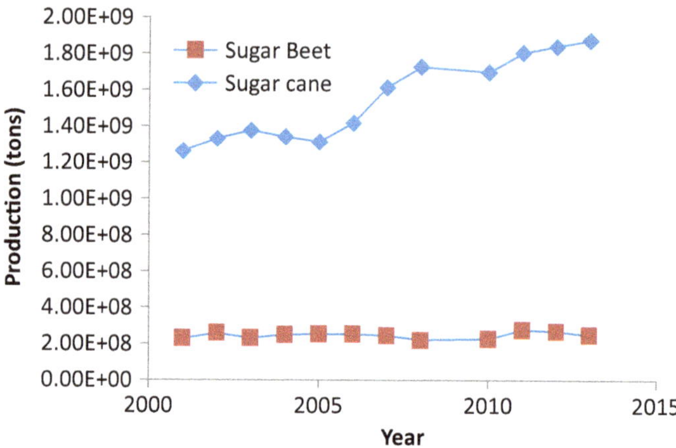

Fig. 24.1 World sugar cane and sugar beet production. (FOAOSTAT data, 2014)

1.3 Uses of Sugar

1.3.1 Technical uses

Chemically, and in every other way, cane sugar and beet sugar are the same. In addition to providing energy for the body and sweetness to foods, sugar performs numerous other roles in the food industry. Sugar is used in baked products where it contributes to the desirable texture of baked goods, and it stabilizes the foam of beaten egg whites. When it caramelizes, it imparts a unique, desirable color and flavor to surfaces of pastries and cakes. It is used in ice cream and other dairy products, in beverages, and in other types of food; it is used in the home, in institutions, and in restaurants for foods and beverages. Sugar is also used in some nonfood products such as pet food and other animal feeds and baits (Jeffery 1993).

One of the most obvious uses of sugar is as a sweetener but, not all sugars are equally sweet (see Table 24.2). Some forms of sugar (lactose) are only 16% as sweet as sucrose. Sugars may also be added for texture and appearance qualities (Table 24.3). Sugar interaction with water can contribute to viscosity, which is important in the consistency, body, and mouth feel of some foods (Table 24.3). Sugar interactions with water also underlie the role as preservative and antimicrobial agent. The availability of water to microorganisms is measured as water activity (Aw) and sugars, with their affinity for water, depress Aw in some foods such as jams and jellies, giving a preservative effect. Sugar also has a high refractive index and is responsible for the shiny appearance in high sugar products such as syrups, jellies, and dried fruits. Sugars may also possess some antioxidant value by virtue of their reducing properties or role as metal chelators though this role has yet to be characterized in more detail.

Table 24.1 Sugar and confectionery manufacturing

NAICS	Sector
31–33	Manufacturing
311	Food manufacturing
3113	Sugar and confectionery product manufacturing
31131	Sugar manufacturing
311313	Beet sugar manufacturing
311314	Cane sugar manufacturing
31134	Nonchocolate confectionery manufacturing
31135	Chocolate and chocolate confectionery manufacturing
311351	Chocolate and confectionery from cacao beans
311352	Confectionery manufacturing from purchased chocolate
325199	Synthetic sweeteners (i.e., sweetening agents) manufacturing
325998	Sugar substitutes (i.e., synthetic sweeteners blended with other ingredients)

Table 24.2 Degree of sweetness of sugars and other sweeteners

Sweetener	Degree of Sweetness
Sucrose	100
Fructose	173
Glucose	74
lactose	16
Maltose	32
Galactose	32
Saccharine	30,000-50,000
Sodium cyclamate	10,000
Neohesperidin dihydrochalcone	1,000,000

Table 24.3 Functional properties of sugars

Sensory function	Physical function
Sweetener	Solubility
Flavor	Freezing point depression
Texture and viscosity	Boiling point elevation
Tenderizer	Crystallization
Appearance	Glass formation
Bulking agent	Increases volume or weight

1.3.2 Consumption trends

The consumption of sugar is regarded as a public health issue, owing to the reported link with chronic illness such as diabetes. Current research indicates that the global sugar consumption rate will grow by about 1.4% per year for a decade (2017-2027). Alongside of nutrition transitions, the rise in sugar consumption is predicted to be higher for developing countries in Asia (China, Indonesia, and Pakistan) and Africa. By contrast, sugar consumption is expected to remain stagnant in those markets where purchases are already high. The per capita sugar consumption was highest for Brazil (67 kg) compared with the moderate sugar intake in countries such as the US (33 kg) to very low levels (<20 kg) in many low-income countries. There are calls to reduce sugar intake in many developed countries (Wolf 2016). Current per capita sugar consumption for some countries are shown Appendix 1 (OECD/ Food and Agricultural Organization 2018).

Table 24.4 World production of sugar (2014)

World Centrifugal Sugar						
	2010/11	2011/12	2012/13	2013/14	May 2014/15	Nov 2014/15
Production (x1000)						
Brazil	38,350	36,150	38,600	37,800	36,800	35,800
India	26,574	28,620	27,337	26,605	27,900	27,250
EU	15,939	18,320	16,655	16,010	16,300	16,300
China	11,199	12,341	14,001	14,263	13,700	13,300
Thailand	9663	10,235	10,024	11,333	11,000	10,200
US	7104	7700	8148	7672	7706	7677
Mexico	5495	5351	7393	6383	6890	6508
Pakistan	3920	4520	5000	5215	4860	4700
Australia	3700	3683	4250	4400	4400	4600
Russia	2996	5545	5000	4400	4400	4200
Guatemala	2048	2499	2778	2852	2850	2850
Indonesia	1770	1830	2300	2300	2500	2500
Philippines	2520	2400	2400	2450	2500	2500
Turkey	2274	2262	2130	2300	2300	2400
Colombia	2280	2270	1950	2300	2300	2300
South Africa	1985	1897	2020	2435	2500	2200
Argentina	2030	2150	2300	1780	2000	2050
Egypt	1830	1980	2000	2013	2050	2050
Cuba	1150	1400	1525	1600	1500	1650
Ukraine	1540	2300	2400	1300	1600	1600
Other	17,822	18,844	19,346	19,599	19,533	19,823
Total	162,189	172,297	177,557	175,010	175,589	172,458

Source: USDA Foreign Agricultural Services (2014). "Sugar: World Markets and Trade, November 20, 2014: World Production, Markets, and Trade Reports." Retrieved Dec 2014, from http://www.fas.usda.gov/data/sugar-world-markets-and-trade

2 Cane Sugar Manufacturing

As noted previously, most of the world's supply of white sugar is from the sugar cane or sugar beet plant. Globally the quantity of sugar cane produced in 2013 was 1887 million metric tons compared with a far smaller amount of 230 metric tons of sugar beet (Fig. 24.1).

2.1 Sugar Cane

The sugar cane plant is a species of giant grass belonging to the genus *Saccharum*. Although nearly all sugar canes are of the same species, differences in growing conditions affect the characteristics of the juices. For example, the sugar content of the juices from sugar cane grown in the tropics is higher than in sugar canes grown in cooler climates. In the US, sugar cane is grown primarily in Louisiana, and some is grown in Florida and Hawaii. Cuba, Puerto Rico, the Virgin Islands, the Philippines, and other countries also produce sugar cane.

Sugar cane is grown by planting cuttings from the stalk, each containing a bud. The length of time that the cane is allowed to grow before harvesting varies in different countries and may be from 7 months to 2 years. The yield of sugar from cane juice is about 14–17%. Sugar cane is harvested by cutting the stalks just above the ground. At this time, the tops of the stalks are cut off, because they contain high concentrations of an enzyme that hydrolyzes and greatly reduces the yield of cane sugar. Also at the time of harvesting, the leaves are stripped from the canes, although they may be burned off prior to harvesting. Parts of the sugar cane other than tops contain some of the enzyme that converts sucrose, so the cane must be processed shortly after harvesting in order to obtain maximum yields.

2.2 Processing Cane Sugar

At processing plants producing raw sugar the cane first passes through shredders and then through three to seven roller mills that press out the juice. After the first pressing, the bagasse (pressed cane) may be mixed with hot water or dilute hot cane juice and again pressed to extract more sugar. Following the final pressing, the bagasse is usually brought directly to the boilers, where it is used as fuel. The wet bagasse is reported to have about one-fourth the fuel value of ordinary fuel oil. However, bagasse is a wood-like fiber, and it is used to manufacture wallboard. It has been the object of study for use in other applications.

The juice, which is dark green in color, has a pH of about 5.2. After extraction, the cane juice is strained to remove pieces of stalk and other detritus, and a mixture of lime powder and water [source of calcium hydroxide, $Ca(OH)_2$] is added to raise the pH. The juice is heated to precipitate and remove impurities. The limed mixture is then held in tanks, where the lime-impurities mixture is allowed to settle out, and the clear juice is separated from the sediment (Fig. 24.2).

The clear juice is then heated (at temperatures below the boiling point of water) under vacuum to evaporate water and concentrate the sugar to the point where there is a mixture of sugar crystals and molasses. More syrup may be added to

Fig. 24.2 Production of raw sugar from sugar cane

Sugar Cane Shred
Press → Residue <Bagasse
Raise pH with Lime
Heat
Allow to Settle in Settling Tank
Bagasse< Usually Used as fuel
Crystallize Clear Juice! under Heat and Vacuum
Centrifuge and Wash → Supernatant
Brown Unpurified Sugar

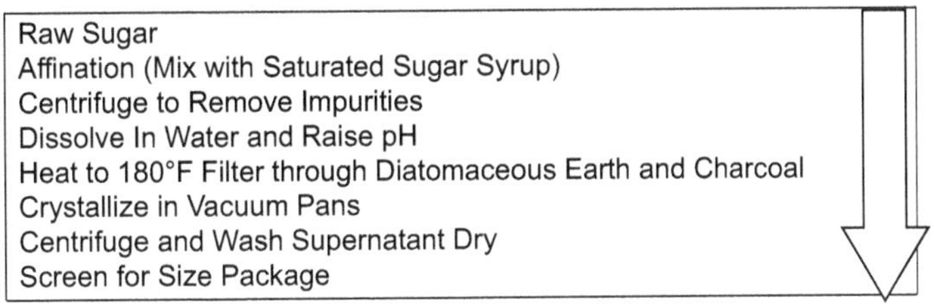

Fig. 24.3 Refining of sugar

Raw Sugar
Affination (Mix with Saturated Sugar Syrup)
Centrifuge to Remove Impurities
Dissolve In Water and Raise pH
Heat to 180°F Filter through Diatomaceous Earth and Charcoal
Crystallize in Vacuum Pans
Centrifuge and Wash Supernatant Dry
Screen for Size Package

the evaporation pans, as the syrup is concentrated. The mixture (called massecuite) is centrifuged to obtain the brown purified sugar. The liquid centrifuged from the sugar crystals still contains dissolved sugar and is returned to the evaporation pans. When the molasses from the centrifuge treatment reaches such a low concentration of sugar that removal of sucrose is uneconomical, it is called blackstrap. This is not discarded, but is not returned to the evaporation pans. Generally, blackstrap is sold to the fermentation industries for the production of rum. Sugar may be purified or refined at the plant that manufactures the raw sugar, but usually the raw sugar is packed in jute bags and shipped to sugar refineries. Higher grades of molasses for domestic use may be made from juice obtained from cane that is not pressurized sufficiently to extract all the liquid. The molasses is heated to 160 °F (71.1 °C) and canned or bottled hot. The containers are sealed, heated in water at 185 to 200 °F (85 to 93.3 °C) for 10 min, and immediately cooled.

2.3 Cane Sugar Refining

Raw sugar delivered to the refinery contains 97–98% sucrose. The first step in refining (Fig. 24.3) consists of mixing the raw sugar with hot saturated sugar syrup. This softens the film of impurities that envelop the sugar crystals. This mixture is then centrifuged, and during centrifugation, the sugar crystals are sprayed with water to remove some of the impurities. The washed sugar crystals are then dissolved in hot water, treated with lime to bring the pH to 7.3 to 7.6, and the temperature is raised to 180 °F (82.2 °C). The coarser colloidal impurities in the hot mixture are filtered out through diatomaceous earth or paper pulp. The coloring material in the sugar solution is removed by filtering the hot liquid through bone charcoal, after which the sugar is crystallized out in vacuum pans and centrifuged to separate it from the liquid. During centrifugation, the crystals are washed with water. They are then dried, screened for crystal size, and packaged.

The finished dried sugar, which has an indefinite shelf life, is made available in different grades. Large-grained sugar is used for the manufacture of candy and other prepared sweet products. Ordinary table sugar is made up of fine-sized grains. Ultrafine sugar grains (for confectioner's sugar) are produced by grinding the crystals in pulverizing hammer mills. To prevent caking in confectioner's sugar, about 3% cornstarch is used. There are other intermediate grades. Sugar is also prepared as cubes or tablets by forming a mixture of sugar crystals and white sugar syrup under pressure, followed by drying. A variety of "soft sugars" ranging in colors from white through various shades of brown, are produced. These sugars are allowed to retain some of the molasses, which provides their unique flavor.

Raw sugar may contain thermophilic bacteria (spore-forming bacteria that grow at high temperatures-as high as 170 °F (76.7 °C). If such bacteria are allowed to grow to high concentrations, they may serve as a source of contamination of such products as canned foods to which sugar is added. The spores of thermophilic bacteria are very difficult to destroy by heat, and canned foods can undergo some deterioration during heat processing. For this reason, during the various filtrations in which sugar solutions are held at high temperatures for comparatively long periods, they should be held at temperatures high enough to prevent the growth of thermophilic bacteria (about 185 °F [85 °C]).

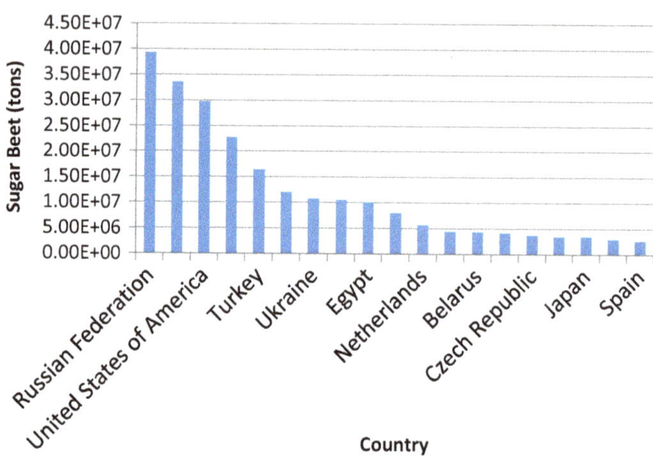

Fig. 24.4 World production of sugar beet in 2013. (FAOSTAT data)

3 Beet Sugar Manufacturing

3.1 Sugar Beet

The sugar beet (*Beta vulgaris)* stores its sugar in the root, unlike the sugar cane, which stores its sugar in the stalk. Another difference between cane sugar production and beet sugar production is that the latter is a continuous operation and does not produce the intermediate raw sugar. Whereas sugar beets contain 16–20% of sugar as sucrose, the yield of sugar per acre is considerably less than that obtained from sugar cane, owing to the quantity of beets or cane harvested per acre.

Sugar beets are planted as seed and require 70 days or more from planting to harvesting. Because the plants are subject to bacterial or mold infection and to infestation with aphids, maggots, and other insects, the plants may need spraying during the growing season. Worldwide the total production quantities for sugar beet did not change drastically from 2001 (230 metric tons) and 2013 (250 metric tons). The top producer nations for sugar beet are the Russian Federation, France and the US (Fig. 24.4). Within the US, beets for sugar are grown mainly in Colorado, California, Michigan, Utah, Idaho, Nebraska, and Montana. Sugar beets are harvested and topped mechanically and brought to the processing plant in bulk by freight cars or trucks. They may be stored outside the processing plant in large piles until treated to extract sugar.

3.2 Processing Beet Sugar

In extracting sugar from beets, the beets are first thoroughly washed to remove mud and stones, and then passed through mechanical slicers that slice them into thin shreds called cossettes. They are then covered with hot water to extract the sugar. Extraction is done even more quickly and effectively by a continuous countercurrent extraction with water. The juice from the extraction process is treated with lime or calcium hydroxide and then with carbon dioxide. The lime removes impurities as a precipitate, and the carbon dioxide is used to precipitate the calcium hydroxide as calcium carbonate ($CaC0_3$). The juice is then filtered and again treated with carbon dioxide to precipitate residual calcium hydroxide. After a second filtration, the extract is treated with sulfur dioxide to bleach out colored components in the liquid. The liquid extract is vacuum concentrated to 60–70% soluble solids and filtered through bone charcoal. Next, the filtered liquid is concentrated in vacuum pans to form crystalline sugar. The sugar is washed as it is centrifuged, dried, screened, and packaged, as in the case of cane sugar. In some manufacturing processes for beet sugar, the liquid from the third centrifugation is used for the manufacture of monosodium glutamate, a flavor enhancer used for many food preparations, including soups, gravies, Chinese foods, and meat dishes. Extracted beet pulp and the tops are dried and used as cattle feed.

Table 24.5 Classification of corn syrups by dextrose equivalent (DE)

Type	DE Value
Low conversion	28–38
Regular conversion	38–48
Intermediate conversion	48–58
High conversion	58–68
Extra high conversion	68–100

4 Other Sugar Sources

Sucrose can be obtained from the sap of a variety of palm trees, one of the most important being the date palm *(Phoenix sylvestris)*. Much of this sucrose is obtained in the Middle East by methods that involve boiling in open kettles, after which there is separation of crystals from molasses, or the unseparated mass may be allowed to set into a whole sugar.

4.1 Maple Syrup

In the northern part of North America, sucrose is obtained from the sap of the hard maple tree *(Acer saccharinum)*. Although the maple syrup sap is largely sucrose, it contains unique impurities that impart to it (when concentrated) a special delicate flavor that makes the natural maple syrup a valuable flavoring for certain preparations. Sucrose can also be produced from the cane of the sorghum plant, which is related to the sugar cane but resembles the corn plant.

4.2 Corn Syrup or Glucose Syrup

Although corn is not a source of sucrose, the value of corn sugar as a sweetener in the food industry makes it worthy of mention. It is noteworthy that the term corn syrup (North America) is often replaced with "glucose syrup" or simply "glucose" in other parts of the world, and chemical glucose is called dextrose in industry. The characteristics of corn syrup and its significance in the food industry have been recently discussed (Hull 2011).

Corn syrup is produced by heating starch in water acidified with hydrochloric acid. In general, the hydrolysis of starch is only partially completed so that the mixture contains some dextrose, maltose, and some longer chains of glucose (dextrin). The extent of starch hydrolysis for different glucose syrups is measured in terms of dextrose equivalents – or the extent to which syrup reacts with Fehling's regent for reducing sugars. A glucose syrup sample that reacts with Fehling's solution to the same extent as an equal weight of dextrose is designated as "dextrose equivalent value" DE equivalent to 100 (Table 24.5). Glucose syrup with DE of 48 is referred to as confectionery grade.

Corn syrups are divided into five commercial classes, depending on their DE value. By this classification scheme, pure dextrose is given a value of 100 DE. In addition to classification by their DE values, syrups are also classified by their solids contents.

A second method is also used for the manufacture of corn sugar is where the hydrolysis is carried out by first heating the starch with acid followed by treating it with an enzyme that hydrolyzes starch. The latter method produces syrup higher in maltose than does the straight-acid hydrolysis method. In this method, the use of amylolytic enzymes (alpha- and beta-amylases) is employed. Beta-amylase is specific in its action, attacking starch molecules at their non-reducing ends and causing progressive breaks in the molecule at 2-carbon intervals, with the resultant release of units of maltose (a 2-carbon sugar). Thus, by this conversion, the corn syrup will have high maltose content and be sweeter compared to the acid-only hydrolyzed corn sugar. Other enzymes, glucosidases, may also be used to supplement either of the two conversion processes. In this case, the syrup will contain significant amounts of glucose (Aiyer 2005) (Table 24.5).

After hydrolysis, the syrup is neutralized by addition of sodium carbonate, then filtered and concentrated to 60% solids, again filtered through bone charcoal, and finally passed through resins (ion exchange) that take out the salt (sodium chloride formed from the acid and sodium carbonate). The corn syrup may be spray or drum dried to about 3% moisture to obtain corn syrup solids. More completely hydrolyzed syrup, after purification, may be concentrated, seeded with fine corn sugar crystals, and crystallized to produce crude corn sugar. Dextrose or corn sugar can be produced in a similar manner from a completely hydrolyzed starch.

In another process, cornstarch is hydrolyzed to produce glucose, and then a percentage of the glucose molecules are converted into fructose. This high-fructose corn syrup (HFCS) has had a significant impact on the soft drink and confection industry. HFCS has been substituted either wholly or partially for sucrose in many formulas, allowing the manufacturers to meet their mammoth product demand with less dependence on the sugar cane industry. Corn syrup, corn sugar, and HFCS are used in bakery products, pharmaceuticals, carbonated beverages, confectioneries, ice cream, jams and jellies, meat products, and dessert powders.

Corn syrup and corn sugar are used largely in the food industry. They are used to supplement sucrose because they are less expensive, while at the same time nearly as effective as sucrose in sweetening characteristics. In addition, they inhibit crystallization of sucrose, especially when their maltose content is high. They are especially useful in the baking and brewing industries because of the quick and complete fermentability of dextrose. Their use in preserves minimizes oxidative discoloration, and other unique properties make them useful in many other applications. The vital role played by glucose syrup in the manufacturing of confectionery is outlined in the next section. Possible health concerns associated with increasing HFCS intake remain controversial (Parker et al. 2010).

4.3 Invert Sugar

Sucrose ($C_{12}H_{22}0_6$) is one of the most important of the naturally occurring disaccharides. It may be hydrolyzed, yielding glucose (dextrose) and fructose (levulose) which are both six-carbon sugars. The mixture of dextrose and levulose is called invert sugar. Invert sugar is made by treating sucrose with an enzyme (invertase) and some acid, whereupon it combines chemically with approximately 5.25% of its weight of water. The temperature must not be so high as to inactivate the enzyme invertase.

Dextrose is not as sweet as sucrose but the levulose is sweeter than sucrose with the result that the final syrup, invert sugar, is slightly sweeter than the syrups having the same concentration of sucrose. Invert sugar has special uses in the food industry. Mixtures of invert sugar and sugar are more soluble than sugar alone. In a proportion of 1:1, a mixture of invert sugar and sugar is more soluble than any other combination of their mixtures.

In candy making, invert sugar is produced directly by heating sugar syrup with acid added to the mix. Confectionery manufacturers prefer cream of tartar (tartaric acid) to promote hydrolysis of sucrose to invert sugar. More frequently, this technology is being replaced by the use of corn syrup. Other sugar products used in candy making include maple sugar, brown sugar (a less refined sugar from the cane sugar refining process), molasses (also from the cane sugar refining process), and honey.

5 Sugar Confectionery

5.1 Global Confectionery Sales

The confectionery sector may be divided into three major categories; (i) products with sugar as the main ingredient, (ii) or those based on chocolate and (iii) a flour based confectionery that are sweetened and baked (Edwards 2003; Hartel et al. 2017). Sugar confectionary products may be further divided into two sub-groups, i.e. those

Table 24.6 Classification of sugar confectionery

Non-crystalline groups	Sugar crystallized products
Boiled sweets	Creams and fudges
Gums and pastilles	Fondants
Jellies	Liquorice sweets.
Marshmallows.	Lozenges
Nougats (the chewy types)	Marshmallow (grained)
Toffees and caramels	Marzipan
	Nougats (short)
	Pralines

Adapted from (Anonymous 1982)

Table 24.7 Global confectionery sales ($bn)

Confectionery types	Sales (US$ billion)
Gum (chewing)	23.0
Chocolate confectionery	88.3
Sugar confectionery	59.0
Caramels and toffees	12.7
Candy (hard boiled sweets)	12.3
Gums and jellies	11.5
Medicated confectionery	8.7
Regular mints	5.9
Other	5.9
Power mints	5.5

Source: Adapted from (Business Insights 2011)

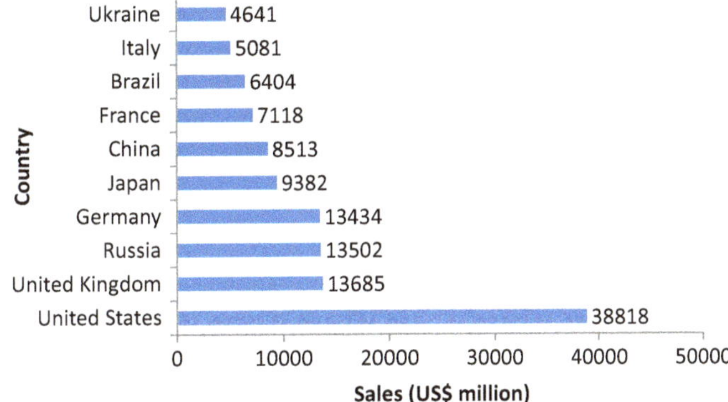

Fig. 24.5 Top 10 Global Market for confectionery 2014

containing crystalline sugar and those in which sugar is not in the crystalline state (Jackson 2000) (Table 24.6).

For 2014 the global sales of confectionery products were estimated at US$200 billion with an annual growth rate average of 2% (Euromonitor International 2014). This figure is slightly higher than predicted 3 years earlier by another market research organization that anticipated the 2014 sales figures US$171 billion (Business Insights 2011). The main confectionery sales categories were chocolate (51%), sugar confectionery (35%), and chewing gums (13.5%). Flour confectioneries are prominent in India and other parts of Asia, but sales of these products are not often captured by current economic metrics.

The US is the largest confectionery market accounting for 19% of global sales. The top-10 nations for confectionary sales account for 67% of market share. The most rapidly growing markets are thought to the BRIC countries (Brazil, Russia, India and China). Some of the largest food corporations are involved in the confectionery sector including; Mars Inc., Cadbury Plc (purchased by Kraft), Nestle' SA, Hershey Co., Kraft Food Plc., Ferrero group, Lotte Group etc. There is a clearly a great deal of economic interest in the confectionery sector (Table 24.7 and Fig. 24.5).

The word candy appears to be "olde" English (circa fifteenth century) apparently derived from *Khandy (Sanskrit)* pointing to the Indian or South Asian origins of the sugar confectionery derived from sugar cane as opposed to honey based sweets of Asia minor. Elsewhere across the world, a variety of acronyms for Candy appear, including terms like sweets, lollies, treats and "toffi".

5.2 Confectionery Art and Science

In the introduction to his well-received book titled "the science of sugar confectionery" the author argues that confectionery manufacturing *was not* (traditionally) a science based industry. However, confectionery was linked historically to pharmaceuticals (Edwards 2000).

In the middle ages, confectionery products were found in the apothecary alongside other medicines, and then later sugar was considered a luxury food items available only for wealthy citizens. Sugar and other ingredients were expensive and exotic items, shipped from faraway places. In renaissance Europe, early confectioners were said to be held in utmost regard and their trade considered "art" in the manner we still talk of culinary art when wishing to emphasize the creative aspects of such professions (Day 2004).

The industrialization of confectionery making in the early nineteenth century is rather well documented and illustrated in the book "Chicago's Sweet Candy History" (Goddard 2012). The confectionery industry is now securely grounded in food manufacturing. Nicholas Appert who developed canning was, initially, a confectioner by profession.

Candy making grew from origins in artisanship–as can be said for many areas of food manufacturing. The modern confectionary industry requires food science and technology competencies. A listing of some important food science and technology concepts for confectionery industry is shown in Table 24.8.

A systematic understanding of confectionery science and technology is a pre-requisite for product innovation, to lower costs, and to enable industry to meet "consumer needs". In line with the focus on carbohydrates in this Chapter, we will have reason to consider the behavior of highly concentrated solutions of sugar in detail later (Ergun et al. 2010).

The behavior of sugar solutions is affected the numbers of dissolved compounds and so appreciation of colligative properties is important. Some candy formulations contain fat, water, sugar and other ingredients so effective mixing, and emulsification may be appropriate (Ergun et al. 2010).

Table 24.8 Some candy science concepts

Some candy science concepts
Water and moisture
Water activity, relative humidity, dew point, Colligative properties, moisture migration
Acidity and pH, buffer action
Emulsions science, Colloids, Surfactants
Material and polymer science
Nucleation and crystallization
Sugar glasses, transition temperatures and their control, Amorphous solids
Colors, dyes and lakes
Food process engineering
Food chemistry, Maillard reaction, Fat deterioration and rancidification
Sugars, fats, additives, general ingredients
High intensity sweeteners
Analytical chemistry and instrumental methods, Polarimetry, Refractometer
Densitometry, Infrared and NMR spectrometry

Adapted from (Edwards 2000; Hartel et al. 2017)

Food dyes and colors are a frequently employed in candy making and some appreciation of the behavior of such substances is essential.

Hard-boiled sweets are described as sugar glasses, which are thermodynamically metastable, super-cooled, solids. The molecules within a glass are trapped by an exceeding high viscosity from the optimum positions required to form a crystalline solid. Glasses are frozen liquids wherein the melting point is above room temperature. The instability of glasses means that they will eventually revert to the more stable crystalline solid on long time-scales. When heated, solid glasses undergo softening (called a glass-to-rubbery state) transition at a characteristic glass transition temperature (Tg).

A glassy material shows large and sudden deviations in physical properties (heat capacity, free volume, viscosity, elasticity et) when the surrounding temperature is raised above the Tg. A common method measuring Tg is using an instrument called a by differential scanning calorimeter (DSC) which detects changes to heat capacity (Cp) or amount of heat needed to raise the temperature of a substance by one degree (Cp; J/°K/g) (Kasapis 2008). The Tg of glassy materials can vary by tens of degrees above or below room temperature depending on moisture content.

Sucrose has a tendency to form crystalline solids. When glucose or fructose is added to sucrose the Tg value declines. Adding monosaccharides to sucrose decreases the crystallizing efficiency of sucrose because differently sized molecules arrange less perfectly within a single crystal matrix (Jeong-Ah et al. 2006). The change in sucrose crystallization behaviour is exploited when glucose syrups is added to sugar ingredient during candy making. As noted above, it is possible also to form invert sugar directly from sucrose by heating in the presence of cream of tartar. Addition or removal of water to sucrose will decrease or increase the Tg value for sucrose, respectively. The value of Tg has been found to provide information related to brittleness, tendency for powders to collapse, stickiness of products, microbiological stability of foods etc. (Ergun et al. 2010).

5.3 Candy Making

Candy making involves manipulating sugar crystals to achieve desired textural effects. This is done by controlling the crystallization of sugar and also sugar/moisture ratio. Examples of sugar-type candies include nougats, fondants, caramels, taffies, and jellies. The ingredient list used in candy making is virtually endless and includes, milk products, egg white, food acids, gums, starches, fats, emulsifiers, flavors, nuts, colors, and more.

The sugar in candies may be crystalline or noncrystalline. The crystals may be large or small and the noncrystalline structure may be glass-like or amorphous. Achieving the desired level of control during the manufacturing process usually means keeping a close eye on the boiling temperature of candy syrup. Softness is another textural quality and soft and hard candies can be made from both crystalline and noncrystalline structures. To achieve softness, higher levels of moisture can be maintained, air can be whipped in, or other ingredients can achieve this textural attribute. Table 24.9 shows the main types of sugar-based candies divided into crystalline and non-crystalline products.

Table 24.9 Some major types of sugar based candies and confectionery

Sugar confectionary	Texture
Type	Creamy or crystalline,
Rock Candy	Large crystals
Fondant, Fudge	Small crystals
Type	Non-crystalline, glassy
Sour Balls, Butterscotch	Hard candies
Peanut Brittles	Brittles
Caramel, Taffy	Chewy candies
Marshmallow, Jellies, Gumdrops	Soft or hard gels, gummy candies

The major sugars used in candy making are sucrose, invert sugar, corn syrups, and HFCS. Starting with a given syrup composition the concentration of sucrose will tend increase "automatically" with cooking time due to moisture loss from the sample. The evaporative loss of water means that syrup shows a continual rise in sugar concentration and this in turn leads to further rise in boiling point (Table 24.10).

In line with colligative properties (see below) the boiling point of a syrup is closely linked with the total solids content (sugar+ glucose syrup etc.) compared to the moisture present. Therefore, it is possible to produce a variety of candy types and textures, ranging from highly crystalline masses to amorphous glasses, if the boiling temperature is monitored and controlled within certain limits (Table 24.10).

To explain why boiling temperature is a good indicator for syrup composition we will borrow from the theory of colligative properties. This useful bit of high-school chemistry, states that the physical properties of liquid water changes predictably, according to the number (not the type) of particles dissolved. The list of colligative properties includes freezing point, boiling point, vapour pressure, relative humidity, osmotic pressure, etc. From common experience, salt depresses the freezing point of water. The theory of colligative properties is demonstrated when salt is added to the driveway or sidewalk in winter.

Interestingly, the boiling point of water increases when sucrose is added. The theory of colligative properties predicts that the boiling point of water will increase with increasing amounts of added

Table 24.10 The boiling point of sugar syrups at sea level relates to composition and candy types

Temperature	Sugar Solids (%)	Candy
Water boils at Sea Level; 212 ° F	0%	Water, Simple sugar syrups
Thread Stage; 215 ° F–234 ° F 101 ° C–112 ° C	80%	Sugar syrup, fruit liqueur and some icings
Soft-Ball Stage; 234 ° F–240 ° F /112 ° C–115 ° C	85%	Fudge, Fondant, pralines, peppermint creams and classic buttercreams
Firm-Ball Stage; 242 ° F–248 ° F /116 ° C–120 ° C	87%	Caramel candies
Hard-Ball Stage; 250 ° F–268 ° F /121 ° C–131 ° C;	92%	Nougat, marshmallows, toffee, gummies, divinity, and rock candy
Soft-Crack Stage; 270 ° F–290 ° F /132 ° C–143 ° C;	95%	Taffy, Butterscotch, Candy apples
Hard-Crack Stage; 300 ° F–310 °F/ 148 ° C–154 ° C;	99%	Brittles, hard candy (lollipops)
320 ° F + / 160 ° C + Sugar (sucrose) melt around 320 ° F and caramelize around 340 ° F.		Caramelizing sugar Thermal Decomposition

Source: Adapted from CraftyBaking, LLC (with permission) http://www.craftybaking.com/howto/candy-sugar-syrup-temperature-chart

sugar. The relation between total solids content and the boiling point of sugar syrup remains true – even as the mixture is heated and moisture is lost from the heated sugar syrup by evaporation. Such background shows why monitoring the boiling point of sugar syrups is such an important activity, and key to producing different types of confectionery.

Appendix 1

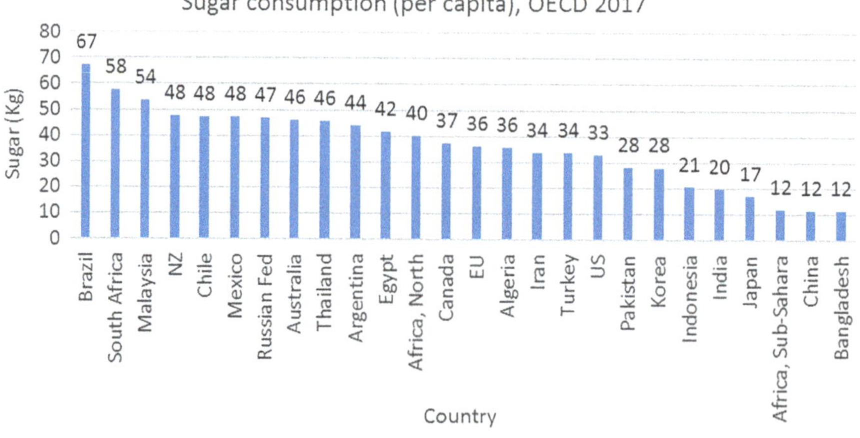

References

Anonymous (1982) General classification of sugar confectionery. Confection Prod 48(5):191–192

Aiyer PV (2005) Amylases and their applications. Afr J Biotechnol 4(13):1525–1529

Business Insights (2011) Future directions in confectionery. Retrieved from http://www.som.cranfield.ac.uk/som/dinamic-content/ki/Businessinsights/BI00049-001/MainDeliverable.pdf

Day I (2004) The art of confectionery. Retrieved from http://www.historicfood.com/The%20Art%20of%20Confectionery.pdf

Edwards WP (2000) The science of sugar confectionery. Retrieved from http://pubs.rsc.org/en/content/ebook/978-1-84755-216-7

Edwards WP (2003) Sweets and candies: sugar confectionery. Encyclopedia of Food Sciences and Nutrition (2nd edn, Caballero B, eds). Academic Press, Oxford, pp 5703–5710. https://doi.org/5710.1016/B5700-5712-227055-X/201173-227051

Ergun R, Lietha R, Hartel RW (2010) Moisture and shelf life in sugar confections. Crit Rev Food Sci Nutr 50(2):162–192. https://doi.org/10.1080/10408390802248833

Euromonitor International (2014) Global confectionery overview: key categories, countries and trends to 2019. Retrieved from http://www.euromonitor.com/Global-confectionery-overview-key-categories-countries-and-trends-to-2019/report

Goddard L (2012) Chicago's sweet candy history. Retrieved from https://books.google.co.uk/books?id=yekGvpKoLWwC

Hartel RW, von Elbe JH, Hofberger R (2017) Confectionery science and technology. Springer, 536pp. https://books.google.com.gh/books?id=QjE5DwAAQBAJ

Hull P (2011) Glucose syrups: technology and applications. Wiley. 392 pp

Jackson EB (1995) Sugar confectionery manufacture. Springer, 400pp. https://books.google.com.gh/books?id=GFw8HEqnLvIC

Jackson B (2000) Fundamentals of sugar confectionery. Manufact Confect 80(8):35–41

Jeffery MS (1993) Key functional properties of sucrose in chocolate and sugar confectionery. Food Technol 47(1):141–144

Jeong-Ah S, Kim SJ, Hyun-Joung K, Yang YS, Hyung Kook K, Yoon-Hwae H (2006) The glass transition temperatures of sugar mixtures. Carbohydr Res 341(15):2516–2520. https://doi.org/10.1016/j.carres.2006.08.014

Kasapis S (2008) Recent advances and future challenges in the explanation and exploitation of the network glass transition of high sugar/biopolymer mixtures. Crit Rev Food Sci Nutr 48(2):185–203. https://doi.org/10.1080/10408390701286025

OECD/Food and Agricultural Organization (2018) OECD-FAO agricultural outlook 2018–2027. Retrieved from, Sugar. https://doi.org/10.1787/agr_outlook-2018-en

Parker K, Salas M, Nwosu VC (2010) High fructose corn syrup: production, uses and public health concerns. Biotechnol Mol Biol Rev 5(5):71–78

Science Direct (2021) Sugar confectionery. From https://www.sciencedirect.com/topics/food-science/sugar-confectionery

USDA Foreign Agricultural Services (2014) Sugar: World markets and trade, November 20, 2014: World production, markets, and trade reports. Retrieved from http://www.fas.usda.gov/data/sugar-world-markets-and-trade

Wolf B (2016) Confectionery and sugar-based foods. Reference module in food science. Elsevier, New York, 4pp. https://doi.org/10.1016/B978-0-08-100596-5.03452-1

25 Foodservices

1 Introduction

1.1 General Principles

United States consumer spending for out-of-home eating now exceeds 50% of total expenditure for food. Eating out continued to gain popularity since the 1960s at which time, it comprised about 20% of total food spend. More money was spent on out-of-home eating from 2013 onwards compared with the expenditure on home cooked food (Saksena et al. 2018; USDA Economic Research Service, 2019, June 4 2020).

In general, out-of-home eating involves purchases from the foodservice sector (NAICS722)[1]. The foodservice businesses include full-service restaurants, limited service eating-places, take-outs and institutional providers. Ready-to-eat food and general sale of all other food from supermarkets and grocery stores comes under food retail (NAICS 455)[2]. The foodservice and food retail sectors are the two main entities that sell food directly to consumers.

In turn, the foodservice and food retail sectors undertake business-to-business (B2B) trade with the wholesale distributers, and ultimately with the food and beverage manufacturing industry (NAICS311 & 312), which is the first relay-point from the farms and fisheries (Fig. 25.1). In other words, Fig. 25.1 is representative of a typical food supply chain for foodservice businesses.

In practice, many procurement channels exist for restaurants including, direct sales and other "farm-to-chef" marketing arrangements (Schmit and Hadcock 2010). The farm-to-restaurant procurement route is attractive, but faces a number barriers related to convenience, ensuring product quality and consistency, limited volume, higher costs, lack of all-year round supplies and time constrains (Schmit and Hadcock 2010). This is an ongoing area of research of civic interest (Sharma et al. 2012).

The food and beverage manufacturing industry (more commonly called the "food processing industry") is up-stream from the foodservice and food retail businesses (Fig. 25.1). The means that foodservice businesses obtain the bulk of their ingredients from the food manufacturers (Fig. 25.1). Moreover, home-prepared meals are derived from the food retailers (supermarket and

[1]The business administration aspects of the foodservice and hospitality industry lies outside of the scope of this chapter and readers wishing to journey in that direction should refer to the many excellent texts available on this topic, e.g. Puckett, R. P. (2012). Foodservice manual for health care institutions, John Wiley & Sons, 592pp.

[2]Definition of food retail – "Industries *in the Food and Beverage Stores subsector usually retail food and beverage merchandise from fixed point-of-sale locations. Establishments in this subsector have special equipment (e.g., freezers, refrigerated display cases, refrigerators) for displaying food and beverage goods. They have staff trained in the processing of food products to guarantee the proper storage and sanitary conditions required by regulatory authority*".

R. Owusu-Apenten, E. R. Vieira, *Elementary Food Science*, Food Science Text Series,
https://doi.org/10.1007/978-3-030-65433-7_25

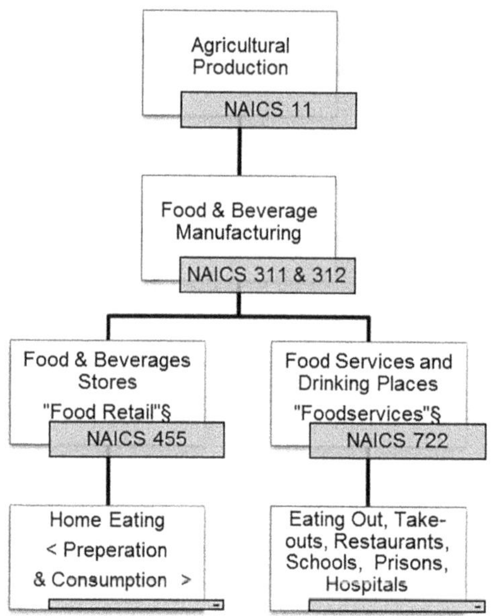

Fig. 25.1 A diagram relating agricultural production (NAICS 11), food and beverage manufacturing (NAICS 311 & NAICS 312), food & beverage stores (NAICS 445), and foodservices/drinking places (NAICS 722)

stores), and the food manufacturers. In support of food literacy (Chap. 6), the foodservice supply chain can expressed as, Agricultural production → Food & beverage manufacturing (food processing) → Wholesale → Foodservice. Similarly, the supply chain for home food preparation involves, Agricultural Production → Food processing industry (aka. Food & beverage manufacturing) → Wholesale / Retail → Home preparation. Clearly, foodservice enterprises combine the functions of food preparation with sales/retail.

Both foodservice and food manufacturers "process" foods for commercial purposes. Foodservice meals are prepared using industry-scale kitchens usually by staff wearing white overalls, soft hats and hairnets. Meanwhile, food-manufacturing operations occur within food factories (processing plants) where staff wear white overalls but with hard hats.

The building premises for foodservice or food process plant are both purpose-built to comply with cGMP. Foodservice commercial kitchens resemble domestic kitchens in terms of feel and look of the premises. Starting from the 1960s, new food production systems (e.g. cook-chill, cook-freeze, souse-vide and steamplicity) were introduced into the foodservice sector thereby blending some of the characteristics of the industrial kitchen and food factory. Other characteristics of foodservice industry such as customization will be discussed later.

The cook-chill system brought elements of foodservices and food science together but the discussions were focused in a narrow area (Dahl and Matthews 1980; Khan and Rao 1983; Rollin and Matthews 1977). The theoretical basis for food technology, using industrial-scale kitchen equipment is not well developed (Parham and Benes 1997; Research Chefs Association 2016). For reasons that are unclear, food science and engineering educational courses over the past 50–60 years were focused on food manufacturing (Floros et al. 2010). The intersection between the foodservice and food science disciplines is a promising area for future applied research.

There are overlaps between the foodservices and the nutrition professions in terms of competencies for management dietitians versus foodservice managers (Oregon State University 2019). Most nutrition programs provide a practical foodservice component with hands on experience using a pilot-scale professional kitchen equipped for meal preparation (NutritionEd.org 2019). By contrast, food science programs require practical experience or training using a food pilot plant (Hartel 2002).

1.2 Characteristics of the Foodservice Sector (NAICS 722)

Food services and drinking places (NAICS 722) is the official designation for businesses commonly described as, foodservices or food services establishments[3]. In the United States (US), magazines, websites and academic literature refer to the "foodservice" (Payne-Palacio and

[3]The two alternative spellings "food service" or foodservice occur in equal proportions in the literature.

Theis 2011a). However, USDA documents refer to "food services" (USDA Economic Research Service 2019a). The designation "food service" is used by the UK and the European Union (Edwards and Causa 2009; Edwards 2013). Some UK trade literature also refers to the "Catering industry" to mean the entire foodservice sector. In the US, the catering industry is a small part of the larger foodservices sector according to NAICS (US Census Bureau 2012; USDA Economic Research Service 2019a, b). In this text, we will refer to foodservices, in line with the style from US academic and trade literature[4].

Foodservice companies are defined as "*establishments that prepare food to customer order for immediate consumption*" (US Census Bureau, 2012). According to the USDA, foodservice outlets are "*...facilities that serve meals and snacks for immediate consumption on site (food away from home)…This category includes full-service restaurants, fast food outlets, caterers, some cafeterias, and other places that prepare, serve, and sell food to the general public for a profit…*" (USDA Economic Research Service 2019a). Other businesses may (or may not) be excluded from foodservice category. Purchasing food from the supermarket (retail) or from a "take-out" can offer similar experiences for the consumer, but these outlets belong to different industry sectors for the purpose of taxation (Edwards and Causa 2009; Edwards 2013).

Preparing food "to order" is a core feature of all foodservices (US Census Bureau 2012; USDA Economic Research Service 2019a). The-made-to order (MTO) style of operation means that product manufacturing starts only once an *order is received from a customer* [5]. One advantage of the MTO supply-chain is that this reduces the quantity of finished stock held by the manufacturer (Asprova Corporation 2019). The MTO process also allows higher degrees of customization, where a consumer determines their preferences based on a set menu of options (Table 25.1) (Anon 2019).

Table 25.1 Some advantages of made-to order manufacturing for foodservices

Made to order (MTO)	Made to stock (MST)
Suited to low volumes of product	Large volume of product
Suited for high value products	Low value product
Highly variable products, bespoke products	Restricted variety of products
Avoids costly storage (of stocks)	Low cost of storage
Enables customizability	Products requiring long lead time

Adapted from Anon (2019)

With the made-to-stock (MTS) supply chain, products are manufactured fully prior to any order being received. The MTS system is suited to highly regular, low cost products, which require a significant amount of time to produce. Adopting the MTS supply chain strategy reduces customer waiting, but also ties up resources in terms of the stock held. A blend of MTS and MTO approaches is thought to be advantageous in some cases. For instance, the manufacturer could assemble, semi-processed ingredients, and then mix these as needed for MTO purposes. In principle, some foodservices operations employ a mix of MTS and MTO systems (cf. Sect. 2) but there are few formal discussions of this topic (Van Donk 2001).

1.3 Classification of Foodservice Businesses

There are essentially four major subcategories of foodservices (Fig. 25.2); full-service restaurants (NAICS7221; NAICS722511), limited service eating-places (NAICS 7222; NAICS 722513) and the special foodservices (NAICS 7223). As shown below (Fig. 25.2) the special foodservices comprise three subcomponents listed officially as, foodservice contractors (NAICS 72231), caterers (NAICS 72232), and providers of mobile foodservices (NAICS 72233). The fourth foodservice category consist of drinking places and retailers of alcoholic drinks (NAICS 7224).

[4]The compound term foodservice and food service both appear within written sources. Food service will used if there is reference to official documents and citations.

[5]An equivalent term appears to be Purchase To Order (PTO). The MTO or PTO concept is applied to vast areas of manufacturing including, the automobile and aviation industries.

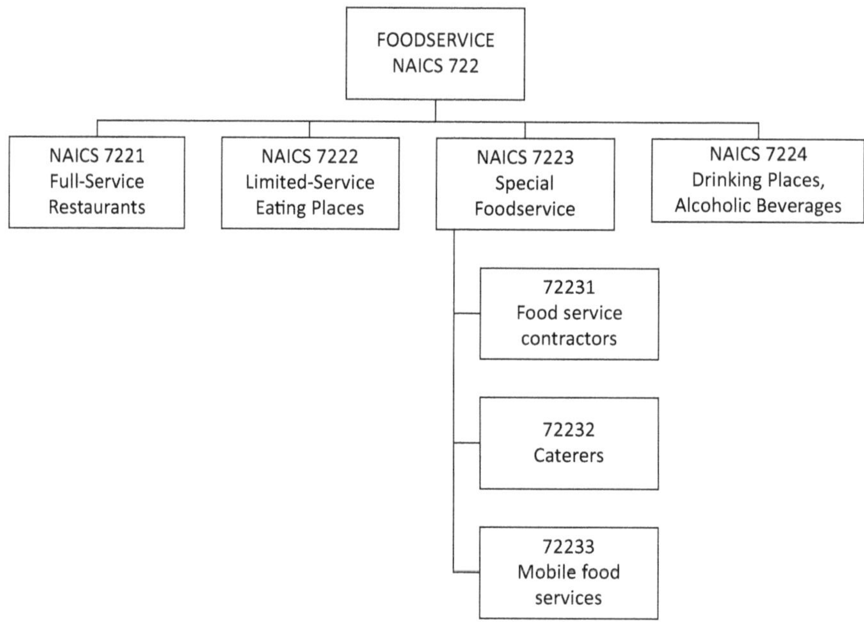

Fig. 25.2 Classification of foodservice industry (NAICSS 7221-4). Post 2017 literature refers to NAICS 722511 for full service restaurants and NAICS 722513 for Limited service eating places

The contract foodservices (NAICS 72231) designation coincides with the "institutional foodservices". However, not all institutional foodservices are contracted or outsourced. Currently, nearly 50% of foodservices are managed by various institutions in-house. The main distinction between contracted and non-contracted foodservices is that the former provide meals to institutions or organizations on a continual basis. However, restaurants provide non-contract foodservices directly to the public, or at one-off events (Tables 25.2 and 25.3).

A significant group of foodservices is described as "non-commercial" meaning they are run by public sector or not-for profit institutions, where foodservice is not the primary business. Commercial foodservices operate with the private sector or else with for-profit businesses such as, airlines and entertainment venues. The different types of foodservices are summarized in Tables 25.2, 25.3 and 25.4.

As noted elsewhere, foodservice enterprises prepare meals, snacks and beverages to order. In addition, food may be consumed on-premises or off-premises. The main differences between a full service restaurant and the limited service eating-places are summarized in Table 25.4. A full service restaurant must provide sit down meals and waiter/waitress service. By contrast, most limited-service eating-places (e.g. Take-outs) offer counter-service and payment but no seating. Other limited service eating-places (Fast-food outlets) offer counter service, seating and disposable cutlery (USDA Economic Research Service, 2019b). The buffet-style "all you can eat" establishment com-

Table 25.2 Characteristics of non-contract foodservice providers

Designation	Non-contract	Clients
Catering Services(Caterers)	One off-events	Weddings, Private parties,
Catering Services(Caterers)	One off-events	Corporate functions
Full-Service Restaurants	Sit-down meals	General public
Limited-Service Restaurants	Sit-down meals	General public
Take-Out, Take-Away	Drop in services	General public
Pubs And Bars	Sit down beverage,& food	General public

Table 25.3 Characteristics of contract foodservices providers

Designation	Description	Clients
Foodservice contractors	On-site Off-site	Airlines Amusement parks Businesses, Casinos, Civic and social organizations Country clubs, and similar recreational facilities Entertainment venues Ferries and cruise ships, Hotels and motels; Museums Sporting venues, Theaters Tourist attractions

Table 25.4 Differences between full service restaurants and limited service eating-places

Index	Full service	Limited service
Waiters/Waitress	Y	N
Ceramic dishware	Y	N
Disposable utensils	N	Y
Counter service	N	Y
Alcohol served	Y	N
Menu	Full	Limited
Convenience	Y	YY
Speed	N	Y

From USDA Economic Research Service (2019b), Yes (Y) or No (N) provision

bine some of characteristics of a full service restaurant (e.g. ceramic ware, metal cutlery, some waiter service) alongside of counter payments (Table 25.4).

1.4 Non-Contract Foodservices

About 80% of foodservice businesses provide meals on a one-off basis (Table 25.2). The non-contract foodservices are privately owned, part of business chains, or part of franchises. Non-contract foodservice establishments include the FSR, limited-service eating-places (unofficially- limited service restaurants), fast food outlets, and take-outs businesses (Lockyer 2016). Some non-contract foodservices serve corporate functions, weddings, and private parties. Finally, a sub-group of non-contract foodservice providers can be found within institutional settings, e.g. staff restaurants located in hospital premises. The non-contract foodservices is also sometimes called the *commercial* foodservice (Table 25.5).

1.5 Contract Foodservices as Institutional Providers

The main differences between contract and non-contract (institutional) foodservices are summarized in Table 25.6. The type of consumer served by foodservice providers are different noticeably. Institutional foodservices have a captive customer base whilst restaurants and the other non-contract providers have constantly changing clientele.

Intuitional foodservices often serve meals within a dining hall setting. Nevertheless, meals can be served also in a hospital ward, prison cell or within a school classroom. By comparison, FSR utilize dedicated dining rooms with multiple seating per room or occasionally private rooms. The range of food available for non-commercial foodservice tend to be more restricted and less diverse compared to FSR. There are usually budgetary constraints to ensure the amount of spend per meal is fixed for hospitals, military, police, schools, correctional institutions and other public sector institutional foodservice providers. However, the nutritional quality may be higher for institutional foodservice providers, which have to meet national nutritional guidelines.

Table 25.5 Non-Contract foodservices

Type	Name	Sales US$ billion	No of locations
LS	Limited Service Eating-Places	245.6	308,682
FSR	Full-service restaurants	171.5	324,351
FSR	Lodgings, FSR	44.7	68,663
Retail	Supermarkets, convenience	24.8	98,994
Retail	Convenience	31.5	151,053
Retail	Other	2	10,055
	Total	520.1 US $billion	1,227,670.0

*US$641 billion including contract. FSR = full service restaurant, LSR = limited service eating-places (LSEP), Adapted from Lockyer (2016)

Table 25.6 Some differences between institutional foodservices and commercial foodservices

Index	Institutional/Contract	Commercial/Non-contract
Clients	Captive/Assured	Non-captive
Meals	All day meals	One meal per day
Meal location	In bed, mess hall	Dining room, cafeteria
Menu type	Cycle menu	Daily menu, ala-carte
Budget/spend per meal	Limited, fixed budget	Customer pricing
CStaff training	Limited	Professionally trained
Special dietary needs	Special dietary needs	General populations
External standards	Nutritional standards	Food safety standards
Payment	Meal package, subsidized	Client payment
Acceptable level of waste	High	Low
Feeding assistance	Sometimes provided	Not available

Adapted from Conner (2014)

1.6 Economic Significance of the Foodservice Sector

Foodservice economic results and general characteristics for 1997-2012 are summarized in Table 25.8. As of 2012, the US foodservices sector consisted of 600,000 enterprises (approx.) and about 10.0 million employees (Table 25.7 & Table 25.8). For the fiscal year 2013, foodservice earnings reached US$600 billion with 4.5% annual growth. By fiscal year 2014, the foodservices revenue was US$731 billion compared with a total of US$1400 billion consumer spending on all foods; as noted previously, out-of-home eating now exceeds home eating (Lockyer, 2016; USDA Economic Research Service, 2019a, 2019b). Overall, about 20% of foodservice earnings is derived from the Institutional entities whilst 80% of the revenue comes from the restaurants (Lockyer, 2016).

For each dollar spent in the restaurant business, about US$2 was generated by allied industry and so the yearly impact of the foodservices sector was estimated (2014) at US$1.8 trillion or 4% of GDP (National Restaurant Association, 2014).

The total number of jobs available from the foodservice sector (2014) exceeded by 10-fold the number of jobs from food manufacturing. Indeed, about 10% of all jobs within the US economy were provided by the foodservice sector (USDA Economic Research Service, 2019a, 2019b)[6,7].

For 2012, the average person spent about $1632 per annum eating out (Table 25.8). Interestingly, the average foodservices estab-

[6]Economic data from two independent US sources are in good agreement, but not identical for reasons discussed in Chap. 1.

[7]For comparative purposes, the UK show1.65 million workers in the foodservice sector (5.7% of the working adults) with 0.41 million in food and beverage manufacturing an overall employment rate of 13.4% in the food and agriculture.

Table 25.7 The US Accommodation and foodservice sector (NAICSS 72)

NAICSS	NAICSS code description	Employees (2012)	Businesses
72----	Accommodation and foodservice (total)	11,556,285	649,011
721	*Accommodation*	1,864,708	63,398
7211	Traveler accommodation	1,815,117	54,126
7212	RV parks and recreational camps	38,273	7189
7213	Rooming and boarding houses	11,318	2083
722	*Foodservice and drinking places*		
7221	Full-service restaurants	4,570,174	227,161
7222	Limited service eating-places	4,106,307	275,939
7223	Special foodservice	668,117	39,654
72231	Foodservice contractors	530,597	26,134
72232	Caterers	129,621	10,815
72233	Mobile foodservice	7899	2705
7224	Drinking places (alc. Beverages)	346,979	42,859
	Foodservice and drinking subtotal	9,691,577	585,613 (¥)

Adapted from (United States Census 2014). (¥) There is a slight disparity between this value and 598,656 businesses (2012) cited in Table 24.8. Post 2017 literature refers to NAICS 722511 for full service restaurants and NAICS 722513 for Limited service eating places

Table 25.8 Characteristics of the US foodservices sector

Category/Year	1997	2002	2005	2012	2007–2012 (§)
Number of establishments	486,906	504,641	571,621	598,656	4.7%
Sales ($ Millions)	251,942	321,401	433,405	512,280	18.2%
Annual payroll ($ Millions)	70,334	92,599	124,434	145,983	17.3%
Total employment	7,754,567	8,307,625	9,630,090	10,057,608	4.4%
Sales / establishment ($1000)	517	637	758	856	12.9%
Sales per employee ($1000)	32	39	45	51	13.2%
Sales per $ of payroll ($)	3.58	3.47	3.48	3.51	.8%
Payroll per employee ($)	9070	11,146	12,921	14,515	12.3%
Employees per establishment	15.93	16.46	16.85	16.80	−.3%
Sales per capita ($)	924	1117	1439	1632	13.4%
Population per establishment	560	570	527	524	−.5%

Adapted from USDA Economic Research Service (2019a), § indicates % change

lishment is considered small to medium enterprise (SME) with an average of 17 employees and a wages of $14,000 per annum (USDA Economic Research Service 2019b). For 2020, spending for out-of-home eating fell to 49.8 %, i.e. slightly below 50 % and could hover around this value for their near future.

NAICS (Sect. 1.2). In addition, the foodservices can be differentiated according to the site of food preparation; (i) conventional foodservice, (ii) centralized foodservice, (iii) ready-prepared foodservice, and (iv) the assembly-serve foodservice (Table 25.9). (National Food Service Management Institute 2002).

2 Choosing Foodservice Systems

This section provides a brief overview of foodservice systems –for food science students. Four classes of foodservices can be identified using

2.1 Conventional and Centralized Foodservices

A conventional food service is one where all activities take place on-site. The food is prepared and served immediately on the premises. However, with

Table 25.9 Characteristics of the four major types of foodservice

Foodservice/Factor	Conventional	Centralized Commissary	Ready-Prepared	Assembly
Customer #	Small	Large	Large	Small/ large
Menu	Broad	Limited	Broad	Broad
Labor	High	Low	Low	Low
Equipment	Low	High	Low	Low
Site #	Few	Many	Few	Few
Transport	No	Yes	Yes	Yes
Food safety	High	High	High	High
Examples	Hospital	Schools, Airlines Restaurants	Hospitals	

Source: adapted from National Food Service Management Institute (2002) and Brown (2005)

a centralized foodservice, food is prepared at a single central facility, refrigerated immediately and then transported to satellite kitchens for reheating and serving. The food must be cooled to 41 °F (5 °C) for cold holding or heated to 135 °F (57 °C) for hot holding. In addition, any food that has been cooked and then cooled must be reheated to 165 °F (74 °C) for 15 seconds to render it suitable for hot holding. Reheating by microwave requires 165 °F (74 °C) for 2 minutes (U.S. Food and Drug Administration 2017). The preceding temperatures represent values that must be attained within the food product and which can be confirmed objectively, by food inspectors (Food Code 2017.p574-581).

The cook-chill and cook-freeze systems, which were introduced from the 1960s and 1970s, are popular with some hospitals (Nettles, et al. 1997; Snyder, 1984; Yang, 1990). The centralized foodservices were more likely for larger hospitals (Mibey, 2002) but less popular with schools (Brown, 2005). One study found that the majority of schools (45%) were using the onsite or conventional foodservice, compared with 12% for a centralized food service. Some 36% of schools used a mix of onsite and centralized foodservices as they provided food to offsite locations. Hot holding was the preferred mode of food delivery for schools (Brown 2005).

2.2 Advantages and Disadvantages of Different Foodservices

The choice of foodservice depends on a variety of local factors (Tables 25.10 and 25.11). Labor availability and cost are seen as important considerations. The centralized foodservice is the least labor intensive. A centralized service allows the sharing of buildings and facilities by many clients. However, the initial investment, and equipment need are also considered higher for centralized provision. On occasion, a hybrid provision may be achieved by incorporated a particular foodservice feature into a different system. For example, recipe standardization, which is a feature of centralized provision, could be useful to deal with inconsistencies experienced at different sites.

2.3 Food Transit through Foodservice Systems

There are ten major unit operations in the typical foodservice (Fig. 25.3); the stages are menu-planning, purchasing, receiving, storing, pre-cooking preparation, cooking, holding, and serving on-site. Centralized foodservices incorporate additional stages e.g. a cook-chill or cook-freeze system. This means food is cooked, and then chilled rapidly to a safe holding temperature, transported and subjected to reheating before serving (Bryan and Lyon 1984).

The challenges faced by all foodservices is that food is highly perishable. The demand for meals is also variable. Peak demand for meals occurs round about traditional meal times, e.g. morning breakfast, mid-day lunch and evening supper. These peak times are followed by periods of low demand. Only some fluctuations in the demand for meals is predictable. Interestingly, a 24/7 service, ensuring the availability of meals throughout the day is the norm now for many non-contract, limited service restaurants. Institutional foodservice providers are

Table 25.10 Some advantages and disadvantages of centralized foodservices

Advantages/Improvements	Disadvantages/Adverse issues
Ingredient control	Automation
Inventory control	Decreased job satisfaction
Labor and food cost	Equipment costs
Larger size	Equipment failure impact
Mechanization	Foodborne outbreak impact
Production scheduling flexibility	High technical knowledge
Purchase power	Large scale production
Quality control – Nutritional value	Need for capital
Quality control – Safety	Perceived food quality
Quality control – Sensory, aesthetic	Recipe standardization
Training & supervision	Transportation costs
Utilization of USDA commodities	
Flexible location	

Adapted from LeBruto and Farsad (1993) and National Food Service Management Institute (2002).

Table 25.11 Some advantages and disadvantages of the Conventional foodservices

Advantages/Improvements	Disadvantages/Adverse issues
Food quality	Labor intensive
Menu flexibility	High labor costs
On-site serving	Product & service consistency
Traditional recipe	Control of costs

Adapted from LeBruto and Farsad (1993) and National Food Service Management Institute (2002).

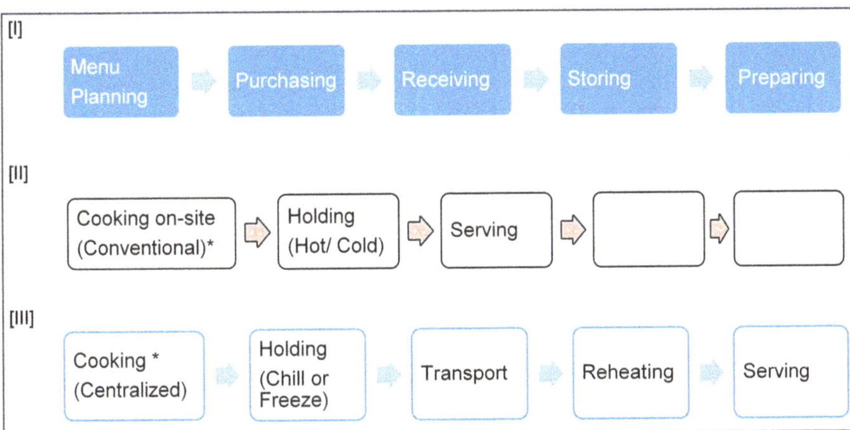

Fig. 25.3 Transit of food through the conventional (I & II) or centralized (I&III) foodservice. (*)The cook step is not used for all menu items. (Adapted from National Food Service Management Institute 2002)

increasingly being required to provide meals on demand for clients, e.g. hospital patients.

2.4 Processing Continuum for the Foodservice

Foodservice meals can be prepared using fresh ingredients as well as secondary ingredients according to a "food process continuum". For example, pizza can be prepared efficiently, starting with the ready-made base, tomato paste, and chopped cheese. Alternatively, one could begin with whole meal flour, fresh tomatoes and non-grated cheese. In this case, the preparation time would be greater. Using partially processed ingredients leads to greater convenience and some cost reductions. There is a third option which is least favorable; these days it would be economically non-viable to make pizza starting from "paleo ingredients", i.e. un-milled wheat, tomatoes and fresh milk to make the cheese toping. The first two economic options show that foodservices meals can be produced using a variety of constituents ranging from fresh produce to partly processed ingredients. At the other extreme, the assembly-serve foodservice employs wholly ready-to-eat dishes, which are re-apportioned prior to serving (National Food Service Management Institute 2002).

2.5 Foodservice Sustainability and Waste Consideration

The rising world population and affluence in former developing countries are the main push factors for more sustainable foodservices and food system as a whole. Switching from an animal-protein dietary pattern, towards more plant-based foods may help protect the environment. A plant-based diet is thought likely to lead to improvements in public health and enhancement of food security (Reynolds et al. 2014) but this thesis is controversial (Drewnowski and Team 2017) (Table 25.12).

Foodservices sustainability is considered important, because of the large economic impact of the sector (4% GDP for the US) and numbers of people involved in this industry (10% of the US workforce). Reducing the carbon footprint, reducing energy use, reducing water use, and improving equipment maintenance to avoid waste are some of strategies suggested for reducing the environmental impact of foodservice operations (Peregrin 2011, 2012). Martin-Rios from the Swiss Hospitality Management School at Lausanne discussed several approaches for decreasing foodservice waste effectively (Martin-Rios et al. 2018).

Table 25.12 Defining sustainability, food systems, and food and nutrition security

Terms	Definition
Sustainable	Meeting the need for people currently living without harming the environment and reducing resources for future populations
Food security	A situation where people have access to foods unimpaired by economic, physical and social factors
Sustainable diet	Where food and diet have low impact on the environment, maintains biodiversity, and is culturally, economically acceptable and fair
Food system	Chain or interrelationships linking areas of activities important for food provision, including agricultural production, processing, packaging, distribution, marketing, consumption and waste disposal
Sustainable food system	A food system meets the need for people currently living but does not harm the environment or resources for future populations

Adapted from Drewnowski and Team (2017)

3 Institutional Foodservices

3.1 Definitions

The institutional foodservices (NAICS 72231) provide food on a contract basis for institutions listed as "*schools, colleges and universities, and hospitals, as well as correctional facilities, public and private cafeterias, nursing homes, and day-care and senior centers*" (Conner 2014) [8]. Some institutional foodservices are managed in-house. Foodservices run by the institution as not-for profit divisions are described as non-commercial foodservices.

It is recommended that institutional foodservices appoint a dietician on their staff to provide expertise in menu planning and meeting nutrition standards, which is required for federal support. Dietitians also play a vital role in clinical nutrition, which means addressing special dietary requirements for patients. An interesting new role for the foodservice dietitian is becoming evident, in private sector organizations and restaurants businesses as nutrition standards become more widespread (Chong et al. 2000) (Table 25.13).

3.2 Foodservice Demographic Considerations

School foodservice programs have a welfare function, to provide nutritionally balanced meals to children, some of whom are from disadvan-

[8]Some non-contract foodservice providers are located on institutional premises, e.g. hospital restaurants that cater for visitors and staff.

Table 25.13 On-site institutional foodservices

Type	Name	Sales US$billon	No of locations
Onsite	Recreation	20.5	62,201
Onsite	Vending	20.3	4011
Onsite	Colleges, Universities	17.4	4706
Onsite	Hospitals	15.6	5686
Onsite	K12-School	15.0	109,629
Onsite	Business & Industry	11.9	11,106
Onsite	Catering	7.8	10,625
Onsite	Senior Living	5.7	34,385
Onsite	Long-term care	3.1	15,632
Onsite	Transportation	2.4	2493
Onsite	Correctional institutions	2.3	4875
Onsite	Military	2.2	523
	Subtotal (§)	124 US$ billion	265,872
LS	Limited Service Eating-Places	245.6	308,682
FSR	Full-Service Restaurants	171.5	324,351
FSR	Lodgings, FSR	44.7	68,663
Retail	Supermarkets, Convenience	24.8	98,994
Retail	Convenience	31.5	151,053
Retail	Other	2	10,055
	Total	644.3 US $billion	1,227,670

(§) Non-commercial or Institutional foodservice subtotal, Adapted from (Lockyer 2016)

taged backgrounds, in order to ensure equal opportunity for education (Sect. 4.1). Children that are food insecure, those from low-income socioeconomic backgrounds, or children from areas of economic deprivation are believed to be at risk from nutrition-related health problems. Food programs provide affordable and nutritional food for school age children. School food programs must also address children with special diets – food allergies and intolerances, medical needs, religion/culture, vegetarian requirements (Food and Nutrition Service USDA 2012).

Foodservice provision for hospital population deals with people who are either ill or else recovering from physical trauma (e.g. broken limbs). Long stay patients in hospitals, the bedridden, and patients suffering from diabetes, hypertension or kidney diseases may have specialized dietary needs. For instance, renal patience require a reduced salt diet. Diabetics may require low sugar diet. Ensuring an appropriate meal planning for the sick requires skills from clinical dietitian (Chong et al. 2000).

Foodservice provision for residents of care homes caters to the elderly and those liable to exhibit age-related changes in physiological function (e.g. decreased taste acuity), impaired cognitive function, and conditions such as Alzheimer's disease or dementia. Compared to the general population, the elderly have impaired immunity and delayed recovery from illness. Elderly people also show increased risk of undernutrition, owing to decreases of appetite and anorexia.

Prison populations show poorer general health due societal inequalities prior to incarceration. Drug/substance abuse, smoking, lack of exercise, poor dietary habits, low literacy, emotional stress, personal identity issues, limited access to food and health services are also prevalent. The prison population is also frequently more diverse ethnically, with a wider range of cultural, personal and religious constraints on food choice. Whilst in prison, the inmates tend to be deprived of personal freedom. Boredom maybe rife and so inmates may consider meal-times to be highlights of the day (Stein 2000).

Foodservices for the military and to a lesser degree, police, para-legal forces, and firemen are for people who are predominantly healthy and fit adults who must engage in intense physical activity.

Such personnel have normal nutrient requirements except under virtual or actual combat conditions. It is accepted that energy intake may be higher for army personnel under combat conditions. Foodservices for the military will usually need to address environmental and logistics constraints; food need to be light and easy to carry.

3.3 Menus Types

The menus served by foodservices tend to be specialized. Care home residents and hospitals tend to have flexible menus, protected mealtimes, 24-h catering and assistance with feeding. The clientele in different institutions have distinct demographic traits that affects the menu served, e.g. age, gender, ethnicity, socioeconomic background, education, occupation (Sect.3.2) (Fig. 25.4).

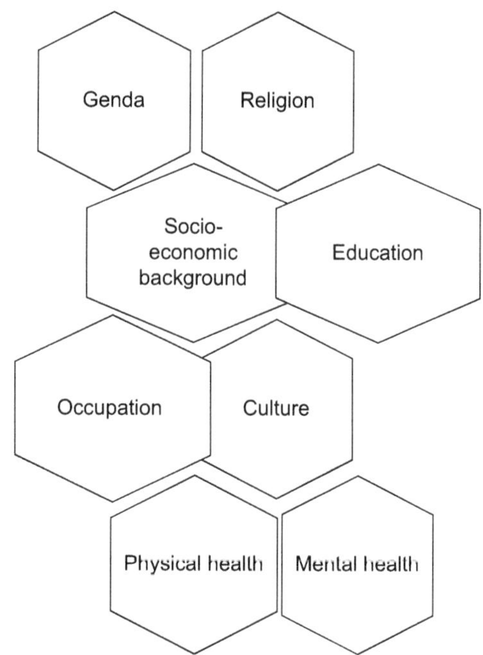

Fig. 25.4 Demographic, health and other considerations likely to affect a foodservice provision

3.4 Consumer Satisfaction with Foodservices

Consumer satisfaction with foodservices is essential for repeat purchases. Hospital patients that were contented with a foodservice provision showed increased food intake, which helps to maintain nutritional status, improve the rate of recovery and reduce the length of stay (Aase 2011). Some survey work suggests that satisfaction is dependent on the type of institution and the kind of study participants (Dall'Oglio et al. 2015; Doyle et al. 2017). Many of the investigations looking at satisfaction with institutional foodservices were about hospital foods (Dall'Oglio et al. 2015).

Another study identified 7 core issues from 26 factors that affect satisfaction with a foodservice provision (Dube et al. 1994b) –see Table 25.14. Another study using elderly patients examined how 6 foodservice variables affected patient satisfaction and isolated 3 areas related to, the taste of food, whether foods were sufficiently cold, and the presentation of meals. The foodservice scored highly for taste, quantity of food and for presentation of meals (Ohara et al. 1997). A number of other influences for satisfaction were recorded for hospital foodservices, including appearance, taste and quality of foods as well as the intervals (days) between menus selection and getting the actual food (Lengyel et al. 2004).

Table 25.14 Variables contributing to hospital foodservice satisfaction

Dube's seven factors	Other factors
Quality of food Timeliness of service Reliability of service Temperature of food Staff attitude (service) Staff attribute (menu) Customization of meals	Appetite Demographic characteristics (gender, age, education, health, beliefs) Time for menu choice Type of diet (normal or therapeutic diet)

From various sources Dube et al. (1994a) and Ohara et al. (1997)

Satisfaction with school foodservices is important for the large numbers of children who

Table 25.15 Variables contributing to school foodservice satisfaction

Factors	Grp.A	Grp.B[a]
Variety of food	0.35	0.40
Attractiveness of presentation	0.15	0.19
Flavor of food	0.12	0.19
Quality of choices	0.09	0.16
Meeting cultural, ethnic needs	0.07	0.08
Staff (courteousness, smile, listening)	0.06	0.12
Special promotions	0.00	0.11
Number of serving lines	0.00	−0.11

[a]Scores for two groups of students, GrpA & Grp B who have (GrpA) or do not have (GrpB) a choice of using the foodservice. A large score (effect size) shows the variable is important. Extracted from (Meyer 2000; Meyer and Conklin 1998)

consume school meals daily to meet nutritional needs (Meyer 2000; Meyer and Conklin 1998). Studies from the University of Mississippi identified some factors that influence school foodservice satisfaction. When 13-19 year old students were surveyed with a 38-item questionnaire, satisfaction with the foodservice provision increased with, a greater variety of food served, attractiveness of food presentations, flavor, and the quality of choices (Table 25.15). Certain behaviors of foodservice staff contributed to satisfaction, which is worth noting as this can be improved by proper training (Meyer 2000; Meyer and Conklin 1998).

The satisfaction ratings for school foodservices were not affected by the time allowed for meals or the dining room temperature. However, inadequate numbers of serving lines affected satisfaction scores adversely. Also, students who identified as having "no choice" about using school foodservices expressed consistently lower satisfaction scores for most attributes. Finally, there was a strong link between satisfaction scores for foodservices and student intention to purchase (Meyer and Conklin 1998). One of the few studies involving middle school and junior school children at the time, found that foodservice satisfaction was largely (39%) determined by 3/13 factors; liking the menu, overall quality of food, and prices (Meyer 2000).

Foodservice satisfaction is one determinant of the overall quality score for an institution. Evidence from hospitals showed that managers did not prioritize foodservice satisfaction compared with other quality measures for an institution (Aase 2011). It is thought that foodservice satisfaction may receive increasing attention in future, in order to facilitate core functions of institutions. For instance, foodservice satisfaction may help achieve nutritional and/ clinical outcomes for schools and hospitals. Researchers from China noted that satisfaction with the school foodservice contributes to student mental attitude, and study effectiveness (Binge et al. 2012). More patient-centered foodservices, private room foodservices, and a more restaurant like experience are just a few of the things suggested for improving hospital foodservice satisfaction (Aase 2011).

3.5 Management Service Companies

Managed-Service Companies (MSC) provide contracted foodservices for institutions [9]. In most cases, the MSC assume responsibilities for management, planning, and purchasing (Table 25.16) as outlined by LeBruto (LeBruto and Farsad 1993). As noted earlier, both in-house and outsourced foodservice provisions co-exist. An average of 47% of all institutional foodservices seem to be contracted but the range of engagement varies by sector (Fleming et al. 2006). For example, 85% of business and industry establishments, 62.5% of universities and colleges, and 45% of hospitals contracted foodservices to MSC (Fig. 25.5).

Some major foodservice companies are multinational corporations such as, the Compass Group North America (USA), Aramark Corporation (USA), and Sodexo Ltd. (France). Typically, such companies provide foodservices and allied solutions across many institutions and industries; schools, universities, hospitals and healthcare, as well as the manufacturing industry.

Some advantages of outsourcing foodservoces to MSC are listed Table 25.17. Ultimately,

[9]Appendix 1 of this chapter lists the top-50 foodservice management companies.

outsourcing may lead to improved savings – provided that there is the necessary economy of scale (LeBruto and Farsad 1993). MSC can achieve greater cost savings owing to bulk buying and their increased negotiation power. The MSC undertake initial foodservice planning, equipment installations, responsibility for human resources, and staff training (Johnson and Chambers 2000; LeBruto and Farsad 1993; Manask et al. 2002). More generally, outsourcing the foodservice function allows an institution to focus on their primary business activity. One incentive for using foodservice management companies is that it allows hospitals and schools to focus on their health or education function, respectively.

Institutions with a small foodservice requirement tend to run these in-house. The availability of staff is another factor that facilitates an in-house foodservice. Currently 85% of retirement centers, 85% correctional institutions, 75% nursing and care homes and about 75% of schools run their foodservices in-house (Fig. 25.5). Common experience suggests that correctional institutions have ready access to a foodservice workforce that can be mobilized following appropriate training. In the case of residential care homes, greater value may be placed on achieving intimate relations with foodservice operators to combat the tendency for malnutrition (see Sect. 4.4). The Table below lists some self-operated health care foodservice systems (Table 25.18).

Table 25.16 Some contractor, client functions in some foodservice

Function	Responsibility
Employment of contractor	Client
Authority of client	Client
Regulatory requirements	Contractor
Foodservice	Contractor
Facilities and equipment	Client
Cleaning responsibilities	Contractor
Health certification, permits	Contractor
Personnel	Contractor
Equal employment opportunity	Contractor
Employee Health examinations	Contractor
Purchasing	Contractor
Inventory of food and supplies	Client
Receipt of Government donated foods & commodities	Client
Licenses, permits, and taxes	Contractor
Insurance, workers compensation	Contractor
Indemnification	Contractor
Fiscal arrangement	Contractor
Accounting	Contractor
Remedies	None applicable
Confidentiality, menu ownership, IP	Contractor

Adapted from LeBruto and Farsad (1993)

Table 25.17 Benefits of foodservice management-service companies

Benefits
Efficient interaction with federal bodies
Efficient business management of internal controls
Increased control of employee benefits
Better reporting to federal bodies
For profit
Disadvantages
Decreased personal relations between consumer and institution
For profit financial goals and impersonal ethos
Restricted access to foodservice facilities by institutional staff
Loss of operational control by institutional management boards

Adopted from LeBruto and Farsad (1993)

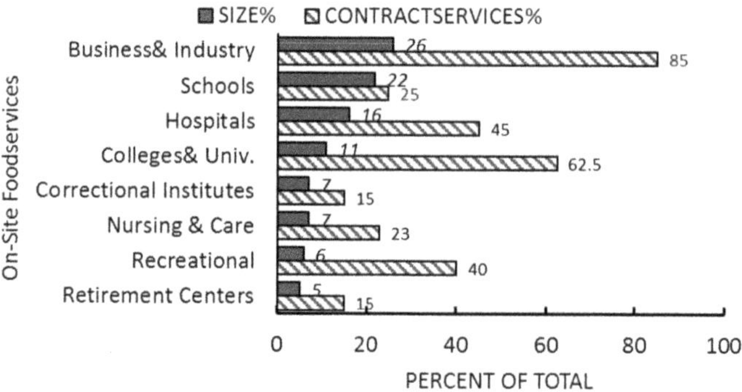

Fig. 25.5 Sizes of institutional foodservices and percentage, contracted. (Drawn suing data from Fleming et al. 2006)

Table 25.18 Example of self-operated healthcare foodservices

Name	Purchases $Million	Sales $Million	Daily transactions	No of beds
Continuum Health Partners (New York City, NY)	10.2	2.7	6050	2322
Massachusetts General Hospital (Boston, MA)	8	16.9	21,000	906
Aurora Health Care (Milwaukee, Wisconsin)	7.5	5.5	6194	1696
UCLA Medical Centre (Los Angeles, California)	5.8	7.8	6800	1005
New York University Hospitals Center	5.5	5.3	3900	1200
Hartford Hospital (Hartford, Connecticut)	4.7	4.2	5175	867
University of North Carolina Hospitals	3.7	5.1	3620	728
Via Christi Regional Medical Center (Wichita)	3.5	4.1	3000	1520
BryanLGH Medical Center (Lincoln, Nebraska)	3.5	2.7	3700	535
Swedish Medical Center (Seattle, Washington)	3.5	3.8	2090	867

Adapted from Fleming et al. (2006)

4 USDA Food Programs

4.1 School Food Programs

The USDA Food and Nutrition Services administers several food programs: the National School Lunch Program (NSLP), school breakfast program and the Summer Food Service Program (Hopkins & Gunther, 2015). The NSLP is accessible to 50 million schoolchildren. Currently, about 31.7 million (66.3%) lunches are provided daily with partial or full reimbursement [10]. Some 17 million breakfasts are served as part of the school brakefast program each day. Each year about 2.9 billion free lunches are served alongside of 1.8 billion partly paid meals. The cost per meal is approximately $2.79 for lunch and $1.80 for breakfast. The total expenses for the NSLP and school brakefast program was US$13.7 billion for fiscal year 2010, and was expected to reach US$ 16.0 billion for fiscal year 2016 (Food and Nutrition Service USDA 2012) (Fig. 25.6).

[10]USDA Food and Nutrition Service (2019). "Nutrition Standards for School Meals." Retrieved Feb, 2019, from https://www.fns.usda.gov/school-meals/nutrition-standards-school-meals.T

4.2 Child and Adult Care Food Program

The USDA program for Adults Requiring Daily Supervision as well as infants and children is called the Child & Adult Care Food Program (CACFP). The program, which started in 1968, currently has 111,400 disabled adult participants and 3.3 million infants and children. CACFP is described as one of the most far reaching of the USDA sponsored programs with annual costs estimated at US$2.0 billion (Institute of Medicine Committee 2011). Participating establishments include (i) after-school children care centers, (ii) homes for children, (iii) after school care for at risk children, (iv) shelters for homeless and (v) adult care centers. Interestingly CACFP enables food provision by State providers with cost reimbursement to those that meet USDA nutrition standards. Like the other USDA programs, CACFP meals must conform to dietary recommendations for Americans. There are requirements, for meals to be practical, that foods are healthful, that equipment and resources are appropriate for the task, and that the foodservice aims for controlled costs.

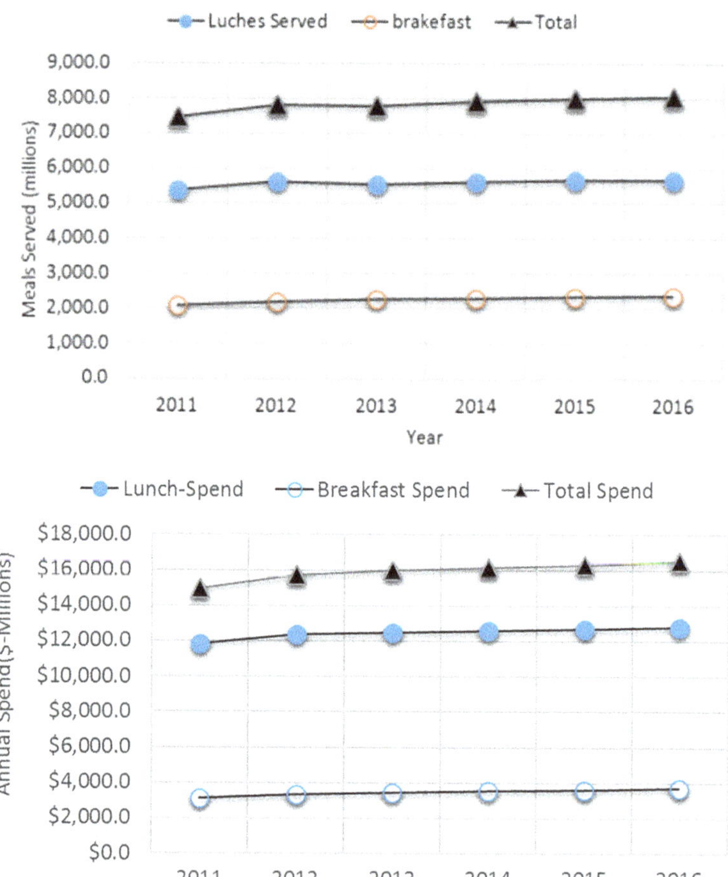

Fig. 25.6 Meals served (top panel) and annual expenditure (lower panel) for the school breakfast program (NSLP) of the US. (Drawn using data from Food and Nutrition Service USDA 2012)

4.3 Nutritional Guidelines for Foodservices

There is a legal requirement for USDA sponsored programs to comply with the "Dietary Guidelines for Americans" [11]. Another legal requirement is that the USDA update nutrition standards based on available evidence derived ultimately from the Institute of Medicine (IOM), food and nutrition board (Food and Nutrition Service USDA 2012) [12]. Some trends and changes of nutrition standards between the years 1946 to 2010 were summarized recently (Hirschman and Chriqui 2013; Hopkins and Gunther 2015). Nutrition standard developed by the IOM are available online (Institute of Medicine Committee 2011) (Table 25.19).

Current nutrition standards require food-based menu planning, and not nutrient based meal planning. This means that meal requirements should be described in terms of real food items, and not isolated nutrients. Providers must refer to real food items and avoid fortification as a strategy for meeting nutrition standards. Some of the other food-based recommendations can be summarized as follows (Food and Nutrition Service USDA 2012; Institute of Medicine Committee 2011):

1. *Fruit as separate item.* One serving of fruits to be provided at lunch or breakfast, either fresh,

[11]Richard B. Russell National School Lunch Act (NSLA) – Section 9(a) (4), 42 U.S.C. 1758(a)(4).

[12]Section 201 of the Healthy, Hunger-Free Kids Act of 2010 (Pub. L. 111–296, HHFKA)

Table 25.19 Summary of USDA Nutrition standards for foodservices

Increase availability
Fruits
Vegetables
Whole grain
Fat-free or low fat milk
Reduce levels
Sodium (sodium chloride)
Saturated fats
Trans-fatty acids
Meat and meat substitutes
Meat
Meat substitutes (tofu)
Other protein sources
Mature beans
Yogurts
Energy/Calories
Mitigate obesity

Adapted from Food and Nutrition Service USDA (2012)

canned, frozen or dried with preference for no-added sugar varieties.

2. *Vegetables as separate item*. Dark green, or yellow/ orange color vegetables or legumes should be served daily.
3. *Whole Grains*. Grain food with 50% whole-grain is recommended at breakfast or lunch with a weekly ration of 8–12 oz. (228–340 g). depending on age.
4. *Meat & Meat substitute*. For lunch or breakfast daily serve lean meat, low salt or cold cut meat is allowed with a portion size of 1-2 oz. (14–56 g). Meat – substitutes (e.g., tofu) and other protein sources (e.g. mature beans, peas or yogurt) can be used.
5. *Low-fat/fat-free fluid milk*. Plain low-fat milk, fat-free milk or fat-free flavored milk is useable, but not low-fat flavored milk, to avoid undue amounts of sugars and saturated fat in the diet.
6. *Nutrient specifications*. Standards focus on specific foods.

6a. *Calories*. Weekly average maximum and minimum caloric intake is indicated for different age groups (K-5, K6-8 & K9-12) to meet energy needs for growth and address issues of childhood obesity.

6b. *Saturated fat*. Dietary saturated fat should not exceed 10% of daily energy intake. No limits are set for total fats.

6c. *Total fats*. No limits for total fat content for school foodservice meals. Instead, low fat was encouraged indirectly, by controlling for total energy, trans-fats and saturated fat content for meals.

6d.*Trans-fatty acids*. There is zero allowance for trans-fatty acids (<0.5 g/serving). The limitation applies to synthetic trans-fats, derived from partially hydrogenated vegetable oil (PHO) associated with margarine and snack food.

6e. *Sodium*. A phased reduction of sodium (chloride) intake is envisaged, reaching eventually 50% of the 2012 average for all schools (1400 mg/lunch). In comparison, the recommended salt intake was 2300 mg/day for the public or 1500 mg/day for at risk groups (African Americans, elderly persons, and hypertensive individuals). Stakeholders raised concerns about the time needed for product reformulation or weather salt substitutes could be used.

7. *Mandatory tracking*. State sponsors for foodservices should review performances every 3 years to ensure compliance with nutrition standards, and the recommended intakes of particular food groups; the numbers of meals at the point of service is checked. Monitoring also requires a formal nutritional analysis to determine calories, saturated fat, trans-fatty acids and sodium intake over a survey period of 2 weeks.
8. *Offer versus serve*. To ensure compliance with recommended dietary patterns, pupils should not decline two food items for lunch and one item for breakfast as served.
9. *Snack*-type foods. A variety of grain or fruit snack type foods are prohibited for reimbursement by federal funds.
10. *Fortification*. Some fortified grain-fruit products are excluded from current school foodservice standards, but ready-to-eat cereals that are a good source of whole grain are allowed.

4.4 Evaluation of School Foodservice Nutrition Programs

USDA meal programs were some of the earliest to be established worldwide (Hirschman and Chriqui 2013). Food programs appeared in other parts of the world since 1946, for example Canada (Godin et al. 2017), Australia, Sweden, the UK (Lucas et al. 2017), diverse countries in Europe and also New Zealand (Micha et al. 2018), Korea (Yoon et al. 2012), Japan (Tanaka and Miyoshi 2012) and India (Chutani 2012) and Vietnam (Le 2012).

Evaluations of the USDA food programmes shows that they had a positive effect on childhood nutrition, particularly for those from a food insecure background. Interventions to improve healthy eating were successful, but not always with high effectiveness. The persistent rise in childhood obesity in the US and other western countries remains a concern in relation to possible effects of food programs. However, such evaluations are just beginning and more research is undoubtedly needed (Hirschman and Chriqui 2013).

5 Food Safety & Hygiene for Foodservices

5.1 Regulatory Requirements for Food Safety

Historically, foodservices practices developed within the restaurants, hotels, and hospitality disciplines (Payne-Palacio and Theis 2015; Payne-Palacio and Theis 2011b; Puckett 2012). Experts believe there is a need to introduce the foodservices sector to some elements of food science, in particular the principles of food safety. On the other hand, food science readers can find more extensive accounts of the foodservices sector within the following textbooks (Drummond and Brefere 2014; Manask et al. 2002; Payne-Palacio and Theis 2015; Payne-Palacio and Theis 2011b; Puckett 2012).

It is a legal requirement that foodservice outlets must provide only safe food for customers. Serving food that is a cause of foodborne illness can lead to death. Unsafe food will have adverse consequences also for the foodservice provider and may lead to legal action. Food safety incorporates food law, policy and food science (food microbiology, engineering, food chemistry etc.). Current food safety regulations worldwide are based on the following premise.

(i) Only safe and wholesome food is marketed,
(ii) Food safety regulations will be science based, transparent and readily accessible / achievable and
(iii) Food safety regulations will be enforceable by independent bodies (e.g. the FDA or EU food safety agency),
(iv) Manufacturers are held liable for non-compliance with food safety legislation.

5.2 Foodborne Disease Outbreaks and Foodservices

A food safety survey from 2016 showed that 85% of all foodborne outbreaks (for the census period) were linked with foodservice businesses. In all, 926 outbreaks of foodborne disease produced over 15,000 hospitalization and 15 deaths (Centers for Disease Control and Prevention, 2018). Some of the highlights of the CDC report are summarized in Fig. 25.7.

Etiology The most common causative agents for foodborne illness were bacterial (55% of cases), viruses (37% of cases), chemical agents (26% of cases) and parasites (1% of cases). The most common single agents were Norovirus (36%) and Salmonella (33%) compared with 4% of foodborne disease identified with E.coli and campylobacter.

Food Agents About 38% of food agents responsible for infections were identified without doubt. This means that 62% of foodborne disease were not ascribed to a single food item (Fig. 25.7).

Fig. 25.7 Foodborne disease incidence in the US, 2016. (Centers for Disease Control and Prevention 2018)

TOTAL – 920 CASES
ILLNESSES 15,202
HOSPITALIZATIONS 950
NUMBER OF DEATHS 15
SINGLE LOCATIONS 86%
MULTIPLE LOCATIONS 6%

Locations
Restaurants 60%
Catering & Banquet 14%
Institutions 5%
Hospital & Homes 1%

Foods
Fish 13%
Chicken 11%
Pork 10%
Dairy 9%

Causative agents
Bacteria 54%
Viruses 38%
Chemical 7%

Type of Foods Risk Animal foods were most associated with foodborne illness (42%), followed by aquatic foods (30%) and plant foods (25%). Specific food/ bacterial pairs were identified, e.g., fish was most frequently linked with ciguatoxn, chicken and pork were associated with salmonella, and dairy was a likely source of campylobacter.

Hospitalization Risk The highest risk of hospitalization due to foodborne disease was due to hepatitis A infection from consuming mollusks. On some occasions hepatitis A infections from fruit were also likely to lead to hospitalization. Salmonella infections from chicken also produced hospitalization.

Death Risk The risks of death from foodborne illness were noted for Hepatitis A infections from mollusks, listeria infection from dairy, and finally salmonella infection from eating beef or chicken.

Poor attitude and a lack of training of some workers probably contributes to foodborne outbreaks. However, in accordance with the principles of Good Manufacturing Practices (GMP) it is the responsibility of the managers to train staff in order to improve food-handling skills (Chap. 17). In addition, restaurants are visited by large numbers of people over short periods, which places heavy demand on the cleaning and food handling efforts.

5.3 Foodservice and retail HACCP

Foodservice (and food retail) businesses consists of nearly 13 million people handling foods and meals in a professional capacity (Chap. 2, Sect. 2.6). In all cases, the requirements for safe food handling is met partly through Good Management Practice (Chap. 10) and personnel training in food hygiene measures as required by the US food code (U.S. Food and Drug Administration 2017).

Currently there is no mandatory requirement for HACCP in foodservice and food retail businesses and for good reason. The FDA notes, *"conventional HACCP cannot apply readily to foodservice businesses, because the operational steps for food preparation are more highly diverse compared to the unit operations used for food manufacturing"*. Therefore, starting from 2006 the FDA suggested a modified HACCP process for foodservices and retail businesses, to deal with one of three food handling scenarios; (a) foods without a cook step, (b) foods with a cook step and one holding step and (c) complex foods with hold - cook - hold steps. Ultimately, the foodservices HACCP process is based on adopting and documenting a time/temperature control for safety scheme that ensures food is kept outside of the 41-135 °F (5 – 57 °C) temperature danger zone. Details of the HACCP process for foodservices and food retail are provided by the FDA (Food and drug Administration 2006; U.S. Food and Drug Administration 2017).

Appendix 1

A1.Top Foodservice Management Companies

Rank	Company	Website
1	Compass Group North America	www.compass-usa.com
2	Aramark	www.aramark.com
3	Sodexo	www.sodexousa.com
4	Delaware North Companies	www.delawarenorth.com
5	Centerplate	www.centerplate.com
6	AVI Foodsystems	www.avifoodsystems.com
7	Thompson Hospitality	www.thompsonhospitality.com
8	Trusthouse Services Group	www.trusthouseservices.com
9	Healthcare Services Group	www.hcsgcorp.com
10	Guest Services, Inc.	www.guestservices.com
11	Ovations Food Services	www.ovationsfoodservices.com
12	Guckenheimer Enterprises	www.guckenheimer.com
13	Legends Hospitality	www.legendshm.com
14	Unidinc Corp.	www.unidinc.com
15	CulinArt, Inc.	www.culinartinc.com
16	Metz Culinary Management	www.metzltd.com
17	Gourmet Services	www.gourmetservicesinc.com
18	Southwest Foodservice Excellence	www.sfellc.org
19	Whitsons Culinary Group	www.whitsons.com
20	Parkhurst Dining	www.parkhurstdining.com
21	The Nutrition Group	www.thenutritiongroup.biz
22	Treat America Food Services	www.treatamerica.com
23	Taher	www.taher.com
24	Thomas Cuisine Management	www.thomascuisine.com
25	Cura Hospitality	www.curahospitality.com
26	Continental Services	www.continentalserv.com
27	Revolution Foods	www.revfoods.com
28	Southern Foodservice	www.southernfoodservice.com
29	Creative Dining Services	www.creativedining.com
30	MMI Dining Systems	www.mmihospitality.com
31	LPM Affiliated Companies dba Epicurean Feast	www.epicureanfeast.com
32	Pomptonian Food Service	www.pomptonian.com
33	Lessing's Food Service Management	www.lessings.com
34	ABM Healthcare Support Services	www.abm.com
35	Lancer Hospitality	www.lancerhospitality.com
36	Epicurean Group	www.epicurean-group.com
37	Food Management Group	www.fmg.com
38	Café Services	www.cafeservices.com
39	Sterling Spoon Culinary Management	www.sterlingspoon.com
40	HHS Culinary & Nutrition Solutions	www.hhs1.com
41	Lakeview Center, Inc. dba Gulf Coast Enterprises	www.gulfcoastenterprises.org
42	Corporate Chefs, Inc.	www.corporatechefs.com
43	Brock & Co.	www.brockco.com
44	Prince Food Systems	www.princefoodsystems.com
45	Quest Food Management Services	www.questfms.com

Rank	Company	Website
46	Food For Thought	www.fftchicago.com
47	Luby's Culinary Services	www.lubyscs.com
48	Food Services, Inc.	www.fsifoodservices.com
49	Lintons Managed Services	www.lintons1.com
50	RMA Hospitality Management	www.restaurantmarketing.us

*Source: Adapted from National Association of College and University Foodservice (2015)

References

Aase S (2011) Hospital foodservice and patient experience: what's new?. J Am Diet Assoc 111(8):1118–+. https://doi.org/10.1016/j.jada.2011.06.017

Anon (2019) Make-to-stock vs. Make-to-order (Push vs. Pull). Retrieved from https://www.youtube.com/watch?time_continue=66&v=HKLpHNK-vS0

Asprova Corporation (2019) MTO (Make to Order). Retrieved from http://www.lean-manufacturing-japan.com/scm-terminology/mto-make-to-order.html

Binge C, Xufen H, Guoying L, Chunyue W, Tingting Y (2012) Impacts of campus foodservice on students' life: an anthropological case study of Shantou University. Int J China Market 2(2):123

Brown DM (2005) Prevalence of food production systems in school foodservice. J Am Diet Assoc 105(8):1261–1265. https://doi.org/10.1016/j.jada.2005.05.010

Bryan FL, Lyon JB (1984) Critical control points of hospital foodservice operations. J Food Protect 47(12):950–963

Centers for Disease Control and Prevention (2018) Surveillance for foodborne disease outbreaks, United States, 2016, Annual Report. In: Atlanta, Georgia: U.S. Department of Health and Human Services, CDC, 2018. Retrieved from https://stacks.cdc.gov/view/cdc/46104

Chong YY, Unklesbay N, Dowdy R (2000) Clinical nutrition and foodservice personnel in teaching hospitals have different perceptions of total quality management performance. J Am Diet Assoc 100(9):1044–1049. https://doi.org/10.1016/s0002-8223(00)00305-9

Chutani AM (2012) School lunch program in India: background, objectives and components. Asia Pac J Clin Nutr 21(1):151–154

Conner DS (2014) Institutional food service. In: Thompson PB, Kaplan DM (eds) Encyclopedia of food and agricultural ethics. Springer, Dordrecht, pp 1258–1263

Dahl CA, Matthews ME (1980) Cook-chill foodservice system with a microwave-oven – thiamin content in portions of beef loaf after microwave-heating. J Food Sci 45(3):608–612. https://doi.org/10.1111/j.1365-2621.1980.tb04112.x

Dall'Oglio I, Nicolò R, Di Ciommo V, Bianchi N, Ciliento G, Gawronski O et al (2015) A systematic review of hospital foodservice patient satisfaction studies. J Acad Nutr Diet 115(4):567–584. https://doi.org/10.1016/j.jand.2014.11.013

Doyle E, Simmance N, Wilding H, Porter J (2017) Systematic review and meta-analyses of foodservice interventions and their effect on nutritional outcomes and satisfaction of adult oncology patients. Nutr Diet 74(2):116–128. https://doi.org/10.1111/1747-0080.12342

Drewnowski A, Team TEI (2017) The Chicago consensus on sustainable food systems science. Front Nutr 4. Retrieved from https://www.ncbi.nlm.nih.gov/pmc/articles/PMC5930345/

Drummond KE, Brefere LM (2014) Nutrition for foodservice and culinary professionals, 8th edn. Wiley, Hoboken

Dube L, Trudeau E, Belanger MC (1994a) Determining the complexity of patient satisfaction with food services. J Am Diet Assoc 94(4):394–+. https://doi.org/10.1016/0002-8223(94)90093-0

Dube L, Trudeau E, Belanger MC (1994b) Determining the complexity of patient satisfaction with foodservices. J Am Diet Assoc 94(4):394–398, 401; quiz 399–400

Edwards JSA (2013) The foodservice industry: eating out is more than just a meal. Food Qual Prefer 27(2):223–229. https://doi.org/10.1016/j.foodqual.2012.02.003

Edwards JS, Causa H (2009) What is food service? J Food Serv 20(1):1–3. Retrieved from https://onlinelibrary.wiley.com/doi/full/10.1111/j.1748-0159.2008.00122.x

Fleming CA, Miller RK, Washington KD (2006) Chapter 21: contract foodservice. In: Restaurant & foodservice market research handbook, 14p. Retrieved from http://search.ebscohost.com/login.aspx?direct=true&db=hjh&AN=22731282&site=ehost-live

Floros, JD, Newsome R, Fisher W, Barbosa-Canovas GV, Hongda C, Dunne CP, German JB, Hall RL, Heldman DR, Karwe MV, Knabel SJ, Labuza TP, Lund DB, Newell-Mcgloughlin M, Robinson JL, Sebranek JG, Shewfelt RL, Tracy WF, Weaver CM, Ziegler GR (2010) Feeding the world today and tomorrow: the importance of food science and technology. Compr Rev Food Sci Food Saf 9(5):572–599. Retrieved from http://www.ift.org/knowledge-center/read-ift-publications/science-reports/~/media/Knowledge%20Center/Science%20Reports/IFTScientificReview_feedingtheworld.pdf, https://onlinelibrary.wiley.com/doi/pdf/10.1111/j.1541-4337.2010.00127.x

Food and drug Administration (2006) Managing food safety: a regulator's manual for applying HACCP principles to risk-based retail and food service inspections and evaluating voluntary food safety management systems. Retrieved from http://www.fda.gov/downloads/Food/GuidanceRegulation/UCM078159.pdf

Food and Nutrition Service USDA (2012) Nutrition standards in the National School Lunch and school breakfast programs. final rule. Fed Regist 77(17):4088–4167

Godin KM, Kirkpatrick SI, Hanning RM, Stapleton J, Leatherdale ST (2017) Examining guidelines for school-based breakfast programs in Canada: a systematic review of the grey literature. Can J Diet Pract Res 78(2):92–100. https://doi.org/10.3148/cjdpr-2016-037

Hartel RW (2002) Core competencies in food science: background information on the development of the IFT education standards. J Food Sci Educ 1(1):3–5. https://onlinelibrary.wiley.com/doi/pdf/10.1111/j.1541-4329.2002.tb00003.x

Hirschman J, Chriqui JF (2013) School food and nutrition policy, monitoring and evaluation in the USA. Public Health Nutr 16(6):982–988. https://doi.org/10.1017/s1368980012004144

Hopkins LC, Gunther C (2015) A historical review of changes in nutrition standards of USDA child meal programs relative to research findings on the nutritional adequacy of program meals and the diet and nutritional health of participants: implications for future research and the summer food service program. Nutrients 7(12):10145–10167. https://doi.org/10.3390/nu7125523

Institute of Medicine Committee (2011) Child and adult care food program aligning dietary guidance for all. Child and adult care food program: aligning dietary guidance for all. Retrieved from https://www.nap.edu/catalog/12959/child-and-adult-care-food-program-aligning-dietary-guidance-for

Johnson BC, Chambers MJ (2000) Foodservice benchmarking: practices, attitudes, and beliefs of foodservice directors. J Am Diet Assoc 100(2):175–180. https://doi.org/10.1016/s0002-8223(00)00056-0

Khan MA, Rao RM (1983) Nutritional quality of meals served by selected foodservices. CRC Crit Rev Food Sci Nutr 19(2):151–171. https://doi.org/10.1080/10408398309527373

Le DS (2012) School meal program in Ho Chi Minh city, Vietnam: reality and future plan. Asia Pac J Clin Nutr 21(1):139–143

LeBruto SM, Farsad B (1993) Contracted school food service: advantages, disadvantages, and political concerns. Hosp Rev 11(1):57–67

Lengyel CO, Smith JT, Whiting SJ, Zello GA (2004) A questionnaire to examine food service satisfaction of elderly residents in long-term care facilities. J Nutr Elder 24(2):5–18. https://doi.org/10.1300/J052v24n02_02

Lockyer S (2016) Foodservice industry overview 2016. Retrieved from https://www.retailstrategies.com/wp-content/uploads/2016/04/CasualDining_RetailStrategies_Webinar2016.pdf

Lucas PJ, Patterson E, Sacks G, Billich N, Evans CEL (2017) Preschool and school meal policies: an overview of what we know about regulation, implementation, and impact on diet in the UK, Sweden, and Australia. Nutrients 9(7). https://doi.org/10.3390/nu9070736

Manask AM, Schechter ME, NetLibrary I (2002) The complete guide to foodservice in cultural institutions: keys to success in restaurants, catering, and special events. Wiley, New York, 304pp

Martin-Rios C, Demen-Meier C, Gössling S, Cornuz C (2018) Food waste management innovations in the foodservice industry. Waste Manag 79:196–206. https://doi.org/10.1016/j.wasman.2018.07.033

Meyer MK (2000) Top predictors of middle/junior high school students' satisfaction with school foodservice and nutrition programs. J Am Diet Assoc 100(1):100–103

Meyer MK, Conklin MT (1998) Variables affecting high school students' perceptions of school foodservice. J Am Diet Assoc 98(12):1424–+. https://doi.org/10.1016/s0002-8223(98)00322-8

Mibey R, Williams P (2002) Food services trends in New South Wales hospitals, 1993–2001. Food Serv Technol 2(2):95–103

Micha R, Karageorgou D, Bakogianni I, Trichia E, Whitsel LP, Story M, Mozaffarian D (2018) Effectiveness of school food environment policies on children's dietary behaviors: a systematic review and meta-analysis. PLoS One 13(3):e0194555. https://doi.org/10.1371/journal.pone.0194555

National Association of College & University Foodservice (2015) Top contract management companies. Retrieved from http://kitchindex.com/contractmgmt.html

National Food Service Management Institute (2002) Introduction to foodservice systems. Retrieved from https://webcache.googleusercontent.com/search?q=cache:-vL1rsF0TG4J:https://slideplayer.com/slide/7383697/+&cd=20&hl=en&ct=clnk&gl=us

National Restaurant Association (2014) Industry impact. Retrieved from http://www.restaurant.org/Industry-Impact/Employing-America/Economic-Engine

Nettles MF, Gregoire MB, Canter DH (1997) Analysis of the decision to select a conventional or cook-chill system for hospital foodservice. J Am Diet Assoc 97(6):626–631

NutritionEd.org (2019) Registered dietitian career overview. Retrieved from https://www.nutritioned.org/registered-dietitian.html

Ohara PA, Harper DW, Kangas M, Dubeau J, Borsutzky C, Lemire N (1997) Taste, temperature, and presentation predict satisfaction with foodservices in a Canadian continuing-care hospital. J Am Diet Assoc 97(4):401–405. https://doi.org/10.1016/s0002-8223(97)00100-4

Oregon State University (2019) Nutrition and foodservice systems: undergraduate options. College of public health and human sciences. Retrieved from https://health.oregonstate.edu/nutrition/nutrition-and-foodservice-systems

Parham ES, Benes BA (1997) Development needs of faculty in foodservice management. J Am Diet Assoc 97(3):262–265. https://doi.org/10.1016/S0002-8223(97)00069-2

Payne-Palacio J, Theis M (2011a) Introduction to foodservice. Pearson Education, 744p

Payne-Palacio J, Theis M (2011b). Introduction to foodservice: United States Edition. Pearson Education, 744p

Payne-Palacio J, Theis M (2015) Foodservice management: principles and practices: Global Edition. Pearson, 529p

Peregrin T (2011) Sustainability in foodservice operations: an update. J Am Diet Assoc 111(9):1286–+. https://doi.org/10.1016/j.jada.2011.07.017

Peregrin T (2012) Sustainability in foodservice operations: an update. J Acad Nutr Diet 112(5 Suppl):S12–S15. https://doi.org/10.1016/j.jand.2012.03.023

Puckett RP (2012) Foodservice manual for health care institutions. Wiley, 592pp

Research Chefs Association (2016) Culinology: the intersection of culinary art and food science Retrieved from http://eu.wiley.com/WileyCDA/WileyTitle/productCd-047048134X.html

Reynolds CJ, Buckley JD, Weinstein P, Boland J (2014) Are the dietary guidelines for meat, fat, fruit and vegetable consumption appropriate for environmental sustainability? A review of the literature. Nutrients 6(6):2251–2265

Rollin J, Matthews ME (1977) Cook/chill foodservice systems: temperature histories of a cooked ground beef product during the chilling process (1). J Food Prot 40(11):782–784. https://doi.org/10.4315/0362-028x-40.11.782

Saksena M, Okrent A, Anekwe TD, Cho C, Dicken C, Effland A, Elitzak H, Guthrie J, Hamrick K, Hyman J, Young Jo, Lin B-H, Mancino L, Mclaughlin PW, Rahkovsky I, Ralston K, Smith TA, Stewart H, Todd JE, Tuttle C (2018) America's eating habits: food away from home. EIB-196 U.S. Department of Agriculture, Economic Research Service, Economic Information Bulletin No. (EIB-196), 172 pp. Retrieved from https://www.ers.usda.gov/publications/pub-details/?pubid=90227

Sharma A, Strohbehn C, Radhakrishna RB, Ortiz A (2012) Economic viability of selling locally grown produce to local restaurants. J Agric Food Syst Commun Dev 3(1):181–198. https://doi.org/10.5304/jafscd.2012.5031.5014

Schmit TM, Hadcock SE (2010) Assessing barriers to expansion of farm-to-chef sales: a case study from upstate. New York, 23pp. https://ageconsearch.umn.edu/record/126973/

Snyder PO, Matthews ME (1984) Microbiological quality of foodservice menu items produced and stored by cook/chill, cook/freeze, cook/hot-hold and heat/serve methods. J Food Prot 47(11):876–885

Stein K (2000) Foodservice in correctional facilities. J Am Diet Assoc 100(5):508–509. https://doi.org/10.1016/s0002-8223(00)00153-x

Tanaka N, Miyoshi M (2012) School lunch program for health promotion among children in Japan. Asia Pac J Clin Nutr 21(1):155–158

United States Census (2014) 2011 county business patterns (NAICS). Retrieved from http://censtats.census.gov/cgi-bin/cbpnaic/cbpsect.pl or http://censtats.census.gov/cgi-bin/cbpnaic/cbpdetl.pl

US Census Bureau (2012) 2012 NAICS: 722 – Food services and drinking places: definition. Retrieved from https://www.census.gov/econ/isp/sampler.php?naicscode=722&naicslevel=3#

U.S. Food and Drug Administration (2017) Food Code (2017). Retrieved from https://www.fda.gov/media/110822/download

USDA Economic Research Service (2019) June 4, 2020). Food service, market segment. Retrieved from https://www.ers.usda.gov/topics/food-markets-prices/food-service-industry/market-segments/

USDA Economic Research Service (2019a) Food service industry. Retrieved from https://www.ers.usda.gov/topics/food-markets-prices/

USDA Economic Research Service (2019b, October 10, 2017) Food service. Market segment. Retrieved from https://www.ers.usda.gov/topics/food-markets-prices/food-service-industry/market-segments/

Van Donk DP (2001) Make to stock or make to order: the decoupling point in the food processing industries. Int J Prod Econ 69(3):297–306

Yang M-F (1990) An assessment of cook-chill foodservice systems. Retrieved from https://ir.library.oregonstate.edu/concern/graduate_thesis_or_dissertations/s7526g09g

Yoon J, Kwon S, Shim JE (2012) Present status and issues of school nutrition programs in Korea. Asia Pac J Clin Nutr 21(1):128–133

Index

R. Owusu-Apenten, E. R. Vieira, *Elementary Food Science*, Food Science Text Series,
https://doi.org/10.1007/978-3-030-65433-7

G

H

I

U

www.ingramcontent.com/pod-product-compliance
Ingram Content Group UK Ltd.
Pitfield, Milton Keynes, MK11 3LW, UK
UKHW050138280726
14058UKWH00006B/700

* 9 7 8 3 0 3 0 6 5 4 3 1 3 *